Lecture Notes in Computer Science   13674

More information about this series at https://link.springer.com/bookseries/558

Shai Avidan · Gabriel Brostow ·
Moustapha Cissé · Giovanni Maria Farinella ·
Tal Hassner (Eds.)

Computer Vision – ECCV 2022

17th European Conference
Tel Aviv, Israel, October 23–27, 2022
Proceedings, Part XIV

 Springer

Editors
Shai Avidan
Tel Aviv University
Tel Aviv, Israel

Gabriel Brostow ⓘ
University College London
London, UK

Moustapha Cissé
Google AI
Accra, Ghana

Giovanni Maria Farinella ⓘ
University of Catania
Catania, Italy

Tal Hassner ⓘ
Facebook (United States)
Menlo Park, CA, USA

ISSN 0302-9743   ISSN 1611-3349 (electronic)
Lecture Notes in Computer Science
ISBN 978-3-031-19780-2   ISBN 978-3-031-19781-9 (eBook)
https://doi.org/10.1007/978-3-031-19781-9

This Springer imprint is published by the registered company Springer Nature Switzerland AG
The registered company address is: Gewerbestrasse 11, 6330 Cham, Switzerland

Foreword

Organizing the European Conference on Computer Vision (ECCV 2022) in Tel-Aviv during a global pandemic was no easy feat. The uncertainty level was extremely high, and decisions had to be postponed to the last minute. Still, we managed to plan things just in time for ECCV 2022 to be held in person. Participation in physical events is crucial to stimulating collaborations and nurturing the culture of the Computer Vision community.

There were many people who worked hard to ensure attendees enjoyed the best science at the 16th edition of ECCV. We are grateful to the Program Chairs Gabriel Brostow and Tal Hassner, who went above and beyond to ensure the ECCV reviewing process ran smoothly. The scientific program includes dozens of workshops and tutorials in addition to the main conference and we would like to thank Leonid Karlinsky and Tomer Michaeli for their hard work. Finally, special thanks to the web chairs Lorenzo Baraldi and Kosta Derpanis, who put in extra hours to transfer information fast and efficiently to the ECCV community.

We would like to express gratitude to our generous sponsors and the Industry Chairs, Dimosthenis Karatzas and Chen Sagiv, who oversaw industry relations and proposed new ways for academia-industry collaboration and technology transfer. It's great to see so much industrial interest in what we're doing!

Authors' draft versions of the papers appeared online with open access on both the Computer Vision Foundation (CVF) and the European Computer Vision Association (ECVA) websites as with previous ECCVs. Springer, the publisher of the proceedings, has arranged for archival publication. The final version of the papers is hosted by SpringerLink, with active references and supplementary materials. It benefits all potential readers that we offer both a free and citeable version for all researchers, as well as an authoritative, citeable version for SpringerLink readers. Our thanks go to Ronan Nugent from Springer, who helped us negotiate this agreement. Last but not least, we wish to thank Eric Mortensen, our publication chair, whose expertise made the process smooth.

October 2022

Rita Cucchiara
Jiří Matas
Amnon Shashua
Lihi Zelnik-Manor

Preface

Welcome to the proceedings of the European Conference on Computer Vision (ECCV 2022). This was a hybrid edition of ECCV as we made our way out of the COVID-19 pandemic. The conference received 5804 valid paper submissions, compared to 5150 submissions to ECCV 2020 (a 12.7% increase) and 2439 in ECCV 2018. 1645 submissions were accepted for publication (28%) and, of those, 157 (2.7% overall) as orals.

846 of the submissions were desk-rejected for various reasons. Many of them because they revealed author identity, thus violating the double-blind policy. This violation came in many forms: some had author names with the title, others added acknowledgments to specific grants, yet others had links to their github account where their name was visible. Tampering with the LaTeX template was another reason for automatic desk rejection.

ECCV 2022 used the traditional CMT system to manage the entire double-blind reviewing process. Authors did not know the names of the reviewers and vice versa. Each paper received at least 3 reviews (except 6 papers that received only 2 reviews), totalling more than 15,000 reviews.

Handling the review process at this scale was a significant challenge. To ensure that each submission received as fair and high-quality reviews as possible, we recruited more than 4719 reviewers (in the end, 4719 reviewers did at least one review). Similarly we recruited more than 276 area chairs (eventually, only 276 area chairs handled a batch of papers). The area chairs were selected based on their technical expertise and reputation, largely among people who served as area chairs in previous top computer vision and machine learning conferences (ECCV, ICCV, CVPR, NeurIPS, etc.).

Reviewers were similarly invited from previous conferences, and also from the pool of authors. We also encouraged experienced area chairs to suggest additional chairs and reviewers in the initial phase of recruiting. The median reviewer load was five papers per reviewer, while the average load was about four papers, because of the emergency reviewers. The area chair load was 35 papers, on average.

Conflicts of interest between authors, area chairs, and reviewers were handled largely automatically by the CMT platform, with some manual help from the Program Chairs. Reviewers were allowed to describe themselves as senior reviewer (load of 8 papers to review) or junior reviewers (load of 4 papers). Papers were matched to area chairs based on a subject-area affinity score computed in CMT and an affinity score computed by the Toronto Paper Matching System (TPMS). TPMS is based on the paper's full text. An area chair handling each submission would bid for preferred expert reviewers, and we balanced load and prevented conflicts.

The assignment of submissions to area chairs was relatively smooth, as was the assignment of submissions to reviewers. A small percentage of reviewers were not happy with their assignments in terms of subjects and self-reported expertise. This is an area for improvement, although it's interesting that many of these cases were reviewers hand-picked by AC's. We made a later round of reviewer recruiting, targeted at the list of authors of papers submitted to the conference, and had an excellent response which helped provide enough emergency reviewers. In the end, all but six papers received at least 3 reviews.

The challenges of the reviewing process are in line with past experiences at ECCV 2020. As the community grows, and the number of submissions increases, it becomes ever more challenging to recruit enough reviewers and ensure a high enough quality of reviews. Enlisting authors by default as reviewers might be one step to address this challenge.

Authors were given a week to rebut the initial reviews, and address reviewers' concerns. Each rebuttal was limited to a single pdf page with a fixed template.

The Area Chairs then led discussions with the reviewers on the merits of each submission. The goal was to reach consensus, but, ultimately, it was up to the Area Chair to make a decision. The decision was then discussed with a buddy Area Chair to make sure decisions were fair and informative. The entire process was conducted virtually with no in-person meetings taking place.

The Program Chairs were informed in cases where the Area Chairs overturned a decisive consensus reached by the reviewers, and pushed for the meta-reviews to contain details that explained the reasoning for such decisions. Obviously these were the most contentious cases, where reviewer inexperience was the most common reported factor.

Once the list of accepted papers was finalized and released, we went through the laborious process of plagiarism (including self-plagiarism) detection. A total of 4 accepted papers were rejected because of that.

Finally, we would like to thank our Technical Program Chair, Pavel Lifshits, who did tremendous work behind the scenes, and we thank the tireless CMT team.

October 2022

Gabriel Brostow
Giovanni Maria Farinella
Moustapha Cissé
Shai Avidan
Tal Hassner

Organization

General Chairs

Rita Cucchiara   University of Modena and Reggio Emilia, Italy
Jiří Matas   Czech Technical University in Prague, Czech
    Republic
Amnon Shashua   Hebrew University of Jerusalem, Israel
Lihi Zelnik-Manor   Technion – Israel Institute of Technology, Israel

Program Chairs

Shai Avidan   Tel-Aviv University, Israel
Gabriel Brostow   University College London, UK
Moustapha Cissé   Google AI, Ghana
Giovanni Maria Farinella   University of Catania, Italy
Tal Hassner   Facebook AI, USA

Program Technical Chair

Pavel Lifshits   Technion – Israel Institute of Technology, Israel

Workshops Chairs

Leonid Karlinsky   IBM Research, Israel
Tomer Michaeli   Technion – Israel Institute of Technology, Israel
Ko Nishino   Kyoto University, Japan

Tutorial Chairs

Thomas Pock   Graz University of Technology, Austria
Natalia Neverova   Facebook AI Research, UK

Demo Chair

Bohyung Han   Seoul National University, Korea

Social and Student Activities Chairs

Tatiana Tommasi   Italian Institute of Technology, Italy
Sagie Benaim   University of Copenhagen, Denmark

Diversity and Inclusion Chairs

Xi Yin   Facebook AI Research, USA
Bryan Russell   Adobe, USA

Communications Chairs

Lorenzo Baraldi   University of Modena and Reggio Emilia, Italy
Kosta Derpanis   York University & Samsung AI Centre Toronto,
                                Canada

Industrial Liaison Chairs

Dimosthenis Karatzas   Universitat Autònoma de Barcelona, Spain
Chen Sagiv   SagivTech, Israel

Finance Chair

Gerard Medioni   University of Southern California & Amazon,
                                USA

Publication Chair

Eric Mortensen   MiCROTEC, USA

Area Chairs

Lourdes Agapito   University College London, UK
Zeynep Akata   University of Tübingen, Germany
Naveed Akhtar   University of Western Australia, Australia
Karteek Alahari   Inria Grenoble Rhône-Alpes, France
Alexandre Alahi   École polytechnique fédérale de Lausanne,
                                Switzerland
Pablo Arbelaez   Universidad de Los Andes, Columbia
Antonis A. Argyros   University of Crete & Foundation for Research
                                and Technology-Hellas, Crete
Yuki M. Asano   University of Amsterdam, The Netherlands
Kalle Åström   Lund University, Sweden
Hadar Averbuch-Elor   Cornell University, USA

Matthijs Douze   Facebook AI Research, USA
Mohamed Elhoseiny   King Abdullah University of Science and
                                   Technology, Saudi Arabia

Sergio Escalera   University of Barcelona, Spain
Yi Fang   New York University, USA
Ryan Farrell   Brigham Young University, USA
Alireza Fathi   Google, USA
Christoph Feichtenhofer   Facebook AI Research, USA
Basura Fernando   Agency for Science, Technology and Research
                                   (A*STAR), Singapore

Vittorio Ferrari   Google Research, Switzerland
Andrew W. Fitzgibbon   Graphcore, UK
David J. Fleet   University of Toronto, Canada
David Forsyth   University of Illinois at Urbana-Champaign, USA
David Fouhey   University of Michigan, USA
Katerina Fragkiadaki   Carnegie Mellon University, USA
Friedrich Fraundorfer   Graz University of Technology, Austria
Oren Freifeld   Ben-Gurion University, Israel
Thomas Funkhouser   Google Research & Princeton University, USA
Yasutaka Furukawa   Simon Fraser University, Canada
Fabio Galasso   Sapienza University of Rome, Italy
Jürgen Gall   University of Bonn, Germany
Chuang Gan   Massachusetts Institute of Technology, USA
Zhe Gan   Microsoft, USA
Animesh Garg   University of Toronto, Vector Institute, Nvidia,
                                   Canada

Efstratios Gavves   University of Amsterdam, The Netherlands
Peter Gehler   Amazon, Germany
Theo Gevers   University of Amsterdam, The Netherlands
Bernard Ghanem   King Abdullah University of Science and
                                   Technology, Saudi Arabia

Ross B. Girshick   Facebook AI Research, USA
Georgia Gkioxari   Facebook AI Research, USA
Albert Gordo   Facebook, USA
Stephen Gould   Australian National University, Australia
Venu Madhav Govindu   Indian Institute of Science, India
Kristen Grauman   Facebook AI Research & UT Austin, USA
Abhinav Gupta   Carnegie Mellon University & Facebook AI
                                   Research, USA
Mohit Gupta   University of Wisconsin-Madison, USA
Hu Han   Institute of Computing Technology, Chinese
                                   Academy of Sciences, China

Ivan Laptev	Inria Paris, France
Laura Leal-Taixé	Technical University of Munich, Germany
Erik Learned-Miller	University of Massachusetts, Amherst, USA
Gim Hee Lee	National University of Singapore, Singapore
Seungyong Lee	Pohang University of Science and Technology, Korea
Zhen Lei	Institute of Automation, Chinese Academy of Sciences, China
Bastian Leibe	RWTH Aachen University, Germany
Hongdong Li	Australian National University, Australia
Fuxin Li	Oregon State University, USA
Bo Li	University of Illinois at Urbana-Champaign, USA
Yin Li	University of Wisconsin-Madison, USA
Ser-Nam Lim	Meta AI Research, USA
Joseph Lim	University of Southern California, USA
Stephen Lin	Microsoft Research Asia, China
Dahua Lin	The Chinese University of Hong Kong, Hong Kong, China
Si Liu	Beihang University, China
Xiaoming Liu	Michigan State University, USA
Ce Liu	Microsoft, USA
Zicheng Liu	Microsoft, USA
Yanxi Liu	Pennsylvania State University, USA
Feng Liu	Portland State University, USA
Yebin Liu	Tsinghua University, China
Chen Change Loy	Nanyang Technological University, Singapore
Huchuan Lu	Dalian University of Technology, China
Cewu Lu	Shanghai Jiao Tong University, China
Oisin Mac Aodha	University of Edinburgh, UK
Dhruv Mahajan	Facebook, USA
Subhransu Maji	University of Massachusetts, Amherst, USA
Atsuto Maki	KTH Royal Institute of Technology, Sweden
Arun Mallya	NVIDIA, USA
R. Manmatha	Amazon, USA
Iacopo Masi	Sapienza University of Rome, Italy
Dimitris N. Metaxas	Rutgers University, USA
Ajmal Mian	University of Western Australia, Australia
Christian Micheloni	University of Udine, Italy
Krystian Mikolajczyk	Imperial College London, UK
Anurag Mittal	Indian Institute of Technology, Madras, India
Philippos Mordohai	Stevens Institute of Technology, USA
Greg Mori	Simon Fraser University & Borealis AI, Canada

Vittorio Murino	Istituto Italiano di Tecnologia, Italy
P. J. Narayanan	International Institute of Information Technology, Hyderabad, India
Ram Nevatia	University of Southern California, USA
Natalia Neverova	Facebook AI Research, UK
Richard Newcombe	Facebook, USA
Cuong V. Nguyen	Florida International University, USA
Bingbing Ni	Shanghai Jiao Tong University, China
Juan Carlos Niebles	Salesforce & Stanford University, USA
Ko Nishino	Kyoto University, Japan
Jean-Marc Odobez	Idiap Research Institute, École polytechnique fédérale de Lausanne, Switzerland
Francesca Odone	University of Genova, Italy
Takayuki Okatani	Tohoku University & RIKEN Center for Advanced Intelligence Project, Japan
Manohar Paluri	Facebook, USA
Guan Pang	Facebook, USA
Maja Pantic	Imperial College London, UK
Sylvain Paris	Adobe Research, USA
Jaesik Park	Pohang University of Science and Technology, Korea
Hyun Soo Park	The University of Minnesota, USA
Omkar M. Parkhi	Facebook, USA
Deepak Pathak	Carnegie Mellon University, USA
Georgios Pavlakos	University of California, Berkeley, USA
Marcello Pelillo	University of Venice, Italy
Marc Pollefeys	ETH Zurich & Microsoft, Switzerland
Jean Ponce	Inria, France
Gerard Pons-Moll	University of Tübingen, Germany
Fatih Porikli	Qualcomm, USA
Victor Adrian Prisacariu	University of Oxford, UK
Petia Radeva	University of Barcelona, Spain
Ravi Ramamoorthi	University of California, San Diego, USA
Deva Ramanan	Carnegie Mellon University, USA
Vignesh Ramanathan	Facebook, USA
Nalini Ratha	State University of New York at Buffalo, USA
Tammy Riklin Raviv	Ben-Gurion University, Israel
Tobias Ritschel	University College London, UK
Emanuele Rodola	Sapienza University of Rome, Italy
Amit K. Roy-Chowdhury	University of California, Riverside, USA
Michael Rubinstein	Google, USA
Olga Russakovsky	Princeton University, USA

Mathieu Salzmann	École polytechnique fédérale de Lausanne, Switzerland
Dimitris Samaras	Stony Brook University, USA
Aswin Sankaranarayanan	Carnegie Mellon University, USA
Imari Sato	National Institute of Informatics, Japan
Yoichi Sato	University of Tokyo, Japan
Shin'ichi Satoh	National Institute of Informatics, Japan
Walter Scheirer	University of Notre Dame, USA
Bernt Schiele	Max Planck Institute for Informatics, Germany
Konrad Schindler	ETH Zurich, Switzerland
Cordelia Schmid	Inria & Google, France
Alexander Schwing	University of Illinois at Urbana-Champaign, USA
Nicu Sebe	University of Trento, Italy
Greg Shakhnarovich	Toyota Technological Institute at Chicago, USA
Eli Shechtman	Adobe Research, USA
Humphrey Shi	University of Oregon & University of Illinois at Urbana-Champaign & Picsart AI Research, USA
Jianbo Shi	University of Pennsylvania, USA
Roy Shilkrot	Massachusetts Institute of Technology, USA
Mike Zheng Shou	National University of Singapore, Singapore
Kaleem Siddiqi	McGill University, Canada
Richa Singh	Indian Institute of Technology Jodhpur, India
Greg Slabaugh	Queen Mary University of London, UK
Cees Snoek	University of Amsterdam, The Netherlands
Yale Song	Facebook AI Research, USA
Yi-Zhe Song	University of Surrey, UK
Bjorn Stenger	Rakuten Institute of Technology
Abby Stylianou	Saint Louis University, USA
Akihiro Sugimoto	National Institute of Informatics, Japan
Chen Sun	Brown University, USA
Deqing Sun	Google, USA
Kalyan Sunkavalli	Adobe Research, USA
Ying Tai	Tencent YouTu Lab, China
Ayellet Tal	Technion – Israel Institute of Technology, Israel
Ping Tan	Simon Fraser University, Canada
Siyu Tang	ETH Zurich, Switzerland
Chi-Keung Tang	Hong Kong University of Science and Technology, Hong Kong, China
Radu Timofte	University of Würzburg, Germany & ETH Zurich, Switzerland
Federico Tombari	Google, Switzerland & Technical University of Munich, Germany

James Tompkin	Brown University, USA
Lorenzo Torresani	Dartmouth College, USA
Alexander Toshev	Apple, USA
Du Tran	Facebook AI Research, USA
Anh T. Tran	VinAI, Vietnam
Zhuowen Tu	University of California, San Diego, USA
Georgios Tzimiropoulos	Queen Mary University of London, UK
Jasper Uijlings	Google Research, Switzerland
Jan C. van Gemert	Delft University of Technology, The Netherlands
Gul Varol	Ecole des Ponts ParisTech, France
Nuno Vasconcelos	University of California, San Diego, USA
Mayank Vatsa	Indian Institute of Technology Jodhpur, India
Ashok Veeraraghavan	Rice University, USA
Jakob Verbeek	Facebook AI Research, France
Carl Vondrick	Columbia University, USA
Ruiping Wang	Institute of Computing Technology, Chinese Academy of Sciences, China
Xinchao Wang	National University of Singapore, Singapore
Liwei Wang	The Chinese University of Hong Kong, Hong Kong, China
Chaohui Wang	Université Paris-Est, France
Xiaolong Wang	University of California, San Diego, USA
Christian Wolf	NAVER LABS Europe, France
Tao Xiang	University of Surrey, UK
Saining Xie	Facebook AI Research, USA
Cihang Xie	University of California, Santa Cruz, USA
Zeki Yalniz	Facebook, USA
Ming-Hsuan Yang	University of California, Merced, USA
Angela Yao	National University of Singapore, Singapore
Shaodi You	University of Amsterdam, The Netherlands
Stella X. Yu	University of California, Berkeley, USA
Junsong Yuan	State University of New York at Buffalo, USA
Stefanos Zafeiriou	Imperial College London, UK
Amir Zamir	École polytechnique fédérale de Lausanne, Switzerland
Lei Zhang	Alibaba & Hong Kong Polytechnic University, Hong Kong, China
Lei Zhang	International Digital Economy Academy (IDEA), China
Pengchuan Zhang	Meta AI, USA
Bolei Zhou	University of California, Los Angeles, USA
Yuke Zhu	University of Texas at Austin, USA

Todd Zickler   Harvard University, USA
Wangmeng Zuo   Harbin Institute of Technology, China

Technical Program Committee

Davide Abati
Soroush Abbasi
  Koohpayegani
Amos L. Abbott
Rameen Abdal
Rabab Abdelfattah
Sahar Abdelnabi
Hassan Abu Alhaija
Abulikemu Abuduweili
Ron Abutbul
Hanno Ackermann
Aikaterini Adam
Kamil Adamczewski
Ehsan Adeli
Vida Adeli
Donald Adjeroh
Arman Afrasiyabi
Akshay Agarwal
Sameer Agarwal
Abhinav Agarwalla
Vaibhav Aggarwal
Sara Aghajanzadeh
Susmit Agrawal
Antonio Agudo
Touqeer Ahmad
Sk Miraj Ahmed
Chaitanya Ahuja
Nilesh A. Ahuja
Abhishek Aich
Shubhra Aich
Noam Aigerman
Arash Akbarinia
Peri Akiva
Derya Akkaynak
Emre Aksan
Arjun R. Akula
Yuval Alaluf
Stephan Alaniz
Paul Albert
Cenek Albl

Filippo Aleotti
Konstantinos P.
  Alexandridis
Motasem Alfarra
Mohsen Ali
Thiemo Alldieck
Hadi Alzayer
Liang An
Shan An
Yi An
Zhulin An
Dongsheng An
Jie An
Xiang An
Saket Anand
Cosmin Ancuti
Juan Andrade-Cetto
Alexander Andreopoulos
Bjoern Andres
Jerone T. A. Andrews
Shivangi Aneja
Anelia Angelova
Dragomir Anguelov
Rushil Anirudh
Oron Anschel
Rao Muhammad Anwer
Djamila Aouada
Evlampios Apostolidis
Srikar Appalaraju
Nikita Araslanov
Andre Araujo
Eric Arazo
Dawit Mureja Argaw
Anurag Arnab
Aditya Arora
Chetan Arora
Sunpreet S. Arora
Alexey Artemov
Muhammad Asad
Kumar Ashutosh

Sinem Aslan
Vishal Asnani
Mahmoud Assran
Amir Atapour-Abarghouei
Nikos Athanasiou
Ali Athar
ShahRukh Athar
Sara Atito
Souhaib Attaiki
Matan Atzmon
Mathieu Aubry
Nicolas Audebert
Tristan T.
  Aumentado-Armstrong
Melinos Averkiou
Yannis Avrithis
Stephane Ayache
Mehmet Aygün
Seyed Mehdi
  Ayyoubzadeh
Hossein Azizpour
George Azzopardi
Mallikarjun B. R.
Yunhao Ba
Abhishek Badki
Seung-Hwan Bae
Seung-Hwan Baek
Seungryul Baek
Piyush Nitin Bagad
Shai Bagon
Gaetan Bahl
Shikhar Bahl
Sherwin Bahmani
Haoran Bai
Lei Bai
Jiawang Bai
Haoyue Bai
Jinbin Bai
Xiang Bai
Xuyang Bai

Yang Bai
Yuanchao Bai
Ziqian Bai
Sungyong Baik
Kevin Bailly
Max Bain
Federico Baldassarre
Wele Gedara Chaminda
    Bandara
Biplab Banerjee
Pratyay Banerjee
Sandipan Banerjee
Jihwan Bang
Antyanta Bangunharcana
Aayush Bansal
Ankan Bansal
Siddhant Bansal
Wentao Bao
Zhipeng Bao
Amir Bar
Manel Baradad Jurjo
Lorenzo Baraldi
Danny Barash
Daniel Barath
Connelly Barnes
Ioan Andrei Bârsan
Steven Basart
Dina Bashkirova
Chaim Baskin
Peyman Bateni
Anil Batra
Sebastiano Battiato
Ardhendu Behera
Harkirat Behl
Jens Behley
Vasileios Belagiannis
Boulbaba Ben Amor
Emanuel Ben Baruch
Abdessamad Ben Hamza
Gil Ben-Artzi
Assia Benbihi
Fabian Benitez-Quiroz
Guy Ben-Yosef
Philipp Benz
Alexander W. Bergman

Urs Bergmann
Jesus Bermudez-Cameo
Stefano Berretti
Gedas Bertasius
Zachary Bessinger
Petra Bevandić
Matthew Beveridge
Lucas Beyer
Yash Bhalgat
Suvaansh Bhambri
Samarth Bharadwaj
Gaurav Bharaj
Aparna Bharati
Bharat Lal Bhatnagar
Uttaran Bhattacharya
Apratim Bhattacharyya
Brojeshwar Bhowmick
Ankan Kumar Bhunia
Ayan Kumar Bhunia
Qi Bi
Sai Bi
Michael Bi Mi
Gui-Bin Bian
Jia-Wang Bian
Shaojun Bian
Pia Bideau
Mario Bijelic
Hakan Bilen
Guillaume-Alexandre
    Bilodeau
Alexander Binder
Tolga Birdal
Vighnesh N. Birodkar
Sandika Biswas
Andreas Blattmann
Janusz Bobulski
Giuseppe Boccignone
Vishnu Boddeti
Navaneeth Bodla
Moritz Böhle
Aleksei Bokhovkin
Sam Bond-Taylor
Vivek Boominathan
Shubhankar Borse
Mark Boss

Andrea Bottino
Adnane Boukhayma
Fadi Boutros
Nicolas C. Boutry
Richard S. Bowen
Ivaylo Boyadzhiev
Aidan Boyd
Yuri Boykov
Aljaz Bozic
Behzad Bozorgtabar
Eric Brachmann
Samarth Brahmbhatt
Gustav Bredell
Francois Bremond
Joel Brogan
Andrew Brown
Thomas Brox
Marcus A. Brubaker
Robert-Jan Bruintjes
Yuqi Bu
Anders G. Buch
Himanshu Buckchash
Mateusz Buda
Ignas Budvytis
José M. Buenaposada
Marcel C. Bühler
Tu Bui
Adrian Bulat
Hannah Bull
Evgeny Burnacv
Andrei Bursuc
Benjamin Busam
Sergey N. Buzykanov
Wonmin Byeon
Fabian Caba
Martin Cadik
Guanyu Cai
Minjie Cai
Qing Cai
Zhongang Cai
Qi Cai
Yancheng Cai
Shen Cai
Han Cai
Jiarui Cai

Bowen Cai
Mu Cai
Qin Cai
Ruojin Cai
Weidong Cai
Weiwei Cai
Yi Cai
Yujun Cai
Zhiping Cai
Akin Caliskan
Lilian Calvet
Baris Can Cam
Necati Cihan Camgoz
Tommaso Campari
Dylan Campbell
Ziang Cao
Ang Cao
Xu Cao
Zhiwen Cao
Shengcao Cao
Song Cao
Weipeng Cao
Xiangyong Cao
Xiaochun Cao
Yue Cao
Yunhao Cao
Zhangjie Cao
Jiale Cao
Yang Cao
Jiajiong Cao
Jie Cao
Jinkun Cao
Lele Cao
Yulong Cao
Zhiguo Cao
Chen Cao
Razvan Caramalau
Marlène Careil
Gustavo Carneiro
Joao Carreira
Dan Casas
Paola Cascante-Bonilla
Angela Castillo
Francisco M. Castro
Pedro Castro

Luca Cavalli
George J. Cazenavette
Oya Celiktutan
Hakan Cevikalp
Sri Harsha C. H.
Sungmin Cha
Geonho Cha
Menglei Chai
Lucy Chai
Yuning Chai
Zenghao Chai
Anirban Chakraborty
Deep Chakraborty
Rudrasis Chakraborty
Souradeep Chakraborty
Kelvin C. K. Chan
Chee Seng Chan
Paramanand Chandramouli
Arjun Chandrasekaran
Kenneth Chaney
Dongliang Chang
Huiwen Chang
Peng Chang
Xiaojun Chang
Jia-Ren Chang
Hyung Jin Chang
Hyun Sung Chang
Ju Yong Chang
Li-Jen Chang
Qi Chang
Wei-Yi Chang
Yi Chang
Nadine Chang
Hanqing Chao
Pradyumna Chari
Dibyadip Chatterjee
Chiranjoy Chattopadhyay
Siddhartha Chaudhuri
Zhengping Che
Gal Chechik
Lianggangxu Chen
Qi Alfred Chen
Brian Chen
Bor-Chun Chen
Bo-Hao Chen

Bohong Chen
Bin Chen
Ziliang Chen
Cheng Chen
Chen Chen
Chaofeng Chen
Xi Chen
Haoyu Chen
Xuanhong Chen
Wei Chen
Qiang Chen
Shi Chen
Xianyu Chen
Chang Chen
Changhuai Chen
Hao Chen
Jie Chen
Jianbo Chen
Jingjing Chen
Jun Chen
Kejiang Chen
Mingcai Chen
Nenglun Chen
Qifeng Chen
Ruoyu Chen
Shu-Yu Chen
Weidong Chen
Weijie Chen
Weikai Chen
Xiang Chen
Xiuyi Chen
Xingyu Chen
Yaofo Chen
Yueting Chen
Yu Chen
Yunjin Chen
Yuntao Chen
Yun Chen
Zhenfang Chen
Zhuangzhuang Chen
Chu-Song Chen
Xiangyu Chen
Zhuo Chen
Chaoqi Chen
Shizhe Chen

Xiaotong Chen
Xiaozhi Chen
Dian Chen
Defang Chen
Dingfan Chen
Ding-Jie Chen
Ee Heng Chen
Tao Chen
Yixin Chen
Wei-Ting Chen
Lin Chen
Guang Chen
Guangyi Chen
Guanying Chen
Guangyao Chen
Hwann-Tzong Chen
Junwen Chen
Jiacheng Chen
Jianxu Chen
Hui Chen
Kai Chen
Kan Chen
Kevin Chen
Kuan-Wen Chen
Weihua Chen
Zhang Chen
Liang-Chieh Chen
Lele Chen
Liang Chen
Fanglin Chen
Zehui Chen
Minghui Chen
Minghao Chen
Xiaokang Chen
Qian Chen
Jun-Cheng Chen
Qi Chen
Qingcai Chen
Richard J. Chen
Runnan Chen
Rui Chen
Shuo Chen
Sentao Chen
Shaoyu Chen
Shixing Chen

Shuai Chen
Shuya Chen
Sizhe Chen
Simin Chen
Shaoxiang Chen
Zitian Chen
Tianlong Chen
Tianshui Chen
Min-Hung Chen
Xiangning Chen
Xin Chen
Xinghao Chen
Xuejin Chen
Xu Chen
Xuxi Chen
Yunlu Chen
Yanbei Chen
Yuxiao Chen
Yun-Chun Chen
Yi-Ting Chen
Yi-Wen Chen
Yinbo Chen
Yiran Chen
Yuanhong Chen
Yubei Chen
Yuefeng Chen
Yuhua Chen
Yukang Chen
Zerui Chen
Zhaoyu Chen
Zhen Chen
Zhenyu Chen
Zhi Chen
Zhiwei Chen
Zhixiang Chen
Long Chen
Bowen Cheng
Jun Cheng
Yi Cheng
Jingchun Cheng
Lechao Cheng
Xi Cheng
Yuan Cheng
Ho Kei Cheng
Kevin Ho Man Cheng

Jiacheng Cheng
Kelvin B. Cheng
Li Cheng
Mengjun Cheng
Zhen Cheng
Qingrong Cheng
Tianheng Cheng
Harry Cheng
Yihua Cheng
Yu Cheng
Ziheng Cheng
Soon Yau Cheong
Anoop Cherian
Manuela Chessa
Zhixiang Chi
Naoki Chiba
Julian Chibane
Kashyap Chitta
Tai-Yin Chiu
Hsu-kuang Chiu
Wei-Chen Chiu
Sungmin Cho
Donghyeon Cho
Hyeon Cho
Yooshin Cho
Gyusang Cho
Jang Hyun Cho
Seungju Cho
Nam Ik Cho
Sunghyun Cho
Hanbyel Cho
Jaesung Choe
Jooyoung Choi
Chiho Choi
Changwoon Choi
Jongwon Choi
Myungsub Choi
Dooseop Choi
Jonghyun Choi
Jinwoo Choi
Jun Won Choi
Min-Kook Choi
Hongsuk Choi
Janghoon Choi
Yoon-Ho Choi

Yukyung Choi
Jaegul Choo
Ayush Chopra
Siddharth Choudhary
Subhabrata Choudhury
Vasileios Choutas
Ka-Ho Chow
Pinaki Nath Chowdhury
Sammy Christen
Anders Christensen
Grigorios Chrysos
Hang Chu
Wen-Hsuan Chu
Peng Chu
Qi Chu
Ruihang Chu
Wei-Ta Chu
Yung-Yu Chuang
Sanghyuk Chun
Se Young Chun
Antonio Cinà
Ramazan Gokberk Cinbis
Javier Civera
Albert Clapés
Ronald Clark
Brian S. Clipp
Felipe Codevilla
Daniel Coelho de Castro
Niv Cohen
Forrester Cole
Maxwell D. Collins
Robert T. Collins
Marc Comino Trinidad
Runmin Cong
Wenyan Cong
Maxime Cordy
Marcella Cornia
Enric Corona
Huseyin Coskun
Luca Cosmo
Dragos Costea
Davide Cozzolino
Arun C. S. Kumar
Aiyu Cui
Qiongjie Cui

Quan Cui
Shuhao Cui
Yiming Cui
Ying Cui
Zijun Cui
Jiali Cui
Jiequan Cui
Yawen Cui
Zhen Cui
Zhaopeng Cui
Jack Culpepper
Xiaodong Cun
Ross Cutler
Adam Czajka
Ali Dabouei
Konstantinos M. Dafnis
Manuel Dahnert
Tao Dai
Yuchao Dai
Bo Dai
Mengyu Dai
Hang Dai
Haixing Dai
Peng Dai
Pingyang Dai
Qi Dai
Qiyu Dai
Yutong Dai
Naser Damer
Zhiyuan Dang
Mohamed Daoudi
Ayan Das
Abir Das
Debasmit Das
Deepayan Das
Partha Das
Sagnik Das
Soumi Das
Srijan Das
Swagatam Das
Avijit Dasgupta
Jim Davis
Adrian K. Davison
Homa Davoudi
Laura Daza

Matthias De Lange
Shalini De Mello
Marco De Nadai
Christophe De
    Vleeschouwer
Alp Dener
Boyang Deng
Congyue Deng
Bailin Deng
Yong Deng
Ye Deng
Zhuo Deng
Zhijie Deng
Xiaoming Deng
Jiankang Deng
Jinhong Deng
Jingjing Deng
Liang-Jian Deng
Siqi Deng
Xiang Deng
Xueqing Deng
Zhongying Deng
Karan Desai
Jean-Emmanuel Deschaud
Aniket Anand Deshmukh
Neel Dey
Helisa Dhamo
Prithviraj Dhar
Amaya Dharmasiri
Yan Di
Xing Di
Ousmane A. Dia
Haiwen Diao
Xiaolei Diao
Gonçalo José Dias Pais
Abdallah Dib
Anastasios Dimou
Changxing Ding
Henghui Ding
Guodong Ding
Yaqing Ding
Shuangrui Ding
Yuhang Ding
Yikang Ding
Shouhong Ding

Haisong Ding
Hui Ding
Jiahao Ding
Jian Ding
Jian-Jiun Ding
Shuxiao Ding
Tianyu Ding
Wenhao Ding
Yuqi Ding
Yi Ding
Yuzhen Ding
Zhengming Ding
Tan Minh Dinh
Vu Dinh
Christos Diou
Mandar Dixit
Bao Gia Doan
Khoa D. Doan
Dzung Anh Doan
Debi Prosad Dogra
Nehal Doiphode
Chengdong Dong
Bowen Dong
Zhenxing Dong
Hang Dong
Xiaoyi Dong
Haoye Dong
Jiangxin Dong
Shichao Dong
Xuan Dong
Zhen Dong
Shuting Dong
Jing Dong
Li Dong
Ming Dong
Nanqing Dong
Qiulei Dong
Runpei Dong
Siyan Dong
Tian Dong
Wei Dong
Xiaomeng Dong
Xin Dong
Xingbo Dong
Yuan Dong

Samuel Dooley
Gianfranco Doretto
Michael Dorkenwald
Keval Doshi
Zhaopeng Dou
Xiaotian Dou
Hazel Doughty
Ahmad Droby
Iddo Drori
Jie Du
Yong Du
Dawei Du
Dong Du
Ruoyi Du
Yuntao Du
Xuefeng Du
Yilun Du
Yuming Du
Radhika Dua
Haodong Duan
Jiafei Duan
Kaiwen Duan
Peiqi Duan
Ye Duan
Haoran Duan
Jiali Duan
Amanda Duarte
Abhimanyu Dubey
Shiv Ram Dubey
Florian Dubost
Lukasz Dudziak
Shivam Duggal
Justin M. Dulay
Matteo Dunnhofer
Chi Nhan Duong
Thibaut Durand
Mihai Dusmanu
Ujjal Kr Dutta
Debidatta Dwibedi
Isht Dwivedi
Sai Kumar Dwivedi
Takeharu Eda
Mark Edmonds
Alexei A. Efros
Thibaud Ehret

Max Ehrlich
Mahsa Ehsanpour
Iván Eichhardt
Farshad Einabadi
Marvin Eisenberger
Hazim Kemal Ekenel
Mohamed El Banani
Ismail Elezi
Moshe Eliasof
Alaa El-Nouby
Ian Endres
Francis Engelmann
Deniz Engin
Chanho Eom
Dave Epstein
Maria C. Escobar
Victor A. Escorcia
Carlos Esteves
Sungmin Eum
Bernard J. E. Evans
Ivan Evtimov
Fevziye Irem Eyiokur
    Yaman
Matteo Fabbri
Sébastien Fabbro
Gabriele Facciolo
Masud Fahim
Bin Fan
Hehe Fan
Deng-Ping Fan
Aoxiang Fan
Chen-Chen Fan
Qi Fan
Zhaoxin Fan
Haoqi Fan
Heng Fan
Hongyi Fan
Linxi Fan
Baojie Fan
Jiayuan Fan
Lei Fan
Quanfu Fan
Yonghui Fan
Yingruo Fan
Zhiwen Fan

Zicong Fan
Sean Fanello
Jiansheng Fang
Chaowei Fang
Yuming Fang
Jianwu Fang
Jin Fang
Qi Fang
Shancheng Fang
Tian Fang
Xianyong Fang
Gongfan Fang
Zhen Fang
Hui Fang
Jiemin Fang
Le Fang
Pengfei Fang
Xiaolin Fang
Yuxin Fang
Zhaoyuan Fang
Ammarah Farooq
Azade Farshad
Zhengcong Fei
Michael Felsberg
Wei Feng
Chen Feng
Fan Feng
Andrew Feng
Xin Feng
Zheyun Feng
Ruicheng Feng
Mingtao Feng
Qianyu Feng
Shangbin Feng
Chun-Mei Feng
Zunlei Feng
Zhiyong Feng
Martin Fergie
Mustansar Fiaz
Marco Fiorucci
Michael Firman
Hamed Firooz
Volker Fischer
Corneliu O. Florea
Georgios Floros

Wolfgang Foerstner
Gianni Franchi
Jean-Sebastien Franco
Simone Frintrop
Anna Fruehstueck
Changhong Fu
Chaoyou Fu
Cheng-Yang Fu
Chi-Wing Fu
Deqing Fu
Huan Fu
Jun Fu
Kexue Fu
Ying Fu
Jianlong Fu
Jingjing Fu
Qichen Fu
Tsu-Jui Fu
Xueyang Fu
Yang Fu
Yanwei Fu
Yonggan Fu
Wolfgang Fuhl
Yasuhisa Fujii
Kent Fujiwara
Marco Fumero
Takuya Funatomi
Isabel Funke
Dario Fuoli
Antonino Furnari
Matheus A. Gadelha
Akshay Gadi Patil
Adrian Galdran
Guillermo Gallego
Silvano Galliani
Orazio Gallo
Leonardo Galteri
Matteo Gamba
Yiming Gan
Sujoy Ganguly
Harald Ganster
Boyan Gao
Changxin Gao
Daiheng Gao
Difei Gao

Chen Gao
Fei Gao
Lin Gao
Wei Gao
Yiming Gao
Junyu Gao
Guangyu Ryan Gao
Haichang Gao
Hongchang Gao
Jialin Gao
Jin Gao
Jun Gao
Katelyn Gao
Mingchen Gao
Mingfei Gao
Pan Gao
Shangqian Gao
Shanghua Gao
Xitong Gao
Yunhe Gao
Zhanning Gao
Elena Garces
Nuno Cruz Garcia
Noa Garcia
Guillermo
    Garcia-Hernando
Isha Garg
Rahul Garg
Sourav Garg
Quentin Garrido
Stefano Gasperini
Kent Gauen
Chandan Gautam
Shivam Gautam
Paul Gay
Chunjiang Ge
Shiming Ge
Wenhang Ge
Yanhao Ge
Zheng Ge
Songwei Ge
Weifeng Ge
Yixiao Ge
Yuying Ge
Shijie Geng

Zhengyang Geng
Kyle A. Genova
Georgios Georgakis
Markos Georgopoulos
Marcel Geppert
Shabnam Ghadar
Mina Ghadimi Atigh
Deepti Ghadiyaram
Maani Ghaffari Jadidi
Sedigh Ghamari
Zahra Gharaee
Michaël Gharbi
Golnaz Ghiasi
Reza Ghoddoosian
Soumya Suvra Ghosal
Adhiraj Ghosh
Arthita Ghosh
Pallabi Ghosh
Soumyadeep Ghosh
Andrew Gilbert
Igor Gilitschenski
Jhony H. Giraldo
Andreu Girbau Xalabarder
Rohit Girdhar
Sharath Girish
Xavier Giro-i-Nieto
Raja Giryes
Thomas Gittings
Nikolaos Gkanatsios
Ioannis Gkioulekas
Abhiram
    Gnanasambandam
Aurele T. Gnanha
Clement L. J. C. Godard
Arushi Goel
Vidit Goel
Shubham Goel
Zan Gojcic
Aaron K. Gokaslan
Tejas Gokhale
S. Alireza Golestaneh
Thiago L. Gomes
Nuno Goncalves
Boqing Gong
Chen Gong

Yuanhao Gong
Guoqiang Gong
Jingyu Gong
Rui Gong
Yu Gong
Mingming Gong
Neil Zhenqiang Gong
Xun Gong
Yunye Gong
Yihong Gong
Cristina I. González
Nithin Gopalakrishnan
    Nair
Gaurav Goswami
Jianping Gou
Shreyank N. Gowda
Ankit Goyal
Helmut Grabner
Patrick L. Grady
Ben Graham
Eric Granger
Douglas R. Gray
Matej Grcić
David Griffiths
Jinjin Gu
Yun Gu
Shuyang Gu
Jianyang Gu
Fuqiang Gu
Jiatao Gu
Jindong Gu
Jiaqi Gu
Jinwei Gu
Jiaxin Gu
Geonmo Gu
Xiao Gu
Xinqian Gu
Xiuye Gu
Yuming Gu
Zhangxuan Gu
Dayan Guan
Junfeng Guan
Qingji Guan
Tianrui Guan
Shanyan Guan

Denis A. Gudovskiy
Ricardo Guerrero
Pierre-Louis Guhur
Jie Gui
Liangyan Gui
Liangke Gui
Benoit Guillard
Erhan Gundogdu
Manuel Günther
Jingcai Guo
Yuanfang Guo
Junfeng Guo
Chenqi Guo
Dan Guo
Hongji Guo
Jia Guo
Jie Guo
Minghao Guo
Shi Guo
Yanhui Guo
Yangyang Guo
Yuan-Chen Guo
Yilu Guo
Yiluan Guo
Yong Guo
Guangyu Guo
Haiyun Guo
Jinyang Guo
Jianyuan Guo
Pengsheng Guo
Pengfei Guo
Shuxuan Guo
Song Guo
Tianyu Guo
Qing Guo
Qiushan Guo
Wen Guo
Xiefan Guo
Xiaohu Guo
Xiaoqing Guo
Yufei Guo
Yuhui Guo
Yuliang Guo
Yunhui Guo
Yanwen Guo

Akshita Gupta
Ankush Gupta
Kamal Gupta
Kartik Gupta
Ritwik Gupta
Rohit Gupta
Siddharth Gururani
Fredrik K. Gustafsson
Abner Guzman Rivera
Vladimir Guzov
Matthew A. Gwilliam
Jung-Woo Ha
Marc Habermann
Isma Hadji
Christian Haene
Martin Hahner
Levente Hajder
Alexandros Haliassos
Emanuela Haller
Bumsub Ham
Abdullah J. Hamdi
Shreyas Hampali
Dongyoon Han
Chunrui Han
Dong-Jun Han
Dong-Sig Han
Guangxing Han
Zhizhong Han
Ruize Han
Jiaming Han
Jin Han
Ligong Han
Xian-Hua Han
Xiaoguang Han
Yizeng Han
Zhi Han
Zhenjun Han
Zhongyi Han
Jungong Han
Junlin Han
Kai Han
Kun Han
Sungwon Han
Songfang Han
Wei Han

Xiao Han
Xintong Han
Xinzhe Han
Yahong Han
Yan Han
Zongbo Han
Nicolai Hani
Rana Hanocka
Niklas Hanselmann
Nicklas A. Hansen
Hong Hanyu
Fusheng Hao
Yanbin Hao
Shijie Hao
Udith Haputhanthri
Mehrtash Harandi
Josh Harguess
Adam Harley
David M. Hart
Atsushi Hashimoto
Ali Hassani
Mohammed Hassanin
Yana Hasson
Joakim Bruslund Haurum
Bo He
Kun He
Chen He
Xin He
Fazhi He
Gaoqi He
Hao He
Haoyu He
Jiangpeng He
Hongliang He
Qian He
Xiangteng He
Xuming He
Yannan He
Yuhang He
Yang He
Xiangyu He
Nanjun He
Pan He
Sen He
Shengfeng He

Songtao He
Tao He
Tong He
Wei He
Xuehai He
Xiaoxiao He
Ying He
Yisheng He
Ziwen He
Peter Hedman
Felix Heide
Yacov Hel-Or
Paul Henderson
Philipp Henzler
Byeongho Heo
Jae-Pil Heo
Miran Heo
Sachini A. Herath
Stephane Herbin
Pedro Hermosilla Casajus
Monica Hernandez
Charles Herrmann
Roei Herzig
Mauricio Hess-Flores
Carlos Hinojosa
Tobias Hinz
Tsubasa Hirakawa
Chih-Hui Ho
Lam Si Tung Ho
Jennifer Hobbs
Derek Hoiem
Yannick Hold-Geoffroy
Aleksander Holynski
Cheeun Hong
Fa-Ting Hong
Hanbin Hong
Guan Zhe Hong
Danfeng Hong
Lanqing Hong
Xiaopeng Hong
Xin Hong
Jie Hong
Seungbum Hong
Cheng-Yao Hong
Seunghoon Hong

Yi Hong
Yuan Hong
Yuchen Hong
Anthony Hoogs
Maxwell C. Horton
Kazuhiro Hotta
Qibin Hou
Tingbo Hou
Junhui Hou
Ji Hou
Qiqi Hou
Rui Hou
Ruibing Hou
Zhi Hou
Henry Howard-Jenkins
Lukas Hoyer
Wei-Lin Hsiao
Chiou-Ting Hsu
Anthony Hu
Brian Hu
Yusong Hu
Hexiang Hu
Haoji Hu
Di Hu
Hengtong Hu
Haigen Hu
Lianyu Hu
Hanzhe Hu
Jie Hu
Junlin Hu
Shizhe Hu
Jian Hu
Zhiming Hu
Juhua Hu
Peng Hu
Ping Hu
Ronghang Hu
MengShun Hu
Tao Hu
Vincent Tao Hu
Xiaoling Hu
Xinting Hu
Xiaolin Hu
Xuefeng Hu
Xiaowei Hu

Yang Hu
Yueyu Hu
Zeyu Hu
Zhongyun Hu
Binh-Son Hua
Guoliang Hua
Yi Hua
Linzhi Huang
Qiusheng Huang
Bo Huang
Chen Huang
Hsin-Ping Huang
Ye Huang
Shuangping Huang
Zeng Huang
Buzhen Huang
Cong Huang
Heng Huang
Hao Huang
Qidong Huang
Huaibo Huang
Chaoqin Huang
Feihu Huang
Jiahui Huang
Jingjia Huang
Kun Huang
Lei Huang
Sheng Huang
Shuaiyi Huang
Siyu Huang
Xiaoshui Huang
Xiaoyang Huang
Yan Huang
Yihao Huang
Ying Huang
Ziling Huang
Xiaoke Huang
Yifei Huang
Haiyang Huang
Zhewei Huang
Jin Huang
Haibin Huang
Jiaxing Huang
Junjie Huang
Keli Huang

Lang Huang
Lin Huang
Luojie Huang
Mingzhen Huang
Shijia Huang
Shengyu Huang
Siyuan Huang
He Huang
Xiuyu Huang
Lianghua Huang
Yue Huang
Yaping Huang
Yuge Huang
Zehao Huang
Zeyi Huang
Zhiqi Huang
Zhongzhan Huang
Zilong Huang
Ziyuan Huang
Tianrui Hui
Zhuo Hui
Le Hui
Jing Huo
Junhwa Hur
Shehzeen S. Hussain
Chuong Minh Huynh
Seunghyun Hwang
Jaehui Hwang
Jyh-Jing Hwang
Sukjun Hwang
Soonmin Hwang
Wonjun Hwang
Rakib Hyder
Sangeek Hyun
Sarah Ibrahimi
Tomoki Ichikawa
Yerlan Idelbayev
A. S. M. Iftekhar
Masaaki Iiyama
Satoshi Ikehata
Sunghoon Im
Atul N. Ingle
Eldar Insafutdinov
Yani A. Ioannou
Radu Tudor Ionescu

Umar Iqbal
Go Irie
Muhammad Zubair Irshad
Ahmet Iscen
Berivan Isik
Ashraful Islam
Md Amirul Islam
Syed Islam
Mariko Isogawa
Vamsi Krishna K. Ithapu
Boris Ivanovic
Darshan Iyer
Sarah Jabbour
Ayush Jain
Nishant Jain
Samyak Jain
Vidit Jain
Vineet Jain
Priyank Jaini
Tomas Jakab
Mohammad A. A. K.
  Jalwana
Muhammad Abdullah
  Jamal
Hadi Jamali-Rad
Stuart James
Varun Jampani
Young Kyun Jang
YeongJun Jang
Yunseok Jang
Ronnachai Jaroensri
Bhavan Jasani
Krishna Murthy
  Jatavallabhula
Mojan Javaheripi
Syed A. Javed
Guillaume Jeanneret
Pranav Jeevan
Herve Jegou
Rohit Jena
Tomas Jenicek
Porter Jenkins
Simon Jenni
Hae-Gon Jeon
Sangryul Jeon

Boseung Jeong
Yoonwoo Jeong
Seong-Gyun Jeong
Jisoo Jeong
Allan D. Jepson
Ankit Jha
Sumit K. Jha
I-Hong Jhuo
Ge-Peng Ji
Chaonan Ji
Deyi Ji
Jingwei Ji
Wei Ji
Zhong Ji
Jiayi Ji
Pengliang Ji
Hui Ji
Mingi Ji
Xiaopeng Ji
Yuzhu Ji
Baoxiong Jia
Songhao Jia
Dan Jia
Shan Jia
Xiaojun Jia
Xiuyi Jia
Xu Jia
Menglin Jia
Wenqi Jia
Boyuan Jiang
Wenhao Jiang
Huaizu Jiang
Hanwen Jiang
Haiyong Jiang
Hao Jiang
Huajie Jiang
Huiqin Jiang
Haojun Jiang
Haobo Jiang
Junjun Jiang
Xingyu Jiang
Yangbangyan Jiang
Yu Jiang
Jianmin Jiang
Jiaxi Jiang

Jing Jiang
Kui Jiang
Li Jiang
Liming Jiang
Chiyu Jiang
Meirui Jiang
Chen Jiang
Peng Jiang
Tai-Xiang Jiang
Wen Jiang
Xinyang Jiang
Yifan Jiang
Yuming Jiang
Yingying Jiang
Zeren Jiang
ZhengKai Jiang
Zhenyu Jiang
Shuming Jiao
Jianbo Jiao
Licheng Jiao
Dongkwon Jin
Yeying Jin
Cheng Jin
Linyi Jin
Qing Jin
Taisong Jin
Xiao Jin
Xin Jin
Sheng Jin
Kyong Hwan Jin
Ruibing Jin
SouYoung Jin
Yueming Jin
Chenchen Jing
Longlong Jing
Taotao Jing
Yongcheng Jing
Younghyun Jo
Joakim Johnander
Jeff Johnson
Michael J. Jones
R. Kenny Jones
Rico Jonschkowski
Ameya Joshi
Sunghun Joung

Felix Juefei-Xu
Claudio R. Jung
Steffen Jung
Hari Chandana K.
Rahul Vigneswaran K.
Prajwal K. R.
Abhishek Kadian
Jhony Kaesemodel Pontes
Kumara Kahatapitiya
Anmol Kalia
Sinan Kalkan
Tarun Kalluri
Jaewon Kam
Sandesh Kamath
Meina Kan
Menelaos Kanakis
Takuhiro Kaneko
Di Kang
Guoliang Kang
Hao Kang
Jaeyeon Kang
Kyoungkook Kang
Li-Wei Kang
MinGuk Kang
Suk-Ju Kang
Zhao Kang
Yash Mukund Kant
Yueying Kao
Aupendu Kar
Konstantinos Karantzalos
Sezer Karaoglu
Navid Kardan
Sanjay Kariyappa
Leonid Karlinsky
Animesh Karnewar
Shyamgopal Karthik
Hirak J. Kashyap
Marc A. Kastner
Hirokatsu Kataoka
Angelos Katharopoulos
Hiroharu Kato
Kai Katsumata
Manuel Kaufmann
Chaitanya Kaul
Prakhar Kaushik

Yuki Kawana
Lei Ke
Lipeng Ke
Tsung-Wei Ke
Wei Ke
Petr Kellnhofer
Aniruddha Kembhavi
John Kender
Corentin Kervadec
Leonid Keselman
Daniel Keysers
Nima Khademi Kalantari
Taras Khakhulin
Samir Khaki
Muhammad Haris Khan
Qadeer Khan
Salman Khan
Subash Khanal
Vaishnavi M. Khindkar
Rawal Khirodkar
Saeed Khorram
Pirazh Khorramshahi
Kourosh Khoshelham
Ansh Khurana
Benjamin Kiefer
Jae Myung Kim
Junho Kim
Boah Kim
Hyeonseong Kim
Dong-Jin Kim
Dongwan Kim
Donghyun Kim
Doyeon Kim
Yonghyun Kim
Hyung-Il Kim
Hyunwoo Kim
Hyeongwoo Kim
Hyo Jin Kim
Hyunwoo J. Kim
Taehoon Kim
Jaeha Kim
Jiwon Kim
Jung Uk Kim
Kangyeol Kim
Eunji Kim

Daeha Kim
Dongwon Kim
Kunhee Kim
Kyungmin Kim
Junsik Kim
Min H. Kim
Namil Kim
Kookhoi Kim
Sanghyun Kim
Seongyeop Kim
Seungryong Kim
Saehoon Kim
Euyoung Kim
Guisik Kim
Sungyeon Kim
Sunnie S. Y. Kim
Taehun Kim
Tae Oh Kim
Won Hwa Kim
Seungwook Kim
YoungBin Kim
Youngeun Kim
Akisato Kimura
Furkan Osman Kınlı
Zsolt Kira
Hedvig Kjellström
Florian Kleber
Jan P. Klopp
Florian Kluger
Laurent Kneip
Byungsoo Ko
Muhammed Kocabas
A. Sophia Koepke
Kevin Koeser
Nick Kolkin
Nikos Kolotouros
Wai-Kin Adams Kong
Deying Kong
Caihua Kong
Youyong Kong
Shuyu Kong
Shu Kong
Tao Kong
Yajing Kong
Yu Kong

Zishang Kong
Theodora Kontogianni
Anton S. Konushin
Julian F. P. Kooij
Bruno Korbar
Giorgos Kordopatis-Zilos
Jari Korhonen
Adam Kortylewski
Denis Korzhenkov
Divya Kothandaraman
Suraj Kothawade
Iuliia Kotseruba
Satwik Kottur
Shashank Kotyan
Alexandros Kouris
Petros Koutras
Anna Kreshuk
Ranjay Krishna
Dilip Krishnan
Andrey Kuehlkamp
Hilde Kuehne
Jason Kuen
David Kügler
Arjan Kuijper
Anna Kukleva
Sumith Kulal
Viveka Kulharia
Akshay R. Kulkarni
Nilesh Kulkarni
Dominik Kulon
Abhinav Kumar
Akash Kumar
Suryansh Kumar
B. V. K. Vijaya Kumar
Pulkit Kumar
Ratnesh Kumar
Sateesh Kumar
Satish Kumar
Vijay Kumar B. G.
Nupur Kumari
Sudhakar Kumawat
Jogendra Nath Kundu
Hsien-Kai Kuo
Meng-Yu Jennifer Kuo
Vinod Kumar Kurmi

Yusuke Kurose
Keerthy Kusumam
Alina Kuznetsova
Henry Kvinge
Ho Man Kwan
Hyeokjun Kweon
Heeseung Kwon
Gihyun Kwon
Myung-Joon Kwon
Taesung Kwon
YoungJoong Kwon
Christos Kyrkou
Jorma Laaksonen
Yann Labbe
Zorah Laehner
Florent Lafarge
Hamid Laga
Manuel Lagunas
Shenqi Lai
Jian-Huang Lai
Zihang Lai
Mohamed I. Lakhal
Mohit Lamba
Meng Lan
Loic Landrieu
Zhiqiang Lang
Natalie Lang
Dong Lao
Yizhen Lao
Yingjie Lao
Issam Hadj Laradji
Gustav Larsson
Viktor Larsson
Zakaria Laskar
Stéphane Lathuilière
Chun Pong Lau
Rynson W. H. Lau
Hei Law
Justin Lazarow
Verica Lazova
Eric-Tuan Le
Hieu Le
Trung-Nghia Le
Mathias Lechner
Byeong-Uk Lee

Chen-Yu Lee
Che-Rung Lee
Chul Lee
Hong Joo Lee
Dongsoo Lee
Jiyoung Lee
Eugene Eu Tzuan Lee
Daeun Lee
Saehyung Lee
Jewook Lee
Hyungtae Lee
Hyunmin Lee
Jungbeom Lee
Joon-Young Lee
Jong-Seok Lee
Joonseok Lee
Junha Lee
Kibok Lee
Byung-Kwan Lee
Jangwon Lee
Jinho Lee
Jongmin Lee
Seunghyun Lee
Sohyun Lee
Minsik Lee
Dogyoon Lee
Seungmin Lee
Min Jun Lee
Sangho Lee
Sangmin Lee
Seungeun Lee
Seon-Ho Lee
Sungmin Lee
Sungho Lee
Sangyoun Lee
Vincent C. S. S. Lee
Jaeseong Lee
Yong Jae Lee
Chenyang Lei
Chenyi Lei
Jiahui Lei
Xinyu Lei
Yinjie Lei
Jiaxu Leng
Luziwei Leng

Jan E. Lenssen
Vincent Lepetit
Thomas Leung
María Leyva-Vallina
Xin Li
Yikang Li
Baoxin Li
Bin Li
Bing Li
Bowen Li
Changlin Li
Chao Li
Chongyi Li
Guanyue Li
Shuai Li
Jin Li
Dingquan Li
Dongxu Li
Yiting Li
Gang Li
Dian Li
Guohao Li
Haoang Li
Haoliang Li
Haoran Li
Hengduo Li
Huafeng Li
Xiaoming Li
Hanao Li
Hongwei Li
Ziqiang Li
Jisheng Li
Jiacheng Li
Jia Li
Jiachen Li
Jiahao Li
Jianwei Li
Jiazhi Li
Jie Li
Jing Li
Jingjing Li
Jingtao Li
Jun Li
Junxuan Li
Kai Li

Kailin Li
Kenneth Li
Kun Li
Kunpeng Li
Aoxue Li
Chenglong Li
Chenglin Li
Changsheng Li
Zhichao Li
Qiang Li
Yanyu Li
Zuoyue Li
Xiang Li
Xuelong Li
Fangda Li
Ailin Li
Liang Li
Chun-Guang Li
Daiqing Li
Dong Li
Guanbin Li
Guorong Li
Haifeng Li
Jianan Li
Jianing Li
Jiaxin Li
Ke Li
Lei Li
Lincheng Li
Liulei Li
Lujun Li
Linjie Li
Lin Li
Pengyu Li
Ping Li
Qiufu Li
Qingyong Li
Rui Li
Siyuan Li
Wei Li
Wenbin Li
Xiangyang Li
Xinyu Li
Xiujun Li
Xiu Li

Xu Li
Ya-Li Li
Yao Li
Yongjie Li
Yijun Li
Yiming Li
Yuezun Li
Yu Li
Yunheng Li
Yuqi Li
Zhe Li
Zeming Li
Zhen Li
Zhengqin Li
Zhimin Li
Jiefeng Li
Jinpeng Li
Chengze Li
Jianwu Li
Lerenhan Li
Shan Li
Suichan Li
Xiangtai Li
Yanjie Li
Yandong Li
Zhuoling Li
Zhenqiang Li
Manyi Li
Maosen Li
Ji Li
Minjun Li
Mingrui Li
Mengtian Li
Junyi Li
Nianyi Li
Bo Li
Xiao Li
Peihua Li
Peike Li
Peizhao Li
Peiliang Li
Qi Li
Ren Li
Runze Li
Shile Li

Sheng Li
Shigang Li
Shiyu Li
Shuang Li
Shasha Li
Shichao Li
Tianye Li
Yuexiang Li
Wei-Hong Li
Wanhua Li
Weihao Li
Weiming Li
Weixin Li
Wenbo Li
Wenshuo Li
Weijian Li
Yunan Li
Xirong Li
Xianhang Li
Xiaoyu Li
Xueqian Li
Xuanlin Li
Xianzhi Li
Yunqiang Li
Yanjing Li
Yansheng Li
Yawei Li
Yi Li
Yong Li
Yong-Lu Li
Yuhang Li
Yu-Jhe Li
Yuxi Li
Yunsheng Li
Yanwei Li
Zechao Li
Zejian Li
Zeju Li
Zekun Li
Zhaowen Li
Zheng Li
Zhenyu Li
Zhiheng Li
Zhi Li
Zhong Li

Zhuowei Li
Zhuowan Li
Zhuohang Li
Zizhang Li
Chen Li
Yuan-Fang Li
Dongze Lian
Xiaochen Lian
Zhouhui Lian
Long Lian
Qing Lian
Jin Lianbao
Jinxiu S. Liang
Dingkang Liang
Jiahao Liang
Jianming Liang
Jingyun Liang
Kevin J. Liang
Kaizhao Liang
Chen Liang
Jie Liang
Senwei Liang
Ding Liang
Jiajun Liang
Jian Liang
Kongming Liang
Siyuan Liang
Yuanzhi Liang
Zhengfa Liang
Mingfu Liang
Xiaodan Liang
Xuefeng Liang
Yuxuan Liang
Kang Liao
Liang Liao
Hong-Yuan Mark Liao
Wentong Liao
Haofu Liao
Yue Liao
Minghui Liao
Shengcai Liao
Ting-Hsuan Liao
Xin Liao
Yinghong Liao
Teck Yian Lim

Che-Tsung Lin
Chung-Ching Lin
Chen-Hsuan Lin
Cheng Lin
Chuming Lin
Chunyu Lin
Dahua Lin
Wei Lin
Zheng Lin
Huaijia Lin
Jason Lin
Jierui Lin
Jiaying Lin
Jie Lin
Kai-En Lin
Kevin Lin
Guangfeng Lin
Jiehong Lin
Feng Lin
Hang Lin
Kwan-Yee Lin
Ke Lin
Luojun Lin
Qinghong Lin
Xiangbo Lin
Yi Lin
Zudi Lin
Shijie Lin
Yiqun Lin
Tzu-Heng Lin
Ming Lin
Shaohui Lin
SongNan Lin
Ji Lin
Tsung-Yu Lin
Xudong Lin
Yancong Lin
Yen-Chen Lin
Yiming Lin
Yuewei Lin
Zhiqiu Lin
Zinan Lin
Zhe Lin
David B. Lindell
Zhixin Ling

Zhan Ling
Alexander Liniger
Venice Erin B. Liong
Joey Litalien
Or Litany
Roee Litman
Ron Litman
Jim Little
Dor Litvak
Shaoteng Liu
Shuaicheng Liu
Andrew Liu
Xian Liu
Shaohui Liu
Bei Liu
Bo Liu
Yong Liu
Ming Liu
Yanbin Liu
Chenxi Liu
Daqi Liu
Di Liu
Difan Liu
Dong Liu
Dongfang Liu
Daizong Liu
Xiao Liu
Fangyi Liu
Fengbei Liu
Fenglin Liu
Bin Liu
Yuang Liu
Ao Liu
Hong Liu
Hongfu Liu
Huidong Liu
Ziyi Liu
Feng Liu
Hao Liu
Jie Liu
Jialun Liu
Jiang Liu
Jing Liu
Jingya Liu
Jiaming Liu

Jun Liu
Juncheng Liu
Jiawei Liu
Hongyu Liu
Chuanbin Liu
Haotian Liu
Lingqiao Liu
Chang Liu
Han Liu
Liu Liu
Min Liu
Yingqi Liu
Aishan Liu
Bingyu Liu
Benlin Liu
Boxiao Liu
Chenchen Liu
Chuanjian Liu
Daqing Liu
Huan Liu
Haozhe Liu
Jiaheng Liu
Wei Liu
Jingzhou Liu
Jiyuan Liu
Lingbo Liu
Nian Liu
Peiye Liu
Qiankun Liu
Shenglan Liu
Shilong Liu
Wen Liu
Wenyu Liu
Weifeng Liu
Wu Liu
Xiaolong Liu
Yang Liu
Yanwei Liu
Yingcheng Liu
Yongfei Liu
Yihao Liu
Yu Liu
Yunze Liu
Ze Liu
Zhenhua Liu

Zhenguang Liu
Lin Liu
Lihao Liu
Pengju Liu
Xinhai Liu
Yunfei Liu
Meng Liu
Minghua Liu
Mingyuan Liu
Miao Liu
Peirong Liu
Ping Liu
Qingjie Liu
Ruoshi Liu
Risheng Liu
Songtao Liu
Xing Liu
Shikun Liu
Shuming Liu
Sheng Liu
Songhua Liu
Tongliang Liu
Weibo Liu
Weide Liu
Weizhe Liu
Wenxi Liu
Weiyang Liu
Xin Liu
Xiaobin Liu
Xudong Liu
Xiaoyi Liu
Xihui Liu
Xinchen Liu
Xingtong Liu
Xinpeng Liu
Xinyu Liu
Xianpeng Liu
Xu Liu
Xingyu Liu
Yongtuo Liu
Yahui Liu
Yangxin Liu
Yaoyao Liu
Yaojie Liu
Yuliang Liu

Yongcheng Liu

Yuan Liu

Yufan Liu

Yu-Lun Liu

Yun Liu

Yunfan Liu

Yuanzhong Liu

Zhuoran Liu

Zhen Liu

Zheng Liu

Zhijian Liu

Zhisong Liu

Ziquan Liu

Ziyu Liu

Zhihua Liu

Zechun Liu

Zhaoyang Liu

Zhengzhe Liu

Stephan Liwicki

Shao-Yuan Lo

Sylvain Lobry

Suhas Lohit

Vishnu Suresh Lokhande

Vincenzo Lomonaco

Chengjiang Long

Guodong Long

Fuchen Long

Shangbang Long

Yang Long

Zijun Long

Vasco Lopes

Antonio M. Lopez

Roberto Javier
   Lopez-Sastre

Tobias Lorenz

Javier Lorenzo-Navarro

Yujing Lou

Qian Lou

Xiankai Lu

Changsheng Lu

Huimin Lu

Yongxi Lu

Hao Lu

Hong Lu

Jiasen Lu

Juwei Lu

Fan Lu

Guangming Lu

Jiwen Lu

Shun Lu

Tao Lu

Xiaonan Lu

Yang Lu

Yao Lu

Yongchun Lu

Zhiwu Lu

Cheng Lu

Liying Lu

Guo Lu

Xuequan Lu

Yanye Lu

Yantao Lu

Yuhang Lu

Fujun Luan

Jonathon Luiten

Jovita Lukasik

Alan Lukezic

Jonathan Samuel Lumentut

Mayank Lunayach

Ao Luo

Canjie Luo

Chong Luo

Xu Luo

Grace Luo

Jun Luo

Katie Z. Luo

Tao Luo

Cheng Luo

Fangzhou Luo

Gen Luo

Lei Luo

Sihui Luo

Weixin Luo

Yan Luo

Xiaoyan Luo

Yong Luo

Yadan Luo

Hao Luo

Ruotian Luo

Mi Luo

Tiange Luo

Wenjie Luo

Wenhan Luo

Xiao Luo

Zhiming Luo

Zhipeng Luo

Zhengyi Luo

Diogo C. Luvizon

Zhaoyang Lv

Gengyu Lyu

Lingjuan Lyu

Jun Lyu

Yuanyuan Lyu

Youwei Lyu

Yueming Lyu

Bingpeng Ma

Chao Ma

Chongyang Ma

Congbo Ma

Chih-Yao Ma

Fan Ma

Lin Ma

Haoyu Ma

Hengbo Ma

Jianqi Ma

Jiawei Ma

Jiayi Ma

Kede Ma

Kai Ma

Lingni Ma

Lei Ma

Xu Ma

Ning Ma

Benteng Ma

Cheng Ma

Andy J. Ma

Long Ma

Zhanyu Ma

Zhiheng Ma

Qianli Ma

Shiqiang Ma

Sizhuo Ma

Shiqing Ma

Xiaolong Ma

Xinzhu Ma

Gautam B. Machiraju
Spandan Madan
Mathew Magimai-Doss
Luca Magri
Behrooz Mahasseni
Upal Mahbub
Siddharth Mahendran
Paridhi Maheshwari
Rishabh Maheshwary
Mohammed Mahmoud
Shishira R. R. Maiya
Sylwia Majchrowska
Arjun Majumdar
Puspita Majumdar
Orchid Majumder
Sagnik Majumder
Ilya Makarov
Farkhod F.
    Makhmudkhujaev
Yasushi Makihara
Ankur Mali
Mateusz Malinowski
Utkarsh Mall
Srikanth Malla
Clement Mallet
Dimitrios Mallis
Yunze Man
Dipu Manandhar
Massimiliano Mancini
Murari Mandal
Raunak Manekar
Karttikeya Mangalam
Puneet Mangla
Fabian Manhardt
Sivabalan Manivasagam
Fahim Mannan
Chengzhi Mao
Hanzi Mao
Jiayuan Mao
Junhua Mao
Zhiyuan Mao
Jiageng Mao
Yunyao Mao
Zhendong Mao
Alberto Marchisio

Diego Marcos
Riccardo Marin
Aram Markosyan
Renaud Marlet
Ricardo Marques
Miquel Martí i Rabadán
Diego Martin Arroyo
Niki Martinel
Brais Martinez
Julieta Martinez
Marc Masana
Tomohiro Mashita
Timothée Masquelier
Minesh Mathew
Tetsu Matsukawa
Marwan Mattar
Bruce A. Maxwell
Christoph Mayer
Mantas Mazeika
Pratik Mazumder
Scott McCloskey
Steven McDonagh
Ishit Mehta
Jie Mei
Kangfu Mei
Jieru Mei
Xiaoguang Mei
Givi Meishvili
Luke Melas-Kyriazi
Iaroslav Melekhov
Andres Mendez-Vazquez
Heydi Mendez-Vazquez
Matias Mendieta
Ricardo A. Mendoza-León
Chenlin Meng
Depu Meng
Rang Meng
Zibo Meng
Qingjie Meng
Qier Meng
Yanda Meng
Zihang Meng
Thomas Mensink
Fabian Mentzer
Christopher Metzler

Gregory P. Meyer
Vasileios Mezaris
Liang Mi
Lu Mi
Bo Miao
Changtao Miao
Zichen Miao
Qiguang Miao
Xin Miao
Zhongqi Miao
Frank Michel
Simone Milani
Ben Mildenhall
Roy V. Miles
Juhong Min
Kyle Min
Hyun-Seok Min
Weiqing Min
Yuecong Min
Zhixiang Min
Qi Ming
David Minnen
Aymen Mir
Deepak Mishra
Anand Mishra
Shlok K. Mishra
Niluthpol Mithun
Gaurav Mittal
Trisha Mittal
Daisuke Miyazaki
Kaichun Mo
Hong Mo
Zhipeng Mo
Davide Modolo
Abduallah A. Mohamed
Mohamed Afham
    Mohamed Aflal
Ron Mokady
Pavlo Molchanov
Davide Moltisanti
Liliane Momeni
Gianluca Monaci
Pascal Monasse
Ajoy Mondal
Tom Monnier

Simone Palazzo
Luca Palmieri
Bowen Pan
Hao Pan
Lili Pan
Tai-Yu Pan
Liang Pan
Chengwei Pan
Yingwei Pan
Xuran Pan
Jinshan Pan
Xinyu Pan
Liyuan Pan
Xingang Pan
Xingjia Pan
Zhihong Pan
Zizheng Pan
Priyadarshini Panda
Rameswar Panda
Rohit Pandey
Kaiyue Pang
Bo Pang
Guansong Pang
Jiangmiao Pang
Meng Pang
Tianyu Pang
Ziqi Pang
Omiros Pantazis
Andreas Panteli
Maja Pantic
Marina Paolanti
Joao P. Papa
Samuele Papa
Mike Papadakis
Dim P. Papadopoulos
George Papandreou
Constantin Pape
Toufiq Parag
Chethan Parameshwara
Shaifali Parashar
Alejandro Pardo
Rishubh Parihar
Sarah Parisot
JaeYoo Park
Gyeong-Moon Park

Hyojin Park
Hyoungseob Park
Jongchan Park
Jae Sung Park
Kiru Park
Chunghyun Park
Kwanyong Park
Sunghyun Park
Sungrae Park
Seongsik Park
Sanghyun Park
Sungjune Park
Taesung Park
Gaurav Parmar
Paritosh Parmar
Alvaro Parra
Despoina Paschalidou
Or Patashnik
Shivansh Patel
Pushpak Pati
Prashant W. Patil
Vaishakh Patil
Suvam Patra
Jay Patravali
Badri Narayana Patro
Angshuman Paul
Sudipta Paul
Rémi Pautrat
Nick E. Pears
Adithya Pediredla
Wenjie Pei
Shmuel Peleg
Latha Pemula
Bo Peng
Houwen Peng
Yue Peng
Liangzu Peng
Baoyun Peng
Jun Peng
Pai Peng
Sida Peng
Xi Peng
Yuxin Peng
Songyou Peng
Wei Peng

Weiqi Peng
Wen-Hsiao Peng
Pramuditha Perera
Juan C. Perez
Eduardo Pérez Pellitero
Juan-Manuel Perez-Rua
Federico Pernici
Marco Pesavento
Stavros Petridis
Ilya A. Petrov
Vladan Petrovic
Mathis Petrovich
Suzanne Petryk
Hieu Pham
Quang Pham
Khoi Pham
Tung Pham
Huy Phan
Stephen Phillips
Cheng Perng Phoo
David Picard
Marco Piccirilli
Georg Pichler
A. J. Piergiovanni
Vipin Pillai
Silvia L. Pintea
Giovanni Pintore
Robinson Piramuthu
Fiora Pirri
Theodoros Pissas
Fabio Pizzati
Benjamin Planche
Bryan Plummer
Matteo Poggi
Ashwini Pokle
Georgy E. Ponimatkin
Adrian Popescu
Stefan Popov
Nikola Popović
Ronald Poppe
Angelo Porrello
Michael Potter
Charalambos Poullis
Hadi Pouransari
Omid Poursaeed

Shraman Pramanick
Mantini Pranav
Dilip K. Prasad
Meghshyam Prasad
B. H. Pawan Prasad
Shitala Prasad
Prateek Prasanna
Ekta Prashnani
Derek S. Prijatelj
Luke Y. Prince
Véronique Prinet
Victor Adrian Prisacariu
James Pritts
Thomas Probst
Sergey Prokudin
Rita Pucci
Chi-Man Pun
Matthew Purri
Haozhi Qi
Lu Qi
Lei Qi
Xianbiao Qi
Yonggang Qi
Yuankai Qi
Siyuan Qi
Guocheng Qian
Hangwei Qian
Qi Qian
Deheng Qian
Shengsheng Qian
Wen Qian
Rui Qian
Yiming Qian
Shengju Qian
Shengyi Qian
Xuelin Qian
Zhenxing Qian
Nan Qiao
Xiaotian Qiao
Jing Qin
Can Qin
Siyang Qin
Hongwei Qin
Jie Qin
Minghai Qin

Yipeng Qin
Yongqiang Qin
Wenda Qin
Xuebin Qin
Yuzhe Qin
Yao Qin
Zhenyue Qin
Zhiwu Qing
Heqian Qiu
Jiayan Qiu
Jielin Qiu
Yue Qiu
Jiaxiong Qiu
Zhongxi Qiu
Shi Qiu
Zhaofan Qiu
Zhongnan Qu
Yanyun Qu
Kha Gia Quach
Yuhui Quan
Ruijie Quan
Mike Rabbat
Rahul Shekhar Rade
Filip Radenovic
Gorjan Radevski
Bogdan Raducanu
Francesco Ragusa
Shafin Rahman
Md Mahfuzur Rahman
    Siddiquee
Hossein Rahmani
Kiran Raja
Sivaramakrishnan
    Rajaraman
Jathushan Rajasegaran
Adnan Siraj Rakin
Michaël Ramamonjisoa
Chirag A. Raman
Shanmuganathan Raman
Vignesh Ramanathan
Vasili Ramanishka
Vikram V. Ramaswamy
Merey Ramazanova
Jason Rambach
Sai Saketh Rambhatla

Clément Rambour
Ashwin Ramesh Babu
Adín Ramírez Rivera
Arianna Rampini
Haoxi Ran
Aakanksha Rana
Aayush Jung Bahadur
    Rana
Kanchana N. Ranasinghe
Aneesh Rangnekar
Samrudhdhi B. Rangrej
Harsh Rangwani
Viresh Ranjan
Anyi Rao
Yongming Rao
Carolina Raposo
Michalis Raptis
Amir Rasouli
Vivek Rathod
Adepu Ravi Sankar
Avinash Ravichandran
Bharadwaj Ravichandran
Dripta S. Raychaudhuri
Adria Recasens
Simon Reiß
Davis Rempe
Daxuan Ren
Jiawei Ren
Jimmy Ren
Sucheng Ren
Dayong Ren
Zhile Ren
Dongwei Ren
Qibing Ren
Pengfei Ren
Zhenwen Ren
Xuqian Ren
Yixuan Ren
Zhongzheng Ren
Ambareesh Revanur
Hamed Rezazadegan
    Tavakoli
Rafael S. Rezende
Wonjong Rhee
Alexander Richard

Christian Richardt
Stephan R. Richter
Benjamin Riggan
Dominik Rivoir
Mamshad Nayeem Rizve
Joshua D. Robinson
Joseph Robinson
Chris Rockwell
Ranga Rodrigo
Andres C. Rodriguez
Carlos Rodriguez-Pardo
Marcus Rohrbach
Gemma Roig
Yu Rong
David A. Ross
Mohammad Rostami
Edward Rosten
Karsten Roth
Anirban Roy
Debaditya Roy
Shuvendu Roy
Ahana Roy Choudhury
Aruni Roy Chowdhury
Denys Rozumnyi
Shulan Ruan
Wenjie Ruan
Patrick Ruhkamp
Danila Rukhovich
Anian Ruoss
Chris Russell
Dan Ruta
Dawid Damian Rymarczyk
DongHun Ryu
Hyeonggon Ryu
Kwonyoung Ryu
Balasubramanian S.
Alexandre Sablayrolles
Mohammad Sabokrou
Arka Sadhu
Aniruddha Saha
Oindrila Saha
Pritish Sahu
Aneeshan Sain
Nirat Saini
Saurabh Saini

Takeshi Saitoh
Christos Sakaridis
Fumihiko Sakaue
Dimitrios Sakkos
Ken Sakurada
Parikshit V. Sakurikar
Rohit Saluja
Nermin Samet
Leo Sampaio Ferraz
    Ribeiro
Jorge Sanchez
Enrique Sanchez
Shengtian Sang
Anush Sankaran
Soubhik Sanyal
Nikolaos Sarafianos
Vishwanath Saragadam
István Sárándi
Saquib Sarfraz
Mert Bulent Sariyildiz
Anindya Sarkar
Pritam Sarkar
Paul-Edouard Sarlin
Hiroshi Sasaki
Takami Sato
Torsten Sattler
Ravi Kumar Satzoda
Axel Sauer
Stefano Savian
Artem Savkin
Manolis Savva
Gerald Schaefer
Simone Schaub-Meyer
Yoni Schirris
Samuel Schulter
Katja Schwarz
Jesse Scott
Sinisa Segvic
Constantin Marc Seibold
Lorenzo Seidenari
Matan Sela
Fadime Sener
Paul Hongsuck Seo
Kwanggyoon Seo
Hongje Seong

Dario Serez
Francesco Setti
Bryan Seybold
Mohamad Shahbazi
Shima Shahfar
Xinxin Shan
Caifeng Shan
Dandan Shan
Shawn Shan
Wei Shang
Jinghuan Shang
Jiaxiang Shang
Lei Shang
Sukrit Shankar
Ken Shao
Rui Shao
Jie Shao
Mingwen Shao
Aashish Sharma
Gaurav Sharma
Vivek Sharma
Abhishek Sharma
Yoli Shavit
Shashank Shekhar
Sumit Shekhar
Zhijie Shen
Fengyi Shen
Furao Shen
Jialie Shen
Jingjing Shen
Ziyi Shen
Linlin Shen
Guangyu Shen
Biluo Shen
Falong Shen
Jiajun Shen
Qiu Shen
Qiuhong Shen
Shuai Shen
Wang Shen
Yiqing Shen
Yunhang Shen
Siqi Shen
Bin Shen
Tianwei Shen

Xi Shen
Yilin Shen
Yuming Shen
Yucong Shen
Zhiqiang Shen
Lu Sheng
Yichen Sheng
Shivanand Venkanna
    Sheshappanavar
Shelly Sheynin
Baifeng Shi
Ruoxi Shi
Botian Shi
Hailin Shi
Jia Shi
Jing Shi
Shaoshuai Shi
Baoguang Shi
Boxin Shi
Hengcan Shi
Tianyang Shi
Xiaodan Shi
Yongjie Shi
Zhensheng Shi
Yinghuan Shi
Weiqi Shi
Wu Shi
Xuepeng Shi
Xiaoshuang Shi
Yujiao Shi
Zenglin Shi
Zhenmei Shi
Takashi Shibata
Meng-Li Shih
Yichang Shih
Hyunjung Shim
Dongseok Shim
Soshi Shimada
Inkyu Shin
Jinwoo Shin
Seungjoo Shin
Seungjae Shin
Koichi Shinoda
Suprosanna Shit

Palaiahnakote
    Shivakumara
Eli Shlizerman
Gaurav Shrivastava
Xiao Shu
Xiangbo Shu
Xiujun Shu
Yang Shu
Tianmin Shu
Jun Shu
Zhixin Shu
Bing Shuai
Maria Shugrina
Ivan Shugurov
Satya Narayan Shukla
Pranjay Shyam
Jianlou Si
Yawar Siddiqui
Alberto Signoroni
Pedro Silva
Jae-Young Sim
Oriane Siméoni
Martin Simon
Andrea Simonelli
Abhishek Singh
Ashish Singh
Dinesh Singh
Gurkirt Singh
Krishna Kumar Singh
Mannat Singh
Pravendra Singh
Rajat Vikram Singh
Utkarsh Singhal
Dipika Singhania
Vasu Singla
Harsh Sinha
Sudipta Sinha
Josef Sivic
Elena Sizikova
Geri Skenderi
Ivan Skorokhodov
Dmitriy Smirnov
Cameron Y. Smith
James S. Smith
Patrick Snape

Mattia Soldan
Hyeongseok Son
Sanghyun Son
Chuanbiao Song
Chen Song
Chunfeng Song
Dan Song
Dongjin Song
Hwanjun Song
Guoxian Song
Jiaming Song
Jie Song
Liangchen Song
Ran Song
Luchuan Song
Xibin Song
Li Song
Fenglong Song
Guoli Song
Guanglu Song
Zhenbo Song
Lin Song
Xinhang Song
Yang Song
Yibing Song
Rajiv Soundararajan
Hossein Souri
Cristovao Sousa
Riccardo Spezialetti
Leonidas Spinoulas
Michael W. Spratling
Deepak Sridhar
Srinath Sridhar
Gaurang Sriramanan
Vinkle Kumar Srivastav
Themos Stafylakis
Serban Stan
Anastasis Stathopoulos
Markus Steinberger
Jan Steinbrener
Sinisa Stekovic
Alexandros Stergiou
Gleb Sterkin
Rainer Stiefelhagen
Pierre Stock

Ombretta Strafforello
Julian Straub
Yannick Strümpler
Joerg Stueckler
Hang Su
Weijie Su
Jong-Chyi Su
Bing Su
Haisheng Su
Jinming Su
Yiyang Su
Yukun Su
Yuxin Su
Zhuo Su
Zhaoqi Su
Xiu Su
Yu-Chuan Su
Zhixun Su
Arulkumar Subramaniam
Akshayvarun Subramanya
A. Subramanyam
Swathikiran Sudhakaran
Yusuke Sugano
Masanori Suganuma
Yumin Suh
Yang Sui
Baochen Sun
Cheng Sun
Long Sun
Guolei Sun
Haoliang Sun
Haomiao Sun
He Sun
Hanqing Sun
Hao Sun
Lichao Sun
Jiachen Sun
Jiaming Sun
Jian Sun
Jin Sun
Jennifer J. Sun
Tiancheng Sun
Libo Sun
Peize Sun
Qianru Sun

Shanlin Sun
Yu Sun
Zhun Sun
Che Sun
Lin Sun
Tao Sun
Yiyou Sun
Chunyi Sun
Chong Sun
Weiwei Sun
Weixuan Sun
Xiuyu Sun
Yanan Sun
Zeren Sun
Zhaodong Sun
Zhiqing Sun
Minhyuk Sung
Jinli Suo
Simon Suo
Abhijit Suprem
Anshuman Suri
Saksham Suri
Joshua M. Susskind
Roman Suvorov
Gurumurthy Swaminathan
Robin Swanson
Paul Swoboda
Tabish A. Syed
Richard Szeliski
Fariborz Taherkhani
Yu-Wing Tai
Keita Takahashi
Walter Talbott
Gary Tam
Masato Tamura
Feitong Tan
Fuwen Tan
Shuhan Tan
Andong Tan
Bin Tan
Cheng Tan
Jianchao Tan
Lei Tan
Mingxing Tan
Xin Tan

Zichang Tan
Zhentao Tan
Kenichiro Tanaka
Masayuki Tanaka
Yushun Tang
Hao Tang
Jingqun Tang
Jinhui Tang
Kaihua Tang
Luming Tang
Lv Tang
Sheyang Tang
Shitao Tang
Siliang Tang
Shixiang Tang
Yansong Tang
Keke Tang
Chang Tang
Chenwei Tang
Jie Tang
Junshu Tang
Ming Tang
Peng Tang
Xu Tang
Yao Tang
Chen Tang
Fan Tang
Haoran Tang
Shengeng Tang
Yehui Tang
Zhipeng Tang
Ugo Tanielian
Chaofan Tao
Jiale Tao
Junli Tao
Renshuai Tao
An Tao
Guanhong Tao
Zhiqiang Tao
Makarand Tapaswi
Jean-Philippe G. Tarel
Juan J. Tarrio
Enzo Tartaglione
Keisuke Tateno
Zachary Teed

Ajinkya B. Tejankar
Bugra Tekin
Purva Tendulkar
Damien Teney
Minggui Teng
Chris Tensmeyer
Andrew Beng Jin Teoh
Philipp Terhörst
Kartik Thakral
Nupur Thakur
Kevin Thandiackal
Spyridon Thermos
Diego Thomas
William Thong
Yuesong Tian
Guanzhong Tian
Lin Tian
Shiqi Tian
Kai Tian
Meng Tian
Tai-Peng Tian
Zhuotao Tian
Shangxuan Tian
Tian Tian
Yapeng Tian
Yu Tian
Yuxin Tian
Leslie Ching Ow Tiong
Praveen Tirupattur
Garvita Tiwari
George Toderici
Antoine Toisoul
Aysim Toker
Tatiana Tommasi
Zhan Tong
Alessio Tonioni
Alessandro Torcinovich
Fabio Tosi
Matteo Toso
Hugo Touvron
Quan Hung Tran
Son Tran
Hung Tran
Ngoc-Trung Tran
Vinh Tran

Phong Tran
Giovanni Trappolini
Edith Tretschk
Subarna Tripathi
Shubhendu Trivedi
Eduard Trulls
Prune Truong
Thanh-Dat Truong
Tomasz Trzcinski
Sam Tsai
Yi-Hsuan Tsai
Ethan Tseng
Yu-Chee Tseng
Shahar Tsiper
Stavros Tsogkas
Shikui Tu
Zhigang Tu
Zhengzhong Tu
Richard Tucker
Sergey Tulyakov
Cigdem Turan
Daniyar Turmukhambetov
Victor G. Turrisi da Costa
Bartlomiej Twardowski
Christopher D. Twigg
Radim Tylecek
Mostofa Rafid Uddin
Md. Zasim Uddin
Kohei Uehara
Nicolas Ugrinovic
Youngjung Uh
Norimichi Ukita
Anwaar Ulhaq
Devesh Upadhyay
Paul Upchurch
Yoshitaka Ushiku
Yuzuko Utsumi
Mikaela Angelina Uy
Mohit Vaishnav
Pratik Vaishnavi
Jeya Maria Jose Valanarasu
Matias A. Valdenegro Toro
Diego Valsesia
Wouter Van Gansbeke
Nanne van Noord

Simon Vandenhende
Farshid Varno
Cristina Vasconcelos
Francisco Vasconcelos
Alex Vasilescu
Subeesh Vasu
Arun Balajee Vasudevan
Kanav Vats
Vaibhav S. Vavilala
Sagar Vaze
Javier Vazquez-Corral
Andrea Vedaldi
Olga Veksler
Andreas Velten
Sai H. Vemprala
Raviteja Vemulapalli
Shashanka
    Venkataramanan
Dor Verbin
Luisa Verdoliva
Manisha Verma
Yashaswi Verma
Constantin Vertan
Eli Verwimp
Deepak Vijaykeerthy
Pablo Villanueva
Ruben Villegas
Markus Vincze
Vibhav Vineet
Minh P. Vo
Huy V. Vo
Duc Minh Vo
Tomas Vojir
Igor Vozniak
Nicholas Vretos
Vibashan VS
Tuan-Anh Vu
Thang Vu
Mårten Wadenbäck
Neal Wadhwa
Aaron T. Walsman
Steven Walton
Jin Wan
Alvin Wan
Jia Wan

Jun Wan
Xiaoyue Wan
Fang Wan
Guowei Wan
Renjie Wan
Zhiqiang Wan
Ziyu Wan
Bastian Wandt
Dongdong Wang
Limin Wang
Haiyang Wang
Xiaobing Wang
Angtian Wang
Angelina Wang
Bing Wang
Bo Wang
Boyu Wang
Binghui Wang
Chen Wang
Chien-Yi Wang
Congli Wang
Qi Wang
Chengrui Wang
Rui Wang
Yiqun Wang
Cong Wang
Wenjing Wang
Dongkai Wang
Di Wang
Xiaogang Wang
Kai Wang
Zhizhong Wang
Fangjinhua Wang
Feng Wang
Hang Wang
Gaoang Wang
Guoqing Wang
Guangcong Wang
Guangzhi Wang
Hanqing Wang
Hao Wang
Haohan Wang
Haoran Wang
Hong Wang
Haotao Wang

Hu Wang
Huan Wang
Hua Wang
Hui-Po Wang
Hengli Wang
Hanyu Wang
Hongxing Wang
Jingwen Wang
Jialiang Wang
Jian Wang
Jianyi Wang
Jiashun Wang
Jiahao Wang
Tsun-Hsuan Wang
Xiaoqian Wang
Jinqiao Wang
Jun Wang
Jianzong Wang
Kaihong Wang
Ke Wang
Lei Wang
Lingjing Wang
Linnan Wang
Lin Wang
Liansheng Wang
Mengjiao Wang
Manning Wang
Nannan Wang
Peihao Wang
Jiayun Wang
Pu Wang
Qiang Wang
Qiufeng Wang
Qilong Wang
Qiangchang Wang
Qin Wang
Qing Wang
Ruocheng Wang
Ruibin Wang
Ruisheng Wang
Ruizhe Wang
Runqi Wang
Runzhong Wang
Wenxuan Wang
Sen Wang

Shangfei Wang
Shaofei Wang
Shijie Wang
Shiqi Wang
Zhibo Wang
Song Wang
Xinjiang Wang
Tai Wang
Tao Wang
Teng Wang
Xiang Wang
Tianren Wang
Tiantian Wang
Tianyi Wang
Fengjiao Wang
Wei Wang
Miaohui Wang
Suchen Wang
Siyue Wang
Yaoming Wang
Xiao Wang
Ze Wang
Biao Wang
Chaofei Wang
Dong Wang
Gu Wang
Guangrun Wang
Guangming Wang
Guo-Hua Wang
Haoqing Wang
Hesheng Wang
Huafeng Wang
Jinghua Wang
Jingdong Wang
Jingjing Wang
Jingya Wang
Jingkang Wang
Jiakai Wang
Junke Wang
Kuo Wang
Lichen Wang
Lizhi Wang
Longguang Wang
Mang Wang
Mei Wang

Min Wang
Peng-Shuai Wang
Run Wang
Shaoru Wang
Shuhui Wang
Tan Wang
Tiancai Wang
Tianqi Wang
Wenhai Wang
Wenzhe Wang
Xiaobo Wang
Xiudong Wang
Xu Wang
Yajie Wang
Yan Wang
Yuan-Gen Wang
Yingqian Wang
Yizhi Wang
Yulin Wang
Yu Wang
Yujie Wang
Yunhe Wang
Yuxi Wang
Yaowei Wang
Yiwei Wang
Zezheng Wang
Hongzhi Wang
Zhiqiang Wang
Ziteng Wang
Ziwei Wang
Zheng Wang
Zhenyu Wang
Binglu Wang
Zhongdao Wang
Ce Wang
Weining Wang
Weiyao Wang
Wenbin Wang
Wenguan Wang
Guangting Wang
Haolin Wang
Haiyan Wang
Huiyu Wang
Naiyan Wang
Jingbo Wang

Jinpeng Wang
Jiaqi Wang
Liyuan Wang
Lizhen Wang
Ning Wang
Wenqian Wang
Sheng-Yu Wang
Weimin Wang
Xiaohan Wang
Yifan Wang
Yi Wang
Yongtao Wang
Yizhou Wang
Zhuo Wang
Zhe Wang
Xudong Wang
Xiaofang Wang
Xinggang Wang
Xiaosen Wang
Xiaosong Wang
Xiaoyang Wang
Lijun Wang
Xinlong Wang
Xuan Wang
Xue Wang
Yangang Wang
Yaohui Wang
Yu-Chiang Frank Wang
Yida Wang
Yilin Wang
Yi Ru Wang
Yali Wang
Yinglong Wang
Yufu Wang
Yujiang Wang
Yuwang Wang
Yuting Wang
Yang Wang
Yu-Xiong Wang
Yixu Wang
Ziqi Wang
Zhicheng Wang
Zeyu Wang
Zhaowen Wang
Zhenyi Wang

Zhenzhi Wang
Zhijie Wang
Zhiyong Wang
Zhongling Wang
Zhuowei Wang
Zian Wang
Zifu Wang
Zihao Wang
Zirui Wang
Ziyan Wang
Wenxiao Wang
Zhen Wang
Zhepeng Wang
Zi Wang
Zihao W. Wang
Steven L. Waslander
Olivia Watkins
Daniel Watson
Silvan Weder
Dongyoon Wee
Dongming Wei
Tianyi Wei
Jia Wei
Dong Wei
Fangyun Wei
Longhui Wei
Mingqiang Wei
Xinyue Wei
Chen Wei
Donglai Wei
Pengxu Wei
Xing Wei
Xiu-Shen Wei
Wenqi Wei
Guoqiang Wei
Wei Wei
XingKui Wei
Xian Wei
Xingxing Wei
Yake Wei
Yuxiang Wei
Yi Wei
Luca Weihs
Michael Weinmann
Martin Weinmann

Congcong Wen
Chuan Wen
Jie Wen
Sijia Wen
Song Wen
Chao Wen
Xiang Wen
Zeyi Wen
Xin Wen
Yilin Wen
Yijia Weng
Shuchen Weng
Junwu Weng
Wenming Weng
Renliang Weng
Zhenyu Weng
Xinshuo Weng
Nicholas J. Westlake
Gordon Wetzstein
Lena M. Widin Klasén
Rick Wildes
Bryan M. Williams
Williem Williem
Ole Winther
Scott Wisdom
Alex Wong
Chau-Wai Wong
Kwan-Yee K. Wong
Yongkang Wong
Scott Workman
Marcel Worring
Michael Wray
Safwan Wshah
Xiang Wu
Aming Wu
Chongruo Wu
Cho-Ying Wu
Chunpeng Wu
Chenyan Wu
Ziyi Wu
Fuxiang Wu
Gang Wu
Haiping Wu
Huisi Wu
Jane Wu

Jialian Wu
Jing Wu
Jinjian Wu
Jianlong Wu
Xian Wu
Lifang Wu
Lifan Wu
Minye Wu
Qianyi Wu
Rongliang Wu
Rui Wu
Shiqian Wu
Shuzhe Wu
Shangzhe Wu
Tsung-Han Wu
Tz-Ying Wu
Ting-Wei Wu
Jiannan Wu
Zhiliang Wu
Yu Wu
Chenyun Wu
Dayan Wu
Dongxian Wu
Fei Wu
Hefeng Wu
Jianxin Wu
Weibin Wu
Wenxuan Wu
Wenhao Wu
Xiao Wu
Yicheng Wu
Yuanwei Wu
Yu-Huan Wu
Zhenxin Wu
Zhenyu Wu
Wei Wu
Peng Wu
Xiaohe Wu
Xindi Wu
Xinxing Wu
Xinyi Wu
Xingjiao Wu
Xiongwei Wu
Yangzheng Wu
Yanzhao Wu

Yawen Wu
Yong Wu
Yi Wu
Ying Nian Wu
Zhenyao Wu
Zhonghua Wu
Zongze Wu
Zuxuan Wu
Stefanie Wuhrer
Teng Xi
Jianing Xi
Fei Xia
Haifeng Xia
Menghan Xia
Yuanqing Xia
Zhihua Xia
Xiaobo Xia
Weihao Xia
Shihong Xia
Yan Xia
Yong Xia
Zhaoyang Xia
Zhihao Xia
Chuhua Xian
Yongqin Xian
Wangmeng Xiang
Fanbo Xiang
Tiange Xiang
Tao Xiang
Liuyu Xiang
Xiaoyu Xiang
Zhiyu Xiang
Aoran Xiao
Chunxia Xiao
Fanyi Xiao
Jimin Xiao
Jun Xiao
Taihong Xiao
Anqi Xiao
Junfei Xiao
Jing Xiao
Liang Xiao
Yang Xiao
Yuting Xiao
Yijun Xiao

Yao Xiao
Zeyu Xiao
Zhisheng Xiao
Zihao Xiao
Binhui Xie
Christopher Xie
Haozhe Xie
Jin Xie
Guo-Sen Xie
Hongtao Xie
Ming-Kun Xie
Tingting Xie
Chaohao Xie
Weicheng Xie
Xudong Xie
Jiyang Xie
Xiaohua Xie
Yuan Xie
Zhenyu Xie
Ning Xie
Xianghui Xie
Xiufeng Xie
You Xie
Yutong Xie
Fuyong Xing
Yifan Xing
Zhen Xing
Yuanjun Xiong
Jinhui Xiong
Weihua Xiong
Hongkai Xiong
Zhitong Xiong
Yuanhao Xiong
Yunyang Xiong
Yuwen Xiong
Zhiwei Xiong
Yuliang Xiu
An Xu
Chang Xu
Chenliang Xu
Chengming Xu
Chenshu Xu
Xiang Xu
Huijuan Xu
Zhe Xu

Jie Xu
Jingyi Xu
Jiarui Xu
Yinghao Xu
Kele Xu
Ke Xu
Li Xu
Linchuan Xu
Linning Xu
Mengde Xu
Mengmeng Frost Xu
Min Xu
Mingye Xu
Jun Xu
Ning Xu
Peng Xu
Runsheng Xu
Sheng Xu
Wenqiang Xu
Xiaogang Xu
Renzhe Xu
Kaidi Xu
Yi Xu
Chi Xu
Qiuling Xu
Baobei Xu
Feng Xu
Haohang Xu
Haofei Xu
Lan Xu
Mingze Xu
Songcen Xu
Weipeng Xu
Wenjia Xu
Wenju Xu
Xiangyu Xu
Xin Xu
Yinshuang Xu
Yixing Xu
Yuting Xu
Yanyu Xu
Zhenbo Xu
Zhiliang Xu
Zhiyuan Xu
Xiaohao Xu

Yanwu Xu
Yan Xu
Yiran Xu
Yifan Xu
Yufei Xu
Yong Xu
Zichuan Xu
Zenglin Xu
Zexiang Xu
Zhan Xu
Zheng Xu
Zhiwei Xu
Ziyue Xu
Shiyu Xuan
Hanyu Xuan
Fei Xue
Jianru Xue
Mingfu Xue
Qinghan Xue
Tianfan Xue
Chao Xue
Chuhui Xue
Nan Xue
Zhou Xue
Xiangyang Xue
Yuan Xue
Abhay Yadav
Ravindra Yadav
Kota Yamaguchi
Toshihiko Yamasaki
Kohei Yamashita
Chaochao Yan
Feng Yan
Kun Yan
Qingsen Yan
Qixin Yan
Rui Yan
Siming Yan
Xinchen Yan
Yaping Yan
Bin Yan
Qingan Yan
Shen Yan
Shipeng Yan
Xu Yan

Yan Yan
Yichao Yan
Zhaoyi Yan
Zike Yan
Zhiqiang Yan
Hongliang Yan
Zizheng Yan
Jiewen Yang
Anqi Joyce Yang
Shan Yang
Anqi Yang
Antoine Yang
Bo Yang
Baoyao Yang
Chenhongyi Yang
Dingkang Yang
De-Nian Yang
Dong Yang
David Yang
Fan Yang
Fengyu Yang
Fengting Yang
Fei Yang
Gengshan Yang
Heng Yang
Han Yang
Huan Yang
Yibo Yang
Jiancheng Yang
Jihan Yang
Jiawei Yang
Jiayu Yang
Jie Yang
Jinfa Yang
Jingkang Yang
Jinyu Yang
Cheng-Fu Yang
Ji Yang
Jianyu Yang
Kailun Yang
Tian Yang
Luyu Yang
Liang Yang
Li Yang
Michael Ying Yang

Yang Yang
Muli Yang
Le Yang
Qiushi Yang
Ren Yang
Ruihan Yang
Shuang Yang
Siyuan Yang
Su Yang
Shiqi Yang
Taojiannan Yang
Tianyu Yang
Lei Yang
Wanzhao Yang
Shuai Yang
William Yang
Wei Yang
Xiaofeng Yang
Xiaoshan Yang
Xin Yang
Xuan Yang
Xu Yang
Xingyi Yang
Xitong Yang
Jing Yang
Yanchao Yang
Wenming Yang
Yujiu Yang
Herb Yang
Jianfei Yang
Jinhui Yang
Chuanguang Yang
Guanglei Yang
Haitao Yang
Kewei Yang
Linlin Yang
Lijin Yang
Longrong Yang
Meng Yang
MingKun Yang
Sibei Yang
Shicai Yang
Tong Yang
Wen Yang
Xi Yang

Xiaolong Yang
Xue Yang
Yubin Yang
Ze Yang
Ziyi Yang
Yi Yang
Linjie Yang
Yuzhe Yang
Yiding Yang
Zhenpei Yang
Zhaohui Yang
Zhengyuan Yang
Zhibo Yang
Zongxin Yang
Hantao Yao
Mingde Yao
Rui Yao
Taiping Yao
Ting Yao
Cong Yao
Qingsong Yao
Quanming Yao
Xu Yao
Yuan Yao
Yao Yao
Yazhou Yao
Jiawen Yao
Shunyu Yao
Pew-Thian Yap
Sudhir Yarram
Rajeev Yasarla
Peng Ye
Botao Ye
Mao Ye
Fei Ye
Hanrong Ye
Jingwen Ye
Jinwei Ye
Jiarong Ye
Mang Ye
Meng Ye
Qi Ye
Qian Ye
Qixiang Ye
Junjie Ye

Sheng Ye
Nanyang Ye
Yufei Ye
Xiaoqing Ye
Ruolin Ye
Yousef Yeganeh
Chun-Hsiao Yeh
Raymond A. Yeh
Yu-Ying Yeh
Kai Yi
Chang Yi
Renjiao Yi
Xinping Yi
Peng Yi
Alper Yilmaz
Junho Yim
Hui Yin
Bangjie Yin
Jia-Li Yin
Miao Yin
Wenzhe Yin
Xuwang Yin
Ming Yin
Yu Yin
Aoxiong Yin
Kangxue Yin
Tianwei Yin
Wei Yin
Xianghua Ying
Rio Yokota
Tatsuya Yokota
Naoto Yokoya
Ryo Yonetani
Ki Yoon Yoo
Jinsu Yoo
Sunjae Yoon
Jae Shin Yoon
Jihun Yoon
Sung-Hoon Yoon
Ryota Yoshihashi
Yusuke Yoshiyasu
Chenyu You
Haoran You
Haoxuan You
Yang You

Quanzeng You
Tackgeun You
Kaichao You
Shan You
Xinge You
Yurong You
Baosheng Yu
Bei Yu
Haichao Yu
Hao Yu
Chaohui Yu
Fisher Yu
Jin-Gang Yu
Jiyang Yu
Jason J. Yu
Jiashuo Yu
Hong-Xing Yu
Lei Yu
Mulin Yu
Ning Yu
Peilin Yu
Qi Yu
Qian Yu
Rui Yu
Shuzhi Yu
Gang Yu
Tan Yu
Weijiang Yu
Xin Yu
Bingyao Yu
Ye Yu
Hanchao Yu
Yingchen Yu
Tao Yu
Xiaotian Yu
Qing Yu
Houjian Yu
Changqian Yu
Jing Yu
Jun Yu
Shujian Yu
Xiang Yu
Zhaofei Yu
Zhenbo Yu
Yinfeng Yu

Zhuoran Yu
Zitong Yu
Bo Yuan
Jiangbo Yuan
Liangzhe Yuan
Weihao Yuan
Jianbo Yuan
Xiaoyun Yuan
Ye Yuan
Li Yuan
Geng Yuan
Jialin Yuan
Maoxun Yuan
Peng Yuan
Xin Yuan
Yuan Yuan
Yuhui Yuan
Yixuan Yuan
Zheng Yuan
Mehmet Kerim Yücel
Kaiyu Yue
Haixiao Yue
Heeseung Yun
Sangdoo Yun
Tian Yun
Mahmut Yurt
Ekim Yurtsever
Ahmet Yüzügüler
Edouard Yvinec
Eloi Zablocki
Christopher Zach
Muhammad Zaigham
  Zaheer
Pierluigi Zama Ramirez
Yuhang Zang
Pietro Zanuttigh
Alexey Zaytsev
Bernhard Zeisl
Haitian Zeng
Pengpeng Zeng
Jiabei Zeng
Runhao Zeng
Wei Zeng
Yawen Zeng
Yi Zeng

Yiming Zeng
Tieyong Zeng
Huanqiang Zeng
Dan Zeng
Yu Zeng
Wei Zhai
Yuanhao Zhai
Fangneng Zhan
Kun Zhan
Xiong Zhang
Jingdong Zhang
Jiangning Zhang
Zhilu Zhang
Gengwei Zhang
Dongsu Zhang
Hui Zhang
Binjie Zhang
Bo Zhang
Tianhao Zhang
Cecilia Zhang
Jing Zhang
Chaoning Zhang
Chenxu Zhang
Chi Zhang
Chris Zhang
Yabin Zhang
Zhao Zhang
Rufeng Zhang
Chaoyi Zhang
Zheng Zhang
Da Zhang
Yi Zhang
Edward Zhang
Xin Zhang
Feifei Zhang
Feilong Zhang
Yuqi Zhang
GuiXuan Zhang
Hanlin Zhang
Hanwang Zhang
Hanzhen Zhang
Haotian Zhang
He Zhang
Haokui Zhang
Hongyuan Zhang

Hengrui Zhang
Hongming Zhang
Mingfang Zhang
Jianpeng Zhang
Jiaming Zhang
Jichao Zhang
Jie Zhang
Jingfeng Zhang
Jingyi Zhang
Jinnian Zhang
David Junhao Zhang
Junjie Zhang
Junzhe Zhang
Jiawan Zhang
Jingyang Zhang
Kai Zhang
Lei Zhang
Lihua Zhang
Lu Zhang
Miao Zhang
Minjia Zhang
Mingjin Zhang
Qi Zhang
Qian Zhang
Qilong Zhang
Qiming Zhang
Qiang Zhang
Richard Zhang
Ruimao Zhang
Ruisi Zhang
Ruixin Zhang
Runze Zhang
Qilin Zhang
Shan Zhang
Shanshan Zhang
Xi Sheryl Zhang
Song-Hai Zhang
Chongyang Zhang
Kaihao Zhang
Songyang Zhang
Shu Zhang
Siwei Zhang
Shujian Zhang
Tianyun Zhang
Tong Zhang

Tao Zhang
Wenwei Zhang
Wenqiang Zhang
Wen Zhang
Xiaolin Zhang
Xingchen Zhang
Xingxuan Zhang
Xiuming Zhang
Xiaoshuai Zhang
Xuanmeng Zhang
Xuanyang Zhang
Xucong Zhang
Xingxing Zhang
Xikun Zhang
Xiaohan Zhang
Yahui Zhang
Yunhua Zhang
Yan Zhang
Yanghao Zhang
Yifei Zhang
Yifan Zhang
Yi-Fan Zhang
Yihao Zhang
Yingliang Zhang
Youshan Zhang
Yulun Zhang
Yushu Zhang
Yixiao Zhang
Yide Zhang
Zhongwen Zhang
Bowen Zhang
Chen-Lin Zhang
Zehua Zhang
Zekun Zhang
Zeyu Zhang
Xiaowei Zhang
Yifeng Zhang
Cheng Zhang
Hongguang Zhang
Yuexi Zhang
Fa Zhang
Guofeng Zhang
Hao Zhang
Haofeng Zhang
Hongwen Zhang

Hua Zhang
Jiaxin Zhang
Zhenyu Zhang
Jian Zhang
Jianfeng Zhang
Jiao Zhang
Jiakai Zhang
Lefei Zhang
Le Zhang
Mi Zhang
Min Zhang
Ning Zhang
Pan Zhang
Pu Zhang
Qing Zhang
Renrui Zhang
Shifeng Zhang
Shuo Zhang
Shaoxiong Zhang
Weizhong Zhang
Xi Zhang
Xiaomei Zhang
Xinyu Zhang
Yin Zhang
Zicheng Zhang
Zihao Zhang
Ziqi Zhang
Zhaoxiang Zhang
Zhen Zhang
Zhipeng Zhang
Zhixing Zhang
Zhizheng Zhang
Jiawei Zhang
Zhong Zhang
Pingping Zhang
Yixin Zhang
Kui Zhang
Lingzhi Zhang
Huaiwen Zhang
Quanshi Zhang
Zhoutong Zhang
Yuhang Zhang
Yuting Zhang
Zhang Zhang
Ziming Zhang

Zhizhong Zhang
Qilong Zhangli
Bingyin Zhao
Bin Zhao
Chenglong Zhao
Lei Zhao
Feng Zhao
Gangming Zhao
Haiyan Zhao
Hao Zhao
Handong Zhao
Hengshuang Zhao
Yinan Zhao
Jiaojiao Zhao
Jiaqi Zhao
Jing Zhao
Kaili Zhao
Haojie Zhao
Yucheng Zhao
Longjiao Zhao
Long Zhao
Qingsong Zhao
Qingyu Zhao
Rui Zhao
Rui-Wei Zhao
Sicheng Zhao
Shuang Zhao
Siyan Zhao
Zelin Zhao
Shiyu Zhao
Wang Zhao
Tiesong Zhao
Qian Zhao
Wangbo Zhao
Xi-Le Zhao
Xu Zhao
Yajie Zhao
Yang Zhao
Ying Zhao
Yin Zhao
Yizhou Zhao
Yunhan Zhao
Yuyang Zhao
Yue Zhao
Yuzhi Zhao

Bowen Zhao
Pu Zhao
Bingchen Zhao
Borui Zhao
Fuqiang Zhao
Hanbin Zhao
Jian Zhao
Mingyang Zhao
Na Zhao
Rongchang Zhao
Ruiqi Zhao
Shuai Zhao
Wenda Zhao
Wenliang Zhao
Xiangyun Zhao
Yifan Zhao
Yaping Zhao
Zhou Zhao
He Zhao
Jie Zhao
Xibin Zhao
Xiaoqi Zhao
Zhengyu Zhao
Jin Zhe
Chuanxia Zheng
Huan Zheng
Hao Zheng
Jia Zheng
Jian-Qing Zheng
Shuai Zheng
Meng Zheng
Mingkai Zheng
Qian Zheng
Qi Zheng
Wu Zheng
Yinqiang Zheng
Yufeng Zheng
Yutong Zheng
Yalin Zheng
Yu Zheng
Feng Zheng
Zhaoheng Zheng
Haitian Zheng
Kang Zheng
Bolun Zheng

Haiyong Zheng
Mingwu Zheng
Sipeng Zheng
Tu Zheng
Wenzhao Zheng
Xiawu Zheng
Yinglin Zheng
Zhuo Zheng
Zilong Zheng
Kecheng Zheng
Zerong Zheng
Shuaifeng Zhi
Tiancheng Zhi
Jia-Xing Zhong
Yiwu Zhong
Fangwei Zhong
Zhihang Zhong
Yaoyao Zhong
Yiran Zhong
Zhun Zhong
Zichun Zhong
Bo Zhou
Boyao Zhou
Brady Zhou
Mo Zhou
Chunluan Zhou
Dingfu Zhou
Fan Zhou
Jingkai Zhou
Honglu Zhou
Jiaming Zhou
Jiahuan Zhou
Jun Zhou
Kaiyang Zhou
Keyang Zhou
Kuangqi Zhou
Lei Zhou
Lihua Zhou
Man Zhou
Mingyi Zhou
Mingyuan Zhou
Ning Zhou
Peng Zhou
Penghao Zhou
Qianyi Zhou

Shuigeng Zhou
Shangchen Zhou
Huayi Zhou
Zhize Zhou
Sanping Zhou
Qin Zhou
Tao Zhou
Wenbo Zhou
Xiangdong Zhou
Xiao-Yun Zhou
Xiao Zhou
Yang Zhou
Yipin Zhou
Zhenyu Zhou
Hao Zhou
Chu Zhou
Daquan Zhou
Da-Wei Zhou
Hang Zhou
Kang Zhou
Qianyu Zhou
Sheng Zhou
Wenhui Zhou
Xingyi Zhou
Yan-Jie Zhou
Yiyi Zhou
Yu Zhou
Yuan Zhou
Yuqian Zhou
Yuxuan Zhou
Zixiang Zhou
Wengang Zhou
Shuchang Zhou
Tianfei Zhou
Yichao Zhou
Alex Zhu
Chenchen Zhu
Deyao Zhu
Xiatian Zhu
Guibo Zhu
Haidong Zhu
Hao Zhu
Hongzi Zhu
Rui Zhu
Jing Zhu

Jianke Zhu
Junchen Zhu
Lei Zhu
Lingyu Zhu
Luyang Zhu
Menglong Zhu
Peihao Zhu
Hui Zhu
Xiaofeng Zhu
Tyler (Lixuan) Zhu
Wentao Zhu
Xiangyu Zhu
Xinqi Zhu
Xinxin Zhu
Xinliang Zhu
Yangguang Zhu
Yichen Zhu
Yixin Zhu
Yanjun Zhu
Yousong Zhu
Yuhao Zhu
Ye Zhu
Feng Zhu
Zhen Zhu
Fangrui Zhu
Jinjing Zhu
Linchao Zhu
Pengfei Zhu
Sijie Zhu
Xiaobin Zhu
Xiaoguang Zhu
Zezhou Zhu
Zhenyao Zhu
Kai Zhu
Pengkai Zhu
Bingbing Zhuang
Chengyuan Zhuang
Liansheng Zhuang
Peiye Zhuang
Yixin Zhuang
Yihong Zhuang
Junbao Zhuo
Andrea Ziani
Bartosz Zieliński
Primo Zingaretti

Nikolaos Zioulis
Andrew Zisserman
Yael Ziv
Liu Ziyin
Xingxing Zou
Danping Zou
Qi Zou

Shihao Zou
Xueyan Zou
Yang Zou
Yuliang Zou
Zihang Zou
Chuhang Zou
Dongqing Zou

Xu Zou
Zhiming Zou
Maria A. Zuluaga
Xinxin Zuo
Zhiwen Zuo
Reyer Zwiggelaar

Contents – Part XIV

Watermark Vaccine: Adversarial Attacks to Prevent Watermark Removal

Xinwei Liu[1,2], Jian Liu[3], Yang Bai[4], Jindong Gu[5], Tao Chen[3],
Xiaojun Jia[1,2(✉)], and Xiaochun Cao[1,6]

[1] SKLOIS, Institute of Information Engineering, CAS, Beijing, China
{liuxinwei,jiaxiaojun}@iie.ac.cn
[2] School of Cyber Security, University of Chinese Academy of Sciences,
Beijing, China
[3] Ant Group, Beijing, China
{rex.lj,boshan.ct}@antgroup.com
[4] Tencent Security Zhuque Lab, Beijing, China
mavisbai@tencent.com
[5] University of Munich, Munich, Germany
[6] School of Cyber Science and Technology, Shenzhen Campus,
Sun Yat-sen University, Shenzhen 518107, China
caoxiaochun@mail.sysu.edu.cn

Abstract. As a common security tool, visible watermarking has been widely applied to protect copyrights of digital images. However, recent works have shown that visible watermarks can be removed by DNNs without damaging their host images. Such watermark-removal techniques pose a great threat to the ownership of images. Inspired by the vulnerability of DNNs on adversarial perturbations, we propose a novel defence mechanism by adversarial machine learning for good. From the perspective of the adversary, blind watermark-removal networks can be posed as our target models; then we actually optimize an imperceptible adversarial perturbation on the host images to proactively attack against watermark-removal networks, dubbed *Watermark Vaccine*. Specifically, two types of vaccines are proposed. Disrupting Watermark Vaccine (DWV) induces to ruin the host image along with watermark after passing through watermark-removal networks. In contrast, Inerasable Watermark Vaccine (IWV) works in another fashion of trying to keep the watermark not removed and still noticeable. Extensive experiments demonstrate the effectiveness of our DWV/IWV in preventing watermark removal, especially on various watermark removal networks. The Code is released in https://github.com/thinwayliu/Watermark-Vaccine.

Keywords: Visible watermark removal · Watermark protection · Adversarial attack

Supplementary Information The online version contains supplementary material available at https://doi.org/10.1007/978-3-031-19781-9_1.

1   Introduction

With the rapid development of digital media and the increasing dependence of deep neural networks (DNNs) on enormous training data, copyright protection attracts great attention especially for image data [45]. Visible watermarking thus becomes an essential technique [3]. It prevents illicit users from obtaining some critical information and using copyrighted high-quality images. As a result, it can reduce illegal theft and play a role in publicity and warning. Mintzer *et al.* [35] posed two characteristics for visible watermark, that is, it can be recognized by human eyes but will not significantly obscure the details.

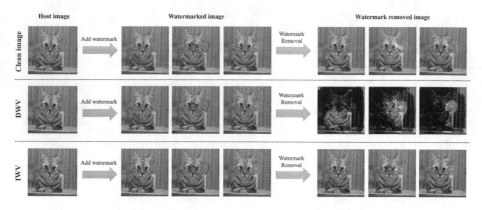

Fig. 1. The protective effects of our watermark vaccines on different watermark patterns or parameters. The current blind watermark-removal technique, such as WDNet [32], can effectively remove the watermarks (**top**). When the host images are equipped with Disrupting Watermark Vaccine (DWV), the watermark-removed images will be ruined (**middle**). However, when the host images are equipped with Inerasable Watermark Vaccine (IWV), the results can not be purified successfully as the host images (**bottom**).

However, visible watermark is in face of security issues as it can be effectively removed by some watermark-removal techniques [2,10,42,46,53]. Among these techniques, some require the location area of the watermark. Huang *et al.* [21] propose to remove the watermark using image inpainting. Park *et al.* [40] propose to formulate it into a feature matching problem. With the rapid development of deep learning, the community has proposed some blind visible watermark-removal DNNs, which can reconstruct watermarked images end-to-end without any information about watermarks [4,8,11,20,29,30,32]. These works usually adopt two-stage strategies. In the first stage, the networks aim to predict the watermark region. After that, in the second stage, they work on recovering the background of such a watermark region. Without strong assumptions, the blind watermark-removal network has become a mainstream method. In Fig. 1(a), we take WDNet [32] as an example to show its performance in both identifying and removing the watermark.

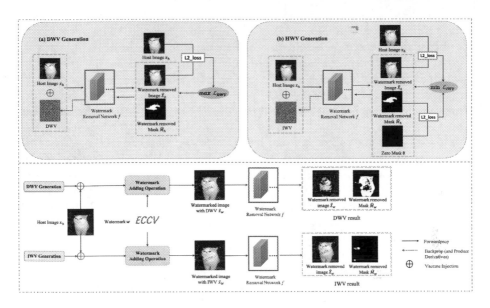

Fig. 2. The overview figure of the generation (on the first row) and application (on the second row) of our proposed watermark vaccine. We propose to maximize $\mathcal{L}_{DWV}$ to generate DWV and minimize $\mathcal{L}_{IWV}$ to generate IWV, as shown in (a) and (b), respectively. Then we first apply DWV/IWV on the host images to generate 'protected host images'. When watermarks are added on those protected host images, they are hard to be normally removed, tending to show either ruined or watermark-preserving results.

Due to these advanced watermark-removal technologies, traditional watermarking methods can no longer effectively protect the copyrights of picture owners. Recently, Khachaturov *et al.* [25] proposed to fool the inpainting-based removal networks to protect the watermark. However, this type of networks are demanding and not widely used. So we focus on preventing the blind watermark removal networks in this paper. Inspired by recent studies on adversary [1,5,17,22,49,56], which show that imperceptible adversarial perturbations can cause some incorrectly outputs for DNNs, 'adversarial for good' is thus a new protection method. To note that, generating a simple adversarial perturbation on the watermarked image could not work directly. Because in real-world scenarios, a watermark is always automatically generated at the last step by the system or website, and the protected image is required to finish uploading before it. More importantly, the watermark is not permanent for one host image. The watermark could be changed with some specific circumstances (such as Enterprise renaming, logo changes, year updates, etc.). So it can be costly to regenerate an adversarial perturbation for the same host image yet with a different watermark. Thus, a universal perturbation can be useful and efficient for watermark protection.

In this paper, we propose a watermark-agnostic perturbation against blind watermark-removal network, dubbed **Watermark Vaccine**, which is injected on host images before adding watermark just like vaccination in reality. Our method is equivalent to a 'double insurance' for copyright protection: the visible watermark serves as a warning and annotating function, telling people not to infringe. The watermark vaccine ensures that the visible watermark won't be removed by blind watermark-removal networks, which can effectively reduce illegal dissemination and other infringements. Then the vaccinated image can protect any watermark from being removed. Specifically, we propose two types of watermark vaccines according to their attack effects: Disrupting Watermark Vaccine (DWV) and Inerasable Watermark Vaccine (IWV). DWV aims to disrupt the watermark-removed image while IWV attempts to still keep the watermark through blind watermark-removal networks. The framework of our proposed watermark vaccine's generation and is shown in Fig. 2. In Fig. 1(b) and (c), we can see that after injected with DWV or IWV, the watermark-removed images will either be ruined or the watermarks are not completely removed. In addition, the masks are disrupted for DWV or are induced elsewhere for IWV. Both results demonstrate the effectiveness of our proposed watermark vaccine in successfully preventing watermark removal.

Our key contributions are summarized as follows:

- We are the first to propose the watermark-agnostic perturbations for blind watermark-removal networks, dubbed *Watermark Vaccine*, to prevent the watermark removal from host images.
- We present two types of effective and powerful watermark vaccines (DWV and IWV), which aim to either disrupt the watermark-removed images or keep the watermarks uncleared respectively.
- We evaluate the effectiveness and universality of two vaccines. The results demonstrate that they generalize well on different watermark patterns, sizes, locations as well as transparencies. In addition, our watermark vaccine can also resist some common image processing operations.

2   Related Work

Visible Watermark Removal. Visible watermark-removal techniques are proposed to evaluate and improve the robustness of visible watermarks at the beginning. In the earlier works [21,40,41], the user's interaction is always required to remove watermark. Namely, they require the location of the watermark and recover that area. However, it can be not practical when processing massive images without location information. In [12] and [15], they assume that the same watermark is added to all host images. Nevertheless, this assumption is also too strong to apply in real scenarios.

With the development of deep learning, neural networks show a great power in computer vision tasks [16,18,27,57]. Several works try to apply neural networks to formulate an end-to-end problem, and there are two popular ways applied to solve it. One way is to directly formulate the watermark removal as an image-to-image translation task [4,29]; the other way is adopting a two-stage strategy to formulate the problem: the first step is to locate the by a mask, and the second step is to recover the background in the watermark area and train a network to solve both at the same time [11,20,30,32]. The latter method was found in experiments can be more effective in watermark removal, so we mainly focus on preventing the second type of networks in this paper.

Adversarial Attacks on Generative Models. Szegedy *et al.* [49] first found and proposed the adversarial examples. In [17], Goodfellow *et al.* proposed the Fast Gradient Sign Method (FGSM), which is a one-step gradient attack. After that, some stronger generation methods are proposed like I-FGSM [28], M-FGSM [13], and Projected Gradient Descent (PGD) [34]. In the black-box setting, adversarial examples can transfer across the different models [14,31,38,39,52] or can be generated by approximating gradients [6,51]. However, most of attack and defense works are about the classification problem [23,24].

Recent works have focused more on attacks on generative models. In [26] and [50], the authors firstly explore adversarial attacks against Variational Autoencoders (VAE) and VAE-GANS. Ruiz *et al.* [43,44] apply transferable adversarial attacks to disrupt facial manipulation systems. From then on some other works [7,47,54,55] adopt adversarial machine learning in translation-based deepfake models. These works show that adversarial attacks can defend against some malicious generative networks to protect users' privacy. Khachaturov *et al.* [25] find that adversarial perturbations can fool an inpainting system into generating a patch similar to random pictures. Although this work can be applied to protect watermarks, the inpainting-based watermarking method needs obtaining a mask in image, and is not practical for big watermarks. Therefore, attacking the inpainting-based watermarking method [25] cannot protect the watermark in natural scenes.

3 Methodology

3.1 Preliminary

The watermark vaccine generation can be formalized as an optimization problem with constraints. We assume a host image as x_h, and the invisible watermark vaccine δ is injected onto it before adding watermark, which is restricted in L_∞ norm bound ϵ. Thus, we get the vaccinated image $\hat{x}_h$ by

$$\hat{x}_h = x_h + \delta$$
$$\|\delta\|_\infty \le \varepsilon. \tag{1}$$

Then we select a watermark sample w from the watermark set W, where $w \in W$. The adding watermark operation can be assumed as g. And g requires some parameters θ to identify the location (p, q), the size (u, v) and the transparency α of the watermark w injected on the host image x_h. So a watermarked image $\hat{x}_w$ with vaccine can be defined as follows,

$$\hat{x}_w = g\left(\hat{x}_h, \omega, \theta\right), \tag{2}$$

where $\theta = (p, q, u, v, \alpha)$. In practice, the parameters p, q, n, m and α are always random. Similarly, the watermarked image without vaccine x_w can also be obtained in the same way. Then we assume the blind watermark-removal network as f, and we can get the watermark removed images and masks of x_w and $\hat{x}_w$ respectively through the network, which is defined as

$$\begin{aligned} X_w, M_w &= f\left(x_w\right), \\ \hat{X}_w, \hat{M}_w &= f\left(\hat{x}_w\right). \end{aligned} \tag{3}$$

Here, we denote the measurement of watermark removal effect as $Q(\cdot)$, and the goal of the vaccine is to degrade the removal effect of watermark removed images by minimizing $Q(f(g(x_h + \delta, w, \theta)))$. In addition, we desire our watermark vaccine to be universal for different watermark patterns, positions, sizes, transparencies, thus the expected $Q(f(g(x_h + \delta, w, \theta)))$ over different watermark w and adding parameters θ is required to make vaccine watermark-agnostic. Our watermark vaccine generation can be formulated as follows,

$$\min_{\delta} \quad \mathbb{E}_{w \sim W} \mathbb{E}_{\theta \sim \Theta} \left[Q(f(g(x_h + \delta, w, \theta)))\right]. \tag{4}$$

Unfortunately, there are two challenges in solving the above optimization problem. First, the effect of watermark removal $Q(\cdot)$ can be customized in a variety of different ways, but it is required to be differentiable during optimization. In addition, the two expectation over W and Θ is hard to optimize by considering the loss of all combinations simultaneously. Although we can refer to 'universal adversarial perturbation' in [19, 36, 37, 48], it is still time-consuming and difficult to obtain the optimal vaccine. To address these issues, we further propose two types of watermark vaccine in the following.

3.2 Disrupting Watermark Vaccine (DWV)

One way to protect the watermark is to disrupt the watermark-removed images, which means that as long as the watermarked image with the watermark vaccine passes through the watermark-removal networks, the output image will be ruined and could never be used. Thus, we call this vaccine as **Disrupting Watermark Vaccine (DWV)**. Next, in order to avoid the two expectations in Eq. 4, we decide to generate the watermark-agnostic vaccine on the host images instead of watermarked images. Therefore, we inject vaccine δ on the clean host image x_h and get the vaccinated image $\hat{x}_h$. After passing through the network f, we get the watermark removed image and watermark removed mask, which is denoted as $\hat{X}_h$ and $\hat{M}_h$, because they are generated on clean host images without a watermark. Such operation can be formulated as,

$$\hat{X}_h, \hat{M}_h = f(\hat{x}_h), \tag{5}$$

Then, we define the $\mathcal{L}_{\mathcal{DWV}}$ to measure the distance between the watermark removed image $\hat{X}_h$ and the clean host image x_h,

$$\mathcal{L}_{\mathcal{DWV}}(x_h, \delta) = \left\| \hat{X}_h - x_h \right\|^2, \tag{6}$$

where the image distance adopt the mean-square error to measure.

The objective here is to maximize $\mathcal{L}_{\mathcal{DWV}}$ such that the watermark removed image is significantly different from the host image. Thus, the watermark removed image of the host image is severely ruined by watermark-removal networks. Naturally, whatever watermark is added onto the image, the benign areas (the areas without the watermark) will be destroyed as well. As a result, the watermark removed images of watermarked images sufficiently deteriorate such that it has to be discarded or such that the modification is perceptually evident The problem can be formally expressed as follows,

$$\begin{aligned} \max_\delta \quad & \mathcal{L}_{\mathcal{DWV}}(x_h, \delta) \\ \text{s.t.} \quad & \hat{x}_h = x_h + \delta \\ & \|\delta\|_\infty \leqslant \varepsilon, \end{aligned} \tag{7}$$

To be consistent with the previous requirements, we restrict the perturbations δ in L_∞ norm bound ϵ. We use projected gradient descent (PGD) [34] to solve the optimization problem in Eq. 8, and we can get the optimal δ according to the following iterative formula,

$$\delta^{t+1} = \mathrm{Proj}\left(\delta^t + \alpha\,\mathrm{sign}\left(\nabla_{\delta^t}\,\mathcal{L}_{\mathcal{DWV}}(x_h, \delta^t)\right)\right), \tag{8}$$

where $\nabla_{\delta^t}\,\mathcal{L}_{\mathcal{DWV}}(x_h, \delta^t)$ is the gradient of the disrupting loss w.r.t δ^t. α is the step size, $\mathrm{Proj}()$ denotes project the δ^t within the norm bound $(-\epsilon, \epsilon)$ and project the $x + \delta^t$ within the valid space $(0, 1)$. In Fig. 2(a), we can see the framework of DWV generation, and at the bottom shows the inference of DWV.

3.3   Inerasable Watermark Vaccine (IWV)

Contrasted with DWV to protect the watermark by ruining the watermark removed images, another solution is to prevent the watermark from being identified and removed. To this end, as an alternative to DWV, we propose another vaccine in this section, **Inerasable Watermark Vaccine (IWV)**. It aims to make the watermarks hard to be detected and removed. As a result, the watermark patterns can not be erased completely on the watermark removed images. Inspired by the Eq. (6), we design the $\mathcal{L}_{\mathcal{IWV}}$ as follow:

$$\mathcal{L}_{\mathcal{IWV}}(x_h, \delta) = \frac{1}{2}\left(\beta \left\| \hat{X}_h - x_h \right\|^2 + \|\hat{M}_h - \mathbf{0}\|^2\right), \tag{9}$$

where $\hat{X}_h$ and $\hat{M}_h$ is the output of the blind watermark-removal network f, x_h is the host image, $\mathbf{0}$ is a zero matrix, which is the same size as the predicted mask. There are two distance terms in the loss $\mathcal{L}_{\mathcal{IWV}}$: image term and mask

---

Algorithm 1: Watermark Vaccine Generation

Input: host image x_h, blind watermark-removal network f, iteration T, step
    size α, perturbation bound ϵ
Output: Host image with watermark vaccine $\hat{x}_h$

1 $\delta \leftarrow 0, \hat{x}_h \leftarrow x_h + \delta$;
2 **for** $i = 1$ *to* T **do**
3   **if** *vaccine is 'DWV'* **then**
4   using Equation (6) to calculate the $\mathcal{L}_{\mathcal{DWV}}$;
5        $\delta \leftarrow \delta + \alpha \, \text{sign}(\nabla_\delta \mathcal{L}_{\mathcal{DWV}}(x_h, \delta))$;
6   **else**
7   using Equation (9) to calculate the $\mathcal{L}_{\mathcal{IWV}}$;
8        $\delta \leftarrow \delta - \alpha \, \text{sign}(\nabla_\delta \mathcal{L}_{\mathcal{IWV}}(x_h, \delta))$;
9   **end**
10    $\hat{x}_h \leftarrow x_h + \text{clip}(\delta, -\epsilon, \epsilon)$;
11 **end**
12 $\hat{x}_h \leftarrow \text{clip}(\hat{x}_h, 0, 1)$;

---

term. The image term is equal to the Eq. (6), and the mask term measures the distance between the predicted mask $\hat{M}_h$ and a zero matrix $\mathbf{0}$. The β is the hyperparameter to balance two loss terms.

Ideally, the predicted image $\hat{X}_h$ should be almost the same as the input image x_h, and the predicted mask $\hat{M}_h$ should be almost black, which means there is no watermark can be detected on $\hat{x}_h$ and the $\mathcal{L}_{\mathcal{IWV}}$ should be very close to 0. However, in reality, these are not 0 and show a large loss actually, as shown in Fig. 2(b). Based on this situation, we decide to minimize the $\mathcal{L}_{\mathcal{IWV}}$ to generate the vaccine that can make the output of the watermark-removal network close to the ideal one. This seems to be a well-intentioned fix for performance, but it is actually superfluous and adversarial. In the test stage, whatever watermark is added to the host image, the IWV will suppress the removal network to recognize it, and the outputs still preserve the watermarks and they tend to be the same as watermarked image inputs. Thus, the IWV generation can be formulated as,

$$\begin{aligned} \min_\delta \;\; & \mathcal{L}_{\mathcal{IWV}}(x_h, \delta) \\ \text{s.t.} \;\; & \hat{x}_h = x_h + \delta \\ & \|\delta\|_\infty \leqslant \varepsilon. \end{aligned} \tag{10}$$

We also restrict the perturbations δ in L_∞ norm bound ϵ, and solve it by projected gradient descent (PGD) [34] again as follows,

$$\delta^{t+1} = \text{Proj}\left(\delta^t - \alpha \, \text{sign}\left(\nabla_{\delta^t} \mathcal{L}_{\mathcal{IWV}}(x_h, \delta)\right)\right). \tag{11}$$

In Fig. 2(b), we can see the framework of IWV generation and the inference of IWV. The pseudocode of our algorithm to generate watermark vaccine including DWV and IWV is shown in Algorithm 1. We use projected gradient descent (PGD) [34] to solve the problem (7) or (10), and then we generate the DWV/IWV. During the inference stage, we inject the DWV and IWV onto host image and add the watermark on it. After that, we evaluate their adversarial results through the watermark-removal networks f and expect to get the damaged or watermark-preserved image. In addition, we theoretically analyze why our vaccines work effectively and give the lower or upper bound of the watermark protection for DWV and IWV in Sect. 1 of the Supplementary Material.

4   Experiments

4.1   Experimental Setups

Datasets. We use the CLWD (Colored Large-scale Watermark Dataset) [32] in our experiments, which contains three parts: watermark-free images, watermarks and watermarked images. We first pretrain the watermark-removal networks using watermarked images in the train set of CLWD. Then in the attack stage, we use the watermark-free images as host images to generate watermark vaccines, and then add the watermarks with generated watermark vaccines. The details about the dataset can be checked in Sect. 2 of the Supplementary Material.

Models Architectures. We choose three advanced blind watermark-removal networks: BVMR [20], SplitNet [11], and WDNet [32]. We train them on the watermarked images of CLWD and save the best checkpoint parameters.

Evaluation Metrics. Following the previous work [11,30,32], Peak Signal-to-Noise Ratio (PSNR), Structural Similarity (SSIM), Root-Mean-Square distance (RMSE), and weighted Root-Mean-Square distance ($RMSE_w$) are adopted as our evaluation metrics. The difference between RMSE and $RMSE_w$ is that $RMSE_w$ only focuses on the watermarked area. For DWV, we specify these metrics as $PSNR^h$, $SSIM^h$, $RMSE^h$, $RMSE_w^h$ compared with host images for a better illustration. The lower $PSNR^h$/$SSIM^h$ or the higher $RMSE^h$/$RMSE_w^h$ mean the worse results of watermark-removal networks thus the better protection performance of proposed DWV. For IWV, we specify the metrics as $PSNR^w$, $SSIM^w$, $RMSE^w$, $RMSE_w^w$ which are compared with watermarked images. Different from DWV, the higher $PSNR^w$/$SSIM^w$ or the lower $RMSE^w$/$RMSE_w^w$ mean that the more excellent performance on keeping the watermarks on thus the better protection performance of proposed IWV.

Attack Parameters. During attack, we empirically set the L_∞ norm bound ϵ as 8/255, which is imperceptible by human eyes. We set the step size α as 2/255, and iteration T as 50. We set the hyperparameter β in Eq. 9 for IWV to 2 initially. We also further discuss the sensitivity of these hyperparameters in the Sect. 5 of Supplemental Material.

4.2   Effectiveness of Watermark Vaccine

We evaluate the effectiveness of DWV and IWV on 10,000 random host-watermark image combinations. We test three models with the same watermark

Fig. 3. Qualitative comparison of DWV and IWV. In each row, we show the watermark removal effects on the images without any vaccine, the images with random noises, the images with DWV, and the images with IWV under the same network.

parameters θ on the same dataset. We compare the clean input and the input with random noise at the same time, which is also restricted in L_∞ norm bound $(-\epsilon, \epsilon)$.

In Fig. 3, we show the qualitative visualization of different networks and their corresponding results. It shows that no matter which network it is, the watermark removed images can be ruined if the host images are injected with DWV, although the watermarks can be successfully removed. For IWV, there are still noticeable all or part of watermarks on the watermark removed images, and other parts of the images are not damaged. On the contrary, the inputs with random noise present no protective effect on the watermark removed results for any watermark-removal network. We show more visualization results in Sect. 3 of the Supplementary Material.

Table 1 demonstrates the quantitative results of watermark vaccines on different watermark-removal networks. In Table 1(a), we can find that random noise could not disrupt the watermark removed image, while the DWV can significantly degrade the quality of watermarked removed images with the lower $PSNR^h/SSIM^h$ and the higher $RMSE^h/RMSE_w^h$ than others. On the other hand, by observing Table 1(b), the watermark removed images with IWV have a better similarity with watermarked input with a little higher PSNR/SSIM. It is noticeable that the $RMSE_w^w$ for IWV is much lower than others, which is to evaluate whether the watermark part is well preserved. Moreover, the above phenomena in different watermark-removal networks tend to be the same, and the quantitative results are consistent with the qualitative visualization.

Although DWV can ruin the watermark removed images, the watermark patterns can also be removed. On the contrary, the watermarks with IWV could be still noticeable on the watermark removed images by human eyes. Therefore, which type of vaccine to choose depends on the need for protection.

Table 1. Impact of the two vaccines on WDNet, BVMR, and SplitNet on the same dataset and with the same parameters θ. The perturbations and random noises are restricted in L_∞ norm bound 8/255. **Clean** denotes the watermarked image with no vaccines, and **RN** denotes the watermarked images with random noise. For DWV, the lower $PSNR^h/SSIM^h$ or the higher $RMSE^h/RMSE_w^h$ the better. For IWV, the higher $PSNR^w/SSIM^w$ or the lower $RMSE^w/RMSE_w^w$ the better. The best-protection results are denoted in boldface.

Metrics	WDNet [32]				BVMR [20]				SplitNet [11]			
	$PSNR^h$	$SSIM^h$	$RMSE^h$	$RMSE_w^h$	$PSNR^h$	$SSIM^h$	$RMSE^h$	$RMSE_w^h$	$PSNR^h$	$SSIM^h$	$RMSE^h$	$RMSE_w^h$
Clean	38.62	0.9946	3.09	16.25	41.96	0.9955	2.09	23.86	42.32	0.9939	2.12	21.86
RN	38.19	0.9938	3.23	17.06	42.48	0.9957	1.98	24.13	42.73	0.9943	2.07	21.33
DWV (ours)	**29.68**	**0.6360**	**8.47**	**41.36**	**29.43**	**0.6462**	**8.68**	**26.85**	**34.12**	**0.8951**	**5.18**	**67.68**

(a) The effect of DWV on different watermark-removal networks.

Metrics	WDNet [32]				BVMR [20]				SplitNet [11]			
	$PSNR^w$	$SSIM^w$	$RMSE^w$	$RMSE_w^w$	$PSNR^w$	$SSIM^w$	$RMSE^w$	$RMSE_w^w$	$PSNR^w$	$SSIM^w$	$RMSE^w$	$RMSE_w^w$
Clean	37.76	0.9788	3.42	52.77	41.88	0.9893	2.13	42.68	40.91	0.9788	2.53	49.67
RN	37.53	0.9755	3.50	52.95	42.59	0.9917	2.00	42.73	41.59	0.9795	2.41	49.29
IWV (ours)	**45.16**	**0.9831**	**2.24**	**28.00**	**43.31**	**0.9926**	**1.86**	**37.42**	**42.79**	**0.9834**	**2.23**	**35.00**

(b) The effect of IWV on different watermark-removal networks.

Table 2. Mean and standard deviation over evaluation metrics of DWV and IWV for random watermark patterns and location parameters.

Metrics	Watermark		Location		Metrics	Watermark		Location	
	Clean	DWV	Clean	DWV		Clean	IWV	Clean	IWV
$PSNR^h$	39.12±0.02	29.40±0.03	40.82±0.04	28.95±0.01	$PSNR^w$	38.42±0.02	47.35±0.21	40.37±0.03	52.30±0.30
$SSIM^h$	0.9957±0.0000	0.6021±0.0028	0.9974±0.0001	0.5288±0.0020	$SSIM^w$	0.9874±0.002	0.9938±0.0003	0.9956±0.0001	0.9981±0.0001
$RMSE^h$	2.85±0.01	8.74±0.02	2.37±0.01	9.15±0.01	$RMSE^w$	3.08±0.01	1.63±0.03	2.48±0.01	1.08±0.03
$RMSE_w^h$	17.15±0.18	42.88±0.77	16.67±0.11	52.02±0.68	$RMSE_w^w$	54.25±0.54	20.67±0.39	48.28±0.38	10.39±0.55

(a) DWV

(b) IWV.

4.3 Universality of Watermark Vaccine

As mentioned in Sect. 3.2 and 3.3, the watermark vaccine we proposed can adapt to different watermarks and parameters and has a good universality. To illustrate this, we investigate the different watermark patterns, sizes, locations and transparency of watermarks for 1,000 host images. The WDNet [32] is selected as the model for an example. Other models can be found in the Supplementary Material. We test every host image with ten random-selected watermark patterns and fix other parameters. Similarly, we test every host image with ten random locations with a fixed watermark and other fixed parameters for the location. To show their universality, we calculate the mean and variance of these results from different settings. Concerning about the watermark size and transparency, we select six sizes: 60×60, 70×70, 80×80, 90×90, 100×100 and six transparency parameters: $\alpha = 0.45, 0.50, 0.55, 0.60, 0.65$. We fix the $\alpha = 0.55$ if the size varies, and fix size $= 80 \times 80$, if the transparency varies. Finally, we calculate their evaluation metrics respectively. The above quantitative results are shown in Table 2 and 3.

In Table 2, compared to the means of the clean input, the means of DWV and IWV show that our watermark vaccines are still effective, and the minor variances of vaccines prove that our vaccines can be universal among different

Table 3. The evaluation metrics for the Clean/DWV/IWV under different size and transparency of watermarks. Each row shows the results under different watermark sizes or different transparencies. The best-attacking results are denoted in boldface.

Metrics	$PSNR^h$		$SSIM^h$		$RMSE^h$		$RMSE^h_w$		$PSNR^w$		$SSIM^w$		$RMSE^w$		$RMSE^w_w$	
Input	Clean	DWV	Clean	DWV	Clean	DWV	Clean	DWV	Clean	IWV	Clean	IWV	Clean	IWV	Clean	IWV
Size = 60	39.91	**29.16**	0.9967	**0.5610**	2.62	**8.95**	18.02	**48.87**	39.36	**50.28**	0.9927	**0.9968**	2.76	**1.31**	51.55	**13.70**
Size = 70	39.53	29.25	0.9962	0.5784	2.72	8.86	17.73	45.59	38.87	50.04	0.9901	0.9958	2.92	1.35	53.31	16.14
Size = 80	39.14	29.25	0.9957	0.5963	2.84	8.78	17.03	41.13	38.39	47.52	0.9868	0.9936	3.09	1.63	55.16	20.71
Size = 90	38.67	29.54	0.9950	0.6261	3.00	8.58	17.02	38.11	37.82	45.69	0.9833	0.9911	3.29	1.90	56.35	26.03
Size = 100	38.25	29.67	0.9945	0.6455	3.15	8.47	16.81	37.32	37.32	43.06	0.9792	0.9864	3.49	2.34	57.29	32.87
$\alpha = 0.45$	39.24	**29.31**	0.9961	**0.5984**	2.81	**8.80**	15.51	**49.44**	38.35	**48.28**	0.9896	**0.9948**	3.10	**1.52**	45.92	**18.00**
$\alpha = 0.50$	39.19	29.43	0.9959	0.6129	2.83	8.70	16.28	44.32	38.36	46.27	0.9883	0.9940	3.10	1.73	50.69	20.67
$\alpha = 0.55$	39.14	29.25	0.9957	0.5963	2.84	8.78	17.03	41.13	38.39	47.52	0.9868	0.9936	3.09	1.63	55.16	20.71
$\alpha = 0.60$	39.08	29.37	0.9955	0.6004	2.86	8.75	17.79	39.26	38.34	47.70	0.9855	0.9934	3.10	1.59	59.44	22.14
$\alpha = 0.65$	38.99	29.32	0.9952	0.6048	2.89	8.79	18.53	39.31	38.27	47.19	0.9841	0.9934	3.13	1.54	62.54	21.40

Table 4. Vaccines Transferability. The columns correspond to the target model, while the rows correspond to the source model. For brevity, we show the $RMSE^h$ for DWV and the $RMSE^w_w$ for IWV.

Target model		WDNet	BVMR	SplitNet	Target model		WDNet	BVMR	SplitNet
Clean		3.14	2.78	2.25	Clean		54.47	43.84	51.10
RN		3.29	2.70	2.22	RN		54.63	43.85	51.15
Source Model	WDNet	**8.30**	2.80	2.34	Source Model	WDNet	**30.14**	43.80	50.80
	BVMR	3.35	**8.61**	2.47		BVMR	54.38	**40.13**	50.93
	SplitNet	3.23	2.79	**5.17**		SplitNet	54.52	43.50	**37.87**

(a) $RMSE^h$ of DWV   (b) $RMSE^w_w$ of IWV

watermark patterns or locations. Table 3 shows that the metrics in each row are not much different, although they are under different sizes and transparencies of the watermark. Hence, the above results indicate that DWV and IWV have good universality, regardless of the watermark pattern, position, size and transparency. The visualization of universality can be seen in the Sect. 6 of Supplementary Material.

Interestingly, according to Table 3, we find that when the size of the watermark becomes larger, the performance of the DWV and IWV has dropped. Moreover, if the transparency parameter α of the watermark becomes larger, the effect of the protection will be worsen, especially for IWV. This phenomenon is consistent with our analysis in Sect. 1 of Supplementary Material, that the watermarking variation $\|w\|$ is one of the factors that determine the effectiveness of watermark protection. The better performance of the watermark vaccine depends on a smaller variation of $\|w\|$. Therefore, it can be a challenge for the copyright owners to choose a suitable size and transparency for the watermark, where a larger and low-transparency watermark is convenient for copyright identification. In comparison, a smaller and high-transparency watermark is more beneficial to protect the watermark vaccine.

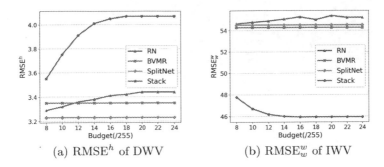

(a) RMSE^h of DWV   (b) RMSE$_w^w$ of IWV

Fig. 4. Testing stacked vaccines on WDNet [32]. For a comparison, we first add a random perturbation baseline and also test the vaccines generated by BVMR and SplitNet respectively. The stacked ones perform clearly the best.

4.4 Transferability of Watermark Vaccine

First, we explore the transferability of our vaccines across different watermark removal networks, and Table 4 shows the results. We find that the vaccines show limited transferability across different watermark removal methods, which is the common problem of the adversarial examples on generative models. The reason may be related to the different procedures and network structures of removal networks. e.g., BVMR [20] is a one-stage method of predicting watermark removed images, while SplitNet [11] and WDNet [32] contain detection, removal and refinement steps. In addition, SplitNet [11] adopts stacked attention-guided ResUNets, but other models do not.

In real world, humans can be protected against different kinds of viruses by inoculating different vaccines. Inspired by this, we study the stacked vaccine assembled by three networks. We test the stacked vaccine on different watermark removal networks (see the partial test results on WDNet in Fig. 4). Compared to the random perturbation baseline and the vaccines generated by other source models, the stacked vaccines perform better under various perturbation budgets. In future work, we will explore how to improve the transferability across different watermark removal networks/frameworks.

4.5 Resistance to Image Processing Operations

In this section, we explore whether our vaccine can resist the common image processing operation. We select two common transformations: JPEG compression [33] and Gaussian blur [9], and take the WDNet as an example for the model. We average the results of 1,000 watermarked images and plot them as Fig. 5. For brevity, we only show the RMSE^h of DWV and the RMSE$_w^w$ of IWV, and other metrics can be found in Sect. 7 of the Supplementary Material.

In Fig. 5(a), as the degree of JPEG compression ratio increases, it shows that the RMSE^h for DWV is declined and higher than random noise at first. Then it gradually rises and approaches the variation of random noise finally. It is possibly

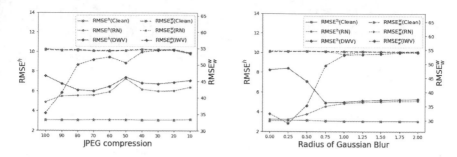

Fig. 5. Effect of two image-based transformation operations (JPEG Compression, Blur) on watermark vaccine. The solid lines show the change of $RMSE^h$, while the dashed lines show the $RMSE_w^w$ change.

because the performance degradation of the watermark vaccine is stronger than the image deterioration at the early stage, and when the degradation is strong enough, the result of DWV is similar to random noise. Regarding the IWV, we can find that the $RMSE_w^w$ of IWV has a sharp rise when the compression ratio increases, then it flattens out. It is worth noting that our watermark vaccines still have effects if the compression ratio is less than 80. The phenomenon in Gaussian blur is quite the same as that in JPEG compression in Fig. 5(b), and our watermark vaccines can resist the blur operation if the radius of Gaussian blur is less than 0.75. Besides the two image processing operations described above, we also consider some other operations that may affect our watermark vaccines, which will be present in the supplementary material. In conclusion, although some image-based transformation operations could reduce the effect of the watermark vaccine if their degradation is too substantial, they could also result in a lower quality of the image. Therefore, to some degree, our watermark vaccine can effectively resist some image processing operations.

5   Conclusions

Watermarking is an important and effective tool to protect copyright yet in face of the watermark removal threat. In this paper, we develop an idea of a watermark vaccine to protect watermarks. Our watermark vaccine is obtained by optimizing adversarial perturbations to attack the blind watermark removal network. Specifically, we propose two types of vaccine, dubbed disrupting watermark vaccine (DWV) and inerasable watermark vaccine (IWV). When malicious removal is presented, DWV will bring catastrophic damage to the host image, while IWV will keep the watermarks still clearly noticeable to human eyes. Both theoretical analysis and empirical experiments show that our vaccines is universal to different watermark patterns, sizes, locations, and transparencies, and they can also resist typical image transformation operations to a certain extent. This work makes the first exploration to protect watermarks from malicious removal.

There is still space to improve our approach, e.g. by improving the transferability of watermark vaccines across target models. We leave further explorations in future work.

Acknowledgment. Supported by the National Key R&D Program of China under (Grant 2019YFB 1406500), Sponsored by Ant Group Security and Risk Management Fund.

References

1. Akhtar, N., Mian, A.: Threat of adversarial attacks on deep learning in computer vision: a survey. IEEE Access **6**, 14410–14430 (2018)
2. Bertalmio, M., Sapiro, G., Caselles, V., Ballester, C.: Image inpainting. In: SIGGRAPH (2000)
3. Braudaway, G.W.: Protecting publicly-available images with an invisible image watermark. In: ICIP (1997)
4. Cao, Z., Niu, S., Zhang, J., Wang, X.: Generative adversarial networks model for visible watermark removal. IET Image Process. **13**(10), 1783–1789 (2019)
5. Carlini, N., Wagner, D.: Towards evaluating the robustness of neural networks. In: IEEE Symposium on Security and Privacy (2017)
6. Chen, P.Y., Zhang, H., Sharma, Y., Yi, J., Hsieh, C.J.: ZOO: zeroth order optimization based black-box attacks to deep neural networks without training substitute models. In: Proceedings of the 10th ACM Workshop on Artificial Intelligence and Security (2017)
7. Chen, Z., Xie, L., Pang, S., He, Y., Zhang, B.: MagDR: mask-guided detection and reconstruction for defending deepfakes. In: CVPR (2021)
8. Cheng, D., et al.: Large-scale visible watermark detection and removal with deep convolutional networks. In: Lai, J.-H., et al. (eds.) PRCV 2018. LNCS, vol. 11258, pp. 27–40. Springer, Cham (2018). https://doi.org/10.1007/978-3-030-03338-5_3
9. Cohen, J., Rosenfeld, E., Kolter, Z.: Certified adversarial robustness via randomized smoothing. In: ICML (2019)
10. Cox, I., Miller, M., Bloom, J., Fridrich, J., Kalker, T.: Digital Watermarking and Steganography. Morgan Kaufmann (2007)
11. Cun, X., Pun, C.M.: Split then refine: stacked attention-guided ResUNets for blind single image visible watermark removal. In: AAAI (2021)
12. Dekel, T., Rubinstein, M., Liu, C., Freeman, W.T.: On the effectiveness of visible watermarks. In: CVPR (2017)
13. Dong, Y., et al.: Boosting adversarial attacks with momentum. In: CVPR (2018)
14. Dong, Y., Pang, T., Su, H., Zhu, J.: Evading defenses to transferable adversarial examples by translation-invariant attacks. In: CVPR (2019)
15. Gandelsman, Y., Shocher, A., Irani, M.: "Double-DIP": unsupervised image decomposition via coupled deep-image-priors. In: CVPR (2019)
16. Goodfellow, I., et al.: Generative adversarial nets. In: NeurIPS (2014)
17. Goodfellow, I.J., Shlens, J., Szegedy, C.: Explaining and harnessing adversarial examples. In: ICLR (2015)
18. He, K., Zhang, X., Ren, S., Sun, J.: Deep residual learning for image recognition. In: CVPR (2016)
19. Hendrik Metzen, J., Chaithanya Kumar, M., Brox, T., Fischer, V.: Universal adversarial perturbations against semantic image segmentation. In: ICCV (2017)

20. Hertz, A., Fogel, S., Hanocka, R., Giryes, R., Cohen-Or, D.: Blind visual motif removal from a single image. In: CVPR (2019)
21. Huang, C.H., Wu, J.L.: Attacking visible watermarking schemes. TMM **6**(1), 16–30 (2004)
22. Jia, X., Wei, X., Cao, X., Han, X.: Adv-watermark: a novel watermark perturbation for adversarial examples. In: ACMMM (2020)
23. Jia, X., Zhang, Y., Wu, B., Ma, K., Wang, J., Cao, X.: LAS-AT: adversarial training with learnable attack strategy. In: CVPR (2022)
24. Jia, X., Zhang, Y., Wu, B., Wang, J., Cao, X.: Boosting fast adversarial training with learnable adversarial initialization. TIP **31**, 4417–4430 (2022)
25. Khachaturov, D., Shumailov, I., Zhao, Y., Papernot, N., Anderson, R.: Markpainting: adversarial machine learning meets inpainting. In: ICML (2021)
26. Kos, J., Fischer, I., Song, D.: Adversarial examples for generative models. In: IEEE Symposium on Security and Privacy Workshops (2018)
27. Krizhevsky, A., Sutskever, I., Hinton, G.E.: ImageNet classification with deep convolutional neural networks. In: NeurIPS (2012)
28. Kurakin, A., Goodfellow, I., Bengio, S., et al.: Adversarial examples in the physical world. In: ICLR Workshop (2017)
29. Li, X., et al.: Towards photo-realistic visible watermark removal with conditional generative adversarial networks. In: Zhao, Y., Barnes, N., Chen, B., Westermann, R., Kong, X., Lin, C. (eds.) ICIG 2019. LNCS, vol. 11901, pp. 345–356. Springer, Cham (2019). https://doi.org/10.1007/978-3-030-34120-6_28
30. Liang, J., Niu, L., Guo, F., Long, T., Zhang, L.: Visible watermark removal via self-calibrated localization and background refinement. In: ACM MM (2021)
31. Lin, J., Song, C., He, K., Wang, L., Hopcroft, J.E.: Nesterov accelerated gradient and scale invariance for adversarial attacks. In: ICLR (2020)
32. Liu, Y., Zhu, Z., Bai, X.: WDNet: watermark-decomposition network for visible watermark removal. In: WACV (2021)
33. Liu, Z., et al.: Feature distillation: DNN-oriented JPEG compression against adversarial examples. In: CVPR (2019)
34. Madry, A., Makelov, A., Schmidt, L., Tsipras, D., Vladu, A.: Towards deep learning models resistant to adversarial attacks. In: ICLR Poster (2018)
35. Mintzer, F., Braudaway, G.W., Yeung, M.M.: Effective and ineffective digital watermarks. In: ICIP (1997)
36. Moosavi-Dezfooli, S.M., Fawzi, A., Fawzi, O., Frossard, P.: Universal adversarial perturbations. In: CVPR (2017)
37. Mopuri, K.R., Uppala, P.K., Babu, R.V.: Ask, acquire, and attack: data-free UAP generation using class impressions. In: Ferrari, V., Hebert, M., Sminchisescu, C., Weiss, Y. (eds.) ECCV 2018. LNCS, vol. 11213, pp. 20–35. Springer, Cham (2018). https://doi.org/10.1007/978-3-030-01240-3_2
38. Papernot, N., McDaniel, P., Goodfellow, I.: Transferability in machine learning: from phenomena to black-box attacks using adversarial samples. arXiv preprint arXiv:1605.07277 (2016)
39. Papernot, N., McDaniel, P., Goodfellow, I., Jha, S., Celik, Z.B., Swami, A.: Practical black-box attacks against machine learning. In: AsiaCCS (2017)
40. Park, J., Tai, Y.W., Kweon, I.S.: Identigram/watermark removal using cross-channel correlation. In: CVPR (2012)
41. Pei, S.C., Zeng, Y.C.: A novel image recovery algorithm for visible watermarked images. IEEE Trans. Inf. Forensics Secur. **1**(4), 543–550 (2006)

42. Qin, C., He, Z., Yao, H., Cao, F., Gao, L.: Visible watermark removal scheme based on reversible data hiding and image inpainting. Sig. Process. Image Commun. **60**, 160–172 (2018)
43. Ruiz, N., Bargal, S.A., Sclaroff, S.: Disrupting deepfakes: adversarial attacks against conditional image translation networks and facial manipulation systems. In: Bartoli, A., Fusiello, A. (eds.) ECCV 2020. LNCS, vol. 12538, pp. 236–251. Springer, Cham (2020). https://doi.org/10.1007/978-3-030-66823-5_14
44. Ruiz, N., Bargal, S.A., Sclaroff, S.: Protecting against image translation deepfakes by leaking universal perturbations from black-box neural networks. arXiv preprint arXiv:2006.06493 (2020)
45. Samuel, S., Penzhorn, W.: Digital watermarking for copyright protection. IEEE Commun. Mag. (2004)
46. Santoyo-Garcia, H., Fragoso-Navarro, E., Reyes-Reyes, R., Sanchez-Perez, G., Nakano-Miyatake, M., Perez-Meana, H.: An automatic visible watermark detection method using total variation. In: IWBF (2017)
47. Segalis, E., Galili, E.: OGAN: disrupting deepfakes with an adversarial attack that survives training. arXiv e-prints (2020)
48. Shafahi, A., Najibi, M., Xu, Z., Dickerson, J., Davis, L.S., Goldstein, T.: Universal adversarial training. In: AAAI (2020)
49. Szegedy, C., et al.: Intriguing properties of neural networks. In: ICLR (2014)
50. Tabacof, P., Tavares, J., Valle, E.: Adversarial images for variational autoencoders. arXiv preprint arXiv:1612.00155 (2016)
51. Uesato, J., O'donoghue, B., Kohli, P., Oord, A.: Adversarial risk and the dangers of evaluating against weak attacks. In: ICML (2018)
52. Xie, C., et al.: Improving transferability of adversarial examples with input diversity. In: CVPR (2019)
53. Xu, C., Lu, Y., Zhou, Y.: An automatic visible watermark removal technique using image inpainting algorithms. In: ICSAI (2017)
54. Yang, C., Ding, L., Chen, Y., Li, H.: Defending against GAN-based deepfake attacks via transformation-aware adversarial faces. In: 2021 International Joint Conference on Neural Networks (IJCNN), pp. 1–8. IEEE (2021)
55. Yeh, C.Y., Chen, H.W., Tsai, S.L., Wang, S.D.: Disrupting image-translation-based deepfake algorithms with adversarial attacks. In: WACV Workshops (2020)
56. Yuan, X., He, P., Zhu, Q., Li, X.: Adversarial examples: attacks and defenses for deep learning. IEEE Trans. Neural Netw. Learn. Syst. **30**(9), 2805–2824 (2019)
57. Zhao, H., Shi, J., Qi, X., Wang, X., Jia, J.: Pyramid scene parsing network. In: CVPR (2017)

Explaining Deepfake Detection
by Analysing Image Matching

Shichao Dong, Jin Wang, Jiajun Liang, Haoqiang Fan, and Renhe Ji[✉]

MEGVII Technology, Beijing, China
{dongshichao,wangjin,liangjiajun,fhq,jirenhe}@megvii.com

Abstract. This paper aims to interpret how deepfake detection models learn artifact features of images when just supervised by binary labels. To this end, three hypotheses from the perspective of image matching are proposed as follows. 1. Deepfake detection models indicate real/fake images based on visual concepts that are neither source-relevant nor target-relevant, that is, considering such visual concepts as artifact-relevant. 2. Besides the supervision of binary labels, deepfake detection models implicitly learn artifact-relevant visual concepts through the FST-Matching (*i.e.* the matching **f**ake, **s**ource, **t**arget images) in the training set. 3. Implicitly learned artifact visual concepts through the FST-Matching in the raw training set are vulnerable to video compression. In experiments, the above hypotheses are verified among various DNNs. Furthermore, based on this understanding, we propose the FST-Matching Deepfake Detection Model to boost the performance of forgery detection on compressed videos. Experiment results show that our method achieves great performance, especially on highly-compressed (*e.g.* c40) videos.

Keywords: Deepfake detection · Image matching · Interpretability

1 Introduction

Recently, deepfake methods [11, 21, 23, 39, 40] have exhibited superior performance in synthesizing realistic faces. Such face forgeries may easily be used by attackers for malicious purposes, causing severe social problems and political threats. To this end, plenty of studies [1, 32] have achieved great success in detecting various manipulated media by simply considering it as a binary classification task. However, understanding how these models learn artifact features of images when just supervised by binary labels (real/fake) is still a challenge to state-of-the-art algorithms.

In this paper, we aim to interpret the success of deepfake detection models from the novel perspective of image matching. We consider the matching images

---

S. Dong and J. Wang—Equal contribution.

Supplementary Information The online version contains supplementary material available at https://doi.org/10.1007/978-3-031-19781-9_2.

S. Avidan et al. (Eds.): ECCV 2022, LNCS 13674, pp. 18–35, 2022.
https://doi.org/10.1007/978-3-031-19781-9_2

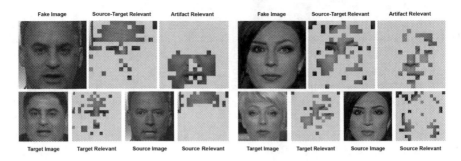

Fig. 1. The relationship between source/target-relevant visual concepts and artifact-relevant visual concepts. Here, visual concepts represent image regions such as eyes, mouths and foreheads of human faces. In this paper, we find that well-trained deepfake detection models mainly consider artifact-relevant visual concepts as neither source-relevant nor target-relevant from the perspective of image matching.

as follows. As shown in Fig. 1, the face of the source image is manipulated with representations of the target image to generate the corresponding fake image. Then the above fake image, source image and target image are considered as the matching images, termed as the FST-Matching. To this end, we design different metrics to quantitatively evaluate the effectiveness of image matching and propose three hypotheses as follows.

Hypothesis 1: Deepfake detection models indicate real/fake images based on visual concepts that are neither source-relevant nor target-relevant, that is, considering such visual concepts as artifact-relevant. In this paper, visual concepts represent the image regions such as the mouths, noses or eyes of human faces. Intuitively, fake images are generated from visual concepts that are either from source images or target images. However, some visual concepts may inevitably be manipulated by deepfake methods, causing them to be different from both source images and target images. Well-trained deepfake detection models are supposed to indicate real/fake images based on both source-irrelevant and target-irrelevant visual concepts.

Hypothesis 2: Besides the supervision of binary labels, deepfake detection models implicitly learn artifact-relevant visual concepts through the FST-Matching in the training set. Intuitively, binary labels are not sufficient enough to accomplish the deepfake detection task. Training images usually contain other artifact-irrelevant visual concepts, such as the identity of images. Such visual concepts may co-appear on certain real/fake images, causing deepfake detection models to learn biased representations of the forgeries. For example, deepfake detection models may infer the results based on the gender of images if real images are all male and fake images are all female. To this end, FST-Matching images are supposed to help deepfake detection models to discard artifact-irrelevant visual concepts and focus on artifact-relevant visual concepts, since they share common artifact-irrelevant visual concepts but are annotated with opposite labels.

Hypothesis 3: Implicitly learned artifact visual concepts through the FST-Matching in the raw training set are vulnerable to the video compression. Deepfake detection models trained on raw images usually suffer from significant performance drop when testing on compressed images [24,32,50]. We assume that it is because the implicit learning of artifact visual concepts through FST-Matching is fragile to the video compression. Specifically, the implicitly learned artifact visual concepts may become indistinguishable from compressed source visual concepts and target visual concepts on fake images due to the compression, causing deepfake detection models to make false predictions.

Methods: To verify the proposed hypotheses, we propose an explanation method based on the Shapley value [35] to interpret the predictions of deepfake detection models with various backbones. The Shapley value was firstly proposed in game theory [35] and is widely used in recent studies [2,27] to interpret the representations inside DNNs. Specifically, the Shapley value unbiasedly estimates the contributions of each player to the total award of the game. It naturally satisfies four properties, *i.e.* the linearity property, the dummy property, the symmetry property, and the efficiency property [41], which ensures its fairness and trustworthiness. Based on the Shapley value, we evaluate the visual concepts on images from the novel perspective of image matching to verify the proposed hypotheses.

Furthermore, during the verification of hypotheses, we surprisingly find the learned source/target visual concepts are more consistent among compressed images than the implicitly learned artifact visual concepts on images. Combined with the understanding of hypothesis 1, we then devise a simple model by disentangling source/target-irrelevant representations from the source/target visual concepts to indicate images (termed as the FST-Matching Deepfake Detection Model), which aims to boost the performance of the forgery detection on compressed videos. Results in our experiments show that such simple architecture achieves great performance, especially on highly compressed (*e.g.* c40) videos.

Contributions: Our contributions can be summarized as follows.

1. We propose a method to interpret the success of deepfake detection models from the novel perspective of image matching, *i.e.* the FST-Matching.
2. Three hypotheses from the perspective of the FST-Matching are proposed and verified, which offers new insights into the task of deepfake detection.
3. We further propose the FST-Matching Deepfake Detection Model to improve the performance on compressed videos.

2   Related Work

2.1   Deepfake Detection

The goal for deepfake detection is to classify the input media as either real or fake. Previous studies of deepfake detection mainly focused on improving the model performance on various datasets. Some methods [1,3,8,30–32] considered it as a binary classification task and directly trained models on the largely-collected dataset, such as Celeb-DF [25], DFDC [9], FF++ [32] and *etc.* These methods achieved great performance on the in-dataset evaluation, *i.e.* testing models on images manipulated by learned deepfake methods. However, these methods often failed to detect unseen datasets with newly proposed deepfake methods. To this end, other studies [20,49,51,54] aim to increase the generalization of deepfake detection models. These methods usually assumed that fake images share common human-perceived artifact representations introduced in the process of deepfake methods, such as blending boundaries [24], geometric features [37] and frequency features [14,22,26,28]. However, such assumptions usually represent human's understanding of artifact representations and may not hold in all real-life scenarios. It still presents continuous challenges to correctly understand the key differences between real and fake images, *i.e.* exploring the essence of the artifact representations on images.

To the best of our knowledge, studies focused on interpreting the learned representations of deepfake detection models are rare. In this paper, we aim to interpret deepfake detection models from the novel perspective of image matching to demonstrate what artifact representations are to deepfake detection models, how they learned artifact representations and how to further boost their performance in real-life scenarios.

2.2 Interpretability of DNNs

Previous studies on the interpretability of DNNs can be roughly divided into two categories. Some studies [10,29,36,42,43,53] focused on semantic explanations for DNNs by visualizing the learned visual concepts. Grad-CAM [34] and Grad-CAM++ [5] explored the attribution maps of input images based on gradient information. Zhou *et al.* [52] visualized the actual receptive fields of various units inside the DNNs. Fong *et al.* [12] explored the relationship between multiple filters and learned semantic visual concepts. Zhang *et al.* proposed to explore the relationships between the learned semantic visual concepts of DNNs via a graph model [47] and a decision tree [48]. However, different from general classification tasks, deepfake detection models aim to learn artifact-relevant visual concepts on images. Such representation is often imperceptible to people, making it difficult to evaluate the correctness of the explanation results derived from the above methods. Moreover, other studies proposed to explain the representations of DNNs mathematically to refrain from human evaluation of semantic representation. To this end, some studies proposed to understand DNNs based on entropy-based methods [7,15]. Some studies explored the representations of DNNs from a game-theoretical view [44–46]. However, although the above methods can be theoretically applied to various types of DNNs, it still remains a challenge to further exploit the explanation results to instruct the learning of specific tasks, such as deepfake detection.

In this paper, we aim to bridge the gap between the general explanation results and learning better deepfake detection models from the novel perspective of image matching. To this end, we designed the FST-Matching Deepfake Detection Model based on our explanation results and further boosted the performance on compressed videos.

3   Algorithms

In this section, given a well-trained deepfake detection model, we aim to interpret its prediction from the novel perspective of image matching. To this end, three hypotheses are proposed. To verify these hypotheses, we propose an explanation method to evaluate the contributions of visual concepts on images based on the Shapley value [35]. Please see supplementary materials for more information about the Shapley value.

3.1   Artifact Representations for Deepfake Detection Models

Hypothesis 1: Deepfake detection models indicate real/fake images based on visual concepts that are neither source-relevant nor target-relevant, that is, considering such visual concepts as artifact-relevant.

In this section, given a well-trained deepfake detection model $v_d(\cdot)$ (also termed as the detection encoder in this paper), we aim to evaluate the learned visual concepts on input images from the perspective of image matching. Specifically, we aim to explore what visual concepts on input images are considered as source-relevant, target-relevant and artifact-relevant. Then, we expect to evaluate the relationship between these visual concepts to verify the hypothesis.

The core challenge is to decide fairly what visual concepts are related to the source, target and artifact representations. Specifically, we do not annotate these visual concepts on images manually since it usually represents human's understanding of artifact representations, rather than the artifact representations inside the models. To this end, we train a source encoder $v_s(\cdot)$ and a target encoder $v_t(\cdot)$ to indicate the source/target-relevant visual concepts on images.

Intuitively, each fake image shares certain common visual concepts with its corresponding source and target image. We believe that when the source encoder v_s classifies each fake image and its corresponding source image as the same category, v_s would tend to focus on source-relevant visual concepts on each fake image. The same way goes for the target encoder v_t. Specifically, we use the additional attribute labels[1] of images to train v_s and v_t for convenience. To train the source/target encoder v_s/v_t, each fake image is considered as the same attribute label as the corresponding source/target image. Each real image is considered as its original attribute label.

We use the Shapley value [35] to evaluate the regional contributions of visual concepts on images to the prediction of each encoder. To reduce the computation cost, we divide the input image into $L \times L$ grids and calculate the contribution of each grid respectively. Let $G = \{g_{11}, g_{12}, ..., g_{LL}\}$ denote the set of all grids. $\phi_{v_d} \in R^{L \times L}, \phi_{v_s} \in R^{L \times L}, \phi_{v_t} \in R^{L \times L}$ represent the contributions of all grids to the prediction of the detection encoder v_d, the source encoder v_s and the target encoder v_t respectively. In this way, ϕ_{v_d}, ϕ_{v_s} and ϕ_{v_t} indicate the artifact,

---

[1] Implemented as the identity labels of images for convenience.

source and target visual concepts on images respectively. More specifically, given $\forall g_{ij} \in G$, it is considered to be artifact-relevant if $\phi_{v_d}(g_{ij}|G) > 0$ and artifact-irrelevant if $\phi_{v_d}(g_{ij}|G) \leq 0$. The same way goes for the source encoder v_s and target encoder v_t.

Based on the grid-level contributions, we propose a metric to evaluate the relationship between the artifact-relevant visual concepts, source-relevant visual concepts and target-relevant visual concepts. According to the hypothesis, deepfake detection models are supposed to consider artifact-relevant visual concepts as neither source-relevant nor target-relevant. Therefore, artifact-relevant visual concepts are supposed to barely have intersections with source/target-relevant visual concepts. To this end, we firstly generate a mask $M_\tau = I(max(\phi_{v_s}, \phi_{v_t}) > \tau)$ to denote the most source/target-relevant visual concepts, where $I(\cdot)$ is the indicator function and τ is a certain threshold. $I(\cdot)$ returns 1 if the condition inside is valid, otherwise $I(\cdot)$ returns 0. The metric is then designed to evaluate the intensities of the intersections between these visual concepts as follows.

$$Q_\tau = \frac{(1 - M_\tau) \cdot \phi_{v_d}}{\sum_{g_{ij} \in G} [1 - M_\tau(g_{ij})]} - \frac{M_\tau \cdot \phi_{v_d}}{\sum_{g_{ij} \in G} M_\tau(g_{ij})} \tag{1}$$

where $\cdot$ denotes the inner product. The first term measures the average intensities of the intersections between source/target-irrelevant visual concepts and artifact-relevant visual concepts. The second term measures the average intensities of the intersections between the source/target-relevant visual concepts and artifact-relevant visual concepts. $Q_\tau > 0$ represents that artifact-relevant visual concepts are more related to source/target-irrelevant visual concepts than the source/target-relevant visual concepts. $Q_\tau < 0$ represents that artifact-relevant visual concepts are less related to source/target-irrelevant visual concepts than the source/target-relevant visual concepts.

3.2   Learning the Artifact Representations

> **Hypothesis 2:** Besides the supervision of binary labels, deepfake detection models implicitly learn artifact-relevant visual concepts through the FST-Matching in the training set.

In this section, to verify the hypothesis, we expect to evaluate how the FST-Matching in the training set affects the learning of deepfake detection models. Specifically, FST-Matching in the training set means that real images contain the corresponding source and target images of fake images. To this end, we train two models with the paired training set and the unpaired training set separately. In the paired training set, the real images are only the corresponding source images and target images of fake images. In the unpaired images, the real images are of the same number as real images in the paired training set but do not correspond to any fake images. Then we compare the ACC, video-level AUC and the proposed metric Q_τ on these two models to evaluate the effectiveness of the FST-Matching.

3.3   Vulnerability of Artifact Representations to Video Compression

Hypothesis 3: Implicitly learned artifact visual concepts through the FST-Matching in the raw training set are vulnerable to the video compression.

In this section, to verify the hypothesis, we aim to measure the stability of implicitly learned artifact visual concepts to the video compression. Note that the detection encoder v_d is firstly trained on raw images and tested on compressed images afterwards. To this end, we design the stability metric to evaluate the changes among artifact visual concepts under the conditions of different compression rates *i.e.* c23, c40. The stability metric is designed as follows.

$$\delta_{v_d} = E_{cmp\in\{c23,c40\}}[cos(\phi_{v_d}^{cmp}, \phi_{v_d}^{raw})] \tag{2}$$

where $\phi_{v_d}^{cmp}$ represents the grids contributions to the predictions of the detection encoder v_d when tested on the compressed images. $\phi_{v_d}^{raw}$ represents the grids contributions tested on the raw images. $cos(\cdot, \cdot)$ denotes the operation of calculating the cosine similarity. A smaller value of $\delta_{v_d} \in [-1, 1]$ indicates that the implicitly learned artifact visual concepts are vulnerable to the compression. Moreover, we also evaluate the stability of the learned source/target visual concepts for source/target encoder v_s/v_t on compressed videos for more comparisons.

3.4   FST-Matching Deepfake Detection Model

Based on the understanding of deepfake detection models from the perspective of FST-Matching, we propose the FST-Matching Deepfake Detection Model to further boost the performance of deepfake detection models on compressed videos. During the verification of hypothesis 3, we surprisingly found that source/target visual concepts learned by the source encoder v_s and the target encoder v_t (*i.e.* ϕ_{v_s} and ϕ_{v_t}) are more consistent than the artifact visual concepts implicitly learned by the detection encoder v_d (*i.e.* ϕ_{v_d}) on compressed images (shown in the bottom of Fig. 2). Inspired by the understanding of hypothesis 1, we believe that directly disentangling source/target-irrelevant representations from source/target visual concepts to indicate images may improve the model performance on compressed videos. Please see supplementary for detailed verification.

The structure of the FST-Matching Deepfake Detection Model is shown in Fig. 2, which aims to classify face forgeries based on source/target-irrelevant visual concepts on images according to hypothesis 1. To this end, we first use the Source Feature Encoder and the Target Feature Encoder to directly learn the source feature $f_s \in R^{B\times C_s}$ and the target feature $f_t \in R^{B\times C_t}$ on images. B indicates the number of input images. C_s and C_t indicate the number of output channels. Then we design the Feature Disentanglement Module to automatically disentangle the source/target-irrelevant feature f_s^{ir}, f_t^{ir} and source/target-relevant feature f_s^r, f_t^r on the channel-level. Similar to [19], we use the channel-wise attention vectors $a_s \in R^{B\times C_s}$ and $a_t \in R^{B\times C_t}$ to disentangle f_s and f_t, which are calculated as follows.

$$a_s = \sigma(MLP(f_s)), \ a_t = \sigma(MLP(f_t)) \tag{3}$$

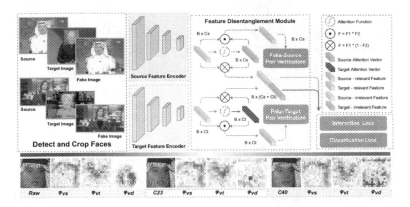

Fig. 2. The FST-Matching Deepfake Detection Model. As shown in the bottom of the figure, we surprisingly find that ϕ_{v_s} and ϕ_{v_t} are more robust to video compression than ϕ_{v_d}. To this end, we use a Source Feature Encoder and a Target Feature Encoder to explicitly learn the source and target representations on images. The Feature Disentanglement Module further extracts source/target-irrelevant representations to indicate the realism of images *i.e.* real or fake.

where MLP denotes the multi-layer perceptron and σ denotes the sigmoid function. In this way, the source and target relevant feature f_s^r, f_t^r are calculated as $f_s^r = a_s \circ f_s$ and $f_t^r = a_t \circ f_t$. The source and target irrelevant feature f_s^{ir}, f_t^{ir} are calculated as $f_s^{ir} = (1 - a_s) \circ f_s$ and $f_t^{ir} = (1 - a_t) \circ f_t$. Here $\circ$ denotes the channel-wise product.

To ensure the effectiveness of the feature disentanglement, we use the Fake-Source Pair Verification module to classify f_s^r as the same attribute label of the source images (See Footnote 1). Similarly, f_t^r is classified as the same attribute label of the target image through the Fake-Target Pair Verification module. f_s^{ir} and f_t^{ir} are then concatenated to predict the final real/fake label of the input image. Let y_s, y_t, y_d denote the source attribute label, target attribute label and forgery detection label of the image. $\hat{y}_s, \hat{y}_t, \hat{y}_d$ denote the predicted source attribute, target attribute and forgery prediction. The classification loss of the FST-Matching Deepfake Detection Model is designed as follows.

$$Loss_{cls} = -E[y_d log\hat{y}_d] - \lambda_s E[y_s log\hat{y}_s] - \lambda_t E[y_t log\hat{y}_t] \qquad (4)$$

Moreover, inspired by [45], we design another loss to further strengthen the interaction between f_s^{ir} and f_t^{ir} for the final prediction. Let $h(\cdot)$ denote the final prediction module. The interaction loss aims to increase the additional award caused by the coalition $[f_s^{ir}, f_t^{ir}]$ *w.r.t.* the sum of the award when f_s^{ir} and f_t^{ir} contribute to the final prediction individually. The interaction loss is designed as follows.

$$Loss_{interaction} = -E[h([f_s^{ir}, f_t^{ir}]) - h([\mathbf{0}, f_t^{ir}]) - h([f_s^{ir}, \mathbf{0}]) + h([\mathbf{0}, \mathbf{0}])] \qquad (5)$$

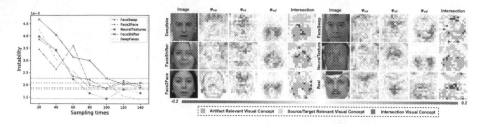

Fig. 3. Instability of the Shapley value (left) and verification of hypothesis 1(right). The left figure shows that as the sampling times increase, the Shapley value becomes stable. The right figure shows the visualization of source, target and artifact visual concepts, *i.e.* ϕ_{v_s}, ϕ_{v_t} and ϕ_{v_d}. Results show that artifact-relevant visual concepts barely have intersections with source/target-relevant visual concepts among various manipulation algorithms, which supports hypothesis 1.

where $\mathbf{0}$ represents the zero vector in the same size with f_s^{ir} and f_t^{ir}. $h([\mathbf{0}, \mathbf{0}]$ represents the basic score when neither f_s^{ir} nor f_t^{ir} contributes to the final prediction. The overall loss is designed as follows.

$$Loss = Loss_{cls} + \lambda_{inter} Loss_{interaction} \tag{6}$$

4   Experiment

4.1   Implementation Details

DNNs and Datasets: To verify the proposed hypotheses, we conduct various experiments on different backbones. Specifically, we used ResNet-18/34 [18] and EfficientNet-b3 [38] as the backbones for the detection encoder v_d, v_s and v_t. Besides, we also used the pre-trained models released in [32] and [49] for the detection encoder v_d for more comparisons with state-of-the-art methods.

We trained and tested our models on the widely-used FF++ [32] dataset. FF++ [32] dataset contains 5000 videos, including 1000 original videos and 4000 fake videos manipulated by different forgery methods, such as Deepfake [11], FaceSwap [21], FaceShifter [23], NeuralTextures [39] and Face2Face [40]. All models were pre-trained on the ImageNet [33] dataset and fine-tuned on FF++ [32]. Moreover, the attribute label of the input image is set as the identity of the image for convenience. Specifically, for the fake image, the source/target encoder is expected to classify the image as the identity of its corresponding source/target image. For the real image, the source encoder and the target encoder are both expected to classify the image as its own original identity.

Implementation of the Shapley Value: The precise calculation of the Shapley value is computationally intolerable. To this end, we used the sampling-based method [4] to approximately calculate the contributions of all the visual concepts. During the sampling process, the unsampled grids of images were set as the baseline value, which is set to be zero in this paper. Moreover, we used the selected scalar before the softmax layer corresponding to the ground truth label of the image as the output score for all the encoders.

4.2   Fairness of the Shapley Value

Accuracy of the Shapley Value. To ensure the stability of the approximated Shapley value, we evaluated the effect of sampling times T *w.r.t* the change of the Shapley value. Specifically, similar to [44], we repeated the sampling procedures [4] two times for the same sampling times T to get ϕ_1 and ϕ_2 respectively. Then we measured the change between ϕ_1 and ϕ_2 *w.r.t* to the sampling times T via the instability metric $\frac{||\phi_1-\phi_2||_2}{||\phi_1+\phi_2||_2}$ among all test images. As shown in Fig. 3, we calculated the instability metric for ResNet18-based ϕ_{v_d} for different sampling times. Results show that when $T \geq 100$, we get the relatively stable Shapley value, which ensures the fairness of our results.

4.3   Verification of Hypotheses

Verification of Hypothesis 1. Hypothesis 1 assumes that well-trained deep-fake detection models indicate images based on neither source-relevant nor target-relevant visual concepts, *i.e.* considering them to be artifact-relevant. In this section, we both qualitatively and quantitatively verify the hypothesis.

For the qualitative analysis, we find that artifact-relevant visual concepts barely have intersections with source/target-relevant visual concepts. In Fig. 3, we showed the visual results of $\phi_{v_s}, \phi_{v_t}, \phi_{v_d}$ and the intersections among the main contributed visual concepts for different manipulation algorithms used in FF++ [32]. For the better visualization, we normalized $\phi_{v_s}, \phi_{v_t}, \phi_{v_d}$ all to the unit vector. The backbone of the detection decoder v_d is ResNet-18 [18]. The source and target relevant visual concepts are denoted based on the mask M_τ. For more clarity, in the column of *Intersection*, we only kept the top highest 30% contributed grids. Results show that deepfake detection models mainly consider artifact-relevant concepts as neither source-relevant nor target-relevant.

For the quantitative analysis, we evaluated the proposed metric Q among various DNNs and manipulation algorithms. In Table 1, we calculated the average value of Q among different thresholds τ for a fair comparison. Specifically, τ was set to different values to keep $\{0.60L^2, 0.65L^2, ..., 0.85L^2, 0.9L^2, 0.95L^2\}$ grids on M_τ respectively. $Q > 0$ represents that the learned artifact-relevant visual concepts are more related to source/target-irrelevant visual concepts than the source/target-relevant visual concepts. Results show that various types of DNNs mainly consider artifact-relevant visual concepts as neither source-relevant nor target-relevant. Moreover, such results are not essentially related to the choices on backbones of v_s and v_t, which further verify the generality of the hypothesis. Note that $Q < 0$ for Xception [32] when tested on images manipulated by FaceShifter [23]. It is because that the originally released pre-trained models Xception in [32] was never trained on forged images of FaceShifter [23] before, thus unable to locate the artifact-relevant visual concepts for FaceShifter [23].

Fig. 4. Verification of hypothesis 2: comparison of the proposed metric Q_τ between models trained on the paired training set and the unpaired train set. The horizontal coordinate represents the percentage of the kept grids in the mask M_τ when setting different thresholds τ. The backbone of the detection encoder is ResNet-18 [18]. Results show that models trained on the paired training set have larger values of Q_τ, showing that FST-Matching helps models to locate artifact-relevant visual concepts.

Table 1. Verification of hypothesis 1: comparison of the proposed metric Q ($\times 10^{-2}$) for different deepfake detection models among various manipulation algorithms. Results show that well-trained deepfake detection models have larger values of Q, which indicates that these models consider source/target-irrelevant visual concepts as artifact-relevant.

Backbone of v_s/v_t	Forgery methods	Backbone of v_d ($Q(\times 10^{-2})$)				
		ResNet-18	ResNet-34	Efficient-b3	MAT [49]	Xception [32]
ResNet-18 [18]	FaceSwap [21]	2.77	2.88	2.02	2.57	3.10
	Face2Face [40]	2.31	2.63	2.08	2.54	2.59
	FaceShifter [23]	2.45	3.22	2.10	2.42	−0.73
	Deepfake [11]	2.53	2.67	2.30	2.79	2.61
	NeuralTexture [39]	2.30	2.67	2.07	2.51	1.00
Efficient-b3 [38]	FaceSwap [21]	2.85	2.99	2.08	2.49	3.20
	Face2Face [40]	2.19	2.63	2.00	2.49	2.61
	FaceShifter [23]	2.38	3.22	2.07	2.33	−0.67
	Deepfake [11]	2.51	2.71	2.17	2.77	2.64
	NeuralTexture [39]	2.32	2.69	2.05	2.47	1.06

Verification of Hypothesis 2. Hypothesis 2 assumes that well-trained deep-fake detection models implicitly learned artifact-relevant visual concepts through the FST-Matching in the training set. To verify the hypothesis, we trained two models of the same backbone on the paired training set and the unpaired training set separately. In the paired training set, real images are only the source and target images corresponding to the fake images. In contrast, the real images in the unpaired training set do not match fake images, but are of the same number as the real images in the paired training set. **Both the paired and unpaired training set are downsampled from FF++ [32] dataset containing only 40 identities of images, which is significantly small compared to the initial 1000 identities in the FF++ [32] dataset.** In this section, we conduct extensive experiments to demonstrate that FST-Matching is crucial for learning deepfake detection models.

Firstly, we compared the ACC and video-level AUC on each trained model. As shown in Table 2, models trained on the paired training set achieved similar

Table 2. Verification of hypothesis 2: performance comparison between models trained on the whole FF++ [32] dataset (denoted as the *Baseline*), the paired training set and the unpaired training set. In the paired training set, real images are the corresponding source and target images of fake images *i.e.* satisfying the FST-Matching. Results show that models trained on the paired training set achieve similar performance to the baseline. Note that paired training set is of a significantly smaller size. Such results demonstrate the effectiveness of the FST-Matching.

Models	Forgery methods	Baseline		Pair		Unpair	
		ACC	AUC	ACC	AUC	ACC	AUC
ResNet-18 [18]	FaceSwap [21]	98.93	100	97.50	99.91	53.93	75.41
	Face2Face [40]	96.79	99.43	97.14	99.27	64.29	85.74
	FaceShifter [23]	99.29	99.99	97.14	99.82	81.07	93.03
	Deepfake [11]	98.21	100	97.50	99.87	69.64	86.51
	NeuralTexture [39]	90.71	98.89	95.71	98.73	60.00	76.60
Efficient-b3 [38]	FaceSwap [21]	100	100	99.64	100	77.50	87.51
	Face2Face [40]	99.29	99.77	99.29	99.72	81.79	93.36
	FaceShifter [23]	99.29	99.93	99.29	99.96	84.29	96.10
	Deepfake [11]	100	100	100	100	85.36	97.81
	NeuralTexture [39]	99.29	99.85	98.93	99.56	82.86	92.30

Table 3. Verification of hypothesis 3: comparisons between the stability metric δ of different visual concepts. The backbones of the source, target and detection encoders are all ResNet-18 [18]. Results show that learned source and target visual concepts are more consistent to video compression than implicitly learned artifact visual concepts.

Visual concept	Forgery methods (δ)				
	FaceSwap	Face2Face	FaceShifter	Deepfake	NeuralTexture
Source	0.73	0.74	0.73	0.74	0.74
Target	0.73	0.76	0.71	0.75	0.76
Artifact (baseline)	0.17	−0.02	0.14	−0.15	−0.14

performance to the baseline models, which are trained on the whole FF++ [32] dataset. Note that the paired training set is significantly smaller than the original FF++ [32] dataset, which demonstrates the importance of FST-Matching in the training set. In contrast, models trained on the unpaired training set, although of the same size as the paired training set, showed apparently worse results. Such results also show that FST-Matching in the training set is of great value to learning deepfake detection models.

Moreover, we compared the proposed metric Q_τ between each trained model as well. To make a fair comparison, we calculated the value of the metric Q_τ of different τ among all the test images. As shown in Fig. 4, models trained on the paired training set have larger values of Q_τ, showing that FST-Matching in the training set effectively helps models to localize source/target-irrelevant visual concepts and consider them as artifact-relevant.

Verification of Hypothesis 3. Hypothesis 3 assumes that the implicitly learned artifact visual concepts through the FST-Matching in the raw training set are vulnerable to the video compression. To verify the hypothesis, we tested the raw-trained models on compressed videos and calculated the proposed metric δ_{v_d} among all test images. For the qualitative analysis, as shown in Fig. 2, raw-trained models indicate compressed images with significantly different visual concepts compared with the raw images. For the quantitative analysis, in Table 3, the calculated $\delta_{v_d} \in [-1, 1]$ is near 0, which also indicates the great change of ϕ_{v_d} under the condition of different compression rate.

Moreover, we also evaluated the stability of the source/target visual concept. Surprisingly, as Fig. 2 and Table 3 show, such learned visual concepts show great consistency to the video compression, compared to the implicitly learned artifact visual concepts. Such results motivate us to improve the model performance on compressed videos by devising a model, which explicitly exploits the FST-Matching in the training set.

Table 4. Performance comparison on compressed videos with state-of-the-art methods. Our method achieves great performance on compressed videos, especially on c40 videos.

Models	Backbone	C23		C40	
		ACC	*AUC*	*ACC*	*AUC*
Steg.Features [13]	–	70.97	–	55.98	–
LD-CNN [8]	–	78.45	–	58.69	–
Face-x-ray [24]	HRNet	–	87.30	–	61.60
MesoNet [1]	Xception	83.10	–	70.47	–
Xception [32]	Xception	92.39	94.86	80.32	81.76
Xception-ELA [16]	Xception	93.86	94.80	79.63	82.90
Xception-PAFilters [6]	Xception	–	–	87.16	90.20
SPSL [26]	Xception	91.50	95.32	81.57	82.82
MAT-Xception [49]	Xception	96.37	98.97	86.95	87.26
MAT-Efficient [49]	Efficient-b4	**97.60**	**99.29**	88.69	90.40
FST-Matching (ours)	ResNet-18	94.52	98.34	**88.92**	**92.02**
	Xception	94.05	98.27	87.38	90.44
	Efficient-b3	95.95	98.75	87.62	90.89
	Efficient-b4	96.19	98.81	88.69	91.27

4.4   FST-Matching Deepfake Detection Model

Performance Comparison on Compressed Videos. In this section, we compared the performance of our model to current state-of-the-art methods. Table 4 shows the performance on compressed videos. Specifically, when aligned with the same backbone of other methods, our model achieved great performance on compressed videos, especially on highly-compressed (*e.g.* c40) videos. Such results also indicate the broad applicability of our method. Meanwhile, note that there still exists a slight performance gap with MAT [49] on c23 in Table 4. Different from our method, MAT [49] designed specific modules to learn the frequency features of images. Such features are widely shown to be effective to enhance the performance of deepfake detection models on compressed videos [14,22,26,28]. To this end, we believe that integrating such features into our model may potentially fill this performance gap. Moreover, since our method is merely the first attempt to exploit our innovative explanation results, we believe that more effective methods could be further inspired based on our study in the future.

Performance Comparison on Raw Videos. In order to have a more comprehensive analysis, we also evaluated our models on raw videos. Results in Table 5 show that our method still performed well on raw images.

Table 5. Evaluation on raw videos.

Models	Backbone	RAW	
		ACC	*AUC*
Face-x-ray [24]	HRNet	–	98.80
MesoNet [1]	Xception	95.23	–
Xception [32]	Xception	**99.26**	99.20
Xception-ELA [16]	Xception	98.57	98.40
MAT-Efficient [49]	Efficient-b4	97.77	99.61
FST-Matching (ours)	ResNet-18	98.14	99.72
	Xception	98.71	99.91
	Efficient-b3	98.93	99.90
	Efficient-b4	99.00	**99.92**

Table 6. Cross-dataset evaluation.

Models	Backbones	Celeb-DF
Xception [32]	Xception	49.03
SPSL [26]	Xception	76.88
MAT [49]	Efficient-b4	68.44
Face-x-ray [24]	HRNet	80.58
FST-Matching (ours)	ResNet-18	86.00
	Xception	88.44
	Efficient-b3	**89.39**
	Efficient-b4	88.13

Table 7. Robustness evaluation to image editing in terms of AUC (%) on FF++.

Method	Saturation	Contrast	Block	Noise	Blur	Pixel	Avg
Xception [32]	99.3	98.6	99.7	53.8	60.2	74.2	81.0
Face-x-ray [24]	97.6	88.5	99.1	49.8	63.8	88.6	81.2
LipForensices [17]	**99.9**	99.6	87.4	73.8	96.1	95.6	92.1
FST-Matching (ours)	99.6	**99.9**	**99.9**	84.8	**99.2**	**98.7**	**97.0**

Evaluation on the Generalization Ability. We conduct another experiment to evaluate the generalization ability of our method. To this end, we followed the same cross-dataset experimental setting in SPSL [26]. Results are shown in Table 6, where the metric is AUC (%). Our models trained on FF++ [32] achieved great performance on Celeb-DF [25], regardless of different backbones.

Robustness to Image Editing Operations. We conduct another experiment to evaluate our method when image editing operations are applied to images.

To this end, we followed the same robustness experiment setting in LipForensics [17]. Results are shown in Table 7, where the metric is AUC (%). Our method also demonstrated great robustness to listed perturbations.

5   Conclusions

In this paper, we interpret the success of deepfake detection models from the novel perspective of image matching. To this end, three hypotheses are proposed and verified among various DNNs, *i.e.* 1. Deepfake detection models indicate real/fake images based on visual concepts that are neither source-relevant nor target-relevant, that is, considering such visual concepts as artifact-relevant. 2. Besides the supervision of binary labels, deepfake detection models implicitly learn artifact-relevant visual concepts through the FST-Matching in the training set. 3. Implicitly learned artifact visual concepts through the FST-Matching in the raw training set are vulnerable to video compression. Based on the understanding, we further propose the FST-Matching Deepfake Detection Model and achieve great performance on the compressed videos. This research provides an opportunity to explore the essence of artifact representation of images and sheds new light on the task of deepfake detection.

References

1. Afchar, D., Nozick, V., Yamagishi, J., Echizen, I.: MesoNet: a compact facial video forgery detection network. In: 2018 IEEE International Workshop on Information Forensics and Security (WIFS), pp. 1–7. IEEE (2018)
2. Ancona, M., Oztireli, C., Gross, M.: Explaining deep neural networks with a polynomial time algorithm for shapley value approximation. In: International Conference on Machine Learning, pp. 272–281. PMLR (2019)
3. Bayar, B., Stamm, M.C.: A deep learning approach to universal image manipulation detection using a new convolutional layer. In: Proceedings of the 4th ACM Workshop on Information Hiding and Multimedia Security, pp. 5–10 (2016)
4. Castro, J., Gómez, D., Tejada, J.: Polynomial calculation of the Shapley value based on sampling. Comput. Oper. Res. **36**(5), 1726–1730 (2009)
5. Chattopadhay, A., Sarkar, A., Howlader, P., Balasubramanian, V.N.: Grad-CAM++: generalized gradient-based visual explanations for deep convolutional networks. In: 2018 IEEE Winter Conference on Applications of Computer Vision (WACV), pp. 839–847. IEEE (2018)
6. Chen, M., Sedighi, V., Boroumand, M., Fridrich, J.: JPEG-phase-aware convolutional neural network for steganalysis of JPEG images. In: Proceedings of the 5th ACM Workshop on Information Hiding and Multimedia Security, pp. 75–84 (2017)
7. Cheng, X., Rao, Z., Chen, Y., Zhang, Q.: Explaining knowledge distillation by quantifying the knowledge. In: Proceedings of the IEEE/CVF Conference on Computer Vision and Pattern Recognition, pp. 12925–12935 (2020)
8. Cozzolino, D., Poggi, G., Verdoliva, L.: Recasting residual-based local descriptors as convolutional neural networks: an application to image forgery detection. In: Proceedings of the 5th ACM Workshop on Information Hiding and Multimedia Security, pp. 159–164 (2017)

9. Dolhansky, B., et al.: The deepfake detection challenge dataset. arXiv e-prints, arXiv-2006 (2020)
10. Dosovitskiy, A., Brox, T.: Inverting visual representations with convolutional networks. In: Proceedings of the IEEE Conference on Computer Vision and Pattern Recognition, pp. 4829–4837 (2016)
11. FaceSwapDevs: Deepfakes (2019). https://github.com/deepfakes/faceswap
12. Fong, R., Vedaldi, A.: Net2Vec: quantifying and explaining how concepts are encoded by filters in deep neural networks. In: Proceedings of the IEEE Conference on Computer Vision and Pattern Recognition, pp. 8730–8738 (2018)
13. Fridrich, J., Kodovsky, J.: Rich models for steganalysis of digital images. IEEE Trans. Inf. Forensics Secur. **7**(3), 868–882 (2012)
14. Gu, Q., Chen, S., Yao, T., Chen, Y., Ding, S., Yi, R.: Exploiting fine-grained face forgery clues via progressive enhancement learning. arXiv preprint arXiv:2112.13977 (2021)
15. Guan, C., Wang, X., Zhang, Q., Chen, R., He, D., Xie, X.: Towards a deep and unified understanding of deep neural models in NLP. In: International Conference on Machine Learning, pp. 2454–2463. PMLR (2019)
16. Gunawan, T.S., Hanafiah, S.A.M., Kartiwi, M., Ismail, N., Za'bah, N.F., Nordin, A.N.: Development of photo forensics algorithm by detecting photoshop manipulation using error level analysis. Indones. J. Electr. Eng. Comput. Sci. **7**(1), 131–137 (2017)
17. Haliassos, A., Vougioukas, K., Petridis, S., Pantic, M.: Lips don't lie: a generalisable and robust approach to face forgery detection. In: Proceedings of the IEEE/CVF Conference on Computer Vision and Pattern Recognition, pp. 5039–5049 (2021)
18. He, K., Zhang, X., Ren, S., Sun, J.: Deep residual learning for image recognition. In: Proceedings of the IEEE Conference on Computer Vision and Pattern Recognition, pp. 770–778 (2016)
19. Hu, J., Shen, L., Sun, G.: Squeeze-and-excitation networks. In: Proceedings of the IEEE Conference on Computer Vision and Pattern Recognition, pp. 7132–7141 (2018)
20. Hu, Z., Xie, H., Wang, Y., Li, J., Wang, Z., Zhang, Y.: Dynamic inconsistency-aware deepfake video detection. In: IJCAI (2021)
21. Kowalski, M.: FaceSwap (2018). https://github.com/MarekKowalski/FaceSwap
22. Li, J., Xie, H., Li, J., Wang, Z., Zhang, Y.: Frequency-aware discriminative feature learning supervised by single-center loss for face forgery detection. In: Proceedings of the IEEE/CVF Conference on Computer Vision and Pattern Recognition, pp. 6458–6467 (2021)
23. Li, L., Bao, J., Yang, H., Chen, D., Wen, F.: FaceShifter: towards high fidelity and occlusion aware face swapping. arXiv preprint arXiv:1912.13457 (2019)
24. Li, L., et al.: Face X-ray for more general face forgery detection. In: Proceedings of the IEEE/CVF Conference on Computer Vision and Pattern Recognition, pp. 5001–5010 (2020)
25. Li, Y., Yang, X., Sun, P., Qi, H., Lyu, S.: Celeb-DF: a large-scale challenging dataset for deepfake forensics. In: Proceedings of the IEEE/CVF Conference on Computer Vision and Pattern Recognition, pp. 3207–3216 (2020)
26. Liu, H., et al.: Spatial-phase shallow learning: rethinking face forgery detection in frequency domain. In: Proceedings of the IEEE/CVF Conference on Computer Vision and Pattern Recognition, pp. 772–781 (2021)
27. Lundberg, S.M., Lee, S.I.: A unified approach to interpreting model predictions. In: Advances in Neural Information Processing Systems, vol. 30 (2017)

28. Luo, Y., Zhang, Y., Yan, J., Liu, W.: Generalizing face forgery detection with high-frequency features. In: Proceedings of the IEEE/CVF Conference on Computer Vision and Pattern Recognition, pp. 16317–16326 (2021)
29. Mahendran, A., Vedaldi, A.: Understanding deep image representations by inverting them. In: Proceedings of the IEEE Conference on Computer Vision and Pattern Recognition, pp. 5188–5196 (2015)
30. Nguyen, H.H., Yamagishi, J., Echizen, I.: Use of a capsule network to detect fake images and videos. arXiv preprint arXiv:1910.12467 (2019)
31. Rahmouni, N., Nozick, V., Yamagishi, J., Echizen, I.: Distinguishing computer graphics from natural images using convolution neural networks. In: 2017 IEEE Workshop on Information Forensics and Security (WIFS), pp. 1–6. IEEE (2017)
32. Rossler, A., Cozzolino, D., Verdoliva, L., Riess, C., Thies, J., Nießner, M.: FaceForensics++: learning to detect manipulated facial images. In: Proceedings of the IEEE/CVF International Conference on Computer Vision, pp. 1–11 (2019)
33. Russakovsky, O., et al.: ImageNet large scale visual recognition challenge. Int. J. Comput. Vis. **115**(3), 211–252 (2015). https://doi.org/10.1007/s11263-015-0816-y
34. Selvaraju, R.R., Cogswell, M., Das, A., Vedantam, R., Parikh, D., Batra, D.: Grad-CAM: visual explanations from deep networks via gradient-based localization. In: Proceedings of the IEEE International Conference on Computer Vision, pp. 618–626 (2017)
35. Shapley, L.S.: A value for n-person games, contributions to the theory of games **2**, 307–317 (1953)
36. Simonyan, K., Vedaldi, A., Zisserman, A.: Deep inside convolutional networks: visualising image classification models and saliency maps. arXiv preprint arXiv:1312.6034 (2013)
37. Sun, Z., Han, Y., Hua, Z., Ruan, N., Jia, W.: Improving the efficiency and robustness of deepfakes detection through precise geometric features. In: Proceedings of the IEEE/CVF Conference on Computer Vision and Pattern Recognition, pp. 3609–3618 (2021)
38. Tan, M., Le, Q.: EfficientNet: rethinking model scaling for convolutional neural networks. In: International Conference on Machine Learning, pp. 6105–6114. PMLR (2019)
39. Thies, J., Zollhöfer, M., Nießner, M.: Deferred neural rendering: image synthesis using neural textures. ACM Trans. Graph. (TOG) **38**(4), 1–12 (2019)
40. Thies, J., Zollhofer, M., Stamminger, M., Theobalt, C., Nießner, M.: Face2Face: real-time face capture and reenactment of RGB videos. In: Proceedings of the IEEE Conference on Computer Vision and Pattern Recognition, pp. 2387–2395 (2016)
41. Weber, R.J.: Probabilistic values for games. In: The Shapley Value. Essays in Honor of Lloyd S. Shapley, pp. 101–119 (1988)
42. Yosinski, J., Clune, J., Nguyen, A., Fuchs, T., Lipson, H.: Understanding neural networks through deep visualization. arXiv preprint arXiv:1506.06579 (2015)
43. Zeiler, M.D., Fergus, R.: Visualizing and understanding convolutional networks. In: Fleet, D., Pajdla, T., Schiele, B., Tuytelaars, T. (eds.) ECCV 2014. LNCS, vol. 8689, pp. 818–833. Springer, Cham (2014). https://doi.org/10.1007/978-3-319-10590-1_53
44. Zhang, D., et al.: Building interpretable interaction trees for deep NLP models. In: Proceedings of the AAAI Conference on Artificial Intelligence, vol. 35, pp. 14328–14337 (2021)
45. Zhang, H., Li, S., Ma, Y., Li, M., Xie, Y., Zhang, Q.: Interpreting and boosting dropout from a game-theoretic view. In: International Conference on Learning Representations (2020)

46. Zhang, H., Xie, Y., Zheng, L., Zhang, D., Zhang, Q.: Interpreting multivariate shapley interactions in DNNs. In: Proceedings of the AAAI Conference on Artificial Intelligence, vol. 35, pp. 10877–10886 (2021)
47. Zhang, Q., Cao, R., Shi, F., Wu, Y.N., Zhu, S.C.: Interpreting CNN knowledge via an explanatory graph. In: Proceedings of the AAAI Conference on Artificial Intelligence, vol. 32 (2018)
48. Zhang, Q., Yang, Y., Ma, H., Wu, Y.N.: Interpreting CNNs via decision trees. In: Proceedings of the IEEE/CVF Conference on Computer Vision and Pattern Recognition (CVPR), June 2019
49. Zhao, H., Zhou, W., Chen, D., Wei, T., Zhang, W., Yu, N.: Multi-attentional deepfake detection. In: Proceedings of the IEEE/CVF Conference on Computer Vision and Pattern Recognition, pp. 2185–2194 (2021)
50. Zhao, T., Xu, X., Xu, M., Ding, H., Xiong, Y., Xia, W.: Learning to recognize patch-wise consistency for deepfake detection. arXiv preprint arXiv:2012.09311 (2020)
51. Zhao, T., Xu, X., Xu, M., Ding, H., Xiong, Y., Xia, W.: Learning self-consistency for deepfake detection. In: Proceedings of the IEEE/CVF International Conference on Computer Vision, pp. 15023–15033 (2021)
52. Zhou, B., Khosla, A., Lapedriza, A., Oliva, A., Torralba, A.: Object detectors emerge in deep scene CNNs. arXiv preprint arXiv:1412.6856 (2014)
53. Zhou, B., Khosla, A., Lapedriza, A., Oliva, A., Torralba, A.: Learning deep features for discriminative localization. In: Proceedings of the IEEE Conference on Computer Vision and Pattern Recognition, pp. 2921–2929 (2016)
54. Zhou, Y., Lim, S.N.: Joint audio-visual deepfake detection. In: Proceedings of the IEEE/CVF International Conference on Computer Vision, pp. 14800–14809 (2021)

FrequencyLowCut Pooling - Plug and Play Against Catastrophic Overfitting

Julia Grabinski[1,2,3]([envelope])[iD], Steffen Jung[4][iD], Janis Keuper[2,3][iD],
and Margret Keuper[1,4][iD]

[1] Visual Computing, Siegen University, Siegen, Germany
[2] Competence Center High Performance Computing, Fraunhofer ITWM,
Kaiserslautern, Germany
`julia.grabinski@itwm.fraunhofer.de`
[3] Institute for Machine Learning and Analytics, Offenburg University,
Offenburg, Germany
[4] Max Planck Institute for Informatics, Saarland Informatics Campus,
Saarbrücken, Germany

Abstract. Over the last years, Convolutional Neural Networks (CNNs) have been the dominating neural architecture in a wide range of computer vision tasks. From an image and signal processing point of view, this success might be a bit surprising as the inherent spatial pyramid design of most CNNs is apparently violating basic signal processing laws, i.e. *Sampling Theorem* in their down-sampling operations. However, since poor sampling appeared not to affect model accuracy, this issue has been broadly neglected until model robustness started to receive more attention. Recent work [18] in the context of adversarial attacks and distribution shifts, showed after all, that there is a strong correlation between the vulnerability of CNNs and aliasing artifacts induced by poor down-sampling operations. This paper builds on these findings and introduces an aliasing free down-sampling operation which can easily be plugged into any CNN architecture: FrequencyLowCut pooling. Our experiments show, that in combination with simple and Fast Gradient Sign Method (FGSM) adversarial training, our hyper-parameter free operator substantially improves model robustness and avoids catastrophic overfitting. Our code is available at https://github.com/GeJulia/flc_pooling.

Keywords: CNNs · Adversarial robustness · Aliasing

1   Introduction

The robustness of convolutional neural networks has evolved to being one of the most crucial computer vision research topics in recent years. While state-of-the-art models provide high accuracy in many tasks, their susceptibility to adversarial attacks [9] and even common corruptions [20] is hampering their

Supplementary Information The online version contains supplementary material available at https://doi.org/10.1007/978-3-031-19781-9_3.

deployment in many practical applications. Therefore, a wide range of publications aim to provide models with increased robustness by adversarial training (AT) schemes [15,43,47], sophisticated data augmentation techniques [35] and enriching the training with additional data [4,16]. As a result, robuster models can be learned with common CNN architectures, yet arguably at a high training cost - even without investigating the reasons for CNN's vulnerability. These reasons are of course multifold, starting with the high dimensionality of the feature space and sparse training data such that models easily tend to overfit [36,44]. Recently, the pooling operation in CNNs has been discussed in a similar context for example in [18] who measured the correlation between aliasing and a network's susceptibility to adversarial attacks. [49] have shown that commonly used pooling operations even prevent the smoothness of image representations under small input translations.

Our contributions are summarized as follows:

- We introduce FrequencyLowCut pooling, ensuring aliasing-free downsampling within CNNs.
- Through extensive experiments with various datasets and architectures, we show empirically that FLC pooling prevents single step AT from catastrophic overfitting, while this is not the case for other recently published improved pooling operations (e.g. [49]).
- FLC pooling is substantially faster, around five times, and easier to integrate than previous AT or defence methods. It provides a hyperparameter-free plug and play module for increased model robustness.

1.1   Related Work

Adversarial Attacks. Adversarial attacks reveal CNNs vulnerabilities to intentional pixel perturbations which are crafted either having access to the full model (so-called white-box attacks) [3,15,27,33,33,37,42] or only having access to the model's prediction on given input images (so-called back-box attacks) [1,7]. The Fast Gradient Sign Method [15], FGSM, is an efficient single step white box attack. More effective methods use multiple optimization steps, e.g. as in the white-box Projected Gradient Descent (PGD) [27] or in black-box attacks such as Squares [1]. AutoAttack [9] is an ensemble of different attacks including an adaptive version of PGD and is widely used to benchmark adversarial robustness because of its strong performance [8]. In relation to image down-sampling, [46] and [30] demonstrate steganography-based attacks on the pre-processing pipeline of CNNs.

Adversarial Training. Some adversarial attacks are directly proposed with a dedicated defence [15,37]. Beyond these attack-specific defences, there are many methods for more general adversarial training (AT) schemes. These typically add an additional loss term which accounts for possible perturbations [12,47] or introduces additional training data [4,39]. Both are combined for example in [43], while [16] use data augmentation which is typically combined with weight averaging [35]. A widely used source for additional training data is *ddpm* [17, 34,35], which contains one million extra samples for CIFAR-10 and is generated with the model proposed by [21]. [17] receive an additional boost in robustness by adding specifically generated images while [34] add wrongly labeled data to the training-set. RobustBench [8] gives an overview and evaluation of a variety of models w.r.t. their adversarial robustness and the additional data used.

A common drawback of all AT methods is the vast increase in computation needed to train networks: large amounts of additional adversarial samples and slower convergence due to the harder learning problem typically increase the training time by a factor between seven and fifteen [27,43,45,47].

Catastrophic Overfitting. AT with single step FGSM is a simple approach to achieve basic adversarial robustness [6,36]. Unfortunately, the robustness of this approach against stronger attacks like PGD is starting to drop again after a certain amount of training epochs. [44] called this phenomenon *catastrophic overfitting*. They concluded that one step adversarial attacks tend to overfit to the chosen adversarial perturbation magnitude (given by ϵ) but fail to be robust against multi-step attacks like PGD. [36] introduced early stopping as a countermeasure. After each training epoch, the model is evaluated on a small portion of the dataset with a multi-step attack, which again increases the computation time. As soon as the accuracy drops compared with a hand selected threshold the model training is stopped. [25] and [41] showed that the observed overfitting is related to the flatness of the loss landscape. They introduced a method to compute the *optimal* perturbation length ϵ' for each image and do single step FGSM training with this optimal perturbation length to prevent catastrophic overfitting. [2] showed that catastrophic overfitting not only occurs in deep neural networks but can also be present in single-layer convolutional neural networks. They propose a new kind of regularization, called GradAlign to improve FGSM perturbations and flatten the loss landscape to prevent catastrophic overfitting.

Anti-aliasing. The problem of aliasing effects in the context of CNN-based neural networks has already been addressed from various angles in literature: [49] improve the shift-invariance of CNNs using anti-aliasing filters implemented as convolutions. [50] further improve shift invariance by using learned instead of predefined blurring filters. [29] rely on the low frequency components of wavelets during pooling operations to reduce aliasing and increase the robustness against common image corruptions. In [22] a depth adaptive blurring filter before pooling as well as an anti-aliasing activation function are used. Anti-aliasing is also relevant in the context of image generation. [24] propose to use blurring filters to remove aliases during image generation in generative adversarial networks (GANs) while [11] and [23] employ additional loss terms in the frequency space to address aliasing. In [18], we empirically showed via a proposed aliasing measure that adversarially robust models exhibit less aliasing in their down-sampling layers than non-robust models. Based on this motivation, we here propose an aliasing-free down-sampling operation that avoids catastrophic overfitting.

2 Preliminaries

2.1 Adversarial Training

In general, AT can be formalized as an optimization problem given by a *min-max* formulation:

$$\min_{\theta} \max_{\delta \in \Delta} L(x + \delta, y; \theta), \tag{1}$$

where we seek to optimize network weights θ such that they minimize the loss L between inputs x and labels y under attacks δ. The maximization over δ can thereby be efficiently performed using the Fast Gradient Sign Method (FGSM), which takes one big step defined by ϵ into the direction of the gradient [15]:

$$x' = x + \epsilon \cdot \text{sign}(\nabla_x L(\theta, x, y)). \tag{2}$$

Specific values of the perturbation size ϵ are usually set to be fractions of eight-bit encodings of the image color channels. A popular choice on the CIFAR-10 [26] dataset is $\epsilon = \frac{8}{255}$ which can be motivated by the human color perception [14]. The Projected Gradient Descent method, PGD, works similar to FGSM but instead of taking one big step in the direction of the gradient with step size ϵ, it iteratively optimizes the adversarial example with a smaller, defined step size α. Random restarts further increase its effectiveness. The final attack is clipped to the maximal step size of ϵ.

$$x'_{N+1} = \text{Clip}_{X,\epsilon}\{x'_N + \alpha \cdot \text{sign}(\nabla_x L(\theta, x, y))\} \tag{3}$$

PGD is one of the strongest attacks, due to its variability in step size and its random restarts. Yet, its applicability for AT is limited as it requires a relatively long optimization time for every example. Additionally, PGD is dependent on several hyperparameters, which makes it even less attractive for training in practice. In contrast, FGSM is fast and straight-forward to implement. Yet, models that use FGSM for AT tend to overfit on FGSM attacks and are not robust to other attacks such as PGD, i.e. they suffer from catastrophic overfitting [44].

2.2 Down-Sampling in CNNs

Independent of their actual network topology, CNNs essentially perform a series of stacked convolutions and non-linearities. Using a vast amount of learnable convolution filters, CNNs are capable of extracting local texture information from all intermediate representations (input data and feature maps). To be able to abstract from this localized spatial information and to learn higher order relations of parts, objects and entire scenes, CNNs apply down-sampling operations to implement a spatial pyramid representation over the network layers.

This down-sampling is typically performed via a convolution with stride greater than one or by so-called pooling layers (see Fig. 1). The most common pooling layers are AveragePooling and MaxPooling. All of these operations are highly sensitive to small shifts or noise in the layer input [5,29,49].

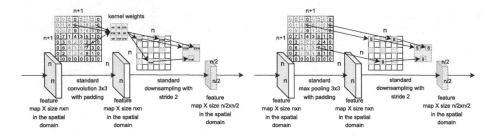

Fig. 1. Standard down-sampling operations used in CNNs. Left: down-sampling via convolution with stride two. First the feature map is padded and the actual convolution is executed. The stride defines the step-size of the kernel. Hence, for stride two, the kernel is moved two spatial units. In practice, this down-sampling is often implemented by a standard convolution with stride one and then discarding every second point in every spatial dimension. Right: down-sampling via MaxPooling. Here the max value for each spatial window location is chosen and the striding is implemented accordingly.

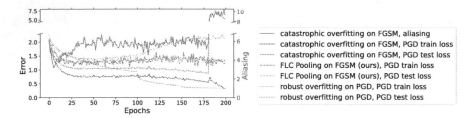

Fig. 2. Examples of AT facing catastrophic overfitting and its relationship to aliasing as well as robust overfitting and our FLC pooling. While FGSM training is prone to catastrophic overfitting, PGD training takes much longer and is also prone to robust overfitting. Our method, FLC pooling, is able to train with the fast FGSM training while preventing catastrophic overfitting.

Aliasing. Common CNNs sub-sample their intermediate feature maps to aggregate spatial information and increase the invariance of the network. However, no aliasing prevention is incorporated in current sub-sampling methods. Concretely, sub-sampling with too low sampling rates will cause pathological overlaps in the frequency spectra (Fig. 3). They arise as soon as the sampling rate is below the double bandwidth of the signal [40] and cause ambiguities: high frequency components can not be clearly distinguished from low frequency components. As a result, CNNs might misconceive local uncorrelated image perturbations as global manipulations. [18] showed that aliasing in CNNs strongly coincides with the robustness of the model. Based on this finding, one can hypothesize that models that overfit to high frequencies in the data tend to be less robust. This thought is also in line with the widely discussed texture bias [13]. To substantiate this hypothesis in the context of adversarial robustness, we investigate and empirically show in Fig. 2 that catastrophic overfitting coincides with increased aliasing during FGSM AT. Based on this observation, we expect networks that sample without aliasing to be better behaved in AT FGSM settings. The FrequencyLowCut pooling, which we propose, trivially fulfills this property.

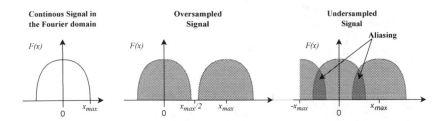

Fig. 3. Aliasing is apparent in the frequency domain. Left: The frequency spectrum of a 1D signal with maximal frequency $x_{\max}$. After down-sampling, replica of the signal appear at a distance proportional to the sampling rate. Center: The spectrum after sampling with a sufficiently large sampling rate. Right: The spectrum after undersampling with aliases due to overlapping replica.

3   FrequencyLowCut Pooling

Several previous approaches such as [49,50] reduce high frequencies in features maps before pooling to avoid aliasing artifacts. They do so by classical blurring operations in the spatial domain. While those methods reduce aliasing, they can not entirely remove it due to sampling theoretic considerations in theory and limited filter sizes in practice (see Appendix A.3 or [14] for details). We aim to perfectly remove aliases in CNNs' down-sampling operations without adding additional hyperparameters. Therefore, we directly address the down-sampling operation in the frequency domain, where we can sample according to the Nyquist rate, i.e. remove all frequencies above $\frac{\text{samplingrate}}{2}$ and thus discard aliases. In practice, the proposed down-sampling operation first performs a Discrete Fourier Transform (DFT) of the feature maps f. Feature maps with height M and width N to be down-sampled are then represented as

$$F(k,l) = \frac{1}{MN} \sum_{m=0}^{M-1} \sum_{n=0}^{N-1} f(m,n) e^{-2\pi j \left(\frac{k}{M} m + \frac{l}{N} n \right)}. \tag{4}$$

In the resulting frequency space representation F (Eq. (4)), all frequencies k, l, with $|k|$ or $|l| > \frac{\text{samplingrate}}{2}$ have to be set to 0 before down-sampling. CNNs commonly down-sample with a factor of two, i.e. sampling rate $= \frac{1}{2}$. Down-sampling thus corresponds to finding $F_d(k,l) = F(k,l)$, $\forall$ frequencies k, l with $|k|, |l| < \frac{1}{4}$. Practically, the DFT(f) returns an array F of complex numbers with size $K \times L = M \times N$, where the frequency $k, l = 0$ is stored in the upper left corner and the highest frequency is in the center. We thus shift the low frequency components into the center of the array via FFT-shift to get F_s and crop the frequencies below the Nyquist frequency

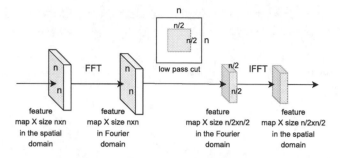

Fig. 4. FrequencyLowCut pooling, the proposed, guaranteed alias-free pooling operation. We first transform feature maps into frequency space via FFT, then crop the low frequency components. The result is transformed back into the spatial domain. This corresponds to a sinc-filtered and down-sampled feature map and is fed into the next convolutional layer.

as $F_{sd} = F_s[K' : 3K', L' : 3L']$ for $K' = \frac{K}{4}$ and $L' = \frac{L}{4}$, for all samples in a batch and all channels in the feature map. After the inverse FFT-shift, we obtain array F_d with size $[\hat{K}, \hat{L}] = [\frac{K}{2}, \frac{L}{2}]$, containing exactly all frequencies below the Nyquist frequency F_d, which we can backtransform to the spatial domain via inverse DFT for the spatial indices $\hat{m} = 0 \ldots \frac{M}{2}$ and $\hat{n} = 0 \ldots \frac{N}{2}$.

$$f_d(\hat{m}, \hat{n}) = \frac{1}{\hat{K}\hat{L}} \sum_{k=0}^{\hat{K}-1} \sum_{l=0}^{\hat{L}-1} F_d(k, l) e^{2\pi j \left(\frac{\hat{m}}{\hat{K}} k + \frac{\hat{n}}{\hat{L}} l \right)}. \tag{5}$$

We thus receive the aliasing-free down-sampled feature map f_d with size $[\frac{M}{2}, \frac{N}{2}]$.

Figure 4 shows this procedure in detail. In the spatial domain, this operation would amount to convolving the feature map with an infinitely large (non-bandlimited) $\text{sinc}(m) = \frac{\sin(m)}{m}$ filter, which can not be implemented in practice.

4   Experiments

4.1   Native Robustness of FLC Pooling

We evaluate our proposed FLC pooling in a standard training scheme with Preact-ResNet-18 (PRN-18) architectures on CIFAR-10 (see Appendix A.1 for details). Table 1 shows that both the decrease in clean accuracy as well as the increase in robustness are marginal compared to the baseline models. We argue that these results are in line with our hypothesis that the removal of aliasing artifacts alone will not lead to enhanced robustness and we need to combine correct down-sampling with AT to compensate for the persisting problems induced by the very high dimensional decision spaces in CNNs.

4.2   FLC Pooling for FGSM Training

In the following series of experiments we apply simple FGSM AT with $\epsilon = \frac{8}{255}$ on different architectures and evaluate the resulting robustness with different

Table 1. Clean training of Preact-ResNet-18 (PRN-18) architectures on CIFAR-10. We compare clean and robust accuracy against FGSM [15] with L_inf, $\epsilon = \frac{8}{255}$, PGD [27] with L_inf, $\epsilon = \frac{1}{255}$ as well as L_2 with $\epsilon = 0.5$ (20 iterations) and common corruptions (CC) [20] (mean over all corruptions and severities).

Method	Clean	FGSM $\epsilon = \frac{8}{255}$	PGD L_inf $\epsilon = \frac{1}{255}$	PGD L_2 $\epsilon = 0.5$	CC
Baseline	**95.08**	34.08	7.15	6.68	74.38
FLC Pooling	94.66	**34.65**	**10.00**	**11.27**	**74.70**

Table 2. FGSM AT of PRN-18 and Wide-ResNet-28-10 (WRN-28-10) architectures on CIFAR-10. Comparison of clean and robust accuracy (high is better) against PGD [27] and AutoAttack [9] on the full dataset with L_inf with $\epsilon = 8/255$ and L_2 with $\epsilon = 0.5$. FGSM test accuracies indicate catastrophic overfitting on the AT data, hence this column is set to gray.

Method	Clean	FGSM $\epsilon = \frac{8}{255}$	PGD L_inf $\epsilon = \frac{8}{255}$	AA L_inf $\epsilon = \frac{8}{255}$	AA L_2 $\epsilon = 0.5$	AA L_inf $\epsilon = \frac{1}{255}$
Preact-ResNet-18						
Baseline: FGSM training	**90.81**	90.37	0.16	0.00	0.01	53.10
Baseline & early stopping	82.88	61.71	11.82	3.76	17.44	72.95
BlurPooling [49]	86.24	78.36	1.33	0.06	1.96	66.88
Adaptive BlurPooling [50]	90.35	77.39	0.23	0.00	0.07	39.00
Wavelet Pooling [28]	85.02	64.16	12.13	5.92	19.65	10.08
FLC Pooling (ours)	84.81	58.25	**38.41**	**36.69**	**55.58**	**80.63**
WRN-28-10						
Baseline: FGSM training	86.67	83.64	1.64	0.09	1.47	59.39
Baseline & early stopping	82.29	56.36	31.26	28.54	46.03	76.87
Blurpooling [49]	91.40	89.44	0.22	0.00	0.00	38.45
Adaptive BlurPooling [50]	91.10	89.76	0.00	0.00	0.00	7.42
Wavelet Pooling [28]	**92.19**	90.85	0.00	0.00	0.00	10.08
FLC Pooling (ours)	84.93	53.81	**39.48**	**38.37**	**52.89**	**80.27**

pooling methods. We compare the models in terms of their clean, FGSM, PGD and AutoAttack accuracy, where the FGSM attack is run with $\epsilon = 8/255$, PGD with 50 iterations and 10 random restarts and $\epsilon = 8/255$ and $\alpha = 2/255$. For AutoAttack, we evaluate the standard L_inf norm with $\epsilon = 8/255$ and a smaller ϵ of $1/255$, as AutoAttack is almost too strong to be imperceptible to humans [31]. Additionally, we evaluate AutoAttack with L_2 norm and $\epsilon = 0.5$.

CIFAR-10. Table 2 shows the evaluation of a PRN-18 as well as a Wide-ResNet-28-10 (WRN-28-10) on CIFAR-10 [26]. For both network architectures, we observe that our proposed FLC pooling is the only method that is able to prevent catastrophic overfitting. All other pooling methods heavily overfit on the FGSM training data, achieving high robustness towards FGSM attacks, but fail to generalize towards PGD or AutoAttack. Our hyper-parameter free approach also outperforms early stopping methods which are additionally suffering from the difficulty that one has to manually choose a suitable threshold in order to maintain the best model robustness.

Table 3. FGSM AT on CINIC-10 for PRN-18 architectures. We compare clean and robust accuracy (higher is better) against PGD [27] as well as AutoAttack [9] on the full dataset with L_{inf} with $\epsilon = 8/255$ and L_2 with $\epsilon = 0.5$. FGSM test accuracies indicate catastrophic overfitting on the AT data, hence this column is set to gray.

Method	Clean	FGSM $\epsilon = \frac{8}{255}$	PGD L_{inf} $\epsilon = \frac{8}{255}$	AA L_{inf} $\epsilon = \frac{8}{255}$	AA L_2 $\epsilon = 0.5$	AA L_{inf} $\epsilon = \frac{1}{255}$
Baseline	87.46	58.83	1.31	0.12	1.55	55.21
Baseline & early stopping	82.79	42.58	27.55	30.76	50.28	**79.88**
Blurpooling [49]	87.13	54.16	1.29	0.20	4.68	70.56
Adaptive BlurPooling [50]	**90.21**	52.27	0.05	0.00	0.01	40.96
Wavelet Pooling [28]	88.81	64.16	1.76	0.12	3.38	66.61
FLC Pooling (ours)	82.56	38.39	**36.28**	**49.61**	**60.51**	78.50

Table 4. FGSM AT on CIFAR-100 for PRN-18 architectures. We compare clean and robust accuracy (higher is better) against PGD [27] and AutoAttack [9] on the full dataset with L_{inf} with $\epsilon = 8/255$ and L_2 with $\epsilon = 0.5$. FGSM test accuracies indicate robustness to training data, so this column is set to gray. Here, none of the models overfit, while FLC pooling still yields best overall robustness.

Method	Clean	FGSM $\epsilon = \frac{8}{255}$	PGD L_{inf} $\epsilon = \frac{8}{255}$	AA L_{inf} $\epsilon = \frac{8}{255}$	AA L_2 $\epsilon = 0.5$	AA L_{inf} $\epsilon = \frac{1}{255}$
Baseline	51.92	23.25	15.41	11.13	25.67	44.53
Baseline & early stopping	52.09	23.34	15.51	10.88	25.78	44.61
Blurpooling [49]	52.68	23.40	16.81	12.43	26.79	45.68
Adaptive BlurPooling [50]	52.08	9.77	18.68	6.05	11.32	21.04
Wavelet Pooling [28]	**55.08**	25.70	18.36	13.76	**27.51**	47.52
FLC Pooling (ours)	54.66	26.82	**19.83**	**15.40**	26.30	**47.83**

CINIC-10. Table 3 shows similar results on CINIC-10 [10]. Our model exhibits no catastrophic overfitting, while previous pooling methods do. It should be noted that CINIC-10 is not officially reported by AutoAttack. This might explain why the accuracies under AutoAttack are higher on CINIC-10 than on CIFAR-10. We assume that AutoAttack is optimized for CIFAR-10 and CIFAR-100 and therefore less strong on CINIC-10.

CIFAR-100. Table 4 shows the results on CIFAR-100 [26], using the same experimental setup as for CIFAR-10 in Table 2. Due to the higher complexity of CIFAR-100, with ten times more classes than CIFAR-10, AT tends to suffer from catastrophic overfitting much later (in terms of epochs) in the training process. Therefore we trained the Baseline model for 300 epochs. While the gap towards the robustness of other methods is decreasing with the amount of catastrophic overfitting, our method still outperforms other pooling approaches in most cases - especially on strong attacks.

ImageNet. Table 5 evaluates our FLC Pooling on ImageNet. We compare against results reported on RobustBench [8], with emphasis on the model by [44] which also uses fast FGSM training. The clean accuracy of our model using

Table 5. Comparison of ResNet-50 models clean and robust accuracy against AutoAttack [9] on ImageNet. We compare against models reported on RobustBench [8].

Method	Clean	PGD L_{inf} $\epsilon = \frac{4}{255}$
Standard [8]	76.52	0.00
FGSM & FLC Pooling (ours)	63.52	27.29
Wong et al., 2020 [44]	55.62	26.24
Robustness lib, 2019 [12]	62.56	29.22
Salman et al., 2020 [38]	64.02	34.96

FLC pooling is about 8% better than the one reached by [44], with a 1% improvement in robust accuracy. All other models are trained with more time consuming methods like PGD (more details can be found in the Appendix A.2).

Analysis. The presented experiments on several datasets and architectures show that baseline FGSM training, as well as other pooling methods, strongly overfit on the adversarial data and do not generalize their robustness towards other attacks. We also show that our FLC pooling sufficiently prevents catastrophic overfitting and is able to generalize robustness over different networks, datasets, and attack sizes in terms of different ϵ-values.

Attack Structures. In Fig. 5, we visualize AutoAttack adversarial attacks. Perturbations created for the baseline trained with FGSM differ substantially from those created for FLC pooling trained with FGSM. While perturbations for the baseline model exhibit high frequency structures, attacks to FLC pooling rather affect the global image structure.

4.3 Training Efficiency

Most AT approaches use adversarial image perturbations during training [15,27, 44]. Thereby the time and memory needed depend highly on the specific attack used to generate the perturbations. Multi-step attacks like PGD [27] require substantially more time than single step attacks like FGSM [44]. TRADES [47] incorporates different loss functions to account for a good trade-off between clean and robust accuracy. With our FLC pooling, we provide a simple and fast method for more robust models. Therefore we compare our method with state-of-the-art training schedules in terms of time needed per epoch when trained in their most basic form. Table 6 shows that FGSM training is fastest. However, FGSM with early stopping is not able to maintain high robustness against AutoAttack [9] due to catastrophic overfitting. PGD training can establish robustness against AutoAttack. It relies on the same training procedure as FGSM but uses expensive multi-step perturbations and thereby increases the computation time by over a factor of four (4.23). For Adversarial Weight Perturbations (AWP) the training time per epoch is over six times (6.57), for TRADES by eight times (8.04) higher. Our FLC pooling increases the training only by a factor of 1.26 while achieving

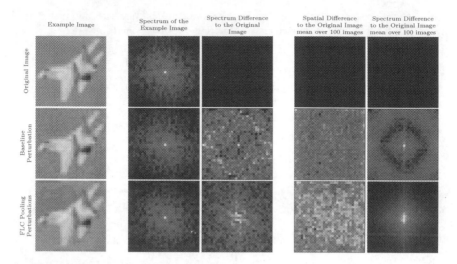

Fig. 5. Spatial and spectral differences of adversarial perturbations created by AutoAttack with $\epsilon = \frac{8}{255}$ on the baseline model as well as our FLC Pooling. On the left side for one specific example of an airplane and on the right side the average difference over 100 images.

a good clean and robust accuracy. When adding additional data like the *ddpm* dataset to the training as it is done in all leading RobustBench [8] models, the training time is increased by a factor of twenty. The *ddpm* dataset incorporates one million extra samples, which is over sixteen times more than the original CIFAR-10 dataset. We report our training times for ImageNet in Appendix A.2 to show that FLC pooling is scalable in terms of practical runtime.

4.4   Black Box Attacks

PGD and AutoAttack are intrinsically related to FGSM. Therefore, to allow for a clean evaluation of the model robustness without bias towards the training scheme, we also evaluate black box attacks. Squares [1], which is also part of the AutoAttack pipeline, adds perturbations in the form of squares onto the image until the label flips. Besides Squares, we evaluate two transferred perturbations. The first perturbation set is constructed through the baseline network which is not robust at all. The second set is constructed from the baseline network which is trained with FGSM and early stopping. We evaluate against different PRN-18 and WRN-28-10 models on CIFAR-10 as well as PRN-18 models on CIFAR-100 provided by RobustBench [8]. Note that all networks marked with * are models which rely on additional data sources such as *ddmp* [21]. Other RobustBench models like [17] rely on training data that is not available anymore such that fair comparison is currently not possible. Arguably, we always expect models to further improve as training data is added.

Table 7 shows that for PRN-18 models our FLC pooling is consistently able to prevent black box attacks better while maintaining clean accuracy compared to

Table 6. Runtime of AT in seconds per epoch over 200 epochs and a batch size of 512 trained with a PRN-18 for training on the original CIFAR-10 dataset without additional data. Experiments are executed on one Nvidia Tesla V100. Evaluation for clean and robust accuracy, higher is better, on AutoAttack [9] with our trained models. The models reported by the original authors may have different numbers due to different hyperparameter selection. The top row reports the baseline without AT.

Method	Seconds per epoch (avg)	Clean Acc	AA Acc
Baseline	14.6 ± 0.1	95.08	0.00
FGSM & early stopping [44]	**27.3 ± 0.1**	82.88	11.82
FGSM & FLC Pooling (ours)	34.5 ± 0.1	**84.81**	38.41
PGD [27]	115.4 ± 0.2	83.11	40.35
Robustness lib [12]	117 ± 19.0	76.37	32.10
AWP [45]	179.4 ± 0.4	82.61	**49.43**
MART [43]	180.4 ± 0.8	55.49	8.63
TRADES [47]	219.4 ± 0.5	81.49	46.91

other robust models from RobustBench. For WRN-28-10 models, we see a clear trend that models trained with additional data can achieve higher robustness. This is expected as wider networks can leverage additional data more effectively. One should note that all of these methods require different training schedules which are at least five times slower than ours and additional data which further increases the training time. For example, incorporating the *ddpm* dataset into the training increases the amount of training time by a factor of twenty. For CIFAR-100 (Table 8) our model is on par with [36].

4.5   Corruption Robustness

To demonstrate that our model generalizes the concept of robustness beyond adversarial examples, we also evaluate it on common corruptions incorporated with CIFAR-C [19]. We compare our model against our baseline as well as other RobustBench [8] models. Similar to the experiments on black box adversarial attacks we distinguish between models using only CIFAR-10 training data and models using extra-data like *ddpm* (marked by *). Table 7 shows that our FLC pooling, when trained only on CIFAR-10, can outperform other adversarially robust models as well as the baseline in terms of robustness against common corruptions for the PRN-18 architecture. As discussed above, WRN-28-10 models are designed to efficiently leverage additional data. As our model is exclusively trained on the clean CIFAR-10 dataset we can not establish the same robustness as other methods on wide networks. However, we can also see a substantial boost in robustness. Table 8 reports the results for CIFAR-100. There we can see that FLC pooling not only boosts clean accuracy but also robust accuracy on common corruptions.

4.6   Shift-Invariance

Initially, anti-aliasing in CNNs has also been discussed in the context of shift-invariance [49]. Therefore, after evaluating our model against adversarial and

Table 7. Robustness against black box attacks on PRN-18 and WRN-28-10 models with CIFAR-10. First against Squares [1] with $\epsilon = 1/255$ and then against perturbations which were created on the baseline network, meaning transferred perturbations (TP), and the baseline model including early stopping (TPE). As well as the accuracy under common corruptions (CC).

Model	Clean	Squares	TP	TPE	CC
Preact-ResNet-18					
Baseline	**90.81**	78.04	0.00	69.33	71.81
FGSM & early stopping	82.88	77.58	77.67	3.76	71.80
FGSM & FLC Pooling (ours)	84.81	**81.40**	**83.64**	**80.49**	**76.15**
Andriushchenko and Flammarion, 2020 [2]	79.84	76.78	78.65	75.06	72.05
Wong et al., 2020 [44]	83.34	80.25	82.03	78.81	74.60
Rebuffi et al., 2021 [35] *	83.53	81.24	82.36	80.28	75.79
WRN-28-10					
Baseline	86.67	76.17	0.09	67.3	77.33
FGSM & early stopping	82.29	78.01	80.8	28.54	72.55
FGSM & FLC Pooling (ours)	84.93	81.06	83.85	72.56	75.44
Carmon et al., 2019 [4] *	**89.69**	**87.70**	**89.12**	**83.55**	81.30
Hendrycks et al., 2019 [20]	87.11	85.02	86.47	80.12	85.02
Wang et al., 2020 [43] *	87.50	85.30	86.74	80.65	**85.30**
Zhang et al., 2021 [48]	89.36	87.45	88.70	83.08	80.11

Table 8. Robustness against black box attacks for PRN-18 on CIFAR-100. First against Squares [1] with $\epsilon = 1/255$ and then against perturbations which were created on the baseline network, meaning transferred perturbations (TP), and the baseline model including early stopping (TPE). As well as the accuracy under common corruptions (CC).

Model	Clean	Squares	TP	TPE	CC
Baseline	51.92	45.74	11.13	23.91	41.22
FGSM & early stopping	52.09	45.75	23.90	10.88	41.15
FGSM & FLC Pooling (ours)	**54.66**	48.85	45.59	45.31	**44.18**
Rice et al., 2020 [36]	53.83	**48.92**	**45.97**	**46.11**	43.48

common corruptions, we also analyze its behavior under image shifts. We compare our model with the baseline as well as the shift-invariant models from [49] and [50].

FLC pooling can outperform all these specifically designed approaches in terms of consistency under shift, while BlurPooling [49] does not outperform the baseline. We assume that BlurPooling is optimized for larger image sizes like ImageNet, 224 by 224 pixels, compared to 32 by 32 pixels for CIFAR-10. The adaptive model from [50] is slightly better than the baseline but can not reach the consistency of our model (Table 9).

Table 9. Consistency of PRN-18 model prediction under image shifts on CIFAR-10. Each model is trained without AT with the same training schedule (see Appendix A.1 for details).

Model	Clean	Consistency under shift
Baseline	94.78	86.48
BlurPooling [49]	**95.04**	86.19
adaptive BlurPooling [50]	94.97	91.47
FLC Pooling (ours)	94.66	**94.46**

5   Discussion and Conclusions

The problem of aliasing in CNNs or GANs has recently been widely discussed [11,23,24]. We contribute to this field by developing a fully aliasing-free down-sampling layer that can be plugged into any down-sampling operation. Previous attempts in this direction are based on blurring before down-sampling. This can help to reduce aliasing but can not eliminate it. With FLC pooling we developed a hyperparameter-free and easy plug-and-play down-sampling which supports CNNs native robustness. Thereby, we can overcome the issue of catastrophic overfitting in single-step AT and provide a path to reliable and fast adversarial robustness. We hope that FLC pooling will be used to evolve to fundamentally improved CNNs which do not need to account for aliasing effects anymore.

A   Appendix

A.1   Training Schedules

CIFAR-10 Adversarial Training Schedule: For our baseline experiments on CIFAR-10, we used the PRN-18 as well as the WRN-28-10 architecture as they give a good trade-off between complexity and feasibility. For the PRN-18 models, we trained for 300 epochs with a batch size of 512 and a circling learning rate schedule with the maximal learning rate 0.2 and minimal learning rate 0. We set the momentum to 0.9 and weight decay to $5e^{-4}$. The loss is calculated via Cross Entropy Loss and as an optimizer, we use Stochastic Gradient Descent (SGD). For the AT, we used the FGSM attack with an ϵ of 8/255 and an α of 10/255 (in Fast FGSM the attack is computed for step size α once and then projected to ϵ). For the WRN-28-10 we used a similar training schedule as for the PRN-18 models but used only 200 epochs and a smaller maximal learning rate of 0.08.

CIFAR-10 Clean Training Schedule: Each model is trained without AT. We used 300 epochs, a batch size of 512 for each training run and a circling learning rate schedule with the maximal learning rate at 0.2 and minimal at 0. We set the momentum to 0.9 and a weight decay to $5e^{-4}$. The loss is calculated via Cross Entropy Loss and as an optimizer, we use Stochastic Gradient Descent (SGD).

CINIC-10 Adversarial Training Schedule: For our baseline experiments on CINIC-10 we used the PRN-18 architecture. We used 300 epochs, a batch size of 512 for each training run and a circling learning rate schedule with the maximal learning rate at 0.1 and minimal at 0. We set the momentum to 0.9 and weight decay to $5e^{-4}$. The loss is calculated via Cross Entropy Loss and as an optimizer, we use Stochastic Gradient Descent (SGD). For the AT, we used the FGSM attack with an epsilon of 8/255 and an alpha of 10/255.

CIFAR-100 Adversarial Training Schedule: For our baseline experiments on CIFAR-100 we used the PRN-18 architecture as it gives a good trade-off between complexity and feasibility. We used 300 epochs, a batch size of 512 for each training run and a circling learning rate schedule with the maximal learning rate at 0.01 and minimal at 0. We set the momentum to 0.9 and a weight decay to $5e^{-4}$. The loss is calculated via Cross Entropy Loss and as an optimizer, we use Stochastic Gradient Descent (SGD). For the AT, we used the FGSM attack with an epsilon of 8/255 and an alpha of 10/255.

ImageNet Adversarial Training Schedule: For our experiment on ImageNet we used the ResNet50 architecture. We trained for 150 epochs with a batch size of 400, and a multistep learning rate schedule with an initial learning rate 0.1, $\gamma = 0.1$, and milestones [30, 60, 90, 120]. We set the momentum to 0.9 and weight decay to $5e^{-4}$. The loss is calculated via Cross Entropy Loss and as an optimizer, we use Stochastic Gradient Descent (SGD). For the AT, we used FGSM attack with an epsilon of 4/255 and an alpha of 5/255.

A.2   ImageNet Training Efficiency

When evaluating practical training times (in minutes) on ImageNet per epoch, we can not see a measurable difference in the costs between a ResNet50 with FLC pooling or strided convolution.

We varied the number of workers for dataloaders with clean training on 4 A-100 GPUs and measured ≈43 m for 12 workers, ≈22 m for 48 workers and ≈18 m for 72 workers for both. FGSM-based AT with the pipeline by [42] takes 1:07 h for both FLC pooling and strided convolutions per epoch. We conclude that training with FLC pooling in terms of practical runtime is scalable (runtime increase in ms-s range) and training times are likely governed by other factors.

The training time of our model should be comparable to the one from Wong et al. [44] while other reported methods have a significantly longer training time. Yet, the clean accuracy of the proposed model using FLC pooling improves about 8% over the one reached by [44], with a 1% improvement in robust accuracy. For example [12] has an increased training time by factor four compared to our model, already on CIFAR10 (see Table 6). This model achieves overall comparable results to ours. The model by Salman et al. [38] is trained with the training schedule from Madry et al. [32] and uses a multi-step adversarial attack for training. Since there is no release of the training script of this model on ImageNet, we can only roughly estimate their training times. Since they adopt the training schedule from Madry et al., we assume a similar training time increase of a factor of four, which is similar to the multi-step times reported for PGD in Table 6.

A.3   Aliasing Free Down-Sampling

Previous approaches like [49,50] have proposed to apply blurring operations before down-sampling, with the purpose of achieving models with improved shift invariance. Therefore, they apply Gaussian blurring directly on the feature maps via convolution. In the following, we briefly discuss why this setting can not guarantee to prevent aliasing in the feature maps, even if large convolutional kernels would be applied, and why, in contrast, the proposed FLC pooling can guarantee to prevent aliasing.

To prevent aliasing, the feature maps need to be band-limited before down-sampling [14]. This band limitation is needed to ensure that after down-sampling no replica of the frequency spectrum overlap (see Fig. 3). To guarantee the required band limitation for sub-sampling with a factor of two to $N/2$ where N is the size of the original signal, one has to remove (reduce to zero) all frequency components above $N/2$.

Spatial Filtering Based Approaches. [49,50] propose to apply approximated Gaussian filter kernels to the feature map. This operation is motivated by the fact that an actual Gaussian in the spatial domain corresponds to a Gaussian in the frequency (e.g. Fourier) domain. As the standard deviation of the Gaussian in the spatial domain increases, the standard deviation of its frequency representation decreases. Yet, the Gaussian distribution has infinite support, regardless of its standard deviation, i.e. the function never actually drops to zero. The convolution in the spatial domain corresponds to the point-wise multiplication in the frequency domain.

Therefore, even after convolving a signal with a perfect Gaussian filter with large standard deviation (and infinite support), all frequency components that were $\neq 0$ before the convolution will be afterwards (although smaller in magnitude). Specifically, the convolution with a Gaussian (even in theoretically ideal settings), can reduce the apparent aliasing but some amount of aliasing will always persist. In practice, these ideal settings are not given: Prior works such as [49,50] have to employ approximated Gaussian filters with finite support (usually not larger than 7×7).

FLC Pooling. Therefore, FLC pooling operates directly in the frequency domain, where it removes all frequencies that can cause aliases.

This operation in the Fourier domain is called the *ideal low pass filter* and corresponds to a point-wise multiplication of the Fourier transform of the feature maps with a rectangular pulse $H(\hat{m}, \hat{n})$.

$$H(\hat{m}, \hat{n}) = \begin{cases} 1 & \text{for all } \hat{m}, \hat{n} \text{ below M}/2 \text{ and N}/2 \\ 0 & \text{otherwise} \end{cases} \tag{6}$$

This trivially guarantees all frequencies above below M/2 and N/2 to be zero.

Could We Apply FLC Pooling as Convolution in the Spatial Domain?
In the spatial domain, the ideal low pass filter operation from above corresponds
to a convolution of the feature maps with the Fourier transform of the rectangular
pulse $H(\hat{m}, \hat{n})$ (by the Convolution Theorem, e.g. [14]). The Fourier transform
of the rectangle function is

$$sinc(m,n) = \begin{cases} \frac{sin(\sqrt{m^2+n^2})}{\sqrt{m^2+n^2}} & m,n \neq 0 \\ 1 & m,n = 0 \end{cases} \tag{7}$$

However, while the ideal low pass filter in the Fourier domain has finite support,
specifically all frequencies above $N/2$ are zero, $sinc(m,n)$ in the spatial domain
has infinite support. Hence, we need an infinitely large convolution kernel to
apply perfect low pass filtering in the spatial domain. This is obviously not
possible in practice. In CNNs the standard kernel size is 3×3 and one hardly
applies kernels larger than 7×7 in CNNs.

A.4   Model Confidences

In Table 10, we evaluate the confidence of model predictions. We compare each
model's confidence on correctly classified clean examples to its respective confidence on wrongly classified adversarial examples. Ideally, the confidence on the
adversarial examples should be lower. The results for the different methods show
that FLC yields comparably high confidence on correctly classified clean examples with a 20% gap in confidence to wrongly classified adversarial examples. In
contrast, the baseline model is highly confident in both cases. Other, even state-
of-the-art robustness models have on average lower confidences but are even
less confident in their correct predictions on clean examples than on erroneously
classified adversarial examples (e.g. MART [43] and PGD [27]). Only the model
from [45] has a trade-off preferable over the one from the proposed, FLC model.

Table 10. Evaluation for clean and robust accuracy, higher is better, on AutoAttack
[9] with our trained models. The models reported by the original authors may have
different numbers due to different hyperparameter selection. We report each models
confidence on their correct predictions on the clean data (Clean Confidence) and the
models confidence on its false predictions due to adversarial perturbations (Perturbation Confidence). The top row reports the baseline without adversarial training.

Method	Clean Acc	AA Acc	Clean confidence	Perturbation confidence
Baseline	95.08	0.00	100.00	97.89
FGSM & early stopping [44]	82.88	11.82	90.50	84.26
FGSM & FLC Pooling (ours)	**84.81**	38.41	**98.84**	70.98
PGD [27]	83.11	40.35	56.58	75.00
Robustness lib [12]	76.37	32.10	95.22	78.91
AWP [45]	82.61	**49.43**	88.83	**37.98**
MART [43]	55.49	8.63	24.44	50.17
TRADES [47]	81.49	46.91	53.94	50.46

A.5   AutoAttack Attack Structure

In the main paper we showed one example of an image optimized by AutoAttack [9] to fool our model and the baseline in Fig. 5. In Fig. 6, we give more examples for better visualisation and comparison.

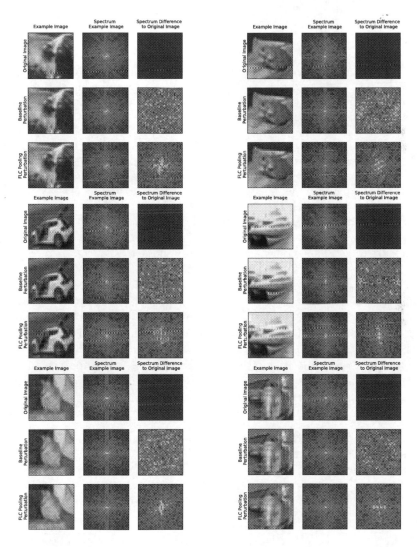

Fig. 6. Spectrum and spectral differences of adversarial perturbations created by AutoAttack with $\epsilon = \frac{8}{255}$ on the baseline model as well as our FLC Pooling. The classes from top left down to the bottom right are: Bird, Frog, Automobile, Ship, Cat and Truck.

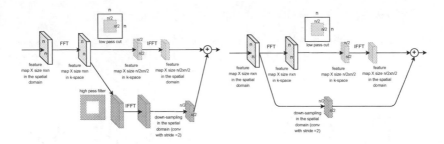

Fig. 7. LC pooling plus, which either includes the original down-sampled signal like it is done traditionally (right) or with the high frequency components filtered by a high pass filter in the Fourier domain and down-sampled in the spatial domain by an identity convolution of stride two (left).

A.6   Ablation Study: Additional Frequency Components

In addition to the low frequency components we tested different settings in which we establish a second path through which we aim to add high frequency or the original information. We either add up the feature maps or contacted them. The procedure of how to include a second path is represented in Fig. 7. One approach is to execute the standard down-sampling and add it to the FLC pooled feature map. The other is to perform a high pass filter on the feature map and down-sample these feature maps. Afterwards, the FLC pooled feature maps as well as the high pass filtered and down-sampled ones are added. With this ablation, we aim to see if we do lose too much through the aggressive FLC pooling and if we would need additional high frequency information which is discarded through the FLC pooling. Table 11 show that we can gain minor points for the clean but not for the robust accuracy. Hence we did not see any improvement in the robustness and an increase in training time per epoch as well as a minor increase in model size, we will stick to the simple FLC pooling.

Table 11. Accuracies for CIFAR-10 Baseline LowCutPooling plus the original or high freqeuncy part of the featuremaps down-sampled in the spatial domain for FGSM Training. We can see that the additional data does not improve the robust accuracy and gives only minor improvement for the clean accuracy. Due to the additional computations necessary for the high frequency /original part we decided to fully discard them and stick to the pure low frequency cutting.

Method	Clean	PGD $L_{\inf}$ $\epsilon = \frac{8}{255}$	Seconds per epoch (avg)	Model size (MB)
FLC pooling	84.81	38.41	34.6 ± 0.1	42.648
FLC pooling + HighPass pooling	85.38	38.02	45.2 ± 0.4	42.652
FLC pooling + Original pooling	85.37	38.30	35.4 ± 0.1	42.652

References

1. Andriushchenko, M., Croce, F., Flammarion, N., Hein, M.: Square attack: a query-efficient black-box adversarial attack via random search. In: Vedaldi, A., Bischof, H., Brox, T., Frahm, J.-M. (eds.) ECCV 2020. LNCS, vol. 12368, pp. 484–501. Springer, Cham (2020). https://doi.org/10.1007/978-3-030-58592-1_29
2. Andriushchenko, M., Flammarion, N.: Understanding and improving fast adversarial training. In: Advances in Neural Information Processing Systems, vol. 33, pp. 16048–16059 (2020)
3. Carlini, N., Wagner, D.: Towards evaluating the robustness of neural networks. In: 2017 IEEE Symposium on Security and Privacy (SP), pp. 39–57. IEEE (2017)
4. Carmon, Y., Raghunathan, A., Schmidt, L., Duchi, J.C., Liang, P.S.: Unlabeled data improves adversarial robustness. In: Advances in Neural Information Processing Systems, vol. 32 (2019)
5. Chaman, A., Dokmanic, I.: Truly shift-invariant convolutional neural networks. In: Proceedings of the IEEE/CVF Conference on Computer Vision and Pattern Recognition, pp. 3773–3783 (2021)
6. Chen, T., Zhang, Z., Liu, S., Chang, S., Wang, Z.: Robust overfitting may be mitigated by properly learned smoothening. In: International Conference on Learning Representations (2021). https://openreview.net/forum?id=qZzy5urZw9
7. Cheng, M., Le, T., Chen, P.Y., Yi, J., Zhang, H., Hsieh, C.J.: Query-efficient hard-label black-box attack: an optimization-based approach. arXiv preprint http://arxiv.org/abs/1807.04457arXiv:1807.04457 (2018)
8. Croce, F., et al.: RobustBench: a standardized adversarial robustness benchmark. arXiv preprint http://arxiv.org/abs/2010.09670arXiv:2010.09670 (2020)
9. Croce, F., Hein, M.: Reliable evaluation of adversarial robustness with an ensemble of diverse parameter-free attacks. In: ICML (2020)
10. Darlow, L.N., Crowley, E.J., Antoniou, A., Storkey, A.J.: CINIC-10 is not ImageNet or CIFAR-10. arXiv preprint arXiv:1810.03505 (2018)
11. Durall, R., Keuper, M., Keuper, J.: Watch your up-convolution: CNN based generative deep neural networks are failing to reproduce spectral distributions (2020)
12. Engstrom, L., Ilyas, A., Salman, H., Santurkar, S., Tsipras, D.: Robustness (python library) (2019). https://github.com/MadryLab/robustness
13. Geirhos, R., Rubisch, P., Michaelis, C., Bethge, M., Wichmann, F.A., Brendel, W.: ImageNet-trained CNNs are biased towards texture; increasing shape bias improves accuracy and robustness. arXiv preprint http://arxiv.org/abs/1811.12231arXiv:1811.12231 (2018)
14. Gonzalez, R.C., Woods, R.E.: Digital Image Processing, 3rd edn. Prentice-Hall Inc. (2006)
15. Goodfellow, I.J., Shlens, J., Szegedy, C.: Explaining and harnessing adversarial examples (2015)
16. Gowal, S., Qin, C., Uesato, J., Mann, T., Kohli, P.: Uncovering the limits of adversarial training against norm-bounded adversarial examples (2021)
17. Gowal, S., Rebuffi, S.A., Wiles, O., Stimberg, F., Calian, D.A., Mann, T.A.: Improving robustness using generated data. In: Advances in Neural Information Processing Systems, vol. 34 (2021)
18. Grabinski, J., Keuper, J., Keuper, M.: Aliasing coincides with CNNs vulnerability towards adversarial attacks. In: The AAAI-2022 Workshop on Adversarial Machine Learning and Beyond (2022). https://openreview.net/forum?id=vKc1mLxBebP

19. Hendrycks, D., Dietterich, T.: Benchmarking neural network robustness to common corruptions and perturbations. In: Proceedings of the International Conference on Learning Representations (2019)

20. Hendrycks, D., Lee, K., Mazeika, M.: Using pre-training can improve model robustness and uncertainty. In: International Conference on Machine Learning, pp. 2712–2721. PMLR (2019)

21. Ho, J., Jain, A., Abbeel, P.: Denoising diffusion probabilistic models. In: Advances in Neural Information Processing Systems, vol. 33, pp. 6840–6851 (2020)

22. Hossain, M.T., Teng, S.W., Sohel, F., Lu, G.: Anti-aliasing deep image classifiers using novel depth adaptive blurring and activation function (2021)

23. Jung, S., Keuper, M.: Spectral distribution aware image generation. In: AAAI (2021)

24. Karras, T., et al.: Alias-free generative adversarial networks. In: Advances in Neural Information Processing Systems, vol. 34 (2021)

25. Kim, H., Lee, W., Lee, J.: Understanding catastrophic overfitting in single-step adversarial training (2020)

26. Krizhevsky, A.: Learning multiple layers of features from tiny images. University of Toronto, May 2012

27. Kurakin, A., Goodfellow, I., Bengio, S.: Adversarial machine learning at scale (2017)

28. Li, Q., Shen, L., Guo, S., Lai, Z.: Wavelet integrated CNNs for noise-robust image classification. In: Proceedings of the IEEE/CVF Conference on Computer Vision and Pattern Recognition, pp. 7245–7254 (2020)

29. Li, Q., Shen, L., Guo, S., Lai, Z.: WaveCNet: wavelet integrated CNNs to suppress aliasing effect for noise-robust image classification. IEEE Trans. Image Process. **30**, 7074–7089 (2021). https://doi.org/10.1109/tip.2021.3101395

30. Lohn, A.J.: Downscaling attack and defense: turning what you see back into what you get (2020)

31. Lorenz, P., Strassel, D., Keuper, M., Keuper, J.: Is robustbench/autoattack a suitable benchmark for adversarial robustness? In: The AAAI-2022 Workshop on Adversarial Machine Learning and Beyond (2022). https://openreview.net/forum?id=aLB3FaqoMBs

32. Madry, A., Makelov, A., Schmidt, L., Tsipras, D., Vladu, A.: Towards deep learning models resistant to adversarial attacks. arXiv preprint arXiv:1706.06083 (2017)

33. Moosavi-Dezfooli, S.M., Fawzi, A., Frossard, P.: DeepFool: a simple and accurate method to fool deep neural networks. In: Proceedings of the IEEE Conference on Computer Vision and Pattern Recognition, pp. 2574–2582 (2016)

34. Rade, R., Moosavi-Dezfooli, S.M.: Helper-based adversarial training: reducing excessive margin to achieve a better accuracy vs. robustness trade-off. In: ICML 2021 Workshop on Adversarial Machine Learning (2021). https://openreview.net/forum?id=BuD2LmNaU3a

35. Rebuffi, S.A., Gowal, S., Calian, D.A., Stimberg, F., Wiles, O., Mann, T.: Fixing data augmentation to improve adversarial robustness (2021)

36. Rice, L., Wong, E., Kolter, Z.: Overfitting in adversarially robust deep learning. In: International Conference on Machine Learning, pp. 8093–8104. PMLR (2020)

37. Rony, J., Hafemann, L.G., Oliveira, L.S., Ayed, I.B., Sabourin, R., Granger, E.: Decoupling direction and norm for efficient gradient-based L2 adversarial attacks and defenses. In: Proceedings of the IEEE/CVF Conference on Computer Vision and Pattern Recognition, pp. 4322–4330 (2019)

38. Salman, H., Ilyas, A., Engstrom, L., Kapoor, A., Madry, A.: Do adversarially robust imagenet models transfer better? In: Advances in Neural Information Processing Systems, vol. 33, pp. 3533–3545 (2020)
39. Sehwag, V., et al.: Improving adversarial robustness using proxy distributions (2021)
40. Shannon, C.: Communication in the presence of noise. Proc. IRE **37**(1), 10–21 (1949). https://doi.org/10.1109/JRPROC.1949.232969
41. Stutz, D., Hein, M., Schiele, B.: Relating adversarially robust generalization to flat minima (2021)
42. Szegedy, C., et al.: Intriguing properties of neural networks. In: International Conference on Learning Representations (2014). http://arxiv.org/abs/1312.6199
43. Wang, Y., Zou, D., Yi, J., Bailey, J., Ma, X., Gu, Q.: Improving adversarial robustness requires revisiting misclassified examples. In: International Conference on Learning Representations (2020). https://openreview.net/forum?id=rklOg6EFwS
44. Wong, E., Rice, L., Kolter, J.Z.: Fast is better than free: revisiting adversarial training. In: International Conference on Learning Representations (2020). https://openreview.net/forum?id=BJx040EFvH
45. Wu, D., Xia, S.T., Wang, Y.: Adversarial weight perturbation helps robust generalization. In: Advances in Neural Information Processing Systems, vol. 33, pp. 2958–2969 (2020)
46. Xiao, Q., Li, K., Zhang, D., Jin, Y.: Wolf in sheep's clothing - the downscaling attack against deep learning applications (2017)
47. Zhang, H., Yu, Y., Jiao, J., Xing, E.P., Ghaoui, L.E., Jordan, M.I.: Theoretically principled trade-off between robustness and accuracy. In: International Conference on Machine Learning (2019)
48. Zhang, J., Zhu, J., Niu, G., Han, B., Sugiyama, M., Kankanhalli, M.: Geometry-aware instance-reweighted adversarial training. In: International Conference on Learning Representations (2021). https://openreview.net/forum?id=iAX0l6Cz8ub
49. Zhang, R.: Making convolutional networks shift-invariant again. In: ICML (2019)
50. Zou, X., Xiao, F., Yu, Z., Lee, Y.J.: Delving deeper into anti-aliasing in convnets. In: BMVC (2020)

TAFIM: Targeted Adversarial Attacks Against Facial Image Manipulations

Shivangi Aneja[1]([✉]), Lev Markhasin[2], and Matthias Nießner[1]

[1] Technical University of Munich, Munich, Germany
shivangi.aneja@tum.de
[2] Sony Europe RDC Stuttgart, Stuttgart, Germany

Abstract. Face manipulation methods can be misused to affect an individual's privacy or to spread disinformation. To this end, we introduce a novel data-driven approach that produces image-specific perturbations which are embedded in the original images. The key idea is that these protected images prevent face manipulation by causing the manipulation model to produce a predefined manipulation target (uniformly colored output image in our case) instead of the actual manipulation. In addition, we propose to leverage differentiable compression approximation, hence making generated perturbations robust to common image compression. In order to prevent against multiple manipulation methods simultaneously, we further propose a novel attention-based fusion of manipulation-specific perturbations. Compared to traditional adversarial attacks that optimize noise patterns for each image individually, our generalized model only needs a single forward pass, thus running orders of magnitude faster and allowing for easy integration in image processing stacks, even on resource-constrained devices like smartphones (Project Page: https://shivangi-aneja.github.io/projects/tafim).

1 Introduction

The spread of disinformation on social media has raised significant public attention in the recent few years, due to its implications on democratic processes and society in general. The emergence and constant improvement of generative models, and in particular face image manipulation methods, has signaled a new possible escalation of this problem. For instance, face-swapping methods [8,37] whose models are publicly accessible can be misused to generate non-consensual synthetic imagery. Other examples include face attribute manipulation methods [9,10,39,41] that change the appearance of real photos, thus generating fake images that might then be used for criminal activities [1]. Although a variety of manipulation tools have been open-sourced, surprisingly only a handful of methods have achieved widespread applicability among users (for details see the supplemental material). One reason is that re-training these methods is not only compute intensive but they

Supplementary Information The online version contains supplementary material available at https://doi.org/10.1007/978-3-031-19781-9_4.

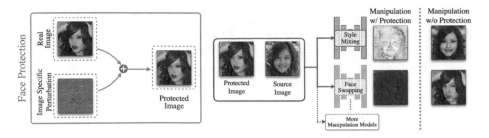

Fig. 1. Left: We propose a novel approach to protect facial images from several image manipulation models simultaneously. We leverage neural network to encode the generation of quasi-imperceptible perturbations for different manipulation models and fuse them together using attention mechanism to generate manipulation-agnostic perturbation. This perturbation, when added to the real image, forces the face manipulation models to produce a predefined manipulation target as output (white/blue image in this case). This is several orders of magnitude faster and can also be used for real-time applications. **Right:** Without any protection applied, manipulation models can be misused to generate fake images for malicious activities. (Color figure online)

also require specialized knowledge and skill sets for training. As a result, most end users only apply easily accessible pre-trained models of a few popular methods. In this work, we exploit these popular manipulation methods and their models which are known in advance and propose targeted adversarial attacks to protect against facial image manipulations.

As powerful face image manipulation tools became easier to use and more widely available, many efforts to detect image manipulations were initiated by the research community [13]. This has led to the task of automatically detecting manipulations as a classification task where predictions indicate whether a given image is real or fake. Several learning-based approaches [2,4,6,11,12,27, 28,36,43,56,62] have shown promising results in identifying manipulated images. Despite the success and high classification accuracies of these methods, they can only be helpful if they are actually being used by the end-user. However, manipulated images typically spread in private groups or on social media sites where manipulation detection is rarely available.

An alternative avenue to detecting manipulations is to prevent manipulations from happening in the first place by disrupting potential manipulation methods [19,44,57–59]. Here, the idea is to disrupt generative neural network models with low-level noise patterns, similar to the ideas of adversarial attacks used in the context of classification tasks [18,48]. Methods optimizing noise patterns for every image from scratch [44,57–59] require several seconds to process a single image. In practice, this slow run time largely prohibits their use on mobile devices (e.g., as part of the camera stack). At the same time, these manipulation prevention methods aim to either disrupt [19,29,44,47] or nullify [58,59] the results of image manipulation models, which makes it difficult to identify which face manipulation technique was used.

To address these challenges, we propose a targeted adversarial attack against face image manipulation methods. More specifically, we introduce a data-driven approach that generates quasi-imperceptible perturbations specific to a given image. Our objective is that when an image manipulation is attempted, a predefined manipulation target is generated as output instead of the originally intended manipulation. In contrast to previous optimization-based approaches, our perturbations are generated by a generalizable conditional model requiring only a few milliseconds for generation. We additionally incorporate a differentiable compression module during training, to achieve robustness against common image processing pipelines. Finally, to handle multiple manipulation models simultaneously, we propose a novel attention-based fusion mechanism to combine model-specific perturbations. In summary, the contributions in the paper are:

- A data-driven approach to synthesize image-specific perturbations that outputs a predefined manipulation target (depending on the manipulation model used), instead of per-image optimization; this is not only significantly faster but also outperforms existing methods in terms of image-to-noise quality.
- Incorporation of differentiable compression during training to achieve robustness to common image processing pipelines.
- An attention-based fusion and refinement of model-specific perturbations to prevent against multiple manipulation models simultaneously.

2   Related Work

Image Manipulation. Recent advances in image synthesis models have made it possible to generate detailed and expressive human faces [7,20–23,38,51] which might be used for unethical activities/frauds. Even more problematic can be the misuse of real face images to synthesize new ones. For instance, face-attribute modification techniques [9,10,39,41] and face-swapping models [8,37] facilitate the manipulation of existing face images. Similarly, facial re-enactment tools [17,24,46,50,60] also use real images/videos to synthesize fake videos.

Facial Manipulation Detection. The increasing availability of these image manipulation models calls for the need to reliably detect synthetic images in an automated fashion. Traditional facial manipulation detection leverages hand-crafted features such as gradients or compression artifacts, in order to find inconsistencies in an image [3,15,31]. While such self-consistency can produce good results, these methods are less accurate than more recent learning-based techniques [5,6,11,12,43], which are able to detect fake imagery with a high degree of confidence. In contrast to detecting forgeries, we aim to prevent manipulations from happening in the first place by rendering the respective manipulation models ineffective by introducing targeted adversarial attacks.

Adversarial Attacks. Adversarial attacks were initially introduced in the context of classification tasks [14,18,35,48] and eventually expanded to semantic segmentation and detection models [16,40,55]. The key idea behind these methods is to make imperceptible changes to an image in order to disrupt the feature extraction of the underlying neural networks. While these methods have achieved great success in fooling state-of-the-art vision models, one significant drawback is that optimizing a pattern for every image individually makes the optimization process quite slow. In order to address this challenge, generic universal image-agnostic noise patterns were introduced [33,34]. This has shown to be effective for misclassification tasks but gives suboptimal results for generative models, as we show in Sect. 4.

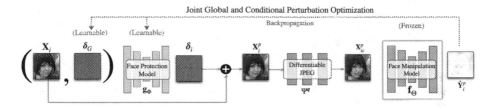

Fig. 2. We first pass the real image $\mathbf{X}_i$ and the global perturbation $\boldsymbol{\delta}_G$ through the face protection model $\mathbf{g}_\Phi$ to generate the image-specific perturbation $\boldsymbol{\delta}_i$. This perturbation is then added to the original image to create the protected image $\mathbf{X}_i^p$. The protected image is then compressed using the differentiable JPEG $\boldsymbol{\Psi}^q$ that generates compressed protected image $\mathbf{X}_{ic}^p$, which is passed through face manipulation model $\mathbf{f}_\Theta$ to generate the manipulated output $\hat{\mathbf{Y}}_i^p$. The output of the face manipulation model is then used to drive the optimization.

Manipulation Prevention. Deep steganography and watermarking techniques [30,49,53,54,57,63] can be used to embed an image-specific watermark to secure an image. For instance, FaceGuard [57] embeds a binary vector to the original image representative of a person's identity and classifies whether the image is fake by checking if the watermark is intact after being used for face manipulation tasks. These methods, however, cannot prevent the manipulation of face images which is the key focus of our work.

Recent works that aim to prevent image manipulations exploit adversarial attack techniques to break image manipulation models. Ruiz et al. [44] disrupt the output of deepfake generation models. Yeh et al. [58,59] aim to nullify the effect of image manipulation models. Other approaches [29,47] aim to disturb the output of face detection and landmark extraction steps, which are usually used as pre-processing by deepfake generation methods. One commonality of these methods is that they optimize a pattern for each image separately which is computationally very expensive, thus having limited applicability for real-world applications like resource-constrained devices. Very recently, Huang et al. [19] proposed a neural network based approach to generate image-specific patterns for low-resolution images, however, they do not consider compression, which is a common practical scenario that can make these generated patterns ineffective (as shown in Sect. 4). Additionally, this method only considers a single manipulation model at a time, thus limiting its applicability to protect against multiple manipulations simultaneously. To this end, we propose (a) a novel data-driven method to generate image-specific perturbations which are robust to compression and (b) fusion of manipulation-specific perturbations. Our method not only require less computational effort compared to existing adversarial attacks works, but can protect against multiple manipulation methods simultaneously.

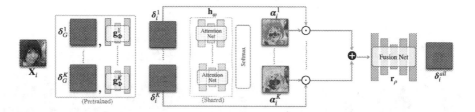

Fig. 3. For a given RGB image $\mathbf{X}_i$, we first use the pre-trained manipulation-specific global noise and protection models $\{\boldsymbol{\delta}_G^k, \mathbf{g}_\Phi^k\}_{k=1}^K$ to generate manipulation-specific perturbations $\{\boldsymbol{\delta}_i^k\}_{k=1}^K$, which are passed into a shared attention backbone h_ω to generate the spatial attention maps $\{\boldsymbol{\alpha}_i^k\}_{k=1}^K$. These attention maps are then combined with manipulation-specific $\{\boldsymbol{\delta}_i^k\}_{k=1}^K$ using channel-wise hadamard product and blended together using addition operation. Finally, the blended perturbation is then refined using FusionNet r_ρ to generate manipulation-agnostic perturbation $\boldsymbol{\delta}_i^{\text{all}}$.

3 Proposed Method

Our goal is to prevent face image manipulations and simultaneously identify which model was used for the manipulation. That is, for a given face image, we aim to find an imperceptible perturbation that disrupts the generative neural network of a manipulation method such that a solid color image is produced as output instead of originally-intended manipulation. Algorithmically, this is a targeted adversarial attack where the predefined manipulation targets make it easy for a human to identify the used manipulation method.

3.1 Method Overview

We consider a setting where we are given K manipulation models $\mathcal{M} = \{\mathbf{f}_\Theta^k\}_{k=1}^K$ where $\mathbf{f}_\Theta^k$ denotes the k-th manipulation model. For a given RGB image $\mathbf{X}_i \in \mathbb{R}^{H \times W \times 3}$ of height H and width W, the goal is to find the optimal perturbation $\boldsymbol{\delta}_i \in \mathbb{R}^{H \times W \times 3}$ that is embedded in the original image $\mathbf{X}_i$ to produce a valid protected image $\mathbf{X}_i^p \in \mathbb{R}^{H \times W \times 3}$. The manipulation model $\mathbf{f}_\Theta^k$, which is parametrized by its neural network weights Θ, is also given as input to the method. Note that we use $\mathbf{f}_\Theta^k$ only to drive the perturbation optimization and do not alter its weights. For the given image $\mathbf{X}_i$, the output synthesized by the manipulation model $\mathbf{f}_\Theta^k$ is denoted as $\hat{\mathbf{Y}}_{ik} \in \mathbb{R}^{H \times W \times 3}$. We define the uniformly-colored predefined manipulation targets for the K manipulation models as $\mathcal{Y} = \{\mathbf{Y}_k^{\text{target}}\}_{k=1}^K$.

In order to protect face images and obtain image perturbations, we propose two main ideas: First, for a given manipulation model $\mathbf{f}_\Theta$, we jointly optimize for a global perturbation pattern $\boldsymbol{\delta}_G \in \mathbb{R}^{H \times W \times 3}$ and a generative neural network

g_{Φ} (parameterized by its weights Φ) to produce image-specific perturbations δ_i. The global pattern δ_G is generalized across the entire data distribution. The generative model g_{Φ} is conditioned on the global perturbation δ_G as well as the real image X_i. Our intuition is that the global perturbation provides a strong prior for the global noise structure, thus enabling the conditional model to produce more effective perturbations. We also incorporate a differentiable JPEG module to ensure the robustness of the perturbations towards compression. This is shown in Fig. 2.

Second, to handle multiple manipulation models simultaneously, we leverage an attention network h_{ω} (parametrized by ω) to first generate attention maps $\{\alpha_k\}_{k=1}^{K}$ for the K manipulation methods, which are then used to refine the model-specific perturbations $\{\delta_i^k\}_{k=1}^{K}$ with an encoder-decoder network denoted as FusionNet r_{ρ} (parametrized by ρ) to generate a single final perturbation δ_i^{all} for the given image X_i. δ_i^{all} can protect the image from the given K manipulation methods simultaneously. An overview is shown in Fig. 3.

3.2   Methodology

We define an optimization strategy where the objective is to find the smallest possible perturbation that achieves the largest disruption in the output manipulation; i.e., where the generated output for the k-th manipulation model is closest to its predefined target image Y_k^{target}. This is explained in detail below.

Joint Global and Conditional Generative Model Optimization. The global perturbation $\delta_G \in \mathbb{R}^{H \times W \times 3}$ is a fixed image-agnostic perturbation shared across the data distribution. The conditional generative neural network model g_{Φ} is a UNet [42] based encoder-decoder architecture. For a given manipulation method, we jointly optimize global perturbation δ_G and the parameters Φ of this conditional model g_{Φ} together in order to generate image-specific perturbations.

$$\delta_G^*, \Phi^* = \text{argmin}_{\delta_G, \Phi} \ \mathcal{L}_k \tag{1}$$

where $\mathcal{L}_k$ refers to overall loss (Eq. 2).

$$\mathcal{L}_k = \left[\sum_{i=1}^{N} \mathcal{L}_i^{\text{recon}} + \lambda \mathcal{L}_i^{\text{perturb}} \right]_k, \tag{2}$$

where the parameter λ regularizes the strength of perturbation added to the real image, N denotes the number of images in the dataset, i denotes the image index and k denotes the manipulation method. $\mathcal{L}_i^{\text{recon}}$ and $\mathcal{L}_i^{\text{perturb}}$ represent reconstruction and perturbation losses for i-th image. The model g_{Φ} is conditioned on the globally-optimized perturbation δ_G as well as the original input image X_i. Conditioning the model g_{Φ} on δ_G facilitates the transfer of global structure from the facial imagery to produce highly-efficient perturbations, i.e., these perturbations are more successful in disturbing manipulation models to produce results close to the manipulation targets. The real image $\mathbf{X}_i$ and global perturbation $\boldsymbol{\delta}_G$ are first concatenated channel-wise, $\widehat{\mathbf{X}}_i = [\mathbf{X}_i, \boldsymbol{\delta}_G]$, to generate a six-channel input $\widehat{\mathbf{X}}_i \in \mathbb{R}^{H \times W \times 6}$.

$\widehat{\mathbf{X}}_i$ is then passed through the conditional model $\boldsymbol{g}_{\boldsymbol{\Phi}}$ to generate image-specific perturbation $\boldsymbol{\delta}_i = \boldsymbol{g}_{\boldsymbol{\Phi}}(\widehat{\mathbf{X}}_i)$. These image-specific perturbations $\boldsymbol{\delta}_i$ are then added to the respective input images $\mathbf{X}_i$ to generate the protected image $\mathbf{X}_i^p$ as

$$\mathbf{X}_i^p = \mathrm{Clamp}_\varepsilon(\mathbf{X}_i + \boldsymbol{\delta}_i). \tag{3}$$

The $\mathrm{Clamp}_\varepsilon(\xi)$ function projects higher/lower values of ξ into the valid interval $[-\varepsilon, \varepsilon]$. Similarly, we generate the protected image using global perturbation $\boldsymbol{\delta}_G$ as

$$\mathbf{X}_i^{Gp} = \mathrm{Clamp}_\varepsilon(\mathbf{X}_i + \boldsymbol{\delta}_G). \tag{4}$$

For the generated conditional and global protected image $\mathbf{X}_i^p$ and $\mathbf{X}_i^{Gp}$ and the given manipulation model $\mathbf{f}_{\boldsymbol{\Theta}}^k$, the reconstruction loss $\mathcal{L}_i^{\mathrm{recon}}$ and perturbation loss $\mathcal{L}_i^{\mathrm{perturb}}$ are formulated as

$$\mathcal{L}_i^{\mathrm{recon}} = \left\| \mathbf{f}_{\boldsymbol{\Theta}}^k(\mathbf{X}_i^p) - \mathbf{Y}_k^{\mathrm{target}} \right\|_2 + \left\| \mathbf{f}_{\boldsymbol{\Theta}}^k(\mathbf{X}_i^{Gp}) - \mathbf{Y}_k^{\mathrm{target}} \right\|_2. \tag{5}$$

$$\mathcal{L}_i^{\mathrm{perturb}} = \left\| \mathbf{X}_i^p - \mathbf{X}_i \right\|_2 + \left\| \mathbf{X}_i^{Gp} - \mathbf{X}_i \right\|_2. \tag{6}$$

Finally, the overall loss can then be written as

$$\mathcal{L}_k = \left[\sum_{i=1}^{N} \left\| \mathbf{f}_{\boldsymbol{\Theta}}^k(\mathbf{X}_i^p) - \mathbf{Y}_k^{\mathrm{target}} \right\|_2 + \left\| \mathbf{f}_{\boldsymbol{\Theta}}^k(\mathbf{X}_i^{Gp}) - \mathbf{Y}_k^{\mathrm{target}} \right\|_2 + \right.$$
$$\left. \lambda \left(\left\| \mathbf{X}_i^p - \mathbf{X}_i \right\|_2 + \left\| \mathbf{X}_i^{Gp} - \mathbf{X}_i \right\|_2 \right) \right]_k. \tag{7}$$

The global perturbation $\boldsymbol{\delta}_G$ is initialized with a random vector sampled from a multivariate uniform distribution, i.e., $\boldsymbol{\delta}_{\mathbf{G}}^0 \sim \mathcal{U}(\mathbf{0}, \mathbf{1})$ and optimized iteratively. Note that $\mathbf{X}_i^{Gp}$ is used only to drive the optimization of $\boldsymbol{\delta}_G$. For further details on the network architecture and hyperparameters, we refer to Sect. 4 and the supplemental material.

Differentiable JPEG Compression. In many practical scenarios, images shared on social media platforms get compressed over the course of transmission. Our initial experiments suggest that protected images $\mathbf{X}_i^p$ generated from the previous steps can easily become ineffective by applying image compression. In order to make our perturbations robust, we propose to incorporate a differentiable JPEG compression into our generative model; i.e., we aim to generate perturbations that still disrupt the manipulation models even if the input is compressed. The actual JPEG compression technique [52] is non-differentiable due to the lossy quantization step (details in supplemental) where information loss happens with the round operation as, $x := \text{round}(x)$. Therefore, we cannot train our protected images against the original JPEG technique. Instead, we leverage continuous and differentiable approximations [25,45] to the rounding operator. For our experiments, we use the sin approximation by Korus et al. [25]

$$x := x - \frac{\sin(2\pi x)}{2\pi}. \tag{8}$$

This differentiable round approximation coupled with other transformations from the actual JPEG technique can be formalized into differentiable JPEG operation. We denote the full differentiable JPEG compression as $\boldsymbol{\Psi}^q$, where q denotes the compression quality.

For training, we first map the protected image $\mathbf{X}_i^p$ to RGB colorspace $[0, 255]$ before applying image compression, obtaining $\widetilde{\mathbf{X}}_i^p$. Next, the image $\widetilde{\mathbf{X}}_i^p$ is passed through differential JPEG layers $\boldsymbol{\Psi}^q$ to generate a compressed image $\widetilde{\mathbf{X}}_{ic}^p$, which is then normalized again as $\mathbf{X}_{ic}^p$ before passing it to the manipulation model $\mathbf{f}_{\boldsymbol{\Theta}}$.

Training with a fixed compression quality ensures robustness to that specific quality but shows limited performance when evaluated with different compression qualities. We therefore, generalize across compression levels by training our model with different compression qualities. Specifically, at each iteration, we randomly sample quality q from a discrete uniform distribution $\mathcal{U}_D(1, 99)$, i.e. $q \sim \mathcal{U}_D(1, 99)$ and compress the protected image $\mathbf{X}_i^p$ at quality level q.

This modifies the reconstruction loss $\mathcal{L}_{\text{recon}}$ as follows

$$\mathcal{L}_i^{\text{recon}} = \left\| \mathbf{f}_{\boldsymbol{\Theta}}^k(\boldsymbol{\Psi}^q(\mathbf{X}_i^p)) - \mathbf{Y}_k^{\text{target}} \right\|_2 + \left\| \mathbf{f}_{\boldsymbol{\Theta}}^k(\mathbf{X}_i^{Gp}) - \mathbf{Y}_k^{\text{target}} \right\|_2 \tag{9}$$

where $\mathbf{X}_{ic}^p = \boldsymbol{\Psi}^q(\mathbf{X}_i^p)$ denotes the compressed protected image. Backpropagating the gradients through $\boldsymbol{\Psi}^q$ during training ensures that the added perturbations survive different compression qualities. At test time, we evaluate results with actual JPEG compression technique instead of approximated/differential used during training to report the results.

Multiple Manipulation Methods. To handle multiple manipulation models simultaneously, we combine model-specific perturbations $\{\boldsymbol{\delta}_i^k\}_{k=1}^K$ obtained previously using $\{\boldsymbol{\delta}_G^k, \mathbf{g}_{\boldsymbol{\Phi}}^k\}_{k=1}^K$ and feed them to our attention network $\mathbf{h}_{\boldsymbol{\omega}}$ (parameterized by $\boldsymbol{\omega}$) and fusion network $\mathbf{r}_{\boldsymbol{\rho}}$ (parameterized by $\boldsymbol{\rho}$) to generate model-agnostic perturbations $\boldsymbol{\delta}_i^{\text{all}}$.

$$\boldsymbol{\omega}^*, \boldsymbol{\rho}^* = \underset{\boldsymbol{\omega}, \boldsymbol{\rho}}{\text{argmin}} \ \mathcal{L}_{\text{all}}. \tag{10}$$

We leverage the pre-trained global pattern and conditional perturbation model pairs $\{\boldsymbol{\delta}_G^k, \mathbf{g}_{\boldsymbol{\Phi}}^k\}_{k=1}^K$ for each of the K different models to generate the final perturbation $\boldsymbol{\delta}_i^{\text{all}}$ for image $\mathbf{X}_i$. More precisely, for the image $\mathbf{X}_i$, we first use the pre-trained $\{\boldsymbol{\delta}_G^k, \mathbf{g}_{\boldsymbol{\Phi}}^k\}_{k=1}^K$ to generate the model-specific perturbations $\{\boldsymbol{\delta}_i^k\}_{k=1}^K$ as:

$$\boldsymbol{\delta}_i^k = \mathbf{g}_{\boldsymbol{\Phi}}^k(\mathbf{X}_i, \boldsymbol{\delta}_G^k). \tag{11}$$

Next, these model-specific perturbations $\{\boldsymbol{\delta}_i^k\}_{k=1}^K$ are fed into attention module $\mathbf{h}_\omega$ coupled with the softmax operation to generate spatial attention maps $\{\boldsymbol{\alpha}_i^k\}_{k=1}^K$ as:

$$\boldsymbol{\alpha}_i^k = \frac{\exp\big(\mathbf{h}_\omega(\boldsymbol{\delta}_i^k, C_k)\big)}{\sum\limits_{k=1}^K \exp\big(\mathbf{h}_\omega(\boldsymbol{\delta}_i^k, C_k)\big)}. \tag{12}$$

where $\boldsymbol{\alpha}_i^k \in \mathbb{R}^{H \times W}$ and C_k refer to class label for the k-th manipulation model. These spatial attention maps are then blended with model-specific perturbations and refined with a fusion network $\mathbf{r}_\rho$ to generate the final perturbation $\boldsymbol{\delta}_i^{\text{all}}$ as:

$$\boldsymbol{\delta}_i^{\text{all}} = \mathbf{r}_\rho\left(\sum_{k=1}^K (\boldsymbol{\alpha}_i^k \odot \boldsymbol{\delta}_i^k)\right) \tag{13}$$

Finally, $\boldsymbol{\delta}_i^{\text{all}}$ is added to the image $\mathbf{X}_i$ to generate the common protected image $\mathbf{X}_i^{\text{all}} = \text{Clamp}_\varepsilon(\mathbf{X}_i + \boldsymbol{\delta}_i^{\text{all}})$ and total loss is formalized as

$$\mathcal{L}_{\text{all}} = \sum_{i=1}^N \left[\sum_{k=1}^K \left(\left\|\mathbf{f}_\Theta^k(\mathbf{X}_i^{\text{all}}) - \mathbf{Y}_k^{\text{target}}\right\|_2\right) + \lambda\left\|\boldsymbol{\delta}_i^{\text{all}}\right\|_2\right]. \tag{14}$$

4   Results

We compare our method against well-studied adversarial attack baselines I-FGSM [26] and I-PGD [32]. To demonstrate our results, we perform experiments with three different models: (1) pSp Encoder [41] which can be used for self-reconstruction and style-mixing (protected with solid white image as manipulation target), and (2) SimSwap [8] for face-swapping (protected with solid blue as manipulation target). (3) StyleClip [39] for text-driven manipulation (protected with solid red as manipulation target). For all these manipulations, we use the publicly available pre-trained models. For pSp encoder, we use a model that is trained for a self-reconstruction task. The same model can also be used for style-mixing to synthesize new images by mixing the latent style features of two images. For style-mixing and face-swapping, protection is applied to the target image. We introduce a custom split on FFHQ [22] for our experiments. We use 10K images for training and 1K images for val and test split each. More details can be found in supplemental. All results are reported on the corresponding test sets for each task respectively.

Experimental Setup. All images are first resized to 256×256 pixels. The global perturbation and conditional model are jointly optimized for 100k iterations with a learning rate of 0.0001 and Adam optimizer. For the protection $\boldsymbol{g}_\Phi$, attention $\boldsymbol{h}_\omega$ and fusion $\boldsymbol{r}_\rho$ network, we use the same UNet-64 encoder-decoder architecture. We use a batch size of 1 for all our experiments. For I-PGD, we use a step size of 0.01. Both I-FGSM and I-PGD are optimized for 100 steps for every image in the test split. More details on training setup and hyperparameters can be found in the supplemental.

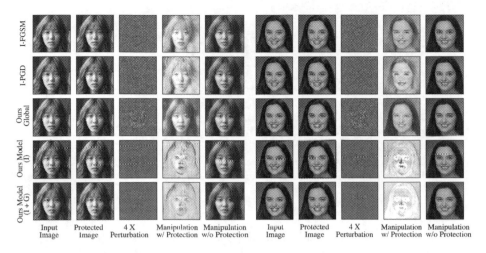

Fig. 4. Comparison on self-reconstruction task with white image as manipulation target. Perturbation enlarged (4×) for better visibility. *Ours Global* refers to the optimized single global perturbation for all the images. *Ours Model (I)* refers to the model conditioned only on real images and *Ours Model (I + G)* refers to the model conditioned on global perturbation and real image, outperforms alternate baselines.

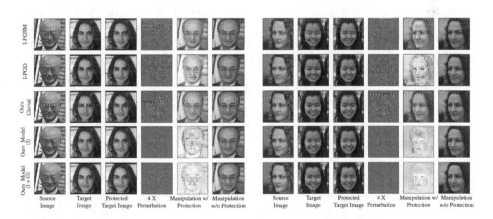

Fig. 5. Comparison on the style-mixing task (white target). The protection is applied to the target image. All methods are trained only for the self-reconstruction task and evaluated on style-mixing. Perturbation enlarged (4×) for better visibility. *Ours Global* refers to the optimized single global perturbation. *Ours Model (I)* refers to the model conditioned only on real images and *Ours Model (I + G)* refers to the model conditioned on global perturbation and real image, outperforms alternate baselines.

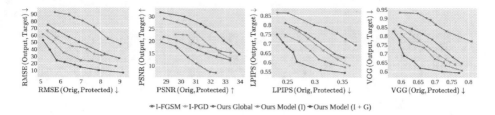

Fig. 6. Comparison with different optimization techniques evaluated on self-reconstruction (white target). We plot the output image quality (y-axis) corresponding to different levels of perturbations added to the image (x-axis). *Orig* and *Protected* refer to the original and protected image. *Output* refers to the output of the manipulation model and *Target* indicates the predefined manipulation target. Note that our method outperforms other baselines at all the different perturbation levels.

Metrics. To evaluate the output quality, we compute relative performance at different perturbation levels, i.e., we plot a graph with the x-axis showing different perturbation levels for the image and the y-axis showing how close is the output of the face manipulation model to the predefined manipulation target. We plot the graph for RMSE, PSNR, LPIPS [61] and VGG loss. In the optimal setting, for a low perturbation in the image, the output should look identical to the manipulation target; i.e., a lower graph is better for RMSE, LPIPS and VGG loss and higher for PSNR.

Baseline Comparisons. To compare our method against other adversarial attack baselines, we first evaluate the results of our proposed method on a single manipulation model without compression; i.e., neither training nor evaluating for JPEG compression. Visual results for self-reconstruction and style mixing are shown in Figs. 4 and 5. The performance graph for different perturbation levels is shown in Fig. 6. We observe that the model conditioned on the global perturbation as well as real images outperforms the model trained only with real images, indicating that the global perturbation provides a strong prior in generating more powerful perturbations.

Runtime Comparison. We compare run-time performance against state-of-the-art in Table 1. I-FGSM [26] and I-PGD [32] optimize for perturbation patterns for each image individually at run time; hence they are orders of magnitude slower than our method that only requires a single forward pass of our conditional generative neural network. Our model takes only 77.89 ± 2.71 ms and 117.0 MB memory to compute the perturbation for a single image on an Intel(R) Xeon(R) W-2133 CPU @ 3.60 GHz. This is an order of magnitude faster compared to per-image methods that are run on GPUs. We believe this makes our method ideally suited to real-time scenarios, even on mobile hardware.

Robustness to JPEG Compression. Next, we investigate the sensitivity of perturbations to different compression qualities. We apply the actual JPEG compression technique to report results. We observe that without training the model against different compression leads to degraded results when evaluated on compressed images, Fig. 7 and 8. Training the model with fixed compression quality makes the perturbation robust to that specific compression quality; however, it fails for other compression levels; see Fig. 9. We therefore train across different compression levels varied during training iterations.

Table 1. Run-time performance (averaged over 10 runs) to generate a perturbation for a single image on the self-reconstruction task. Our method runs an order of magnitude faster than existing works that require per-image optimization. All timings are measured on an Nvidia Titan RTX 2080 GPU.

Method	Time
I-FGSM [26]	17517.71 ms (±124.08 ms)
I-PGD [32]	17523.01 ms (±204.15 ms)
Ours	**10.66 ms** (±0.21 ms)

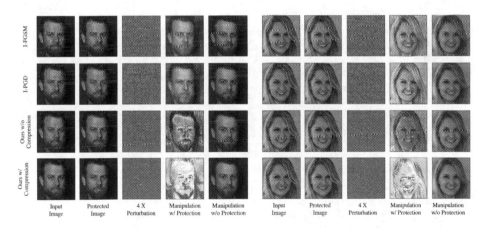

Fig. 7. Qualitative comparison in the presence of JPEG compression (white target). Methods trained without compression struggle; in contrast, our model trained with compression is able to produce perturbations that are robust to compression. *Ours w/ Compression* refers to the model trained with random compression. *Ours w/o Compression* refers to model trained without compression. Compression is applied on the protected images. All methods are evaluated at compression quality C-80.

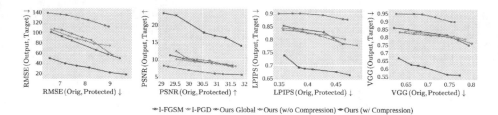

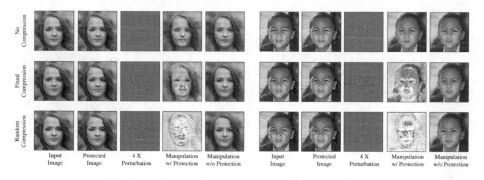

Fig. 8. Performance comparison in the presence of JPEG compression. Our method without differentiable JPEG training manages to disrupt the model; however, training with random compression levels significantly outperforms the uncompressed baselines. All methods are evaluated at compression quality C-80.

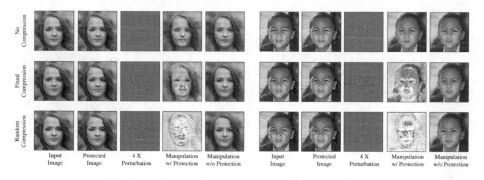

Fig. 9. Comparison for our method trained without compression, fixed compression, and random compression for self-reconstruction task (white target). The fixed compression model was trained with compression quality C-80. All methods are evaluated on compression quality C-30. The randomly compressed model outperforms both fixed and no compression models.

Multiple Manipulation Models. We leverage manipulation-specific perturbations as priors and combine them using attention-based fusion to generate a single perturbation to protect against multiple manipulations at the same time. As a baseline, we also compare against a model trained directly for all manipulation methods combined without manipulation-specific priors or attention. We notice that this setup is unable to produce optimal perturbations due to absence of prior information from manipulation-specific perturbations which provide the most optimal perturbations to produce predefined manipulation targets. We color-code the manipulation targets with different colors for different manipulation models. This protection technique has an advantage over simple disruption since it gives more information about which technique was used to manipulate the image. We conduct experiments with three different state-of-the-art methods: pSp [41] with solid white image as the manipulation target, SimSwap [8] with solid blue image as target and StyleClip [39] with solid red as target image. Visual results and performance graph comparison are shown in Fig. 10. Our combined model with attention produces more effective results than without attention baseline, and without any significant degradation compared to manipulation-specific baselines when handling multiple manipulation at the same time.

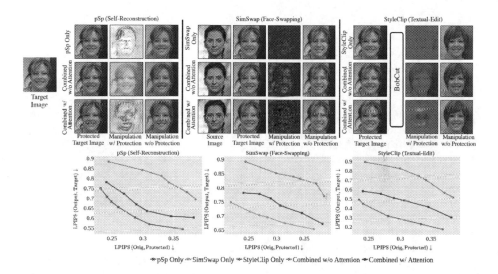

Fig. 10. Visual results (top) and performance graph (bottom) for multiple targets simultaneously. *pSp Only*, *SimSwap Only*, and *StyleClip Only* refer to the individual protection models trained only for the respective manipulations. *Combined w/o Attention* refers to a model trained directly for all manipulation methods combined. *Combined w/ Attention* refers to our proposed attention-based fusion approach. Our proposed attention model performs much better the no attention baseline, and is comparable to individual models.

5   Conclusion

In this work, we proposed a data-driven approach to protect face images from potential popular manipulations. Our method can both prevent and simultaneously identify the manipulation technique by generating the predefined manipulation target as output. In comparison to existing works, our method not only run orders of magnitude faster, but also achieves superior performance; i.e., with smaller perturbations of a given input image, we can achieve larger disruptions in the respective manipulation methods. In addition, we proposed an end-to-end compression formulation to make the perturbation robust to compression. Furthermore, we propose a new attention-based fusion approach to handle multiple manipulations simultaneously. We believe our generalized, data-driven method takes an important step towards addressing the potential misuse of popular face image manipulation techniques.

Acknowledgment. This work is supported by a TUM-IAS Rudolf Mößbauer Fellowship, the ERC Starting Grant Scan2CAD (804724), and Sony Semiconductor Solutions Corporation. We would also like to thank Angela Dai for video voice over.

References

1. Maisy Kinsley fake account. https://twitter.com/sokane1/status/1111023838467362816. Accessed 27 Mar 2019
2. Afchar, D., Nozick, V., Yamagishi, J., Echizen, I.: Mesonet: a compact facial video forgery detection network (2018). https://doi.org/10.1109/wifs.2018.8630761
3. Agarwal, S., Farid, H.: Photo forensics from jpeg dimples. In: 2017 IEEE Workshop on Information Forensics and Security (WIFS), pp. 1–6 (2017). https://doi.org/10.1109/WIFS.2017.8267641
4. Agarwal, S., Farid, H., Gu, Y., He, M., Nagano, K., Li, H.: Protecting world leaders against deep fakes. In: Proceedings of the IEEE Conference on Computer Vision and Pattern Recognition (CVPR) Workshops, p. 8. IEEE, Long Beach (2019)
5. Agarwal, S., Farid, H., Gu, Y., He, M., Nagano, K., Li, H.: Protecting world leaders against deep fakes. In: CVPR Workshops (2019)
6. Aneja, S., Nießner, M.: Generalized Zero and few-shot transfer for facial forgery detection. ArXiv preprint arXiv:2006.11863 (2020)
7. Brock, A., Donahue, J., Simonyan, K.: Large scale GAN training for high fidelity natural image synthesis. In: International Conference on Learning Representations (2019). https://openreview.net/forum?id=B1xsqj09Fm
8. Chen, R., Chen, X., Ni, B., Ge, Y.: Simswap: an efficient framework for high fidelity face swapping. In: MM 2020: The 28th ACM International Conference on Multimedia, pp. 2003–2011. ACM (2020)
9. Choi, Y., Choi, M., Kim, M., Ha, J.W., Kim, S., Choo, J.: Stargan: unified generative adversarial networks for multi-domain image-to-image translation. In: Proceedings of the IEEE Conference on Computer Vision and Pattern Recognition (2018)
10. Choi, Y., Uh, Y., Yoo, J., Ha, J.W.: Stargan v2: diverse image synthesis for multiple domains. In: Proceedings of the IEEE Conference on Computer Vision and Pattern Recognition (2020)
11. Cozzolino, D., Rössler, A., Thies, J., Nießner, M., Verdoliva, L.: Id-reveal: identity-aware deepfake video detection. In: Proceedings of the IEEE/CVF International Conference on Computer Vision (ICCV), pp. 15108–15117 (2021)
12. Cozzolino, D., Thies, J., Rössler, A., Riess, C., Nießner, M., Verdoliva, L.: Forensic-transfer: Weakly-supervised domain adaptation for forgery detection. arXiv (2018)
13. Dolhansky, B., Bitton, J., Pflaum, B., Lu, J., Howes, R., Wang, M., Ferrer, C.C.: The deepfake detection challenge (dfdc) dataset (2020)
14. Dong, Y., et al.: Boosting adversarial attacks with momentum. In: 2018 IEEE/CVF Conference on Computer Vision and Pattern Recognition, pp. 9185–9193 (2018)
15. Ferrara, P., Bianchi, T., De Rosa, A., Piva, A.: Image forgery localization via fine-grained analysis of CFA artifacts. IEEE Trans. Inf. Forensics Secur. **7**(5), 1566–1577 (2012). https://doi.org/10.1109/TIFS.2012.2202227
16. Fischer, V., Kumar, M.C., Metzen, J.H., Brox, T.: Adversarial examples for semantic image segmentation (2017)

17. Gafni, G., Thies, J., Zollhöfer, M., Nießner, M.: Dynamic neural radiance fields for monocular 4D facial avatar reconstruction. In: Proceedings of the IEEE/CVF Conference on Computer Vision and Pattern Recognition (CVPR), pp. 8649–8658 (2021)
18. Goodfellow, I.J., Shlens, J., Szegedy, C.: Explaining and harnessing adversarial examples (2015)
19. Huang, Q., Zhang, J., Zhou, W., WeimingZhang, Yu, N.: Initiative defense against facial manipulation (2021)
20. Karras, T., Aila, T., Laine, S., Lehtinen, J.: Progressive growing of GANs for improved quality, stability, and variation. In: International Conference on Learning Representations (2018). https://openreview.net/forum?id=Hk99zCeAb
21. Karras, T., et al.: Alias-free generative adversarial networks. In: Proceedings of NeurIPS (2021)
22. Karras, T., Laine, S., Aila, T.: A style-based generator architecture for generative adversarial networks (2019)
23. Karras, T., Laine, S., Aittala, M., Hellsten, J., Lehtinen, J., Aila, T.: Analyzing and improving the image quality of stylegan. In: Proceedings of the IEEE/CVF Conference on Computer Vision and Pattern Recognition (CVPR) (2020)
24. Kim, H., et al.: Deep video portraits. ACM Trans. Graph. (TOG) (2018)
25. Korus, P., Memon, N.: Content authentication for neural imaging pipelines: end-to-end optimization of photo provenance in complex distribution channels. In: Proceedings of the IEEE/CVF Conference on Computer Vision and Pattern Recognition (CVPR) (2019)
26. Kurakin, A., Goodfellow, I.J., Bengio, S.: Adversarial machine learning at scale. ArXiv abs/1611.01236 (2017)
27. Li, L., et al.: Face x-ray for more general face forgery detection. In: Proceedings of the IEEE/CVF Conference on Computer Vision and Pattern Recognition, pp. 5001–5010 (2020)
28. Li, Y., Lyu, S.: Exposing deepfake videos by detecting face warping artifacts (2018)
29. Li, Y., Yang, X., Wu, B., Lyu, S.: Hiding faces in plain sight: disrupting AI face synthesis with adversarial perturbations. ArXiv abs/1906.09288 (2019)
30. Luo, X., Zhan, R., Chang, H., Yang, F., Milanfar, P.: Distortion agnostic deep watermarking. In: 2020 IEEE/CVF Conference on Computer Vision and Pattern Recognition (CVPR), pp. 13545–13554 (2020)
31. Lyu, S., Pan, X., Zhang, X.: Exposing region splicing forgeries with blind local noise estimation. Int. J. Comput. Vision $110(2)$, 202–221 (2014)
32. Madry, A., Makelov, A., Schmidt, L., Tsipras, D., Vladu, A.: Towards deep learning models resistant to adversarial attacks. ArXiv abs/1706.06083 (2018)
33. Metzen, J.H., Kumar, M.C., Brox, T., Fischer, V.: Universal adversarial perturbations against semantic image segmentation (2017). https://arxiv.org/abs/1704.05712
34. Moosavi-Dezfooli, S.M., Fawzi, A., Fawzi, O., Frossard, P.: Universal adversarial perturbations. In: Proceedings of the IEEE Conference on Computer Vision and Pattern Recognition (CVPR) (July 2017)
35. Moosavi-Dezfooli, S.M., Fawzi, A., Frossard, P.: Deepfool: a simple and accurate method to fool deep neural networks. In: 2016 IEEE Conference on Computer Vision and Pattern Recognition (CVPR), pp. 2574–2582 (2016)
36. Nguyen, H.H., Yamagishi, J., Echizen, I.: Capsule-forensics: using capsule networks to detect forged images and videos. In: ICASSP 2019–2019 IEEE International Conference on Acoustics, Speech and Signal Processing (ICASSP) (2019). https://doi.org/10.1109/icassp.2019.8682602

37. Nirkin, Y., Keller, Y., Hassner, T.: FSGAN: subject agnostic face swapping and reenactment. In: Proceedings of the IEEE International Conference on Computer Vision, pp. 7184–7193 (2019)
38. Park, T., Liu, M.Y., Wang, T.C., Zhu, J.Y.: Gaugan: semantic image synthesis with spatially adaptive normalization. In: ACM SIGGRAPH 2019 Real-Time Live! SIGGRAPH 2019. Association for Computing Machinery, New York (2019)
39. Patashnik, O., Wu, Z., Shechtman, E., Cohen-Or, D., Lischinski, D.: Styleclip: text-driven manipulation of stylegan imagery. In: Proceedings of the IEEE/CVF International Conference on Computer Vision (ICCV), pp. 2085–2094 (2021)
40. Poursaeed, O., Katsman, I., Gao, B., Belongie, S.: Generative adversarial perturbations. In: Proceedings of the IEEE Conference on Computer Vision and Pattern Recognition, pp. 4422–4431 (2018)
41. Richardson, E., et al.: Encoding in style: a stylegan encoder for image-to-image translation. In: IEEE/CVF Conference on Computer Vision and Pattern Recognition (CVPR) (2021)
42. Ronneberger, O., Fischer, P., Brox, T.: U-net: convolutional networks for biomedical image segmentation. In: Navab, N., Hornegger, J., Wells, W.M., Frangi, A.F. (eds.) MICCAI 2015. LNCS, vol. 9351, pp. 234–241. Springer, Cham (2015). https://doi.org/10.1007/978-3-319-24574-4_28
43. Rössler, A., Cozzolino, D., Verdoliva, L., Riess, C., Thies, J., Nießner, M.: Face-Forensics++: learning to detect manipulated facial images. In: International Conference on Computer Vision (ICCV) (2019)
44. Ruiz, N., Bargal, S.A., Sclaroff, S.: Disrupting deepfakes: adversarial attacks against conditional image translation networks and facial manipulation systems (2020)
45. Shin, R.: Jpeg-resistant adversarial images (2017)
46. Siarohin, A., Lathuilière, S., Tulyakov, S., Ricci, E., Sebe, N.: First order motion model for image animation. In: Conference on Neural Information Processing Systems (NeurIPS) (2019)
47. Sun, P., Li, Y., Qi, H., Lyu, S.: Landmark breaker: obstructing deepfake by disturbing landmark extraction. In: 2020 IEEE International Workshop on Information Forensics and Security (WIFS), pp. 1–6 (2020). https://doi.org/10.1109/WIFS49906.2020.9360910
48. Szegedy, C., et al.: Intriguing properties of neural networks (2014)
49. Tancik, M., Mildenhall, B., Ng, R.: Stegastamp: invisible hyperlinks in physical photographs. In: IEEE Conference on Computer Vision and Pattern Recognition (CVPR) (2020)
50. Thies, J., Zollhöfer, M., Stamminger, M., Theobalt, C., Nießner, M.: Face2face: real-time face capture and reenactment of rgb videos. In: Proceedings of Computer Vision and Pattern Recognition (CVPR). IEEE (2016)
51. Thies, J., Zollhöfer, M., Nießner, M.: Deferred neural rendering: image synthesis using neural textures. ACM Trans. Graph. (TOG) **38**, 1–12 (2019)
52. Wallace, G.K.: The jpeg still picture compression standard. IEEE Trans. Cons. Electron. **38**(1), xviii–xxxiv (1992)
53. Wang, R., Juefei-Xu, F., Luo, M., Liu, Y., Wang, L.: Faketagger: robust safeguards against deepfake dissemination via provenance tracking (2021)
54. Wengrowski, E., Dana, K.: Light field messaging with deep photographic steganography. In: Proceedings of the IEEE Conference on Computer Vision and Pattern Recognition, pp. 1515–1524 (2019)

55. Xie, C., Wang, J., Zhang, Z., Zhou, Y., Xie, L., Yuille, A.: Adversarial examples for semantic segmentation and object detection. In: International Conference on Computer Vision. IEEE (2017)
56. Yang, X., Li, Y., Lyu, S.: Exposing deep fakes using inconsistent head poses. In: ICASSP 2019–2019 IEEE International Conference on Acoustics, Speech and Signal Processing (ICASSP) (2019). https://doi.org/10.1109/icassp.2019.8683164
57. Yang, Y., Liang, C., He, H., Cao, X., Gong, N.Z.: Faceguard: proactive deepfake detection (2021)
58. Yeh, C.Y., Chen, H., Tsai, S.L., Wang, S.D.: Disrupting image-translation-based deepfake algorithms with adversarial attacks. In: 2020 IEEE Winter Applications of Computer Vision Workshops (WACVW), pp. 53–62 (2020)
59. Yeh, C.Y., Chen, H.W., Shuai, H.H., Yang, D.N., Chen, M.S.: Attack as the best defense: Nullifying image-to-image translation gans via limit-aware adversarial attack. In: Proceedings of the IEEE/CVF International Conference on Computer Vision (ICCV), pp. 16188–16197 (2021)
60. Zakharov, E., Shysheya, A., Burkov, E., Lempitsky, V.: Few-shot adversarial learning of realistic neural talking head models (2019)
61. Zhang, R., Isola, P., Efros, A.A., Shechtman, E., Wang, O.: The unreasonable effectiveness of deep features as a perceptual metric. In: CVPR (2018)
62. Zhou, P., Han, X., Morariu, V.I., Davis, L.S.: Two-stream neural networks for tampered face detection (2017). https://doi.org/10.1109/cvprw.2017.229
63. Zhu, J., Kaplan, R., Johnson, J., Fei-Fei, L.: Hidden: hiding data with deep networks. In: ECCV (2018)

FingerprintNet: Synthesized Fingerprints for Generated Image Detection

Yonghyun Jeong[1], Doyeon Kim[2], Youngmin Ro[3], Pyounggeon Kim[4,5], and Jongwon Choi[4(✉)]

[1] Clova, NAVER, Seongnam-si, South Korea
yonghyun.jeong@navercorp.com
[2] LINE Plus, Seongnam-si, South Korea
doyeon.k@linecorp.com
[3] Department of Artificial Intelligence, University of Seoul, Seoul, Korea
youngmin.ro@uos.ac.kr
[4] Department of Advanced Imaging, Chung-Ang University, Seoul, Korea
{trytty,choijw}@cau.ac.kr
[5] Samsung SDS, Seongnam-si, South Korea

Abstract. While recent advances in generative models benefit the society, the generated images can be abused for malicious purposes, like fraud, defamation, and false news. To prevent such cases, vigorous research is conducted on distinguishing the generated images from the real ones, but challenges still remain with detecting the unseen generated images outside of the training settings. To overcome this problem, we analyze the distinctive characteristic of the generated images called 'fingerprints,' and propose a new framework to reproduce diverse types of fingerprints generated by various generative models. By training the model with the real images only, our framework can avoid data dependency on particular generative models and enhance generalization. With the mathematical derivation that the fingerprint is emphasized at the frequency domain, we design a generated image detector for effective training of the fingerprints. Our framework outperforms the prior state-of-the-art detectors, even though only real images are used for training. We also provide new benchmark datasets to demonstrate the model's robustness using the images of the latest anti-artifact generative models for reducing the spectral discrepancies.

1 Introduction

Based on the recent enhancement of the generative models, such as Generative Adversarial Networks (GAN) [19], it has become easy to obtain high-quality synthesized images [32,33]. Many recent generative models can even transform the target images to include the specific properties of the users' choices [9,10,46,

Supplementary Information The online version contains supplementary material available at https://doi.org/10.1007/978-3-031-19781-9_5.

58,59]. However, with technological improvement, the risk of maliciously abusing such images also rises, such as fraud, defamation, and fake news [36,38,44]. To prevent such cases, it is important to distinguish between the real images and the generated images [48].

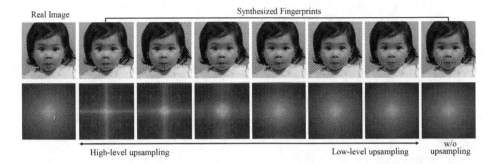

Fig. 1. The synthesized fingerprints varying by the level of upsampling process. Using the real images from FFHQ [32] as shown at the upper-left corner, the autoencoders can reconstruct images with various levels of upsampling processes, as shown in the columns of two to eight from the left. Their average 2D spectra in the second row show diverse synthesized fingerprints varying by the level of upsampling

Many recently generated image detectors have advanced to find the distinguishable features resulting during the image generation process [48]. For example, the checkerboard traces discovered in the frequency-level generated images are called the '*fingerprints*,' which are created during the upsampling estimation of the generator [6,14,15,18]. Unfortunately, the appearance of the frequency-level fingerprints varies by the generative models and also by the object categories. Thus, when tested with the generated images of the unseen GAN models or object categories, the generated image detectors inevitably suffer from a performance decline [20]. In addition, recent studies have advanced to reduce the aliasing effect that occurs during the upsampling process of CNN in order to generate more realistic images. Such effort makes it challenging for the detectors to distinguish the fingerprints in generated images.

To overcome the issues, we suggest *FingerprintNet* composed of a fingerprint generator and a generated image detector. The overall framework utilizes the real images only and ignores the generated images by specific GAN models for training. Instead, the fingerprint generator reconstructs the real images to insert the general fingerprints to cover various GAN models. To synthesize the general fingerprints, we employ three mechanisms for the fingerprint generator, which include a random layer selection, a multi-kernel deconvolution layer, and a feature blender. Since the number of upsampling operations affects the appearance of the fingerprints as shown in Fig. 1, we control the number of upsampling operations of the fingerprint generator by applying the random layer selection. The multi-kernel deconvolution layer and the feature blender are employed to handle the diverse fingerprints due to various kernel sizes and the diverse amplitude of the fingerprints, respectively.

By using the real images and the images reconstructed from the fingerprint generator, we train the generated image detector to distinguish the generated images from the real ones. We mathematically derive the reasons why the fingerprint can be easily detected in the frequency-level domain, and accordingly, the input of our GAN detector is the magnitude spectrum to increase the detection performance. Our method is tested with various real-world scenarios to validate state-of-the-art generalization ability of our model in detecting even the unseen GAN models and object categories. Especially, our self-supervised detector shows a similar performance as the supervised detector when detecting the images generated from the recent generative model, such as Fréchet Inception Distance (FID) [21] and the anti-aliasing GAN models [7, 29].

We can summarize our contributions as follows:

- Unlike the previous GAN detectors dependent on the specific GAN models, our model utilizes the self-supervised training method to obtain generalized detection ability and avoid data dependency.
- We propose a network that can generate various fingerprints, and a new way to train the detector by adjusting the amplitude of various fingerprints or perturbations.
- We provide a comprehensive analysis including visualizations and derivations on the artificial fingerprints observed in the frequency domain.
- We offer an extended benchmark dataset including the images generated by the latest anti-aliasing GAN models, which validate the state-of-the-art performance of our framework even for the unseen GAN models.

2   Related Work

We explore the previous literature on GAN image detection and the recent methods on GAN image creation, which have evolved to be more challenging to detect.

2.1   Generated Image Detection

The pixel-level characteristics in the GAN-based generated images can be used to identify the generated images. The identifiers can be referred to as the 'artifacts,' which are created due to upsampling process of the generator in GANs [6,14, 20]. Some studies analyze the inconsistencies in blocking artifacts from JPEG compression [50,53], or demosaicing artifacts created by a color filter array [13, 17]. Other image-based detection methods include an adaptable autoencoder-based neural network architecture for new target domains [12] and cross-model manipulation detection, such as JPEG and blur [52]. Recently, [49] proposed LRNet for detecting deepfakes based on temporal modeling.

Many generated image detectors focus on the unique patterns in the frequency spectra. [35] analyzed the artifacts in the spatial, frequency domain with the variance of the prediction residue, while [23] suggested Fast Fourier Transform [11] to distinguish image manipulations, such as JPEG compression. Also, a study by [43] employed frequency-based, GAN-specific detection using the artificial fingerprints, while [3] proposed a manipulation localization using the frequency domain correlation to find the forged areas. [18] analyzed the GAN-based artifacts using Discrete Cosine Transform [1], and [56] suggested studying the artifacts induced by the up-sampler of GANs. Also, [14,15] suggested exploiting the spectral distortions via Azimuthal integration. Recently, [27] suggested using bilateral high-pass filters for generalized detection, and [28] utilized the frequency-level perturbations for robust deepfake detection. Also, [57] proposed a new multi-attentional deepfake detection network, while [41] presented a spatial-phase shallow learning method for detecting artifacts of face forgeries.

Similar to our study, [26,56] utilized the autoencoder to reconstruct the generated images. However, they ignored the difference of fingerprints among the various GAN models, while our study employs the additional mechanism to obtain the generality for the unseen GAN models.

2.2 Advancement in Generative Models

Recently, generative models have become capable of creating images without the synthesized traces thanks to anti-aliasing, which refers to reducing the effect of artifacts in the generated images. Since it has now become more challenging to distinguish the generated images, it is important to analyze the latest generative models to upgrade the current detecting technologies. One of the popular anti-aliasing methods is to apply blur after deconvolution [30,32,33] and employ interpolation instead of the deconvolution layer [4,10]. Recently, applying kaiser filter to the activation function is newly proposed for anti-aliasing by Karras et al. [31]. Generative models besides GANs have also advanced to generate high-quality images. An example is DDPM [22], which uses diffusion probabilistic models based on denoising score matching. Recently, ILVR [8] proposed a method to guide and condition the generative process of DDPM. Another example of high-quality generative models is NVAE [51], a deep hierarchical VAE using depth-wise separable convolutions and batch normalization. Some studies [5,29] focused on reducing the spectral discrepancies in the spatial and spectral domains to obtain the anti-artifact characteristics. For example, SSD-GAN [7] enhanced GAN models to alleviate the loss of spectral information to generate the exact details of real images.

Since new manipulation methods quickly emerge, it is impractical to constantly update the detector's training in a supervised way [2]. Instead, it is much more practical to improve the generalization ability of generated image detectors. To improve the issue, some studies [2,12,24] adopted transfer learning, which utilizes a pre-trained model for another task using less amount of data. Recently, [34] proposed a method to perform domain adaptation on deepfake detection using transfer learning. However, transfer learning requires the pre-obtained knowledge on which GAN model is used for image synthesis, which makes it difficult to utilize for generalization of generated image detectors.

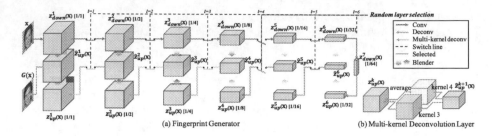

Fig. 2. The overall architecture of FingerprintNet. (a) depicts an example in which l is selected to 4 by *the random layer selection* requiring three times of upsampling. (b) shows details of the multi-kernel deconvolution layer

3 Fingerprint Generator

The purpose of the fingerprint generator is to mimic various kinds of fingerprints based on the reconstruction of the real images. The previous literature [14,52] mentioned that the fingerprints are created in generated images due to the upsampling process of the generative models. Based on the findings, we have developed the fingerprint generator using the autoencoder, which is designed to contain a number of upsampling process. The number of upsampling processes and the kernel size only affect the frequency of the fingerprints. Thus, to generate diverse kinds of fingerprints, we design the fingerprint generator to include the additional modules, such as blending of features, random selection of layers, as well as the multi-kernel deconvolution layers.

3.1 Overall Architecture

Figure 2 shows the overall architecture of the fingerprint generator, which is composed of 7 blocks each containing the upsampling and downsampling convolution layers. First, the input image is turned into a small resolution by the convolution layers consecutively applied for image compression. The first layer with stride 1 is exempt, but all the other layers are required to compress the images with the stride of 2. Then, the resolution of the feature map becomes 1/64 of the original input size at the last layer. The compressed feature maps from k-th convolution layer can be defined as $z^k_{down}(\mathbf{x})$ for a given input image $\mathbf{x}$. Also, every convolution layer is paired with a ReLU activation function, which is applied after the convolution.

Then, in order to restore the feature map's resolution, we consecutively apply the deconvolution layers to the compressed features. Except for stride 1's last deconvolution layer to reconstruct the original image, every deconvolution layer enlarges the feature map's resolution twice as big as stride 2. The deconvolution layers consist of a variety of layers as follows: in the order of a transposed convolution, a batch normalization, a blur kernel for anti-aliasing, and a ReLU activation function layers. The k-th deconvolution layer's feature map is defined as $z^{k-1}_{up}(\mathbf{x})$. For easy understanding, we designate the same index of k to the same size of feature maps of $z_{down}^k(\mathbf{x})$ and $z_{up}^k(\mathbf{x})$. The fingerprint generator's output image is defined as $G(\mathbf{x})$ for the input of $\mathbf{x}$.

3.2   Training Loss

The fingerprint generator's training set is composed of only the real images. Also, additional to the conventional reconstruction loss of the autoencoder, the fingerprint generator's training loss takes the similarity losses to decrease the disparities between the feature maps of the corresponding deconvolution and convolution layers. Therefore, the fingerprint generator's training loss can be represented as:

$$\mathcal{L}(\mathbf{X}_r, G) = \mathbb{E}_{\mathbf{x} \sim \mathbf{X}_r}\left[||\mathbf{x} - G(\mathbf{x})||_2 + \sum_{k=1}^{6}||z_{down}^k(\mathbf{x}) - z_{up}^k(\mathbf{x})||_2\right], \qquad (1)$$

where $\mathbf{X}_r$ indicates the training set composed of the real images. Based on the additional loss term using the latent feature maps, we can take $z_{down}^k(\mathbf{x})$ as the artifact-free version of the feature map of $z_{up}^k(\mathbf{x})$. These characteristics of the feature maps in the fingerprint generator are utilized for the feature blender, which is explained in Sect. 3.5.

3.3   Random Layer Selection

To handle the different numbers of upsampling operations in various GAN models, we employ the random layer selection in the fingerprint generator. At every training iteration, the module of the random layer selection randomly selects one value l from $\{1, 2, 3, 4, 5, 6\}$ according to the uniform distribution. Then, instead of using the entire layers, $z_{down}^l(\mathbf{x})$ is fed into l-th deconvolution layer to estimate $z_{up}^l(\mathbf{x})$. We set the feature maps from the remaining convolution and deconvolution layers with the indices larger than l as zero. Since l-th convolution and deconvolution layers have the equivalent resolutions and channel sizes, we can use the same weight parameters of the layers even after the random layer selection. As a result, while the original architecture of the fingerprint generator remains, the number of upsampling operations can vary to generalize the appearance of fingerprints in the reconstructed image.

3.4   Multi-kernel Deconvolution Layer

To consider the difference of fingerprints generated by various kernel sizes, we employ multiple kernels in the respective deconvolution layer. Among the various kernel sizes, we focus on the difference between the even and odd sizes of the kernels. Due to the constant stride size of 2 for each deconvolution layer, when estimating z_{up}^k, kernels of even sizes overlap by an even number of pixels, whereas kernels of odd sizes overlap by an odd number of pixels. Thus, instead of employing numerous kernel sizes, we use only two kernels where the sizes are set to 3 and 4, respectively. Especially, the two kernel sizes of 3 and 4 are conventionally used in most of the GAN models [9,10,30,32,59]. Then, one deconvolution layer contains two kernels, which are respectively applied to the input feature maps in parallel. Finally, z_{up}^k is obtained by estimating the average of the two feature maps resulting from the two kernels.

3.5   Feature Blender

The feature blender is employed to consider the various amplitude of fingerprints from the GAN models. Since the amplitude of fingerprints can be dependent on the input images, the feature blender augments the training samples by blending $z_{down}^k(\mathbf{x})$ and $z_{up}^k(\mathbf{x})$. According to our training loss (Eq. 1), the feature maps of the corresponding indices (*i.e.* $z_{down}^k(\mathbf{x})$ and $z_{up}^k(\mathbf{x})$) are trained to be similar to each other. Then, due to the absence of upsampling operations to estimate $z_{down}^k(\mathbf{x})$, $z_{down}^k(\mathbf{x})$ can be seen as the artifact-free feature map that is similar to $z_{up}^k(\mathbf{x})$. Thus, by blending the two feature maps, we can reduce the effect of fingerprints of $z_{up}^k(\mathbf{x})$ even while preserving its semantic information.

By using the feature blender, k-th deconvolution layer is fed by the blended feature map of $\hat{z}_{up}^k(\mathbf{x})$ instead of $z_{up}^k(\mathbf{x})$ that is the original feature map from the leading deconvolution layer. The blended feature map is obtained as follows:

$$\hat{z}_{up}^k(\mathbf{x}) = \mu_k z_{down}^k(\mathbf{x}) + (1 - \mu_k)z_{up}^k(\mathbf{x}), \tag{2}$$

where μ_k is a value randomly sampled from a Beta distribution of $\alpha = 1$ and $\beta = 1$. The value of μ_k is sampled repeatedly at every deconvolution layer and every training iteration. Thus, the fingerprint generator can generate the various amplitudes of fingerprints only by the unified model.

3.6   Fingerprint Generation

After the training of the fingerprint generator, we build the synthetic dataset containing the generated images from the fingerprint generator. During the dataset generation, we fix the indices l of the random layer selection by 1, 2, and 6. Thus, the generated images of our synthetic dataset contain various types of fingerprints. When $l = 2$ or $l = 6$, the fingerprints appear at $G(\mathbf{x})$ due to the upsampling operations, which can be used as an important characteristic to distinguish the generated images. Meanwhile, the generated images with $l = 1$ have had no upsampling operation and thus support robustness on the anti-aliasing GAN models. To improve robustness to various GAN models, the multi-kernel deconvolution layer and the feature blender remain in the dataset generation. Even though multiple images can be generated through the randomness of the feature blender, when one real image is given, we generate only three images respectively for one index l of the random layer selection. Thus, in the synthetic dataset, the quantity of generated images is three times that of real images.

4 Generated Image Detector

To classify the generated images from the real images, we utilize the additional CNN model, which is called the generated image detector. Before explaining the details of the generated image detector, we first derive mathematically the reason why the fingerprint of the generated images becomes distinctive in the frequency-level domain as referred by many studies [14,18]. Based on the derivation, we also utilize the frequency spectrum as the input of the generated image detector. To further improve the robustness of the generated image detector, we employ the mechanism of mixed batch during its training.

4.1 Effect of Frequency-Level Input

In this paper, we derive mathematically the reason why the fingerprints become distinctive in the frequency spectrum. A number of studies have confirmed that the images generated by GAN models contain the fingerprints appearing as the unique patterns in the frequency domain and utilizing those fingerprints can be the key to robust detection of the generated images [14,18]. As shown in Fig. 1, the artificial fingerprints generated by the deconvolution layer are easily discovered in the 2D spectra. To ease the derivation, we consider the 1-D sequence in the following derivations.

As shown in Fig. 1, the fingerprints appear quite impulse train in the frequency spectrum. Thus, when a_m and T represent a scale factor for m-th impulse sequence and the period between the impulse sequences, respectively, we can represent the frequency-level fingerprints by the weighted impulse train as:

$$\mathcal{F}\{\mathbf{g}\}[k] = \sum_{m=-M}^{m-M} a_m \delta[k - mT], \tag{3}$$

where $m \in \{-M, ..., M\}$, $\mathbf{g}$ represents the pixel-level fingerprints, and $\mathcal{F}$ and $\delta[k]$ are the Fourier transformation function and an unit impulse sequence of the frequency component k, respectively.

Then, to acquire the pixel-level artifacts, we estimate the inverse Fourier transformation of the frequency-level fingerprints as follows:

$$\mathbf{g}[n] = \frac{2}{N} \sum_{m=0}^{M} |a_m| \cos\left(\frac{2\pi mT}{N} n + \alpha\right), \tag{4}$$

where N is the length of entire sequence and $\alpha = \arctan\left(\texttt{Im}\{a_m\}/\texttt{Re}\{a_m\}\right)$. The detailed derivation is given in Appendix A.

From the derivation, we can obtain two interesting characteristics of the pixel-domain artifacts. First, the fingerprints in the frequency domain are easier to discover because of their composition in the impulse train format, unlike the pixel-level artifacts based on the smooth trigonometric functions. Second, since M is smaller than $N/2$, we can find that the magnitude of the fingerprint in pixel-level domain cannot be larger than that in frequency-level domain according to the inequality of $g[n] < 2M|a_m|/N$ derived from Eq. 4. Thus, when we transform the input images into the frequency-level domain, the fingerprints can be emphasized to be detected easily. Therefore, we also employ the Fourier transform for the input of the detector.

4.2   Architecture of Detector

The next step after training the artificial fingerprint generator is the training of the generated image detector to discern between the generated images and the real images. As illustrated in Fig. 1 and the derivation in Sect. 4.1, it is effective to utilize the frequency-level analysis to investigate the artificial fingerprints. Therefore, we employ Fast Fourier Transform (FFT) [11] to transform the generated image $\hat{x} \in \hat{X} = \{G(x)|\forall x \in X\}$ of the artificial fingerprint generator into a 2D spectrum.

Our detector is based on ResNet-50 [39] for a fair comparison with the previous research [14,18,52]. In order to train the detector, we procure the generated images by reconstructing the real images from the training dataset. The training of the generated image detector can be challenging due to the unbalancing issue arising from the three generated images from one real image. To solve the issue, we have changed the sampling probability to extract the real images three times as much of the reconstructed images in a mini-batch.

4.3   Training Method with Mixed Batch

The generated image detector's training dataset can be divided into two categories: the generated images and real images. We sample an equal number of generated and real images to make one mini-batch. Then, we mix the sampled images instead of utilizing them directly to lessen data reliance on the category of real images, which is defined as *mixed images*. Also, the mixed images may minimize the noisy information in the generated images, improving the detector's tolerance against high-quality images from contemporary GAN models.

Every sample from the mini-batch is replaced with mixed samples, as stated by $\tilde{\mathbf{S}}$. $\tilde{\mathbf{Y}}$ stands for the labels that belong to the samples of $\tilde{\mathbf{S}}$. We assign 1 for real, and 0 for generated images. First, two images are randomly chosen from a mini-batch $\mathbf{S} = \{X_r, X_g\}$, which are indicated by $\mathbf{s}_i$ and $\mathbf{s}_j$, when we denote the sets of real images by $\mathbf{X}_r$ and generated images by $\mathbf{X}_g$. The mixed sample $\tilde{\mathbf{s}}_{(i,j)}$ and its label $\tilde{y}_{i,j}$ are retrieved by as follows:

$$\tilde{\mathbf{s}}_{(i,j)} = \lambda \mathbf{s}_i + (1 - \lambda)\mathbf{s}_j, \qquad \tilde{y}_{(i,j)} = y_i y_j, \qquad (5)$$

where λ is a mixing scale factor randomly selected from a Beta distribution with $a = 1$ and $b = 1$, and y_i and y_j are labels for $\mathbf{s}_i$ and $\mathbf{s}_j$, respectively. Only when the two real images are blended, we regard the mixed image to be the real image. Then, for each $\tilde{\mathbf{s}}_{(i,j)}$ and $\tilde{y}_{(i,j)}$, $\tilde{\mathbf{S}}$ and $\tilde{\mathbf{Y}}$ are established. The mixing mechanism may seem similar to Mixup method [55], but the notable difference of our work is the designing method of the augmented labels. While Mixup utilizes the augmented labels by integrating the original labels with the scales of the mixed samples, our feature blender considers the samples mixed with any fake images as the perfect fake images.

Then, we train the generated image detector (C) with a softmax cross-entropy loss, which can be represented as follows:

$$\mathcal{L}_C(\tilde{\mathbf{S}}) = \mathbb{E}_{(\tilde{s},\tilde{y})\sim(\tilde{\mathbf{S}},\tilde{\mathbf{Y}})}[CE(\tilde{C}(\mathbf{s}),\tilde{y})], \tag{6}$$

where the softmax cross-entropy loss between the predictions of $\hat{y}$ and its associated ground-truth y is denoted by $CE(\hat{y}, y)$. Also, the training datasets are additionally supplemented with augmentation using JPEG compression and blur as provided in [52].

5   Experimental Results

5.1   Dataset

Through experiments, we compare the performance of each network based on the same data. Since the training settings have a strong impact on the analysis of the generated image detector, we adopt the same training settings as ProGAN [30] and utilize the real horse images of LSUN [54]. In contrast, the comparing models are trained with the 20 categories of ProGAN and the 20 categories of LSUN, which were used to train ProGAN. Also, for evaluation, we utilize the benchmark dataset [52] used for assessment of the generated image detector. The benchmark dataset includes several well-known unconditional GAN models including ProGAN [30], StyleGAN [32], and StyleGAN2 [33], and also a conditional GAN model, such as BigGAN [4]. We also employ the image-to-image translation models for testing, including CycleGAN [59], StarGAN [9], and Gau-GAN [45]. We utilize various GANs with human faces and various objects, and the real images used to train the GANs, including CelebA-HQ [37], CelebA [42], COCO [40], LSUN [54], and ImageNet [47].

Additionally, for evaluations, we utilize the recent GAN models that can generate images with spectral distributions similar to the real images, as well as state-of-the-art score-based generative models and variational autoencoders (VAE). For training, LSUN [54] and FFHQ [32] are used. For evaluations, we utilize the generative models in spatial and spectral domains including SSD-GAN [7], and SpectralGAN [29]. Also, we include the most recent score-based unconditional GAN, DDPM [22] and its conditional model, ILVR [8], as well as the most advanced unconditional VAE, NVAE [51], and state-of-the-art faceswap-based model, FICGAN [25].

5.2   Evaluation Metrics

For performance comparison, we employ the accuracy and average precision [16], which are the metrics commonly used in this field of study. To compare the generalization performance, we follow the suggestion of Wang [52] to use JPEG compression, which is known as the most effective method to test the generalization performance. Also, for the frequency-level analysis, we compare with Frank [18], Durall [14], and Jeong [27]. To evaluate cross-category performance, we compare with a self-supervised model [56].

5.3 Generalization Performance of Our Detector

To show the generalization ability of our detector, we perform two types of evaluations, which include the cross-category performance and the cross-model performance.

Cross-Model Performance

Table 1 shows the results of the first experiment to test the cross-model performance of the generated image detectors. We compare with the previous studies used for the comparison of cross-model performance: Wang [52], Frank [18], Durall [14], and Jeong [27]. Each of them is trained using 1 to 20 categories generated by ProGAN [30], and tested with the generated images of seven other generative models. In contrast, our self-supervised generated image detector is trained with real *horse* images only. Even with the seriously limited setting where no generated images of GAN models are used, our generated image detector achieves the highest accuracy and average precision.

To show the component-wise effectiveness of our framework, we perform the ablation studies in the cross-model experiments. As shown in the bottom section of Table 1, the averaged performance dramatically drops when only one of the components is missing, which verifies the importance of the respective component to cover the various types of generative models. For the first row of our ablation tests (*w Mixup*), we use the Mixup method [55] to replace our training loss, which supports the effectiveness of our novel training loss to recognize the subtle artifacts and improve detection accuracy. Especially, when the random selection module is removed, the amount of performance decline is substantial, which indicates the importance of considering the various numbers of upsampling operations to obtain diverse fingerprints. Based on the discovery, we can conclude that it is necessary to diversify the number of upsampling operations for diversity in generated fingerprints.

Cross-Category Performance.

We conduct a cross-category experiment to compare accuracy in the same test settings as [56]. Using the generated images of the same GAN model, we train the generated image detectors with only one object category and test with the entire object categories to evaluate the generalization performance. Table 2 shows the test results using the 6 classes (apple, horse, orange, summer, winter, and zebra) of the generated image detectors trained with each category of CycleGAN [59]. Compared to the existing model trained in a self-supervised manner [26,56], our model shows superior performance in generalized detection.

Table 1. Cross-model performance with ablation study.

Model	# of class(real/fake)	StyleGAN [32] Acc.	A.P.	StyleGAN2 [33] Acc.	A.P.	BigGAN [4] Acc.	A.P.	CycleGAN [59] Acc.	A.P.	StarGAN [9] Acc.	A.P.	GauGAN [45] Acc.	A.P.	Mean Acc.	A.P.	Min Acc.	A.P.
Wang [52]	(1, 1)	51.6	73.9	52.2	77.8	52.1	69.5	71.4	90.1	58.0	83.7	60.0	92.8	57.6	81.3	51.6	69.5
Durall [14]		64.1	58.6	69.3	62.9	55.4	52.9	69.6	62.8	95.4	91.5	57.5	54.0	68.6	63.8	55.4	52.9
Frank [18]		68.5	80.7	60.8	77.3	72.1	63.0	57.6	56.6	80.1	76.3	74.0	95.5	68.9	74.9	57.6	56.6
Jeong [27]		66.9	72.1	64.7	73.8	80.2	83.9	66.4	82.6	90.4	99.4	82.8	96.2	75.2	84.7	64.7	72.1
Wang [52]	(2, 2)	52.8	82.8	75.7	96.6	51.6	70.5	58.6	81.5	51.2	74.3	53.6	86.6	57.3	82.1	51.2	70.5
Durall [14]		63.5	58.1	68.7	62.4	56.4	53.5	63.5	58.2	89.8	83.1	56.5	53.5	66.4	61.5	56.4	53.5
Frank [18]		70.8	83.8	61.2	75.6	74.9	76.2	74.8	76.8	91.7	97.5	89.2	98.4	77.1	84.7	61.2	75.6
Jeong [27]		71.6	74.1	77.0	81.1	82.6	80.6	86.0	86.6	93.8	80.8	69.6	90.8	80.1	82.3	69.6	74.1
Wang [52]	(4, 4)	63.8	91.4	76.4	97.5	52.9	73.3	72.7	88.6	63.8	90.8	63.9	92.2	65.6	89.0	52.9	73.3
Durall [14]		63.9	58.4	69.0	62.7	58.5	54.7	69.6	63.1	99.0	98.1	57.0	53.8	69.5	65.1	57.0	53.8
Frank [18]		72.2	82.1	64.2	80.1	68.9	82.4	53.7	66.2	89.1	99.2	65.3	90.3	68.9	83.4	53.7	66.2
Jeong [27]		76.9	75.1	76.2	74.7	84.9	81.7	81.9	78.9	94.4	94.4	65.5	94.0	80.0	83.1	65.5	74.7
Wang [52]	(20, 20)	71.4	96.3	67.5	93.4	60.9	83.3	83.8	94.3	84.5	93.6	79.3	98.1	74.6	93.2	60.9	83.3
Durall [14]		64.7	59.0	69.2	62.9	59.4	55.3	66.9	60.9	98.5	97.1	57.2	53.9	69.3	64.9	57.2	53.9
Frank [18]		81.8	91.7	71.4	93.0	76.0	87.8	62.8	77.3	96.9	99.4	73.9	93.1	77.1	90.4	62.8	77.3
Jeong [27]		73.0	83.9	62.7	75.9	78.1	94.8	60.5	85.6	100.0	100.0	68.7	97.4	73.8	89.6	60.5	75.9
w Mix up	(1, 0)	69.0	81.4	68.2	80.6	79.2	94.1	62.7	84.2	98.8	100.0	69.5	89.1	74.6	88.2	62.7	80.6
w/o Similar loss		71.3	81.3	76.6	87.6	76.9	89.6	59.4	95.8	99.1	99.3	65.7	96.8	74.8	91.7	59.4	81.3
w/o Rand. select.		56.4	53.2	57.9	77.3	54.2	69.6	51.9	41.5	86.4	89.6	53.1	75.9	60.0	67.9	51.9	41.5
w/o Multi. kernel		68.5	82.0	71.5	88.8	67.2	91.8	62.6	82.1	98.3	99.7	59.9	78.2	71.3	87.1	59.9	78.2
w/o Mixed batch		69.0	81.4	68.2	80.6	79.2	94.1	62.7	84.2	98.8	100.0	69.5	89.1	74.6	88.2	62.7	80.6
w/o Feat. Blender		78.6	89.7	73.5	88.7	73.9	86.3	63.0	88.8	98.9	99.8	61.7	91.4	74.9	90.8	61.7	**86.3**
w/o FFT		92.1	97.4	89.1	95.9	66.8	65.7	64.0	74.7	99.3	100.0	58.1	63.8	78.2	82.9	58.1	63.8
Ours		74.1	85.3	89.5	96.1	85.0	94.8	71.2	96.9	99.9	100.0	75.9	90.9	**82.6**	**94.0**	**71.2**	85.3

Table 2. Comparison result with self-supervised manner.

Model	Train category						Mean
	Apple	Horse	Orange	Summer	Winter	Zebra	
AutoGAN [56]	76.1	97.4	67.7	97.2	68.1	78.6	80.9
SelfDetector [26]	78.7	95.3	78.3	90.8	80.8	97.7	86.9
Ours	96.1	95.8	88.4	95.0	91.1	96.3	**93.8**

5.4   Generalization for Recent Generative Models

According to [7, 29], the spectral distributions of images are known to vary by the last de-convolution layer, and it can decline the performance of generated image detectors. Based on that, we assess the model's robustness on the synthesized images of anti-artifact generative methods to reduce the spectral discrepancies. For training of each generative model, the real horse images of LSUN [54] are utilized, as in Sect. 5.3. The left section of Table 3 shows the performance of each detector when evaluated with the generated images of the anti-artifact models. Our model and Frank [18] show the most superior performance compared to others. Jeong [27] shows declined performance due to its high-pass filter, since the high-frequency components are modified in the generated images of the anti-artifact models. Also, Durall [14] also suffers from declined performance due to the reduced spectral discrepancies in frequency distributions of images.

Technological advancement in generative models has not only affected GANs but also the score-based diffusion probabilistic models and variational autoencoders. Thus, we additionally evaluate the performance of state-of-the-art generative methods, including DDPM [22], ILVR [8], NVAE [51], and FICGAN [25]. DDPM is the most well-known diffusion probabilistic model, and ILVR is a conditioning method for DDPM. Also, NVAE is the most recent unconditional variational autoencoder for high fidelity synthesized images, while FICGAN is a face-swapping method for high-quality deepfake images. The right section of the Table 3 shows the performance of each model evaluated with the images generated by state-of-the-art generative models. Our detector achieves stable performance even with the face-swap model, FICGAN. Since other models trained in a supervised manner focus on the distributions of GAN training, they suffer from a decline in performance when tested with non-GAN generative models with different distributions, such as DDPM and VAE.

5.5   Color Manipulation Performance

We conduct an experiment to evaluate the detector's robustness on color manipulated images, using the same settings of the color manipulation experiments of Jeong [27]. First, we resize the images from 1024×1024 to 256×256, then modify colors for assessment. Manipulations in hue, brightness, saturation, gamma, and contrast modify the overall distribution of images to make challenging conditions

Table 3. Robustness to anti-artifact GANs and SOTA models.

Model	# of class (real/fake)	Test models																
		Anti-artifact GANs						State-of-the-art generative models										
		SSD-GAN [7]		SpectralGAN [29]		Mean		DDPM [22]		ILVR [8]		NVAE [51]		FICGAN [25]		Mean	
		Acc.	A.P.	Acc.	A.P	Acc.	A.P.	Acc.	A.P.	Acc.	A.P.	Acc.	A.P.	Acc.	A.P.	Acc.	A.P.
Wang [52]	(1, 1)	50.3	95.3	50.0	95.6	50.2	95.5	49.3	31.8	49.3	32.0	49.7	33.3	49.3	32.6	49.4	32.4
Durall [14]		53.1	51.7	73.2	68.0	63.2	59.9	49.9	49.9	55.6	53.0	56.1	53.2	54.7	52.4	54.1	52.1
Frank [18]		96.2	99.6	96.2	99.7	96.2	99.7	68.7	82.6	70.8	84.3	78.0	89.2	76.9	88.9	73.6	86.3
Jeong [27]		89.5	95.0	89.0	89.4	89.3	92.2	78.7	86.5	79.8	87.8	81.2	86.4	89.4	96.2	82.3	89.2
Wang [52]	(20, 20)	75.8	84.6	87.9	89.5	81.9	87.1	68.4	79.2	75.8	84.6	71.0	80.9	76.6	86.5	73.0	82.8
Durall [14]		68.7	62.4	44.5	48.3	56.6	55.4	57.5	54.1	57.6	54.1	57.7	54.2	57.0	53.7	57.5	54.0
Frank [18]		96.2	100.0	96.4	100.0	96.3	100.0	88.4	93.2	86.5	93.6	92.4	96.2	82.7	92.2	**87.5**	93.8
Jeong [27]		84.2	99.9	84.2	99.5	84.2	99.7	83.5	93.4	83.0	92.4	82.0	91.5	83.6	94.1	83.0	92.9
Ours	(1, 0)	96.4	99.2	98.5	99.9	**97.5**	99.6	78.7	92.3	84.0	95.4	91.4	97.1	92.6	98.5	86.7	**95.8**

Table 4. Color manipulation performance.

Model	Original		Hue		Brightness		Saturation		Gamma		Contrast		Mean	
	Acc.	A.P.	Acc.	A.P.	Acc.	A.P.	Acc.	A.P.	Acc.	A.P.	Acc.	A.P.	Acc.	A.P.
Wang [52]	99.9	100.0	73.9	81.3	61.8	74.7	74.3	84.4	70.2	83.2	66.6	79.7	74.5	83.9
Frank [18]	95.2	96.5	85.5	97.2	84.2	97.2	91.2	98.0	85.4	97.4	84.3	96.7	87.6	97.2
Durall [14]	86.2	93.4	86.2	81.9	85.9	81.9	86.2	81.9	85.1	80.8	85.2	81.2	85.8	83.5
Jeong [27]	97.0	98.1	92.0	97.8	92.0	97.9	91.9	96.7	91.7	96.8	92.4	98.1	92.8	97.6
Ours	97.4	99.8	97.1	99.8	89.4	96.6	94.1	99.2	96.9	99.9	89.6	97.7	**94.1**	**98.8**

for detectors to work [27]. The hue factor is the amount of shift in the hue channel by 0.2, while brightness, saturation, gamma, and contrast are adjusted by 1.3, respectively. Table 4 indicates the variance in detecting performance when images are manipulated and the characteristics of the artifacts have changed. For a fair comparison, we apply the supervised learning to train the detector based on ProGAN [30] face and FFHQ [32] as in [27], and do not apply the center crop. The experimental results validate that the frequency-based methods [14,18,27] including ours are more robust to color manipulations compared to image-based method [52].

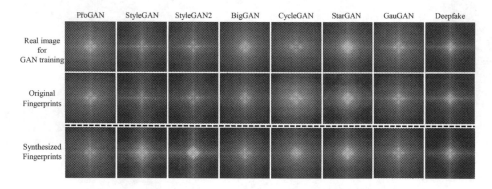

Fig. 3. The averaged spectra of the real images, images from GAN models, and images from fingerprint generator. The first row shows the averaged spectra of the real images used for training the GAN models, while the middle row shows those of the generated images from the GAN models. The last row shows the averaged spectra where we can obtain the highest resemblance between the spectra of the generated images and the synthesized fingerprints

5.6   Visualization

Figure 3 shows the resemblance between the reconstructed average 2D spectra by adjusting the level of the upsampling process and those generated by the actual GAN models. By adjusting the level of downsampling in autoencoders, we can observe the close resemblance among the reconstructed patterns generated by each GAN model in FFT. Also, we can confirm that the transposed convolution-based GANs, including StyleGAN, StyleGAN2, CycleGAN, and StarGAN, generate more distinct fingerprints, which are close to the spectrum of the reconstructed data with a high level of upsampling. From the visualization, we can confirm that the fingerprints from our fingerprint generator can be effective for the training of the generated image detector. We provide every visualization result in Appendix B.

6   Conclusion

We propose a novel framework composed of a fingerprint generator and a generated image detector for robust generalization. First, we analyze the diverse types of fingerprints in generated images and develop a fingerprint generator, which can synthesize and insert the fingerprints on real images for high-quality training data. Based on the analysis, we newly introduce a training method using real images only for generalized detection and validate its efficacy through robust performance of our model. Surpassing others trained in a supervised manner, our model achieves impressive performance in zero-shot learning, even when tested with unseen categories and GAN models. Also, we include the most recent anti-artifact generative models for evaluation and verify our model's consistent performance. We hope that the suggested framework can be enhanced in the future to manage the unexpected developments of new generative models by using the extra modules to address the additional properties of their fingerprints.

Acknowledgments. It was supported by Samsung SDS and Institute of Information & communications Technology Planning & Evaluation (IITP) grant funded by the Korea government (MSIT) (2021-0-01341, Artificial Intelligence Graduate School Program(Chung-Ang University); 2021-0-01778, Development of Human Image Synthesis and Discrimination Technology Below the Perceptual Threshold; 2021-0-02067, Next Generation AI for Multi-purpose Video Search).

References

1. Ahmed, N., Natarajan, T., Rao, K.R.: Discrete cosine transform. IEEE Trans. Comput. **100**(1), 90–93 (1974)
2. Aneja, S., Nießner, M.: Generalized zero and few-shot transfer for facial forgery detection. arXiv preprint arXiv:2006.11863 (2020)
3. Bappy, J.H., Simons, C., Nataraj, L., Manjunath, B., Roy-Chowdhury, A.K.: Hybrid LSTM and encoder-decoder architecture for detection of image forgeries. IEEE Trans. Image Process. **28**(7), 3286–3300 (2019)
4. Brock, A., Donahue, J., Simonyan, K.: Large scale GAN training for high fidelity natural image synthesis. In: International Conference on Learning Representations (2019). https://openreview.net/forum?id=B1xsqj09Fm
5. Chandrasegaran, K., Tran, N.T., Cheung, N.M.: A closer look at Fourier spectrum discrepancies for CNN-generated images detection. In: Proceedings of the IEEE/CVF Conference on Computer Vision and Pattern Recognition (2021)

6. Chen, S., Yao, T., Chen, Y., Ding, S., Li, J., Ji, R.: Local relation learning for face forgery detection. arXiv preprint arXiv:2105.02577 (2021)
7. Chen, Y., Li, G., Jin, C., Liu, S., Li, T.: SSD-GAN: measuring the realness in the spatial and spectral domains. In: Proceedings of the AAAI Conference on Artificial Intelligence (2021)
8. Choi, J., Kim, S., Jeong, Y., Gwon, Y., Yoon, S.: ILVR: conditioning method for denoising diffusion probabilistic models. In: IEEE International Conference on Computer Vision (2021)
9. Choi, Y., Choi, M., Kim, M., Ha, J.W., Kim, S., Choo, J.: StarGAN: unified generative adversarial networks for multi-domain image-to-image translation. In: IEEE Conference on Computer Vision and Pattern Recognition (2018)
10. Choi, Y., Uh, Y., Yoo, J., Ha, J.W.: StarGAN v2: diverse image synthesis for multiple domains. In: IEEE Conference on Computer Vision and Pattern Recognition (2020)
11. Cooley, J.W., Lewis, P.A., Welch, P.D.: The fast Fourier transform and its applications. IEEE Trans. Educ. **12**(1), 27–34 (1969)
12. Cozzolino, D., Thies, J., Rössler, A., Riess, C., Nießner, M., Verdoliva, L.: Forensic-Transfer: weakly-supervised domain adaptation for forgery detection. arXiv (2018)
13. Dirik, A.E., Memon, N.: Image tamper detection based on demosaicing artifacts. In: 2009 16th IEEE International Conference on Image Processing, pp. 1497–1500 (2009)
14. Durall, R., Keuper, M., Keuper, J.: Watch your up-convolution: CNN based generative deep neural networks are failing to reproduce spectral distributions. In: IEEE Conference on Computer Vision and Pattern Recognition, Seattle, WA, United States (2020)
15. Durall, R., Keuper, M., Pfreundt, F.J., Keuper, J.: Unmasking deepfakes with simple features. arXiv preprint arXiv:1911.00686 (2019)
16. Everingham, M., Gool, L.V., Williams, C.K.I., Winn, J., Zisserman, A.: The pascal visual object classes (VOC) challenge. Int. J. Comput. Vis. **88**, 303–338 (2010). https://doi.org/10.1007/s11263-009-0275-4
17. Ferrara, P., Bianchi, T., De Rosa, A., Piva, A.: Image forgery localization via fine-grained analysis of CFA artifacts. IEEE Trans. Inf. Forensics Secur. **7**(5), 1566–1577 (2012)
18. Frank, J., Eisenhofer, T., Schönherr, L., Fischer, A., Kolossa, D., Holz, T.: Leveraging frequency analysis for deep fake image recognition. In: International Conference on Machine Learning, pp. 3247–3258. PMLR (2020)
19. Goodfellow, I., et al.: Generative adversarial nets. In: Advances in Neural Information Processing Systems, pp. 2672–2680 (2014)
20. Gragnaniello, D., Cozzolino, D., Marra, F., Poggi, G., Verdoliva, L.: Are GAN generated images easy to detect? A critical analysis of the state-of-the-art. arXiv preprint arXiv:2104.02617 (2021)
21. Heusel, M., Ramsauer, H., Unterthiner, T., Nessler, B., Hochreiter, S.: GANs trained by a two time-scale update rule converge to a local Nash equilibrium. In: Advances in Neural Information Processing Systems (2017)
22. Ho, J., Jain, A., Abbeel, P.: Denoising diffusion probabilistic models. In: Neural Information Processing Systems (NeurIPS) (2020)
23. Huang, D.Y., Huang, C.N., Hu, W.C., Chou, C.H.: Robustness of copy-move forgery detection under high JPEG compression artifacts. Multimed. Tools Appl. **76**(1), 1509–1530 (2017). https://doi.org/10.1007/s11042-015-3152-x

24. Jeon, H., Bang, Y.O., Kim, J., Woo, S.: T-GD: transferable GAN-generated images detection framework. In: International Conference on Machine Learning, pp. 4746–4761. PMLR (2020)
25. Jeong, Y., et al.: FICGAN: facial identity controllable GAN for de-identification. arXiv preprint arXiv:2110.00740 (2021)
26. Jeong, Y., Kim, D., Kim, P., Ro, Y., Choi, J.: Self-supervised GAN detector. arXiv preprint arXiv:2111.06575 (2021)
27. Jeong, Y., Kim, D., Min, S., Joe, S., Gwon, Y., Choi, J.: BiHPF: bilateral high-pass filters for robust deepfake detection. arXiv preprint arXiv:2109.00911 (2021)
28. Jeong, Y., Kim, D., Ro, Y., Choi, J.: FrePGAN: robust deepfake detection using frequency-level perturbations. arXiv preprint arXiv:2202.03347 (2022)
29. Jung, S., Keuper, M.: Spectral distribution aware image generation. In: Proceedings of the AAAI Conference on Artificial Intelligence (2021)
30. Karras, T., Aila, T., Laine, S., Lehtinen, J.: Progressive growing of GANs for improved quality, stability, and variation. In: International Conference on Learning Representations (2018). https://openreview.net/forum?id=Hk99zCeAb
31. Karras, T., et al.: Alias-free generative adversarial networks. In: Proceedings of the Neural Information Processing Systems (NeurIPS) (2021)
32. Karras, T., Laine, S., Aila, T.: A style-based generator architecture for generative adversarial networks. In: IEEE Conference on Computer Vision and Pattern Recognition, pp. 4401–4410 (2019)
33. Karras, T., Laine, S., Aittala, M., Hellsten, J., Lehtinen, J., Aila, T.: Analyzing and improving the image quality of StyleGAN. CoRR abs/1912.04958 (2019)
34. Kim, M., Tariq, S., Woo, S.S.: FReTAL: generalizing deepfake detection using knowledge distillation and representation learning. In: Proceedings of the IEEE/CVF Conference on Computer Vision and Pattern Recognition, pp. 1001–1012 (2021)
35. Kirchner, M.: Fast and reliable resampling detection by spectral analysis of fixed linear predictor residue. In: ACM Workshop on Multimedia and Security, pp. 11–20 (2008)
36. Kwon, P., You, J., Nam, G., Park, S., Chae, G.: KoDF: a large-scale Korean deepfake detection dataset. arXiv preprint arXiv:2103.10094 (2021)
37. Lee, C.H., Liu, Z., Wu, L., Luo, P.: MaskGAN: towards diverse and interactive facial image manipulation. In: IEEE Conference on Computer Vision and Pattern Recognition (2020)
38. Lee, S., Tariq, S., Shin, Y., Woo, S.S.: Detecting handcrafted facial image manipulations and GAN-generated facial images using Shallow-FakeFaceNet. Appl. Soft Comput. **105**, 107256 (2021)
39. Li, Y., Lyu, S.: Exposing deepfake videos by detecting face warping artifacts. In: IEEE Conference on Computer Vision and Pattern Recognition Workshops (2019)
40. Lin, T.-Y., et al.: Microsoft COCO: common objects in context. In: Fleet, D., Pajdla, T., Schiele, B., Tuytelaars, T. (eds.) ECCV 2014. LNCS, vol. 8693, pp. 740–755. Springer, Cham (2014). https://doi.org/10.1007/978-3-319-10602-1_48
41. Liu, H., et al.: Spatial-phase shallow learning: rethinking face forgery detection in frequency domain. In: Proceedings of the IEEE/CVF Conference on Computer Vision and Pattern Recognition, pp. 772–781 (2021)
42. Liu, Z., Luo, P., Wang, X., Tang, X.: Deep learning face attributes in the wild. In: International Conference on Computer Vision, December 2015
43. Marra, F., Gragnaniello, D., Verdoliva, L., Poggi, G.: Do GANs leave artificial fingerprints? In: IEEE Conference on Multimedia Information Processing and Retrieval, pp. 506–511. IEEE (2019)

44. Nguyen, T.T., Nguyen, C.M., Nguyen, D.T., Nguyen, D.T., Nahavandi, S.: Deep learning for deepfakes creation and detection. arXiv preprint arXiv:1909.11573 (2019)

45. Park, T., Liu, M.Y., Wang, T.C., Zhu, J.Y.: Semantic image synthesis with spatially-adaptive normalization. In: Proceedings of the IEEE/CVF Conference on Computer Vision and Pattern Recognition, pp. 2337–2346 (2019)

46. Pidhorskyi, S., Adjeroh, D.A., Doretto, G.: Adversarial latent autoencoders. In: Proceedings of the IEEE/CVF Conference on Computer Vision and Pattern Recognition, pp. 14104–14113 (2020)

47. Russakovsky, O., et al.: ImageNet large scale visual recognition challenge. Int. J. Comput. Vis. **115**(3), 211–252 (2015). https://doi.org/10.1007/s11263-015-0816-y

48. Sun, K., et al.: Domain general face forgery detection by learning to weight (2021)

49. Sun, Z., Han, Y., Hua, Z., Ruan, N., Jia, W.: Improving the efficiency and robustness of deepfakes detection through precise geometric features. In: Proceedings of the IEEE/CVF Conference on Computer Vision and Pattern Recognition, pp. 3609–3618 (2021)

50. Tralic, D., Petrovic, J., Grgic, S.: JPEG image tampering detection using blocking artifacts. In: International Conference on Systems, Signals and Image Processing, pp. 5–8. IEEE (2012)

51. Vahdat, A., Kautz, J.: NVAE: a deep hierarchical variational autoencoder. In: Neural Information Processing Systems (NeurIPS) (2020)

52. Wang, S.Y., Wang, O., Zhang, R., Owens, A., Efros, A.A.: CNN-generated images are surprisingly easy to spot...for now. In: IEEE Conference on Computer Vision and Pattern Recognition (2020)

53. Ye, S., Sun, Q., Chang, E.C.: Detecting digital image forgeries by measuring inconsistencies of blocking artifact. In: IEEE International Conference on Multimedia and Expo, pp. 12–15. IEEE (2007)

54. Yu, F., Zhang, Y., Song, S., Seff, A., Xiao, J.: LSUN: construction of a large-scale image dataset using deep learning with humans in the loop. arXiv preprint arXiv:1506.03365 (2015)

55. Zhang, H., Cisse, M., Dauphin, Y.N., Lopez-Paz, D.: mixup: beyond empirical risk minimization (2018)

56. Zhang, X., Karaman, S., Chang, S.F.: Detecting and simulating artifacts in GAN fake images. In: 2019 IEEE International Workshop on Information Forensics and Security (WIFS), pp. 1–6. IEEE (2019)

57. Zhao, H., Zhou, W., Chen, D., Wei, T., Zhang, W., Yu, N.: Multi-attentional deepfake detection. In: Proceedings of the IEEE/CVF Conference on Computer Vision and Pattern Recognition, pp. 2185–2194 (2021)

58. Zhu, J., Shen, Y., Zhao, D., Zhou, B.: In-domain GAN inversion for real image editing. In: Vedaldi, A., Bischof, H., Brox, T., Frahm, J.-M. (eds.) ECCV 2020. LNCS, vol. 12362, pp. 592–608. Springer, Cham (2020). https://doi.org/10.1007/978-3-030-58520-4_35

59. Zhu, J.Y., Park, T., Isola, P., Efros, A.A.: Unpaired image-to-image translation using cycle-consistent adversarial networks. In: IEEE International Conference on Computer Vision (2017)

Detecting Generated Images by Real Images

Bo Liu⑩, Fan Yang⑩, Xiuli Bi$^{(\boxtimes)}$⑩, Bin Xiao$^{(\boxtimes)}$⑩, Weisheng Li⑩,
and Xinbo Gao⑩

Chongqing University of Posts and Telecommunications, Chongqing, China
S200201074@stu.cqupt.edu.cn
{boliu,bixl,xiaobin,liws,gaoxb}@cqupt.edu.cn

Abstract. The widespread of generative models have called into question the authenticity of many things on the web. In this situation, the task of image forensics is urgent. The existing methods examine generated images and claim a forgery by detecting visual artifacts or invisible patterns, resulting in generalization issues. We observed that the noise pattern of real images exhibits similar characteristics in the frequency domain, while the generated images are far different. Therefore, we can perform image authentication by checking whether an image follows the patterns of authentic images. The experiments show that a simple classifier using noise patterns can easily detect a wide range of generative models, including GAN and flow-based models. Our method achieves state-of-the-art performance on both low- and high-resolution images from a wide range of generative models and shows superior generalization ability to unseen models. The code is available at https://github.com/Tangsenghenshou/Detecting-Generated-Images-by-Real-Images.

Keywords: Image forensics · Forgery detection · Image noise · Frequency domain analysis · GAN · Generated images

1   Introduction

Can you find out the fake images in Fig. 1? The answer is that all the images are fake. The popularity of deep neural networks has driven the rapid development of synthesis technology. Various mind-boggling technologies have entered our lives, from image editing to composite scenes, from face attribute tampering to face-swapping. For example, in the GPU technology conference hosted by Jen-Hsun Huang at NVIDIA in 2021, the video and Jen-Hsun Huang himself were synthesized, successfully fooling most people and bringing the image forgery to the limelight. Meanwhile, the concerns about image synthesis technology are growing as making global tampering becomes very easy. In particular, impressive progress has been made on generative models such as Generative Adversarial Networks(GAN) [1] and its variants. Examples include conditional GANs such as CycleGAN [2] based on unpaired data, StarGAN [3] that uses a generator and

© The Author(s), under exclusive license to Springer Nature Switzerland AG 2022
S. Avidan et al. (Eds.): ECCV 2022, LNCS 13674, pp. 95–110, 2022.
https://doi.org/10.1007/978-3-031-19781-9_6

Fig. 1. Which pictures are real and which are fake?

a discriminator to learn mappings between multiple domains, and GauGAN [4] that uses spatially adaptive normalization; unconditional GANs such as Big-GAN [5] based on orthogonal regularization, ProGAN [6] using feature vector normalization of pixels, and StyleGAN [7] using nonlinear mapping networks and an improved version of StyleGAN2 [8]. The other generative models, such as HiSD [9] based on hierarchical style decoupling, and the flow model Glow [10] based on reversible 1×1 convolution, can also produce high-quality generated images. Currently, many generated images can deceive the human eyes. Therefore it is urgent to pay more attention to image forensics. This paper proposed a detection method to expose globally tampered images yielded by generative models.

Generated image detection methods can be divided into two main categories: artifacts detection and data-driven approaches. The former detects artifacts in the spatial domain in generated images left by the upsampling components of networks or the periodical signals in the frequency domain. They are effective for most of the generated images in low quality by checking the traces generated by conditional GANs during upsampling. However, they become ineffective to unconditional GANs with high image quality. The data-driven approaches learn a large number of real and fake images, making the classifier learn the common features in GAN-generated images. However, the classifier is susceptible to unseen models and therefore does not generalize well as it is impossible to learn the common features shared by all generative models. A generic data-driven-based approach is introduced by Wang et al. [11]. However, such methods pay attention to the characteristics of generated images, resulting in generalization issues. We perform forgery detection from the perspective of real images. Specifically, we learn the shared properties of real images so that the detection network can work across various generative models, even with unseen models.

In this paper, we rethink the relationship between real and generated images. Analysis shows that real images possess spatial and frequency domain features not presented in generated images. This discrepancy can be observed in the representations of the image noise under the high-dimensional spatial mapping of the neural network, which we call the Learned Noise Patterns (LNP). We used a network to classify real and generated images with the help of LNP. Using LNP can effectively suppress the high-frequency information of images and reduce the influence of image semantics on classification. In order to make full use of the information in the LNP, we utilized the amplitude and phase spectrum of images along with the LNP so that the network uses the spatial and frequency domain features.

To sum up, this paper proposes a method to detect generated images. The main contributions of this paper can be summarised as follows:

• Our frequency domain analysis of the noise patterns of real images reveals its consistency in real images, while the generated images are far different.

• We discriminate the generated images by their inconsistent noise patterns to real images rather than detecting the artifacts or patterns of generated images.

• The proposed detection method achieves the SOTA performance in publicly available datasets and shows superior generalization ability to unseen models.

2   Related Work

Existing methods for detecting generated images can be classified into image artifacts detection and data-driven approaches. For those focus image artifacts, Dang et al. [12] found that the spatial information of the tampered region is important, and the tampered region is located by estimating the attention map of a particular image. Liu et al. [13] proposed GramNet, proving that CNNs consider texture as an important factor while finding that the texture statistics of real and false images differ significantly. Zhao et al. [14] used the attention mechanism to improve detection performance by extracting texture information at shallow and locating forgery at deep levels. Zhang et al. [15] introduce a generator that simulates sampling artifacts on several common GANs and demonstrates superior performance in the frequency domain by learning to classify sampling artifacts on GANs in both the spatial and frequency domains. It is argued in [16] that local information is easier to extract helpful information than global information. Frank et al. [17] demonstrate that upsampling operations in the pipeline cause artifacts in GAN-generated images, and the detection is performed using the DCT transform. Durall et al. [18] show that commonly used up-sampling operations (deconvolution or transposed convolution) make such models fail to reproduce the spectral distribution of the training data correctly.

For data-driven methods, Wang et al. [11] directly trained ResNet50 as the classifier with a large number of real images and ProGAN-generated images, which can be well generalized to the detection of different generative models using global information. On this basis, Gragnaniello et al. [19] used the modified ResNet50 network with two fewer down-sampling layers to improve the detection performance but significantly increase the training time.

Image noise is widely utilized in local tampering and source device identification. Each device will have its specific fingerprints left on the shooting process, which is also caused by imperfections in the manufacturing process of the device, and this pattern is called the PRNU noise pattern. Therefore, the equipment identification can be performed based on these fingerprints, e.g. [20]. Based on the specific properties of PRNU, Davide et al. [21] introduce a method

Fig. 2. (a) row represents the real and generated images, and (b) row represents their LNP. The first, third and fifth columns are real images. The second, fourth and sixth columns are images generated by StarGAN, StyleGAN and StyleGAN2, respectively. The red box indicates that the generated images show the grid effect. (Color figure online)

that learns the camera noise by denoising the network for local forgery detection. Ghosh et al. [22] extract noise fingerprints to identify real patches and forged patches for local tampering detection.

However, the previous approaches, whether detecting artifacts or by data-driven, look for fingerprints left by generative models, resulting in lower versatility. Instead, we focus on learning the common properties of real images to avoid generalization issues.

3   Method

3.1   Learned Noise Patterns (LNP)

Although the generated images from early GAN models are easy to detect, with the development of unconditional GAN such as StyleGAN and stylegan2, the current GAN-generated images have become more and more realistic. As shown in Fig. 2(a), we can hardly distinguish between StyleGAN (column 4), Style-GAN2 (column 6), and the real images (columns 3 and 5) with naked eyes. It is necessary to extract the discriminative features of the images to amplify their differences.

In the imaging process, a camera converts photons into electrons, and then the signal goes through components such as digital-to-analog converters. As the photons enter a camera, the incident light intensity at various places in a real image will show no regularity. Therefore, the pixel values do not change periodically for most real images.

In the pipeline of GANs for generating images, papers such as [15] have introduced the artifacts in the generated images due to up-sampling operations. However, in unconditional GANs with high generation quality, the artifacts are

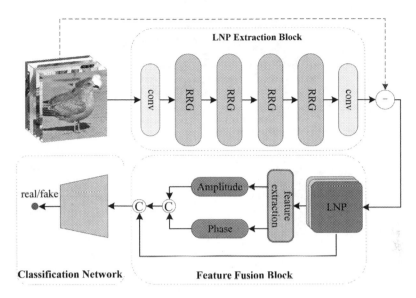

Fig. 3. The structure of the image verification network. ©indicates concatenation

not apparent in the spatial domain. Moreover, a large amount of semantic information in the spatial domain interferes with the classifier's performance. Existing methods directly use images for classification. Although good results can be achieved after extensive training, their generalization performance has room to improve. For example, for the popular generative models, detection results are not satisfied (Table 3). It is because the classifier focuses too much on artifacts in fake images, but different generative models produce different artifacts. In order to discriminate generated images from real images, we should find a feature or a pattern shared only by real images.

The exclusive pattern of real images can be extracted in image noise space, and neural networks can learn this pattern. For real images, the smooth regions show different patterns depending on the light intensity, as in column 1, column 3, and column 5 in Fig. 2(b). However, in the images generated by the GANs, the smooth regions exhibit checkerboard patterns, exhibiting periodicity, as in columns 2, 4, and 6 in Fig. 2(b).

Our goal is to find common properties in real images, so we do not need semantic information of images. A denoising network takes a set of noisy images and outputs a set of clean images after denoising. Therefore, the denoising network can maintain detailed information such as the edge texture of the original image. Then we can use the original image minus the denoised image to obtain the noise pattern without semantic interference, and we name it Learned Noise Patterns (LNP). The denoising network can be described as

$$Dst = F(Src(x, y)), \tag{1}$$

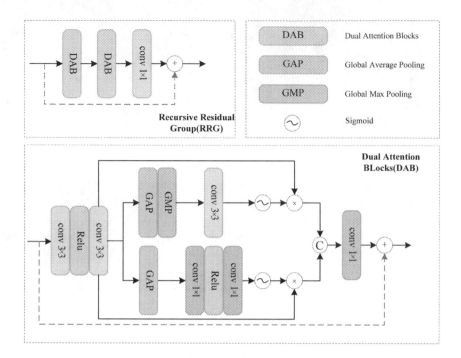

Fig. 4. The structure of the RRG module

where $Src(x, y)$ denotes the input noisy image, and $F(\cdot)$ denotes the denoising network, and Dst denotes the final clean image. We use the result of $Src(x, y) - Dst$ as LNP.

The early denoising networks, such as DNCNN, add Gaussian white noise (AWGN) to images to form training data. It is superficial and very different from the real world since there are not just AWGNs in real images. To simulate the real-world scene, [23] uses an RGB image to construct its RAW image, adding noise to the RAW image and then converting the RAW image to an RGB image to simulate the process of a real camera shot. To extract more real noise patterns, we used CycleISP [23] denoising network (LNP extraction block in Fig. 3) which was trained on real image dataset and synthetic dataset. We can then use this denoising network to extract LNP from images. For an image $I_{in}(x, y)$, $1 \leq x \leq M, 1 \leq y \leq N$, where M and N are the size of that image, it will be processed as

$$M_0 = K_3(I_{in}(x, y)), \tag{2}$$

where K_3 denotes a 3×3 convolution, and M_0 contains multiple feature maps with low-level features. Then, we used the Recursive Residual Group module (RRG) (Fig. 4) to further process the features. The RRG module is composed of two Dual Attention Blocks (DABs). Each DAB calibrates the features by two types of channel attention and spatial attention. This process can be expressed as

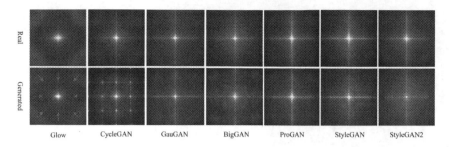

Fig. 5. The amplitude spectrum is plotted by averaging all the original or generated images for each GAN model from the dataset provided in [11]. The top indicates the average amplitude spectrum of LNP of the real images in their dataset, and the bottom shows the average amplitude spectrum of the LNP of the generated images from each model. The red arrows indicate peaks. (Color figure online)

$$M_1 = RRG(RRG(RRG(RRG(M_0)))). \tag{3}$$

Finally, the three-channel feature map M_2 can be obtained by $M_2 = K_3(M_1)$. The extracted LNP is $I_{LNP} = -M_2$.

3.2 LNP Amplitude Spectrum

The LNP characteristics of real images are not fully shown in the spatial domain. For a better exploration of the LNP, we analyzed its amplitude spectrum. For an image of M × N size, its two-dimensional discrete Fourier transform can be described as

$$F(u,v) = \frac{1}{MN} \sum_{x=0}^{M-1} \sum_{y=0}^{N-1} f(x,y)e^{-i2\pi ux/M}e^{-i2\pi vy/N}. \tag{4}$$

where $F(u,v)$ denotes the frequency component at frequency domain (u,v) and $f(x,y)$ is the gray value at point (x,y) in the spatial domain of a channel of the input image. The high frequency corresponds to the part of the image where the pixel value changes drastically. And the low frequency corresponds to the flat area of the image. The amplitude spectrum A in the frequency domain can be expressed as

$$A(u,v) = \sqrt{R^2(u,v) + I^2(u,v)}. \tag{5}$$

where $R(u,v)$ and $I(u,v)$ denote the real and imaginary parts of $F(x,y)$, respectively.

Figure 5 shows the averaged amplitude spectrum of LNP of all images from each generative model, including the generated and real images provided by [11]. For GauGAN [4], BigGAN [5], ProGAN [6], the networks use the nearest neighbor interpolation for upsampling with a period of 4. In contrast, StyleGAN [7] and StyleGAN2 [8] use bilinear interpolation for up-sampling, and we can see

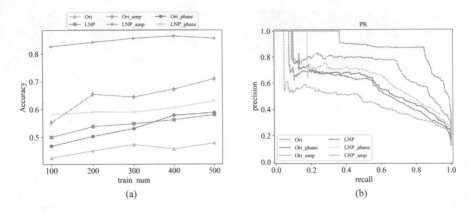

Fig. 6. Multiple classifications with SVM to discern original images, CycleGAN [2], StarGAN [3], GauGAN [4], BigGAN [5], ProGAN [6], StyleGAN [7] and StyleGAN2 [8], where the test sets include 500 images. The horizontal coordinate of plot (a) indicates the number of training images, and the vertical coordinates indicate the test accuracy. (b) represents the PR curves for 500 images in each training set

Table 1. Accuracy in the CycleGAN and StarGAN datasets using OC-SVM in the amplitude spectrum of the original images compared to the LNP amplitude spectrum

Dataset	CycleGAN [2]			StarGAN [3]		
Method	Ori_amp	LNP_amp	Gain	Ori_amp	LNP_amp	Gain
Accuracy	47.80%	73.80%	+26.0%	49.90%	98.80%	+48.90%

that the periodicity on the amplitude spectrum of their LNP is 8. For Cycle-GAN [2], upsampling is performed using deconvolution, and the LNP amplitude spectrum has strong vibrations with a period of 4. Glow [10] uses linear interpolation, and its LNP amplitude spectral period is also 4. Since the generated image has a prominent periodicity, the original image can easily be distinguished for lacking grid artifacts. For real images, their LNP are very similar in the frequency domain. Therefore, we can distinguish generated images by learning the special properties of real images.

To demonstrate the discriminative ability of LNP, we used eight datasets of real images, CycleGAN [2], StarGAN [3], GauGAN [4], BigGAN [5], ProGAN [6], StyleGAN [7] and StyleGAN2 [8] for multi-classification. Figure 6(a) shows that LNP has better classification results than using the original images, and using the amplitude spectrum of LNP has a great improvement compared to using the original images. Figure 6(b) shows better performance in PR curves.

To demonstrate the superiority of using LNP compared to the original images, we trained one-class SVM (OC-SVM) by the amplitude spectrum of the real images only. Table 1 shows that our method achieves good performance on

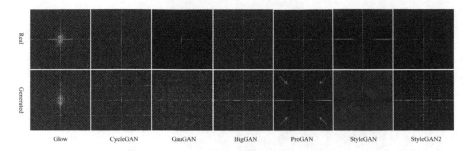

Fig. 7. The phase spectrum is plotted by averaging all the original and generated images for each GAN model from the dataset provided in [11]. The top indicates the average phase spectrum of LNP of the real images in their dataset, and the bottom shows the average phase spectrum of the LNP of the generated images in the dataset for the model. The red arrows indicate peaks. (Color figure online)

the one-class classification task that only learns the amplitude spectrum features from real images. More experimental details are in Subsect. 4.3.

3.3 LNP Phase Spectrum

The phase spectrum can be described as

$$\phi(u, v) = \arctan[\frac{I(u, v)}{R(u, v)}]. \tag{6}$$

The frequency spectrum does not contain all the information in the frequency domain. Take the basic sine wave for example, the different phases determine the position of the wave. In addition to the frequency spectrum (amplitude spectrum), we also included the phase spectrum. Neural networks are more concerned with pixel information and will learn more information about the amplitude spectrum but lack the ability to learn structural information directly [24]. The phase spectrum contains more structural information in the image. Thus, we can fully use the image information by using the phase spectrum. As in Fig. 7, we can find that the LNP of real images in the phase spectrum is similar to the amplitude spectrum. Both have similar characteristics. The dataset used for the real images of the Glow model is Celeba-HQ, which used post-processing such as face alignment and cropping on the Celeba dataset. Therefore, the LNP of its real image is slightly different from the rest of the images. In general, The phase spectrum of the LNP exhibits a grid effect and is therefore also periodic as well. We have verified in Fig. 6 that the LNP phase spectrum is easier to extract useful information than the original image phase spectrum. The accuracy of using the LNP phase spectrum and PR curve is better than using the original image phase spectrum.

Table 2. The specifications of test datasets

Generative model type	Generative model	Source	Nums
Conditional model	StarGAN [3]	CelebA	4k
	CycleGAN [2]	Style/object transfer	2.6k
	GauGAN [4]	COCO	10k
	HiSD [9]	ClelebA	4k
Unconditional model	BigGAN [5]	ImageNet	8k
	ProGAN [6]	LSUN	8k
	Glow [10]	CelebA-HQ	2k
	StyleGAN [7](LR)	LSUN	12k
	StyleGAN [7](HR)	FFHQ	5k
	StyleGAN2 [8](LR)	LSUN	16k
	StyleGAN2 [8](HR)	FFHQ	2k

3.4 LNP Network

The above analysis shows that LNP has a good discriminative ability. Compared to solely using real images for training, the amplitude spectrum information can improve the classification performance of the network, and the phase spectrum provides more contour information in the frequency domain. Therefore, we started from the perspective of real images and found the commonality that real images have. We built the network architecture in Fig. 3 by making the LNP blend with its amplitude spectrum and phase spectrum.

4 Experiments

4.1 Datasets

We used 20 classes of images provided in [11], which contain 362K real images with 362K images generated by ProGAN [6] as the training set, 4k images generated by ProGAN [6] with 4k real images as the validation set. For a fair comparison, we evaluated the publicly available dataset in [11], the GAN-generated image dataset[1], and the face dataset[2]. These include conditional generative models (StarGAN [3], CycleGAN [2], GauGAN [4]), and unconditional generative models (BigGAN [5], ProGAN [6], StyleGAN [7], StyleGAN2 [8]). In order to fully validate the effectiveness of our method, we added non-GAN generative models to our test set: HiSD [9] and Glow [10]. The details of each generative model are shown in Table 2. The StyleGAN [7], StyleGAN2 [8] dataset contains low resolution (LR) images (256×256 resolution) and high resolution (HR) images (1024×1024 resolution), where the HR images are selected from the

---

[1] http://www.grip.unina.it/download/DoGANs/.
[2] http://www.seeprettyface.com/information.html.

FFHQ face dataset. The real images for the Glow [10] model were selected from the Celeba-HQ dataset. For the HiSD [9] model we used the officially published pre-trained model[3] without any post-processing.

Table 3. The comparison of the accuracy with the state-of-the-art methods. We used only ProGAN on both the training and validation sets

Method	LR									HR		AVG
	Cycle GAN	Star GAN	Gau GAN	Big GAN	Pro GAN	HiSD	Glow	Style GAN	Style GAN2	Style GAN	Style GAN2	
AutoGAN-Spec(19')	75.3	81.2	73.4	74.9	76.9	69.3	49.5	59.7	53.3	84.9	84.2	71.2
DCT-CNN(20')	67.8	49.7	51.7	42.6	57.4	56.0	47.6	60.1	55.9	53.2	52.1	54.0
Wang(20')	83.9	90.9	77.0	75.7	99.9	80.2	27.0	91.6	90.9	89.8	88.5	81.4
Gragniello(21')	71.9	**100**	56.5	68.9	**100**	**98.4**	40.9	88.7	**98.9**	95.6	96.6	83.3
Ours	**91.6**	**100**	**79.7**	**88.1**	99.1	95.9	**80.0**	**96.0**	92.3	**99.1**	**98.8**	**92.8**

Table 4. The comparison of AP with the state-of-the-art methods

Method	LR									HR		AVG
	Cycle GAN	Star GAN	Gau GAN	Big GAN	Pro GAN	HiSD	Glow	Style GAN	Style GAN2	Style GAN	Style GAN2	
AutoGAN-Spec(19')	83.3	81.4	78.7	71.6	85.9	73.6	40.1	60.7	55.1	92.4	91.5	74.0
DCT-CNN(20')	50.5	38.8	48.3	42.6	47.3	31.6	53.6	43.0	42.5	55.4	39.9	44.9
Wang(20')	91.5	98.1	79.1	77.3	100	89.8	33.2	98.5	99.1	96.2	99.6	87.5
Gragniello(21')	79.1	**100**	60.0	67.4	100	100	33.7	98.9	**100**	99.9	**99.9**	85.4
Ours	**98.1**	**100**	**83.3**	**95.2**	100	100	**68.6**	**99.6**	98.9	**100**	99.5	**94.8**

4.2   Setup

In our experiments, ResNet50 pre-trained in ImageNet was used. Training was performed using the Adam training optimizer with $\beta_1 = 0.9$ and $\beta_2 = 0.999$ with a batch size of 256 and an initial learning rate of 1e-4. It is worth noting that if the validation set accuracy does not rise within five epochs, the learning rate decays by a factor of ten, with a minimum learning rate of 1e-6. The validation set was indirectly involved in the training, so only ProGAN [6] was used for the validation set. Our model did not see images from other generative models during the training period except for ProGAN. All training processes were implemented on an NVIDIA Tesla V100 (32G) GPU.

4.3   Comparisons

We utilized OC-SVM to discriminate generated images using the amplitude spectrum of original images and the LNP amplitude spectrum (Table 1). We used 1000 real images from the StarGAN dataset as the training set, while 1000 fake

---

[3] https://github.com/imlixinyang/HiSD.

Table 5. The comparison of F1-Score with the state-of-the-art methods

Method	LR									HR		AVG
	Cycle GAN	Star GAN	Gau GAN	Big GAN	Pro GAN	HiSD	Glow	Style GAN	Style GAN2	Style GAN	Style GAN2	
AutoGAN-Spec(19')	0.750	0.815	0.733	0.752	0.763	0.694	0.326	0.598	0.492	0.848	0.850	0.693
DCT-CNN(20')	0.708	0.660	0.606	0.553	0.606	0.689	0.542	0.644	0.617	0.180	0.173	0.544
Wang (20')	0.830	0.905	0.758	0.762	0.999	0.814	0.412	0.922	0.911	0.889	0.869	0.825
Gragnaniello(21')	0.623	1.0	0.561	0.574	**1.0**	**0.981**	0.147	0.897	**0.987**	0.954	0.965	0.790
Ours	**0.914**	**1.0**	**0.765**	**0.877**	0.991	0.961	**0.795**	**0.961**	0.928	**0.991**	**0.988**	**0.925**

images and 1000 real images as the test set. In the CycleGAN dataset, 500 real images were used as the training set, and 500 fake and 500 real images were used as the test set. The experimental results show that our method can extract more useful information than the original images. Moreover, our method can effectively distinguish real images from fake images based on the common attributes of real images.

We compared our method with three state-of-the-art deep learning methods for generated image detection: Zhang et al. [15], Frank et al. [17], Wang et al. [11], and Gragnaniello et al. [19]. Tables 3, 4 and 5 report the accuracy, AP and $F1$ values with the threshold of 0. For the others methods, we chose the best result in three experiments. [19] removed two down-sampling layers in ResNet50, so the training and testing time is ten times larger than our method. Our method achieves excellent performance on LR images and good generalization performance. Our experiments show that we have good performance not only on the GAN-generated images but also on other generative models, such as the flow-based models (Glow [10]). Our average accuracy across all models is over 90%. In the HR test set, we used real images different from the LR ones to avoid reusing data. On HR images, an average accuracy of over 98% was achieved, with an increase of around 10% compared to the rest of the methods.

4.4   Ablation Study

LNP Extraction Block. In order to provide a more suitable LNP Extraction Block, five different models were compared, including CycleISP [23], DNCNN [25], CBDNet [26], DeamNet [27] and InvDN [28]. We compared the accuracy of the five models on the test set, trained in line with Sect. 4.3. The results are shown in Table 6. We found that the information extracted by CycleISP is more favorable and can significantly improve the experimental results.

Feature Fusion Block. To evaluate the necessity of the individual components of our model, we used accuracy and mAP on both LR images and HR images. Detection results are presented in Table 7. We first evaluated the performance of using the LNP alone, which performs well. We then used the amplitude spectrum for single and three channels and the phase spectrum for single and three channels for the input to the classification network. The experimental results

Table 6. The performance of different denoising network in the LNP Extraction Block

Method	LR			HR		
	ACC	mAP	F1	ACC	mAP	F1
DNCNN [25]	78.2	90.3	82.7	75.9	87.2	0.808
CBDNet [26]	80.5	85.4	80.8	93.3	97.3	0.927
DeamNet [27]	81.2	88.7	84.8	91.5	98.3	0.926
InvDNInvDN [28]	76.6	83.7	81.2	54.6	74.8	0.682
CycleISP [23]	**91.4**	**93.7**	**91.0**	**98.9**	**99.7**	**0.989**

Table 7. The ablation study of the Feature Fusion Block

LNP	Amp(1 channel)	Amp(3 channels)	Phase(1 channel)	Phase(3 channels)	LR		HR	
					ACC	mAP	ACC	mAP
✓					87.2	92.9	97.6	**99.7**
✓	✓				84.3	91.8	89.8	99.4
✓		✓			88.1	93.3	97.8	**99.7**
✓			✓		85.3	91.8	97.3	**99.7**
✓				✓	89.3	92.2	95.9	99.2
✓	✓			✓	84.6	90.1	76.4	94.7
✓	✓		✓		**92.8**	**94.8**	**98.9**	**99.7**

show an improvement in the results using LNP and three-channel amplitude spectra and LNP and three-channel phase spectra. Using LNP, three-channel amplitude spectrum, three-channel phase spectrum combining the results into nine channels has a significant degradation. This is because the number of channels in the network does not change in ResNet50, but the proportion of LNP is reduced. Using the LNP, single-channel amplitude spectrum, and single-channel phase spectrum combined into a 5-channel feature ensures the dominance of the LNP. It allows the network to learn useful information about the amplitude and phase information.

Classification Backbone. To verify the effectiveness of the different classification networks, we conducted experiments using VGG16, VGG19, ResNet18, ResNet34, and ResNet50. Figure 8 shows our results. The VGG model has more parameters than the ResNet model and therefore is more accurate on LR images than ResNet18 and ResNet34. Since the VGG model has a fully connected layer, we cropped to 224^2 on HR images. As the number of layers on the network deepens, the results are optimal on ResNet50.

4.5   Robustness

In real-world scenes, images are subjected to various post-processing processes, such as blurring and cropping. We test our method against post-processing,

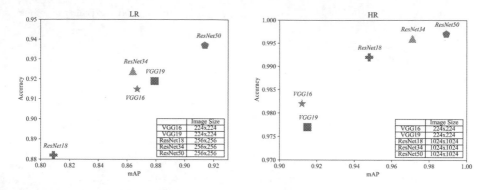

Fig. 8. The performance of different backbones in LR and HR

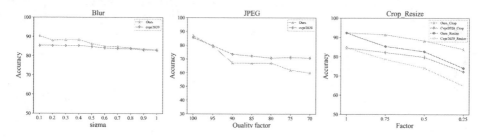

Fig. 9. The robustness of our model compared to Wang(CVPR20') [11]

including Gaussian blurring (sigma: 0.1–1), JPEG quality factors (70–100), image cropping, and resizing (cropping/scaling factor: 0.25–1). We randomly selected 1000 images in each generative model dataset for robustness experiments. Figure 9 shows the robustness results of our comparison with [11]. Our model is better when blurring, cropping, and resizing. However, the results are lower in the JEPG compression case. The JPEG scheme generates multiple peaks in the frequency domain, similar to the periodicity in generated images. Therefore our method does not work well in the JPEG case. In the following work, we will solve this problem.

5   Conclusions

In this paper, we detect generated images using LNP of real images. We demonstrated that the LNP of real images are very similar in amplitude and phase spectrum, while the LNP of generated images is far different. Experimental results show that the method outperforms existing methods in image authentication. The superior generalization ability of the proposed method allows use in realistic scenes, even with future unseen models.

References

1. Goodfellow, I., et al.: Generative adversarial nets. Adv. Neural Inf. Process. Syst. **27** (2014)
2. Zhu, J.Y., Park, T., Isola, P., Efros, A.A.: Unpaired image-to-image translation using cycle-consistent adversarial networks. In: Proceedings of the IEEE International Conference on Computer Vision, pp. 2223–2232 (2017)
3. Choi, Y., Choi, M., Kim, M., Ha, J.W., Kim, S., Choo, J.: Stargan: unified generative adversarial networks for multi-domain image-to-image translation. In: Proceedings of the IEEE Conference on Computer Vision and Pattern Recognition, pp. 8789–8797 (2018)
4. Park, T., Liu, M.Y., Wang, T.C., Zhu, J.Y.: Semantic image synthesis with spatially-adaptive normalization. In: Proceedings of the IEEE/CVF Conference on Computer Vision and Pattern Recognition, pp. 2337–2346 (2019)
5. Brock, A., Donahue, J., Simonyan, K.: Large scale GAN training for high fidelity natural image synthesis. arXiv preprint arXiv:1809.11096 (2018)
6. Karras, T., Aila, T., Laine, S., Lehtinen, J.: Progressive growing of gans for improved quality, stability, and variation. arXiv preprint arXiv:1710.10196 (2017)
7. Karras, T., Laine, S., Aila, T.: A style-based generator architecture for generative adversarial networks. In: Proceedings of the IEEE/CVF Conference on Computer Vision and Pattern Recognition, pp. 4401–4410 (2019)
8. Karras, T., Laine, S., Aittala, M., Hellsten, J., Lehtinen, J., Aila, T.: Analyzing and improving the image quality of stylegan. In Proceedings of the IEEE/CVF Conference on Computer Vision and Pattern Recognition, pp. 8110–8119 (2020)
9. Li, X., et al.: Image-to-image translation via hierarchical style disentanglement. In: Proceedings of the IEEE/CVF Conference on Computer Vision and Pattern Recognition, pp. 8639–8648 (2021)
10. Kingma, D.P., Dhariwal, P.: Glow: generative flow with invertible $1{\times}1$ convolutions. arXiv preprint arXiv:1807.03039 (2018)
11. Wang, S.Y., Wang, O., Zhang, R., Owens, A., Efros, A.A.: Cnn-generated images are surprisingly easy to spot... for now. In: Proceedings of the IEEE/CVF Conference on Computer Vision and Pattern Recognition, pp. 8695–8704 (2020)
12. Dang, H., Liu, F., Stehouwer, J., Liu, X., Jain, A.K.: On the detection of digital face manipulation. In: Proceedings of the IEEE/CVF Conference on Computer Vision and Pattern recognition, pp. 5781–5790 (2020)
13. Liu, Z., Qi, X., Torr, P.H.S.: Global texture enhancement for fake face detection in the wild. In: Proceedings of the IEEE/CVF Conference on Computer Vision and Pattern Recognition, pp. 8060–8069 (2020)
14. Zhao, H., Zhou, W., Chen, D., Wei, T., Zhang, W., Yu, N.: Multi-attentional deepfake detection. In: Proceedings of the IEEE/CVF Conference on Computer Vision and Pattern Recognition, pp. 2185–2194 (2021)
15. Zhang, X., Karaman, S., Chang, S.F.: Detecting and simulating artifacts in gan fake images. In: 2019 IEEE International Workshop on Information Forensics and Security (WIFS), pp. 1–6. IEEE (2019)
16. Chai, L., Bau, D., Lim, S.-N., Isola, P.: What makes fake images detectable? understanding properties that generalize. In: Vedaldi, A., Bischof, H., Brox, T., Frahm, J.-M. (eds.) ECCV 2020. LNCS, vol. 12371, pp. 103–120. Springer, Cham (2020). https://doi.org/10.1007/978-3-030-58574-7_7
17. Frank, J., Eisenhofer, T., Schönherr, L., Fischer, A., Kolossa, D., Holz, T.: Leveraging frequency analysis for deep fake image recognition. In: International Conference on Machine Learning, pp. 3247–3258. PMLR (2020)

18. Durall, R., Keuper, M., Keuper, J.: Watch your up-convolution: Cnn based generative deep neural networks are failing to reproduce spectral distributions. In: Proceedings of the IEEE/CVF Conference on Computer Vision and Pattern Recognition, pp. 7890–7899 (2020)

19. Gragnaniello, D., Cozzolino, D., Marra, F., Poggi, G., Verdoliva, L.: Are gan generated images easy to detect? a critical analysis of the state-of-the-art. In: 2021 IEEE International Conference on Multimedia and Expo (ICME), pp. 1–6. IEEE (2021)

20. Cozzolino, D., Marra, F., Gragnaniello, D., Poggi, G., Verdoliva, L.: Combining prnu and noiseprint for robust and efficient device source identification. EURASIP J. Inf. Secur. **2020**(1), 1–12 (2020)

21. Cozzolino, D., Verdoliva, L.: Noiseprint: a cnn-based camera model fingerprint. IEEE Trans. Inf. Forensics Secur. **15**, 144–159 (2019)

22. Ghosh, A., Zhong, Z., Cruz, S., Veeravasarapu, S., Singh, M., Boult, T.E.: Infoprint: information theoretic digital image forensics. In: 2020 IEEE International Conference on Image Processing (ICIP), pp. 638–642. IEEE (2020)

23. Zamir, S.W., et al.: Cycleisp: real image restoration via improved data synthesis. In: Proceedings of the IEEE/CVF Conference on Computer Vision and Pattern Recognition, pp. 2696–2705 (2020)

24. Chen, G., Peng, P., Ma, L., Li, J., Du, L., Tian, Y.: Amplitude-phase recombination: Rethinking robustness of convolutional neural networks in frequency domain. In: Proceedings of the IEEE/CVF International Conference on Computer Vision, pp. 458–467 (2021)

25. Zhang, K., Zuo, W., Chen, Y., Meng, D., Zhang, L.: Beyond a gaussian denoiser: residual learning of deep cnn for image denoising. IEEE Trans. Image Process. **26**(7), 3142–3155 (2017)

26. Guo, S., Yan, Z., Zhang, K., Zuo, W., Zhang, L.: Toward convolutional blind denoising of real photographs. In: Proceedings of the IEEE/CVF Conference on Computer Vision and Pattern Recognition, pp. 1712–1722 (2019)

27. Ren, C., He, X., Wang, C., Zhao, Z.: Adaptive consistency prior based deep network for image denoising. In: Proceedings of the IEEE/CVF Conference on Computer Vision and Pattern Recognition, pp. 8596–8606 (2021)

28. Liu, Y., et al.: Invertible denoising network: a light solution for real noise removal. In: Proceedings of the IEEE/CVF Conference on Computer Vision and Pattern Recognition, pp. 13365–13374 (2021)

An Information Theoretic Approach for Attention-Driven Face Forgery Detection

Ke Sun[1,3], Hong Liu[2], Taiping Yao[3], Xiaoshuai Sun[1,4(✉)], Shen Chen[3], Shouhong Ding[3(✉)], and Rongrong Ji[1,4]

[1] Media Analytics and Computing Lab, Department of Artificial Intelligence, School of Informatics, Xiamen University, Xiamen 361005, China
xssun@xmu.edu.cn
[2] National Institute of Informatics, Tokyo, Japan
[3] Youtu Lab, Tencent, Shanghai, China
skjack@stu.xmu.edu.cn
[4] Institute of Artificial Intelligence, Xiamen University, Xiamen 361005, China

Abstract. Recently, Deepfake arises as a powerful tool to fool the existing real-world face detection systems, which has received wide attention in both academia and society. Most existing forgery face detection methods use heuristic clues to build a binary forgery detector, which mainly takes advantage of the empirical observation based on abnormal texture, blending clues, or high-frequency noise, etc.. However, heuristic clues only reflect certain aspects of the forgery, which might lead to model bias or sub-optimization. Our recent observations indicate that most of the forgery clues are hidden in the informative region, which can be measured quantitatively by the classic information maximization theory. Motivated by this, we make the first attempt to introduce the self-information metric to enhance the feature representation for forgery detection. The proposed metric can be formulated as a plug-and-play block, termed self-information attention (SIA) module, which can be integrated with most of the top-performance deep models to boost their detection performance. The SIA module can explicitly help the model locate the informative regions and recalibrate channel-wise feature responses, which improves both model's performance and generalization with few additional parameters. Extensive experiments on several large-scale benchmarks demonstrate the superiority of the proposed method against the state-of-the-art competitors.

Keywords: Face forgery detection · Information maximization · Attention mechanism

1 Introduction

Recently, face forgery generation methods have received lots of attention in the computer vision community [12,21,40,46,49,53], which may cause severe trust

---

Supplementary Information The online version contains supplementary material available at https://doi.org/10.1007/978-3-031-19781-9_7.

S. Avidan et al. (Eds.): ECCV 2022, LNCS 13674, pp. 111–127, 2022.
https://doi.org/10.1007/978-3-031-19781-9_7

issues and seriously disturb the social order. For example, the producer can forge the video of world leaders to influence or manipulate politics and social sentiment. Even worse, these fake videos are of high quality and can be easily generated by open-source codes like DeepFaceLab. Therefore, it is urgent to develop effective face forgery detection methods to mitigate malicious abuse of face forgery.

A simple way is to model face forgery detection as a binary classification problem [1,12,16,44]. Basically, a pretrained convolutional neural network (CNN) is used to distinguish the authenticity of the input face, which is a golden standard in Deepfake-Detection-Challenge [12]. However, the generated fake faces become more and more authentic, which means the differences between real and fake faces are more subtle. Although CNN models possess discriminative features, it is still hard to directly use such models to capture those forgery clues in a unified framework, resulting in unsatisfactory performance.

To tackle this issue, many heuristic methods [7,11,17,23,32,45] usually use the prior knowledge or observed clues to learn more discriminative features, which are instrumental in distinguishing real and fake faces. For example, F3-Net [39] learns forgery patterns with the awareness of frequency, Gram-Net [32] leverages global image texture representations for robust fake image detection, and Face X-ray [27] takes advantage of blending boundary for a forged image to enhance the performance. Though these methods can help to improve the performance, these heuristic methods lack unified theoretical support and only reflect certain aspects of the face forgery, leading to model bias or suboptimization.

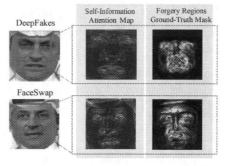

Fig. 1. Visualization of self-information map and manipulation ground truth mask for forgery faces by different manipulations (Deepfakes and FaceSwap). The self-information map is calculated by Eq. 1. The ground truth mask is generated by subtracting the forged faces and the corresponding real faces with some morphological transformations

To address this issue, we revisit face forgery detection from a new perspective, *i.e.*, face forgery is highly associated with high-information content. Inspired by [5,6], we make the first attempt to introduce self-information as a theoretic guidance to improve the discriminativeness of the model. Specially, self-information can be easily defined by the current or surrounding regions [43], where a high-information region is significantly different from their neighborhoods that can reflect the amount of information of the image content. Moreover, we find that most existing clues are always in high self-information regions. For example, due to the instability of the generative model, some abnormal textures always appear in forgery faces. These high-frequency artifacts are often very different from the surrounding facial features or skin, where the self-information can highlight these clues. Another example is blending artifacts. Face x-ray [27] demonstrate that the forged boundary is widely existed in forgery faces because of blending operation. The skin color or texture difference between real and forgery part enlarge the self-information in blending artifacts regions. Motivated by this observation, we design a novel self-information attention module called Self-Information Attention (SIA), which calculates pixel-wise self-information and uses it as a spatial attention map to capture more subtle abnormal clues. Additionally, the SIA module calculates the average self-information of each channel's feature map and uses it as attention weights to select the most informative feature map. As shown in Fig. 1, the self-information map of the original image highlight the same region as the ground truth mask, which indicates its effectiveness in face forgery detection task.

We conduct our experiments on several widely-used benchmarks. And experimental results show that our proposed method significantly outperforms the state-of-the-art competitors. Particularly, the proposed SIA module can be flexibly plugged into most CNN architectures with little parameter increase. Our main contributions can be summarized as follows:

- We propose a new perspective for face forgery detection based on information theory, where self-information is introduced as a theoretic guidance for detection models to capture more critical forgery cues.
- We specially design a novel attention module based on self-information, which helps the model capture more informative regions and learn more discriminative features. Besides, the SIA attention can be plugged into most existing 2D CNNs with negligible parameter increase.
- Extensive experiments and visualizations demonstrate that our method can achieve consistent improvement over multiple competitors with a comparable amount of parameters.

2   Related Work

2.1   Forgery Face Manipulation

Face forgery generation methods have a security influence on scenarios related to identity authentication, which achieve more and more attention in computer vision communities. In particular, *deepfakes* is the first deep learning based face identity swap method [49], which uses two Autoencoders to simulate changes in facial expressions. The other stream of research is to design GAN based models [4,14,15,21] for generating entire fake faces. Recently, graphics-based approaches are widely used for identity transfer, which are more stable compared with deep learning based approaches. For instance, Face2Face [50] is can operate face swap using only an RGB camera in real-time. Averbuch-Elor *et al.* [3] proposed a reenactment method that deforms the target image to match the expressions of the source face. NeuralTextures [48] renders a fake face via computing reenactment result with neural texture. Kim *et al.* [25] combined image-to-image translation network with computer graphics renderings to convert face attributes. These forgery methods focus on manipulate high-information areas and may leave some high-frequency subtle clues, thus we introduce self-information learning to assist in identifying forged faces.

2.2   Face Forgery Detection

To detect the authenticity of input faces, early works usually extract low-level features such as RGB patterns [36], inconsistency of JPEG compression [2], visual artifacts [35]. More recently, binary convolution neural network has been widely used to this task [12] and achieve better performance. However, directly using vanilla CNN tend to extract semantic information while may ignore the subtle and local forgery patterns [54]. Thus, some heuristic methods are proposed, which leverage observation or prior knowledge to help model to mine the forgery pattern. For instance, Face X-ray [27] is supervised by the forged boundary. F3-Net [39] leverage frequency clues as to the auxiliary to RGB features. Local-Relation [9] measures the similarity between features of local regions based on the observation of inconsistency between forgery parts and real parts. However, these methods still cannot cover all the forgery clues, leading to sub-optimal performance. Thus, we introduce self-information to help the model capture informative region adaptively. In addition, our proposed method only contains a few parameters and can serve as a plug-and-play module upon several backbones.

3   Proposed Method

3.1   Preliminaries

Problem Formulation. Many works [24] have been proposed to identify a given face, real or fake, but most of them are based on experimental observations that show the remarkable difference between real and fake faces. Recent work [35] found that these observations belong to the discriminative artifacts clues that are subtle but abnormal compared with their neighborhoods, because of the generative model's instability and the imperfection of the blending methods. On the other hand, existing models just consider one or a small number of these different clues, which are integrated into the vanilla CNN, leading to bias or sub-optimization model. These raise a natural question that, *is there a metric that can adaptively capture differential information?* To answer this question, this paper focuses on the information theory and uses classical self-information to adaptively qualify the saliency clues.

Self-information Analysis. The self-information is a metric of the information content related to the outcome of a random variable [5], which is also called *surprisal, i.e.,* it can reflect the surprise of an event's outcome. Given a random variable X with probability mass function P_X, the self-information of X as outcome x is $I_X(x) = -log(P_X(x))$. As a result, we can derive that the smaller its probability, the higher the self-information it has. That is, the more different the region from its neighboring patches, the more self-information it contains. Inspired by [5,43], self-information can apply to the joint likelihood of statistics in a local neighborhood of the current patch, which provides a transformation between probability and the degree of information inherent in the

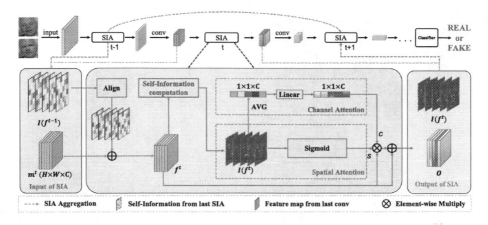

Fig. 2. The overview of our face forgery detection framework with Self-information attention (SIA) module. The SIA module is embedded in the middle layer of CNN. The orange dotted block means the channel attention part, while the blue dotted block represents the spatial attention part. The details of self-information is shown in Fig. 3 (Best viewed in color) (Color figure online)

local statistics. For face forgery detection, the heuristic unusual forgery clues (such as high-frequency noise, blending boundary, abnormal textures, *etc.*) are hidden in the high-information context. Therefore, it is intuitive to introduce the self-information metric into face forgery detection to help model additively learn high-information features.

3.2 Overall Framework

In this paper, we design a new attention mechanism, that is based on the self-information metric, which could highlight the manipulated regions. We call this newly defined model as Self-Information Attention (SIA) module, whose overview framework is shown in the Fig. 2. In particular, the proposed SIA module mainly contains three key parts: 1) **Self-Information Computation**: To capture the high-information content region, we calculate the self-information from the input feature map and output a new discriminative attention map. 2) **Self-Information based Dual Attention**: To maximize the ability of using self-information by backbone model, the self-information from the input feature map would be used on both channel-wise attention and spatial-wise attention. 3) **Self-Information Aggregation**: Motivated by [19,54], we densely forward all previous self-information feature maps to the current SIA block, which is to preserve the detail area to the greatest extent.

3.3 Self-information Computation

Let $f^t \in R^{C \times H \times W}$ denotes the input of the t-th SIA module with C channels and spatial shape of $H \times W$, where $f_k^t(i,j)$ denotes the k-th channel's

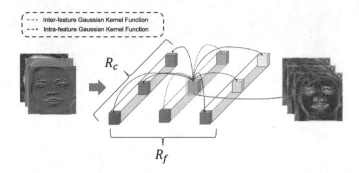

Fig. 3. Visualization of Self-Information Computation. R_f denotes the local receptive filed region, and R_c is the channel offset region (Best viewed in color)

pixel of f^t located by the coordinate (i, j). As mentioned before [5], the self-information can be approximated by the joint probability distribution of the current pixel together with its neighborhoods with Gaussian kernel function. Different with previous work [5], we consider the self-information through two orthogonal dimensions, one is to find the neighborhoods in the spatial dimension, and the other is to search the neighborhoods in channel dimension.

We define the spatial space intra-feature self-information as:

$$I_{\text{intra}}(f_k^t(i,j)) = -\log \sum_{m,n \in R_f} e^{-\frac{||f_k^t(i,j) - f_k^t(i+m, j+n)||^2}{2h^2}}, \tag{1}$$

where R_f are the local receptive filed region near the pixel (i, j), m and n are the pixel indexes in the R_f, and h is the bandwidth.

When the neighborhoods are located in the channel dimension, we define the self-information in channel as inter-feature self-information I_{inter}, which is shown as:

$$I_{\text{inter}}(f_k^t(i,j)) = -\log \sum_{s \in R_c} e^{-\frac{||f_k^t(i,j) - f_{k+s}^t(i,j)||^2}{2h^2}}, \tag{2}$$

where s is the index of the channel offset region R_c. The inter-feature self-information could help us avoid some observation noise that exists in the channels.

As a result, the whole self-information $I(f_k)$ can be formulated as:

$$I(f_k^t(i,j)) = I_{\text{intra}}(f_k^t(i,j)) + \lambda I_{\text{inter}}(f_k^t(i,j)), \tag{3}$$

where λ is the weight parameter that balance the importance of the inter-feature self-information. The Fig. 3 illustrates the computation of self-information.

3.4 Self-information Based Dual Attention

We propose a new dual attention model, where the saliency is qualified by the self-information measure [6]. Inspired by [20], we consider the saliency features through spatial dimension and channel dimension.

Spatial-Wise Attention Module. We introduce a spatial attention module based on self-information, as the flowchart shown in Fig. 2. In detail, we calculate each pixel's self-information features $I(f_k^t)$ via Eq. 3. Then, we use the Sigmoid function to normalize such features and output the self-information based spatial attention map. Finally, we perform an element-wise multiply operation with the input feature f_k^t. The whole formulation of the Spatial-wise Attention Module is shown as follows:

$$s_k = \text{Sigmoid}(I(f_k^t)) * f_k^t. \tag{4}$$

This attention map focuses on the high-information region with little parameter improvement, which can adaptively enhance many artifact subtle clues, such as blending boundary and high-frequency noise. For more details please refer to the Sect. 4.

Channel-Wise Attention Module. Apart from the spatial attention module, we further introduce the channel attention module, which pipeline is illustrated in Fig. 2. Similar to [20], we calculate the average self-information of channel feature maps and generate channel-wise statistical feature c_k for the k-th element of f^t as follows:

$$c_k = \frac{1}{H \times W} \sum_{i=1}^{H} \sum_{j=1}^{W} I(f_k^t). \tag{5}$$

To mitigate the problem of training stability, we opt to employ a simple linear transform with sigmoid activation on vector $c = \{c_1, c_2...c_C\}$ as channel attention c':

$$c' = \text{Sigmoid}(Wc), \tag{6}$$

where $W \in R^{C \times C}$ is the linear function. This module could improve the self-information of the feature map that contains high-information, which helps locate the saliency in explicit contents.

Dual Attention Module Embedded in CNN. Finally, we combine the two attention modules mentioned above and perform an element-wise sum operation between the processed attention map and f_k to output a residual error feature $o_k \in R^{C \times H \times W}$, which formulated as:

$$O_k = c'_k * s_k + f_k^t. \tag{7}$$

The proposed SIA module is a flexible module, which can be easily inserted into any CNN-based architecture. Also, we can also flexibly choose the spatial attention module and the channel attention module. The SIA module does not increase many parameters yet can enhance the performance of the model.

3.5 Self-information Aggregation

General CNN such as EfficientNet [47] usually use down-sampling operations to reduce the parameters and expand the receptive field, which tend to eliminate subtle clues with high information content in face forgery detection task.

To overcome this problem, inspired by [22], we design a self-information aggregation operation, cascading different levels of SIA modules via self-information attention map. Thus the local and subtle forgery clues can be preserved. As shown in Fig. 2, we add the attention map of the previous stage with the current input feature map to preserve the shallow high informative texture. Due to the different sizes of attention maps at different levels, we use 1×1 convolution to align the number of channels and use the interpolation method to align the size of the feature map. This alignment operation could be presented as the function Align. As a result, the t-th input feature f^t can be defined as:

$$f^t = \sum_{i=1}^{t-1} \text{Align}_i(I(f^i)) + m^t, \qquad (8)$$

where m^t is the feature map adjacent to the t-th SIA module.

3.6   Loss Function

We use the Cross-entropy as loss function, which is defined as:

$$L_{ce} = -\frac{1}{n} \sum_{i=1}^{n} y_i \log(\hat{y}_i) + (1 - y_i) \log(1 - \hat{y}_i), \qquad (9)$$

where n is the number of images, $\hat{y}_i$ is the prediction of the i-th fake image, and y_i is the label of the sample.

4   Experiment

In this section, we evaluate the proposed SIA module against some state-of-the-art face forgery detection methods $[1, 8, 10, 11, 13, 18, 27, 34, 39, 47, 54]$ and some attention techniques $[20, 32, 42]$. We explore the robustness under unseen manipulation methods and conduct some ablation studies, and further give some visualization results.

4.1   Experimental Setup

Datasets. We conduct our experiments on several challenging dataset to evaluate the effectiveness of our proposed method. **FaceForensics++** [40] is a large-scale deepfake detection dataset containing $1,000$ videos, in which 720 videos are used for training and the rest 280 videos are used for validation or testing. There are four different face synthesis approaches, including two graphics-based methods (*Face2Face* and *FaceSwap*) and two learning-based approaches (*DeepFakes* and *NeuralTextures*). The videos in FaceForensics++ have two kinds of video quality: high quality (C23) and low quality (C40). **Celeb-DF** [30] is another widely-used Deepfakes dataset, which contains 590 real videos and $5,639$ fake videos. In Celeb-DF, the DeepFake videos are generated by swapping faces for each pair of the 59 subjects. Following the prior works $[39, 47, 54]$, we use the

Table 1. Comparison on FaceForensics++ dataset in terms of ACC and AUC with different qualities (HQ and LQ). The highest results are highlighted in bold. The F3-Net use 0.5 as a threshold

Table 2. Cross-dataset evaluation from FF++ (LQ) to deepfake class of FF++ and Celeb-DF in terms of AUC. The highest results are highlight in bold

Methods	ACC (LQ)	AUC (LQ)	ACC (HQ)	AUC (HQ)
MesoNet [1]	70.47	–	83.10	–
Face X-ray [27]	–	61.60	–	87.40
Xception [10]	86.86	89.30	95.73	96.30
Xception-ELA [18]	79.63	82.90	93.86	94.80
Xception-PAF [8]	87.16	90.20	–	–
Two Branch [34]	86.34	86.59	86.34	86.59
EfficientNet-B4 [47]	86.95	88.91	96.63	99.18
F3-Net [39]	86.89	93.30	97.31	98.10
MAT [54]	88.69	90.40	97.60	99.29
SPSL [31]	81.57	82.82	91.50	95.32
RFM [51]	87.06	89.83	95.69	98.79
Freq-SCL [26]	89.00	92.39	96.69	99.28
Ours	**90.23**	**93.45**	**97.64**	**99.35**

Methods	FF++	Celeb-DF
Two-stream [55]	70.10	53.80
Meso4 [1]	83.00	53.60
FWA [29]	80.10	56.90
DSP-FWA [29]	93.00	64.60
Xception [40]	95.50	65.50
EN-b4 [12]	96.39	71.10
Multi-task [37]	76.30	54.30
Capeule [38]	96.60	57.50
SMIL [28]	96.80	56.30
Two Branch [34]	93.18	73.41
MAT [54]	96.41	72.50
GFF [33]	95.73	74.12
SPSL [31]	96.91	76.88
Ours	**96.94**	**77.35**

multi-task cascaded CNNs to extract faces, and we randomly select 50 frames from each video to construct the training set and test set. **WildDeepfake** [56] is a recently released forgery face dataset that contains 3805 real face sequences and 3509 fake face sequences, which is obtained from the internet. Therefore, wild deepfake has a variety of synthesis methods, backgrounds, and ids.

Evaluation Metrics. We apply accuracy score (ACC) and area under the receiver operating characteristic curve (AUC) as our basic evaluation metrics.

Implementation Details. We use EfficientNet-b4 [47] pretrained on the ImageNet as our backbones, which are widely used in face forgery detection. The backbone contains seven layers, and we put our proposed SIA module in the output of layer1, layer2, and layer4. This is due to the shallow and middle layers contain low-level and middle-level features, which reflect the subtle artifact clues well. We resize each input face to 299×299. The hyperparameters λ in Eq. 3 is set to 0.5. We use Adam optimizer to train the network's parameters, where the weight decay is equal to $1e - 5$ with betas of 0.9 and 0.999. The initial learning rate is set to 0.001, and we use StepLR scheduler with 5 step-size decay and gamma is set to 0.1. The batch size is set to 32.

4.2 Experimental Results

Intra-dataset Testing. We evaluate the performance under two quality settings on FaceForensics++. Note that the results of F3-Net use a threshold of

Table 3. Performance on Celeb-DF and WildDeepfake datasets in terms of ACC and AUC

Method	Celeb-DF		WildDeepfake	
	ACC	AUC	ACC	AUC
Xception [10]	97.90	99.73	77.25	86.76
EfficientNet-B4 [47]	97.63	99.20	81.63	90.36
RFM [51]	97.96	99.94	77.38	83.92
F3-Net [39]	95.95	98.93	80.66	87.53
MAT [54]	97.84	99.81	82.86	90.71
Ours	**98.48**	**99.96**	**83.95**	**91.34**

Table 4. ACC of different pretrained backbones on FaceForensics++ HQ, FaceForensics++ LQ and Celeb-DF datasets

Backbone	FF++ (HQ)	FF++ (LQ)	Celeb-DF
EffecinetNet-b0	96.32	85.66	97.81
EffecinetNet-b0+Ours	**96.95**	**87.03**	**98.37**
XceptionNet	95.73	86.86	97.90
XceptionNet+ours	**96.85**	**87.30**	**98.03**
MobileNet-v2	95.58	85.19	97.41
MobileNet-v2+Ours	**97.05**	86.78	**98.22**

0.5. The overall results in Table 1 show that the proposed method obtains state-of-the-art performance on the both high-quality and low-quality settings. Compared Freq-SCL [26] and SPSL [31] leverage frequency clues as to the auxiliary to RGB features, both of which convert RGB image into the frequency domain together with a dual-stream framework. Both two methods boost performance via input perspective, but our method takes consideration of promoting representation learning. Compared with the recent attention based Multi-attentional [54], our method achieves better performance. This is because that SIA gives more accurate guidance for the attention mechanism on both channel-wise and spatial-wise dimensions, which provide more adaptive information for the model.

To further demonstrate the effectiveness of our method, we evaluate the SIA module on two famous forgery datasets: Celeb-DF and WildDeepfake. The results are shown in Table 3. We can observe that our SIA outperforms all comparison methods. Specifically, compared with F3-Net which requires 80M parameters and 21G macs, our EN-b4+SIA only contains 35M parameters and 6.05G macs and achieves about 3% improvement on both two datasets. In addition, we evaluate the proposed SIA module on DFDC [12] dataset and achieve SOTA performance with 82.31% in terms of ACC and 90.96% in terms of AUC. Due to the page limit, we put the results in the supplementary material.

Cross-Dataset Testing. To further demonstrate the generalization of SIA, we conduct cross-dataset evaluations. Specifically, following the setting of [34], we train our model on FF++ (LQ) and test it on Deepfakes class and Celeb-DF. The quantitative results are shown in Table 2, we can observe that our method obtain state-of-the-art performance especially in cross-database setting. Our SIA outperforms by 4% and 2% in terms of AUC compared with the recent SPSL and GFF on cross-dataset setting and achieve slight improvement on the intra-dimain setting. The reason for the improvement is that our module guide the backbone focuses on the informative subtle details which are commonly present on all forgery face.

Dependency on Backbone. The proposed SIA module is a plug-and-play block, which can be embedded in any deep learning based model. Therefore, we verify the effectiveness of the SIA module by using different backbones. We select

Table 5. Quantitative results on Celeb-DF and FaceForensics++ dataset with different qualities (HQ and LQ). The compared methods are all plug-and-play attention modules. The last column represents the parameter increase after adding the corresponding module. The highest results are highlighted in bold

Method	FF++ (HQ)			FF++ (LQ)			Celeb-DF			Parameter increase
	ACC	AUC	EER	ACC	AUC	EER	ACC	AUC	EER	
Baseline+Selayer	97.05	99.20	4.04	89.05	90.99	19.01	98.00	99.65	2.15	45k
Baseline+NL	96.79	99.16	4.20	89.55	91.34	19.50	97.89	99.61	2.20	764k
Baseline+Selayer+NL	97.21	99.25	3.95	89.78	91.56	18.98	97.80	99.68	2.04	811k
Baseline+GSA	94.68	97.92	7.03	88.92	89.03	22.32	97.88	99.63	2.63	1024k
Ours	**97.64**	**99.35**	**3.83**	**90.23**	**93.75**	**18.57**	**98.48**	**99.96**	**1.87**	42k

the EffecinetNet-b0, MobileNet-v2, and XceptionNet as other backbones, and we evaluate the results on FaceForensics++ and Celeb-DF. All the SIA module is embedded in the first and middle layer. For instance, for the EffiecientNet-b0, we put out the module after the 2th, 3th and 5th MBConvBlock. For the XceptionNet, our module is inserted between 3th block and 4 block. As for the MobileNet-v2, the module is embedded after the 3th and 7th InvertedResidual block. The result is shown in Table 4. We find that our methods do improve the network performance regardless of the types of backbones, which proves the flexibility and generality of our method.

Compared with Attention Methods. We compare the proposed method with several classical attention-based methods to show the effectiveness of self-information in this task: **(1) Baseline**: The EffecinetNet-b4 pretrained on the ImageNet. **(2) Baseline+SE-layer** [20]: The channel attention module. **(3) Baseline+Non-local** [52]: Non-local attention has been used in deepfake detection [32]. Here we use the Gaussian embedded version with both batchnorm layer and sub-sample strategy. **(4) Baseline+SE-layer+Non-local**: We use the SE-layer and Non-local to realize both channel attention and spatial attention module. **(5) Baseline+GSA** [42]: GSA is the state-of-the-art attention module that considers both pixel content and spatial information. Here the number heads are set to 8, and the dimensional key is set to 64.

The comparison results are reported in Table 5. The results show that our proposed SIA module outperforms all the reference methods on both two benchmarks. Specifically, after adding our SIA module on the baseline, the performance has about 1.5% ACC improvement with little parameters increase. This reflects that the self-information does fit for face forgery detection task.

4.3   Ablation Study

Impact of Different Components. To further explore the impact of different components of SIA module, we split each part separately for experimental verification. The ACC and AUC results on FF++ (LQ) are shown in Table 6. The results demonstrate that the three key components have a positive effect

Table 6. Abalation study on FaceForensics++(LQ) dataset

Spatial	Channel	Aggregation	ACC	AUC
✓			89.65	91.95
	✓		89.83	91.34
✓	✓		89.90	92.43
✓	✓	✓	90.23	93.75

Table 7. Comparative experiment of module insertion position

Layers	ACC on FF++	AUC on FF++
L1, L2, L3	89.56	92.73
L1, L2, L4	**90.23**	**93.75**
L3, L4, L5	89.14	91.15
L1, L4, L6	88.35	90.21

on the performance, all of which are necessary for the face forgery detection. Among them, spatial attention has a relatively large impact on performance, which demonstrate the importance of capturing high-information regions for the face forgery task.

Impact of Embedding Layer. We further conduct some ablation experiments to explore the effect of insertion place of our method. The attention module is embedded in different layers of EffiecientNet-b4 and tested on the FaceForensics++ LQ dataset. The results on the left of Table 7 show that the best performance is achieved when the attention module is embedded in layer1, layer 2 and layer4, which is in the shallow and middle of the backbone. The SIA module is derived from the theory of self-information (SI), which is usually built on the shallow structural and textural features. Therefore, it is intuitive to insert the SIA module in the shallow layers, which helps enhance SI. In the middle layers, SIA module helps reduce the global inconsistency bringing from long-range forgery patterns and pass the useful local and subtle forgery information via the self-information aggregation scheme. However, in deeper layers, down-sampling operation will neglect many local and subtle forgery information, in which SI can hardly find useful cues for forgery detection. In sum, it is natural and reasonable to plug SIA into either shallow or middle layers (L1, L2, L4), and our experiments indeed had verified this.

4.4   Visualization and Analysis

Analysis on SIA Module. To analysis our attention module, we visualize the feature maps from different channels sorted by channel-wise attention weight and the highest weight channel's SIA map. Figure 4 shows the result (all visualizations are colored according to the normalized feature map). We can observe that the channels with high self-information contain more local high-frequency clues and subtle details, while the lower ones have more semantic information and smoother clues which is less helpful for the face forgery detection task. In addition, the self-information based attention map enhances high-information areas such as mouth, eyes, high-frequency textures, and blending boundary, while weakening repetitive low-frequency areas. These visualizations demonstrate that our SIA module can effectively mine the informative channels and subtle clues, which are critical for performance improvement.

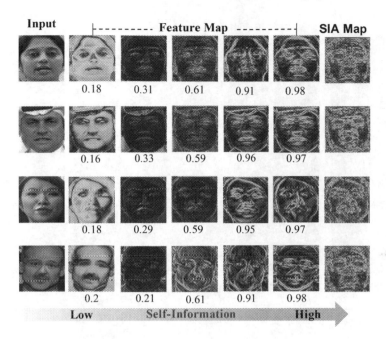

Fig. 4. Visualization of our channel-wise attention scores and their corresponding SIA maps. The feature maps are sorted according to the attention weights from low to high. The last column shows the SIA map calculated by the highest channel score feature map (Best viewed in color)

Visualization of Grad-CAM. We apply Grad-CAM [41] and Guided Grad-CAM tools to the baseline model and our model, which are widely-used methods to explain the attention of deep neural networks. The Grad-CAM can identify the regions that the network considers import, while Guided Grad-CAM can reflect more details of activation. Through Fig. 5, we can observe that our module helps the network to capture more subtle artifacts compared with the baseline backbone. The red circle indicates the obvious high-information forgery details. We also find that the baseline model ignores these artifacts (white circle) while our SIA module helps networks pay more attention to these clues. For example, the forgery face in the fourth line has an obvious blending boundary, but the baseline CAM does not pay attention to this area. After going through our SIA module, the network clearly focuses on this high-information area. Furthermore, the activation area of the guided grad-cam is larger than the baseline, because our module help the network enhances the most informative channel.

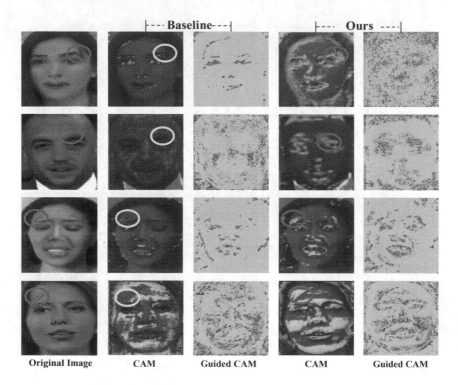

Fig. 5. Grad CAM and Guided Grad CAM on baseline model and our proposed model (layer 1 of EffiecientNet-B4). The red circles indicate obvious clues that are ignored by previous approach but well captured by our method (Best viewed in color) (Color figure online)

5   Conclusion

In this work, we propose an information theoretic framework with self-Information Attention (SIA) for effective face forgery detection. The proposed SIA module has a strong theoretic basis, which leads to an effective and interpretable method that can achieve superior face forgery detection performance with negligible parameter increase. Specially, self-information of each feature map is extracted as the bases of dual attention to help model capture the informative regions which contains critical forgery clues. Experiments on several datasets demonstrate the effectiveness of our method.

Future Work. Currently, we only evaluate our SIA module on the RGB domain. In future work, we will evaluate it in the frequency domain to further demonstrate its effectiveness and generality.

Acknowledgments. This work was supported by the National Science Fund for Distinguished Young Scholars (No. 62025603), the National Natural Science Foundation of China (No. U21B2037, No. 62176222, No. 62176223, No. 62176226, No. 62072386, No. 62072387, No. 62072389, and No. 62002305), Guangdong Basic and Applied Basic Research Foundation (No. 2019B1515120049, and the Natural Science Foundation of Fujian Province of China (No. 2021J01002).

References

1. Afchar, D., Nozick, V., Yamagishi, J., Echizen, I.: MesoNet: a compact facial video forgery detection network. In: WIFS, pp. 1–7. IEEE (2018)
2. Agarwal, S., Farid, H.: Photo forensics from JPEG dimples. In: WIFS, pp. 1–6. IEEE (2017)
3. Averbuch-Elor, H., Cohen-Or, D., Kopf, J., Cohen, M.F.: Bringing portraits to life. ACM Trans. Graph. (TOG) **36**(6), 1–13 (2017)
4. Brock, A., Donahue, J., Simonyan, K.: Large scale GAN training for high fidelity natural image synthesis. arXiv preprint arXiv:1809.11096 (2018)
5. Bruce, N., Tsotsos, J.: Saliency based on information maximization. In: Advances in Neural Information Processing Systems, pp. 155–162 (2005)
6. Bruce, N., Tsotsos, J.: Attention based on information maximization. J. Vis. **7**(9), 950–950 (2007)
7. Cao, J., Ma, C., Yao, T., Chen, S., Ding, S., Yang, X.: End-to-end reconstruction-classification learning for face forgery detection. In: Proceedings of the IEEE/CVF Conference on Computer Vision and Pattern Recognition, pp. 4113–4122 (2022)
8. Chen, M., Sedighi, V., Boroumand, M., Fridrich, J.: JPEG-phase-aware convolutional neural network for steganalysis of JPEG images. In: Proceedings of the 5th ACM Workshop on Information Hiding and Multimedia Security, pp. 75–84 (2017)
9. Chen, S., Yao, T., Chen, Y., Ding, S., Li, J., Ji, R.: Local relation learning for face forgery detection. In: AAAI (2021)
10. Chollet, F.: Xception: deep learning with depthwise separable convolutions. In: CVPR, pp. 1251–1258 (2017)
11. Cozzolino, D., Poggi, G., Verdoliva, L.: Recasting residual-based local descriptors as convolutional neural networks: an application to image forgery detection. In: Proceedings of the 5th ACM Workshop on Information Hiding and Multimedia Security, pp. 159–164 (2017)
12. Dolhansky, B., et al.: The deepfake detection challenge dataset. arXiv preprint arXiv:2006.07397 (2020)
13. Fridrich, J., Kodovsky, J.: Rich models for steganalysis of digital images. IEEE Trans. Inf. Forensics Secur. **7**(3), 868–882 (2012)
14. Gonzalez-Sosa, E., Fierrez, J., Vera-Rodriguez, R., Alonso-Fernandez, F.: Facial soft biometrics for recognition in the wild: recent works, annotation, and cots evaluation. IEEE Trans. Inf. Forensics Secur. **13**(8), 2001–2014 (2018)
15. Goodfellow, I., et al.: Generative adversarial nets. In: NeurIPS, pp. 2672–2680 (2014)
16. Gu, Q., Chen, S., Yao, T., Chen, Y., Ding, S., Yi, R.: Exploiting fine-grained face forgery clues via progressive enhancement learning. In: AAAI, vol. 36, pp. 735–743 (2022)
17. Gu, Z., et al.: Spatiotemporal inconsistency learning for deepfake video detection. In: ACM MM, pp. 3473–3481 (2021)

18. Gunawan, T.S., Hanafiah, S.A.M., Kartiwi, M., Ismail, N., Za'bah, N.F., Nordin, A.N.: Development of photo forensics algorithm by detecting photoshop manipulation using error level analysis. Indones. J. Electr. Eng. Comput. Sci. **7**(1), 131–137 (2017)

19. Guo, Z., Yang, G., Chen, J., Sun, X.: Fake face detection via adaptive manipulation traces extraction network. Comput. Vis. Image Underst. **204**, 103170 (2021)

20. Hu, J., Shen, L., Sun, G.: Squeeze-and-excitation networks. In: CVPR, pp. 7132–7141 (2018)

21. Huang, D., De La Torre, F.: Facial action transfer with personalized bilinear regression. In: Fitzgibbon, A., Lazebnik, S., Perona, P., Sato, Y., Schmid, C. (eds.) ECCV 2012. LNCS, vol. 7573, pp. 144–158. Springer, Heidelberg (2012). https://doi.org/10.1007/978-3-642-33709-3_11

22. Huang, G., Liu, Z., Van Der Maaten, L., Weinberger, K.Q.: Densely connected convolutional networks. In: CVPR, pp. 4700–4708 (2017)

23. Huang, Y., et al.: FakePolisher: making deepfakes more detection-evasive by shallow reconstruction. arXiv preprint arXiv:2006.07533 (2020)

24. Juefei-Xu, F., Wang, R., Huang, Y., Guo, Q., Ma, L., Liu, Y.: Countering malicious deepfakes: survey, battleground, and horizon. arXiv preprint arXiv:2103.00218 (2021)

25. Kim, H., et al.: Deep video portraits. ACM Trans. Graph. (TOG) **37**(4), 1–14 (2018)

26. Li, J., Xie, H., Li, J., Wang, Z., Zhang, Y.: Frequency-aware discriminative feature learning supervised by single-center loss for face forgery detection. In: CVPR, pp. 6458–6467 (2021)

27. Li, L., et al.: Face X-ray for more general face forgery detection. In: CVPR, pp. 5001–5010 (2020)

28. Li, X., et al.: Sharp multiple instance learning for deepfake video detection. In: ACM MM, pp. 1864–1872 (2020)

29. Li, Y., Lyu, S.: Exposing deepfake videos by detecting face warping artifacts. arXiv preprint arXiv:1811.00656 (2018)

30. Li, Y., Yang, X., Sun, P., Qi, H., Lyu, S.: Celeb-DF: a new dataset for deepfake forensics. arXiv preprint arXiv:1909.12962 (2019)

31. Liu, H., et al.: Spatial-phase shallow learning: rethinking face forgery detection in frequency domain. In: CVPR, pp. 772–781 (2021)

32. Liu, Z., Qi, X., Torr, P.H.: Global texture enhancement for fake face detection in the wild. In: CVPR, pp. 8060–8069 (2020)

33. Luo, Y., Zhang, Y., Yan, J., Liu, W.: Generalizing face forgery detection with high-frequency features. In: CVPR, pp. 16317–16326 (2021)

34. Masi, I., Killekar, A., Mascarenhas, R.M., Gurudatt, S.P., AbdAlmageed, W.: Two-branch recurrent network for isolating deepfakes in videos. In: Vedaldi, A., Bischof, H., Brox, T., Frahm, J.-M. (eds.) ECCV 2020. LNCS, vol. 12352, pp. 667–684. Springer, Cham (2020). https://doi.org/10.1007/978-3-030-58571-6_39

35. Matern, F., Riess, C., Stamminger, M.: Exploiting visual artifacts to expose deepfakes and face manipulations. In: WACVW, pp. 83–92. IEEE (2019)

36. McCloskey, S., Albright, M.: Detecting GAN-generated imagery using color cues. arXiv preprint arXiv:1812.08247 (2018)

37. Nguyen, H.H., Fang, F., Yamagishi, J., Echizen, I.: Multi-task learning for detecting and segmenting manipulated facial images and videos. arXiv preprint arXiv:1906.06876 (2019)

38. Nguyen, H.H., Yamagishi, J., Echizen, I.: Capsule-forensics: using capsule networks to detect forged images and videos. In: ICASSP, pp. 2307–2311. IEEE (2019)

39. Qian, Y., Yin, G., Sheng, L., Chen, Z., Shao, J.: Thinking in frequency: face forgery detection by mining frequency-aware clues. In: Vedaldi, A., Bischof, H., Brox, T., Frahm, J.-M. (eds.) ECCV 2020. LNCS, vol. 12357, pp. 86–103. Springer, Cham (2020). https://doi.org/10.1007/978-3-030-58610-2_6
40. Rossler, A., Cozzolino, D., Verdoliva, L., Riess, C., Thies, J., Nießner, M.: Face-Forensics++: learning to detect manipulated facial images. In: ICCV, pp. 1–11 (2019)
41. Selvaraju, R.R., Cogswell, M., Das, A., Vedantam, R., Parikh, D., Batra, D.: Grad-CAM: visual explanations from deep networks via gradient-based localization. In: ICCV, pp. 618–626 (2017)
42. Shen, Z., Bello, I., Vemulapalli, R., Jia, X., Chen, C.H.: Global self-attention networks for image recognition. arXiv preprint arXiv:2010.03019 (2020)
43. Shi, B., Zhang, D., Dai, Q., Wang, J., Zhu, Z., Mu, Y.: Informative dropout for robust representation learning: a shape-bias perspective. In: ICML, vol. 1 (2020)
44. Stehouwer, J., Dang, H., Liu, F., Liu, X., Jain, A.: On the detection of digital face manipulation. In: CVPR (2019)
45. Sun, K., et al.: Domain general face forgery detection by learning to weight. In: AAAI, vol. 35, pp. 2638–2646 (2021)
46. Sun, K., Yao, T., Chen, S., Ding, S., Li, J., Ji, R.: Dual contrastive learning for general face forgery detection. In: AAAI, vol. 36, pp. 2316–2324 (2022)
47. Tan, M., Le, Q.V.: EfficientNet: rethinking model scaling for convolutional neural networks. In: ICML (2019)
48. Thies, J., Zollhöfer, M., Nießner, M.: Deferred neural rendering: image synthesis using neural textures. ACM Trans. Graph. (TOG) **38**(4), 1–12 (2019)
49. Thies, J., Zollhöfer, M., Nießner, M., Valgaerts, L., Stamminger, M., Theobalt, C.: Real-time expression transfer for facial reenactment. ACM Trans. Graph. **34**(6), 183-1 (2015)
50. Thies, J., Zollhofer, M., Stamminger, M., Theobalt, C., Nießner, M.: Face2Face: real-time face capture and reenactment of RGB videos. In: CVPR, pp. 2387–2395 (2016)
51. Wang, C., Deng, W.: Representative forgery mining for fake face detection. In: CVPR, pp. 14923–14932 (2021)
52. Wang, X., Girshick, R., Gupta, A., He, K.: Non-local neural networks. In: CVPR, pp. 7794–7803 (2018)
53. Wang, X., Yao, T., Ding, S., Ma, L.: Face manipulation detection via auxiliary supervision. In: Yang, H., Pasupa, K., Leung, A.C.-S., Kwok, J.T., Chan, J.H., King, I. (eds.) ICONIP 2020. LNCS, vol. 12532, pp. 313–324. Springer, Cham (2020). https://doi.org/10.1007/978-3-030-63830-6_27
54. Zhao, H., Zhou, W., Chen, D., Wei, T., Zhang, W., Yu, N.: Multi-attentional deepfake detection. In: CVPR (2021)
55. Zhou, P., Han, X., Morariu, V.I., Davis, L.S.: Two-stream neural networks for tampered face detection. In: CVPRW, pp. 1831–1839. IEEE (2017)
56. Zi, B., Chang, M., Chen, J., Ma, X., Jiang, Y.G.: WildDeepfake: a challenging real-world dataset for deepfake detection. In: ACM MM, pp. 2382–2390 (2020)

Exploring Disentangled Content Information for Face Forgery Detection

Jiahao Liang[1], Huafeng Shi[2], and Weihong Deng[1(✉)]

[1] Beijing University of Posts and Telecommunications, Beijing, China
{jiahao.liang,whdeng}@bupt.edu.cn
[2] SenseTime Research, Hefei, China
shihuafeng1@sensetime.com

Abstract. Convolutional neural network based face forgery detection methods have achieved remarkable results during training, but struggled to maintain comparable performance during testing. We observe that the detector is prone to focus more on content information than artifact traces, suggesting that the detector is sensitive to the intrinsic bias of the dataset, which leads to severe overfitting. Motivated by this key observation, we design an easily embeddable disentanglement framework for content information removal, and further propose a *Content Consistency Constraint* (C^2C) and a *Global Representation Contrastive Constraint* (GRCC) to enhance the independence of disentangled features. Furthermore, we cleverly construct two unbalanced datasets to investigate the impact of the content bias. Extensive visualizations and experiments demonstrate that our framework can not only ignore the interference of content information, but also guide the detector to mine suspicious artifact traces and achieve competitive performance.

Keywords: Face forgery detection · Content information · Disentangled representation

1 Introduction

With the incredible success of deep learning, numerous techniques for forgery have emerged, such as Deepfakes [14], Face2Face [42], and FaceSwap [30]. Due to the extremely low barriers and easy accessibility, generative techniques are gradually being misused [1,2].

To defend against, face forgery detection has attracted increasing attention. Early works [10,17,22,32,47] used hand-crafted facial features (*e.g.*, eyes blinking, head poses, lip movements, *etc.*) to capture some visual artifacts and inconsistencies resulting from the forgery generation process. Meanwhile, some works [11,24,35] explored PPG signals representing heart rate information. Later, learning-based methods [7,13,45,46,48] have made significant progress. Nevertheless, these methods are vulnerable to image compression or noise interference. Frank *et al.* [15] found that, compared to the time domain, mining

ⓒ The Author(s), under exclusive license to Springer Nature Switzerland AG 2022
S. Avidan et al. (Eds.): ECCV 2022, LNCS 13674, pp. 128–145, 2022.
https://doi.org/10.1007/978-3-031-19781-9_8

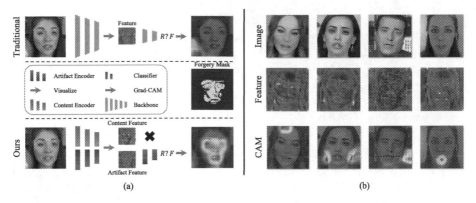

Fig. 1. (a) Unlike the traditional methods (upper), we propose a disentanglement framework (lower) for content information removal. Grad-CAM [39] shows that the traditional detector is distracted by the red object, while our method still mine suspicious artifact traces and the activation region almost consistent with the mask. (b) Visualization of the image (first row), traditional detector's (Xception) features (second row) and Grad-CAM [39] (third row).

forgery information in the frequency domain can still maintain satisfactory results even under severe compression [15,20,25,31,36].

We observe that most methods [31,35,36,53] perform admirably in in-dataset evaluations but struggle to maintain comparable results in cross-domain evaluations, which inspire us to conduct an in-depth analysis of the previous method. Existing methods take it for granted that after proper training, the detector will selectively grasp artifact traces as the basis for authenticity judgment. However, the visualization (shown in Fig. 1(b)) illustrates that the feature of the detector remains recognizable content clues, and the detector is prone to overfitting to small local regions, or even focusing only on content information outside the face region.

Based on this key observation, we conjecture that detectors may no longer mine hard-to-capture artifact traces, and instead overfit certain non-artifact (*i.e.*, content) information, thus leading to the failure of cross-domain evaluations.

Therefore, we propose an easily embeddable framework for disentangling content features and artifact features, and only the disentangled artifact features for face forgery detection, thus ignoring the interference of content information. A brief comparison between the traditional methods and our framework is sketched in Fig. 1(a).

However, most disentanglement methods [26,33,51] consider only the completeness of features, but do not explore the independence of disentangled features in-depth, which leads to the failure of the face forgery detection (see Table 4). To enhance it, we propose a *Content Consistency Constraint* (C^2C) to ensure that the disentangled features contain the corresponding information and a *Global Representation Contrastive Constraint* (GRCC) to further ensure the purity of the disentangled features, which helps our disentanglement framework to achieve competitive performance. Furthermore, we cleverly construct two unbalanced datasets based on the FaceForensics++ [38] to investigate the impact of content bias, and further demonstrate that our framework can ignore the interference of the content bias. Notably, our framework is easily embeddable, we embed some backbones into our framework for extensive evaluations and ablation experiments, and experimental results demonstrate the effectiveness and generalization capability of our framework in face forgery detection.

The contributions of this paper could be summarized as three-fold:

- To the best of our knowledge, we are the first to explore the impact of content information on the generalization performance of face forgery detection, and cleverly construct two unbalanced datasets to further investigate the impact of content bias, which brings a novel perspective for this field.
- We design an easily embeddable disentanglement framework for content information removal, and further propose a *Content Consistency Constraint* (C^2C) and *Global Representation Contrastive Constraint* (GRCC) to enhance the independence of disentangled features.
- Extensive visualizations and experiments demonstrate that our framework can not only ignore the interference of content information, but also guide the detector to mine suspicious artifact traces and achieve competitive performance in face forgery detection.

2 Related Works

2.1 Forgery Detection

Benefiting from the great progress of GAN, forgery techniques, especially for faces, have been incredibly advanced. To avoid its illegal use, researchers have explored forgery detection extensively [4,21,28,40,54].

Later, various learning-based methods [7,13,45,46,48] demonstrated significant improvements. In addition, some works [3,8,27] suggested that shallow local texture details and correlations between local regions of the face can better reflect forgery information. However, almost all of these CNN-based methods only utilize spatial domain information (*i.e.*, RGB, YUV, HSV), and therefore the performance is sensitive to the quality and distribution of the dataset. To counter it, some works [15,20,25,31,36] transformed images into the frequency domain by DCT transform and analyzed the frequency domain statistics, achieving satisfactory results even with severe compression. Recent attempts to boost the generalization of face forgery detection by extending the activated attention region of the network. Zhao *et al.* [53] proposed multiple spatial attention heads to guide the network focus on different local regions. Wang *et al.* [44] encouraged detectors to dig deeper into previously overlooked regions by masking the sensitive facial regions. Although these CNN-based methods significantly enhance the feature extraction capability of the detector, due to the neglect of the content bias implied in the features, the detector is trapped in the intrinsic bias of the dataset, thus hindering the improvement of cross-domain generalization performance.

2.2   Disentangled Representation

Disentangled representation learning is to decompose complex dimensional coupled information into simple features with a strong distinguishing ability [6].
DR-GAN [43] disentangled the face into identity and pose features for synthesizing faces in arbitrary poses to aid in recognition. Niu *et al.* [33] proposed a
cross-verified feature disentangling strategy with robust multi-task physiological
measurements. Zhang *et al.* [52] also adopted a similar structure to disentangle
pose and appearance features from gait videos. In the field of face anti-spoofing,
Zhang *et al.* [51] decompose the facial image into content features and liveness
features and introduced LBP map, depth map as auxiliary supervision. Liu *et*
al. [26] proposed a new adversarial learning framework to separate the spoof
trace into a hierarchical combination of multi-scale patterns. In this paper, we
further propose a *Content Consistency Constraint* (C^2C) and *Global Representation Contrastive Constraint* (GRCC) to enhance the independence of disentangled features. And the disentanglement framework only serves as an underlying
architecture, a detailed ablation analysis can be found in Sect. 4.3.

3   Methods

3.1   Motivation

Consider a forged image, which consists of artifact traces and content information, where the content information can be subdivided into identity information
and background information. The only difference between the forgery image and
the real image is the presence of artifact traces, which is the basis for the detector
to determine the authenticity.

   We observe that most detectors perform admirably in in-dataset evaluations
but struggle to maintain comparable results in cross-dataset evaluation. For
further exploration, we visualize the features of the middle layer of the detector
and the Grad-CAM [39]. The visualization results (see Fig. 1(b)) illustrate that
the feature of the detector remains recognizable content information, and the
detector is prone to overfitting to small local regions (DF, NT), or even focusing
on content information outside the face region (FS).

   Based on this key observation, we conjecture that with the weak constraint of
binary labels alone, detectors may no longer mine hard-to-capture artifact traces,
and instead overfit certain non-artifact information (*i.e.*, content information),
thus failing in cross-dataset evaluations.

   Therefore, we propose an embeddable disentanglement framework that disentangles content and artifact features, and the artifact features are used for
forgery detection, thus eliminating the interference of content information.

3.2   Basic Disentanglement Framework

We assume that the high-dimensional latent representation of an image consists
of content and artifact features. The main purpose is to disentangle them and
use the disentangled artifact features for subsequent detection.

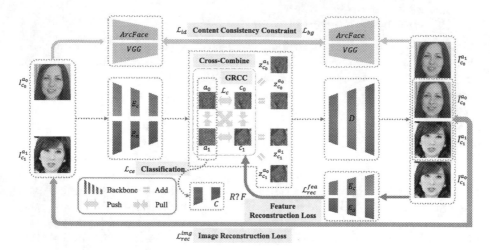

Fig. 2. The overview framework of our method. The input of our network is a pair of images. First, we use artifact encoder E_a and the content encoder E_c to disentangle the content and artifact features, respectively. Then, we feed the artifact features a_0 and a_1 to the classifier C for detection to compute $\mathcal{L}_{ce}$. Next, *Global Representation Contrastive Constraint* (GRCC) is used to compute $\mathcal{L}_c$, and the artifact and content features are cross-combined to get latent representation z_c^a and then reconstruct the images. Finally, the reconstruction loss $\mathcal{L}_{rec}^{img}$, $\mathcal{L}_{rec}^{fea}$ and the *Content Consistency Constraint* (C²C) loss $\mathcal{L}_{id}$, $\mathcal{L}_{env}$ are calculated to ensure the completeness and independence of the disentangled features.

The disentanglement framework mainly consists of two independent encoders E_c and E_a, for extracting content and artifact features, respectively, a decoder D for the reconstruction of the images, and a classifier C for face forgery detection. Among them, we use the front and back parts of the backbone as the artifact encoder E_a and the classifier C, and the artifact encoder E_a and the content encoder E_c have the same structure, but the parameters are not shared.

Specifically, as shown in Fig. 2, with pairwise input images $I_{c_0}^{a_0}$, $I_{c_1}^{a_1}$, where a_0, a_1 and c_0, c_1 denotes the corresponding artifact and content features of the image, respectively. It is worth noting that one of the images is real and the other one is fake. We first use the content encoder E_c and the artifact encoder E_a to get the content features c_0, c_1 and the artifact features a_0, a_1, and the formula is as follow:

$$c_i = E_c(I_{c_i}^{a_i}), a_i = E_a(I_{c_i}^{a_i}), \tag{1}$$

where i denotes the index of feature.

Self-reconstruction. Then element-wise addition is applied to the content and artifact features encoded from the same image to obtain the high-dimensional latent representation features of the image, *i.e.*, $z_{c_i}^{a_i} = a_i + c_i$. Next, $z_{c_i}^{a_i}$ is fed into the decoder D to reconstruct the corresponding original image $\tilde{I}_{c_i}^{a_i}$, and the formula is as follow:

$$\tilde{I}_{c_i}^{a_i} = D(z_{c_i}^{a_i}). \tag{2}$$

Cross-Reconstruction. Moreover, we **cross-combine** content and artifact features from different images to obtain the high-dimensional latent representation features, *i.e.*, $z_{c_{1-i}}^{a_i} = a_i + c_{1-i}$. Also, $z_{c_{1-i}}^{a_i}$ is fed into the decoder D to reconstruct the image $\tilde{I}_{c_{1-i}}^{a_i}$, and the formula is as follow:

$$\tilde{I}_{c_{1-i}}^{a_i} = D(z_{c_{1-i}}^{a_i}). \tag{3}$$

Reconstruction Loss. The decoder D should effectively reconstruct the original image to ensure the completeness of the high-dimensional latent representation feature, so the image reconstruction loss is formulated as:

$$\mathcal{L}_{rec}^{img} = \sum_{i=0}^{1} \left\| I_{c_i}^{a_i} - \tilde{I}_{c_i}^{a_i} \right\|_1. \tag{4}$$

Image reconstruction loss ensures that the reconstructed image and the original image are consistent at the pixel level. In addition, the encoded features of the reconstructed image should still be consistent with the reconstructed features, so we introduce a feature reconstruction loss:

$$\mathcal{L}_{rec}^{fea} = \sum_{i=0}^{1} (\left\| E_c(\tilde{I}_{c_i}^{a_i}) - c_i \right\|_1 + \left\| E_a(\tilde{I}_{c_i}^{a_i}) - a_i \right\|_1 \\ + \left\| E_c(\tilde{I}_{c_{1-i}}^{a_i}) - c_{1-i} \right\|_1 + \left\| E_a(\tilde{I}_{c_{1-i}}^{a_i}) - a_i \right\|_1). \tag{5}$$

3.3 Enhanced Independence of Disentangled Features

Although reconstruction loss can guarantee the completeness of features for the combination of content and artifact features. However, there are still two elements that cannot be guaranteed: (i) Whether the encoders can selectively disentangle features (*i.e.*, whether the disentangled features contain the corresponding information). (ii) Whether the disentangled features contain **only** the corresponding information. We are keenly aware that the key to successful disentangling lies in the establishment of these two conditions, which is proved by subsequent ablation study (Sect. 4.3). Unfortunately, none of the previous related methods [26,33,51] have explored the independence of features in depth. We propose a *Content Consistency Constraint* (C^2C) and a *Global Representation Contrastive Constraint* (GRCC) to further enhance the independence of disentangled features.

Content Consistency Constraint. In cross-reconstruction, content features should determine the background and face ID information of the reconstructed image. Specifically, the cross-reconstructed image $\tilde{I}_{c_i}^{a_{1-i}}$ should have the same content attributes as the origin image $I_{c_i}^{a_i}$ that encodes the content features c_i. As we mentioned before, content features consist of background and face ID, so the *Content Consistency Constraint* (C^2C) can be formulated as:

$$\text{Content}(\widetilde{I}_{c_i}^{a_{1-i}}) = \text{Content}(I_{c_i}^{a_i}),$$

$$\Updownarrow$$

$$\text{Identity}(\widetilde{I}_{c_i}^{a_{1-i}}) = \text{Identity}(I_{c_i}^{a_i}),$$

$$\text{Backgroud}(\widetilde{I}_{c_i}^{a_{1-i}}) = \text{Backgroud}(I_{c_i}^{a_i}), \tag{6}$$

based on this prior condition, we adopt the identity preservation loss $\mathcal{L}_{id}$ and the content perception loss $\mathcal{L}_{bg}$ to preserve the content attributes of the cross-reconstructed images. It is formulated as:

$$\mathcal{L}_{id} = 1 - \cos(\text{ArcFace}(\widetilde{I}_{c_i}^{a_{1-i}}), \text{ArcFace}(I_{c_i}^{a_i})),$$

$$\mathcal{L}_{bg} = \left\| \text{VGG}(\widetilde{I}_{c_i}^{a_{1-i}}) - \text{VGG}(I_{c_i}^{a_i}) \right\|_1, \tag{7}$$

where $\text{ArcFace}(\cdot)$ and $\text{VGG}(\cdot)$ represents a pretrained VGG network and a pretrained ArcFace network, respectively, $\cos(\cdot, \cdot)$ represents the cosine similarity of two vectors. Here $\text{VGG}(\cdot)$ is considered to extract high-level semantic features, and since artifacts are mainly concentrated in low-level texture details [27,53], the extracted content features is pure and does not contain artifact information.

Global Representation Contrastive Constraint. Artifact features and content features should be two fundamentally distinct spaces. In other words, artifact features and content features can be regarded as two different classes, and the inter-classes feature distance should be much larger than the intra-class feature distance. Specifically, we regard the intra-class features as positive pairs and inter-class features as negative pairs, and adopt the contrastive learning protocol to further eliminate the possible overlap of content features and artifact features. Inspired by [27], we take the Gram matrix of content and artifact features as a global and distinctive representation:

$$\mathbf{G} = (F_i^T F_j)_{n \times n} = \begin{bmatrix} F_1^T F_1 & \cdots & F_1^T F_n \\ \vdots & \ddots & \vdots \\ F_n^T F_1 & \cdots & F_n^T F_n \end{bmatrix}, \tag{8}$$

where F denotes the feature, and n denotes the channel of the feature. For feature distance measurement, we adopt the cosine distance, where closer features render larger scores. Finally, we take the advantage of the InfoNCE [34] to construct a *Global Representation Contrastive Constraint* (GRCC) between the artifact and content features:

$$\mathcal{L}_c = -\log\left[\frac{\exp(d(\mathbf{G}_{a_0}, \mathbf{G}_{a_1}))}{\exp(d(\mathbf{G}_{a_0}, \mathbf{G}_{a_1})) + \sum_{i=0}^{1} \exp(d(\mathbf{G}_{a_i}, \mathbf{G}_{c_{1-i}}))}\right]$$

$$- \log\left[\frac{\exp(d(\mathbf{G}_{c_0}, \mathbf{G}_{c_1}))}{\exp(d(\mathbf{G}_{c_0}, \mathbf{G}_{c_1})) + \sum_{i=0}^{1} \exp(d(\mathbf{G}_{a_i}, \mathbf{G}_{c_{1-i}}))}\right], \tag{9}$$

where $\mathbf{G}_{a_i}$ and $\mathbf{G}_{c_i}$ represent the flattened vector of the gram matrix of a_i and c_i, respectively, and $d(\cdot, \cdot)$ represents the cosine similarity.

3.4   Overall Loss

The final loss function of the training process is the weighted sum of the above loss functions.

$$\mathcal{L} = \mathcal{L}_{ce} + \lambda_1\mathcal{L}_{rec}^{img} + \lambda_2\mathcal{L}_{rec}^{fea} + \lambda_3\mathcal{L}_{id} + \lambda_4\mathcal{L}_{bg} + \lambda_5\mathcal{L}_c, \tag{10}$$

where $\mathcal{L}_{ce}$ denotes the cross entropy loss, λ_1, λ_2, λ_3, λ_4, λ_5 are the weights for balancing the loss.

4   Experiments

4.1   Experimental Setting

Datasets. To validate the effectiveness of our method, we choose the most widely used benchmark FaceForensics++ (FF++) [38] for training. It contains 1 real sub-dataset and 4 fake sub-datasets, *i.e.*, Deepfakes (DF) [14], Face2Face (FF) [42], FaceSwap (FS) [30] and NeuralTextures (NT) [41]. Each sub-dataset contains 1,000 videos, and we follow the official standard by using 720 videos for training, 140 videos for validation, and 140 videos for testing, and we adopt the LQ version by default and specify the version otherwise. Celeb-DF [23] uses 59 celebrity interview videos on YouTube as the original videos. In total, 590 real videos and 5,639 DeepFakes videos are included.

Metrics. We apply the accuracy score (ACC), equal error rate (EER), and the area under the receiver operating characteristic (ROC) Curve (AUC) as our evaluation metrics. For a comprehensive evaluation of performance, we also report the true detection rate (TDR) for a given false detection rate (FDR).

Implementation Details. For data preprocessing, we only resize the facial images into a fixed size of 224×224. For training, we set the size of the mini-batch to 128, and the ratio of real and fake images to $1 : 1$. We use Adam [19] as our optimizer with an initial learning rate of 0.001 and a half decay every 5000 iters. The maximum iters number is 30000. And we set λ_1 to λ_5 in Eq. 10 as 1, 0.01, 1, 0.01 and 0.01. All the code is based on the PyTorch framework and trained with NVIDIA GTX 1080Ti.

4.2   Evaluations

To evaluate our method comprehensively, in this section, we perform in-dataset, cross-method and cross-dataset evaluation to demonstrate the generalizability and robustness of our method.

In-Dataset Evaluation. In-Dataset evaluation reflects the ability of the network to fit the distribution of the dataset, as shown in Table 1. In general, with the help of our framework, the performance of both detectors and mainstream networks has been improved in different degrees, which fully proves the effectiveness and adaptability of our framework, Among them, our methods (ResNest-50 + Ours) achieve the state of the art on Deepfakes and NeuralTextures. Notably,

Table 1. In-Dataset evaluation (ACC (%)) on FF++ (LQ). We combine each forgery and real dataset in pairs to construct four sub-datasets, and evaluate the corresponding performance. AVG: the average performance of the four sub-datasets. Noting that results for some methods are from [36]. After embedding into our framework, all detectors achieve considerable performance gains and even outperform other methods.

Method	DF	FF	FS	NT	AVG
Steg. Features [16]	67.00	48.00	49.00	56.00	55.00
LD-CNN [12]	75.00	56.00	51.00	62.00	61.00
C-Conv [5]	87.00	82.00	74.00	74.00	79.25
CP-CNN [37]	80.00	62.00	59.00	59.00	65.00
MesoNet [3]	90.00	83.00	83.00	75.00	82.75
F^3-Net [36]	96.81	94.01	<u>95.85</u>	79.36	91.51
Gram-Net [27]	95.12	88.01	93.34	76.12	88.15
+ Ours	**95.67**	**89.06**	**94.01**	**76.96**	**88.93**
RFM [44]	95.42	91.24	93.60	79.83	90.02
+ Ours	**95.92**	**92.27**	**93.97**	**80.14**	**90.58**
ResNet-50 [18]	95.23	87.79	92.34	76.28	87.91
+ Ours	**95.43**	**88.94**	**93.99**	**77.19**	**88.89**
Xception [9]	95.36	91.94	93.55	78.32	89.79
+ Ours	**96.50**	**93.62**	**94.76**	**79.02**	**90.98**
ResNest-50 [49]	95.98	92.16	93.13	78.22	89.87
+ Ours	**98.95**	**94.32**	**94.56**	<u>**80.46**</u>	**92.10**

Table 2. Cross-Method evaluation (AUC (%)) on FF++ (C40). We adopt Xception [9], which is widely used in face forgery detection, as a baseline for comparison on FF++. Specifically, we use one of the sub-datasets for training, and the rest for testing.

Train set	Method	Test set (AUC(%))				
		DF	FF	FS	NT	AVG
DF	Xception	99.21	58.81	64.79	59.69	70.63
	+ Ours	**99.22**	**60.18**	**68.19**	**61.17**	**72.19**
FF	Xception	66.39	95.40	56.58	57.59	68.99
	+ Ours	**67.13**	**96.07**	**61.36**	**59.98**	**71.14**
FS	Xception	80.00	56.65	94.55	53.42	71.16
	+ Ours	**82.68**	**56.77**	**94.76**	**54.23**	**72.11**
NT	Xception	**69.94**	**67.88**	57.59	86.72	**70.53**
	+ Ours	68.39	65.40	**58.34**	**87.89**	70.01

for Deepfakes, we outperform the F^3-Net [36] and baseline by 2.14% and 2.97% in terms of ACC score. Although our best performance is still slightly worse than F^3-Net on FaceSwap, it is understandable because our method does not pursue a magical modification of the network architecture.

Cross-Method Evaluation. Forgery techniques are constantly iterating, and we need to address not only existing forgery methods, but also the most cutting-edge ones. Table 2 shows our method is superior to the baseline in most cases, but the performance of both methods will drop greatly in cross-method evaluation, which is inevitable, because the extremely strong feature extraction capability of convolutional networks leads to the overfitting of detectors. Our method only mitigates the degree of overfitting to a certain extent, but does not significantly improve the generalization performance.

Table 3. Cross-Dataset evaluation on Celeb-DF (AUC (%)) by training on FF++-DF (ACC (%)). Our method outperforms all the methods with the same backbone (Xception) and achieves the best performance with the backbone of ResNest-50.

BackBone	Method	FF++-DF (Train)	Celeb-DF (Test)
Xception	F^3-Net [36]	97.97	65.17
Efficient-B4	Zhao *et al.* [53]	-	67.44
HRNet	Face X-ray [31]	-	74.76
Xception	SPSL [25]	96.91	76.88
-	Chen *et al.* [8]	98.84	78.26
ResNet-18	Gram-Net [27]	95.12	67.14
	+ Ours	**95.67**	**74.04**
Xception	RFM [44]	95.42	67.21
	+ Ours	**95.92**	**74.44**
ResNet-50	ResNet-50 [18]	95.23	66.84
	+ Ours	**95.43**	**74.71**
Xception	Xception [9]	95.36	65.50
	+ Ours	**96.50**	**76.91**
ResNest-50	ResNest-50 [49]	95.98	68.00
	+ Ours	**98.95**	**82.38**

Cross-Dataset Evaluation. Due to the differences in raw data and experimental details, there can be huge gaps in the distribution between different datasets corresponding to even the same method. As shown in Table 3, regardless of the method, the performance drops significantly when testing on the Celeb-DF dataset, which implies that the difference in the distribution of different datasets for the same method does exist. With the assistance of our framework, the performance of each backbone on FF++ is slightly improved, but the improvement on Celeb-DF is significant. Specifically, our method (ResNest-50+Ours) has a

Table 4. Ablation study on the FF++-DF and Celeb-DF. "Basic" represents the basic disentanglement framework.

Method	Basic	C^{2}C	GRCC	FF++-DF	Celeb-DF
Xception				95.36	65.50
Variant A	✓			94.55	65.11
Variant B	✓	✓		95.73	70.08
Variant C	✓		✓	96.33	72.57
Variant D	✓	✓	✓	**96.50**	**76.91**

Table 5. Results (Δ_{AUC} (%)) of image- and feature-level data augmentation study.

Method	Augmentation			Dataset (Δ_{AUC} (%))	
	Erasing [55]	H-Flip	Mixup [50]	FF++-DF	Celeb-DF
Image	✓			−2.13	**+0.98**
Feature	✓			**+0.34**	+0.94
Image		✓		−0.10	+0.23
Feature		✓		**+0.22**	**+3.62**
Image			✓	−0.85	+3.01
Feature			✓	**−0.07**	**+3.57**

14.38% improvement on Celeb-DF, while the improvement on FF++-DF is only 2.97%. Furthermore, our method (ResNest-50+Ours) outperforms the state-of-the-art results (Chen *et al.* [8]) by 4.12% in terms of AUC score. Among the methods using Xception as the backbone, our method also surpasses others.

4.3   Ablation Study

We perform several ablations to better understand the contributions of each component in our method, the experimental results and visualizations are shown in Table 4 and Fig. 3, respectively.

From the comparison of Variant A and Baseline, we can find that the performance of face forgery detection does not increase but decreases (0.81%) by simply introducing the disentanglement framework. Furthermore, we add *Content Consistency Constraint* (C^{2}C) and *Global Representation Contrastive Constraint* (GRCC) separately, with 4.97% and 7.46% improvement in terms of AUC, respectively, which proves the effectiveness of the enhanced independence of disentangled features. While the performance increases by 11.80% after combining these two, which indicates the two can play a mutually reinforcing role. Overall, C^{2}C and GRCC play a dominant role as the key core of our method.

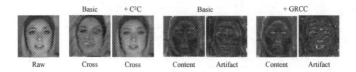

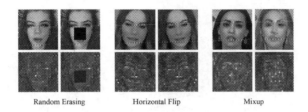

Fig. 3. Visualization of the ablation study, which illustrates the impact of C^2C on the reconstructed images and GRCC on the disentangled features, respectively. "Raw" represents the raw image, and "Cross" represents the cross-reconstruction image.

Fig. 4. Visualization of the image- (1st row) and feature-level (2nd row) augmentation.

4.4 Augmentation Study of Disentangled Features

Our framework first disentangles content features and artifact features from the images, and then uses the artifact features for subsequent detection. It is natural to guess that compared to image-level data augmentation, directly performing data augmentation on artifact features may achieve better performance. To validate it, we select common data augmentation methods such as Random Erasing [55], Horizontal Flip, and Mixup [50] to experiment, the details of the augmentation are shown in Fig. 4. It is worth noting that data augmentation is not used in other experiments.

It can be seen from Table 5 that the performance improvement in the cross-dataset evaluation is greater than in-dataset evaluation. Our explanation is that in the in-dataset evaluation, the distributions of the train and test sets are close, and data augmentation disrupts the consistency of the distribution of the train and test set, resulting in little performance improvement or even reduction. For cross-dataset evaluation, data augmentation can enhance the diversity of the train set, and then pull the distribution between the train and test sets. In addition, the performance improvement of data augmentation at the feature-level is significantly better than that at the image-level, which implies the effectiveness of our disentanglement framework.

4.5 Investigation of Intrinsic Content Bias

To investigate the impact of intrinsic content bias within the dataset on the performance of face forgery detection, we cleverly construct two unbalanced datasets based on the FF++ dataset, the *Identity Unbalanced* dataset and the *Background Unbalanced* dataset.

Table 6. Comparison of our framework with baseline methods on the identity and background unbalanced dataset.

Method	ID unbalanced dataset				BG unbalanced dataset			
	ACC	AUC	EER	$TDR_{0.1}$	ACC	AUC	EER	$TDR_{0.1}$
Gram-Net [27]	89.85	96.49	10.14	89.70	79.84	87.83	20.19	66.10
+ Ours	**94.80**	**99.48**	**3.619**	**98.90**	**94.00**	**98.93**	**5.764**	**96.70**
RFM [44]	90.34	95.34	9.232	90.93	85.09	92.33	14.89	79.58
+ Ours	**95.49**	**99.11**	**4.102**	**97.90**	**95.02**	**98.34**	**4.839**	**96.93**
ResNet-50 [18]	89.61	96.46	10.35	89.30	80.88	91.29	17.17	72.30
+ Ours	**95.39**	**99.54**	**3.571**	**99.00**	**94.46**	**98.76**	**5.524**	**97.20**
Xception [9]	91.06	96.91	8.967	91.50	84.39	92.42	17.17	78.10
+ Ours	**95.85**	**99.32**	**3.434**	**99.10**	**95.14**	**98.71**	**4.762**	**97.00**
ResNest-50 [49]	89.89	97.54	8.507	92.70	81.28	93.68	14.34	79.60
+ Ours	**95.58**	**99.61**	**3.190**	**98.90**	**94.56**	**98.48**	**5.479**	**96.90**

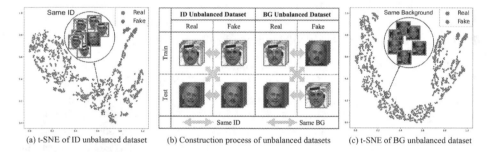

(a) t-SNE of ID unbalanced dataset   (b) Construction process of unbalanced datasets   (c) t-SNE of BG unbalanced dataset

Fig. 5. (b) The construction process of unbalanced datasets. (a)(c) t-SNE feature visualization of the Xception network on the ID and BG unbalanced dataset.

We conduct comparative experiments on these datasets, and the experimental results are shown in Table 6. We can find that the performance on the ID and BG unbalanced datasets suffers a huge drop, which indicates that the existence of the intrinsic bias does interfere with the optimization of the detector. In contrast, our framework can maintain a high performance even on the unbalanced dataset by stripping the content features and thus eliminating the interference of content bias. Furthermore, compared with the ID unbalanced dataset, the performance degradation on the BG unbalanced dataset is more serious.

For a more intuitive understanding of the impact of content bias, we also visualize the t-SNE [29] feature spaces of the Xception network on the ID unbalanced dataset (Fig. 5(a)) and BG unbalanced dataset (Fig. 5(c)). We can observe that some samples with similar content information tend to cluster together, in other words, the distance between some samples with similar content information is much smaller than the distance between samples with similar forgery methods, which reveals that the content bias induce the detector to use content information for discrimination instead of artifact traces.

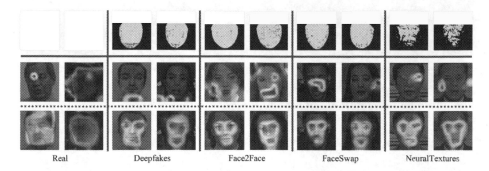

Real   Deepfakes   Face2Face   FaceSwap   NeuralTextures

Fig. 6. Visualization of forgery mask (first row), Xception's (second row) and ours (third row) Grad-CAM on five sub-datasets of FF++. The activation region of our method is comprehensive and almost consistent with the forgery mask.

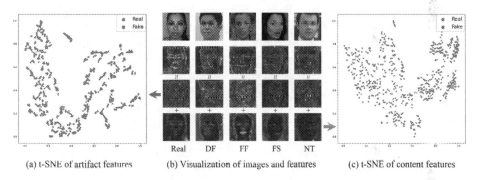

(a) t-SNE of artifact features   (b) Visualization of images and features   (c) t-SNE of content features

Fig. 7. (b) Visualization of the image (first row), traditional detector's (Xception) features (second row), ours disentangled artifact (third row) and content features (fourth row). (a)(c) t-SNE visualization of artifact features and content features.

4.6   Visualization

To more intuitively demonstrate the effectiveness of our method, we visualize the Grad-CAM [39] of the baseline and our method, respectively, and the forgery mask, as shown in Fig. 6. Grad-CAM shows that the baseline is prone to over-fitting to small local regions or focusing on content noise outside the forgery region. In contrast, the activation region of our method is comprehensive and almost consistent with the forgery mask. Such visualization results also explain the motivation of this paper: without additional constraints, the detector has difficulty in mining suspicious artifact regions thorough weak supervision of labels only, and easily falls into content bias, which leads to overfitting or even misleading the direction of optimization. Instead, our goal is to remove the interference of content bias by an pre-disentanglement framework, and guide the detector to mine suspicious artifact trace.

As shown in the Fig. 7(b), traditional methods seek to allocate more attention to the face region, which improve the fitting ability but also exacerbated the overfitting of content bias within the dataset. Instead, we separate content features to eliminate misleading content information, guide the detector to pay attention to suspicious artifact traces, and strengthen the generalization capability fundamentally. Furthermore, Fig. 7(a)(c) demonstrate that the disentangled artifact features are discriminative for forgery detection, while the content features do not, which also validates the validity of our motives.

5   Conclusion

In this paper, we observe that detectors may no longer mine hard-to-capture artifact traces, and instead overfit certain content information, thus leading to the failure of generalization, which brings a novel perspective for face forgery detection. Motivated by this key observation, we design an easily embeddable disentanglement framework for content information removal, and further propose a *Content Consistency Constraint* (C^2C) and a *Global Representation Contrastive Constraint* (GRCC) to enhance the independence of disentangled features. Furthermore, we cleverly construct two unbalanced datasets to investigate the impact of the content bias. Extensive visualizations and experiments demonstrate our framework can not only ignore the interference of content bias but also guide the detector to mine suspicious artifact traces and achieve competitive performance in face forgery detection.

Acknowledgments. This work was supported by National Key R&D Program of China (2019YFB1406504).

References

1. Deepfake porn is ruining women's lives. http://www.technologyreview.com/2021/02/12/1018222/deepfake-revenge-porn-coming-ban. Accessed 15 Aug 2021
2. A voice deepfake was used to scam a ceo out of $243,000. http://www.forbes.com/sites/jessedamiani/2019/09/03/a-voice-deepfake-was-used-to-scam-a-ceo-out-of-243000. Accessed 02 Jan 2021
3. Afchar, D., Nozick, V., Yamagishi, J., Echizen, I.: Mesonet: a compact facial video forgery detection network. In: 2018 IEEE International Workshop on Information Forensics and Security (WIFS), pp. 1–7. IEEE (2018)
4. Asnani, V., Yin, X., Hassner, T., Liu, X.: Reverse engineering of generative models: Inferring model hyperparameters from generated images. arXiv preprint arXiv:2106.07873 (2021)
5. Bayar, B., Stamm, M.C.: A deep learning approach to universal image manipulation detection using a new convolutional layer. In: Proceedings of the 4th ACM Workshop on Information Hiding and Multimedia Security, pp. 5–10 (2016)
6. Bengio, Y., Courville, A., Vincent, P.: Representation learning: a review and new perspectives. IEEE Trans. Pattern Anal. Mach. Intell. **35**(8), 1798–1828 (2013)
7. Chai, L., Bau, D., Lim, S.-N., Isola, P.: What makes fake images detectable? understanding properties that generalize. In: Vedaldi, A., Bischof, H., Brox, T., Frahm, J.-M. (eds.) ECCV 2020. LNCS, vol. 12371, pp. 103–120. Springer, Cham (2020). https://doi.org/10.1007/978-3-030-58574-7_7

8. Chen, S., Yao, T., Chen, Y., Ding, S., Li, J., Ji, R.: Local relation learning for face forgery detection. In: Proceedings of the AAAI Conference on Artificial Intelligence, vol. 35, pp. 1081–1088 (2021)
9. Chollet, F.: Xception: deep learning with depthwise separable convolutions. In: Proceedings of the IEEE Conference on Computer Vision and Pattern Recognition, pp. 1251–1258 (2017)
10. Chugh, K., Gupta, P., Dhall, A., Subramanian, R.: Not made for each other-audio-visual dissonance-based deepfake detection and localization. In: Proceedings of the 28th ACM International Conference on Multimedia, pp. 439–447 (2020)
11. Ciftci, U.A., Demir, I., Yin, L.: Fakecatcher: Detection of synthetic portrait videos using biological signals. IEEE Trans. Pattern Anal. Mach. Intell. (2020)
12. Cozzolino, D., Poggi, G., Verdoliva, L.: Recasting residual-based local descriptors as convolutional neural networks: an application to image forgery detection. In: Proceedings of the 5th ACM Workshop on Information Hiding and Multimedia Security, pp. 159–164 (2017)
13. Dang, H., Liu, F., Stehouwer, J., Liu, X., Jain, A.K.: On the detection of digital face manipulation. In: Proceedings of the IEEE/CVF Conference on Computer Vision and Pattern recognition, pp. 5781–5790 (2020)
14. deepfakes: Deepfakes. https://github.com/deepfakes. Accessed 18 Aug 2021
15. Frank, J., Eisenhofer, T., Schönherr, L., Fischer, A., Kolossa, D., Holz, T.: Leveraging frequency analysis for deep fake image recognition. In: International Conference on Machine Learning, pp. 3247–3258. PMLR (2020)
16. Fridrich, J., Kodovsky, J.: Rich models for steganalysis of digital images. IEEE Trans. Inf. Forensics Secur. 7(3), 868–882 (2012)
17. Haliassos, A., Vougioukas, K., Petridis, S., Pantic, M.: Lips don't lie: a generalisable and robust approach to face forgery detection. In: Proceedings of the IEEE/CVF Conference on Computer Vision and Pattern Recognition, pp. 5039–5049 (2021)
18. He, K., Zhang, X., Ren, S., Sun, J.: Deep residual learning for image recognition. In: Proceedings of the IEEE Conference on Computer Vision and Pattern Recognition, pp. 770–778 (2016)
19. Kingma, D.P., Ba, J.: Adam: A method for stochastic optimization. arXiv preprint arXiv:1412.6980 (2014)
20. Li, J., Xie, H., Li, J., Wang, Z., Zhang, Y.: Frequency-aware discriminative feature learning supervised by single-center loss for face forgery detection. In: Proceedings of the IEEE/CVF Conference on Computer Vision and Pattern Recognition, pp. 6458–6467 (2021)
21. Li, L., Bao, J., Zhang, T., Yang, H., Chen, D., Wen, F., Guo, B.: Face x-ray for more general face forgery detection. In: Proceedings of the IEEE/CVF Conference on Computer Vision and Pattern Recognition. pp. 5001–5010 (2020)
22. Li, Y., Chang, M.C., Lyu, S.: In ictu oculi: Exposing ai created fake videos by detecting eye blinking. In: 2018 IEEE International Workshop on Information Forensics and Security (WIFS), pp. 1–7. IEEE (2018)
23. Li, Y., Yang, X., Sun, P., Qi, H., Lyu, S.: Celeb-df: a large-scale challenging dataset for deepfake forensics. In: Proceedings of the IEEE/CVF Conference on Computer Vision and Pattern Recognition, pp. 3207–3216 (2020)
24. Liang, J., Deng, W.: Identifying rhythmic patterns for face forgery detection and categorization. In: 2021 IEEE International Joint Conference on Biometrics (IJCB), pp. 1–8. IEEE (2021)
25. Liu, H., et al.: Spatial-phase shallow learning: rethinking face forgery detection in frequency domain. In: Proceedings of the IEEE/CVF Conference on Computer Vision and Pattern Recognition, pp. 772–781 (2021)

26. Liu, Y., Stehouwer, J., Liu, X.: On disentangling spoof trace for generic face anti-spoofing. In: Vedaldi, A., Bischof, H., Brox, T., Frahm, J.-M. (eds.) ECCV 2020. LNCS, vol. 12363, pp. 406–422. Springer, Cham (2020). https://doi.org/10.1007/978-3-030-58523-5_24

27. Liu, Z., Qi, X., Torr, P.H.: Global texture enhancement for fake face detection in the wild. In: Proceedings of the IEEE/CVF Conference on Computer Vision and Pattern Recognition, pp. 8060–8069 (2020)

28. Luo, Y., Zhang, Y., Yan, J., Liu, W.: Generalizing face forgery detection with high-frequency features. In: ProceeRings of the IEEE/CVF Conference on Computer Vision and Pattern Recognition, pp. 16317–16326 (2021)

29. Van der Maaten, L., Hinton, G.: Visualizing data using t-sne. J. Mach. Learn. Res. **9**(11) (2008)

30. MarekKowalski: Fakeswap.   htpps://github.com/MarekKowalski/FaceSwap. Accessed 18 Aug 2021

31. Masi, I., Killekar, A., Mascarenhas, R.M., Gurudatt, S.P., AbdAlmageed, W.: Two-branch recurrent network for isolating deepfakes in videos. In: Vedaldi, A., Bischof, H., Brox, T., Frahm, J.-M. (eds.) ECCV 2020. LNCS, vol. 12352, pp. 667–684. Springer, Cham (2020). https://doi.org/10.1007/978-3-030-58571-6_39

32. Matern, F., Riess, C., Stamminger, M.: Exploiting visual artifacts to expose deep-fakes and face manipulations. In: 2019 IEEE Winter Applications of Computer Vision Workshops (WACVW), pp. 83–92. IEEE (2019)

33. Niu, X., Yu, Z., Han, H., Li, X., Shan, S., Zhao, G.: Video-based remote physiolog-ical measurement via cross-verified feature disentangling. In: Vedaldi, A., Bischof, H., Brox, T., Frahm, J.-M. (eds.) ECCV 2020. LNCS, vol. 12347, pp. 295–310. Springer, Cham (2020). https://doi.org/10.1007/978-3-030-58536-5_18

34. Oord, A.v.d., Li, Y., Vinyals, O.: Representation learning with contrastive predic-tive coding. arXiv preprint arXiv:1807.03748 (2018)

35. Qi, H., Guo, Q., Juefei-Xu, F., Xie, X., Ma, L., Feng, W., Liu, Y., Zhao, J.: Deep-rhythm: Exposing deepfakes with attentional visual heartbeat rhythms. In: Pro-ceedings of the 28th ACM International Conference on Multimedia, pp. 4318–4327 (2020)

36. Qian, Y., Yin, G., Sheng, L., Chen, Z., Shao, J.: Thinking in frequency: face forgery detection by mining frequency-aware clues. In: Vedaldi, A., Bischof, H., Brox, T., Frahm, J.-M. (eds.) ECCV 2020. LNCS, vol. 12357, pp. 86–103. Springer, Cham (2020). https://doi.org/10.1007/978-3-030-58610-2_6

37. Rahmouni, N., Nozick, V., Yamagishi, J., Echizen, I.: Distinguishing computer graphics from natural images using convolution neural networks. In: 2017 IEEE Workshop on Information Forensics and Security (WIFS), pp. 1–6. IEEE (2017)

38. Rossler, A., Cozzolino, D., Verdoliva, L., Riess, C., Thies, J., Nießner, M.: Face-forensics++: learning to detect manipulated facial images. In: Proceedings of the IEEE/CVF International Conference on Computer Vision, pp. 1–11 (2019)

39. Selvaraju, R.R., Cogswell, M., Das, A., Vedantam, R., Parikh, D., Batra, D.: Grad-cam: Visual explanations from deep networks via gradient-based localization. In: Proceedings of the IEEE International Conference on Computer Vision, pp. 618–626 (2017)

40. Sun, K., Liu, H., Ye, Q., Liu, J., Gao, Y., Shao, L., Ji, R.: Domain general face forgery detection by learning to weight. In: Proceedings of the AAAI Conference on Artificial Intelligence, vol. 35, pp. 2638–2646 (2021)

41. Thies, J., Zollhöfer, M., Nießner, M.: Deferred neural rendering: image synthesis using neural textures. ACM Trans. Graph. (TOG) **38**(4), 1–12 (2019)

42. Thies, J., Zollhofer, M., Stamminger, M., Theobalt, C., Nießner, M.: Face2face: real-time face capture and reenactment of rgb videos. In: Proceedings of the IEEE Conference on Computer Vision and Pattern Recognition, pp. 2387–2395 (2016)

43. Tran, L., Yin, X., Liu, X.: Disentangled representation learning gan for pose-invariant face recognition. In: Proceedings of the IEEE Conference on Computer Vision and Pattern Recognition, pp. 1415–1424 (2017)

44. Wang, C., Deng, W.: Representative forgery mining for fake face detection. In: Proceedings of the IEEE/CVF Conference on Computer Vision and Pattern Recognition, pp. 14923–14932 (2021)

45. Wang, R., Juefei-Xu, F., Ma, L., Xie, X., Huang, Y., Wang, J., Liu, Y.: Fakespotter: A simple yet robust baseline for spotting ai-synthesized fake faces. In: Bessiere, C. (ed.) Proceedings of the Twenty-Ninth International Joint Conference on Artificial Intelligence, IJCAI-20. pp. 3444–3451. International Joint Conferences on Artificial Intelligence Organization (7 2020). https://doi.org/10.24963/ijcai.2020/476. doi.org/10.24963/ijcai.2020/476, main track

46. Wang, S.Y., Wang, O., Zhang, R., Owens, A., Efros, A.A.: Cnn-generated images are surprisingly easy to spot... for now. In: Proceedings of the IEEE/CVF Conference on Computer Vision and Pattern Recognition, pp. 8695–8704 (2020)

47. Yang, X., Li, Y., Lyu, S.: Exposing deep fakes using inconsistent head poses. In: ICASSP 2019–2019 IEEE International Conference on Acoustics, Speech and Signal Processing (ICASSP), pp. 8261–8265. IEEE (2019)

48. Yu, N., Davis, L.S., Fritz, M.: Attributing fake images to gans: learning and analyzing gan fingerprints. In: Proceedings of the IEEE/CVF International Conference on Computer Vision, pp. 7556–7566 (2019)

49. Zhang, H., Wu, C., Zhang, Z., Zhu, Y., Lin, H., Zhang, Z., Sun, Y., He, T., Mueller, J., Manmatha, R., et al.: Resnest: Split-attention networks. arXiv preprint arXiv:2004.08955 (2020)

50. Zhang, H., Cisse, M., Dauphin, Y.N., Lopez-Paz, D.: mixup: beyond empirical risk minimization. In: International Conference on Learning Representations (2018)

51. Zhang, K.-Y., et al.: Face anti-spoofing via disentangled representation learning. In: Vedaldi, A., Bischof, H., Brox, T., Frahm, J.-M. (eds.) ECCV 2020. LNCS, vol. 12364, pp. 641–657. Springer, Cham (2020). https://doi.org/10.1007/978-3-030-58529-7_38

52. Zhang, Z., et al.: Gait recognition via disentangled representation learning. In: Proceedings of the IEEE/CVF Conference on Computer Vision and Pattern Recognition, pp. 4710–4719 (2019)

53. Zhao, H., Zhou, W., Chen, D., Wei, T., Zhang, W., Yu, N.: Multi-attentional deepfake detection. In: Proceedings of the IEEE/CVF Conference on Computer Vision and Pattern Recognition, pp. 2185–2194 (2021)

54. Zhao, T., Xu, X., Xu, M., Ding, H., Xiong, Y., Xia, W.: Learning to recognize patch-wise consistency for deepfake detection. arXiv preprint arXiv:2012.09311 (2020)

55. Zhong, Z., Zheng, L., Kang, G., Li, S., Yang, Y.: Random erasing data augmentation. In: Proceedings of the AAAI Conference on Artificial Intelligence, vol. 34, pp. 13001–13008 (2020)

RepMix: Representation Mixing for Robust Attribution of Synthesized Images

Tu Bui[1]([✉])[iD], Ning Yu[2], and John Collomosse[1,3]

[1] University of Surrey, Guildford, UK
t.v.bui@surrey.ac.uk
[2] Salesforce Research, Palo Alto, USA
ning.yu@salesforce.com
[3] Adobe Research, San Jose, USA
collomos@adobe.com

Abstract. Rapid advances in Generative Adversarial Networks (GANs) raise new challenges for *image attribution*; detecting whether an image is synthetic and, if so, determining which GAN architecture created it. Uniquely, we present a solution to this task capable of 1) matching images invariant to their semantic content; 2) robust to benign transformations (changes in quality, resolution, shape, etc.) commonly encountered as images are re-shared online. In order to formalize our research, a challenging benchmark, Attribution88, is collected for robust and practical image attribution. We then propose RepMix, our GAN fingerprinting technique based on representation mixing and a novel loss. We validate its capability of tracing the provenance of GAN-generated images invariant to the semantic content of the image and also robust to perturbations. We show our approach improves significantly from existing GAN fingerprinting works on both semantic generalization and robustness. Data and code are available at https://github.com/TuBui/image_attribution.

Keywords: GAN Fingerprinting · Image attribution · Fake image detection · Dataset benchmarking

1 Introduction

Generative imagery is transforming creative practice through intuitive tools that enable controllable and high quality image synthesis. The photo-realism achievable by recent Generative Adversarial Networks (GANs) is often indistinguishable from real imagery [41]; it is difficult for a lay user to tell if an image is synthetic, or to tell images generated by one GAN from those generated by another. Yet, understanding the provenance of visual media has never been more important – to help ensure creative rights, and to mitigate the spread of misinformation

Supplementary Information The online version contains supplementary material available at https://doi.org/10.1007/978-3-031-19781-9_9.

due to abuses of GAN technology. In the near future, parameterizable generative imagery may even begin to challenge or replace traditional stock photography. Tools to trace an image to the GAN that created it are urgently needed to ensure the authenticity and proper attribution of images shared online.

Recent work has already shown initial success at detecting synthetic imagery [22,50,51] and attribution of generative imagery [2,17,56] ('GAN fingerprinting') to a GAN source. Particularly, Wang et al. [51] suggest that today GANs share some common technical flaws that could be easily distinguished from real images. However, image attribution is generally more challenging than synthesis detection due to the diversity in GAN classes; also it is inconclusive what sort of fingerprint a GAN model leaves in its output imagery. Existing image attribution methods, despite reporting near-saturated performance, have two setbacks. First, they mostly focus on attributing images to specific GAN models, which is impractical because a single change in training data, training metaparameters (e.g. learning rate, optimizer, training iterations ...) or even random seed results in a different GAN model [56]. It would be more practical to attribute synthetic imagery to the underlining GAN architecture rather than specific GAN models. Second, the effects of perturbations on synthetic images are largely underestimated. Current works often experiment with few image transformations such as blurring, JPEG compression, random crop [17,51,56] which does not reflect the real-life perturbations that online imagery is subjected through redistribution. Such perturbations could deteriorate GAN fingerprint which is reported to lay between the medium and high frequency bands in an image [63].

The foremost contribution of this paper is a solid benchmark for image attribution, where a GAN class is represented by several GAN models trained on different semantic datasets, and images are subjected to various sources of perturbations. We then propose a novel method to robustly determine the fakeness of an image, and if so, which GAN architecture was used. Both our benchmark and proposed method address two key limitations of existing approaches:

1. Semantic generalization. Existing GAN fingerprinting methods trained on images of a particular class of object (e.g. faces) typically fail on images of other object classes. This is because prior works focus on attribution to one of several GAN models seen at training time. Uniquely, we address the new problem of attributing images of unseen semantic class to the GAN *architecture* that created them. In doing so, we formalize a new problem (attribution to GAN architecture rather than model), and propose a novel representation mix-up training strategy so as to equip GAN fingerprinting with semantic generalization over unseen models producing images containing unseen object classes.

2. Robustness to benign transformation. Images often undergo non-editorial (benign) transformations, such as quality, resolution, or format change as they are redistributed online [8,11,40]. Existing GAN fingerprinting techniques exploit artifacts in the GAN generated images in the pixel domain [56] or frequency domain [17] that are removed or corrupted via redistribution process, causing attribution to fail. In some cases, GANs are actively trained to introduce such artifacts. We employ a contrastive training strategy to enable our GAN attribution model to discriminate GAN architectures passively, based upon artifacts that are seldom removed via benign transformation upon images.

2   Related Work

Generative Adversarial Networks (GANs) [19] have shown outstanding performance in many downstream image synthesis tasks: photo-real blending and in-painting [54], super-resolution [31], facial portrait generation [13], manipulation [43], and texture synthesis [42,55]. GANs have been also applied to bridge multiple modalities such as geometry [1], audio [48], or sketch [36]. Our work focuses upon unconditional GANs [5,6,26–29,39] to avoid introducing additional constraints when producing synthetic images.

Content provenance explores the attribution of media to a trusted source (*e.g.* a database or blockchain [9,10]). Image provenance systems typically rely upon embedded metadata [3,11], watermarking [4,14,21,44] or perceptual hashing [12,32,34,64] to perform visual search robust to the kinds of non-editorial transformation encountered online. Some methods are trained to fail in the presence of digital manipulation [40], whilst others are explicitly trained to match such content and highlight any manipulation [7,8] between the query and matched original. Regardless of applications, robustness and generalization are crucial for content provenance. This is usually addressed via data manipulation (augmentation, data mixing, adversarial attack), implicit representation learning (kernel methods, disentanglement) or explicit learning strategy (ensemble, meta-learning) [49]. In this aspect, RepMix can be considered as a blend of data manipulation (new data is created by mixing existing data points) and representation learning (mixing is performed at feature level).

Digital forensics methods detect and localize image manipulations in the 'blind' *i.e.* without a comparator. The recent 'deep fake detection challenge' (DFDC) [16] identified several approaches to detect GAN generated images or image regions, either upon its statistical properties [52,62] or current limitations of GAN methods (*e.g.* human blinking [33]). Our approach contributes most directly to this area, seeking to determine both the presence, and the source of, synthetic imagery. As such we are aligned with recent GAN fingerprinting work. Prior work has explored this problem mainly for facial images, seeking to identify the model [15,17,56] or the architecture and metaparameters [2]. All these works are passive; the practicality of GAN identification is limited by reliance upon fragile signals within an image that are easily destroyed by benign transformation. In order to mitigate this, Yu *et al.* instead propose to modify the GAN training to inject a robust fingerprint into the synthetic image [57,58]. However such approaches require active participation of the GAN creator, and all remain limited to images of a single semantic class. Our fingerprinting approach is passive and robust to both unseen semantic classes and benign transformation, presenting a further step toward practical GAN attribution in the wild.

Most of the above approaches attribute images towards specific GAN models. Although Ding *et al.* [15] attempts to learn an architecture-specific attributor, their work only covers GAN models of different training seeds. Reverse Engineering [2] shows that GAN architecture parameters could be traced even for unseen GAN models, however such fine-grain attributions mean each GAN class is represented by 1 GAN model; and the robustness of the model is still inconclusive. Recently, Girish *et al.* [18] proposes to automatically discover a new GAN cluster for unseen synthesized images, at the cost of iterative evolution of the attributor. While we share a similar goal with [18] in term of architecture-specific attribution, our work scope limits at a closed world problem (*i.e.* attribution on a fixed set of GAN classes), instead we focus on the generalization on unseen semantic and transformations.

Fig. 1. Illustrating the construction of Attribution88; a new dataset and benchmark that we contribute for synthetic image detection and attribution.

3   The Attribution88 Benchmark

The most popular attribution dataset in literature is introduced by Yu *et al.* [56], containing 5 classes (Real + 4 GANs) of a single semantic object. Each GAN class is represented by one GAN model, thus the learned fingerprint could be entangled with semantic features. This is also not an absolute benchmark since only the GAN models are released (rather than the synthesized images) and there is not a fixed train/test split. Existing approaches [2,17,56] report different results on this dataset, even for the common baselines. Additionally, the reported performance is near saturated. It is important to have a fixed and more challenging benchmark for image attribution. The new benchmark should have GAN classes tied to the GAN design/architecture rather than specific GAN models, meaning images from the same GAN class could come from different model training instances. While we could simply vary the training random seeds (*e.g.* [15]) or other metaparameters to create different model instances of a same GAN, we leave the configuration of these parameters of each GAN model fixed to recommended settings for optimal generative quality. Instead, for each GAN class, we train multiple models on different sets of image objects (semantics). The new benchmark is more challenging as attribution must be agnostic to semantics.

We introduce **Attribution88** - a new dataset made of 8 generator classes and 11 semantics (Fig. 1). We start with 5 generator classes (Real, Progan [26], Cramergan [5], Mmdgan [6], Sngan [39]) as proposed in [56], then add 3 most recent classes of the StyleGAN family (Stylegan [28], Stylegan2 [29] and Stylegan3 [27]). For semantics, we choose 10 objects and scenes from the LSUN dataset [53] plus the popular CelebA face dataset [35]. We note that CelebA is structurally aligned and well curated as compared with other semantic sets, but it is widely used for image attribution/synthesis and adds diversity to our benchmark. For each semantic set, we randomly select 100k images for training the 7 GAN models above, and a disjoint 12k images to serve as *real* images for the attribution task. We use pretrained GAN models when available, otherwise they are trained from scratch using public code, outputting 128×128 images (more details in Sup.Mat).

Next, we generate 100K images per GAN model, resulting in 7.7M synthesized images. Since some images have visible artifacts, we clean them to improve challenge and quality by first extracting perceptual features (of synthesized and real images) using InceptionV3 [61]. We then use K-Means (k = 100) to cluster the synthesized images, determine the closest real image for each, and sort the synthesized images according the distance to its closest real image. We then pick top-k (k = 120) images in each group, assuming the images closest to a real one have the highest quality. This process helps retain a balance between diversity and realism of images. Overall, we obtain 12K images for each of 8 generator sources (Real plus 7 GANs) and 11 semantics, totally ∼1M images. We further partition each set to 10K training, 1K validation and 1K test images. In our experiments, we expose only 6 semantics (*CelebA Face, Bedroom, Airplane, Classroom, Cow, Church Outdoor*) in training and evaluate on all test images (including 5 unseen semantic classes: *Bridge, Bus, Sheep, Kitchen, Cat*).

Perturbations. Images circulated online are subjected to benign perturbations, from mild transformations such as image resizing to strong ones like noises and enhancement effects. It is important to be robust against these. To this end, we employ ImageNet-C [24], a popular benchmark for evaluating classification robustness. ImageNet-C contains 19 common types of corruption, including various additive noises, blurring and effects, each has 5 different corruption levels. Similar to [24], we only expose 15 transformations to training while the test set is subjected to all possible transformations.

4   Methodology

Synthetic image attribution is a classification problem [2,17,22,56]. In our case, the classes correspond to the GAN architectures from which the images are generated. Unlike semantic classification which relies on discriminative features of salient objects, the features useful for image attribution are often subtle and may deteriorate due to noise or other image perturbations [22,56]. In order to learn an attribution model robust against (even unseen) semantics and perturbations, we propose RepMix - a simple feature mixing mechanism to synthesize

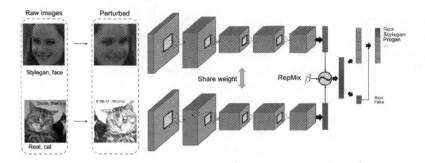

Fig. 2. CNN architecture of our image attribution model. A pair of images is passed though the earlier layers of the CNN model, gets mixed in the RepMix layer before passing to later layers. Training is regulated by a compound loss (see Sect. 4.2).

new data from interpolation between existing data points, then learn to predict the mixing ratio. Figure 2 shows an overview of our approach. Our key technical advancements include (1) the RepMix layer that performs feature mixing between generator classes and (2) the compound loss to predict the mixing ratio for classification.

4.1   Representation Mixing (RepMix) Layer

Suppose we have a training set $\mathcal{X} = \{(\mathbf{x}_i, s_i, y_i), i = 1, 2, ...\}$ where an image $\mathbf{x}_i$ has semantic label $s_i \in \mathcal{S}$ and source label $y_i \in \mathcal{Y}$ (which includes real and a set of GAN source labels). Our goal is to learn mapping $\mathbf{x}_i$ to y_i agnostic to s_i.

Given a training image pair $\mathbf{x}_i$ and $\mathbf{x}_j$ which could either share or differ in source and semantic labels, we first project both images to an intermediate feature space using a nonlinear mapping function f_e:

$$\mathbf{u}_i = f_e(\mathbf{x}_i); \qquad \mathbf{u}_j = f_e(\mathbf{x}_j) \tag{1}$$

where $f_e(.)$ could be the earlier layers of a CNN module. The intermediate representations are input to our RepMix layer:

$$\mathbf{u} = M_\beta(\mathbf{u}_i, \mathbf{u}_j) := \alpha * \mathbf{u}_i + (1 - \alpha) * \mathbf{u}_j \tag{2}$$

with random weight α generated from a certain distribution (here we draw α from a beta distribution[1], $\alpha \sim \text{Beta}(\beta, \beta)$).

Next, the mixed feature map $\mathbf{u}$ is projected into the output via a second mapping function (*e.g.* the later layers of the CNN module):

$$\mathbf{z} = f_l(\mathbf{u}) \in \mathbb{R}^D \tag{3}$$

where D is the output dimension (D = 256 in our work). We call $\mathbf{z}$ the embedding space as it directly precedes the objective function (Subsect. 4.2).

---

[1] https://en.wikipedia.org/wiki/Beta_distribution.

From an implementation perspective, RepMix is portable and can be inserted anywhere in any existing CNN architecture. Since it has no learnable parameters, it introduces minimal overhead at training time. And since it is used for training only, it can be removed during inference (equivalent to duplicating $\mathbf{x}_i$ to make $\mathbf{x}_j$ with the same semantic and source label). We consider RepMix an extension of MixUp and related work [23,25,59,60] regarding the idea of mixing features. The difference is that existing work performs mixing in the raw image space, while RepMix performs at an intermediate layer. We argue that image attribution relies on subtle artifacts on an image (instead of salient objects) to distinguish real from fake as well as classifying different GAN sources. These useful artifacts could be overwritten or canceled out if images are mixed at pixel level, reducing overall performance (see Sect. 5).

4.2 Compound Loss

To attribute an image to its source, existing works [2,17,56] treat the class *real* the same way as other GAN classes prior to modeling classification with a cross-entropy loss. In fact, there is a hierarchical structure in our problem: an image can be either real or fake, if it is fake then it is synthesized from one of the GAN generators. Additionally, real images have a different distribution than GAN synthesized images (see Sect. 5.7), therefore should be treated differently. To this end, we proposed a compound loss that takes into account real/fake detection and attribution at the same time.

We first detect the proportion of realness and fakeness scores in the mix up:

$$z_{\text{real}} = \mathbf{W}_{\text{real}}^T \mathbf{z}; \qquad z_{\text{fake}} = \mathbf{W}_{\text{fake}}^T \mathbf{z} \qquad \in \mathbb{R} \tag{4}$$

$$\bar{z}_{\text{real}} = \frac{e^{z_{\text{real}}}}{e^{z_{\text{real}}} + e^{z_{\text{fake}}}}; \qquad \bar{z}_{\text{fake}} = \frac{e^{z_{\text{fake}}}}{e^{z_{\text{real}}} + e^{z_{\text{fake}}}} \tag{5}$$

$$L_{\text{det}} = -\big(\alpha(1 - y_i^*) + (1-\alpha)(1 - y_j^*)\big)\log(\bar{z}_{\text{real}}) \tag{6}$$

$$- \frac{1}{|\mathcal{Y}| - 1}\big(\alpha y_i^* + (1-\alpha)y_j^*\big)\log(\bar{z}_{\text{fake}}) \tag{7}$$

where $\mathbf{W}_{\text{real}}, \mathbf{W}_{\text{fake}} \in \mathbb{R}^{D \times 1}$ are learnable filters, and pseudo label $y_i^* = 0$ if $\mathbf{x}_i$ is real, otherwise 1 (same for y_j^*). This detection loss essentially measures the weighted cross entropy between real and fakeness of each image in the mix. Since there are generally more fake images than real in the training set, the fake term is scaled down by the number of GAN sources accordingly.

The actual attribution task is performed via another cross-entropy loss, taking into account the real/fake-ness score:

$$\mathbf{z}_{\text{attr}} = \mathbf{W}_{\text{attr}}^T \mathbf{z} + \mathbf{b} \in \mathbb{R}^{|\mathcal{Y}|} \tag{8}$$

$$\hat{\mathbf{z}}_{\text{attr}} = \begin{cases} z_{\text{attr}}^{(y_{\text{real}})} * \bar{z}_{\text{real}} \\ z_{\text{attr}}^{(c)} * \bar{z}_{\text{fake}} & \forall c \in \mathcal{Y}\backslash\{y_{\text{real}}\} \end{cases} \tag{9}$$

$$L_{\text{attr}} = -\alpha\log\Big(\frac{e^{\hat{z}_{\text{attr}}^{(y_i)}}}{\sum_k e^{\hat{z}_{\text{attr}}^{(y_k)}}}\Big) - (1-\alpha)\log\Big(\frac{e^{\hat{z}_{\text{attr}}^{(y_j)}}}{\sum_k e^{\hat{z}_{\text{attr}}^{(y_k)}}}\Big) \tag{10}$$

where $\mathbf{W}_{attr} \in \mathbb{R}^{D \times |\mathcal{Y}|}$ and $\mathbf{b}$ are learnable weight and bias of a fully connected layer to linearly map our embedding $\mathbf{z}$ to the attribution logits. (c) indicates the c-th element of the logit vector. Finally, the total loss is sum of the two above losses $L_{total} = L_{det} + L_{attr}$.

5 Experiments

5.1 Training Details

We use the Resnet50 architecture as the backbone for our RepMix model, with the final N-way classification layer replaced by a FC layer producing the 256-D latent code, followed by our compound loss (Subsect. 4.2). Our RepMix layer is inserted at the first FC layer for optimal performance (c.f. Subsect. 5.6), with $\beta = 0.4$. Image pairs are randomly sampled from the training data, regardless of generator class and semantics. We do not enforce any constraint on sampling the image pairs to maximize all possible source/semantic combinations. During training we resize images to 256×256 and augment with random crop to 224×224, horizontal flip followed by a random *seen* ImageNet-C perturbation with activation probability of 95%. We train our attribution models for maximum 30 epochs, with Adam optimizer and initial learning rate 1e−4, step decaying with $\gamma = 0.85$ and early stopping based on validation accuracy.

Table 1. Performance of RepMix and other baselines on a control set that mimics Yu *et al.* [56] settings, and Attribution88 test set. Yu[†] *et al.* refers to the implementation using the original public code

	1 Sem., Clean			Attribution88		
	Det. Acc. ⇑	Attr. Acc. ⇑	Attr. NMI ⇑	Det. Acc. ⇑	Attr. Acc. ⇑	Attr. NMI ⇑
RepMix	**1.0000**	**0.9994**	**0.9975**	**0.9745**	**0.8207**	**0.6679**
Yu *et al.* [56] (reimp.)	0.9910	0.9838	0.9458	0.9306	0.6784	0.4666
Yu[†] *et al.* [56]	0.9888	0.9844	0.9455	0.9190	0.6322	0.4028
DCT-CNN [17]	0.9922	0.9838	0.9526	0.9001	0.6447	0.4061
Reverse Eng. [2]	0.9976	0.9960	0.9834	0.8665	0.5637	0.3653
EigenFace [47]	0.8262	0.6538	0.4515	0.7829	0.1515	0.0034
PRNU [38]	0.8544	0.8482	0.7389	0.7845	0.1252	0.0003

5.2 Baseline Comparison

We compare our method with 5 baselines: (i) Yu *et al.* [56] attributes images via a simple fingerprinting CNN model; (ii) DCT-CNN [17] classifies images in the frequency space; (iii) Reverse Engineering [2] models GAN architecture details such as number of layers and loss types to assist attribution; (iv) EigenFace [47] builds an Eigen model for each class and classify an image based on its maximum correlation with each model; (v) PRNU [38] is similar to EigenFace but works on noise fingerprints of each class instead. The baseline models are trained using

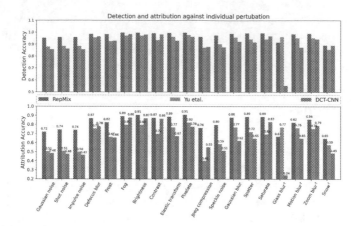

Fig. 3. Detection and attribution performance of our proposed RepMix method vs. two baselines [17,56] in the presence of different benign perturbations of the image

public code with the same data augmentation techniques as in the proposed method. We also provide our re-implementation of Yu *et al.*'s approach. More details on the baseline implementation are in the Sup.Mat.

To validate our training of the baselines and the GAN models, we also perform comparison on a replica of Yu *et al.* [56] dataset, denoted as *1 Sem., Clean*. Specifically, we adopt their data cleaning method, use 5 classes (1 real and 4 GANs) as stated in [56] and without any ImageNet-C perturbation. The only difference is that we use our trained GAN models and we apply random crop and horizontal flip as the minimal augmentation during training and test.

Evaluation Metrics. We report standard classification accuracy and Normalized Mutual Information (NMI) score [18] that measures the dependence between the prediction and the target. Since *real* is one of the target classes, we are also interested in an auxiliary metric, detection accuracy, which is the proportion of images being correctly classified as *real* or *not-real*.

Table 1 compares the performance of RepMix against baselines. The performance on the control set is comparable with existing work [2,17,56], with near-saturated accuracy on the deep learning approaches. Reverse Engineering is the highest scored baseline, next is DCT-CNN [17] which performs slightly better than Yu *et al.* [56]. RepMix achieves perfect detection accuracy and the best attribution accuracy and NMI. However, the baselines underperform on Attribution88. The frequency-based methods (DCT-CNN, Reverse Engineering) under-perform the pixel-based ones (Yu *et al.*). The complexity of our benchmark also causes the shallow methods to either fail completely (PRNU [38]) or just above random prediction (EigenFace [47]). We attribute these changes to the diversity of data (including unseen semantics) and severity of the perturbations. RepMix performs with 4% and 14% higher accuracy than the closest baseline on the detection and attribution scores.

5.3   Robustness Against Individual Perturbation

To analyze the effects of individual perturbation on attribution performance, we evaluate RepMix and the closest competitors, Yu et al. [56] and DCT-CNN [17] on Attribution88 with ImageNet-C perturbations applied on test images (Fig. 3). JPEG compression and additive noise hinders the performance most significantly, especially on the two baselines, while other perturbation sources that transform blocks of neighboring pixels but do not replace them (e.g. blurring) have less severe effects. DCT-CNN is particularly vulnerable to glass blurring. Performance on seen and unseen perturbations is comparable, indicating generalization of our models when being exposed to a large enough sources of augmentations during training. Additionally, detection performance is more robust than attribution, with detection standard deviation of 2.8% across all perturbations versus 8.0% attribution for RepMix (3.7% vs. 12.1% for Yu et al. method; 9.3% vs. 16.8% for DCT-CNN).

Table 2. Attribution errors caused by adversarial attacks on the Attribution88 test set at different levels of max perturbation ϵ. Lower is better

Methods	$\epsilon = 2/255$	$\epsilon = 4/255$	$\epsilon - 8/255$	$\epsilon = 16/255$	$\epsilon = 24/255$	$\epsilon = 32/255$
RepMix	0.1509	0.1952	0.2454	0.3008	0.3333	0.3572
Yu et al. [56]	0.2113	0.2709	0.3328	0.3945	0.4303	0.4534
DCT-CNN [17]	0.1545	0.2190	0.2831	0.3375	0.3642	0.3812

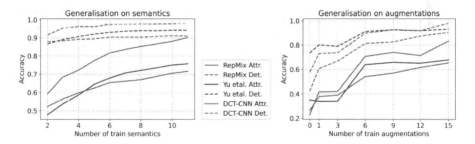

Fig. 4. RepMix performance versus number of (left) semantics and (right) augmentations seen during training.

5.4   Generalization on Semantic and Perturbation

We evaluate the generalization properties of RepMix, Yu et al. and DCT-CNN approaches under the circumstance of limited training data and data augmentation. Figure 4 (left) depicts detection and attribution performance when the models are exposed to increasing number of semantics during training. We evaluate on the full Attribution88 test set. All 3 detection curves stabilize quite early with RepMix consistently maintaining a 3% gap above other two methods. On attribution performance, the more training data leads to more rewarding results, with RepMix having better generalization capability, scoring from 59% accuracy at 2 seen semantics to 90% when all 11 semantics are exposed during training.

Figure 4 (right) shows a similar trend as the number of data augmentation methods increases. We fix the number of training semantics at 6, and increase the number of augmentation methods from 0 to 15, and test on a held-out test set of 4 unseen perturbations. The overall trend is a boost in performance when exposing the models to more perturbations during training, with RepMix gaining more generalization power beyond 15 perturbations.

5.5   Robustness Against Adversarial Attacks

Adversarial attacks introduce to an image a subtle layer of noise which is invisible to the naked eye but enough to change the prediction results of a model. Adversarial attacks work by diverting the gradient w.r.t input image toward the most plausible class other than the groundtruth. Repmix enforces a linear inter-class interpolation in the intermediate feature space, therefore is robust to adversarial attacks by design. To verify this, we perform untargeted whitebox attacks on Repmix, Yu et $al.$ and DCT-CNN models using the I-FGSM method [20]. We use 20 iterations of I-FGSM for every image in the Attribution88 test set and stochastic gradient ascend for optimization. Table 2 shows the attribution errors, which is the difference in attribution accuracy before and after adversarial attacks, at different noise levels. Although all methods suffer a performance drop and the severity is higher at higher noise tolerant levels ($i.e.$ ϵ), RepMix is more robust than the other two approaches. At max perturbation $\epsilon = 32/255$, Rep-Mix accuracy is 2x higher than Yu et $al.$ and DCT-CNN (46.35% vs. 22.49% for Yu et $al.$, and 26.34% for DCT-CNN). Interestingly, DCT-CNN [17] has better resistance than Yu et $al.$ [56], probably because an images in frequency spectrum are visually more monotonous and alike than in the pixel domain thus would require more efforts (aka. iterations) from I-FGSM for a successful attack.

Table 3. Ablation study of RepMix exploring performance at attribution and detection whilst removing different design components, and alternate backbone choices

	Detection Acc. ⇑	Attribution Acc. ⇑	Attribution NMI ⇑
All	**0.9426**	**0.7400**	**0.5546**
w/o compound loss	0.9364	0.7204	0.5280
w/o RepMix	0.9296	0.7188	0.5205
w/o RepMix+Compound loss	0.9283	0.7129	0.5167
w/o augmentation	0.7044	0.2762	0.0856
Different backbones			
VGG16	0.9493	0.7150	0.5315
AlexNet	0.8818	0.5280	0.2817

5.6   Ablation Study

Table 3 shows the performance of RepMix when removing one or several of its components or changing the backbone architecture. Without loss of generality we train and test our ablated models on a subset of Attribution88, with all 8 source classes but 2 semantics during training, and test on 4 semantics (2 seen and 2 unseen). Removing either RepMix layer or compound loss or both results in a drop in performance of all metrics. It can be seen that the compound loss does not benefit only the *real* class (small drop in detection accuracy when removing it), but the whole attribution (2% drop). Finally, removing all ImageNet-C perturbations (leave only random crop and horizontal flip as the data augmentation method) significantly decreases the performance, even causes misleading real/fake detection (detection accuracy below random guess). We also replace Resnet50 with AlexNet [30] and VGG16 [46]. AlexNet leads to a significant performance drop, with NMI score reduced by a half. VGG16 has comparable detection accuracy, but 2.5% lower attribution score. More backbone experiments can be found on Sup.Mat.

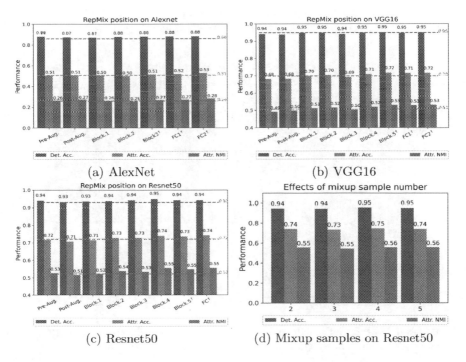

(a) AlexNet   (b) VGG16

(c) Resnet50   (d) Mixup samples on Resnet50

Fig. 5. Effect of RepMix on different layers of (a) AlexNet, (b) VGG16 and (c) Resnet50. Dashed lines refer to baselines without mixing. † indicates the mixing is performed on 1-D feature map (either after Global Average Pooling or FC layer). (d) - The number of mix-up samples have marginal effect on performance of Resnet50

RepMix Position. We experiment with different positions of the RepMix layer in Resnet50, VGG16, and AlexNet. RepMix can be applied to input images at pixel level (equivalent to MixUp [60]), before data augmentation (Pre-Aug.) or after it (Post-Aug.). Within the CNN layers, we insert RepMix after every pooling or FC layer. Figure 5 shows a similar trend across the three networks. Mixing images at pixel level does not improve performance; meaningful subtle artifacts are lost. Post-Aug mixing has the worst score since the image is exposed to double corruption. RepMix is more beneficial at the later layers of the networks, benefiting less on 2D feature maps and more on global representation (FC features). This can be seen from Fig. 6, where the attention heatmap covers larger areas. In Fig. 8, semantic clusters appear even at the embedding layer. However, the GAN classification loss ensures semantic features are weaker at the later layers while the GAN class signal is stronger. Thus, mixing representations at later layers is more beneficial.

Number of Mixup Samples. We test with increasing number of samples to be mixed in RepMix layers. The beta distribution now becomes the Dirichlet distribution to accommodate more than two samples in a mixing group. Figure 5 (d) shows that increasing number of mixing samples has marginal boost in performance, with 1% improvement at 4 mixing samples at most.

Fig. 6. GradCAM visualization on unseen-semantic test images showing the visual artifacts contributing most signficantly to the GAN classification decision.

5.7   Further Analysis

Real Versus Other Classes. We observe that the detection of real images is fairly robust to training data and perturbations and across various ablation settings (c.f. Sects. 5.2–5.6. This interesting behavior is further demonstrated in

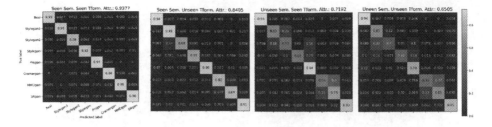

Fig. 7. Confusion matrix of RepMix on seen/unseen semantic classes and on seen/unseen classes of image transformation applied to the test images.

Fig. 7, where class real has the highest score and also appears the most consistent across the seen/unseen semantics and perturbations.

To understand this behavior, we visualize the image regions that contribute the most to the prediction of our model using GradCAM [45]. Figure 6 shows examples of GradCAM heatmaps for several images of *real* and other GAN classes, from both seen and unseen semantics as well as perturbations. For GAN classes, the heatmaps tend to highlight the edge regions which are often more resilient to perturbation attacks. For real images, GradCAM heatmap also focuses on background objects. We therefore reason that real images have a different distribution from synthesized images particularly because they have vivid background, which often attracts the attention of our attribution model.

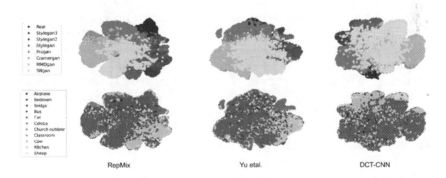

Fig. 8. t-SNE visualization of Attribution88 test set using features extracted from RepMix (left) or Yu *et al.* (middle) and DCT-CNN approach

t-SNE Visualization. We visualize the embedding space **z** of RepMix computed on the Attribution88 test set using t-SNE [37] 2D projection, and compare it with Yu *et al.* approach. Figure 8 shows RepMix has better class separation and semantic fusion than Yu *et al.* Nevertheless, both approaches have a mixed region in the middle of the t-SNE plots where classes are not well separated, which illustrates the challenge of the Attribution88 benchmark.

Limitations. Figure 9 shows examples where RepMix fails, often due to excessive perturbation that distort finer details of an image, narrowing the gap between real/synthesis and between different GAN classes. Another case shown is mis-classification between the three StyleGAN due to architectural similarity.

Fig. 9. Examples of attribution failure. For each inset, left: raw image, middle: image after perturbation, right: GradCAM heatmap justifying its (wrong) prediction

6   Conclusion

We introduce a challenging image attribution benchmark, Attribution88, for detecting and tracing images to the originating GAN architecture, rather than the GAN model. We present a novel GAN fingerprinting technique that introduces strong zero-shot generalization to unseen semantic classes and unseen transformations, in contrast to prior work that generalizes poorly beyond a single class (*e.g.* faces) even if trained with sight of those classes [17,56]. We demonstrate detection accuracy of 97% and attribution accuracy of 82% on this new benchmark, without introducing any change to the GAN training process (per [58]). Our method is particularly robust to detecting real images, by exploiting an unique feature that current GAN methods have not been able to fabricated. Future work could scale our experiments to even broader classes of GAN including conditional GAN frameworks, although we do not believe such experiments necessary to demonstrate the value of benchmark or contrastive training and mix-up strategy in enabling class generalization for GAN attribution.

Acknowledgments. This work was supported by EPSRC DECaDE Grant Ref EP/T022485/1.

References

1. Ashual, O., Wolf, L.: Specifying object attributes and relations in interactive scene generation. In: Proceedings of the ICCV, pp. 4561–4569 (2019)
2. Asnani, V., Yin, X., Hassner, T., Liu, X.: Reverse engineering of generative models: Inferring model hyperparameters from generated images. arXiv preprint arXiv:2106.07873 (2021)

3. Aythora, J., et al.: Multi-stakeholder media provenance management to counter synthetic media risks in news publishing. In: Proceedings of the International Broadcasting Convention (IBC) (2020)
4. Baba, S., Krekor, L., Arif, T., Shaaban, Z.: Watermarking scheme for copyright protection of digital images. IJCSNS **9**(4) (2019)
5. Bellemare, M.G., et al.: The cramer distance as a solution to biased wasserstein gradients. arXiv preprint arXiv:1705.10743 (2017)
6. Bińkowski, M., Sutherland, D.J., Arbel, M., Gretton, A.: Demystifying mmd gans. In: Proceedings of the ICLR (2018)
7. Black, A., Bui, T., Jenni, S., Swaminathan, V., Collomosse, J.: Vpn: Video provenance network for robust content attribution. In: Proceedings of the CVMP, pp. 1–10 (2021)
8. Black, A., Bui, T., Jin, H., Swaminathan, V., Collomosse, J.: Deep image comparator: learning to visualize editorial change. In: Proceedings of the CVPR, pp. 972–980 (2021)
9. Bui, T., et al.: Archangel: tamper-proofing video archives using temporal content hashes on the blockchain. In: Proceedings of the CVPR WS (2019)
10. Bui, T., et al.: Tamper-proofing video with hierarchical attention autoencoder hashing on blockchain. IEEE Trans. Multimedia **22**(11), 2858–2872 (2020)
11. (CAI), C.A.I.: Setting the standard for content attribution. Technical report, Adobe Inc. (2020)
12. Cao, Z., Long, M., Wang, J., Yu, P.S.: Hashnet: deep learning to hash by continuation. In: Proceedings of the CVPR, pp. 5608–5617 (2017)
13. Chen, A., Liu, R., Xie, L., Chen, Z., Su, H., Yu, J.: Sofgan: a portrait image generator with dynamic styling. ACM Trans. Graphics (TOG) **41**(1), 1–26 (2022)
14. Devi, P., Venkatesan, M., Duraiswamy, K.: A fragile watermarking scheme for image authentication with tamper localization using integer wavelet transform. J. Comput. Sci. **5**(11), 831–837 (2019)
15. Ding, Y., Thakur, N., Li, B.: Does a gan leave distinct model-specific fingerprints? In: Proceedings of the BMVC (2021)
16. Dolhansky, B., Bitton, J., Pflaum, B., Lu, J., Howes, R., Wang, M., Ferrer, C.C.: The deepfake detection challenge (DFDC) dataset. CoRR abs/2006.07397 (2020), arxiv.org/abs/2006.07397
17. Frank, J., Eisenhofer, T., Schönherr, L., Fischer, A., Kolossa, D., Holz, T.: Leveraging frequency analysis for deep fake image recognition. In: Proceedings of the ICML, pp. 3247–3258. PMLR (2020)
18. Girish, S., Suri, S., Rambhatla, S.S., Shrivastava, A.: Towards discovery and attribution of open-world gan generated images. In: Proceedings of the ICCV, pp. 14094–14103 (2021)
19. Goodfellow, I., et al.: Generative adversarial nets. NeurIPS 27 (2014)
20. Goodfellow, I.J., Shlens, J., Szegedy, C.: Explaining and harnessing adversarial examples. arXiv preprint arXiv:1412.6572 (2014)
21. Hameed, K., Mumtax, A., Gilani, S.: Digital image watermarking in the wavelet transform domain. WASET **13**, 86–89 (2006)
22. He, Y., Yu, N., Keuper, M., Fritz, M.: Beyond the spectrum: Detecting deepfakes via re-synthesis. In: Proceedings of the IJCAI-21, pp. 2534–2541. International Joint Conferences on Artificial Intelligence Organization (2021)
23. Hendrycks, D., et al.: The many faces of robustness: a critical analysis of out-of-distribution generalization. In: Proceedings of the ICCV, pp. 8340–8349 (2021)
24. Hendrycks, D., Dieterich, T.: Benchmarking neural network robustness to common corruptions and perturbations. In: Proceedings of the ICLR (2018)

25. Hendrycks, D., Mu, N., Cubuk, E.D., Zoph, B., Gilmer, J., Lakshminarayanan, B.: Augmix: a simple data processing method to improve robustness and uncertainty. In: Proceedings of the ICLR (2019)
26. Karras, T., Aila, T., Laine, S., Lehtinen, J.: Progressive growing of gans for improved quality, stability, and variation. In: Proceedings of the ICLR (2018)
27. Karras, T., Aittala, M., Laine, S., Härkönen, E., Hellsten, J., Lehtinen, J., Aila, T.: Alias-free generative adversarial networks. NeurIPS **34** (2021)
28. Karras, T., Laine, S., Aila, T.: A style-based generator architecture for generative adversarial networks. In: Proceedings of the CVPR, pp. 4401–4410 (2019)
29. Karras, T., Laine, S., Aittala, M., Hellsten, J., Lehtinen, J., Aila, T.: Analyzing and improving the image quality of stylegan. In: Proceedings of the CVPR, pp. 8110–8119 (2020)
30. Krizhevsky, A., Sutskever, I., Hinton, G.E.: Imagenet classification with deep convolutional neural networks. NeurIPS **25** (2012)
31. Ledig, C., et al.: Photo-realistic single image super-resolution using a generative adversarial network. In: Proceedings of the CVPR, pp. 4681–4690 (2017)
32. Li, W., Wang, S., Kang, W.C.: Feature learning based deep supervised hashing with pairwise labels. In: Proceedings of the IJCAI, pp. 1711–1717 (2016)
33. Li, Y., Ching, M.C., Lyu, S.: In ictu oculi: Exposing ai created fake videos by detecting eye blinking. In: Proceedings of the IEEE WIFS (2018)
34. Liu, H., Wang, R., Shan, S., Chen, X.: Deep supervised hashing for fast image retrieval. In: Proceedings of the CVPR, pp. 2064–2072 (2016)
35. Liu, Z., Luo, P., Wang, X., Tang, X.: Deep learning face attributes in the wild. In: Proceedings of the ICCV, pp. 3730–3738 (2015)
36. Lu, Y., Wu, S., Tai, Y.-W., Tang, C.-K.: Image generation from sketch constraint using contextual GAN. In: Ferrari, V., Hebert, M., Sminchisescu, C., Weiss, Y. (eds.) ECCV 2018. LNCS, vol. 11220, pp. 213–228. Springer, Cham (2018). https://doi.org/10.1007/978-3-030-01270-0_13
37. Van der Maaten, L., Hinton, G.: Visualizing data using t-sne. J. Mach. Learn. Res. **9**(11) (2008)
38. Marra, F., Gragnaniello, D., Verdoliva, L., Poggi, G.: Do gans leave artificial fingerprints? In: Proceedings of the MIPR, pp. 506–511. IEEE (2019)
39. Miyato, T., Kataoka, T., Koyama, M., Yoshida, Y.: Spectral normalization for generative adversarial networks. In: International Conference on Learning Representations (2018)
40. Nguyen, E., Bui, T., Swaminathan, V., Collomosse, J.: Oscar-net: object-centric scene graph attention for image attribution. In: Proceedings of the ICCV, pp. 14499–14508 (2021)
41. Nightingale, S.J., Farid, H.: Ai-synthesized faces are indistinguishable from real faces and more trustworthy. Proc. Natl. Acad. Sci. **119**(8) (2022)
42. Park, T., Liu, M.Y., Wang, T.C., Zhu, J.Y.: Semantic image synthesis with spatially-adaptive normalization. In: Proceedings of the CVPR, pp. 2337–2346 (2019)
43. Perarnau, G., Van De Weijer, J., Raducanu, B., Álvarez, J.M.: Invertible conditional gans for image editing. arXiv preprint arXiv:1611.06355 (2016)
44. Profrock, D., Schlauweg, M., Muller, E.: Content-based watermarking by geometric wrapping and feature- based image segmentation. In: Proceedings of the SITIS, pp. 572–581 (2006)
45. Selvaraju, R.R., Cogswell, M., Das, A., Vedantam, R., Parikh, D., Batra, D.: Gradcam: Visual explanations from deep networks via gradient-based localization. In: Proceedings of the ICCV, pp. 618–626 (2017)

46. Simonyan, K., Zisserman, A.: Very deep convolutional networks for large-scale image recognition. In: Proceedings of the ICLR (2015)
47. Sirovich, L., Kirby, M.: Low-dimensional procedure for the characterization of human faces. Josa a **4**(3), 519–524 (1987)
48. Wan, C.H., Chuang, S.P., Lee, H.Y.: Towards audio to scene image synthesis using generative adversarial network. In: Proceedings of the ICASSP, pp. 496–500. IEEE (2019)
49. Wang, J., et al.: Generalizing to unseen domains: a survey on domain generalization. IEEE Trans. Knowl. Data Eng. (2022)
50. Wang, R., et al.: Fakespotter: a simple yet robust baseline for spotting ai-synthesized fake faces. In: Proceedings of the IJCAI, pp. 3444–3451 (2021)
51. Wang, S.Y., Wang, O., Zhang, R., Owens, A., Efros, A.A.: Cnn-generated images are surprisingly easy to spot... for now. In: Proceedings of the CVPR, pp. 8695–8704 (2020)
52. Wu, Y., AbdAlmageed, W., Natarajan, P.: Mantra-net: Manipulation tracing network for detection and localization of image forgeries with anomalous features. In: Proceedings of theCVPR, pp. 9543–9552 (2019)
53. Yu, F., Seff, A., Zhang, Y., Song, S., Funkhouser, T., Xiao, J.: Lsun: construction of a large-scale image dataset using deep learning with humans in the loop. arXiv preprint arXiv:1506.03365 (2015)
54. Yu, J., Lin, Z., Yang, J., Shen, X., Lu, X., Huang, T.S.: Generative image inpainting with contextual attention. In: Proceedings of the CVPR, pp. 5505–5514 (2018)
55. Yu, N., Barnes, C., Shechtman, E., Amirghodsi, S., Lukac, M.: Texture mixer: A network for controllable synthesis and interpolation of texture. In: Proceedings of the CVPR, pp. 12164–12173 (2019)
56. Yu, N., Davis, L.S., Fritz, M.: Attributing fake images to gans: Learning and analyzing gan fingerprints. In: Proc. ICCV. pp. 7556–7566 (2019)
57. Yu, N., Skripniuk, V., Abdelnabi, S., Fritz, M.: Artificial fingerprinting for generative models: Rooting deepfake attribution in training data. In: Proceedings of the ICCV, pp. 14448–14457 (2021)
58. Yu, N., Skripniuk, V., Chen, D., Davis, L., Fritz, M.: Responsible disclosure of generative models using scalable fingerprinting (2022)
59. Yun, S., Han, D., Oh, S.J., Chun, S., Choe, J., Yoo, Y.: Cutmix: Regularization strategy to train strong classifiers with localizable features. In: Proceedings of the ICCV, pp. 6023–6032 (2019)
60. Zhang, H., Cisse, M., Dauphin, Y.N., Lopez-Paz, D.: mixup: beyond empirical risk minimization. In: Proceedings of the ICLR (2018)
61. Zhang, R., Isola, P., Efros, A.A., Shechtman, E., Wang, O.: The unreasonable effectiveness of deep features as a perceptual metric. In: Proceedings of the CVPR, pp. 586–595 (2018)
62. Zhang, X., Sun, Z.H., Karaman, S., Chang, S.: Discovering image manipulation history by pairwise relation and forensics tools. IEEE J. Selected Topics Signal Process. **14**(5), 1012–1023 (2020)
63. Zhang, X., Karaman, S., Chang, S.F.: Detecting and simulating artifacts in gan fake images. In: IEEE WIFS, pp. 1–6. IEEE (2019)
64. Zhu, H., Long, M., Wang, J., Cao, Y.: Deep hashing network for efficient similarity retrieval. In: Proceedings of the AAAI (2016)

Totems: Physical Objects for Verifying Visual Integrity

Jingwei Ma[1]([✉]), Lucy Chai[2], Minyoung Huh[2], Tongzhou Wang[2],
Ser-Nam Lim[3], Phillip Isola[2], and Antonio Torralba[2]

[1] University of Washington, Seattle, USA
`jingweim@uw.edu`
[2] MIT, Cambridge, USA
[3] Meta AI, New York, USA

Abstract. We introduce a new approach to image forensics: placing physical refractive objects, which we call totems, into a scene so as to protect any photograph taken of that scene. Totems bend and redirect light rays, thus providing multiple, albeit distorted, views of the scene within a single image. A defender can use these distorted totem pixels to detect if an image has been manipulated. Our approach unscrambles the light rays passing through the totems by estimating their positions in the scene and using their known geometric and material properties. To verify a totem-protected image, we detect inconsistencies between the scene reconstructed from totem viewpoints and the scene's appearance from the camera viewpoint. Such an approach makes the adversarial manipulation task more difficult, as the adversary must modify both the totem and image pixels in a geometrically consistent manner without knowing the physical properties of the totem. Unlike prior learning-based approaches, our method does not require training on datasets of specific manipulations, and instead uses physical properties of the scene and camera to solve the forensics problem.

1 Introduction

As new technologies for photo manipulation become readily accessible, it is vital to maintain our ability to tell apart real images from fake ones. Yet, the realm of current image verification methods is mostly *passive*. From the point of view of a person who wants to be protected from adversarial manipulation, they must trust that the downstream algorithms can recognize the subtle cues and artifacts left behind by the manipulation process.

How can we give a person *active* control overmaintaining image integrity? Imagine if one could place a "signature" in a scene before the photograph is taken. Then, the verification process simply becomes the task of checking

Supplementary Information The online version contains supplementary material available at https://doi.org/10.1007/978-3-031-19781-9_10.

Fig. 1. We envision a setup in which physical objects, called totems, are placed into a scene to protect against adversarial manipulations. From a single camera capture, the totems provide alternative distorted views of the image, which allows us to reconstruct the underlying 3D scene. The reconstruction is then used to highlight potential image manipulations by comparing the scene viewed from the totems to the image observed by the camera.

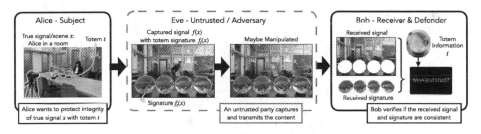

Fig. 2. Totems as a general method to protect a piece of signal (of Alice). Even if the middle person (Eve) can tamper the content, they can't easily edit both the signal and the totem signature consistently. A defender (Bob) can thus detect such manipulations.

whether the signature matches the image content. Inspired by the movie *Inception*, where characters use unique *totems* to distinguish between the real world and the fabricated world, we propose to use physical objects as totems that determine the authenticity of the scene. In our setting, a *totem* is a refractive object that, when placed in a scene, displays a distorted version of the scene on its surface (a *totem view*). After a photo is taken, the defender can check whether the scene captured by the camera is consistent with the totem's appearance.

However, decoding the totems and verifying consistency from a single image presents several challenges. The image of the scene observed through the totem depends on: the totem's physical properties (*e.g.*, index of refraction (IoR)), the totem's position from the camera, and the scene geometry. We assume that the defender *exclusively* knows the physical properties of the totem as the "key" needed to unscramble the totem views. The critical assumption here is that, from the adversary's perspective, it is extremely difficult to manipulate the image and the totem in a geometrically consistent manner without having access to the totem's properties. For the defender, knowing the totem's physical properties makes it possible to estimate the totem's position within the scene and to check for consistency between the scene and the totems. In fact, this process can be bolstered by further complicating the adversary's job, using either multiple totems or a single totem with complicated facets that act to "encrypt" the scene.

More generally, the totems can be viewed as signatures that protect the identity of a piece of signal/information, as shown in Fig. 2. In cryptography, digital signatures also act as a form of active defense, where a message is always sent together with a signature, used to verify both message integrity and sender identity (*e.g.*, the signature can be private-key encoding) [6]. The totems are conceptually similar to such signatures, but are also fundamentally different in that (1) they give control to the subject rather than the party who captures/transmits the photo and (2) they are physical objects and fill the "analog hole" [10], the phenomenon that digital protections become invalid once the content is converted to analog form (*e.g.*, via printing).

In this work, we explore an initial realization of a totem-based verification system where we use simple spherical totems. Our contributions can be summarized as follows:

- We propose an *active* image verification pipeline, in which we can place physical objects called totems within a scene to certify image integrity.
- Totems create multiple, distorted projections of a scene within a single photo. Without access to totem physical properties, it becomes difficult to manipulate the scene in a way that is consistent across all totems.
- For verification, we undo the distortion process and infer the scene geometry from *sparse* totem views and *unknown* totem poses by jointly learning to reconstruct the totem pixels and optimizing the totem pose.
- Comparing the scene reconstructed from the totems and the pixels of the camera viewpoint enables us to detect manipulation from a single image.
- Our work is an initial step towards hardware and geometry-driven approaches to detecting manipulations. The proposed framework is not specific to the scene reconstruction method we test in this paper, and may become more robust as more advanced scene reconstruction methods become available.

2   Related Work

Learning Properties of Refractive Objects. Compared to opaque materials, refractive and transparent objects pose a unique set of challenges in vision and graphics due to their complex interactions with light rays. For example, reconstructing the shape of refractive objects requires multiple views of the refractive object from different viewing angles and makes several assumptions about the environment and capture setup, such as a moving camera and parametric shape formulation [5], known IoR and correspondences between scene 3D points and image pixels [14], or known object maps, environment map, and IoR [18]. Another line of work models the light transport function in refractive objects. These approaches also involve placing planar backgrounds behind the object and capturing multiple viewpoints, which then allow for environment matting to compose a refractive object with different backgrounds [34,45]. In our work, we assume that the knowledge about the totem is asymmetrical between the defender and adversary. The defender knows full totem specifications such as IoR, shape, and size, but the adversary must guess these parameters. Thus, the defender is better able to model the light refraction process through the totems compared to the adversary. Further, we do not require a specific capture environment beyond a single image taken with multiple totems visible in the scene.

Accidental Cameras. Oftentimes, the objects around us can form subtle, unexpected cameras. However, decoding the image from these non-traditional cameras is much more challenging than reading directly from a camera sensor. We are inspired by the classic work of Torralba and Freeman [30], which uses shadows within a room to recover a view of the world outside. By observing changes in the indirect illumination within a room, it is then possible to infer properties such as the motion of people outside of the camera frame [4,29]. Using specular objects as distorted mirrors, Park et al. [24] recover the environment by looking at an RGB-D sequence of a shiny input object, while Zhang et al. [42] decode images from the reflection pattern of randomly oriented small reflective planes comprised from glitter particles. Information about scenes is also inadvertently contained in sparse image descriptors, such as those from a Structure-from-Motion (SfM) point cloud, and can be used to render the scene from a novel viewpoint [25]. With respect to transparent or semi-transparent objects, decoding textures such as water droplets on a glass surface can reveal the structure of the room behind it [12], even if the glass is intentionally obscured using that texture [28]. These hidden cameras have serious implications towards privacy, but here we leverage totems as a hidden camera for an alternative purpose – verifying the integrity of possibly manipulated images using a multiview consistency check.

Detecting Image Manipulations. There are numerous ways to edit an image from its original state, warranting a large collection of works that identify artifacts left behind by various manipulation strategies. A number of manipulation pipelines involve modifying only part of an image (e.g. cut-and-paste operations and facial identity manipulations [7,27]), and thus detection approaches can either directly identify the blending [15] or warping artifacts [17,32] or leverage consistency checks between different parts of the image to locate the modified region [8,11,20,35,43,44]. Our approach intends to build the consistency check into captured image, rather than using a learned pipeline. Other cues for detection include subtle traces left behind by the camera or postprocessing procedures [2,13,26], or human biometric signals [3,16,36]. With the rise of image synthesis techniques and image manipulation using deep neural networks, it has been shown that the architectures of these networks also leave detectable traces [19,33,37,41], and that image generators can also reflect signatures embedded in the training data [38]. Another way to verify image integrity is to assume that we have multiple viewpoints of the same scene, captured at the same time; then detecting inconsistencies among these different viewpoints can signal potential manipulations [31,40]. Our setup is most similar to these latter approaches, but we relax the assumption of having multiple cameras and instead use refractive totems to obtain multiple projections of the scene within a single image. Moreover, using irregular totems as distorted lenses may further increase the difficulty of successful manipulations.

Digitial Signatures, Cryptography, and Physical One-Way Functions.
Our general Totem framework (see Sect. 3.2) is conceptually similar to digital
signature schemes in cryptography [6]. Both ideas add a certificate/signature to a
message, which is then used by the recipient to verify message integrity. However,
as mentioned in the introduction, totems are physical objects that give control
to the subject (rather than photographer/transmitter) and fill the "analog hole"
[10]. Additionally, unlike digital signatures, they are not restricted to a particular
sender and thus does not verify sender identity. This paper focuses particularly
on an instantiation of this general framework in the visual domain, where views
via distorted lenses are assumed to be hard to manipulate. Also utilizing physical
behaviors of complex material, Pappu et al. [23] showed promising results in
creating physical processes with cryptographical properties, termed physical one-
way functions, for their easiness to evaluate and hardness to invert. While such
processes are complex and not readily suited for image verification, they may
have potential implications in future totem geometry designs.

3   The Totem Verification Framework

3.1   Physical Totems for Image Verification

In image manipulation, the adversary modifies the content of a single image
with the intent that a viewer would infer a different scene than the one originally
depicted by the unmodified image. For instance, the perpetrator of a crime might
edit themselves out of a photo of the crime scene.

Easy access to modern image-processing software and deep image manipulations has significantly lowered the barrier to making such realistic attacks on a
single image [1,7]. Is there really no hope to defend against such attacks and protect the integrity of photographs we have taken? In this framework, we propose
a potential solution.

We argue that a single camera image is a rather vulnerable format as it only
represents *one* view of the underlying scene that we want to verify. Moreover,
the camera's mapping from scene to the image is well-understood, so it is rather
easy to infer the scene and edit the image. But what if the defender also receives
other views of the same scene, where the lenses are customized such that the
mappings from scene to such views are only known to the defender? Indeed, it
would be harder for an adversary to provide such a set of views that *consistently*
represent a different scene.

After all, simply obtaining multiple photos of the same scene is no easy
feat, often requiring coordination among multiple cameras, let alone requiring
custom lenses. In comparison, most current image-hosting services only ask for a
single-view image, which can be captured by everyday devices such as cellphones.
Therefore, we desire a set-up where:

- The content includes various distorted views of the scene from different spatial
  locations and angles,
- The distortions (*i.e.*, lens properties) are only known to the defender,

- The process does not require significant equipment investment, high skill, or an obscure content format.

Essentially, such a mechanism can be accessible to common users who create and upload visual content without adding much complications to their workflow.

Towards these goals, we propose placing small refractive objects in the scene and capturing the image as usual. The appearance of these objects in the camera image essentially forms small lenses of the same scene from different locations and angles (see Fig. 1). Such objects can be custom designed and mass-produced with simple materials (*e.g.*, glass) to be widely and cheaply available. Therefore, with the same imaging devices and file format, the uploaded image now itself contains multiple (distorted) views of the same scene. Unlike traditional digital signatures in which the photographer generates a certificate as a post hoc procedure, we use a single certificate (*e.g.*, totems) – often owned by the subject or defender – that is used across all photos, giving an active control to the subject that is getting photographed.

With exclusive knowledge of these objects' physical properties, the defender may then extract such multiple views and check if they, and the rest of the normal camera view, can form a consistent 3D scene. The physical laws and specific properties of these objects place various constraints on these views. The defender may check specific constraints, or attempt a multiview 3D scene reconstruction, as we explore in Sect. 4. Notably, such procedures are not available to adversaries who do not have access to the detailed object properties.

3.2   The General Totem Framework

The above image verification procedure is an instance of a more general approach (Fig. 2). There, we assumed that physical rendering through custom "lenses" is a process that is hard to manipulate but easy to verify. In general, similar processes can be used for verifying the integrity of various data modalities.

Setting. A true signal x (*e.g.*, a 3D scene) is conveyed via a compressed format $y = f(x)$ (*e.g.*, a 2D image). An adversary may manipulate $y \to y'$ such that receiver of y' believes that it represents some different signal x' with $y' = f(x')$ (*e.g.*, editing an image to depict a different scene).

Defense. To detect such attacks, a defender provides a *totem* t and requires every submission of $y = f(x)$ to be accompanied with $f_t(x)$ as a certificate of x generated with t (*e.g.*, totem views of the scene). Now, an adversarial attack will have to manipulate $(f(x), f_t(x))$ to $(f(x'), f_t(x'))$ (without having x'). This would be a much harder task if f_t is sufficiently complex, because the adversary now have to forge a consistent pair $(f(x'), f_t(x'))$, without knowledge of f_t's internal logic. On the other hand, the defender, with access to detailed knowledge of t and f_t, can verify if they have received a consistent pair. The extended submission $(f(x), f_t(x))$ ideally should be easy to create (with x and t) and not require much more overhead (than just $f(x)$).

The specific totems (and corresponding method of verification) depend on the type of signal. In the present paper, we focus on visual signals, where we can use the physical laws of rendering/imaging. Similarly, distorted audio reflectors could be used as totems for protecting recorded speech. Informally,

1. Consistency should be easy to verify and difficult to fake (described above);
2. The totem-generated certificate $f_t(x)$ should be impacted by as many properties of the true signal x as possible, so that a large portion of x is protected.

Relation with Cryptography. The Totem framework has a cryptographical flavor, where the totem represents a function that is easy for the defender to work with (*e.g.*, can invert), but hard to manipulate for the adversary. In fact, when each user is given a special totem, our framework is conceptually similar to cryptographical digital signature schemes, where a message is always sent together with a signature, used to verify message integrity and sender identity (*e.g.*, the signature can be private-key encoding). However, unlike digital signatures, we need not design user-specific totems, and a totem holder can protect their signals even when a third party captures and transmits them.

4   Method

For verification with the Totem framework, geometric consistency can be checked in various ways. Here we describe a specific procedure we use in this work, which verifies 3D consistency via scene reconstruction with neural radiance fields [21]. Specifically, the proposed method verifies the geometric consistency of a totem-protected image with the following 2 steps: (1) reconstructing the camera viewpoint from the provided totem views and (2) running a patch-wise comparison between the image and the reconstruction.

Here we focus on spherical totems, which demonstrate the potential of this framework and avoid costly manufacturing of geometrically-complex totems. However, the method is not fundamentally limited in these aspects, as we discuss in details in Sect. 5.3.

4.1   Scene Reconstruction from Totem Views

Image Formation Process. A totem-protected image is composed of image pixels $f(x)$ and distorted totem pixels $f_t(x)$. Image pixels capture the scene light rays directly passing through the camera optical system and display the scene as it appears to the naked eye. For totem pixels, the scene light rays first scatter through the refractive totems and then pass through the camera, implying that rays corresponding to two neighboring totem pixels may come from drastically different parts of the scene, depending on the complexity of the totem surface geometry. Regarding scene reconstruction, while traditional stereo methods suffice for simple totem views (*e.g.*, radial distortion), they do not generalize to more distorted totem views. For this reason, we choose to model the scene as a neural radiance field [21] and reconstruct by rendering from the camera viewpoint (Fig. 4).

A radiance field F_Θ represents a scene with a 5-dimensional plenoptic function that queries a spatial location and viewing direction and outputs its radiance and density. The color of an image pixel is rendered by querying 3D points along the corresponding scene light ray and computing the expected radiance given a distribution based on point density and occlusion. The mapping from image pixels to scene rays is simply the pinhole model, but for totem pixels, we need to compute the refracted ray which depends on totem-to-camera pose and totem properties such as 3D shape and index of refraction (IoR). With access to the totem 3D shape, we first register totem pose, obtain surface normals, and compute the mapping analytically. For simplicity, we ignore reflections on the totems and assume that the refraction process does not change light intensity.

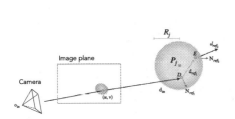
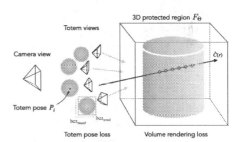

Fig. 3. Camera-ray refraction: Using totem's geometrical properties, we compute the resultant ray direction of an image pixel that passes through the totem. For a spherical totem, this involves two refractions dependent on totem pose P_j.

Fig. 4. Reconstruction pipeline: Using the refracted ray directions from the totems, we learn a radiance field F_Θ to reconstruct the 3D scene. We jointly optimize the totem positions P_j with radiance field to improve the scene reconstruction.

Pixel-to-Ray Mapping. Given an image $\mathcal{I}$ and assuming a set of spherical totems $\mathcal{J}$ indexed $j = \{1 \ldots |\mathcal{J}|\}$, with center positions P_j relative to the camera, radii R_j, and IoR n_j, and the index of refraction of air $n_{air} = 1$, we first compute the mapping from a totem pixel in the image $\mathcal{I}_{u,v}$ to the scene light ray corresponding to refraction through the totem $\mathbf{r}_{out} = \mathbf{o}_{out} + \mathbf{d}_{out} * t$, where $\mathbf{o}_{out}$ is the ray origin and $\mathbf{d}_{out}$ the ray direction.

Following Fig. 3, we begin with a ray $\mathbf{r}_{in} = \mathbf{o}_{in} + \mathbf{d}_{in} * t$ corresponding to pixel (u, v) in the image, and compute point D the first intersection of $\mathbf{r}_{in}$ with the totem, $\mathbf{N}_{ref_1}$ the surface normal at D, and $\mathbf{d}_{ref_1}$ the refracted direction:

$$D = \mathsf{intersect}(P_j, R_j, \mathbf{d}_{in}, \mathbf{o}_{in}), \tag{1}$$

$$\mathbf{N}_{ref_1} = \overrightarrow{P_j D}/\|\overrightarrow{P_j D}\|_2, \tag{2}$$

$$\mathbf{d}_{ref_1} = \mathsf{refract}(n_j, n_{air}, \mathbf{N}_{ref_1}, \mathbf{d}_{in}). \tag{3}$$

Next, we compute the second intersection point E where the ray exits the totem, corresponding surface normal $\mathbf{N}_{ref_2}$, and ray exit direction $\mathbf{d}_{ref_2}$:

$$E = \mathsf{intersect}(P_j, R_j, \mathbf{d}_{ref_1}, D), \tag{4}$$

$$\mathbf{N}_{ref_2} = \overrightarrow{P_jE}/|\overrightarrow{P_jE}\|_2, \tag{5}$$

$$\mathbf{d}_{ref_2} = \mathsf{refract}(n_i, n_{air}, \mathbf{N}_{ref_2}, \mathbf{d}_{ref_1}). \tag{6}$$

We provide the formulas for intersect and refract in supplementary material. We obtain the resulting ray direction $\mathbf{r}_{out} = \mathbf{o}_{out} + \mathbf{d}_{out} * t$ with $\mathbf{o}_{out} = E$ and $\mathbf{d}_{out} = \mathbf{d}_{ref_2}$. $\mathbf{r}_{out}$ is also refered to as $\mathbf{r}$ below.

Joint Optimization of Radiance Field and Totem Position. A key part of the totem framework is determining the positions of the totem centers P_j relative to the camera. While we know P_j in simulator settings, it is necessary to estimate the totem positions in order to operate on real-world images. We assume that a binary mask M_j for each totem is known, which could be annotated by the defender. We first initialize each P_j by projecting the boundary pixels of the annotated binary mask into camera rays, forming a cone. Using the known totem radius R_j, we derive the totem's initial position by fitting a circle corresponding to the intersection of the cone and the spherical totem (details in supplementary).

We then jointly optimize the neural radiance function along with the totem position P_j. We use a photometric loss on F_Θ to reconstruct the color $\mathbf{C}(\mathbf{r})$ of totem pixels in the image corresponding to refracted totem rays $\mathbf{r}$ in batch $\mathcal{R}$ (see [21] for construction of predicted color $\hat{\mathbf{C}}(\mathbf{r})$):

$$\mathcal{L}_{rec} = \sum_{\mathbf{r} \in \mathcal{R}} \left\| \hat{\mathbf{C}}(\mathbf{r}) - \mathbf{C}(\mathbf{r}) \right\|_2^2. \tag{7}$$

However, as minimizing per-pixel loss can cause P_j to deviate far from the initial totem masks, we additionally regularize the optimized P_j using an IoU loss. Given the current totem positions P_j and radius R_j, we again compute the circle formed by the intersection of the totem with the cone of camera rays (see supplementary material for exact derivations). We sample a set of 3D points lying along this circle $\mathcal{X} = \{X_1 \ldots X_n\}$ and project them to 2D image coordinates using camera intrinsics K and depth d_i:

$$(u_i, v_i, 1) = \frac{KX_i}{d_i}. \tag{8}$$

We compute the bounding box of these projected 2D image coordinates:

$$\mathrm{box}_{pred} = \left(\min_i(u_i),\ \max_i(u_i),\ \min_i(v_i),\ \max_i(v_i) \right). \tag{9}$$

We then apply the IoU loss (Jaccard index [9]) over box_{pred} and box_{mask}, the bounding box from the totem binary masks M_j, with the overall loss objective:

$$\mathcal{L} = \lambda * \mathcal{L}_{rec} + \mathcal{L}_{IoU}, \tag{10}$$

where we use $\lambda = 10$. See additional training details in the supplementary.

4.2   Manipulation Detection

As the totems do not uniformly cover the entire scene observed by the camera, we first construct a confidence map of the region intersected by multiple totems. For each totem ray $\mathbf{r}_{out}$, we sample points along the ray $\mathbf{r}_{out}(t) = \mathbf{o}_{out} + \mathbf{d}_{out} * t$ and query F_Θ to obtain their weight contribution $w(\mathbf{r}_{out}(t))$ to the resultant color. We then construct a 3D point cloud where each point X_p is the point with the highest weight along a totem ray:

$$X_p = \mathbf{r}_{out}(\arg\max_t w\,(\mathbf{r}_{out}(t))). \tag{11}$$

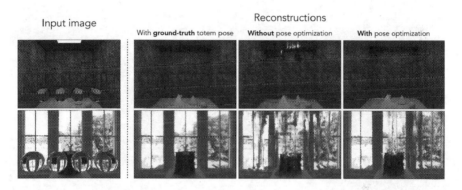

Fig. 5. **Pose optimization on synthetic data:** We start with a setup of spherical totems placed in a simulated environment. We find that four totems is sufficient to recover the scene when the totem pose is accurate; i.e., the ground-truth totem pose used to render the scene. Using the initial estimated totem position leads to artifacts in the reconstructed camera viewpoint, while allowing the totem poses to update while learning the scene representation improves the reconstruction.

We project these points to 2D using Eq. 8. We accumulate the number of points from the point cloud that project to each pixel, apply a box filter of width 30 pixels, and threshold boxes with more than 10% accumulated points. We take a convex hull around this thresholded region and call it the *protected region*. Intuitively, this identifies the part of the scene that is adequately visible within the totems.

Within the protected region, we generate a heatmap for potential manipulations by comparing the scene reconstructed using the totems and F_Θ to the pixels visible in the image $\mathcal{I}$. We use a patch-wise L1 error metric:

$$\mathsf{L}_1(i,j) = \sum_{\substack{|k_i|<K \\ |k_j|<K}} |\mathcal{I}(i+k_i, j+k_j) - \hat{\mathbf{C}}(i+k_i, j+k_j)|, \tag{12}$$

with patch size $K = 64$, where $\hat{\mathbf{C}}(i,j)$ refers to the color along the camera ray corresponding to pixel (i,j). In addition to patch-wise L1, we also use LPIPS, a learned perceptual patch similarity metric [39], on the same patch size.

5   Results

5.1   Data Collection

Synthetic Images. We first demonstrate our method in a simulated setting, where we know all ground-truth information about the totems and camera. We generate the data with Mitsuba2 [22], a differentiable rendering system. We try two settings: (1) we set up a room similar to a Cornell box with random wallpapers and random geometric objects on the floor (2) we generate the room using an environment map (Fig. 5-left). We then place refractive spheres in between the camera and the scene to form the totem views.

Fig. 6. Reconstruction results on real images: For *real images*, we do not know the ground-truth totem positions and must rely on our position estimates. We find that jointly optimizing the pose of the totems together with the scene reconstruction better recovers the scene geometry and obtains a closer match to the input image than using the initial totem pose estimates. We conduct reconstruction on indoor and outdoor scenes under a range of lighting conditions.

Real Images. To demonstrate our framework in more realistic settings, we take pictures using a Canon EOS 5D Mark III camera and place four physical totems at arbitrary positions in front of the scene (Fig. 7-left). We obtain the totem size and IoR from manufacturer specifications and manually annotate the totem masks in the image. Camera intrinsic parameters are obtained via a calibration sequence. We correct radial distortion in the collected images to better approximate a pinhole camera model when computing refracted ray directions.

Image Manipulations. We conduct manipulations to create inconsistencies between the observed scene and the totem views. We locally modify the image by inserting randomly colored patches, adding people by image splicing, removing people with Photoshop Content Aware Fill (CAF), or shifting people in both image and totems to the same reference position. Note that we do not consider manipulations where totems are entirely removed; in that case we consider the image no longer verifiable.

5.2   Decoding the Scene from Totem Views

Reconstructing a Simulated Scene. We conduct initial experiments in simulated scenes to validate components of our learning framework when ground-truth totem parameters are known. Figure 5 shows the reconstruction of the scene using (1) the known totem positions for reconstruction as the oracle (2) only the initial estimate of totem positions derived from annotated totem masks, and (3) jointly estimating the totem position and scene radiance field. We find that small changes in the totem position greatly impact the reconstruction quality; therefore relying on the initial totem position estimate alone leads to sub-optimal reconstruction of the camera viewpoint. The reconstruction improves when allowing the totem positions to update during learning, while using the oracle ground truth totem position obtains the best reconstruction (Table 1). However, ground truth totem positions are only available in simulators, so we must estimate these positions when using real images. In supplementary material, we conduct additional experiments on the number of totems required to

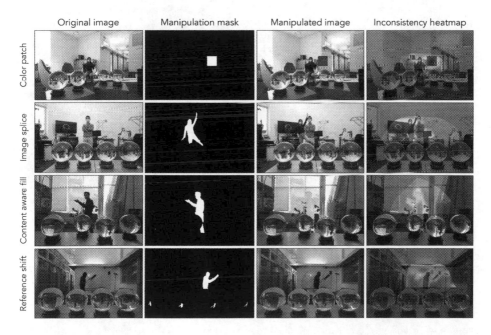

Fig. 7. Detection results: We manipulate scenes by adding random color patches, inserting people with image splicing, removing people with Photoshop (CAF), and shifting people in both camera and totem views to the same reference point (*e.g.* right edge of the wooden table). By comparing the manipulated image and the scene reconstruction, we obtain an inconsistency heatmap over regions of possible manipulation.

reconstruct the scene and find empirically that four totems leads to a reasonable balance between reconstruction quality and visibility of the scene.

Reconstructing Scenes from Real Images. Similar to the simulated environment, we set up four totems in a room in front of the subject, and jointly optimize for reconstruction and totem position. We find that the joint optimization procedure yields a better scene reconstruction when viewed from the camera (Fig. 6). On un-manipulated images, joint optimization decreases the L1 error of the reconstructed scene to the ground truth from 0.15 to 0.11 (Table 1).

Table 1. Camera view reconstruction comparisons: We measure L1 and LPIPS [39] distance of the camera view reconstruction using the learned scene representation.

Dataset	Totem Optimization	Reconstruction		Pose
		L1	LPIPS	L1
Box	✗	0.057	0.658	**0.008**
Box	✓	0.054	0.645	0.108
Box	Oracle	**0.047**	**0.625**	–
Env map	✗	0.173	0.617	0.060
Env map	✓	0.103	0.520	**0.027**
Env map	Oracle	**0.040**	**0.476**	–
Real	✗	0.149	0.644	–
Real	✓	**0.109**	**0.586**	–

Table 2. Detection comparisons: Patch-wise mAP on various image manipulations. Compared to [11] and [35], our method is based on geometric reconstructions and is therefore robust to different manipulation types and image processing.

Method			CAF	Splice	Color
Self-consistency [11]			0.037	0.037	0.801
ManTra-Net [35]			0.151	0.295	0.181
Ours w/o totem opt.	+L1		0.485	0.401	0.944
	+LPIPS		0.489	0.449	0.954
Ours with totem opt.	+L1		0.554	0.638	**0.961**
	+LPIPS		**0.666**	**0.739**	0.946

Detecting Image Manipulations. We next investigate the ability to detect manipulations after reconstructing the scene from only the totem views. We experiment with a patch-wise L1 distance metric (Eq. 12) and a perceptual distance metric [39] to measure the difference between the camera viewpoint and the scene reconstruction, yielding a heatmap over the potentially manipulated area. Qualitative examples are shown in Fig. 7 and we quantify the detection performance in Table 2 by normalizing the heatmap and computing average precision over these patches. While our method relies on 3D geometric consistency obtained from a single image, we compare to an image splice detection method [11], but we note that such learning-based methods tend to fail on setups outside of the training distribution and where manipulations involve parts of two images with the same camera metadata. We also compare to Wu et al. [35] with downsampled images due to GPU memory explosion. The low final mAP is partly due to the method's sensitivity to the exact compression artifacts, which can be affected by any processing (e.g., resizing).

5.3   Potential Avenues for a More General Method

In our current method as described in Sect. 4, we make several simplifying assumptions about the totems and scene. These assumptions are not fundamental limitations of the framework, as they can potentially be addressed by, for example, high-precision totem manufacture/measurement, advanced reconstruction method not requiring known camera intrinsics and/or robust to more diverse totem placements, etc. We briefly discuss these limitations below. Please refer to the FAQ in the supplementary materials for more discussions.

Reconstruction. For high reconstruction quality, the current method is best suited when scene is clearly visible in multiple totem views, so that the neural radiance field fitting has more training samples (pixels) and is more stable. Therefore, detection is more difficult when totem placements only show the manipulated parts in a rather small view (e.g., totems far away from camera) or a highly distorted view (e.g., totems near the sides of the image). See the supplementary material for examples and analysis on totem placement (including number of totems and totem positions). Based on neural radiance fields, the reconstruction is also not fast enough for real-time detection and assumes known camera intrinsics. As scene reconstruction research continues to develop, we believe these limitations will be become much less relevant.

Totem Design. Our experiments use spherical totems. They can be readily purchased online and have known geometry, which allow us to much more easily experiment and analyze this new verification strategy. However, our verification framework, specifically the reconstruction component, is not limited to this one geometry. It directly generalizes to various totem designs as long as the totem geometry and physical properties can be computed/measured. For example, Zhang et al. [42] demonstrates reconstructions under a complex scrambling of light rays once the input-output ray mapping is known.

Much more can be explored in designing totems for ease to use and effectiveness towards forensics tasks. For example, totems can be made more compact, and therefore more portable and less visible to the adversary (totem identities are unknown to the adversary a priori). Totems with complex geometric design exhibit less interpretable distortion patterns and can contain multiple distorted views of the scene simultaneously, which makes it more difficult to achieve geometrically-consistent manipulation and reduces the number of totems required during scene setup.

6   Conclusion

We design a framework for verifying image integrity by placing physical *totems* into the scene, thus encrypting the scene content as a function of the totem geometry and material. By comparing the scene viewed from the camera to the distorted versions of the scene visible from the totems, we can identify the presence of image manipulations from a single reference image. Our approach decodes the distorted totem views by first estimating the totem positions, computing the refracted ray directions, and using the resultant rays to fit a scene radiance field. Furthermore, we show that it is possible to fit this 3D scene representation using sparse totem views, and that jointly optimizing the totem positions and the scene representation improves the reconstruction result. While we assume spherical totems in this work, an avenue for future exploration would be to extend the approach to more complex totems such as those with more complex shapes or randomly oriented microfacets, thus creating a stronger encryption function.

Acknowledgments. This work was supported by Meta AI. We thank Wei-Chiu Ma, Gabe Margolis, Yen-Chen Lin, Xavier Puig, Ching-Yao Chuang, Tao Chen, and Hyojin Bahng for helping us with data collection. LC is supported by the National Science Foundation Graduate Research Fellowship under Grant No. 1745302 and Adobe Research Fellowship.

References

1. Adobe Inc.: Adobe Photoshop. www.adobe.com/products/photoshop.html
2. Agarwal, S., Farid, H.: Photo forensics from jpeg dimples. In: 2017 IEEE Workshop on Information Forensics and Security (WIFS), pp. 1–6. IEEE (2017)
3. Agarwal, S., Farid, H., Gu, Y., He, M., Nagano, K., Li, H.: Protecting world leaders against deep fakes. In: CVPR workshops. vol. 1 (2019)
4. Aittala, M., et al.: Computational mirrors: blind inverse light transport by deep matrix factorization. arXiv preprint arXiv:1912.02314 (2019)
5. Ben-Ezra, M., Nayar, S.: What does motion reveal about transparency? In: Proceedings Ninth IEEE International Conference on Computer Vision, vol. 2, pp. 1025–1032 (2003)
6. Diffie, W., Hellman, M.: New directions in cryptography. IEEE Trans. Inf. Theory **22**(6), 644–654 (1976)
7. Dolhansky, B., Howes, R., Pflaum, B., Baram, N., Ferrer, C.C.: The deepfake detection challenge (dfdc) preview dataset. arXiv preprint arXiv:1910.08854 (2019)
8. Fu, H., Cao, X.: Forgery authentication in extreme wide-angle lens using distortion cue and fake saliency map. IEEE Trans. Inf. Forensics Secur. **7**, 1301–1314 (2012). https://doi.org/10.1109/TIFS.2012.2195492
9. Girshick, R., Donahue, J., Darrell, T., Malik, J.: Rich feature hierarchies for accurate object detection and semantic segmentation. In: Proceedings of the IEEE Conference on Computer Vision and Pattern Recognition, pp. 580–587 (2014)
10. Haber, S., Horne, B., Pato, J., Sander, T., Tarjan, R.E.: If piracy is the problem, is DRM the answer? In: Becker, E., Buhse, W., Günnewig, D., Rump, N. (eds.) Digital Rights Management. LNCS, vol. 2770, pp. 224–233. Springer, Heidelberg (2003). https://doi.org/10.1007/10941270_15
11. Huh, M., Liu, A., Owens, A., Efros, A.A.: Fighting fake news: Image splice detection via learned self-consistency (2018)
12. Iseringhausen, J., et al.: 4D imaging through spray-on optics. ACM Trans. Graph. (Proc. SIGGRAPH 2017) **36**(4), 35:1–35:11 (2017)
13. Johnson, M.K., Farid, H.: Exposing digital forgeries through chromatic aberration. In: Proceedings of the 8th Workshop on Multimedia and Security, pp. 48–55 (2006)
14. Kutulakos, K., Steger, E.: A theory of refractive and specular 3d shape by light-path triangulation, vol. 76, pp. 1448–1455, January 2005. https://doi.org/10.1109/ICCV.2005.26

15. Li, L., et al.: Face x-ray for more general face forgery detection. In: Proceedings of the IEEE/CVF Conference on Computer Vision and Pattern Recognition, pp. 5001–5010 (2020)
16. Li, Y., Chang, M.C., Lyu, S.: In ictu oculi: Exposing ai created fake videos by detecting eye blinking. In: 2018 IEEE International Workshop on Information Forensics and Security (WIFS), pp. 1–7. IEEE (2018)
17. Li, Y., Lyu, S.: Exposing deepfake videos by detecting face warping artifacts. arXiv preprint arXiv:1811.00656 (2018)
18. Li, Z., Yeh, Y.Y., Chandraker, M.: Through the looking glass: Neural 3d reconstruction of transparent shapes. In: Proceedings of the IEEE/CVF Conference on Computer Vision and Pattern Recognition, pp. 1262–1271 (2020)
19. Marra, F., Gragnaniello, D., Verdoliva, L., Poggi, G.: Do gans leave artificial fingerprints? In: 2019 IEEE Conference on Multimedia Information Processing and Retrieval (MIPR), pp. 506–511. IEEE (2019)
20. Mayer, O., Stamm, M.C.: Exposing fake images with forensic similarity graphs. IEEE J. Sel. Top. Sig. Process. **14**(5), 1049–1064 (2020)
21. Mildenhall, B., Srinivasan, P.P., Tancik, M., Barron, J.T., Ramamoorthi, R., Ng, R.: Nerf: representing scenes as neural radiance fields for view synthesis (2020)
22. Nimier-David, M., Vicini, D., Zeltner, T., Jakob, W.: Mitsuba 2: a retargetable forward and inverse renderer. ACM Trans. Graph. (TOG) **38**(6), 1–17 (2019)
23. Pappu, R., Recht, B., Taylor, J., Gershenfeld, N.: Physical one-way functions. Science **297**(5589), 2026–2030 (2002)
24. Park, J.J., Holynski, A., Seitz, S.M.: Seeing the world in a bag of chips. In: Proceedings of the IEEE/CVF Conference on Computer Vision and Pattern Recognition, pp. 1417–1427 (2020)
25. Pittaluga, F., Koppal, S.J., Kang, S.B., Sinha, S.N.: Revealing scenes by inverting structure from motion reconstructions. In: Proceedings of the IEEE/CVF Conference on Computer Vision and Pattern Recognition, pp. 145–154 (2019)
26. Popescu, A.C., Farid, H.: Exposing digital forgeries in color filter array interpolated images. IEEE Trans. Signal Process. **53**(10), 3948–3959 (2005)
27. Rossler, A., Cozzolino, D., Verdoliva, L., Riess, C., Thies, J., Nießner, M.: Faceforensics++: learning to detect manipulated facial images. In: Proceedings of the IEEE/CVF International Conference on Computer Vision, pp. 1–11 (2019)
28. Shan, Q., Curless, B., Kohno, T.: Seeing through obscure glass. In: Daniilidis, K., Maragos, P., Paragios, N. (eds.) ECCV 2010. LNCS, vol. 6316, pp. 364–378. Springer, Heidelberg (2010). https://doi.org/10.1007/978-3-642-15567-3_27
29. Sharma, P., et al.: What you can learn by staring at a blank wall. In: Proceedings of the IEEE/CVF International Conference on Computer Vision, pp. 2330–2339 (2021)
30. Torralba, A., Freeman, W.T.: Accidental pinhole and pinspeck cameras: Revealing the scene outside the picture. In: 2012 IEEE Conference on Computer Vision and Pattern Recognition, pp. 374–381. IEEE (2012)
31. Tursman, E., George, M., Kamara, S., Tompkin, J.: Towards untrusted social video verification to combat deepfakes via face geometry consistency. In: Proceedings of the IEEE/CVF Conference on Computer Vision and Pattern Recognition Workshops, pp. 654–655 (2020)
32. Wang, S.Y., Wang, O., Owens, A., Zhang, R., Efros, A.A.: Detecting photoshopped faces by scripting photoshop (2019)
33. Wang, S.Y., Wang, O., Zhang, R., Owens, A., Efros, A.A.: CNN-generated images are surprisingly easy to spot... for now (2020)

34. Wexler, Y., Fitzgibbon, A., Zisserman, A.: Image-based environment matting. In: SIGGRAPH 2002 (2002)
35. Wu, Y., AbdAlmageed, W., Natarajan, P.: Mantra-net: Manipulation tracing network for detection and localization of image forgeries with anomalous features. In: CVPR (2019)
36. Yang, X., Li, Y., Lyu, S.: Exposing deep fakes using inconsistent head poses. In: ICASSP 2019–2019 IEEE International Conference on Acoustics, Speech and Signal Processing (ICASSP), pp. 8261–8265. IEEE (2019)
37. Yu, N., Davis, L.S., Fritz, M.: Attributing fake images to gans: learning and analyzing gan fingerprints. In: Proceedings of the IEEE/CVF International Conference on Computer Vision, pp. 7556–7566 (2019)
38. Yu, N., Skripniuk, V., Abdelnabi, S., Fritz, M.: Artificial fingerprinting for generative models: rooting deepfake attribution in training data. In: Proceedings of the IEEE/CVF International Conference on Computer Vision, pp. 14448–14457 (2021)
39. Zhang, R., Isola, P., Efros, A.A., Shechtman, E., Wang, O.: The unreasonable effectiveness of deep features as a perceptual metric. In: Proceedings of the IEEE Conference on Computer Vision and Pattern Recognition, pp. 586–595 (2018)
40. Zhang, W., Cao, X., Qu, Y., Hou, Y., Zhao, H., Zhang, C.: Detecting and extracting the photo composites using planar homography and graph cut. IEEE Trans. Inf. Forensics Secur. **5**, 544–555 (2010). https://doi.org/10.1109/TIFS.2010.2051666
41. Zhang, X., Karaman, S., Chang, S.F.: Detecting and simulating artifacts in gan fake images. In: WIFS (2019)
42. Zhang, Z., Isola, P., Adelson, E.H.: Sparklevision: Seeing the world through random specular microfacets. In: Proceedings of the IEEE Conference on Computer Vision and Pattern Recognition Workshops, pp. 1–9 (2015)
43. Zhao, T., Xu, X., Xu, M., Ding, H., Xiong, Y., Xia, W.: Learning self-consistency for deepfake detection. In: Proceedings of the IEEE/CVF International Conference on Computer Vision, pp. 15023–15033 (2021)
44. Zhou, P., Han, X., Morariu, V.I., Davis, L.S.: Two-stream neural networks for tampered face detection. In: 2017 IEEE Conference on Computer Vision and Pattern Recognition Workshops (CVPRW), pp. 1831–1839. IEEE (2017)
45. Zongker, D., Werner, D., Curless, B., Salesin, D.: Environment matting and compositing. In: Proceedings of SIGGRAPH 1999, pp. 205–214, June 1999. https://doi.org/10.1145/311535.311558

Dual-Stream Knowledge-Preserving Hashing for Unsupervised Video Retrieval

Pandeng Li[1] , Hongtao Xie[1](✉) , Jiannan Ge[1] , Lei Zhang[2] ,
Shaobo Min[3] , and Yongdong Zhang[1]

[1] University of Science and Technology of China, Hefei, China
{lpd,gejn}@mail.ustc.edu.cn, {htxie,zhyd73}@ustc.edu.cn
[2] Kuaishou Technology, Beijing, China
zhanglei06@kuaishou.com
[3] Tencent Data Platform, Shenzhen, China
bobmin@tencent.com

Abstract. Unsupervised video hashing usually optimizes binary codes by learning to reconstruct input videos. Such reconstruction constraint spends much effort on frame-level temporal context changes without focusing on video-level global semantics that are more useful for retrieval. Hence, we address this problem by decomposing video information into reconstruction-dependent and semantic-dependent information, which disentangles the semantic extraction from reconstruction constraint. Specifically, we first design a simple dual-stream structure, including a temporal layer and a hash layer. Then, with the help of semantic similarity knowledge obtained from self-supervision, the hash layer learns to capture information for semantic retrieval, while the temporal layer learns to capture the information for reconstruction. In this way, the model naturally preserves the disentangled semantics into binary codes. Validated by comprehensive experiments, our method consistently outperforms the state-of-the-arts on three video benchmarks.

Keywords: Unsupervised video retrieval · Dual-stream hashing

1 Introduction

In view of the explosive growth of informative media (*i.e.,* videos) [24,29,44,46,57], the efficient large-scale retrieval system [3,11,19,52,53] has become an urgent requirement in the real world. Video retrieval system needs to understand the semantic similarity information implicit in videos [55], which can be found by comparing the real-valued features in the last layer of deep networks. Unfortunately, these massive amounts of features take up large storage space [6] and seriously affect the retrieval speed. As a key building block of search algorithms, hashing [12], can alleviate the above issue by compressing high dimensional features into compact binary codes. However, the abundant content and

---

Supplementary Information The online version contains supplementary material available at https://doi.org/10.1007/978-3-031-19781-9_11.

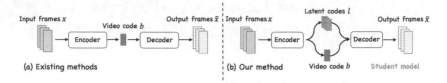

Fig. 1. (a) Existing methods usually optimize binary codes by using an encoder-decoder architecture to reconstruct the visual information of input frames. (b) Our method utilizes latent codes in the student model to model temporal changes required for the reconstruction task, thereby allowing binary codes to focus on global semantics.

Fig. 2. Essential and superfluous information for the semantic video retrieval task.

temporal dynamics of videos make it difficult for binary codes to preserve the similarity structure of the real-valued feature space [23,45]. Besides, compared to image datasets, the manual annotation and pre-training costs of standard large-scale video data are very high [10]. Therefore, unsupervised video hashing has intrigued many researchers in practice [21,22,37,56].

As shown in Fig. 1 (a), existing unsupervised video hashing methods usually optimize binary codes by using an encoder-decoder architecture to reconstruct the visual information of input frames. For example, Zhang *et al.* [56] employ an encoder-decoder Recurrent Neural Networks (RNNs) [33] to capture the temporal nature of videos for binary codes. Later, Li *et al.* [20] introduce Variational Auto-Encoders (VAE) [17] to learn a probabilistic latent code of video variations. However, these binary codes are forced to independently satisfy the goal of video reconstruction, which may be sub-optimal for semantic retrieval due to the heterogeneity of two tasks [13] (*i.e.*, the retrieval and reconstruction tasks).

Specifically, given raw videos as input, existing models tend to compress the information that is essential to reconstruction but may be superfluous for similarity search. This argument can be proved in Information Bottleneck (IB) [42] from an information-theoretic perspective. IB models the information flow [50] from input x to the target $\tilde{x}$ through latent variable b (*e.g.*, binary codes), where the optimal b should contain the minimal sufficient information to predict $\tilde{x}$ but discards all superfluous information in x that is irrelevant for $\tilde{x}$. This provides an optimization principle that maximizes the mutual information $I(b; \tilde{x})$ between the latent variable and the target, and simultaneously constrains $I(x; b)$ small. In the above existing hashing methods, maximizing $I(b; \tilde{x})$ corresponds to minimizing the reconstruction error. The reconstruction-essential information [31] may be the sequence of actions, constant changes, etc., which involves more fine-grained temporal understanding. However, as shown in Fig. 2, binary codes require more attention to global semantic concepts like "biking" or "cat" for ranking videos. Because, the retrieval goal is not to retain all information of the original video data, but to preserve the discriminative similarity information.

Based on the above discussion, we propose a novel Dual-stream Knowledge-Preserving Hashing (DKPH) framework to obtain semantic binary codes by decomposing video information into semantic-dependent and reconstruction-dependent information. As shown in Fig. 3, DKPH fully releases the potential of semantic learning via teacher-student optimization: (1) the student model designs a simple but effective dual-stream structure to disentangle the semantic extraction from reconstruction constraint on a single binary code; (2) the teacher model refines the semantic similarity knowledge to further guide the meaningful information decomposition in the student model.

More concretely, the dual-stream structure contains a parallel temporal layer and hash layer. The temporal layer tries to capture reconstruction-dependent information by learning dynamic frame-level features, while the hash layer focuses on the semantic-dependent part from a global video-level perspective. To achieve the above goal, a teacher model is trained in a self-supervised manner to construct a Gaussian-adaptive similarity graph, which captures the inherent similarity relations between samples. This relation knowledge is preserved into the student hash layer to generate semantic-dependent discriminative binary codes.

Contributions. (1) We propose a novel framework, DKPH, to fully release the potential of semantic learning on binary codes and may shed critical insights for the retrieval community. To our best knowledge, our method is the first work that explores the task heterogeneity in video hashing. (2) A simple but effective dual-stream structure is developed to decompose video information, which can generate semantic-dependent discriminative binary codes by preserving the semantic similarity knowledge from the proposed Gaussian-adaptive similarity graph. (3) Extensive experiments demonstrate that DKPH outperforms state-of-the-art video hashing models on FCVID, ActivityNet and YFCC datasets.

2   Related Work

Unsupervised Hashing. Unsupervised hashing aims to learn hash functions that compress data points into binary codes, which are built on training data without manual annotations. Iterative quantization (ITQ) [12] is a traditional representative method that directly explores the minimum quantization error by learning an optimal rotation of principal component directions. However, non-deep image hashing methods only seek a single linear projection, resulting in poor generalization. Then, Deep Hashing (DH) [9] uses a deep neural network to learn binary codes via multiple hierarchical non-linear transformations.

Due to the explosive growth of short videos, some works [21,36] also focus on video hashing. Multiple Feature Hashing (MFH) [36] mines local structural information while ignoring inter-frame temporal consistency [51]. Later, a series of methods based on encoder-decoder structure have become mainstream methods for video hashing. For example, Self-Supervised Temporal Hashing (SSTH) [56] employs an encoder-decoder RNNs to capture the temporal nature of videos.

Li *et al.* [18] jointly model static visual appearance and temporal pattern into binary codes via two special reconstruction losses. Unsupervised Deep Video Hashing (UDVH) [45] emphasizes balancing dimensional variation for each binary representation. Self Supervised Video Hashing (SSVH) [37] attempts more powerful Bi-LSTM to model more granular inter-frame dependencies. Despite a similar network architecture to SSVH, Neighborhood Preserving Hashing (NPH) [21] encodes the neighborhood-dependent video content as a binary code. Bidirectional Transformer Hashing (BTH) [22] introduces the BERT architecture [7] in NLP to explore inter-frame correlations, and achieves excellent results. However, these video hashing methods fail to consider the heterogeneity between reconstruction and retrieval tasks for optimizing binary codes. Recently, Shen *et al.* [34] propose twin bottlenecks to extract continuous features, but the similarity optimization process for binary codes is still implicit and heavily depends on the reconstruction effects. Besides, more efficient sample relations have not been explored, which affects the semantic discriminative of binary codes.

Knowledge Distillation. [14,32] first propose to transfer knowledge from teacher models to student models through the soft outputs or intermediate layer features. Recently, Knowledge Distillation (KD) is extended to training deep networks in generations and [1,25] find that KD can refine ground truth labels. In unsupervised video hashing, to preserve and distill the semantic knowledge, we refine pre-trained CNN features to visual embeddings in the teacher model, which can further construct an efficient similarity graph for training student model.

3   Method

3.1   Problem Definition

We introduce some notations and the problem definition of unsupervised video hashing. Generally, learning hash functions is considered in an unsupervised manner from a training set of N video data points $\mathcal{V} = \{v_i\}_{i=1}^N \in \mathbb{R}^{N \times M \times D}$, where each $v_i = [x_1, \cdots, x_M] \in \mathbb{R}^{M \times D}$ is a CNN feature set, M is the number of frames, and D is the feature dimension of each frame. DKPH aims to learn nonlinear hash functions based on transformer blocks that map each video data point v_i into a K-dimensional Hamming space $b_i \in \{-1, 1\}^K$, which needs to keep relative semantic similarity between videos.

3.2   Network Overview

DKPH consists of a teacher model Ω^T and a student model Ω^S. As shown in Fig. 3, Ω^T is a common encoder-decoder architecture that can exchange inter-frame information through transformers to obtain long-term semantic knowledge. Ω^S is a dual-stream encoder-decoder architecture that can disentangle the semantic extraction and reconstruction constraint on a single binary code to better capture the semantic information transmitted by Ω^T. In this section, we introduce three key sub-networks: transformer encoder, hash layer and temporal layer, where the structure of transformer encoder is the same in Ω^T and Ω^S.

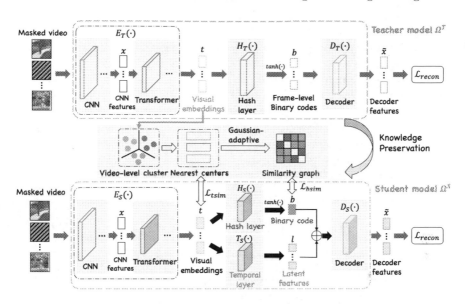

Fig. 3. The proposed DKPH framework which involves (1) training teacher model Ω^T and student model Ω^S in an unsupervised manner, (2) distilling semantic knowledge from Ω^T to guide the information decomposition in Ω^S.

Transformer Encoder. To model long-term semantic correlation in videos, we first employ transformer blocks to handle the pre-processing CNN frame features. Each transformer encoder block has a multi-head self-attention and a feed-forward layer. Different from splitting images into several tokens in ViT [8], we treat frame features $[\boldsymbol{x}_i^1, \cdots, \boldsymbol{x}_i^M]$ as token units, which contain rich visual content information. Besides, to learn the ordering information of each frame inside the original video, we follow the standard procedure in ViT by adding trainable positional encoding embeddings $\mathbf{E}_{pos}$. Thus, the input video matrix $\mathbf{X}_i$ is defined as follows:

$$\mathbf{X}_i = [\boldsymbol{x}_i^1, \cdots, \boldsymbol{x}_i^M] + \mathbf{E}_{pos}. \tag{1}$$

Given the input matrix $\mathbf{X}_i$, we calculate queries $\mathbf{Q}_i$, keys $\mathbf{K}_i$ and values $\mathbf{V}_i$ as follows: $\mathbf{Q}_i = \mathbf{X}_i \mathbf{W}_i^Q$, $\mathbf{K}_i = \mathbf{X}_i \mathbf{W}_i^K$, $\mathbf{V}_i = \mathbf{X}_i \mathbf{W}_i^V$, where $\mathbf{W}_i^Q$, $\mathbf{W}_i^K$ and $\mathbf{W}_i^V$ are linear projections with an output of d dimensions. Then the self-attention outputs can be calculated by

$$\mathrm{Att}(\mathbf{Q}_i, \mathbf{K}_i, \mathbf{V}_i) = \mathrm{softmax}\left(\mathbf{Q}_i \mathbf{K}_i^T / \sqrt{d}\right) \mathbf{V}_i^T. \tag{2}$$

Finally, these frame token units undergo multiple informative interactions, which are transformed into a sequence of visual embeddings $[\boldsymbol{t}_i^1, \cdots, \boldsymbol{t}_i^M]$.

Hash Layer. As shown in the teacher model Ω^T of Fig. 3, the intuitive approach [22] is to directly reduce the visual embedding dimension of each frame through linear mapping $[\hat{\boldsymbol{t}}_i^1, \cdots, \hat{\boldsymbol{t}}_i^M] = H_T\left([\boldsymbol{t}_i^1, \cdots, \boldsymbol{t}_i^M]\right) \in \mathbb{R}^{M \times K}$, and then binarize them to obtain frame-level binary codes $[\boldsymbol{b}_i^1, \cdots, \boldsymbol{b}_i^M] \in \{-1,1\}^{M \times K}$. However, according to the settings of existing methods [22] in the testing phase, Ω^T needs to average $[\boldsymbol{b}_i^1, \cdots, \boldsymbol{b}_i^M]$ to obtain a real-valued code, which is binarized to video-level binary code for retrieval. This leads to two issues: (1) there is a quantization error between real-valued codes and binary codes, resulting in a sub-optimal solution; (2) when $+1$ and -1 numbers of frame binary codes are the same, $\{-1, 0, 1\}^K$ may be generated, which violates the principle of hashing.

Therefore, we directly concatenate the frame visual features $[\boldsymbol{t}_i^1, \cdots, \boldsymbol{t}_i^M]$ of the video from a global perspective in Ω^S, and then extract a real-valued code through the Fully Connected (FC) layer:

$$\hat{\boldsymbol{t}}_i = H_S \left(Concat([\boldsymbol{t}_i^1, \cdots, \boldsymbol{t}_i^M]) \right) \in \mathbb{R}^K. \tag{3}$$

Finally, we can obtain a K-bit binary code:

$$\boldsymbol{b}_i = \mathrm{sgn} \left(\tanh(\hat{\boldsymbol{t}}_i) \right) \in \{-1, 1\}^K. \tag{4}$$

Besides, to avoid the discrete optimization problem [5], we follow [15] for backpropagating gradients. In this way, the encoder-decoder methods [37,56] can compress the visual information as much as possible. However, to meet the goal of video reconstruction, the compression process may contain lots of retrieval-superfluous information, which affects the discriminativeness of binary codes.

Temporal Layer. To alleviate the task heterogeneity problem, a simple but effective dual-stream structure is introduced to decompose video information in Ω^S. Specifically, we design a temporal layer T_S parallel to the hash layer H_S in the dual-stream structure. T_S directly reduces the dimension of frame visual features $[\boldsymbol{t}_i^1, \cdots, \boldsymbol{t}_i^M]$ to obtain frame-level latent features via FC:

$$[\boldsymbol{l}_i^1, \cdots, \boldsymbol{l}_i^M] = T_S \left([\boldsymbol{t}_i^1, \cdots, \boldsymbol{t}_i^M] \right) \in \mathbb{R}^{M \times K}. \tag{5}$$

The temporal layer attempts to model complex information such as dynamic temporal changes via the reconstruction constraint, while for the hash layer, we will design similarity constraints to guide the flow of semantic information. Next, we will introduce how to perform dual-stream reconstruction and similarity knowledge preservation respectively.

3.3   Dual-Stream Reconstruction Learning

Existing video hashing works usually design the reconstruction task to compress visual information into binary codes. Inspired by masked language modeling in BERT [7,43], Ω^T [22] exploits the visual cloze task to optimize transformer blocks and capture inter-frame correlations, which randomly masks the input frame features as tokens and reconstructs the masked tokens in the decoder. In this way, frame-level binary codes in Ω^T can retain all the essential information for reconstruction, rather than retrieval.

To avoid this issue in Ω^S, we first mix $[\boldsymbol{l}_i^1, \cdots, \boldsymbol{l}_i^M]$ and $\boldsymbol{b}_i$ derived in Eq. 4 and Eq. 5, and then leverage the FC layer to reconstruct:

$$[\tilde{\boldsymbol{x}}_i^1, \cdots, \tilde{\boldsymbol{x}}_i^M] = D_S \left([\boldsymbol{l}_i^1 + \boldsymbol{b}_i, \cdots, \boldsymbol{l}_i^M + \boldsymbol{b}_i] \right) \in \mathbb{R}^{M \times D}. \tag{6}$$

Then, we can use the mean square error loss to measure the difference between CNN features $[\boldsymbol{x}_i^1, \cdots, \boldsymbol{x}_i^M]$ and decoder features $[\tilde{\boldsymbol{x}}_i^1, \cdots, \tilde{\boldsymbol{x}}_i^M]$ in $\boldsymbol{\Omega}^S$:

$$\mathcal{L}_{recon} = \frac{1}{DNM} \sum_{i=1}^{N} \sum_{m=1}^{M} \|\boldsymbol{x}_i^m - \tilde{\boldsymbol{x}}_i^m\|_2^2. \tag{7}$$

Please note that, in $\boldsymbol{\Omega}^T$, the definition of $\mathcal{L}_{recon}$ is the same as Eq. 7, but teacher decoder features can only be generated from frame-level binary codes $[\boldsymbol{b}_i^1, \cdots, \boldsymbol{b}_i^M]$, where the code length is fixed to 128 in experiments.

3.4 Semantic Knowledge Preservation

Using the frame-level reconstruction task alone does not make the two layers perform the desired role, so we further guide H_S to learn video-level semantic similarity information. For unsupervised learning, some image hashing works [49] prove that neighborhood structures learned from original CNN features can capture the similarity relations between samples. However, this strategy is time-consuming due to building a similarity graph for all samples directly, and has lots of noisy predictions, which confuses the learning of hash functions. Benefiting from the teacher-student distillation framework in Fig. 3, we construct a Gaussian-adaptive similarity graph from $\boldsymbol{\Omega}^T$ that captures the inherent semantic relations by estimating positives and hard negatives of training videos. These relations can guide H_S to generate discriminative binary codes and maintain the neighborhood structure in Hamming space.

Specifically, we first warm up $\boldsymbol{\Omega}^T$ with the reconstruction task $\mathcal{L}_{recon}$, and exploit the transformer visual embeddings $[\boldsymbol{t}_i^1, \cdots, \boldsymbol{t}_i^M]$ instead of the high dimensional CNN frame features to mine similarity relations. Although visual embeddings may contain redundant information due to $\mathcal{L}_{recon}$, they model inter-frame correlations compared to CNN features, which are helpful for mining long-term semantic concepts. To obtain video-level graph, we average $[\boldsymbol{t}_i^1, \cdots, \boldsymbol{t}_i^M]$ to video embedding $\bar{\boldsymbol{t}}_i$. Then, to solve the time-consuming problem, we follow [26] to use the neighbor graph between each video point $\bar{\boldsymbol{t}}_i$ and the cluster center $\{\boldsymbol{c}_i\}_{i=1}^{N_c}$ of video points to approximate similarity relations between $\bar{\boldsymbol{t}}_i$ and $\{\bar{\boldsymbol{t}}_i\}_{i=1}^{N}$, where N_c is the number of K-means clustering center. For each $\bar{\boldsymbol{t}}_i$, we calculate p nearest centers $\{\boldsymbol{c}_{ij}\}_{j=1}^{p}$, and the similarity matrix $\boldsymbol{Z} \in \mathbb{R}^{N \times N_c}$ is expressed as:

$$\boldsymbol{Z}_{ij} = \frac{\exp\left(-\|\bar{\boldsymbol{t}}_i, \boldsymbol{c}_{ij}\|_2/\alpha\right)}{\sum_{j'=1}^{p} \exp\left(-\|\bar{\boldsymbol{t}}_i, \boldsymbol{c}_{ij'}\|_2/\alpha\right)}, \tag{8}$$

where α is a bandwidth parameter. Note that the similarity values between $\bar{\boldsymbol{t}}_i$ and corresponding $N_c - p$ non-nearest centers in $\boldsymbol{Z}$ are set to 0, for simplicity. Finally, an approximate graph adjacency $\boldsymbol{A} \in \mathbb{R}^{N \times N}$ is calculated as: $\boldsymbol{A} = \boldsymbol{Z}\Lambda^{-1}\boldsymbol{Z}^T$, where $\Lambda = \text{diag}(\boldsymbol{Z}^T\boldsymbol{1}) \in \mathbb{R}^{N_c \times N_c}$. However, $\boldsymbol{A}$ may still be a noisy similarity signal, where the nearest center number p greatly affects the prediction quality. To avoid this dilemma, the existing work [22] builds multiple large adjacency

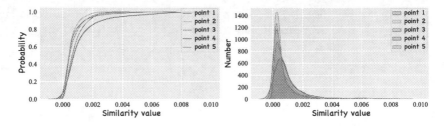

Fig. 4. The cumulative distribution and corresponding histogram distribution of similarity values for 5 video points in $\boldsymbol{A}$ on the FCVID dataset [16], where $p = 10$.

matrices to vote for credible sample relations, but it requires careful parameter tuning and takes up huge storage space on large video datasets.

Different from [22], we develop the Gaussian-adaptive similarity graph inspired by SSDH [48], which requires only one matrix to estimate more efficient sample relations. Compared to SSDH, our novelty lies in building a graph based on each sample and mining hard negative samples. Specifically, we first investigate the cumulative distribution and corresponding histogram of similarity values for each video point in $\boldsymbol{A}$. For better visualization, we randomly select the similarity values corresponding to 5 video points on the FCVID dataset [16], and use kernel density estimation [2] to simulate the real distribution curve in Fig. 4. Observing the cumulative distribution shows that the similarity values between most graph nodes are relatively small, while the histogram of the similarity value corresponding to each video point tends to a Gaussian distribution. This shows from the real data that it is very noisy to directly treat all the similarity signals in $\boldsymbol{A}$ as positive samples. To ensure high confidence in the supervision signal, we adaptively obtain positive samples for each video point. For the video point $\boldsymbol{v}_i$, the mean and standard deviation of similarity values between the nodes can be expressed as μ_i and ϵ_i, then we take the positive sample estimator as $PT_i = \mu_i + \lambda_1 * \epsilon_i$. Some metric learning works [28,58] show that hard negative samples are beneficial to model, so we add negative sample estimator $NT_i = \mu_i - \lambda_2 * \epsilon_i$ to mine hard negative samples for training. In this way, the Gaussian-adaptive graph adjacency matrix can be expressed as:

$$\hat{\boldsymbol{A}}_{ij} = \begin{cases} 1, & \text{if } \boldsymbol{A}_{ij} \geq PT_i \\ -1, & \text{if } NT_i < \boldsymbol{A}_{ij} < \mu_i \,. \\ 0, & \text{otherwise} \end{cases} \tag{9}$$

To preserve the similarity graph relations mined in $\boldsymbol{\Omega}^T$ for binary codes, we design a binary structure similarity loss:

$$\mathcal{L}_{bsim} = \frac{1}{N} \sum_{\{i,j\} \in \mathcal{S}} |\hat{\boldsymbol{A}}_{ij}|(\hat{\boldsymbol{A}}_{ij} - \frac{1}{K}\boldsymbol{b}_i\boldsymbol{b}_j^T)^2, \tag{10}$$

where $\mathcal{S}$ is the equal sampling strategy that samples positive or negative pairs with probability 0.5 based on $\hat{\boldsymbol{A}}$. Finally, we can obtain discriminative codes.

Furthermore, some works [32] argue that the middle layer of the teacher network can serve as a hint to the corresponding layer of the student network, thereby improving the effect of semantic knowledge transfer. Therefore, we consider aligning the visual embeddings between Ω^S and Ω^T. Inspired by [39,41], we design a visual embedding similarity loss:

$$\mathcal{L}_{tsim} = \frac{1}{N} \sum_{\{i,j\}\in\mathcal{S}} \left\|\bar{t}_i - c_{i1}\right\|_2^2 + \eta|\hat{A}_{ij}|(1 - \hat{A}_{ij})\left[\left\|\bar{t}_i - c_{i1}\right\|_2^2 - \left\|\bar{t}_i - c_{j1}\right\|_2^2 + \beta\right]_+,$$

(11)

where $\bar{t}_i$ is the mean visual embedding of i-th video in Ω^S, c_{i1} or c_{j1} is 1-NN nearest center of corresponding teacher visual embedding, η controls the balance and $[x]_+$ means the hinge function $max(0, x)$, which makes $\bar{t}_i$ closer to c_{i1} than negative pair c_{j1} by a fixed margin $\beta = 1.0$.

3.5 Overall Learning

The overall training objectives of Ω^T and Ω^S are as follows:

$$\begin{aligned} \mathcal{L}_{\text{teacher}} &= \mathcal{L}_{\text{recon}} , \\ \mathcal{L}_{\text{student}} &= \mathcal{L}_{\text{recon}} + \gamma_1\mathcal{L}_{\text{bsim}} + \gamma_2\mathcal{L}_{\text{tsim}} , \end{aligned}$$

(12)

where γ_1 and γ_2 relatively weight the losses.

4 Experimental Results

4.1 Datasets, Metrics and Implementation Details

Datasets. We run experiments on three popular video datasets. **FCVID** [16] is a web video dataset consisting of 91,223 YouTube videos annotated into 239 categories. It covers a wide range of topics, with the majority of them being real-world events such as "biking", "making coffee" and "yoga". The dataset is evenly split into training and testing partitions with 45,585 and 45,600 videos. We use the testing partition as the query set and retrieval database. **ActivityNet** [4] consists of 20K YouTube videos annotated with 200 class descriptions. As the testing set labels are not publicly available, the evaluation is performed on the validation set. Following [21], we pick 9,722, 1,000 and 3,760 videos as training set, query set and retrieval database, respectively. **YFCC** [40] is a massive dataset from the Yahoo Webscope program containing 0.8M videos. We randomly select 409,788 unlabeled videos for training and 101,256 labeled videos with 80 semantic concepts [47] for testing. In these labeled videos, we sample 1000 videos as the query set and the remaining ones as retrieval database.

Metrics. We measure the retrieval performance with standard metrics in information retrieval, including Mean Average Precision at top-k retrieved results (MAP@k) and Precision-Recall (PR) curves.

Implementation Details. Our experiments are based on the Pytorch framework [30]. In the video encoding process, we uniformly sample 25 frames from

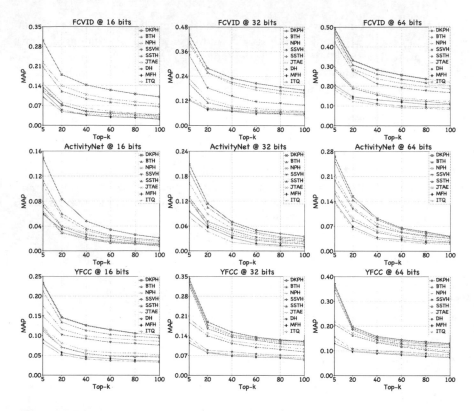

Fig. 5. Performance comparison (*w.r.t.* MAP@k) of DKPH and SOTA methods.

each video and use VGG-16 pretrained on Imagenet [35] to extract frame-wise features. To ensure fair comparison [22], we use a single transformer block with a single attention head as the transformer encoder. For the teacher model Ω^T, we warm up 60, 300 and 200 epochs on FCVID, ActivityNet and YFCC, respectively. Considering the trade-off of visual information compression loss and inter-frame correlations, we set the dimension of the visual embeddings and binary codes to 256 and 128 in Ω^T. In the graph construction process, the number of clustering center N_c is set as 2,000, 1,000 and 2,000 on FCVID, ActivityNet and YFCC respectively. We employ Adam optimizer to train the model with a mini-batch size of 256 and train the student model Ω^S for 48 epochs, where the initial learning rate is 5×10^{-4}. The default hyper-parameters setting is: $\lambda_1 = 2, \lambda_2 = 1, \eta = 0.1, \gamma_1 = 0.11, \gamma_2 = 0.9$. In the testing phase, we only use the student model Ω^S, where the lengths of binary codes are 16, 32 and 64.

4.2   Comparisons with State-of-the-Art (SOTA) Methods

To prove the effectiveness of DKPH, we compare the retrieval performance with two image hashing methods: ITQ [12], DH [9], and six SOTA video hashing methods: MFH [36], SSTH [56], JTAE [18], SSVH [37], NPH [21], and BTH [22].

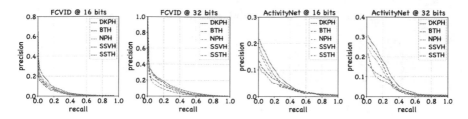

Fig. 6. Performance comparison (*w.r.t.* PR curves) of DKPH and SOTA methods.

Table 1. The impact of different frame feature encoders with or without the dual-stream structure on FCVID. Some abbreviations: TF-Transformer, D-Dual-stream.

Method	16 bits				32 bits				64 bits			
	k = 5	k = 20	k = 60	k = 100	k = 5	k = 20	k = 60	k = 100	k = 5	k = 20	k = 60	k = 100
CNN [54]	0.229	0.116	0.080	0.065	0.395	0.242	0.172	0.140	0.460	0.294	0.207	0.171
CNN [54]+D	0.273	0.152	0.106	0.084	0.407	0.252	0.175	0.147	0.464	0.308	0.223	0.189
LSTM [38]	0.227	0.114	0.077	0.062	0.393	0.240	0.168	0.139	0.457	0.291	0.210	0.174
LSTM [38]+D	0.272	0.150	0.104	0.083	0.404	0.248	0.172	0.146	0.462	0.301	0.224	0.192
TF [7]	0.235	0.122	0.083	0.069	0.421	0.252	0.172	0.143	0.477	0.313	0.238	0.202
TF [7]+D	0.297	0.174	0.120	0.097	0.441	0.275	0.203	0.171	0.494	0.331	0.255	0.228

Figure 5 shows the MAP@K results on three datasets. Compared with SOTA methods, DKPH achieves the best results on three video datasets. Specifically, we obtain 0.7%–8.6% MAP@5 gains for various bits, which demonstrates the efficiency of DKPH. We owe the great advantage of DKPH over these two methods [21, 22] to the full use of dual-stream structure and Gaussian-adaptive similarity graph. Note that the model performance gaps are larger at 16 bits, as we expected. Because the amount of information carried by the binary code is limited by the length. Therefore, in low-bit scenarios, the impact of task heterogeneity will be more serious, leading to inferior results from existing methods [22].

Furthermore, we examine DKPH with PR curves on FCVID and ActivityNet in Fig. 6. DKPH delivers higher precision than SOTA methods at the same recall rate, and improves more significantly at low recall requirements. This illustrates that the model is suitable for real-world video retrieval systems, as people tend to focus more on results with high accuracy rather than finding all similar results.

4.3   Ablation Study

To provide further insight into DKPH, we conduct critical ablation studies.
Analysis of the Dual-Stream Structure with Different Encoders. DKPH employs a transformer encoder and a dual-stream structure to generate binary codes. Thus, we explore the impact of different frame feature encoders (CNN [54], LSTM [38], and transformer [7]) with or without the dual-stream structure in Table 1. Specifically, based on the dual-stream structure, CNN, LSTM and transformer obtain 0.4%–6.2% MAP@5 gains at different bits on FCVID. The following advantages can be clearly observed: (1) the dual-stream structure is

Table 2. Contributions of different modules on FCVID.

Method	16 bits				32 bits				64 bits			
	k = 5	k = 20	k = 60	k = 100	k = 5	k = 20	k = 60	k = 100	k = 5	k = 20	k = 60	k = 100
DKPH+A	0.202	0.119	0.080	0.062	0.399	0.244	0.170	0.139	0.458	0.294	0.211	0.180
DKPH+M	0.269	0.139	0.091	0.075	0.421	0.258	0.176	0.151	0.484	0.321	0.243	0.204
DKPH+T	0.228	0.135	0.090	0.073	0.419	0.251	0.172	0.147	0.463	0.298	0.218	0.181
DKPH-D	0.235	0.122	0.083	0.069	0.421	0.252	0.172	0.143	0.477	0.313	0.238	0.202
DKPH-DR	0.276	0.159	0.111	0.092	0.435	0.267	0.189	0.158	0.487	0.326	0.245	0.218
DKPH-TS	0.158	0.088	0.067	0.051	0.311	0.123	0.078	0.058	0.345	0.169	0.118	0.103
DKPH-Lb	0.174	0.093	0.072	0.058	0.322	0.142	0.080	0.062	0.366	0.187	0.123	0.107
DKPH-Lt	0.209	0.121	0.098	0.080	0.405	0.239	0.177	0.141	0.432	0.276	0.205	0.162
DKPH	**0.297**	**0.174**	**0.120**	**0.097**	**0.441**	**0.275**	**0.203**	**0.171**	**0.494**	**0.331**	**0.255**	**0.228**

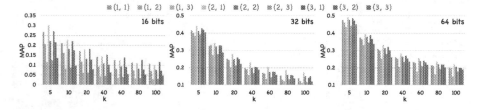

Fig. 7. The MAP@k scores with various configurations about the positive estimator factor and the negative estimator factor (λ_1, λ_2) on FCVID.

a **general-purpose and important design** that consistently improves three encoders, especially at low bits; (2) the transformer outperforms CNN and LSTM due to its strong ability to model long-term inter-frame correlations.

Analysis of Model Components. We compare DKPH with the following variations: (1) DKPH+A. The Gaussian-adaptive graph adjacency matrix $\hat{A}$ is replaced by A; (2) DKPH+M. $\hat{A}$ is replaced by multiple matrices [22]; (3) DKPH+T. We replace the dual-stream structure with twin bottlenecks designed for image hashing [34]; (4) DKPH-D. The dual-stream structure is removed; (5) DKPH-DR. Both the dual-stream structure and $\mathcal{L}_{recon}$ in Ω^S are removed; (6) DKPH-TS. We remove the teacher-student distillation strategy and only use $\mathcal{L}_{recon}$; (7) DKPH-Lb. We remove the binary structure similarity loss $\mathcal{L}_{bsim}$; (8) DKPH-Lt. We remove the visual embedding similarity loss $\mathcal{L}_{tsim}$. Table 2 shows the performance of DKPH and its variations at different bits on FCVID, and proves that each module significantly contributes to the final result.

We have the following observations. First, reasonable mining of positive and hard negative pairs helps discriminate binary codes. DKPH adaptively explores the similarity relations of each video point through the sample estimators. However, in DKPH+M, the multiple matrices strategy consumes more time and space resources, and requires careful adjustment of matrix parameters, which cannot achieve optimal results. Second, twin bottlenecks (DKPH+T) are still difficult to replace the dual-stream structure designed for video hashing. There are two reasons: (1) the mechanism of twin bottlenecks is to learn better reconstructed images to feedback binary codes, which cannot exhibit the advantages

Table 3. Cross-dataset MAP@20 results when training on FCVID and test on YFCC at 64 bits. The blue number indicates the performance drop compared with training and testing both on YFCC (black number).

Method	SSTH [56]	SSVH [37]	NPH [21]	BTH [22]	DKPH
MAP@20	0.155 (−6.3%)	0.173 (−7.8%)	0.180 (−6.0%)	0.191 (−5.7%)	0.199 (−2.8%)

Table 4. The MAP@k results of latent features l and binary codes b at 16 bits.

Method	k = 5	k = 20	k = 60
Only l	0.098	0.077	0.065
Only b	0.297	0.174	0.120

Table 5. The reconstruction errors in a category.

Method	Error
DKPH	0.4849
DKPH$_f$	0.9527

Table 6. The reconstruction errors in test set.

Method	Error
DKPH	0.5586
Remove b	0.5615
Remove l	0.9769

of video information decomposition; (2) twin bottlenecks generate frame-level binary codes, resulting in quantization errors during testing. Third, DKPH-D and DKPH-DR explore the effects of task heterogeneity, which suggests a conflict between $\mathcal{L}_{recon}$ and similarity learning in existing methods [22,37]. We decouple the tasks, which allows binary codes to retain useful information and avoids the conflict. Fourth, results in DKPH-Lb yield an excessive drop due to the lack of similarity guidance, where $\mathcal{L}_{bsim}$ is the core loss of information decomposition.

Hyperparameter Analysis. We investigate various configurations about the positive and negative estimator factors (λ_1, λ_2), as shown in Fig. 7. From this experiment, we find that as λ_1 grows (*i.e.*, PT_i grows), the performance increases at first and reaches the best results, then decreases as a whole. A small PT_i may cause the model to be trained on more noisy signals, while a large PT_i may not fully exploit the underlying positive similarity relations. Moreover, λ_2 has a greater impact on model performance than λ_1, reflecting the vital contribution of hard negative samples to the model.

4.4 Further Analysis

Cross-Dataset Evaluation Comparisons. To investigate the generalization of DKPH for cross-dataset retrieval, we train various methods on FCVID and test on YFCC in Table 3, which shows MAP@20 results for cross-dataset retrieval at 64 bits. DKPH can not only achieve SOTA in the single-dataset setting, but also the performance drop (−2.8%) is the lowest in the cross-dataset setting. This may be because binary codes focus more on semantic concepts rather than the underlying reconstruction information, which ensures good transferability and generalization of DKPH when retrieving unknown datasets.

Information Decomposition Analysis. Table 4 shows the MAP@k results of latent features l and binary codes b at 16 bits on FCVID. Results of Only l are much lower than those of Only b, which indicate that l may not have

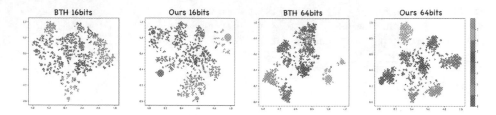

Fig. 8. t-SNE visualizations [27] of BTH and DKPH. Videos are randomly sampled on FCVID database, and samples with different labels are marked with different colors.

enough semantics to support the retrieval task. Next, we examine the effect of dual-stream features for reconstruction at 16 bits on FCVID. In Table 5, we randomly input a category of test videos and then calculate the mean square error between $[\boldsymbol{x}_i^1, \cdots, \boldsymbol{x}_i^M]$ and $[\tilde{\boldsymbol{x}}_i^1, \cdots, \tilde{\boldsymbol{x}}_i^M]$. When we replace $\boldsymbol{l}$ with fixed values (*i.e.,*the mean of latent features), the reconstruction error in DKPH$_f$ increases by 96.5%. In Table 6, we directly remove $\boldsymbol{b}$ or $\boldsymbol{l}$ for reconstruction and calculate errors in all test videos. Removing $\boldsymbol{b}$, the error increases by 0.52%, while Removing $\boldsymbol{l}$ increases the error by 74.9%. Tables 5 and 6 prove that $\boldsymbol{l}$, rather than $\boldsymbol{b}$, contains sufficient essential information (dynamic changes) for reconstruction.

Qualitative Results. Figure 8 shows the t-SNE visualization [27] of binary codes learned by BTH and DKPH. To facilitate the observation, we randomly sample 8 categories of videos twice on 16 bits and 64 bits, respectively, to obtain binary codes. At 16 bits, there is a clear distinction between most categories in our model. In particular, t-SNE embeddings of DKPH in some categories (*e.g.,* 0, 2, 6) can be mapped onto a small circle. This proves that DKPH pays more attention to the learning of global semantics and binary codes of a category are almost very close in Hamming space, so the phenomenon of t-SNE embedding aggregation occurs. At 64 bits, t-SNE embeddings of our model in different categories are well separated, which proves the good discriminativeness.

5   Conclusion

We propose a novel unsupervised video hashing framework, DKPH, to tackle the task heterogeneity problem. Firstly, we design the dual-stream structure to decompose video information, which disentangles the semantic extraction from reconstruction constraint. Then, a Gaussian-adaptive similarity graph is developed to explore the semantic similarity knowledge between samples. With the help of this knowledge, the hash layer in the dual-stream structure can further generate discriminative semantic binary codes. In this paper, we hope not only to present insights into the importance of information decomposition but also to facilitate future work that advances video hashing by solving design flaws rather than mostly trial and error.

Acknowledgements. This work is supported by the National Nature Science Foundation of China (62121002, 62022076, U1936210), the Fundamental Research Funds for the Central Universities under Grant WK3480000011, the Youth Innovation Promotion Association Chinese Academy of Sciences (Y2021122). We acknowledge the support of GPU cluster built by MCC Lab of Information Science and Technology Institution, USTC.

References

1. Bagherinezhad, H., Horton, M., Rastegari, M., Farhadi, A.: Label refinery: improving imagenet classification through label progression. In: AAAI (2021)
2. Botev, Z.I., Grotowski, J.F., Kroese, D.P.: Kernel density estimation via diffusion. Ann Stat (2010)
3. Brown, A., Xie, W., Kalogeiton, V., Zisserman, A.: Smooth-AP: smoothing the path towards large-scale image retrieval. In: Vedaldi, A., Bischof, H., Brox, T., Frahm, J.-M. (eds.) ECCV 2020. LNCS, vol. 12354, pp. 677–694. Springer, Cham (2020). https://doi.org/10.1007/978-3-030-58545-7_39
4. Caba Heilbron, F., Escorcia, V., Ghanem, B., Carlos Niebles, J.: Activitynet: a large-scale video benchmark for human activity understanding. In: CVPR (2015)
5. Cao, Z., Long, M., Wang, J., Yu, P.S.: Hashnet: deep learning to hash by continuation. In: ICCV (2017)
6. Cui, Q., Jiang, Q.-Y., Wei, X.-S., Li, W.-J., Yoshie, O.: ExchNet: a unified hashing network for large-scale fine-grained image retrieval. In: Vedaldi, A., Bischof, H., Brox, T., Frahm, J.-M. (eds.) ECCV 2020. LNCS, vol. 12348, pp. 189–205. Springer, Cham (2020). https://doi.org/10.1007/978-3-030-58580-8_12
7. Devlin, J., Chang, M.W., Lee, K., Toutanova, K.: Bert: Pre-training of deep bidirectional transformers for language understanding. In: NAACL (2019)
8. Dosovitskiy, A., et al.: An image is worth 16x16 words: transformers for image recognition at scale. In: ICLR (2021)
9. Erin Liong, V., Lu, J., Wang, G., Moulin, P., Zhou, J.: Deep hashing for compact binary codes learning. In: CVPR (2015)
10. Gabeur, V., Sun, C., Alahari, K., Schmid, C.: Multi-modal transformer for video retrieval. In: Vedaldi, A., Bischof, H., Brox, T., Frahm, J.-M. (eds.) ECCV 2020. LNCS, vol. 12349, pp. 214–229. Springer, Cham (2020). https://doi.org/10.1007/978-3-030-58548-8_13
11. Ge, J., Xie, H., Min, S., Zhang, Y.: Semantic-guided reinforced region embedding for generalized zero-shot learning. In: AAAI (2021)
12. Gong, Y., Lazebnik, S., Gordo, A., Perronnin, F.: Iterative quantization: a procrustean approach to learning binary codes for large-scale image retrieval. TPAMI (2012)
13. Guo, M., Haque, A., Huang, D.-A., Yeung, S., Fei-Fei, L.: Dynamic task prioritization for multitask learning. In: Ferrari, V., Hebert, M., Sminchisescu, C., Weiss, Y. (eds.) ECCV 2018. LNCS, vol. 11220, pp. 282–299. Springer, Cham (2018). https://doi.org/10.1007/978-3-030-01270-0_17
14. Hinton, G., Vinyals, O., Dean, J., et al.: Distilling the knowledge in a neural network. arXiv preprint arXiv:1503.02531 (2015)
15. Hubara, I., Courbariaux, M., Soudry, D., El-Yaniv, R., Bengio, Y.: Binarized neural networks. NeurIPS (2016)

16. Jiang, Y.G., Wu, Z., Wang, J., Xue, X., Chang, S.F.: Exploiting feature and class relationships in video categorization with regularized deep neural networks. TPAMI (2017)
17. Kingma, D.P., Welling, M.: Auto-encoding variational bayes. In: ICLR (2014)
18. Li, C., Yang, Y., Cao, J., Huang, Z.: Jointly modeling static visual appearance and temporal pattern for unsupervised video hashing. In: CIKM (2017)
19. Li, P., Li, Y., Xie, H., Zhang, L.: Neighborhood-adaptive structure augmented metric learning. In: AAAI (2022)
20. Li, S., Chen, Z., Li, X., Lu, J., Zhou, J.: Unsupervised variational video hashing with 1d-cnn-lstm networks. TMM (2019)
21. Li, S., Chen, Z., Lu, J., Li, X., Zhou, J.: Neighborhood preserving hashing for scalable video retrieval. In: ICCV (2019)
22. Li, S., Li, X., Lu, J., Zhou, J.: Self-supervised video hashing via bidirectional transformers. In: CVPR (2021)
23. Liong, V.E., Lu, J., Tan, Y.P., Zhou, J.: Deep video hashing. TMM (2016)
24. Liu, B., Yeung, S., Chou, E., Huang, D.-A., Fei-Fei, L., Niebles, J.C.: Temporal modular networks for retrieving complex compositional activities in videos. In: Ferrari, V., Hebert, M., Sminchisescu, C., Weiss, Y. (eds.) ECCV 2018. LNCS, vol. 11207, pp. 569–586. Springer, Cham (2018). https://doi.org/10.1007/978-3-030-01219-9_34
25. Liu, Q., Xie, L., Wang, H., Yuille, A.L.: Semantic-aware knowledge preservation for zero-shot sketch-based image retrieval. In: ICCV (2019)
26. Liu, W., Wang, J., Kumar, S., Chang, S.F.: Hashing with graphs. In: ICML (2011)
27. Van der Maaten, L., Hinton, G.: Visualizing data using t-sne. JMLR (2008)
28. Milbich, T., et al.: DiVA: diverse visual feature aggregation for deep metric learning. In: Vedaldi, A., Bischof, H., Brox, T., Frahm, J.-M. (eds.) ECCV 2020. LNCS, vol. 12353, pp. 590–607. Springer, Cham (2020). https://doi.org/10.1007/978-3-030-58598-3_35
29. Min, S., Yao, H., Xie, H., Wang, C., Zha, Z.J., Zhang, Y.: Domain-aware visual bias eliminating for generalized zero-shot learning. In: CVPR (2020)
30. Paszke, A., et al.: Pytorch: an imperative style, high-performance deep learning library. NeurIPS (2019)
31. Qiu, Z., Su, Q., Ou, Z., Yu, J., Chen, C.: Unsupervised hashing with contrastive information bottleneck. In: IJCAI (2021)
32. Romero, A., Ballas, N., Kahou, S.E., Chassang, A., Gatta, C., Bengio, Y.: Fitnets: hints for thin deep nets. In: ICLR (2015)
33. Rumelhart, D.E., Hinton, G.E., Williams, R.J.: Learning representations by back-propagating errors. nature (1986)
34. Shen, Y., et al.: Auto-encoding twin-bottleneck hashing. In: CVPR (2020)
35. Simonyan, K., Zisserman, A.: Very deep convolutional networks for large-scale image recognition. In: ICLR (2015)
36. Song, J., Yang, Y., Huang, Z., Shen, H.T., Hong, R.: Multiple feature hashing for real-time large scale near-duplicate video retrieval. In: ACM MM (2011)
37. Song, J., Zhang, H., Li, X., Gao, L., Wang, M., Hong, R.: Self-supervised video hashing with hierarchical binary auto-encoder. TIP (2018)
38. Srivastava, N., Mansimov, E., Salakhudinov, R.: Unsupervised learning of video representations using lstms. In: ICML (2015)
39. Su, S., Zhang, C., Han, K., Tian, Y.: Greedy hash: Towards fast optimization for accurate hash coding in cnn. In: NeurIPS (2018)
40. Thomee, B., et al.: The new data and new challenges in multimedia research. arXiv preprint arXiv:1503.01817 (2015)

41. Tian, K., Zhou, S., Guan, J.: Deepcluster: a general clustering framework based on deep learning. In: ECML (2017)
42. Tishby, N., Zaslavsky, N.: Deep learning and the information bottleneck principle. In: ITW (2015)
43. Wang, Y., Xie, H., Fang, S., Wang, J., Zhu, S., Zhang, Y.: From two to one: a new scene text recognizer with visual language modeling network. In: ICCV (2021)
44. Wang, Y., Xie, H., Zha, Z.J., Xing, M., Fu, Z., Zhang, Y.: Contournet: taking a further step toward accurate arbitrary-shaped scene text detection. In: CVPR (2020)
45. Wu, G., et al.: Unsupervised deep video hashing via balanced code for large-scale video retrieval. TIP (2018)
46. Wu, W., et al.: End-to-end video text spotting with transformer. arXiv preprint arXiv:2203.10539 (2022)
47. Xiao, J., Hays, J., Ehinger, K.A., Oliva, A., Torralba, A.: Sun database: large-scale scene recognition from abbey to zoo. In: CVPR (2010)
48. Yang, E., Deng, C., Liu, T., Liu, W., Tao, D.: Semantic structure-based unsupervised deep hashing. In: IJCAI (2018)
49. Yang, E., Liu, T., Deng, C., Liu, W., Tao, D.: Distillhash: unsupervised deep hashing by distilling data pairs. In: CVPR (2019)
50. Yang, K., Zhou, T., Tian, X., Tao, D., et al.: Class-disentanglement and applications in adversarial detection and defense. NeurIPS (2021)
51. Ye, G., Liu, D., Wang, J., Chang, S.F.: Large-scale video hashing via structure learning. In: ICCV (2013)
52. Yu, T., Yang, Y., Li, Y., Liu, L., Fei, H., Li, P.: Heterogeneous attention network for effective and efficient cross-modal retrieval. In: SIGIR (2021)
53. Yu, T., Yuan, J., Fang, C., Jin, H.: Product quantization network for fast image retrieval. In: Ferrari, V., Hebert, M., Sminchisescu, C., Weiss, Y. (eds.) ECCV 2018. LNCS, vol. 11205, pp. 191–206. Springer, Cham (2018). https://doi.org/10.1007/978-3-030-01246-5_12
54. Yue-Hei Ng, J., Hausknecht, M., Vijayanarasimhan, S., Vinyals, O., Monga, R., Toderici, G.: Beyond short snippets: deep networks for video classification. In: CVPR (2015)
55. Zhang, B., Hu, H., Sha, F.: Cross-modal and hierarchical modeling of video and text. In: Ferrari, V., Hebert, M., Sminchisescu, C., Weiss, Y. (eds.) ECCV 2018. LNCS, vol. 11217, pp. 385–401. Springer, Cham (2018). https://doi.org/10.1007/978-3-030-01261-8_23
56. Zhang, H., Wang, M., Hong, R., Chua, T.S.: Play and rewind: Optimizing binary representations of videos by self-supervised temporal hashing. In: ACM MM (2016)
57. Zhang, X., Zhang, T., Hong, X., Cui, Z., Yang, J.: Graph wasserstein correlation analysis for movie retrieval. In: Vedaldi, A., Bischof, H., Brox, T., Frahm, J.-M. (eds.) ECCV 2020. LNCS, vol. 12370, pp. 424–439. Springer, Cham (2020). https://doi.org/10.1007/978-3-030-58595-2_26
58. Zhao, Y., Jin, Z., Qi, G., Lu, H., Hua, X.: An adversarial approach to hard triplet generation. In: Ferrari, V., Hebert, M., Sminchisescu, C., Weiss, Y. (eds.) ECCV 2018. LNCS, vol. 11213, pp. 508–524. Springer, Cham (2018). https://doi.org/10.1007/978-3-030-01240-3_31

PASS: Part-Aware Self-Supervised Pre-Training for Person Re-Identification

Kuan Zhu[1,2] ⓘ, Haiyun Guo[1,2(✉)] ⓘ, Tianyi Yan[1,2] ⓘ, Yousong Zhu[1] ⓘ, Jinqiao Wang[1,2,3] ⓘ, and Ming Tang[1,2] ⓘ

[1] National Laboratory of Pattern Recognition, Institute of Automation, Chinese Academy of Sciences, Beijing, China
[2] School of Artificial Intelligence, University of Chinese Academy of Sciences, Beijing, China
[3] Peng Cheng Laboratory, Shenzhen, China
{kuan.zhu,haiyun.guo,tianyi.yan,yousong.zhu,jqwang,tangm}@nlpr.ia.ac.cn

Abstract. In person re-identification (ReID), very recent researches have validated pre-training the models on unlabelled person images is much better than on ImageNet. However, these researches directly apply the existing self-supervised learning (SSL) methods designed for image classification to ReID without any adaption in the framework. These SSL methods match the outputs of *local* views (e.g., red T-shirt, blue shorts) to those of the *global* views at the same time, losing lots of details. In this paper, we propose a ReID-specific pre-training method, Part-Aware Self-Supervised pre-training (PASS), which can generate part-level features to offer fine-grained information and is more suitable for ReID. PASS divides the images into several local areas, and the *local* views randomly cropped from each area are assigned a specific learnable [PART] token. On the other hand, the [PART]s of all local areas are also appended to the *global* views. PASS learns to match the outputs of the *local* views and *global* views on the same [PART]. That is, the learned [PART] of the *local* views from a local area is only matched with the corresponding [PART] learned from the *global* views. As a result, each [PART] can focus on a specific local area of the image and extracts fine-grained information of this area. Experiments show PASS sets the new state-of-the-art performances on Market1501 and MSMT17 on various ReID tasks, e.g., vanilla ViT-S/16 pre-trained by PASS achieves 92.2%/90.2%/88.5% mAP accuracy on Market1501 for supervised/UDA/USL ReID. Our codes are available at https://github.com/CASIA-IVA-Lab/PASS-reID.

Keywords: Person re-identification · Self-supervised pre-training · Local representations

Supplementary Information The online version contains supplementary material available at https://doi.org/10.1007/978-3-031-19781-9_12.

1 Introduction

Person re-identification (ReID) aims to associate the person images captured by different cameras. Limited by the scale of the labeled ReID datasets, most existing methods first pre-train the backbone networks (e.g., ResNet [17], ViT [10]) on ImageNet [8] and then fine-tune them on person ReID datasets to boost the performance [18,22,25]. However, it is arguable whether using ImageNet for pre-training is optimal as there exist large domain gaps between ImageNet and person ReID data, e.g., 1) ImageNet-1K contains a thousand kinds of objects while ReID datasets only contain persons; 2) the model pre-trained on ImageNet will focus on category-level differences, losing lots of rich visual information. Therefore, the fine-grained identity information, which is preferred by ReID, can not be provided by pre-training on ImageNet [11,23].

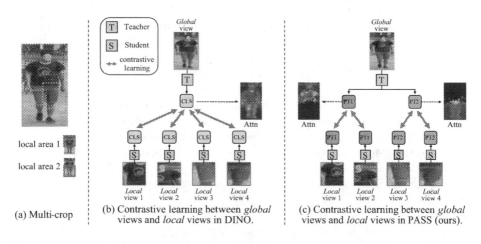

Fig. 1. (a) Multi-crop [1] is a data augmentation strategy which uses a mix of views with different resolutions. (b) DINO [2] learns to match the outputs of all the *local* views and *global* views, losing lots of details. (c) In PASS, the *local* views cropped from each local area are assigned a specific [PART], e.g., the *local* views cropped from the upper local area, *local* view 1 and 2, are assigned [PART]$_1$; the *local* views cropped from the lower local area, *local* view 3 and 4, are assigned [PART]$_2$. The *global* views are appended with all the [PART]s. PASS learns to match the predictions on the same [PART] (i.e., [PART]$_1$ or [PART]$_2$) of *local* views and *global* views, **respectively**

To bridge the gap between pre-training and fine-tuning datasets for better ReID models, Fu et al. [11] propose the first large scale unlabeled person ReID dataset "LUPerson" and demonstrate that unsupervised pre-training the models on LUPerson is quite better than supervised pre-training on ImageNet. Luo et al. [23] further investigate that DINO [2] algorithm with Transformer architecture (ViT [10]) obtains the best ReID performance among the existing self-supervised learning (SSL) methods and network architectures. However, these works [11,23] directly apply the existing SSL methods, that are proposed for image classification, to ReID and do not make any adaption in the SSL framework. Another gap will appear when employing these SSL methods to ReID, which is shown in

Fig. 1(b). DINO matches the outputs of all the *local* views with the *global* views in the same feature space. It is unreasonable to make the outputs of red T-shirt and blue shorts both match the output of the global image at the same time. To compromise on this matching, only the shared features of different views are retrained and lots of local detailed information are removed in the learned features. However, as fine-grained information has been verified to be crucial to describe a person image [21,25,28], it is necessary to design a pre-training method that can extract as many fine-grained clues as possible for ReID.

In this paper, we propose a ReID-specific pre-training method based on Transformer, Part-Aware Self-Supervised pre-training (PASS), by which part-level features can be automatically extracted to offer fine-grained information for person ReID. PASS uses the learning paradigm of knowledge distillation to match the outputs of the teacher network and student network. It first divides the image into several fixed overlapping local areas and randomly crops *local* views from these local areas. The *global* views are randomly cropped from the whole image with higher resolution. All views are passed through the student while only *global* views are passed through the teacher. For simplicity, we only illustrate the comparison between the outputs of *local* views passed through student and *global* views passed through teacher in Fig. 1(c). Before passing through the student, the *local* views cropped from each local area are assigned a specific learnable [PART] token, which is used to learn the local representation. All these [PART]s are also appended to the *global* views and fed to the teacher to learn local features from the whole image. PASS learns to match the corresponding [PART]s of *local* views and *global* views, where the [PART]s of different areas are not compared. Take Fig. 1(c) as an example, the *local* views cropped from the upper local area, *local* views 1 and 2, are assigned [PART]$_1$; the *local* views cropped from the lower local area, *local* views 3 and 4, are assigned [PART]$_2$. The predictions on the [PART]$_1$s of *local* views 1 and 2 are only compared with those on the [PART]$_1$s of the *global* views and so does the [PART]$_2$s.

In the student, the *local* views assigned to each [PART] are cropped from a specific local area, thus the [PART]s can focus on different areas. PASS uses the student to update the teacher, which can guarantee [PART]s in teacher also focus on different local areas and learns fine-grained information. In pre-training, the student learns to match the output of the teacher on the same [PART], which can guarantee each [PART] learns a robust local representation from the *local* views cropped in its corresponding local area. In fine-tuning, all [PART]s are appended to the input image and each [PART] automatically learns the local representation for a specific area.

We summarize the contributions of this work as: (i) In this paper, we propose the ReID-specific pre-training method, Part-Aware Self-Supervised pre-training (PASS), which is more suitable for ReID with part-level features offering fine-grained information. It is worth noting that we do not add any complex module to extract part-level features but only use several learnable tokens. (ii) The pre-trained ViT backbone can be fine-tuned on various ReID downstream tasks, i.e., supervised learning, unsupervised domain adaptation (UDA), and unsupervised learning (USL). PASS helps the ViTs set the new state-of-the-art performance on Market-1501 [34] and MSMT17 [29], e.g., vanilla ViT-S pre-trained by PASS achieves 92.2%/90.2%/88.5% Rank-1 and 69.1%/49.1%/41.0% mAP accuracy on Market1501 and MSMT17 for supervised/UDA/USL ReID, respectively.

2   Related Work

2.1   Self-supervised Learning

Self-supervised learning (SSL) methods aim to learn discriminative representations from large scale unlabeled data. Recently, contrastive learning methods have made remarkable achievements [1,2,4,5,14,16,31], significantly reducing the gap with supervised pre-training. MoCo series [4,5], which are developed from Momentum Contrast, treat the augmentations of a sample as positive pairs and all other samples as negative pairs. Ge et al. [14] propose a new paradigm, BYOL, where the online network predicts the output representation of the target network on the same image under a different augmented view, removing the need of negative pairs. DINO [2] further improves BYOL by using a centering and sharpening of the momentum teacher outputs to avoid model collapse. Besides, DINO adopts the augmentation strategy of multi-crop [1] to conduct the comparison between the representations of *global* views and *local* views. Xie et al. [31] propose MoBY which is a Transformer-specific method and combines MoCo with BYOL. Among these methods, DINO algorithm, plus the ViT architecture, can achieve the best performance on ReID [23]. Therefore, we aim to improve DINO to better adapt to the ReID tasks and obtain higher performance.

2.2   Person Re-identification

Part-Based Person ReID. Employing part-level features for person image description can offer fine-grained information and has been verified as beneficial for person re-identification. Many efforts have been made to develop the part-based person re-ID to boost the state-of-the-art performance [15,25,28,35 37]. PCB [25] first directly partitions the person images into fixed horizontal stripes and extract stripe-based part features. MGN [28] enhances the robustness by dividing images into stripes of different granularities and designs overlap between stripes. SPReID [21] uses a pre-trained human semantic parsing model to provide the mask of body parts to extract part features. ISP [36] proposes to automatically locate both human parts and non-human ones at pixel-level by iterative clustering. TransReID [19], which is a Transformer-specific method, also designs to extract the part-level features by re-arranging the patch embeddings and re-grouping them. All these methods show extracting part-level features to offer fine-grained information is of vital importance for person ReID.

Unsupervised Pre-training for Person ReID. Recently, more and more works find that there exists large domain gaps between ImageNet and person ReID data [11,23], which makes pre-training ReID models on unlabelled person images is much better than pre-training on ImageNet. Fu et al. [11] first focus on this problem and propose a large scale unlabeled person ReID dataset "LUPerson". They also validate that unsupervised pre-training on LUPerson can improve the ReID performance compared with ImageNet-1k pre-training. Luo et al. [23] further investigate several self-supervised learning methods and backbone networks. The results show that DINO [2] with ViT [10] obtains the best performance. [23] also proposes a data filtering method and IBN-based convolution stem for ViT architecture. However, apart from some adaptions in data augmentation or hyper-parameters, they do not specifically improve these pre-training methods for person ReID. The gaps will appear when applying the existing SSL methods proposed for image classification to ReID as shown in Fig. 1. In this paper, we propose a ReID-specific self-supervised pre-training method, PASS, which makes the model learn to extract part-level features in pre-training and automatically extract part-level features in fine-tuning.

3   Method

3.1   Preliminaries

Vision Transformer. We briefly describe the mechanism of the Vision Transformer (ViT) [10,26] here, and please refer to [26] for details about Transformers and to [10] for its adaptation to images. The ViT architecture takes a grid of non-overlapping contiguous image patches of resolution $N \times N$ as input. Typically, we use $N = 16$ ("/16") in this paper. The patches are then mapped to a sequence of patch embeddings by a trainable linear projection. Some extra learnable tokens are appended to the sequence, e.g., class token [CLS] [9,10] and part token [PART] [37]. The role of these tokens is to aggregate information from the patch sequence. [CLS] is proposed to learn a global representation for the input image and [PART] aims to extract part-level feature. There is no structural difference between [CLS] and [PART], while PASS will make them learn different-level features during training.

Student network and Teacher Network. The framework used for this work, PASS, shares a similar overall structure as the popular self-supervised approach, DINO [2]. The overview of PASS is illustrated in Fig. 2, where the learning paradigm of knowledge distillation is used. PASS contains a student network and a teacher network, and they share the same architecture. PASS trains the student f_{θ_s} to match the output of the teacher f_{θ_t}, which are parameterized by θ_s and θ_t respectively. Given an input image x, both networks predict probability distributions of K dimensions on the appended learnable tokens (i.e., [PART] and [CLS]) by projection heads. The probability P is obtained by normalizing the output of the network f with a softmax function. Specifically,

$$P_s^g(x)^{(t)} = \frac{\exp\left(f_{\theta_s}^{cls}(x)^{(t)}/\tau_s\right)}{\sum_{k=1}^{K} \exp\left(f_{\theta_s}^{cls}(x)^{(k)}/\tau_s\right)}, \tag{1}$$

where $f_{\theta_s}^{cls}(x)$ is the predicted distribution on [CLS], and t, k are the indexes of vector components. $\tau_s > 0$ is a temperature parameter that controls the sharpness of the output distribution. The similar formulas hold for P_t^g, $P_s^{l_i}$, $P_t^{l_i}$, where l_i means predicting on the ith [PART].

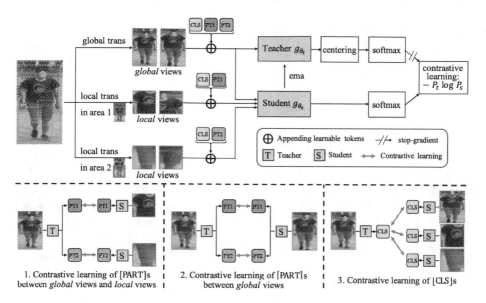

Fig. 2. The overview of PASS. We illustrate PASS in the case of $L = 2$ for simplicity. The global random transformation (global trans) crops out the *global* views from the whole images and the two local random transformations (local trans) crop out *local* views from two local areas, respectively. The *local* views from each local area are assigned a specific [PART]. The [PART]s of all local areas are also appended to the *global* views. [CLS] is also appended to all the views. In pre-training, all views pass through the student while only the *global* views pass through the teacher. Each network predicts several K dimensional features on [CLS] and [PART]s by the projection heads, and the features are normalized with a temperature softmax over the feature dimension. Then the similarities between the same [PART]/[CLS] output by student and teacher are measured with cross-entropy losses. We apply a stop-gradient operator on the teacher to propagate gradients only through the student. The teacher parameters are updated with an exponential moving average (ema) of the student parameters

The teacher f_{θ_t} is not pre-defined but built by means of the past iterations of the student f_{θ_s}. We freeze the teacher network over an epoch and use an exponential moving average on the student weights [2,16]. The update rule is:

$$\theta_t \leftarrow \lambda\theta_t + (1 - \lambda)\theta_s, \qquad (2)$$

with λ following a cosine schedule from 0.996 to 1 during training. The output of the teacher network is centered with a mean calculated over the batch to avoid collapse [2].

3.2   Part-Aware Self-supervised Pre-training

Given an input image x, PASS constructs a set of different views which includes different distorted views, or crops, of x. This set contains M *global* views, x_m^g, randomly cropped from the whole image, and $L \times J$ *local* views of smaller resolution randomly cropped from L local areas of the image, where the jth cropped view of the ith local area is denoted as $x_j^{l_i}, (i \in \{1, ..., L\}, j \in \{1, ..., J\})$.

All views are passed through the student while only the *global* views are passed through the teacher. For a *global* view, all learnable tokens, i.e., [CLS] and [PART]s, are appended to the sequence of patch embeddings. While for a *local* view cropped from the ith local area, only [CLS] and the ith part token [PART]$_i$ are appended to the patch embeddings. The projection heads are added to these extra learnable tokens in both teacher and student, and output the predicted probability distributions P. Specifically, given an input image x, its mth *global* view x_m^g is appended with [CLS] and all the [PART]s, and is passed through both the teacher and student. The predicted distributions by teacher and student are $P_t^g(x_m^g)$, $\{P_t^{l_i}(x_m^g)|i \in \{1, ..., L\}\}$ and $P_s^g(x_m^g)$, $\{P_s^{l_i}(x_m^g)|i \in \{1, ..., L\}\}$, respectively. Its *local* view $x_j^{l_i}$ is only appended with [CLS] and [PART]$_i$, and then is only passed through the student. The student predicts two probability distributions on [CLS] and [PART]$_i$: $P_s^g(x_j^{l_i})$ and $P_s^{l_i}(x_j^{l_i})$.

PASS learns to match the predictions of the same learnable tokens by cross-entropy loss. Specifically, **for** [PART]$_i$, the student outputs two types of predictions: (1) the prediction on [PART]$_i$ appended to $x_j^{l_i}$: $P_s^{l_i}(x_j^{l_i})$. (2) the prediction on [PART]$_i$ appended to x_m^g: $P_s^{l_i}(x_m^g)$. While the teacher only predicts the latter one and outputs $P_t^{l_i}(x_m^g)$. The student learns to match the output distributions of teacher by minimizing the cross-entropy loss w.r.t. the parameters of the student θ_s:

$$\min_{\theta_s} \left\{ \sum_{m=1}^{M} \sum_{j=1}^{J} H\left(P_t^{l_i}(x_m^g), P_s^{l_i}(x_j^{l_i})\right) + \sum_{m_1=1}^{M} \sum_{m_2=1}^{M} H\left(P_t^{l_i}(x_{m_1}^g), P_s^{l_i}(x_{m_2}^g)\right) \right\} \tag{3}$$

where $m_1 \neq m_2$ and $H(a, b) = -a \log b$. The two terms are corresponding to the contrastive learning 1 and 2 in the second row of Fig. 2. PASS applies this optimization to all the [PART]s. That is, any i in $\{1, ..., L\}$ is applicable in Eq. 3.

In the student, the *local* views assigned to each [PART] are randomly cropped from a specific local area, thus the [PART]s can focus on different areas. PASS uses the student to update the teacher, which can guarantee [PART]s in teacher also focus on different local areas and learns fine-grained information. In pre-training, the student learns to match the output of the teacher on the same [PART], which can guarantee each [PART] learns a robust local representation from the *local* views cropped in its corresponding local area.

For the [CLS], the predicted distributions on all the views are matched by:

$$\min_{\theta_s} \left\{ \sum_{m=1}^{M} \sum_{i=1}^{L} \sum_{j=1}^{J} H\left(P_t^g(x_m^g), P_s^g(x_j^{l_i})\right) + \sum_{m_1=1}^{M} \sum_{m_2=1}^{M} H\left(P_t^g(x_{m_1}^g), P_s^g(x_{m_2}^g)\right) \right\} \tag{4}$$

where $m_1 \neq m_2$. This formula makes the [CLS] learn the "local-to-global" correspondences and robust to the difficult scenarios, e.g., occlusion and incorrect detection, which are also why we still append [CLS] to the *local* views.

3.3   Fine-Tuning

Pre-trained by PASS, the [CLS] of ViT backbone has the capability to learn the global description of the input image, and the [PART]s can automatically focus on local areas and extract local representations. Therefore, in fine-tuning, we only need to append the [CLS] and all the [PART]s to the embedding sequence of the input image, and they can automatically focus on different-level features (we do not fixed them in fine-tuning). The teacher network is used for fine-tuning.

Supervised ReID. We concatenate the output [CLS] and the mean of output [PART]s, denoted by $[\overline{\text{Part}}]$, as the representation of the input image. The ReID head [22] is attached to the concatenated feature. Most of the training hyper-parameters are borrowed from the baseline of TransReID [19]. That is, **none** of the overlapping patch embedding, jigsaw patch module, or side information embedding is used in our fine-tuning. The commonly used cross-entropy loss and triplet loss with hard sample mining [20] are adopted to train our model, of which the cross-entropy loss is calculated as:

$$L_{cls} = -\log \mathcal{P}([\text{CLS}] \copyright [\overline{\text{Part}}]), \tag{5}$$

where $\mathcal{P}(\mathbf{x})$ is the probability of $\mathbf{x}$ belonging to its ground truth identity and $\copyright$ means concatenating.

The triplet loss is calculated as:

$$L_{tri} = [d_p - d_n + \alpha]_+, \tag{6}$$

where d_p and d_n are feature distances from positive pair and negative pair, respectively. α is the margin of triplet loss. $[\cdot]_+$ equals to $max(\cdot, 0)$. Therefore, the overall objective function for our model is:

$$L = L_{cls} + L_{tri}. \tag{7}$$

In the testing phase, [CLS] and $[\overline{\text{Part}}]$ are concatenated to represent a person image.

UDA/USL ReID. We follow most of the settings in C-Contrast [7] to conduct our experiments on UDA/USL ReID. Before training in an unsupervised manner on target datasets, UDA ReID need to be first pre-trained on source datasets while USL ReID does not need. Apart from this, UDA ReID and USL ReID share the same training settings. First, all the training images pass through the network to obtain the training data features. Then, clustering is conducted on these features to generate pseudo labels. By averaging the features with the same pseudo labels, the cluster prototypes are obtained. Next in training, the contrastive loss between the output features and the cluster prototypes are computed to optimize the network. More details can be found in [7].

4   Experiments

4.1   Implementation Details

Datasets. LUPerson [11], which contains 4.18 M unlabeled human images collected from 50,534 online videos, are used for pre-training. To evaluate the performance on ReID tasks, we conduct the fine-tuning experiments on two widely used benchmarks, i.e., Market-1501 [34] and MSMT17 [29], which contain 32,668 images of 1501 identities and 126,441 images of 4,101 identities, respectively. In fine-tuning, the images are resized to 256×128 unless mentioned otherwise. Following common practices, the cumulative matching characteristics (CMC) and the mean average precision (mAP) are used for evaluation.

Pre-training. For pre-training on LUPerson with PASS, the model is trained on $8 \times$A100 GPUs for 100 epochs, which costs about 60 h for ViT-S and 120 h for ViT-B. The divided local areas are with overlap to guarantee that the *local* view can be cropped from anywhere in the image. The *global* views are resized to 256×128 and the *local* views are resized to 96×48. Similar to DINO, we set $M = 2$ and $J = \lceil \frac{9}{L} \rceil$, e.g., $J = 3$ when $L = 3$ and $J = 5$ when $L = 2$.

Supervised ReID. To fine-tune the pre-trained Transformer, we use most of the training strategies of the baseline in TransReID [19], which means none of the overlapping patch embedding, jigsaw patch module, or side information embedding is used here. We set the learning rate to $lr = 0.0004 \times \frac{batchsize}{64}$ and warm up the model by 20 epochs [23]. The α in triplet loss is set to 0.3.

UDA/USL ReID. The ViTs are trained for 50 epochs and SGD is used. The initial learning rate is 3.5e-4 and is multiplied by 0.1 every 20 epochs. Each mini-batch contains 256 images of 32 persons, i.e., each ID contains 8 images.

4.2   Comparison with State-of-the-Art Methods

Supervised ReID. We compare PASS to some outstanding state-of-the-art methods on supervised ReID in Table 1. Compared with the methods pre-trained on ImageNet, our method significantly outperforms the existing best methods without adding any complex module to the backbone network, e.g., our ViT-B↑384 obtains 93.3%/74.3% mAP and 96.9%/89.7% Rank-1 accuracy on Market1501/MSMT17, outperforming the state-of-the-art results by 4.8%/8.5% and 1.2%/4.6%, respectively. It is noted that AAformer [37] and TransReID are pre-trained on *ImageNet*-21K which is much larger than LUPerson.

Compared with the self-supervised methods pre-trained on LUPerson, PASS also shows considerable superiority, e.g., our ViT-S↑384 obtains 92.6%/71.7% mAP and 96.8%/87.9% Rank-1 accuracy on Market1501/MSMT17 datasets, surpassing the existing best method (CFS [23]) by 1.1%/2.9% and 0.8%/1.8%, respectively. DINO can be regarded as the baseline of our method, and the results validate the remarkable effectiveness of using [PART] to offer fine-grained

Table 1. Comparison with the state-of-the-art methods of supervised ReID. Methods in the 1st group are pre-trained on ImageNet. Methods in the 2nd group are self-supervised methods pre-trained on LUPerson. Most of the self-supervised methods share the same fine-tuning settings as ours. The last group is our method. The results with underline are the best in their groups. "TransReID⁻" means side information and overlapping patches are removed for a fair comparison

Methods	Backbone	Market1501		MSMT17	
		mAP	Rank-1	mAP	Rank-1
BOT [22]	R50	85.9	94.5	50.2	74.1
MGN [28]	R50↑384	87.5	95.1	63.7	<u>85.1</u>
SCSN [6]	R50↑384	<u>88.5</u>	<u>95.7</u>	58.5	83.8
ABDNet [3]	R50↑384	88.3	95.6	60.8	82.3
AAformer [37]	ViT-B↑384	87.7	95.4	63.2	83.6
TransReID⁻ [19]	ViT-B	87.4	94.6	63.6	82.5
TransReID⁻ [19]	ViT-B↑384	87.6	94.6	<u>65.8</u>	84.4
MoCoV2 [4]	ViT-S	72.1	87.6	27.8	47.4
MoCoV2 [11]	MGN↑384	91.0	<u>96.4</u>	65.7	85.5
MoCoV3 [5]	ViT-S	82.2	92.1	47.4	70.3
MoBY [31]	ViT-S	84.0	92.9	50.0	73.2
DINO [2]	ViT-S	90.3	95.4	64.2	83.4
DINO + CFS [23]	ViT-S	91.0	96.0	66.1	84.6
DINO + CFS [23]	ViT-S↑384	<u>91.5</u>	96.0	<u>68.8</u>	<u>86.1</u>
PASS (ours)	ViT-S	92.2	96.3	69.1	86.5
PASS (ours)	ViT-S↑384	92.6	96.8	71.7	87.9
PASS (ours)	ViT-B	93.0	96.8	71.8	88.2
PASS (ours)	ViT-B↑384	**93.3**	**96.9**	**74.3**	**89.7**

Table 2. Comparison with the state-of-the-art methods of UDA ReID. Methods in the 1st group are pre-trained on ImageNet. Methods in the 2nd group are self-supervised methods pre-trained on LUPerson, and are fine-tuned with C-Contrast [7]

Methods	Backbone	MSMT2Market		Market2MSMT	
		mAP	Rank-1	mAP	Rank-1
DG-Net++ [38]	R50	64.6	83.1	22.1	48.4
MMT [12]	R50	75.6	83.9	24.0	50.1
SPCL [13]	R50	77.5	89.7	26.8	53.7
MCRN [30]	R50	-	-	32.8	<u>64.4</u>
C-Contrast [7]	R50	<u>82.4</u>	<u>92.5</u>	<u>33.4</u>	60.5
MoCoV2 [4]	R50	85.1	94.4	28.3	53.8
DINO [2]	ViT-S	88.5	95.0	43.9	67.7
DINO + CFS [23]	ViT-S	<u>89.4</u>	<u>95.4</u>	<u>47.4</u>	<u>70.8</u>
PASS (ours)	ViT-S	**90.2**	**95.8**	**49.1**	**72.7**

local information in PASS. Besides, Table 1 also shows that self-supervised pre-training on LUPerson is much more effective than supervised pre-training on ImageNet. The ICS module in [23] adds extra IBN [24] layers to the ViT backbone and increases the overall complexity, thus is not compared here for fairness.

UDA ReID. Some latest UDA-ReID methods are compared in the Table 2. Our ViT-S outperforms the existing best method by considerable margins. Specifically, ViT-S obtains 90.2%/49.1% (+0.8%/+1.7%) mAP and 95.8%/72.7% (+0.4%/+1.9%) Rank-1 accuracy on MSMT17 → Market1501 and Market1501 → MSMT17 respectively, which are already comparable to many supervised methods. Besides, PASS also surpasses its baseline method (DINO) by large margins, which validates local details are also vital for unsupervised learning in ReID.

USL ReID. We list some of the latest USL-ReID methods in Table 3, which shows our method outperforms the existing best method by 0.7% (94.9% *vs.* 94.2%) and 0.6% (67.0% *vs.* 66.4%) on Rank-1 accuracy on Market and MSMT17, respectively. The results validate PASS can provide a better initialization to the ViT model on USL ReID.

Table 3. Comparison with the state-of-the-art methods of USL ReID. Methods in the 1st group are pre-trained on ImageNet. Methods in the 2nd group are self-supervised methods pre-trained on LUPerson, and are fine-tuned with C-Contrast [7]

Methods	Backbone	Market1501		MSMT17	
		mAP	Rank-1	mAP	Rank-1
MMCL [27]	R50	45.5	80.3	11.2	35.4
HCT [33]	R50	56.4	80.0	-	-
IICS [32]	R50	72.9	89.5	26.9	52.4
MCRN [30]	R50	80.8	92.5	31.2	63.6
C-Contrast [7]	R50	82.6	93.0	33.1	63.3
MoCoV2 [4]	R50	84.0	93.4	31.4	58.8
DINO [2]	ViT-S	87.8	94.4	38.4	63.8
DINO + CFS [23]	ViT-S	88.2	94.2	40.9	66.4
PASS (ours)	ViT-S	**88.5**	**94.9**	**41.0**	**67.0**

4.3   Ablation Studies

The Choices of L. We first investigate the most suitable division strategy for PASS to divide the person images into several local areas. A *local* view accounts for up to 40% of the whole image, thus when $L = 2$, each local area should occupy 70% of the image (the same width as the original image and 70% height) to guarantee that the *local* view can be cropped from almost anywhere in the image. For the same reason, when $L = 3$, each local area occupies 50% of the image. We also follow MGN [28] to conduct the ablation experiments which use $L = \{2, 3\}$, where two kinds of division granularity are used at the same time. The results in Table 4 show that increasing the number of local areas does not always bring positive feedback, and PASS with $L = 3$ obtains the best performance. Therefore, we recommend dividing the image into 3 local areas with each area occupying 50% of the image.

Table 4. Ablation studies about the number of divided local areas L on supervised ReID

L	Backbone	Market1501		MSMT17	
		mAP	Rank-1	mAP	Rank-1
2	ViT-S	91.7	96.1	67.7	85.6
3	ViT-S	**92.2**	**96.3**	**69.1**	86.5
4	ViT-S	91.9	96.1	68.1	86.0
5	ViT-S	90.9	95.8	67.3	85.4
2,3	ViT-S	92.0	96.3	68.7	**86.6**

Feature Fusion. Next, we conduct experiments to investigate how to effectively integrate the output global feature and local features on supervised/UDA/USL ReID, which are shown in Table 5 and Table 6, respectively. One way is directly concatenating all these features, which is used by TransReID [19]. However, this way increases the dimension of feature embedding several times and cannot obtain the best performance in our experiments. We also conduct the ablation study which uses the mean feature or concatenates [CLS] and [Part]. The results show concatenating [CLS] and [Part] can obtain the best performance on supervised ReID. It is also worth noting that the strategy of "mean feature", which does not increase the feature dimension, already outperforms the existing state-of-the-art methods, some of which increase the feature dimension several times [19,28], by considerable margins. On UDA/USL ReID, MSMT17 dataset prefers the "mean feature" while Market1501 prefers concatenating [CLS] and [Part], which may be related to their data distributions.

Visualization. To further give an intuitive illustration of the effectiveness of PASS, we conduct visualization experiments to show the focus areas of [CLS] and [PART]s in the ViT backbone pre-trained by PASS. As illustrated in Fig. 3, given an input image, [CLS] focuses on the whole images to extract the global feature, and the [PART]s focus on different local areas, i.e., upper-body, waist, and legs, to extract part-level features, which can obtain more fine-grained information. More satisfactory, [PART]s can locate on specific semantic parts rather than simply locating different positions. Take the bottom-left sample in Fig. 3 as an example, when a man carries a box and the box blocks his upper body, the first [PART] does not focus on the box but focus on the visible part of his upper body. This makes the learned feature more robust as the man may carry boxes with different colors in other images. This visualization also validates PASS can well handle the occlusion scenarios. It is worth noting that we do not add any extra module to the ViT backbone to achieve this, but only pre-train the ViT backbone with PASS.

Table 5. Ablation studies about feature fusion on supervised ReID. ⓒ means concatenating. C is the number of channels. The first row concatenates all these features. The second row uses the mean feature. The third row concatenates [CLS] and [$\overline{\text{Part}}$]

Method	Dim	Backbone	Market1501		MSMT17	
			mAP	Rank-1	mAP	Rank-1
[CLS] ⓒ $\frac{[\text{PART}]_1}{L}$ ⓒ... $\frac{[\text{PART}]_L}{L}$	$(L+1)\times C$	ViT-S	90.8	96.1	67.7	85.2
		ViT-B	92.2	96.3	71.3	87.4
$\frac{1}{2}$([CLS]+[$\overline{\text{Part}}$])	C	ViT-S	92.0	**96.3**	68.7	86.3
		ViT-B	92.5	96.7	71.5	87.8
[CLS] ⓒ [$\overline{\text{Part}}$]	2C	ViT-S	**92.2**	**96.3**	**69.1**	**86.5**
		ViT-B	**93.0**	**96.8**	**71.8**	**88.2**

Table 6. Ablation studies about feature fusion on UDA/USL ReID. The backbone is ViT-S and ⓒ means concatenating. The first row concatenates all these features. The second row uses the mean feature. The third row concatenates [CLS] and [$\overline{\text{Part}}$]

Method	UDA reID				USL reID			
	MS2MA		MA2MS		Market1501		MSMT17	
	mAP	R-1	mAP	R-1	mAP	R-1	mAP	R-1
[CLS] ⓒ $\frac{[\text{PART}]_1}{L}$ ⓒ... $\frac{[\text{PART}]_L}{L}$	90.0	95.6	47.1	71.1	88.4	94.6	39.4	64.9
$\frac{1}{2}$([CLS]+[$\overline{\text{Part}}$])	89.8	95.5	**49.1**	**72.7**	**88.6**	94.6	**41.0**	**67.0**
[CLS] ⓒ [$\overline{\text{Part}}$]	**90.2**	**95.8**	44.9	69.4	88.5	**94.9**	36.6	62.5

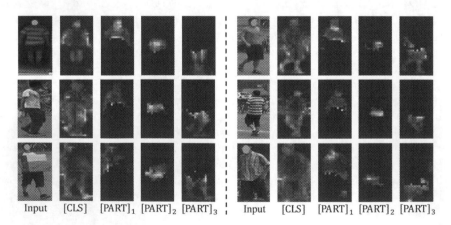

Input [CLS] [PART]₁ [PART]₂ [PART]₃ ¦ Input [CLS] [PART]₁ [PART]₂ [PART]₃

Fig. 3. Visualization of attention maps of [CLS] and [PART]s in the last self-attention layer of ViT pre-trained by PASS. For each [PART], the patches that have the maximal similarity with it among all the [PART]s are visualized. The examples in the last row show the visualization on occlusion and bad detection scenarios

Ranking List. Finally, we compare the ranking results of ViT pre-trained by PASS with DINO (baseline) in Fig. 4. *The first example shows even the person takes different boxes with different colors, our model can still find him in the massive images, which validates our model is not affected by the obstructions and focuses on the details of visible parts.* The other samples show the strong ability of our model in discriminating the extremely similar samples through the identifiable tiny clues, e.g., patterns on clothes. All these samples validate the superiority of PASS over the baseline method DINO.

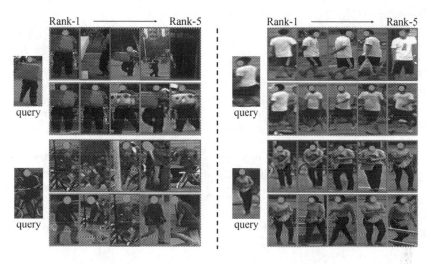

Fig. 4. The ranking lists of models pre-trained by DINO and PASS. For each query, the ranking list of the first row is from DINO and that of the second row is from PASS. The results show the PASS is largely superior to DINO in finding the fine-grained information of local details

5   Conclusion

In this paper, we propose a ReID-specific self-supervised pre-training method, Part-Aware Self-Supervised pre-training (PASS). Pre-trained by PASS, the models can automatically extract part-level features from the input images and offer more fine-grained information. Experimental results validate the powerful performance of PASS in both supervised ReID and UDA/USL ReID.

Acknowledgement. This work was supported by National Key R&D Program of China under Grant No. 2021ZD0110403, National Natural Science Foundation of China (No. 62002356, 61976210, 62002357, 62076235), Open Research Projects of Zhejiang Lab (No. 2021KH0AB07).

References

1. Caron, M., Misra, I., Mairal, J., Goyal, P., Bojanowski, P., Joulin, A.: Unsupervised learning of visual features by contrasting cluster assignments. arXiv preprint arXiv:2006.09882 (2020)
2. Caron, M., Touvron, H., Misra, I., Jégou, H., Mairal, J., Bojanowski, P., Joulin, A.: Emerging properties in self-supervised vision transformers. arXiv preprint arXiv:2104.14294 (2021)
3. Chen, T., Ding, S., Xie, J., Yuan, Y., Chen, W., Yang, Y., Ren, Z., Wang, Z.: Abd-net: attentive but diverse person re-identification. In: Proceedings of the IEEE International Conference on Computer Vision (2019)
4. Chen, X., Fan, H., Girshick, R., He, K.: Improved baselines with momentum contrastive learning. arXiv preprint arXiv:2003.04297 (2020)
5. Chen, X., Xie, S., He, K.: An empirical study of training self-supervised vision transformers. In: Proceedings of the IEEE/CVF International Conference on Computer Vision, pp. 9640–9649 (2021)
6. Chen, X., Fu, C., Zhao, Y., Zheng, F., Song, J., Ji, R., Yang, Y.: Salience-guided cascaded suppression network for person re-identification. In: Proceedings of the IEEE/CVF Conference on Computer Vision and Pattern Recognition, pp. 3300–3310 (2020)
7. Dai, Z., Wang, G., Zhu, S., Yuan, W., Tan, P.: Cluster contrast for unsupervised person re-identification. arxiv 2021. arXiv preprint arXiv:2103.11568 (2021)
8. Deng, J., Dong, W., Socher, R., Li, L.J., Li, K., Fei-Fei, L.: Imagenet: a large-scale hierarchical image database. In: 2009 IEEE Conference on Computer Vision and Pattern Recognition, pp. 248–255. IEEE (2009)
9. Devlin, J., Chang, M.W., Lee, K., Toutanova, K.: Bert: pre-training of deep bidirectional transformers for language understanding. In: Proceedings of NAACL-HLT, pp. 4171–4186 (2019)
10. Dosovitskiy, A., Beyer, L., Kolesnikov, A., Weissenborn, D., Zhai, X., Unterthiner, T., Dehghani, M., Minderer, M., Heigold, G., Gelly, S., et al.: An image is worth 16×16 words: Transformers for image recognition at scale. arXiv preprint arXiv:2010.11929 (2020)
11. Fu, D., et al.: Unsupervised pre-training for person re-identification. In: Proceedings of the IEEE/CVF Conference on Computer Vision and Pattern Recognition, pp. 14750–14759 (2021)
12. Ge, Y., Chen, D., Li, H.: Mutual mean-teaching: Pseudo label refinery for unsupervised domain adaptation on person re-identification. arXiv preprint arXiv:2001.01526 (2020)
13. Ge, Y., Zhu, F., Chen, D., Zhao, R., Li, H.: Self-paced contrastive learning with hybrid memory for domain adaptive object re-id. arXiv preprint arXiv:2006.02713 (2020)
14. Grill, J.B., Strub, F., Altché, F., Tallec, C., Richemond, P., Buchatskaya, E., Doersch, C., Avila Pires, B., Guo, Z., Gheshlaghi Azar, M., et al.: Bootstrap your own latent-a new approach to self-supervised learning. Adv. Neural. Inf. Process. Syst. **33**, 21271–21284 (2020)
15. Guo, H., Wu, H., Zhao, C., Zhang, H., Wang, J., Lu, H.: Cascade attention network for person re-identification. In: 2019 IEEE International Conference on Image Processing (ICIP), pp. 2264–2268. IEEE (2019)
16. He, K., Fan, H., Wu, Y., Xie, S., Girshick, R.: Momentum contrast for unsupervised visual representation learning. In: Proceedings of the IEEE/CVF Conference on computer Vision and Pattern Recognition, pp. 9729–9738 (2020)

17. He, K., Zhang, X., Ren, S., Sun, J.: Deep residual learning for image recognition. In: Proceedings of the IEEE Conference on Computer Vision and Pattern Recognition, pp. 770–778 (2016)
18. He, L., Liao, X., Liu, W., Liu, X., Cheng, P., Mei, T.: Fastreid: A pytorch toolbox for general instance re-identification. arXiv preprint arXiv:2006.02631 (2020)
19. He, S., Luo, H., Wang, P., Wang, F., Li, H., Jiang, W.: Transreid: Transformer-based object re-identification. In: Proceedings of the IEEE International Conference on Computer Vision (2021)
20. Hermans, A., Beyer, L., Leibe, B.: In defense of the triplet loss for person re-identification. arXiv preprint arXiv:1703.07737 (2017)
21. Kalayeh, M.M., Basaran, E., Gökmen, M., Kamasak, M.E., Shah, M.: Human semantic parsing for person re-identification. In: Proceedings of the IEEE Conference on Computer Vision and Pattern Recognition, pp. 1062–1071 (2018)
22. Luo, H., Gu, Y., Liao, X., Lai, S., Jiang, W.: Bag of tricks and a strong baseline for deep person re-identification. In: Proceedings of the IEEE Conference on Computer Vision and Pattern Recognition Workshops, pp. 0–0 (2019)
23. Luo, H., Wang, P., Xu, Y., Ding, F., Zhou, Y., Wang, F., Li, H., Jin, R.: Self-supervised pre-training for transformer-based person re-identification. arXiv preprint arXiv:2111.12084 (2021)
24. Pan, X., Luo, P., Shi, J., Tang, X.: Two at once: Enhancing learning and generalization capacities via ibn-net. In: Proceedings of the European Conference on Computer Vision (ECCV), pp. 464–479 (2018)
25. Sun, Y., Zheng, L., Yang, Y., Tian, Q., Wang, S.: Beyond part models: Person retrieval with refined part pooling (and a strong convolutional baseline). In: Proceedings of the European Conference on Computer Vision (ECCV), pp. 480–496 (2018)
26. Vaswani, A., et al.: Attention is all you need. In: Proceedings of the 31st International Conference on Neural Information Processing Systems, pp. 6000–6010 (2017)
27. Wang, D., Zhang, S.: Unsupervised person re-identification via multi-label classification. In: Proceedings of the IEEE/CVF Conference on Computer Vision and Pattern Recognition, pp. 10981–10990 (2020)
28. Wang, G., Yuan, Y., Chen, X., Li, J., Zhou, X.: Learning discriminative features with multiple granularities for person re-identification. In: 2018 ACM Multimedia Conference on Multimedia Conference, pp. 274–282. ACM (2018)
29. Wei, L., Zhang, S., Gao, W., Tian, Q.: Person transfer gan to bridge domain gap for person re-identification. In: Proceedings of the IEEE Conference on Computer Vision and Pattern Recognition, pp. 79–88 (2018)
30. Wu, Y., et al.: Multi-centroid representation network for domain adaptive person re-id. In: Proceedings of the AAAI Conference on Artificial Intelligence (2022)
31. Xie, Z., et al.: Self-supervised learning with swin transformers. arXiv preprint arXiv:2105.04553 (2021)
32. Xuan, S., Zhang, S.: Intra-inter camera similarity for unsupervised person re-identification. In: Proceedings of the IEEE/CVF Conference on Computer Vision and Pattern Recognition, pp. 11926–11935 (2021)
33. Zeng, K., Ning, M., Wang, Y., Guo, Y.: Hierarchical clustering with hard-batch triplet loss for person re-identification. In: Proceedings of the IEEE/CVF Conference on Computer Vision and Pattern Recognition, pp. 13657–13665 (2020)
34. Zheng, L., Shen, L., Tian, L., Wang, S., Wang, J., Tian, Q.: Scalable person re-identification: a benchmark. In: Computer Vision, IEEE International Conference on Computer Vision, pp. 1116–1124 (2015)

35. Zhu, K., Guo, H., Liu, S., Wang, J., Tang, M.: Learning semantics-consistent stripes with self-refinement for person re-identification. IEEE Trans. Neural Networks Learn. Syst., 1–12 (2022). https://doi.org/10.1109/TNNLS.2022.3151487
36. Zhu, K., Guo, H., Liu, Z., Tang, M., Wang, J.: Identity-guided human semantic parsing for person re-identification, pp. 346–363 (2020)
37. Zhu, K., et al.: AAformer: Auto-Aligned transformer for person re-identification. arXiv preprint arXiv:2104.00921 (2021)
38. Zou, Y., Yang, X., Yu, Z., Kumar, B., Kautz, J.: Joint disentangling and adaptation for cross-domain person re-identification. In: European Conference on Computer Vision. pp. 87–104. Springer (2020)

Adaptive Cross-domain Learning for Generalizable Person Re-identification

Pengyi Zhang[1], Huanzhang Dou[1], Yunlong Yu[2(✉)], and Xi Li[1,3,4(✉)]

[1] College of Computer Science and Technology,
Zhejiang University, Hangzhou, China
{pyzhang,hzdou,xilizju}@zju.edu.cn
[2] College of Information Science and Electronic Engineering,
Zhejiang University, Hangzhou, China
yuyunlong@zju.edu.cn
[3] Shanghai Institute for Advanced Study, Zhejiang University, Hangzhou, China
[4] Shanghai AI Laboratory, Hangzhou, China
https://github.com/peterzpy/ACL-DGReID

Abstract. Domain Generalizable Person Re-Identification (DG-ReID) is a more practical ReID task that is trained from multiple source domains and tested on the unseen target domains. Most existing methods are challenged for dealing with the shared and specific characteristics among different domains, which is called the domain conflict problem. To address this problem, we present an Adaptive Cross-domain Learning (ACL) framework equipped with a CrOss-Domain Embedding Block (CODE-Block) to maintain a common feature space for capturing both the domain-invariant and the domain-specific features, while dynamically mining the relations across different domains. Moreover, our model adaptively adjusts the architecture to focus on learning the corresponding features of a single domain at a time without interference from the biased features of other domains. Specifically, the CODE-Block is composed of two complementary branches, a dynamic branch for extracting domain-adaptive features and a static branch for extracting the domain-invariant features. Extensive experiments demonstrate that the proposed approach achieves state-of-the-art performances on the popular benchmarks. Under Protocol-2, our method outperforms previous SOTA by 7.8% and 7.6% in terms of mAP and rank-1 accuracy.

Keywords: Adaptive cross-domain learning · Common feature space · Dynamic network · Domain conflict · Domain generalizable person re-identification

1   Introduction

Recently, Domain Generalizable Person Re-Identification (DG-ReID), a more practical ReID task that aims at identifying samples from unseen domains

Supplementary Information The online version contains supplementary material available at https://doi.org/10.1007/978-3-031-19781-9_13.

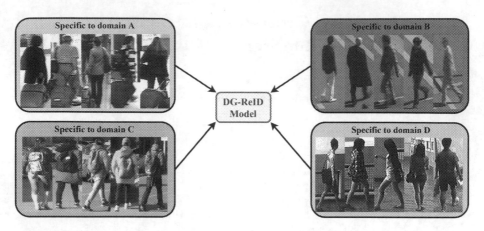

Fig. 1. Examples of multiple person re-identification domains. Different domains are biased on their specific characteristics (e.g., hue, illumination, resolution, clothing style, and carrying objects apart from the common characteristics).

without domain adaptation, has attracted increasing attention due to the conventional ReID approaches suffering from significant performance degradation on the unseen domains. To address this task, some efforts [5,24,25,40,41,43,51,61] propose to jointly train a static ReID model with multiple seen domains and then directly apply it to the unseen domains. However, as shown in Fig. 1, there is a huge gap between different domains (e.g., illumination, hue, resolution, clothing style, and carrying objects), where each domain has its specific characteristics apart from the common characteristics. During joint training, the specific characteristics of one domain may be useless or even interfere with the learning of other domains, which is called the *domain conflict problem* [27,28,45].

To mitigate the domain conflict problem, the early approaches [21,38,57,60] focus on modeling the invariant features between different domains with disentanglement learning or meta-learning. However, these methods do not consider the diverse and complementary information of domain-specific features enough, which may limit the generalization capability on the unseen target domain [63]. Recently, in addition to modeling invariant features, some approaches [2,42,59] also focus on capturing the domain-specific features via designing specific expert networks for the individual domains. Such ways of maintaining a specific space for each domain help reduce the cross-domain interference during training but hardly transfer the specific knowledge from different spaces and capture the specific features from unseen domains. Besides, these methods are challenged by the problem of linear increasing costs, since the number of individual feature spaces must be consistent with the number of domains.

To overcome the above limitations, in this paper, we propose to improve the existing methods from two perspectives: maintaining a common space for both domain-invariant and domain-specific features, and adaptively capturing features through the dynamically adjusted architecture. First, maintaining a common space could capture the relations between different domains, thus benefiting in the knowledge transfer across different domains and avoiding the redundant modeling among domains. Second, capturing the domain-adaptive features through adaptive dynamic architectures enables the model to learn specific features for each novel domain and alleviate the conflicts among domains.

For this purpose, we propose a novel framework called **A**daptive **C**ross-domain **L**earning (ACL) to dynamically capture the adaptive features with both invariant and specific features for each domain. Specifically, we design a **CrO**ss-**D**omain **E**mbedding Block (CODE-Block), which is composed of three components, i.e., a dynamic branch, a static branch, and a fusion module. The dynamic branch is designed with a series of parallel feature embedding networks to reduce the cross-domain interference, each of which captures either the fine-grained domain-invariant or domain specific features, and a domain-aware adapter to produce meta-weights for adaptively capturing domain-adaptive information. The static branch serves as a complement to the dynamic branch to additionally extract the domain-invariant features among domains and makes the whole training process more stable. Following [53], the static branch adopts a style normalization layer for filtering out domain-specific contrast information. By doing this, our framework could focus on learning the corresponding features of a single domain at a time without interference from other domain-specific features. As a result, the domain conflict issue would be alleviated.

In a nutshell, our highlights include:

- To the best of our knowledge, we are the first to adopt the dynamic network to adaptively learn features from different domains for tackling the domain conflict problem in DG-ReID.
- We develop a novel framework equipped with a CODE-Block to adaptively capture both the domain-invariant and the domain-specific features in a common feature space, and aggregate them through a domain-aware adapter to adaptively learn different domains.
- Extensive experiments demonstrate the effectiveness of our framework under multiple testing protocols. The proposed approach obtains 7.8% and 7.6% improvements in terms of mAP and rank-1 accuracy over the SOTA method on the average results of four large-scale benchmarks under Protocol-2.

2   Related Work

2.1   Domain Generalization

Domain Generalization [1,7,15,20,29,31,34,36,44,55,64] is a solution to the potential domain shift problem in practice, which can be categorized into two types of solutions. 1) Representation learning. These methods reduce the discrepancy between different domains by disentanglement learning or probability distribution alignment. For instance, [54] learns a feature transformation to minimize the variances of the class-conditional distributions among multiple source domains for all classes. 2) Domain augmentation. These methods generate more cross-domain samples to regularize the model for avoiding overfitting and improving generalization ability. For example, [39] trains a domain classifier and use its adversarial gradients to generate the cross-domain samples.

Although previous DG methods obtain remarkable performance on closed-set tasks (e.g., classification), they are based on the premise that all the training and testing domains share the same label space. However, there is no overlap of identities under the retrieval tasks like ReID. As a result, it is difficult to perform well when directly applying these methods to ReID.

2.2 Domain Generalizable Person Re-identification

With the application of ReID technology in practice, several DG methods tailored for ReID [5,24,25,40,41,43,51,61] have been proposed recently. There are two main categories of DG-ReID methods. 1) Domain-invariant feature modeling [4,18,19,21,37,38,50,57,60]. These methods aim to learn the shared features among multiple source domains, which could reduce the biases in the single domain and obtain a more robust generalization capability. Specifically, these methods could be achieved by domain-adversarial learning [38], feature disentanglement learning [57], joint training with meta-learning [60], or some normalization-based strategies [37]. 2) Domain-specific feature modeling [2,6,42,56,59]. These methods aim to leverage the diversity of each source domain for complementary learning through modeling the relevance between seen source domains and unseen target domains. Typically, these methods are implemented through Mixture-of-Expert (MoE), where each expert extracts domain-specific features from the corresponding domain. As for the relevance between different domains, it can be calculated using the similarity between features of different domains [56] or the similarity between the IN/BN statistics of different domains [2], as well as directly predicted through a specific network [59].

Compared with the previous methods, our approach can model both the invariant/specific features in a common feature space. Besides, our dynamic network with the cross-domain embedding can better model the cross-domain relevance and reduce the redundant modeling for each domain than previous MoE-based methods.

2.3 Dynamic Neural Networks

Dynamic neural networks [58], including dynamic architecture [13,16,22,65] and dynamic parameters [8,9,11,33,35,52], are wildly used in many fields for their satisfactory representation capability, adaptiveness, and generality. These methods usually adjust the model architectures to allocate appropriate computation based on each sample, which is more adaptive to the specific sample. Motivated by this, we build a dynamic network to solve the domain conflicts in DG-ReID through domain-adaptive architecture adjusting, then learn the meta-knowledge among different domains implicitly for better generalization performance.

3 Adaptive Cross-domain Learning Framework

In this work, we present a novel **A**daptive **C**ross-domain **L**earning (ACL) framework to mitigate the domain conflict issue for DG-ReID via adaptively modeling

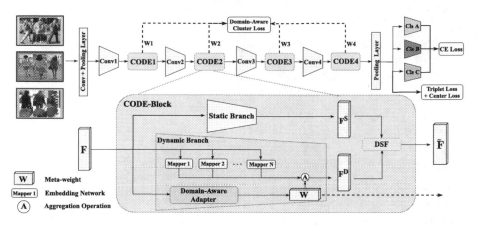

Fig. 2. Illustration of our method. Each conv block means a stage of convolution blocks without the last one. We replace the last convolutional block in each stage with our CODE-Block. In the CODE-Block, we extract a dynamic domain-adaptive feature F^D and a static domain-invariant feature F^S, then we fuse these two features through a dynamic-static fusion module (DSF). Notably, to reduce the domain conflicts, we calculate the cross-entropy loss for each domain by individual classifiers, respectively.

both the common features and specific features from different input domains in a common feature space with a dynamic architecture. In this section, we first give an overview of ACL and then introduce the core component, a novel **CrOss-Domain Embedding** Block (CODE-Block), followed by the detailed objective functions and the optimization process of our ACL framework.

3.1 Overview

As illustrated in Fig. 2, our ACL framework is designed by plugging the CODE-Block into the different layers of the feature extractor architecture, respectively for extracting different levels of semantics (e.g., hue, and contrast in the low-level, and carrying objects, viewpoint, and clothing style in the high-level). To reduce the computation cost, the CODE-Block is plugged into the architecture by replacing the final convolutional blocks at each stage.

3.2 Cross-domain Embedding Block

As shown in Fig. 2, the CODE-Block is designed for modeling both the domain-invariant/specific features from different domains, which consists of a dynamic branch, a static branch, and a fusion module. Let $F \in \mathbb{R}^{H \times W \times C}$ be the input feature map of CODE-Block, where H, W, and C indicate the height, width, and the number of channels, respectively. The CODE-Block is processed as follows:

1) Dynamic branch extracts fine-grained features $\{F_i^D\}_{i=1}^N$ for capturing domain-invariant or domain-specific features through several parallel embedding networks, where N is the number of embedding networks. Then discriminative

domain-adaptive features $F^D \in \mathbb{R}^{H \times W \times C}$ are aggregated with the guidance of the meta-weights $W \in \mathbb{R}^N$ generated by domain-aware adapter; 2) Static branch extracts the robust domain-invariant features $F^S \in \mathbb{R}^{H \times W \times C}$ from the input feature F; 3) Fusion module integrates the F^S from static branch and the F^D from dynamic branch into the final output feature $\tilde{F} \in \mathbb{R}^{H \times W \times C}$.

Dynamic Branch. Dynamic Branch is the core design of the CODE-Block, which consists of several parallel *embedding networks* to model a common feature space for all domains and a *domain-aware adapter* to guide the combination of the discriminative domain-adaptive features for each domain.

1) Embedding Networks. Due to the fact that different domains significantly differ in hue, carrying objects, resolution, and many other principles, it is hard to capture good features for the target domains with a single network, thus we design a series of parallel networks to capture both domain-invariant and domain-specific features in an implicit way, each of which aims at capturing domain-invariant features or domain-specific features for each domain.

Specifically, we build N parallel embedding networks (i.e., bottleneck block in ResNet-IBN but with different initialization). For a light computation cost, we reduce the number of channels for intermediate features to $1/N$. The fine-grained features $\{F_i^D\}_{i=1}^N$ from i-th embedding network are obtained with:

$$F_i^D = Embed_i(F), \; s.t. \; i \in [1 \cdots N], \tag{1}$$

where $Embed_i$ indicates the i-th embedding network. Note that all the fine-grained features from different embedding networks are embedded into the same space, which is the so-called common feature space.

Parallel embedding networks could capture the fine-grained domain-invariant and domain-specific features as the bases for the following combination. In contrast to previous methods [2,56,59] that model the features for each domain using domain-level individual experts, our strategy simultaneously learns domain-invariant and domain-specific features in multiple parallel embedding networks, thus reducing the risk of cross-domain interference and redundant modeling.

2) Domain-Aware Adapter. To adaptively capture the features for each domain, as shown in Fig. 3, the Domain-Aware Adapter produces meta-weights with a weight generator for aggregating the fine-grained features extracted from all embedding networks. The weight generator takes the original features as input and produces the meta-weights. To further enhance the sample-adaptive capability, we propose to generate the parameters of Domain-Aware Adapter with a meta-parameter generator, which is formulated with:

$$W_g = \delta(W_2 \delta(W_1 pool(F))), \tag{2}$$

which consists of a global average pooling layer followed by two FC layers that are parameterized by $W_1 \in \mathbb{R}^{\frac{C}{r} \times C}$ and $W_2 \in \mathbb{R}^{NC \times \frac{C}{r}}$ with a reduction ratio r that is set to 16, and the $\delta(\cdot)$ indicates the ReLU activation function.

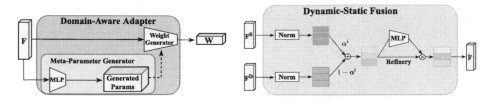

Fig. 3. Architecture of the domain-aware adapter.

Fig. 4. Architecture of the fusion module. Best viewed in color. (Color figure online)

To this end, the meta-weights are obtained with $W = \sigma(W_g pool(F))$, where W has the dimension of $\mathbb{R}^N$ and $\sigma(\cdot)$ denotes the Softmax function. Through the domain-aware combination of the cross-domain features, we integrate the final domain-adaptive feature with $F^D = \sum_{i=1}^{N} W_i F_i^D$.

Static Branch. Apart from the dynamic branch that dynamically extracts discriminative domain-adaptive features for each domain, we additionally maintain a static branch to extract the domain-invariant features for robust lower-bound performance. Specifically, the Static Branch has the same architecture as a conventional bottleneck block in ResNet, but we replace the batch normalization (BN) layers in the bottleneck block with the instance normalization (IN) layers. IN can be regarded as a style normalization [53], which filters the specific features in different domains to reduce the discrepancy of samples, has been widely used in many previous methods [37,50]. Therefore, we could statically extract the robust domain-invariant features for stabilizing and complementing the adaptive features extracted from the dynamic branch.

Fusion Module. Once the dynamic domain-adaptive feature F^D and static domain-invariant feature F^S are extracted, we further develop a fusion module called Dynamic-Static Fusion (DSF) to integrate these two features for complementation. As shown in Fig. 4, DSF first respectively normalizes the two input features, then sums up them with an optimized weight vector $\alpha \in \mathbb{R}^C$, finally refines the fused feature using channel attention [17], which is formulated as:

$$\tilde{F} = \alpha * Norm_S(F^S) + (1 - \alpha) * Norm_D(F^D),$$
$$\tilde{F} = \tilde{F} * Sigmoid(W_2^r \delta(W_1^r pool(\tilde{F}))), \tag{3}$$

where $*$ notes the elementwise multiplication, $\delta(\cdot)$ denotes the ReLU function, W_1^r and W_2^r are FC layers parameterized with $\mathbb{R}^{\frac{C}{r} \times C}$ and $\mathbb{R}^{C \times \frac{C}{r}}$, respectively.

3.3 Objective Function

To train the ACL framework, four loss functions are applied: 1) Domain-Aware Cluster Loss $\mathcal{L}_{cluster}$. To make the generated meta-weights being aware of different domains, we adopt a domain-level cluster regularization to increase the inter-domain variances, while narrowing down the intra-domain variances. Given the predicted meta-weights of all these modules $W \in \mathbb{R}^{M*N}$, where M is the number of CODE-Block and N is the number of embedding networks in a CODE-Block, the intra-domain loss $\mathcal{L}_{intra}$ and the inter-domain loss $\mathcal{L}_{inter}$ are calculated as follows,

$$\mathcal{L}_{intra} = \frac{1}{K \times L} \sum_{i=1}^{K} \sum_{j=1}^{L} [\||W_{i,j} - \overline{W_{i,:}}\||_2 - m_1]_+^2,$$

$$\mathcal{L}_{inter} = \frac{1}{K(K-1)} \sum_{i=1}^{K} \sum_{j \neq i}^{K} [m_2 - ||\overline{W_{i,:}} - \overline{W_{j,:}}||_2]_+^2,$$

(4)

where K is the number of domains, m_1 and m_2 are the margins, L is the number of instances for the corresponding domain, $W_{i,j}$ means the j-th instance of the domain i, and $\overline{W_{i,:}}$ means the cluster center of the domain i. As a result, the final cluster loss $\mathcal{L}_{cluster}$ can be aggregated as follows,

$$\mathcal{L}_{cluster} = \mathcal{L}_{intra} + \mathcal{L}_{inter}.$$

(5)

2) Cross-Entropy Loss $\mathcal{L}_{ce}$, Triplet Loss $\mathcal{L}_{tri}$, and Center Loss $\mathcal{L}_{cntr}$. These three losses follow the standard formulation in previous works [30]. Notably, to further reduce the conflicts between different domains, we calculate the cross-entropy loss for each domain, respectively.

3.4   Optimization Process

To improve the generalization capability on the unseen target domains, we adopt a meta-learning algorithm to simulate the unseen domain scenarios following [37]. As shown in Algorithm 1, the training process of our ACL framework is divided into two stages, a basic training process to train the whole network, and a meta-learning process for optimizing the domain-aware adapter module.

For the first basic training stage, the objective function $\mathcal{L}_{basic}$ consists of cross-entropy loss, triplet loss, center loss, and our cluster loss,

$$\mathcal{L}_{basic}(\mathcal{X}; \theta; \phi) = \mathcal{L}_{ce} + \mathcal{L}_{tri} + \mathcal{L}_{cntr} + \mathcal{L}_{cluster},$$

(6)

where all the parameters in the whole network (i.e., θ for the domain-aware adapter and ϕ for others) will be updated by optimizing these loss functions. For the second meta-learning stage, we follow the process and objective functions of [37], which adds two cluster losses $\mathcal{L}_{scat}$ and $\mathcal{L}_{shuf}$ for a more discriminative feature representation. Specifically, $\mathcal{L}_{scat}$ is to make the feature distribution more compact in each domain and $\mathcal{L}_{shuf}$ is to avoid the model collapse for each domain. We split the K seen domains into $K - 1$ meta-train domains and a meta-test domain, respectively. Then we calculate the objective function $\mathcal{L}_{mtr}$ for the meta-training stage to calculate the temporary optimized parameters θ' for domain-aware adapter as follows,

$$\mathcal{L}_{mtr}(\mathcal{X}; \theta; \phi) = \mathcal{L}_{tri} + \mathcal{L}_{cluster} + \mathcal{L}_{scat} + \mathcal{L}_{shuf}.$$

(7)

Algorithm 1: Training for ACL framework

Input: Source domains $\mathcal{D} = \{\mathcal{D}_1, \mathcal{D}_2, \cdots, \mathcal{D}_K\}$; Learning rate γ; MaxIters; Meta-learning Frequency $\mathcal{F}$.

Output: Trained network $F(\theta; \phi)$ with domain-aware adapter θ and other parameters ϕ.

1   **for** *iter in MaxIters* **do**

2   **Basic Training Stage**:

3   Sample a mini-batch $\mathcal{X}_B$ from $\mathcal{D}$.

4        $\mathcal{L}_{basic}(\mathcal{X}_B; \theta; \phi) = \mathcal{L}_{ce} + \mathcal{L}_{tri} + \mathcal{L}_{cntr} + \mathcal{L}_{cluster}$

5        $(\theta, \phi) \leftarrow (\theta - \gamma\nabla_\theta\mathcal{L}_{basic}(\mathcal{X}_B; \theta; \phi), \phi - \gamma\nabla_\phi\mathcal{L}_{basic}(\mathcal{X}_B; \theta; \phi))$

6   **Meta-Learning Stage**:

7   **if** iter % $\mathcal{F}$ == 0:

8   Split $\mathcal{D}$ as $(\mathcal{D}_{mtr} \cap \mathcal{D}_{mte} = \varnothing, \mathcal{D}_{mtr} \cup \mathcal{D}_{mte} = \mathcal{D})$

9   **Meta-Training**:

10   Sample a mini-batch $\mathcal{X}_S$ from $\mathcal{D}_{mtr}$.

11        $\mathcal{L}_{mtr}(\mathcal{X}_S; \theta; \phi) = \mathcal{L}_{tri} + \mathcal{L}_{cluster} + \mathcal{L}_{scat} + \mathcal{L}_{shuf}$

12        $\theta' = \theta - \gamma\nabla_\theta\mathcal{L}_{mtr}(\mathcal{X}_S; \theta; \phi)$

13   **Meta-Testing**:

14   Sample a mini-batch $\mathcal{X}_T$ from $\mathcal{D}_{mte}$.

15        $\mathcal{L}_{mte}(\mathcal{X}_T; \theta'; \phi) = \mathcal{L}_{tri} + \mathcal{L}_{cntr} + \mathcal{L}_{cluster}$

16        $\theta = \theta - \gamma\nabla_\theta\mathcal{L}_{mte}(\mathcal{X}_T; \theta'; \phi)$

With the temporary parameters θ', we can obtain the final optimized parameters θ by optimizing the loss for the meta-testing stage, which is formulated by

$$\mathcal{L}_{mte}(\mathcal{X}; \theta'; \phi) = \mathcal{L}_{tri} + \mathcal{L}_{cntr} + \mathcal{L}_{cluster}. \tag{8}$$

4   Experiments

4.1   Implementation Details

We use ResNet50 with IBN [51] pretrained on ImageNet as our backbone. Similar to previous methods [12,14], we set the stride of the last layer as 1. Images are resized to 256×128 and the training batch size is set to 64, including 32 identities and two instances for each identity. For data augmentation, we use random horizontal flipping, random cropping, color jittering, and auto augmentation [26]. We optimize the model using the SGD optimizer with a momentum of 0.9 and weight decay of 5e-4 for 60 epochs, and the warmup strategy is used in the first 10 epochs. The margin for intra-domain loss m_1 and inter-domain loss m_2 are set to 0.1 and 0.3, respectively. The meta-learning frequency $\mathcal{F}$ is set to 3. The initial learning rate is set to 4e-2, which is cosine decayed to 4e-5 at the final iteration. We conduct all the experiments with PyTorch on four 1080Ti GPUs.

4.2   Datasets and Evaluation Settings

Datasets. We conduct experiments on several person re-identification benchmarks: Market1501 [23], MSMT17 [27], CUHK02 [48], CUHK03 [47], CUHK-SYSU [46], PRID [32], GRID [3], VIPeR [10], and iLIDs [49]. For CUHK03, we use

Table 1. Comparison with state-of-the-art methods under Protocol-1. We achieve better generalization performance even with fewer data for training (i.e., without DukeMTMC dataset). '*' denotes the re-implementation for the work under the new protocol, based on the author's code on Github. Best results are highlighted in **bold**.

Method	Reference	Source	Target								Average	
			PRID		GRID		VIPeR		iLIDs			
			mAP	R1	mAP	R1	mAP	R1	mAP	R1	mAP	R1
DIMN [18]	CVPR19	M+D+C2 +C3+CS	52.0	39.2	41.1	29.3	60.1	51.2	78.4	70.2	57.9	47.5
SNR [50]	CVPR20		66.5	52.1	47.7	40.2	61.3	52.9	89.9	84.1	66.4	57.3
RaMoE [59]	CVPR21		67.3	57.7	54.2	46.8	64.6	56.6	**90.2**	**85.0**	69.1	61.5
DMG-Net [56]	CVPR21		68.4	60.6	56.6	51.0	60.4	53.9	83.9	79.3	67.3	61.2
QAConv$_{50}$* [24]	ECCV20	M+C2+C3+CS	62.2	52.3	57.4	48.6	66.3	57.0	81.9	75.0	67.0	58.2
M^{3}L* [60]	CVPR21		64.3	53.1	55.0	44.4	66.2	57.5	81.5	74.0	66.8	57.3
MetaBIN* [37]	CVPR21		70.8	61.2	57.9	50.2	64.3	55.9	82.7	74.7	68.9	60.5
Ours		M+C2+C3+CS	**73.4**	**63.0**	**65.7**	**55.2**	**75.1**	**66.4**	86.5	81.8	**75.2**	**66.6**

the 'labeled' data following [59,66]. We do not use the DukeMTMC [62] dataset since its privacy issues. The statistics of these datasets can be seen in supplementary material.

For simplicity, we denote Market1501, MSMT17, CUHK02, CUHK03, CUHK-SYSU as M, MS, C2, C3, and CS in the following.

Evaluation Settings. The mean Average Precision (mAP) and Cumulative Matching Characteristics (CMC) are used for evaluation. There are three evaluation protocols following [2,21,59].

For Protocol-1, the model is trained by both the training and testing data in M+C2+C3+CS datasets and then tested on four small datasets (i.e., PRID, GRID, VIPeR, and iLIDs), respectively. Since some of these datasets have no official split for probe and gallery sets, we randomly split the probe/gallery sets for ten times and report the averaged evaluation performance as the final result. For Protocol-2 and Protocol-3, we orderly use one of the datasets' data for testing (i.e., only the testing data) and the remaining datasets (Protocol-2 only uses training data, whereas Protocol-3 uses both training and testing data of source domains) for training the model. Notably, all ablation studies below are conducted under Protocol-2. More details can be seen in supplementary material.

4.3 Comparison with the State-of-the-Arts

Comparison with DG-ReID Methods Under Protocol-1. We compare our method with previous DG-ReID methods under Protocol-1, which are tested on four datasets (i.e., PRID, GRID, VIPeR, and iLIDs). As shown in Table 1, our method could outperform previous methods by a large margin.

Table 2. Comparison with state-of-the-art methods under Protocol-2 and Protocol-3. Four large-scale datasets are involved in the leave-one-out setting.

Setting	Method	Reference	M+MS+CS→C3		M+CS+C3→MS		MS+CS+C3→M		Average	
			mAP	R1	mAP	R1	mAP	R1	mAP	R1
Protocol-2	SNR* [50]	CVPR20	8.9	8.9	6.8	19.9	34.6	62.7	16.8	30.5
	QAConv$_{50}^*$ [24]	ECCV20	25.4	24.8	16.4	45.3	63.1	83.7	35.0	51.3
	M^{3}L* [60]	CVPR21	34.2	34.4	16.7	37.5	61.5	82.3	37.5	51.4
	MetaBIN* [37]	CVPR21	28.8	28.1	17.8	40.2	57.9	80.1	34.8	49.5
	Ours		**41.2**	**41.8**	**20.4**	**45.9**	**74.3**	**89.3**	**45.3**	**59.0**
Protocol-3	SNR* [50]	CVPR20	17.5	17.1	7.7	22.0	52.4	77.8	25.9	39.0
	QAConv$_{50}^*$ [24]	ECCV20	32.9	33.3	17.6	46.6	66.5	85.0	39.0	55.0
	M^{3}L* [60]	CVPR21	35.7	36.5	17.4	38.6	62.4	82.7	38.5	52.6
	MetaBIN* [37]	CVPR21	43.0	43.1	18.8	41.2	67.2	84.5	43.0	56.3
	Ours		**49.4**	**50.1**	**21.7**	**47.3**	**76.8**	**90.6**	**49.3**	**62.7**

Comparison with DG-ReID Methods Under Protocol-2 and Protocol-3. We also conduct the experiments under the leave-one-out setting with different amounts of data. As shown in Table 2, under these two protocols with large-scale datasets, our method still maintains the superiority in learning a more generalizable feature representation from multiple domains.

4.4   Ablation Study

Effectiveness of the Static/Dynamic Branches. As we have stated above, our CODE-Block consists of two branches, a static branch to extract the domain-invariant features and a dynamic branch to extract the domain-adaptive features. As shown in Table 3, we explore the effects of these two branches to demonstrate the effectiveness of our framework. 'Baseline' indicates a vanilla model without our CODE-Block module, which follows the same training setting as our basic training stage. Our training pipeline shows a great improvement, which could achieve a considerable performance even with just the dynamic branch adopted, showing the effectiveness of the adaptive feature modeling. And the performance can be further improved with the complement of the static branch.

Effectiveness of the Domain-Aware Adapter. Our domain-aware adapter adopts the *meta-parameter generator* to predict the parameter weights of the weight generator for adaptively generate *domain-aware meta-weights* of each sample. The effectiveness of these components is explored in Table 4, respectively.

We first compare our pipeline with a standard MLP-style weight generator, which also consists of two FC layers (i.e., $W_1 \in \mathbb{R}^{\frac{C}{r} \times C}$ and $W_2 \in \mathbb{R}^{N \times \frac{C}{r}}$) but directly predicts the combination weights. Compared to this counterpart, our pipeline can be more adaptive to each sample, which better reduces the conflicts among domains. Moreover, the meta-parameter generator can be regarded as a meta-learning process [9], which further enhances the generalization capability of our model. Therefore, our pipeline surpasses the standard MLP by a considerable margin (i.e., 1.7% mAP and 2.0% rank-1 accuracy).

Table 3. Ablation study on the effectiveness of the static and dynamic branches.

Method	M+MS+CS→C3		M+CS+C3→MS		MS+CS+C3→M		Average	
	mAP	R1	mAP	R1	mAP	R1	mAP	R1
Baseline	34.1	34.4	17.3	42.7	69.2	87.0	40.2	54.7
Static branch	35.4	35.1	17.2	41.7	69.4	87.8	40.7	54.9
Dynamic branch	37.5	37.7	18.7	43.4	72.2	88.2	42.8	56.4
Static+Dynamic (ACL)	**41.2**	**41.8**	**20.4**	**45.9**	**74.3**	**89.3**	**45.3**	**59.0**

Table 4. Ablation study on the effectiveness of the meta-parameter generator and the generated domain-aware meta-weights of domain-aware adapter (DAA).

Method	M+MS+CS→C3		M+CS+C3→MS		MS+CS+C3→M		Average	
	mAP	R1	mAP	R1	mAP	R1	mAP	R1
Standard MLP	38.9	37.8	19.3	44.6	72.6	88.7	43.6	57.0
Meta-Parameter (ACL)	**41.2**	**41.8**	**20.4**	**45.9**	**74.3**	**89.3**	**45.3**	**59.0**
Identity weight	37.9	38.2	19.4	45.3	72.9	88.7	43.4	57.4
Softmax w/o DAA	40.0	39.4	19.6	45.2	73.0	89.0	44.2	57.9
Softmax w/ DAA (ACL)	**41.2**	**41.8**	**20.4**	**45.9**	**74.3**	**89.3**	**45.3**	**59.0**

As shown in Table 4, we demonstrate the effectiveness of our domain-aware meta-weights by comparing it with the naive ensemble model (i.e., the sub mapping features are average summed). Compared with this variant, our domain-aware meta-weights could fully utilize the fine-grained modeling of cross-domain embedding without conflicts among domains. Besides, we also demonstrate the effectiveness of our domain-aware cluster loss for the meta-weights. As shown in Table 4, with the domain-level cluster loss, we could explicitly model the relationship between weights and the domain, and improve the averaged performance by 1.1% mAP and 1.1% rank-1 accuracy.

Effectiveness of the Fusion Module. As shown in Table 5, we demonstrate the effectiveness of the three components in the DSF module, respectively. Without the fusion module, fusing the two branches features using direct summation helps but improves marginally. With the proposed normalization layer, weighted summation layer, and the final refinery channel attention layer, these two features can be fused better and achieve a better generalization performance. As a result, with the fusion module, the averaged performance is improved by 1.8% mAP and 2.3% rank-1 accuracy, respectively.

Table 5. Ablation study on the effectiveness of the fusion module.

Method	M+MS+CS→C3		M+CS+C3→MS		MS+CS+C3→M		Average	
	mAP	R1	mAP	R1	mAP	R1	mAP	R1
W/O DSF	38.4	37.7	19.4	44.2	72.6	88.2	43.5	56.7
+Norm	39.3	38.6	20.0	45.3	72.8	89.0	44.0	57.6
+Weighted sum	40.0	40.4	20.1	45.6	74.1	89.1	44.7	58.4
+Refinery (ACL)	41.2	41.8	20.4	45.9	74.3	89.3	45.3	59.0

Table 6. Ablation study on the effects of different embedding stages for the CODE-Block. 'Stage-1' notes that only embed it in the first stage of the backbone network.

Method	M+MS+CS→C3		M+CS+C3→MS		MS+CS+C3→M		Average	
	mAP	R1	mAP	R1	mAP	R1	mAP	R1
Stage-1	37.1	35.9	18.2	42.4	72.7	88.2	42.7	55.5
Stage-2	37.4	37.1	19.0	43.6	72.8	87.9	43.1	56.2
Stage-3	37.8	36.8	19.0	43.8	73.0	88.1	43.3	56.2
Stage-4	38.1	38.2	19.2	44.2	73.5	88.3	43.6	56.9
Stages-all (ACL)	41.2	41.8	20.4	45.9	74.3	89.3	45.3	59.0

Effects of Different Stages for Embedding CODE-Block. As an embedding module, CODE-Block can be plugged into multiple stages of the backbone network. In this part, we explore the effects of different embedding stages. Specifically, we compare our framework with four variants embedding CODE-Block in different stages. As shown in Table 6, embedding in higher semantic level turns to have a better performance. Moreover, the combination of multiple CODE-Blocks from different semantic levels could better model the features across domains and obtain a better generalization performance.

The Number of Embedding Networks for CODE-Block. As shown in Fig. 5, we analyze the effects of the number N of parallel embedding networks in a CODE-Block. In general, the more embedding networks embedded, the more diverse and fine-grained features can be modeled, but there will be more optimization challenges at the same time. As a result, we choose to use four embedding networks in each CODE-Block for better generalization performance.

4.5   Visualization

To further demonstrate the effectiveness of our domain-aware adapter, we visualize the distribution of generated meta-weights from different domains. As shown in Fig. 6, the meta-weights can be well clustered into their corresponding domains while also having good diversity in each domain cluster.

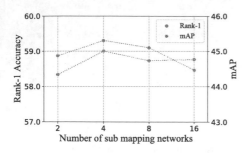
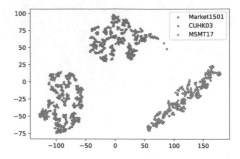

Fig. 5. The averaged performance under Protocol-2 with the different number of parallel embedding networks for the CODE-Block. Best viewed in color. (Color figure online)

Fig. 6. The t-SNE visualization of meta-weights generated by the domain-aware adapter for target data from different domains. Best viewed in color. (Color figure online)

5   Discussion

Since the ReID system may be abused and violate people's privacy, governments and officials must create regulations and legislation for governing the use of the ReID system. However, person ReID is an established computer vision problem with known benchmarks. The research for ReID technology under authorized datasets should not be forbidden for the ethical implications.

6   Conclusion

In this paper, we have proposed a generalizable framework, called Adaptive Cross-Domain Learning (ACL) for tackling the problem of domain generalizable person re-identification (DG-ReID). Our framework is equipped with a novel embedding CrOss-Domain Embedding Block (CODE-Block) that enables the model to adaptively adjust the architecture and capture the adaptive features with both the domain-invariant and domain-specific features in a common feature space, which could further integrate the cross-domain relevance. Specifically, two branches have been designed to extract the domain-adaptive and domain-invariant features, then we introduced a fusion module to integrate these two features for complementation. Extensive experiments have shown the effectiveness of our framework, which achieves state-of-the-art generalization performance.

Acknowledgement. This work is supported in part by Zhejiang Provincial Natural Science Foundation of China under Grant LR19F020004, National Key Research and Development Program of China under Grant 2020AAA0107400, National Natural Science Foundation of China under Grant U20A20222, Key R&D Program of Zhejiang Province, China (2021C01119), the National Natural Science Foundation of China under Grant (62002320, U19B2043).

References

1. Khosla, A., Zhou, T., Malisiewicz, T., Efros, A.A., Torralba, A.: Undoing the damage of dataset bias. In: Fitzgibbon, A., Lazebnik, S., Perona, P., Sato, Y., Schmid, C. (eds.) ECCV 2012. LNCS, vol. 7572, pp. 158–171. Springer, Heidelberg (2012). https://doi.org/10.1007/978-3-642-33718-5_12

2. Boqiang, X., Jian, L., Lingxiao, H., Zhenan, S.: Meta: Mimicking embedding via others' aggregation for generalizable person re-identification. arXiv preprint arXiv:2112.08684 (2021)

3. Change, L.C., Tao, X., Shaogang, G.: Time-delayed correlation analysis for multi-camera activity understanding. IJCV **90**(1), 106–129 (2010)

4. Chen, P., et al.: Dual distribution alignment network for generalizable person re-identification. In: AAAI, pp. 1054–1062 (2021)

5. Luo, C., Song, C., Zhang, Z.: Generalizing person re-identification by camera-aware invariance learning and cross-domain mixup. In: Vedaldi, A., Bischof, H., Brox, T., Frahm, J.-M. (eds.) ECCV 2020. LNCS, vol. 12360, pp. 224–241. Springer, Cham (2020). https://doi.org/10.1007/978-3-030-58555-6_14

6. Ci-Siang, L., Yuan-Chia, C., Frank, W.Y.C.: Domain generalized person re-identification via cross-domain episodic learning. In: ICPR, pp. 6758–6763 (2021)

7. Da, L., Yongxin, Y., Yi-Zhe, S., M, H.T.: Learning to generalize: Meta-learning for domain generalization. In: AAAI, pp. 3490–3497 (2018)

8. David, H., Andrew, D., V, L.Q.: Hypernetworks. arXiv preprint arXiv:1609.09106 (2016)

9. Dominic, Z., von Oswald, J., Seijin, K., João, S., Grewe, B.F.: Meta-learning via hypernetworks. In: NeurIPS (2020)

10. Gray, D., Tao, H.: Viewpoint invariant pedestrian recognition with an ensemble of localized features. In: Forsyth, D., Torr, P., Zisserman, A. (eds.) ECCV 2008. LNCS, vol. 5302, pp. 262–275. Springer, Heidelberg (2008). https://doi.org/10.1007/978-3-540-88682-2_21

11. Feihu, Z., W, W.B.: Supplementary meta-learning: towards a dynamic model for deep neural networks. In: ICCV, pp. 4344–4353 (2017)

12. Fengliang, Q., Bo, Y., Leilei, C., Hongbin, W.: Stronger baseline for person re-identification. arXiv preprint arXiv:2112.01059 (2021)

13. Gao, H., Shichen, L., der Maaten Laurens, V., Q, W.K.: Condensenet: an efficient densenet using learned group convolutions. In: CVPR, pp. 2752–2761 (2018)

14. Hao, L., et al.: A strong baseline and batch normalization neck for deep person re-identification. TMM **22**(10), 2597–2609 (2019)

15. Haoliang, L., Jialin, P.S., Shiqi, W., C, K.A.: Domain generalization with adversarial feature learning. In: CVPR, pp. 5400–5409 (2018)

16. Hubara, I., Courbariaux, M., Soudry, D., El-Yaniv, R., Bengio, Y.: Binarized neural networks. In: NeurIPS, pp. 4107–4115 (2016)

17. Jie, H., Li, S., Gang, S.: Squeeze-and-excitation networks. In: CVPR, pp. 7132–7141 (2018)

18. Song, J., Yang, Y., Song, Y.Z., Xiang, T., Hospedales, T.M.: Generalizable person re-identification by domain-invariant mapping network. In: CVPR, pp. 719–728 (2019)

19. Kaiwen, Y., Xinmei, T.: Domain-class correlation decomposition for generalizable person re-identification. arXiv preprint arXiv:2106.15206 (2021)

20. Kaiyang, Z., Ziwei, L., Yu, Q., Tao, X., Change, L.C.: Domain generalization: a survey. arXiv preprint arXiv:2103.02503 (2021)

21. Kecheng, Z., Jiawei, L., Wei, W., Liang, L., Zheng-jun, Z.: Calibrated feature decomposition for generalizable person re-identification. arXiv preprint arXiv:2111.13945 (2021)
22. Lanlan, L., Jia, D.: Dynamic deep neural networks: optimizing accuracy-efficiency trade-offs by selective execution. In: AAAI, pp. 3675–3682 (2018)
23. Liang, Z., Liyue, S., Lu, T., Shengjin, W., Jingdong, W., Qi, T.: Scalable person re-identification: a benchmark. In: CVPR, pp. 1116–1124 (2015)
24. Liao, S., Shao, L.: Interpretable and generalizable person re-identification with query-adaptive convolution and temporal lifting. In: Vedaldi, A., Bischof, H., Brox, T., Frahm, J.-M. (eds.) ECCV 2020. LNCS, vol. 12356, pp. 456–474. Springer, Cham (2020). https://doi.org/10.1007/978-3-030-58621-8_27
25. Lingxiao, H., et al.: Semi-supervised domain generalizable person re-identification. arXiv preprint arXiv:2108.05045 (2021)
26. Lingxiao, H., Xingyu, L., Wu, L., Xinchen, L., Peng, C., Tao, M.: Fastreid: a pytorch toolbox for general instance re-identification. arXiv preprint arXiv:2006.02631 (2020)
27. Longhui, W., Shiliang, Z., Wen, G., Qi, T.: Person transfer gan to bridge domain gap for person re-identification. In: CVPR, pp. 79–88 (2018)
28. Lu, Y., Lingqiao, L., Yunlong, W., Peng, W., Yanning, Z.: Multi-domain joint training for person re-identification. arXiv preprint arXiv:2201.01983 (2022)
29. M, C.F., Antonio, D., Silvia, B., Barbara, C., Tatiana, T.: Domain generalization by solving jigsaw puzzles. In: CVPR, pp. 2229–2238 (2019)
30. Mang, Y., Jianbing, S., Gaojie, L., Tao, X., Ling, S., CH, H.S.: Deep learning for person re-identification: a survey and outlook. arXiv preprint arXiv:2001.04193 (2020)
31. Maniyar, U., Joseph, K.J., Deshmukh, A.A., Dogan, Ü., Balasubramanian, V.N.: Zero-shot domain generalization. arXiv preprint arXiv:2008.07443 (2020)
32. Hirzer, M., Beleznai, C., Roth, P.M., Bischof, H.: Person re-identification by descriptive and discriminative classification. In: Heyden, A., Kahl, F. (eds.) SCIA 2011. LNCS, vol. 6688, pp. 91–102. Springer, Heidelberg (2011). https://doi.org/10.1007/978-3-642-21227-7_9
33. Misha, D., Babak, S., Laurent, D., Marc'Aurelio, R., Nando, D.F.: Predicting parameters in deep learning. In: NeurIPS, pp. 2148–2156 (2013)
34. Muhammad, G., Bastiaan, K.W., Mengjie, Z., David, B.: Domain generalization for object recognition with multi-task autoencoders. In: ICCV, pp. 2551–2559 (2015)
35. Ma, N., Zhang, X., Huang, J., Sun, J.: WeightNet: revisiting the design space of weight networks. In: Vedaldi, A., Bischof, H., Brox, T., Frahm, J.-M. (eds.) ECCV 2020. LNCS, vol. 12360, pp. 776–792. Springer, Cham (2020). https://doi.org/10.1007/978-3-030-58555-6_46
36. Riccardo, V., Vittorio, M.: Addressing model vulnerability to distributional shifts over image transformation sets. In: ICCV, pp. 7980–7989 (2019)
37. Seokeon, C., Taekyung, K., Minki, J., Hyoungseob, P., Changick, K.: Meta batch-instance normalization for generalizable person re-identification. In: CVPR, pp. 3425–3435 (2021)
38. Lin, S., Li, C.T., Kot, A.C.: Multi-domain adversarial feature generalization for person re-identification. TIP 30, 1596–1607 (2020)
39. Shankar, S., Piratla, V., Chakrabarti, S., Chaudhuri, S., Jyothi, P., Sarawagi, S.: Generalizing across domains via cross-gradient training. In: ICLR (2018)
40. Shengcai, L., Ling, S.: Graph sampling based deep metric learning for generalizable person re-identification. arXiv preprint arXiv:2104.01546 (2021)

41. Shengcai, L., Ling, S.: Transmatcher: deep image matching through transformers for generalizable person re-identification. In: NeurIPS (2021)
42. Shijie, Y., et al.: Multiple domain experts collaborative learning: multi-source domain generalization for person re-identification. arXiv preprint arXiv:2105.12355 (2021)
43. Shiyu, X., Shiliang, Z.: Intra-inter camera similarity for unsupervised person re-identification. In: CVPR, pp. 11926–11935 (2021)
44. Wang, S., Yu, L., Li, C., Fu, C.-W., Heng, P.-A.: Learning from extrinsic and intrinsic supervisions for domain generalization. In: Vedaldi, A., Bischof, H., Brox, T., Frahm, J.-M. (eds.) ECCV 2020. LNCS, vol. 12354, pp. 159–176. Springer, Cham (2020). https://doi.org/10.1007/978-3-030-58545-7_10
45. Tong, X., Hongsheng, L., Wanli, O., Xiaogang, W.: Learning deep feature representations with domain guided dropout for person re-identification. In: CVPR, pp. 1249–1258 (2016)
46. Tong, X., Shuang, L., Bochao, W., Liang, L., Xiaogang, W.: End-to-end deep learning for person search. arXiv preprint arXiv:1604.01850 (2016)
47. Wei, L., Rui, Z., Tong, X., Xiaogang, W.: Deepreid: deep filter pairing neural network for person re-identification. In: CVPR, pp. 152–159 (2014)
48. Wei, L., Xiaogang, W.: Locally aligned feature transforms across views. In: CVPR, pp. 3594–3601 (2013)
49. Wei-Shi, Z., Shaogang, G., Tao, X.: Associating groups of people. In: BMVC, pp. 1–11 (2009)
50. Xin, J., Cuiling, L., Wenjun, Z., Zhibo, C., Li, Z.: Style normalization and restitution for generalizable person re-identification. In: CVPR, pp. 3143–3152 (2020)
51. Pan, X., Luo, P., Shi, J., Tang, X.: Two at once: enhancing learning and generalization capacities via IBN-Net. In: Ferrari, V., Hebert, M., Sminchisescu, C., Weiss, Y. (eds.) ECCV 2018. LNCS, vol. 11208, pp. 484–500. Springer, Cham (2018). https://doi.org/10.1007/978-3-030-01225-0_29
52. Jia, X., De Brabandere, B., Tuytelaars, T., Gool, L.V.: Dynamic filter networks, pp. 667–675 (2016)
53. Xun, H., Serge, B.: Arbitrary style transfer in real-time with adaptive instance normalization. In: Proceedings of the IEEE International Conference on Computer Vision, pp. 1501–1510 (2017)
54. Ya, L., Mingming, G., Xinmei, T., Tongliang, L., Dacheng, T.: Domain generalization via conditional invariant representations. In: AAAI, pp. 3579–3587 (2018)
55. Li, Y., Tian, X., Gong, M., Liu, Y., Liu, T., Zhang, K., Tao, D.: Deep domain generalization via conditional invariant adversarial networks. In: Ferrari, V., Hebert, M., Sminchisescu, C., Weiss, Y. (eds.) ECCV 2018. LNCS, vol. 11219, pp. 647–663. Springer, Cham (2018). https://doi.org/10.1007/978-3-030-01267-0_38
56. Yan, B., et al.: Person30k: a dual-meta generalization network for person re-identification. In: CVPR, pp. 2123–2132 (2021)
57. Yi-Fan, Z., et al.: Learning domain invariant representations for generalizable person re-identification. arXiv preprint arXiv:2103.15890 (2021)
58. Yizeng, H., Gao, H., Shiji, S., Le, Y., Honghui, W., Yulin, W.: Dynamic neural networks: a survey. TPAMI (2021)
59. Yongxing, D., Xiaotong, L., Jun, L., Zekun, T., Ling-Yu, D.: Generalizable person re-identification with relevance-aware mixture of experts. In: CVPR, pp. 16145–16154 (2021)
60. Yuyang, Z., et al.: Learning to generalize unseen domains via memory-based multi-source meta-learning for person re-identification. In: CVPR, pp. 6277–6286 (2021)

61. Zhang, E., et al.: One for more: Selecting generalizable samples for generalizable reid model. In: AAAI, pp. 3324–3332 (2021)
62. Zhedong, Z., Liang, Z., Yi, Y.: Unlabeled samples generated by gan improve the person re-identification baseline in vitro. In: ICCV, pp. 3754–3762 (2017)
63. Zhou, K., Yang, Y., Qiao, Y., Xiang, T.: Domain adaptive ensemble learning. TIP **30**, 8008–8018 (2021)
64. Zhou, K., Yang, Y., Qiao, Y., Xiang, T.: Domain generalization with mixstyle. In: ICLR (2021)
65. Zhuang, L., Jianguo, L., Zhiqiang, S., Gao, H., Shoumeng, Y., Changshui, Z.: Learning efficient convolutional networks through network slimming. In: ICCV, pp. 2736–2744 (2017)
66. Zhun, Z., Liang, Z., Donglin, C., Shaozi, L.: Re-ranking person re-identification with k-reciprocal encoding. In: CVPR, pp. 1318–1327 (2017)

Multi-query Video Retrieval

Zeyu Wang$^{(\boxtimes)}$, Yu Wu , Karthik Narasimhan , and Olga Russakovsky

Princeton University, Princeton, USA
{zeyuwang,yuwu,karthikn,olgarus}@cs.princeton.edu

Abstract. Retrieving target videos based on text descriptions is a task of great practical value and has received increasing attention over the past few years. Despite recent progress, imperfect annotations in existing video retrieval datasets have posed significant challenges on model evaluation and development. In this paper, we tackle this issue by focusing on the less-studied setting of multi-query video retrieval, where multiple descriptions are provided to the model for searching over the video archive. We first show that multi-query retrieval task effectively mitigates the dataset noise introduced by imperfect annotations and better correlates with human judgement on evaluating retrieval abilities of current models. We then investigate several methods which leverage multiple queries at training time, and demonstrate that the multi-query inspired training can lead to superior performance and better generalization. We hope further investigation in this direction can bring new insights on building systems that perform better in real-world video retrieval applications (Code is available at https://github.com/princetonvisualai/MQVR).

Keywords: Video retrieval · Multi-query · Evaluation

1 Introduction

With the vast amount of videos available online and new ones being generated and uploaded everyday, an efficient video search engine is of great practical value. The core of a video search engine is the retrieval task, which deals with finding the best matching videos based on the user's text query [10,16,33,37,39]. With collective efforts over the years, current state-of-the-art methods [3,17, 36] have achieved reasonable performance on several different video retrieval benchmarks [4,9,28,43,55].

Despite these advances, existing works focus mostly on experimenting under the single-query setting, *i.e.*, retrieving target videos given a *single* text description as input. A major issue with this paradigm is the problem of under-specification. As most existing video retrieval benchmarks are based on datasets collected for other tasks (*e.g.*, video captioning) [9,28,43,50,55], lots of text descriptions are of low-quality and unsuited for video retrieval. Figure 1 top

Supplementary Information The online version contains supplementary material available at https://doi.org/10.1007/978-3-031-19781-9_14.

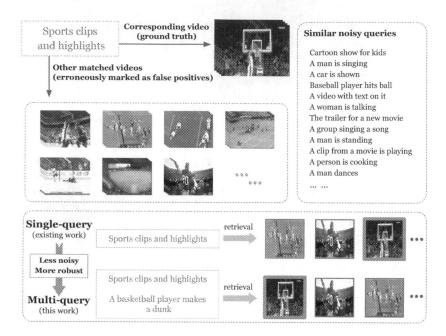

Fig. 1. Existing video retrieval works focus on the single-query setting, where a single text description is input for retrieving target video. However, due to imperfect annotations in the datasets (too general descriptions that match many different videos, shown on top), retrieval model may be unfairly penalized for not finding the single target video. To counteract this under-specification issue, in this work we tackle the previously less-studied *multi-query* setting (bottom), which provides a less noisy benchmarking of retrieval abilities of current models.

shows some examples of noisy annotations from one of the most widely-used video retrieval datasets MSR-VTT [55]. One can notice that these annotations are very general and can be perfectly matched to lots of videos. This can cause trouble during evaluation as all matched videos would be given similar matching scores and essentially ranked randomly, thus resulting in a wrong estimate of the model's true performance. One way to solve such problem potentially is to build better datasets with cleaner annotations. However, it is extremely hard and laborious to build a "perfect" dataset and many of the most widely used ones are all with different imperfections [1,13]. Especially, latest models are more and more often being trained with large-scale web-scraped data [24,42], which inevitably contains various kind of noise.

In this work, we propose the task of *multi-query video retrieval* (MQVR) where multiple text queries are provided to the model simultaneously (Fig. 1 bottom). It addresses the noise effects introduced by imperfect annotations in a simple and effective way. Intuitively, it is less likely that there are many videos in the dataset *all* matching *all* the provided queries. As a result, MQVR can produce a more accurate evaluation of the model's performance compared to a

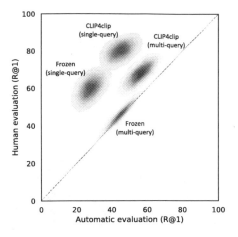
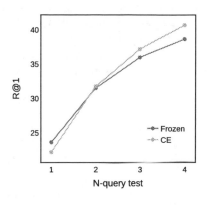

Fig. 2. Left: Density plot of human evaluation vs. automatic evaluation for two video retrieval models [8,30] on MSR-VTT [55]. Single-query retrieval is done by selecting the video with the highest similarity with the input query; for multi-query retrieval, the average similarity with all queries is used. Human evaluation (obtained by manual inspection of the top retrieved video) is compared to automatic evaluation (a binary indicator of whether the input queries are the corresponding annotations in the dataset for the top retrieved video). The results are computed on 100 retrieval samples, and the the plot is generated with Gaussian kernel density estimation on 10,000 samples generated with bootstrapping. Multi-query automatic evaluation shows significantly better correlation with human judgement compared to the single-query counterpart. **Right**: Relative performance of models [8,33] could change when evaluated with a different number of queries, suggesting that multi-query setting may reveal important insights about the retrieval potential of different models. The two models depicted are collaborative experts (CE) [33] and Frozen [8]; both models are trained from scratch on MSR-VTT [55] for a fair comparison.

single-query retrieval benchmark. We verify this intuition empirically by inspecting 100 retrieval results, and manually labeling whether the top retrieved video matches the provided query or queries. As shown in Fig. 2 left, single-query evaluation has a large gap with human judgement: automated methods unfairly penalize models for retrieving videos which actually match the provided query (as per human assessment) but are not the single video intended to be retrieved by the benchmark. Multi-query evaluation significantly reduces the gap by making the retrieval task less under-specified, so the model can better retrieve the target video (higher automatic evaluation score) and by making it less likely for other videos to match input queries. Going further, in Fig. 2 right we show that this has practical consequences as the relative ordering of two retrieval models changes under the single-query vs. multi-query evaluation settings.

Equipped with these insights, we re-purpose existing video retrieval datasets and perform an extensive investigation of state-of-the-art models adapted towards handling multiple queries. We first experiment with different ways of using multiple queries only during inference, and then explore ways of learning retrieval models explicitly with multiple queries provided at training time. Our experiments over three different datasets demonstrate that MQVR more optimally utilizes the capabilities of modern retrieval systems, and can provide new insights for building better retrieval models. Especially, we find that a multi-query inspired training can lead to representations with superior performance and better generalization.

To summarize, we make the following contributions:

- We extensively investigate multi-query video retrieval (MQVR) – where multiple text queries are provided to the model – over multiple video retrieval datasets (MSR-VTT [55], MSVD [9], VATEX [50]).
- We find that dedicated multi-query training methods can provide large gains (up to 21% in R@1) over simply treating each query independently and combining their similarity scores. We also investigate several architectural changes such as localized and contextualized weighting mechanisms which further improve MQVR performance.
- To facilitate future research in this area, we further propose a new metric based on the area under the curve for MQVR with varying number of queries, which is complementary to standard metrics used in single-query retrieval.
- Finally, we also demonstrate that multi-query training methods can be utilized to benefit standard single-query retrieval, and that the multi-query trained representations have better generalization than the single-query trained counterpart.

2   Related Work

Video-Text Representation Learning. The multi-modal learning for videos [2,5,7,14,19,40,45,53,54,56] has been receiving increasing attention in recent years. Among them, learning to generate better video and language representation [19,48,57,59] is one of the most attractive topics. Recent works leverage large-scale vision-language datasets for model pretraining [8,27,31,37]. Howto100M [38] is a large-scale video-text pretraining dataset, which takes texts obtained from narrated videos using automatic speech recognition (ASR) as the supervision for video-text representation learning. Miech *et al.* [37] then proposed a multiple instance learning approach derived from noise contrastive estimation (MIL-NCE) to learn from these noisy instructional video-text pairs. Lei *et al.* [29] proposed ClipBERT which leverages sparse sampling to train the model in an end-to-end approach. Recently, CLIP (Contrastive Language-Image Pre-training) [42] has shown great success by pre-training on large-scale webly-supervised image and text pairs. The learned image and language representations are proved to be helpful in most downstream vision-language tasks.

Text-to-Video Retrieval. To search videos by unconstrained text input, both video and text modalities should be embedded into a shared representation space for further similarity matching. Early works [20,33,37,39,49] used pre-trained models to extract representations from multi-modal data (RGB, motion, audio). Mithun *et al.* [39] proposed to use hard negative mining to improve the training for text and video joint modeling. Liu *et al.* [33] leverage seven modalities (*e.g.*, speech content, scene texts, faces) to build the video representation. Miech *et al.* [37] further proposed a strong joint embedding framework based on mixture-of-expert features. Gabeur *et al.* [20] introduce multi-modal transformer to jointly encode those different modalities with attentions. Recently, there are some works [17,21,30,36] based on the large-scale pretrained vision-language models (*e.g.*, CLIP [42]), which achieve superior performance compared to previous training-from-scratch models. However, these works are evaluated using single-query benchmarks, suffering from the query quality issue. Differently, we study the multi-query video retrieval in this paper, where the input is a set of multiple queries.

Multi-query Retrieval. Many previous studies on multi-query retrieval are in the uni-modal domain, such as content-based image retrieval [18,25,47], landmark retrieval [51,52], or person retrieval using multiple images or video frames [32,58]. These methods take multiple images as input queries where each image is obtained from different angles, distances, conditions of the same objects or scenes [47,52]. Instead of getting multiple queries from the user, works on query expansion [6,11,12,22,23,26,41] take a single query as input and further extend it to multiple queries by pre-processing (*e.g.*, synonym replacement) or using relevant candidates produced during an initial ranking. In this work, we utilize multi-query in the task of text-to-video retrieval to address challenges brought by imperfect annotations in the datasets.

3 Multi-query Video Retrieval

In this section, we first formally define the setting of MQVR, then delineate the various methods experimented to handle multiple queries, and finally introduce a new evaluation metric for MQVR.

3.1 Setting

Video retrieval[1] is a task of searching videos given text descriptions [33,37,39]. Formally, given a video database $\mathcal{V}$ composed of n videos, $\mathcal{V} = \{v_1, v_2, ..., v_n\}$, and a text description q_i of video v_i, the goal is to successfully retrieve v_i from $\mathcal{V}$ based on q_i. This is the setting most existing video retrieval benchmarks adopt [4,28,43,55]. We refer to it as single-query setting, since one caption is used for retrieval during evaluation. Specifically, the goal is to learn a model $\mathcal{M}^{single}$ that can evaluate the similarity between any query-video pair, such

---

[1] Throughout this paper, we use the term 'video retrieval' to refer the specific task of text-to-video retrieval.

that $\mathcal{M}^{single}(q, v) \in \mathbb{R}$ reflects how the query matches with the video. And a perfect retrieval would score the matching pair higher than non-matching pairs,

$$\forall j, j \neq i, \quad \mathcal{M}^{single}(q_i, v_i) > \mathcal{M}^{single}(q_i, v_j). \tag{1}$$

However, this single-query setting might be problematic when the input description has low quality, for example, being too general or abstract that can perfectly describe different videos in the database. This can pose serious trouble during evaluation. Due to the fact that existing video retrieval benchmarks are mostly based on datasets collected for other tasks (*e.g.*, video captioning) [9,28,43,50,55], such low-quality queries are prevalent during model evaluation (as shown in Fig. 1).

To tackle this issue, we study the multi-query retrieval setting where more than one query is available during retrieval (Fig. 1 bottom). Under similar notation, the task of multi-query retrieval is to retrieve a target video v_i from the video database $\mathcal{V}$ based on multiple descriptions $Q_i = \{q_i^1, q_i^2, ..., q_i^k\}$ of it. And similarly we would like to learn a model $\mathcal{M}^{multi}$ that correctly retrieves the target video,

$$\forall j, j \neq i, \quad \mathcal{M}^{multi}(Q_i, v_i) > \mathcal{M}^{multi}(Q_i, v_j). \tag{2}$$

3.2   Methods

Broadly, we divide the methods into two categories: the first category of methods extends the models trained with single-query to multi-query evaluation in a post-hoc fashion without any retraining, and the second one has dedicated modifications for multi-query retrieval during training.

Post-hoc Inference Methods. The models trained using single-query can be easily adapted when multiple queries are available, just by considering multiple queries separately and then aggregating the results.

Similarity Aggregation (SA). A simple form is to take the mean of the similarity scores evaluated between each query and the video as the final multi-query similarity, which is then used to rank videos.

$$\mathcal{M}_{SA}^{multi}(Q, v) = \frac{1}{k} \sum_{i=1}^{k} \mathcal{M}^{single}(q_i, v). \tag{3}$$

Rank Aggregation (RA). Unlike *similarity aggregation* which aggregates the raw similarity score of each query, *rank aggregation* aggregates the retrieval results instead. Denote $\mathcal{R}(v|q, \mathcal{M}^{single}) \in \{1, 2, ..., n\}$ as the rank of v among all videos in the candidate pool based on the evaluation of similarity to q according to $\mathcal{M}^{single}$ (smaller rank means more compatible with the query), then the multi-query similarity can be calculated as:

$$\mathcal{M}_{RA}^{multi}(Q, v) = -\frac{1}{k} \sum_{i=1}^{k} \mathcal{R}(v|q, \mathcal{M}^{single}). \tag{4}$$

Note that the overall similarity score here doesn't have a well-defined quantitative meaning as in the **SA** case, and just serves to order different videos.

Multi-query Training Methods. Unlike post-hoc methods that essentially deal with the multi-query problem in a "single-query way", dedicated training modifications might be helpful if multiple queries are available during the training phase. This is already provided by many standard benchmarks [9,50,55], but to the best of the authors' knowledge, most existing works only treat the descriptions independently and adopt the single-query training, where a training text-video pair is composed of a video and one description, $i.e.$, (q_i, v_i) for a positive sample, (q_i, v_j) for a negative sample. However, as we will show later, this is not the best choice and that multi-query training, $i.e.$, using (Q_i, v_i) as a positive pair and (Q_i, v_j) as a negative pair, where $Q_i = \{q_i^1, q_i^2, ..., q_i^k\}$ contains more than one descriptions, can provide a large gain.

To facilitate the discussion, we first denote the similarity score of a query-video pair as $\mathcal{M}^{single}(q, v) = \mathbf{S}\left(f(q), g(v)\right)$. Here f denotes the feature extraction network of query sentence which maps the input query to a high dimensional vector $f(q) \in \mathbb{R}^m$. Similarly, g, the video feature extractor, maps input video to the same embedding space $g(v) \in \mathbb{R}^m$. $\mathbf{S}$ is the metric that measures the similarity between $f(q)$ and $g(v)$ ($e.g.$, the cosine similarity).

Mean Feature (MF). A naive way of combining multiple queries during training is to take the mean of their features to be the final query feature, with corresponding similarity score calculated as:

$$\mathcal{M}_{MF}^{multi}(Q, v) = \mathbf{S}\left(\frac{1}{k}\sum_{i=1}^{k} f(q_i),\ g(v)\right). \tag{5}$$

We will show in the next section that just by making this simple modification, the trained model can outperform the post-hoc methods by a large margin. However, a potential drawback of *mean feature* is that each query contributes equally to the result regardless of their quality. Thus we make a natural extension and further investigate **weighted feature (WF)** methods,

$$\mathcal{M}_{WF}^{multi}(Q, v) = \mathbf{S}\left(\sum_{i=1}^{k} \alpha_i f(q_i),\ g(v)\right), \tag{6}$$

where $\sum_{i=1}^{k} \alpha_i = 1$. We experiment with several ways to generate the weights α.

Text-to-Text Similarity Weighting (TS-WF). It is desirable for a query to contain complementary information about the target video that is not captured by other queries. On contrary, a query is not informative if it only contains redundant information already captured by others. This inspires a parameter-free method to evaluate the informativeness of one query by comparing the similarity of it to other queries. Specifically, the informativeness of query q_i is computed as $\mathbf{I}(q_i) = -\sum_{j \neq i} \mathbf{S}(f(q_i), f(q_j))$, where $q_i, q_j \in Q$. Notice the minus sign here means the informative queries should be different from others. Finally, the weights can be computed by taking softmax among $I(q)$s.

Localized Weight Generation (LG-WF). The weights can also be learned using a separate network and trained end-to-end with other parameters. We experiment with a multi-layer perceptron (MLP) that maps extracted text features to a scalar. The MLP is shared across all queries and each query is processed separately. The resulting weights are normalized with softmax.

Contextualized Weight Generation (CG-WF). Instead of computing weight for every query individually, *CG-WF* attends to other queries when generating each weight. Specifically, we experiment with a transformer-based attention network, where all query features are first input to a transformer to generate contextualized features, and then a MLP head is used to map the contextualized features to scalars. Softmax is used to compute the final normalized weights like in the *localized weight generation* case.

3.3   Evaluation

We adopt the evaluation metrics widely used in the standard single-query retrieval setting and compute the standard R@K (recall at rank K, higher the better), median rank and mean rank (lower the better) in our experiments. Additionally, specific to multi-query evaluation, we would also like to compare models tested under varying number of query inputs (*e.g.,* Fig. 5 in the experiments) for a more robust evaluation of the models' generalization ability. To facilitate future research in this subject, we propose an area under the curve (AUC) metric. Specifically, $\mathbf{AUC}_n^{R@K}$ is defined as the normalized area under the curve value of R@K with test query-input varying from 1 to n,

$$\mathbf{AUC}_n^{R@K} = \frac{\mathbf{AUC}([R@K_1, R@K_2, ...R@K_n])}{n-1}, \tag{7}$$

where $R@K_m$ is the R@K value when evaluating with m queries as input.

4   Experiments

We first describe in Sect. 4.1 the architecture backbones on which we build our multi-query retrieval experiment. Then specific experiment settings and implementation details are described in Sect. 4.2 and finally the experiment results are shown in Sect. 4.3.

4.1   Architecture Backbones

CLIP4Clip [36]. CLIP [42] has been shown to learn transferable features to benefit lots of downstream tasks including video action recognition, OCR, etc.

Recently, such trend has also been demonstrated for video retrieval. Among multiple models proposed in this line of work [3,17,36], we adopt the CLIP4Clip for its simplicity. For our experiment, we use the publicly released "ViT-B/32" checkpoint for initialization of the CLIP model.

Frozen [8]. This model proposed by Bain *et al.* utilizes a transformer-based architecture [15,46], which is composed of space-time self-attention blocks. For our experiments, we use the checkpoint provided by the original authors which is pretrained on Conceptual Captions [44] and WebVid-2M [8].

4.2 Experimental Setup

Datasets. We conduct our experiments on three datasets, which have multiple descriptions available for each video clip.

- **MSR-VTT** [55] is one of the most widely used datasets for video retrieval. It contains 10K video clips gathered from YouTube and each clip is annotated with 20 natural language descriptions. We follow the previous works to use 9K videos for training and 1K videos for testing.
- **MSVD** [9] is a dataset initially collected for translation and paraphrase evaluation. It contains 1970 videos, each with multiple descriptions in several different languages. We only use the English descriptions in our experiments. Following previous works, we use 1200 videos for training, 100 videos for validation and 670 videos for testing.
- **VATEX** [50] is a large-scale multilingual video description dataset. It contains 34991 videos, each annotated with ten English and ten Chinese captions. We only keep the English annotations and use the standard split with 25991 videos for training, 3000 for validation and 6000 for testing.

Implementation Details. All the models are trained with 8-frame inputs and a batch size of 48 for 30 epochs. We adopt the cross entropy loss over similarity scores and a softmax temperature of 0.05 as used in [8]. AdamW [35] optimizer is used with a cosine learning rate scheduler and a linear warm up of 5 epochs [34]. The max learning rate is 3e-5 and 3e-6 for Frozen and CLIP4Clip respectively. We set query number to five for our multi-query experiments. Specifically, all available captions for each video are used during training, but for each training instance during forward pass, a subset of random five descriptions are input to the multi-query model. During test, we evaluate the N-query performance by sampling N query captions per video. However, to avoid selection bias, we repeat the evaluation for a hundred times (each time selects a different subset of N captions as queries for every video) and report the mean as the final results.

4.3 Results

Table 1 summarizes the evaluation results on the test split of the MSR-VTT, MSVD and VATEX datasets. Due to space limit, we only show the CLIP4Clip

Table 1. Performance of different methods on MSR-VTT, MSVD and VATEX. The baseline is trained and evaluated with one query. Others are evaluated with five-query input. RA, SA are trained with one query. MF, TS-WF, LG-WF, and CG-WF are trained with five-query input. All numbers are the average over 100 evaluations with different query samples. Recall numbers are reported in percent.

	MSR-VTT [55] (CLIP4Clip)					MSR-VTT (Frozen)				
	R@1 ↑	R@5 ↑	R@10 ↑	MdR ↓	MnR ↓	R@1 ↑	R@5 ↑	R@10 ↑	MdR ↓	MnR ↓
Baseline	41.5	69.4	79.3	2.0	15.7	31.9	61.1	72.6	3.0	23.2
RA	56.4	84.5	92.1	1.0	4.3	44.6	76.2	85.2	2.0	7.0
SA	68.4	92.1	97.0	1.0	2.4	55.3	85.2	92.5	1.0	4.2
MF	71.3	93.5	97.6	1.0	2.2	59.6	88.0	93.9	1.0	4.0
TS-WF	72.6	**94.2**	**97.8**	1.0	2.1	**60.6**	**88.5**	**94.2**	1.0	3.8
LG-WF	72.6	94.1	97.7	1.0	2.1	59.6	87.6	93.8	1.0	3.9
CG-WF	**73.1**	94.1	**97.8**	1.0	2.1	59.9	88.1	94.0	1.0	3.7

	MSVD [9] (CLIP4Clip)					VATEX [50] (CLIP4Clip)				
	R@1 ↑	R@5 ↑	R@10 ↑	MdR ↓	MnR ↓	R@1 ↑	R@5 ↑	R@10 ↑	MdR ↓	MnR ↓
Baseline	43.8	74.0	83.0	2.0	10.9	33.3	63.5	76.1	3.0	17.9
RA	47.3	79.9	89.8	2.0	4.8	43.6	75.5	86.1	2.0	7.9
SA	57.6	87.1	93.7	1.0	3.4	47.3	77.9	87.9	2.0	6.7
MF	59.6	88.3	94.2	1.0	3.3	57.2	84.7	92.0	1.0	4.9
TS-WF	60.4	88.8	94.5	1.0	3.3	58.4	85.3	92.5	1.0	4.6
LG-WF	60.2	88.0	94.2	1.0	3.2	58.5	85.2	92.4	1.0	4.8
CG-WF	**61.7**	**89.5**	**94.9**	1.0	3.0	**59.0**	**85.6**	**92.7**	1.0	4.7

numbers for MSVD and VATEX datasets. Please see the Appendix for the full results of Frozen model. Since the experiment findings are similar across both models, we will focus on the CLIP4Clip model during the following discussion.

(R1) Multi-query vs. Single-query. The first row of Table 1 shows the result for the single-query baseline, where the model is both trained and evaluated with one query. The rest of the rows show the results for five-query evaluation. With five queries available, all models see a big boost in terms of retrieval accuracy compared to single-query. While it's anticipated that multi-query models should perform better than the single-query counterpart, such big improvement is still striking, and shows that single-query evaluation could underestimate the real retrieval abilities of models.

(R2) Post-hoc Aggregation Methods. Comparing two post-hoc methods which simply extend the pretrained single-query model to the multi-query setting, it's clear that *similarity aggregation* outperforms *rank aggregation* by a large margin. On CLIP4Clip, the R@1 is improved from 56.4% to 68.4% for MSR-VTT, from 47.3% to 57.6% for MSVD, and from 43.6% to 47.3% for VATEX, respectively. This is not surprising as the rank provided by a low-quality query is noisy and would thus drag down the final rank when combining with ranks provided by the other queries. While the similarity generated by a low-quality

query is also noisy, its value tends to be small and is typically dominated by the similarity scores provided by the high-quality queries.

(R3) Multi-query Training Methods. It's clear from Table 1 that multi-query training methods with ad-hoc training modifications improve a lot over the post-hoc methods. Just by simply feeding five captions and taking the mean of the encoded features as the final text features during training (*mean feature*), R@1 of CLIP4Clip model can be improved by 2.9% (from 68.4% to 71.3%), 2.0% (from 57.6% to 59.6%) and 9.9% (from 47.3% to 57.2%) on MSR-VTT, MSVD and VATEX, respectively. We anticipate that such improvement can be mainly attributed to the de-noising effect of multi-query training. As the ranking loss acts on the combined features, the part of loss that tries to push apart the false-negative text-video pairs (due to general descriptions that match with more than one videos) is lessened, thus avoiding potential over-fitting and providing more robust features (additional evidence is also discussed in **R5**).

Even though the *mean feature* training is already a very strong baseline, additional weighting heuristics still manage to introduce further improvements. On the CLIP4Clip model, with the best-performing *weighted feature* training with *contextualized weight generation*, R@1 can be improved over *mean feature* by 1.8% (from 71.3% to 73.1%), 2.1% (from 59.6% to 61.7%) and 1.8% (from 57.2% to 59.0%) points across the datasets. To show that the learned weights can correctly capture the relative quality of the queries, we compute the average weights given to different queries ordered by their quality (we rank the quality of queries by their single-query retrieval result, 1 is the best, and 5 is the worst). Figure 3 shows the result, and it's clear that the generated weights can correctly reflect the quality of the query. The *contextualized weight generation* works better than the *localized weight generation*. This is expected as instead of trying to learn a standalone quality predictor, the former lessens the task by learning to predict the relative quality. Especially when training data is small, *e.g.*, in MSVD, *localized weight generation* would have a hard time learning the quality and give more spread weights across different queries. Figure 4 shows several qualitative examples on MSR-VTT for a model trained with *contextualized weight generation* (more examples shown in the Appendix). While some of the queries are of low-quality, the model can correctly attend to the better ones and achieve good overall retrieval accuracy.

(R4) Evaluation with Varying Number of Queries. Table 1 shows the result where evaluation is performed under the same number of queries as multi-query models are trained (five-query training and five-query test). It is then natural to question whether such result can be generalized across varying number of queries. Figure 5 shows R@1 curves when the same models are tested under different number of queries and Table 2 summarizes the results in the form of the proposed $\mathbf{AUC}_n^{R@1}$ metric. First, the performance of all methods improves with more queries available. The curve increases rapidly at beginning and gradually saturates as more queries provides marginally additional information. Second, the five-query trained models outperform the single-query trained models across different number of testing queries except one, with dominating scores

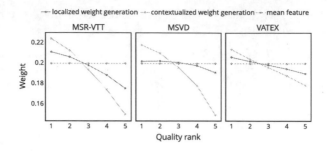

Fig. 3. Averaged weights generated to five queries, which are ranked by their single-query retrieval results. 1 is the best and 5 is the worst. The average is taken over (test split size) * 100 instances.

Video	Query	Weight	Single-query rank	Multi-query rank
	A man is busy in cooking in the kitchen	0.36	14	
				5
	Person is mixing the ingredients and preparing the dessert	**0.64**	8	
	A clip from a video game	0.19	13	
				1
	A girl is talking about having a barn in the sims	**0.81**	1	
	A man is singing	0.19	30	
	A band is performing on stage	0.32	9	2
	Two men playing guitar on stage	**0.49**	1	
	A man is playing the guitar	0.19	12	
	A band is playing on a stage	0.35	3	1
	A man playing a bass guitar and a man playing on a keyboard	**0.46**	1	

Fig. 4. Qualitative examples of a weighted model (CG-WF) on MSR-VTT. It correctly attends to the disrciminative descriptions.

for $\mathbf{AUC}_3^{R@1}$, $\mathbf{AUC}_5^{R@1}$ and $\mathbf{AUC}_{10}^{R@1}$. This shows that multi-query training indeed learns better features suited for multi-query evaluation. While single-query trained models maintain the lead at single-query test, the differences are very small. Comparing *similarity aggregation* with *contextualized weight generalization*, the $R@1_1$ (single-query R@1) is 41.5% vs. 41.0% for MSR-TT, 43.8% vs. 43.6% for MSVD, and 33.3% vs. 32.8% for VATEX respectively.

To get the best of both worlds, we further conduct an experiment with a combination of single-query and five-query *mean feature* training, *i.e.*, some training pairs contain one video and one caption while others contain one video and five captions (as with all other experiments, all available captions are used during training, but for each training instance, a random sample of one or five captions is used as input). The dash lines in Fig. 5 show the results. Rather surprisingly, the *combination training* achieves the best single-query performance and outper-

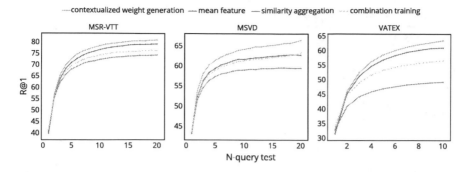

Fig. 5. R@1 performance for SA, MF, CG-WF and *combination training* when evaluated with varying number of queries.

Table 2. Multi-query test results summarized with proposed $\mathbf{AUC}_n^{R@1}$ metric.

		$R@1_1$	$\mathbf{AUC}_3^{R@1}$	$\mathbf{AUC}_5^{R@1}$	$\mathbf{AUC}_{10}^{R@1}$
MSR-VTT	SA	41.5	54.0	59.9	66.1
	MF	40.3	54.4	61.3	68.8
	CG-WF	41.0	**55.5**	**62.7**	**70.4**
	Cmb (MF)	**41.9**	54.6	60.8	67.5
MSVD	SA	43.8	50.6	53.5	56.4
	MF	43.5	51.3	54.8	58.3
	CG-WF	43.6	**52.6**	**56.4**	**60.2**
	Cmb (MF)	**44.0**	51.2	54.5	57.9
VATEX	SA	33.3	40.1	43.1	46.2
	MF	31.8	43.7	49.2	55.1
	CG-WF	32.8	**44.6**	**50.4**	**56.8**
	Cmb (MF)	**34.4**	43.5	47.6	52.1

forms single-query training entirely. This shows that **the de-noising effect of multi-query training can also be utilized to improve standard single-query retrieval**.

(R5) Training with Different Number of Queries. To understand the effect of number of queries (N) during training, we plot the performance of different models in Fig. 6a. We observe that the performance for all models increases at first, followed by a decrease (sometimes to a score even lower than N=1) as more queries are added. We hypothesize that this may be due to over-smoothing effect in the training process when too many queries are used, which causes the model to not learn discriminative representations for each individual query.

To further demonstrate the utility of the MQVR framework in learning robust representations, we perform a zero-shot transfer evaluation on VATEX dataset using models trained on MSR-VTT (Table 6b). Quite promisingly, we

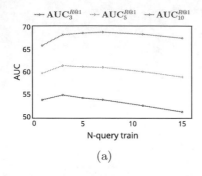

Training	$R@1_1$	$\mathbf{AUC}_3^{R@1}$	$\mathbf{AUC}_5^{R@1}$	$\mathbf{AUC}_{10}^{R@1}$
VATEX 1q	33.3	40.1	43.1	46.2
MSR-VTT 1q	26.5	32.2	34.7	37.3
MSR-VTT 2q	27.0	33.2	36.0	38.9
MSR-VTT 3q	**27.1**	33.8	36.9	40.0
MSR-VTT 5q	**27.1**	34.7	38.2	41.9
MSR-VTT 7q	26.6	**34.8**	**38.7**	**42.8**

(a)   (b)

Fig. 6. Evaluation of *mean feature* model trained on MSR-VTT with different number of queries. (a) In-domain test on MSR-VTT test set. (b) Out of domain zero-shot test on VATEX test set. "nq" means n-query training.

observe that multi-query training (MSR-VTT 5q, MSR-VTT 7q) outperforms their single-query counterparts (MSR-VTT 1q) on both single-query and multi-query evaluations, achieving 0.6% higher single-query R@1 and 2.6–5.5 points higher AUC, respectively. The latter results especially are only a few points below the AUC performance of an in-domain VATEX 1q model, demonstrating that there is a strong transfer and indicating that **multi-query training can lead to better generalization**.

5   Conclusion

In this work, we study the multi-query retrieval problem where multiple descriptions are available for retrieving target videos. We argue that this previously less-studied setting is of practical value both because it can provide significant improvement on retrieval accuracy through incorporating information from multiple queries and that it addresses challenges in model training and evaluation introduced by imperfect descriptions. We then investigate several multi-query training methods and propose a new evaluation metric dedicated for this setting. With extensive experiments, we demonstrate that multi-query inspired training methods can provide superior performance and better generalization. We believe further investigation on the issue will benefit the field and bring new insights to building better retrieval systems for real-world applications.

Acknowledgements. This material is based upon work supported by the National Science Foundation under Grant No. 2107048. Any opinions, findings, and conclusions or recommendations expressed in this material are those of the author(s) and do not necessarily reflect the views of the National Science Foundation. We would like to thank members of the Princeton Visual AI Lab (Jihoon Chung, Zhiwei Deng, William Yang and others) for their helpful comments and suggestions.

References

1. Agrawal, A., Batra, D., Parikh, D., Kembhavi, A.: Don't just assume; look and answer: Overcoming priors for visual question answering. In: CVPR (2018)
2. Alayrac, J.B., et al.: Self-supervised multimodal versatile networks. In: NeurIPS (2020)
3. Andrés Portillo-Quintero, J., Ortiz-Bayliss, J.C., Terashima-Marín, H.: A straightforward framework for video retrieval using CLIP. arXiv:2102.12443 (2021)
4. Anne Hendricks, L., Wang, O., Shechtman, E., Sivic, J., Darrell, T., Russell, B.: Localizing moments in video with natural language. In: ICCV (2017)
5. Anne Hendricks, L., Wang, O., Shechtman, E., Sivic, J., Darrell, T., Russell, B.: Localizing moments in video with natural language. In: ICCV (2017)
6. Arandjelović, R., Zisserman, A.: Three things everyone should know to improve object retrieval. In: CVPR (2012)
7. Arandjelovic, R., Zisserman, A.: Look, listen and learn. In: ICCV (2017)
8. Bain, M., Nagrani, A., Varol, G., Zisserman, A.: Frozen in time: a joint video and image encoder for end-to-end retrieval. arXiv:2104.00650 (2021)
9. Chen, D., Dolan, W.B.: Collecting highly parallel data for paraphrase evaluation. In: ACL (2011)
10. Cheng, X., Lin, H., Wu, X., Yang, F., Shen, D.: Improving video-text retrieval by multi-stream corpus alignment and dual softmax loss. arXiv:2109.04290 (2021)
11. Chum, O., Mikulik, A., Perdoch, M., Matas, J.: Total recall II: query expansion revisited. In: CVPR (2011)
12. Chum, O., Philbin, J., Sivic, J., Isard, M., Zisserman, A.: Total recall: automatic query expansion with a generative feature model for object retrieval. In: ICCV (2007)
13. Chung, J., Wuu, C., Yang, H., Tai, Y.-W., Tang, C.-K.: Haa500: human-centric atomic action dataset with curated videos. In: ICCV (2021)
14. Dong, J., et al.: Dual encoding for video retrieval by text. In: PAMI (2021)
15. Dosovitskiy, A., et al.: An image is worth 16x16 words: transformers for image recognition at scale. arXiv:2010.11929 (2020)
16. Dzabraev, M., Kalashnikov, M., Komkov, S., Petiushko, A.: MDMMT: multidomain multimodal transformer for video retrieval. In: CVPR (2021)
17. Fang, H., Xiong, P., Xu, L., Chen, Y.: CLIP2Video: mastering video-text retrieval via image CLIP. arXiv:2106.11097 (2021)
18. Fernando, B., Tuytelaars, T.: Mining multiple queries for image retrieval: on-the-fly learning of an object-specific mid-level representation. In: CVPR (2013)
19. Gabeur, V., Nagrani, A., Sun, C., Alahari, K., Schmid, C.: Masking modalities for cross-modal video retrieval. In: WACV (2022)
20. Gabeur, V., Sun, C., Alahari, K., Schmid, C.: Multi-modal transformer for video retrieval. In: Vedaldi, A., Bischof, H., Brox, T., Frahm, J.-M. (eds.) ECCV 2020. LNCS, vol. 12349, pp. 214–229. Springer, Cham (2020). https://doi.org/10.1007/978-3-030-58548-8_13
21. Gao, Z., Liu, J., Chen, S., Chang, D., Zhang, H., Yuan, J.: CLIP2TV: an empirical study on transformer-based methods for video-text retrieval. arXiv:2111.05610 (2021)
22. Gordo, A., Almazan, J., Revaud, J., Larlus, D.: End-to-end learning of deep visual representations for image retrieval. IJCV (2017)

23. Gordo, A., Radenovic, F., Berg, T.: Attention-based query expansion learning. In: Vedaldi, A., Bischof, H., Brox, T., Frahm, J.-M. (eds.) ECCV 2020. LNCS, vol. 12373, pp. 172–188. Springer, Cham (2020). https://doi.org/10.1007/978-3-030-58604-1_11

24. Goyal, P., et al.: Self-supervised pretraining of visual features in the wild. arXiv:2103.01988 (2021)

25. Huang, S., Hang, H.M.: Multi-query image retrieval using CNN and SIFT features. In: APSIPA ASC (2017)

26. Imani, A., Vakili, A., Montazer, A., Shakery, A.: Deep neural networks for query expansion using word embeddings. In: ECIR (2019)

27. Kamath, A., Singh, M., LeCun, Y., Synnaeve, G., Misra, I., Carion, N.: MDETR-modulated detection for end-to-end multi-modal understanding. In: ICCV (2021)

28. Krishna, R., Hata, K., Ren, F., Fei-Fei, L., Carlos Niebles, J.: Dense-captioning events in videos. In: ICCV (2017)

29. Lei, J., Li, L., Zhou, L., Gan, Z., Berg, T.L., Bansal, M., Liu, J.: Less is more: clipBERT for video-and-language learning via sparse sampling. In: CVPR (2021)

30. Li, G., He, F., Feng, Z.: A CLIP-Enhanced method for video-language understanding. arXiv:2110.07137 (2021)

31. Li, X., et al.: OSCAR: object-semantics aligned pre-training for vision-language tasks. In: Vedaldi, A., Bischof, H., Brox, T., Frahm, J.-M. (eds.) ECCV 2020. LNCS, vol. 12375, pp. 121–137. Springer, Cham (2020). https://doi.org/10.1007/978-3-030-58577-8_8

32. Liu, T., Lin, Y., Du, B.: Unsupervised person re-identification with stochastic training strategy. TIP (2022)

33. Liu, Y., Albanie, S., Nagrani, A., Zisserman, A.: Use what you have: video retrieval using representations from collaborative experts. arXiv:1907.13487 (2019)

34. Loshchilov, I., Hutter, F.: SGDR: stochastic gradient descent with warm restarts. arXiv:1608.03983 (2016)

35. Loshchilov, I., Hutter, F.: Decoupled weight decay regularization. arXiv:1711.05101 (2017)

36. Luo, H., Ji, L., Zhong, M., Chen, Y., Lei, W., Duan, N., Li, T.: CLIP4clip: an empirical study of CLIP for end to end video clip retrieval. arXiv:2104.08860 (2021)

37. Miech, A., Laptev, I., Sivic, J.: Learning a text-video embedding from incomplete and heterogeneous data. arXiv:1804.02516 (2018)

38. Miech, A., Zhukov, D., Alayrac, J.B., Tapaswi, M., Laptev, I., Sivic, J.: Howto100M: learning a text-video embedding by watching hundred million narrated video clips. In: ICCV (2019)

39. Mithun, N.C., Li, J., Metze, F., Roy-Chowdhury, A.K.: Learning joint embedding with multimodal cues for cross-modal video-text retrieval. In: ICMR (2018)

40. Patrick, M., et al.: Support-set bottlenecks for video-text representation learning. arXiv:2010.02824 (2020)

41. Radenović, F., Tolias, G., Chum, O.: Fine-tuning CNN image retrieval with no human annotation. TPAMI (2018)

42. Radford, A., et al.: Learning transferable visual models from natural language supervision. arXiv:2103.00020 (2021)

43. Rohrbach, A., Rohrbach, M., Tandon, N., Schiele, B.: A dataset for movie description. In: CVPR (2015)

44. Sharma, P., Ding, N., Goodman, S., Soricut, R.: Conceptual captions: a cleaned, hypernymed, image alt-text dataset for automatic image captioning. In: ACL (2018)

45. Sun, C., Myers, A., Vondrick, C., Murphy, K., Schmid, C.: VideoBERT: a joint model for video and language representation learning. In: ICCV (2019)
46. Vaswani, A., et al.: Attention is all you need. In: NeurIPS (2017)
47. Vural, C., Akbacak, E.: Deep multi query image retrieval. SPIC (2020)
48. Wang, L., Li, Y., Lazebnik, S.: Learning deep structure-preserving image-text embeddings. In: CVPR (2016)
49. Wang, X., Zhu, L., Yang, Y.: T2VLAD: global-local sequence alignment for text-video retrieval. In: CVPR (2021)
50. Wang, X., Wu, J., Chen, J., Li, L., Wang, Y.F., Wang, W.Y.: VATEX: a large-scale, high-quality multilingual dataset for video-and-language research. In: ICCV (2019)
51. Wang, Y., Lin, X., Wu, L., Zhang, W.: Effective multi-query expansions: robust landmark retrieval. In: ACMMM (2015)
52. Wang, Y., Lin, X., Wu, L., Zhang, W.: Effective multi-query expansions: collaborative deep networks for robust landmark retrieval. TIP (2017)
53. Wu, Y., Jiang, L., Yang, Y.: Switchable novel object captioner. TPAMI (2022)
54. Wu, Y., Yang, Y.: Exploring heterogeneous clues for weakly-supervised audio-visual video parsing. In: CVPR (2021)
55. Xu, J., Mei, T., Yao, T., Rui, Y.: MSR-VTT: a large video description dataset for bridging video and language. In: CVPR (2016)
56. Yu, Y., Kim, J., Kim, G.: A joint sequence fusion model for video question answering and retrieval. In: Ferrari, V., Hebert, M., Sminchisescu, C., Weiss, Y. (eds.) ECCV 2018. LNCS, vol. 11211, pp. 487–503. Springer, Cham (2018). https://doi.org/10.1007/978-3-030-01234-2_29
57. Zhang, B., Hu, H., Sha, F.: Cross-modal and hierarchical modeling of video and text. In: Ferrari, V., Hebert, M., Sminchisescu, C., Weiss, Y. (eds.) ECCV 2018. LNCS, vol. 11217, pp. 385–401. Springer, Cham (2018). https://doi.org/10.1007/978-3-030-01261-8_23
58. Zheng, L., et al.: MARS: a video benchmark for large-scale person re-identification. In: Leibe, B., Matas, J., Sebe, N., Welling, M. (eds.) ECCV 2016. LNCS, vol. 9910, pp. 868–884. Springer, Cham (2016). https://doi.org/10.1007/978-3-319-46466-4_52
59. Zhu, L., Yang, Y.: ActBERT: learning global-local video-text representations. In: CVPR (2020)

Hierarchical Average Precision Training for Pertinent Image Retrieval

Elias Ramzi[1,2(✉)] ⓘ, Nicolas Audebert[1] ⓘ, Nicolas Thome[1,3] ⓘ,
Clément Rambour[1] ⓘ, and Xavier Bitot[2]

[1] CEDRIC, Conservatoire National des Arts et Métiers, Paris, France
{elias.ramzi,nicolas.audebert,nicolas.thome,clement.rambour}@cnam.fr
[2] Coexya, Paris, France
xavier.bitot@coexya.eu
[3] Sorbonne Université, CNRS, ISIR, 75005 Paris, France

Abstract. Image Retrieval is commonly evaluated with Average Precision (AP) or Recall@k. Yet, those metrics, are limited to binary labels and do not take into account errors' severity. This paper introduces a new hierarchical AP training method for pertinent image retrieval (HAPPIER). HAPPIER is based on a new $\mathcal{H}$-AP metric, which leverages a concept hierarchy to refine AP by integrating errors' importance and better evaluate rankings. To train deep models with $\mathcal{H}$-AP, we carefully study the problem's structure and design a smooth lower bound surrogate combined with a clustering loss that ensures consistent ordering. Extensive experiments on 6 datasets show that HAPPIER significantly outperforms state-of-the-art methods for hierarchical retrieval, while being on par with the latest approaches when evaluating fine-grained ranking performances. Finally, we show that HAPPIER leads to better organization of the embedding space, and prevents most severe failure cases of non-hierarchical methods. Our code is publicly available at https://github.com/elias-ramzi/HAPPIER.

Keywords: Hierarchical Image Retrieval · Hierarchical Average Precision · Ranking

1 Introduction

Image Retrieval (IR) consists in ranking images with respect to a query by decreasing order of visual similarity. IR methods are commonly evaluated using Recall@k (R@k) or Average Precision (AP). Because those metrics are non-differentiable, a rich literature exists on finding adequate surrogate loss functions to optimize them with deep learning, with tuple-wise losses [26,31,38–40],

Supplementary Information The online version contains supplementary material available at https://doi.org/10.1007/978-3-031-19781-9_15.

HAPPIER

Fig. 1. Proposed HAPPIER framework for pertinent image retrieval. Standard ranking metrics based on binary labels, *e.g.* Average Precision (AP), assign the same score to the bottom and top row rankings (0.9). We introduce the $\mathcal{H}$-AP metric based on non-binary labels, that takes into account mistakes' severity. $\mathcal{H}$-AP assigns a smaller score to the bottom row (0.68) than the top one (0.94). HAPPIER maximizes $\mathcal{H}$-AP during training and thus explicitly supports to learn rankings similar to the top one, in contrast to binary ranking losses.

proxy based losses [11,34,37,41] and direct AP optimization methods [2,6,23,27, 28,30].

These metrics are only defined for binary ($\oplus/\ominus$) labels, which we denote as *fine-grained labels*: an image is negative as soon as it is not strictly similar to the query. Binary metrics are by design unable to take into account the severity of the mistakes in a ranking. On Fig. 1, some negative instances are "less negative" than others, *e.g.* given the "Brown Bear" query, "Polar bear" is more relevant than "Butterfly". However, AP is 0.9 for both the top and bottom rankings. Consequently, training on binary metrics (*e.g.* AP or R@k) develops no incentive to produce ranking such as the top row, and often produces rankings similar to the bottom one. To address this problem, we introduce the HAPPIER method dedicated to Hierarchical Average Precision training for Pertinent ImagE Retrieval. HAPPIER provides a smooth training objective, amenable to gradient descent, which explicitly takes into account the severity of mistakes when evaluating rankings.

Our first contribution is to define a new Hierarchical AP metric ($\mathcal{H}$-AP) that leverages the hierarchical tree between concepts and enables a fine weighting between errors in rankings. As shown in Fig. 1, $\mathcal{H}$-AP assigns a larger score (0.94) to the top ranking than to the bottom one (0.68). We show that $\mathcal{H}$-AP provides a consistent generalization of AP for the non-binary setting. We also introduce our HAPPIER$_F$ variant, giving more weights to fine-grained levels of the hierarchy. Since $\mathcal{H}$-AP, like AP, is a non-differentiable metric, our second contribution is to use HAPPIER to directly optimize $\mathcal{H}$-AP by gradient descent. We carefully design a smooth surrogate loss for $\mathcal{H}$-AP that has strong theoretical guarantees and is an upper bound of the true loss. We then define an additional clustering loss to support having a consistency between partial and global rankings.

We validate HAPPIER on six IR datasets, including three standard datasets (Stanford Online Products [22] and iNaturalist-base/full [36]), and three recent hierarchical datasets (DyML [32]). We show that, when evaluating on hierarchical metrics (*e.g.* $\mathcal{H}$-AP), HAPPIER outperforms state-of-the-art methods for fine-grained ranking [27, 34, 40, 41], the baselines and the latest hierarchical method of [32], and only slightly under-performs *vs.* state-of-the-art IR methods at the fine-grained level (*e.g.* AP, R@1). HAPPIER$_F$ performs on par on fine-grained metrics while still outperforming fine-grained methods on hierarchical metrics.

2   Related Work

2.1   Image Retrieval and Ranking

The Image Retrieval community has designed several families of methods to optimize metrics such as AP and R@k. Methods that relies on tuplet-wise losses, like pair losses [15, 26], triplet losses [40], or larger tuplets [20, 31, 38] learn comparison relations between instances. Methods using proxies have been introduced to lower the computational complexity of tuplet based training [11, 21, 34, 37, 41]: they learn jointly a deep model and weight matrix that represent proxies using a cross-entropy based loss. Proxies are approximations of the original data points that should belong to their neighbourhood. Finally, there also has been large amounts of work dedicated to the direct optimization of the AP during training by introducing differentiable surrogates [2, 6, 23, 27, 28, 30], so that models are optimized on the same metric they are evaluated on. However, nearly all of these methods only consider binary labels: two instances are either the same (positive) or different (negative), leading to poor performance when multiple levels of hierarchy are considered.

2.2   Hierarchical Predictions and Metrics

There has been a recent regain of interest in Hierarchical Classification [1, 8, 12] with the introductions of methods based either on a hierarchical softmax function or on multiple classifiers. It is considered that learning from hierarchical relations between labels leads to more robust models that make "better mistakes" [1]. Yet, hierarchical classification means that labels are known in advance and are identical in the train and test sets. This is called a *closed set* setting. However, Hierarchical Image Retrieval does not fall into this framework. Standard IR protocols consider the *open set* paradigm to better evaluate the generalization abilities of learned models: the retrieval task at test time pertains to labels that were not present in the train set, making classification poorly suited to IR.

Meanwhile, the broader Information Retrieval community has been using datasets where documents can be more or less relevant depending on the query and the user making the request [16, 19]. Instead of the mere positive/negative dichotomy, each instance has a continuous score quantifying its relevance to the query. To quantify the quality of their retrieval engine, Information Retrieval researchers have long used ranking based metrics, such as the NDCG [10, 17], that penalize mistakes differently based on whether they occur at the top or the bottom of the ranking and whether wrong documents still have some marginal relevance or not. Average Precision is also used as a retrieval metric [18] and has even been given probabilistic interpretations based on how users interact with the system [13]. Several works have investigated how to optimize those metrics during the training of neural networks, *e.g.* using pairwise losses [4] and later using smooth surrogates of the NDCG in LambdaRank [5], SoftRank [33], ApproxNDCG [25] and Learning-To-Rank [3]. These works however focused on NDCG, the most popular metric for information retrieval, and are without any theoretical guarantees: the surrogates are approximations of the NDCG but not *lower bounds*, *i.e.* their maximization does not imply improved performances during inference.

An additional drawback of this literature is that NDCG does not relate easily to average precision [14], which is the most common metric in image retrieval. Fortunately, there have been some works done to extend AP in a graded setting where relevance between instances is not binary [13, 29]. The graded Average Precision from [29] is the closest to our goal as it leverages SoftRank for direct optimization on non-binary relevance judgements, although there are significant shortcomings. There is no guarantee that the SoftRank surrogate actually minimizes the graded AP, it requires to annotate datasets with pairwise relevances which is unpractical for large scale settings and was only applied to small-scale corpora of a few thousands documents, compared to the hundred thousands of images in IR.

Recently, the authors of [32] introduced three new hierarchical benchmarks datasets for image retrieval, in addition to a novel hierarchical loss CSL. CSL extends proxy-based triplet losses to the hierarchical setting and tries to structure the embedding space in a hierarchical manner. However, this method faces the same limitation as the usual triplet losses: minimizing CSL does not explicitly optimize a well-behaved hierarchical evaluation metric, *e.g.* $\mathcal{H}$-AP. We show experimentally that our method HAPPIER significantly outperforms CSL [32] both on hierarchical metrics and AP-level evaluations.

3   HAPPIER Model

We detail HAPPIER our Hierarchical Average Precision training method for Pertinent ImagE Retrieval. We first introduce the Hierarchical Average Precision, $\mathcal{H}$-AP in Sect. 3.1, that leverages a hierarchical tree (Fig. 2a) of labels. It is based on the hierarchical rank, $\mathcal{H}$-rank, and evaluates rankings so that more relevant instances are ranked before less relevant ones (Fig. 2b). We then show how to directly optimize $\mathcal{H}$-AP by stochastic gradient descent (SGD) using HAPPIER in Sect. 3.2. Our training objective combines a carefully designed smooth upper bound surrogate loss for $\mathcal{L}_{\mathcal{H}\text{-AP}} = 1 - \mathcal{H}$-AP and a clustering loss $\mathcal{L}_{\text{clust.}}$ that supports consistent rankings.

Context. Let us consider a retrieval set $\Omega = \{x_j\}_{j \in [\![1;N]\!]}$ composed of N instances. For a query[1] $q \in \Omega$, we aim to order all $x_j \in \Omega$ so that more relevant (*i.e.* similar) instances are ranked before less relevant instances.

In our hierarchical setting, the relevance of an instance x_j is non-binary. We assume that we have access to a hierarchical tree defining semantic similarities between concepts as in Fig. 2a. For a query q, we leverage this knowledge to partition the set of retrieved instances into $L + 1$ disjoint subsets $\{\Omega^{(l)}\}_{l \in [\![0;L]\!]}$.

---

[1] For the sake of readability, our notations are given for a single query. During training, HAPPIER optimizes our hierarchical retrieval objective by averaging several queries.

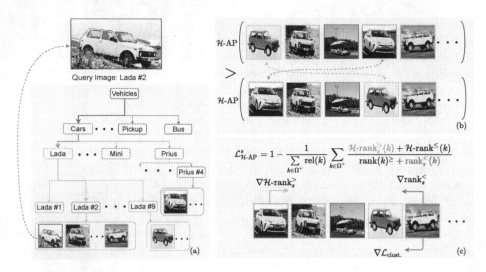

Fig. 2. HAPPIER leverages a hierarchical tree representing the semantic similarities between concepts in (a) to introduce a new hierarchical metric, $\mathcal{H}$-AP in Eq. (3), see (b). $\mathcal{H}$-AP exploits the hierarchy to weight rankings' inversion: given the query image of a "Lada #2", $\mathcal{H}$-AP penalizes an inversion with a "Lada #9" less than with a "Prius #4". To directly train models with $\mathcal{H}$-AP, we carefully study the structure of the problem and introduce the $\mathcal{L}^s_{\mathcal{H}\text{-AP}}$ loss in Eq. (5), which provides a smooth upper bound of $\mathcal{L}_{\mathcal{H}\text{-AP}}$, see (c). We also train HAPPIER with the $\mathcal{L}_{\text{clust.}}$ loss in Eq. (6) to enforce the partial ordering in stochastic optimization to mach the global ones. (Color figure online)

$\Omega^{(L)}$ is the subset of the most similar instances to the query (*i.e.* fine-grained level): for $L = 3$ and a "Lada #2" query, $\Omega^{(3)}$ are the images of the same "Lada #2" (green), see Fig. 2. The set $\Omega^{(l)}$ for $l < L$ contains instances with smaller relevance with respect to the query: $\Omega^{(2)}$ in Fig. 2 is the set of "Lada" that are not "Lada #2" (blue) and $\Omega^{(1)}$ is the set of "Cars" that are not "Lada" (orange). We also define $\Omega^- := \Omega^{(0)}$ as the set of negative instances, *i.e.* the set of vehicles that are not "Cars" (in red) in Fig. 2 and $\Omega^+ = \bigcup_{l=1}^{L} \Omega^{(l)}$. Each instance k of $\Omega^{(l)}$ is thus associated a value through the *relevance function* denoted as **rel(k)** [16].

To rank the instances $x_j \in \Omega$ with respect to the query $\boldsymbol{q}$, we compute cosine similarities in an embedding space. More precisely, we extract embedding vectors using a deep neural network $\boldsymbol{f}$ parameterized by $\boldsymbol{\theta}$, $v_j = f_\theta(x_j)$, and compute the cosine similarity between the query and every image $s_j = f_\theta(q)^T v_j$. Images are then ranked by decreasing cosine similarity score. We learn the parameters $\boldsymbol{\theta}$ of the network with HAPPIER, our framework to directly minimize $\mathcal{L}_{\mathcal{H}\text{-AP}}(\theta) = 1 - \mathcal{H}\text{-AP}(\theta)$. This enforces a ranking where the instances with the highest cosine similarity scores belong to $\Omega^{(L)}$, then $\Omega^{(L-1)}$ *etc.*. and the items with the lowest cosine similarity belong to Ω^-.

3.1 Hierarchical Average Precision

Average Precision (AP) is the most common metric in Image Retrieval. AP evaluates a ranking in a binary setting: for a given query, each instance is either

Fig. 3. Given a "Lada #2" query, the top inversion is less severe than the bottom one. Indeed on the top row instance 1 is semantically closer to the query – as it is a "Lada"– than instance 3 on the bottom row. Indeed instance 3's closest common ancestor with the query, "Cars", is farther in the hierarchical tree (see Fig. 2a). Because of that $\mathcal{H}$-rank(2) is greater on the top row (5/3) than on the bottom row (4/3), leading to a greater $\mathcal{H}$-AP in Fig. 2b for the top row.

positive or negative. It is computed as the average of precision at each rank n over the positive set $AP = \frac{1}{|\Omega^+|} \sum_{n=1}^{N} Prec(n)$. Previous works have written the AP using the ranking operator [2] as in Eq. (1). The rank for an instance k is written as a sum of Heaviside (step) function H [25]: this counts the number of instances j ranked before k, *i.e.* that have a higher cosine similarity ($s_j > s_k$). rank$^+$ is the rank among the positive instances, *i.e.* restricted to Ω^+.

$$AP = \frac{1}{|\Omega^+|} \sum_{k \in \Omega^+} \frac{\text{rank}^+(k)}{\text{rank}(k)}, \text{ with } \begin{cases} \text{rank}(k) = 1 + \sum_{j \in \Omega} H(s_j - s_k) \\ \text{rank}^+(k) = 1 + \sum_{j \in \Omega^+} H(s_j - s_k) \end{cases} \tag{1}$$

Extending AP to Hierarchical Image Retrieval. We propose an extension of AP that leverages non-binary labels. To do so, we extend the concept of rank$^+$ to the hierarchical case with the concept of hierarchical rank, $\mathcal{H}$-rank:

$$\mathcal{H}\text{-rank}(k) = \text{rel}(k) + \sum_{j \in \Omega^+} \min(\text{rel}(k), \text{rel}(j)) \cdot H(s_j - s_k) . \tag{2}$$

Intuitively, $\min(\text{rel}(k), \text{rel}(j))$ corresponds to seeking the closest ancestor shared by instance k and j with the query in the hierarchical tree. As illustrated in Fig. 3, $\mathcal{H}$-rank induces a smoother penalization for instances that do not share the same fine-grained label as the query but still share some coarser semantics, which is not the case for rank$^+$.

From $\mathcal{H}$-rank in Eq. (2) we define the Hierarchical Average Precision, $\mathcal{H}$-AP:

$$\mathcal{H}\text{-AP} = \frac{1}{\sum_{k \in \Omega^+} \text{rel}(k)} \sum_{k \in \Omega^+} \frac{\mathcal{H}\text{-rank}(k)}{\text{rank}(k)} \tag{3}$$

Equation (3) extends the AP to non-binary labels. We replace rank$^+$ by our hierarchical rank $\mathcal{H}$-rank and the normalization term $|\Omega^+|$ is replaced by $\sum_{k \in \Omega^+} \text{rel}(k)$, which both represent the "sum of positives", see more details in supplementary A.2.

$\mathcal{H}$-AP extends the desirable properties of the AP. It evaluates the quality of a ranking by: i) penalizing inversions of instances that are not ranked in decreasing order of relevances with respect to the query, ii) giving stronger emphasis to inversions that occur at the top of the ranking. Finally, we can observe that, by this definition, $\mathcal{H}$-AP is equal to the AP in the binary setting ($L = 1$). This makes $\mathcal{H}$-AP a *consistent generalization* of AP (details in supplementary A.2).

Relevance Function Design. The relevance rel(k) defines how "similar" an instance $k \in \Omega^{(l)}$ is to the query q. While rel(k) might be given as input in Information Retrieval datasets [9,24], we need to define it based on the hierarchical tree in our case. We want to enforce the constraint that the relevance decreases when going up the tree, *i.e.* rel(k) > rel(k') for $k \in \Omega^{(l)}$, $k' \in \Omega^{(l')}$ and $l > l'$. To do so, we assign a total weight of $(l/L)^\alpha$ to each semantic level l, where $\alpha \in \mathbb{R}^+$ controls the decrease rate of similarity in the tree. For example for $L = 3$ and $\alpha = 1$, the total weights for each level are 1, $\frac{2}{3}$, $\frac{1}{3}$ and 0. The instance relevance rel(k) is normalized by the cardinal of $\Omega^{(l)}$:

$$\text{rel}(k) = \frac{(l/L)^\alpha}{|\Omega^{(l)}|} \text{ if } k \in \Omega^{(l)} \tag{4}$$

Other definitions fulfilling the decreasing similarity behaviour in the tree are possible. An interesting option for the relevance enables to recover a weighted sum of AP, denoted as $\sum w\text{AP} := \sum_{l=1}^{L} w_l \cdot \text{AP}^{(l)}$ (supplementary A.2), *i.e.* the weighted sum of AP is a particular case of $\mathcal{H}$-AP.

We set $\alpha = 1$ in Eq. (4) for the $\mathcal{H}$-AP metric and in our main experiments. Setting α to larger values supports better performances on fine-grained levels as their relevances will relatively increase. This variant is denoted HAPPIER$_F$ and discussed in Sect. 4.

3.2   Direct Optimization of $\mathcal{H}$-AP

$\mathcal{H}$-AP in Eq. (3) involves the computation of $\mathcal{H}$-rank and rank, which are non-differentiable due to the summing of Heaviside step functions. We thus introduce a smooth approximation of $\mathcal{H}$-AP to obtain a surrogate loss amenable to gradient descent, which fulfils theoretical guarantees for proper optimization.

Re-writing. $\mathcal{H}$-AP In order to design our surrogate loss for $\mathcal{L}_{\mathcal{H}\text{-AP}} = 1 - \mathcal{H}\text{-AP}$, we decompose $\mathcal{H}$-rank and rank into two quantities. Denoting $\mathcal{H}$-rank$^>(k)$ (resp. $\mathcal{H}$-rank$^\leq(k)$) as the restriction of $\mathcal{H}$-rank to instances of strictly higher relevances (resp. lower or equal), we can see that $\mathcal{H}$-rank(k) = $\mathcal{H}$-rank$^>(k)$ + $\mathcal{H}$-rank$^\leq(k)$. The rank can be decomposed in a similar fashion: rank(k) = rank$^\geq(k)$+rank$^<(k)$ where < (resp. $\geq$) denotes the restriction to instances of strictly lower relevances (resp. higher or equal). The $\mathcal{L}_{\mathcal{H}\text{-AP}}$ can be rewritten as follow:

$$\mathcal{L}_{\mathcal{H}\text{-AP}} = 1 - \frac{1}{\sum_{k \in \Omega^+} \text{rel}(k)} \sum_{k \in \Omega^+} \frac{\mathcal{H}\text{-rank}^>(k) + \mathcal{H}\text{-rank}^\leq(k)}{\text{rank}^\geq(k) + \text{rank}^<(k)}. \tag{5}$$

We choose to optimize over $\mathcal{H}$-rank$^>$ and rank$^<$ in Eq. (5). We maximize $\mathcal{H}$-rank$^>$ to enforce that the kth instance must decrease in cosine similarity score if it is ranked before another instance of higher relevance ($\nabla\mathcal{H}$-rank$^>$ in Fig. 2 enforces the blue instance to be ranked after the green one as it is less relevant to the query). We minimize rank$^<$ to encourage the kth instance to increase in cosine similarity score if it is ranked after one or more instances of lower relevance (∇ rank$^<$ in Fig. 2 enforces that the last green instance moves before less relevant instances). Optimizing both those terms leads to a decrease in $\mathcal{L}_{\mathcal{H}\text{-AP}}$. On the other hand, we purposely do not optimize the two remaining $\mathcal{H}$-rank$^\leq(k)$ and rank$^\geq(k)$ terms, since this could harm training performances as explained in supplementary A.3.

Upper Bound of $\mathcal{L}_{\mathcal{H}\text{-AP}}$. Based on the previous analysis, we now design our surrogate loss $\mathcal{L}_{\mathcal{H}\text{-AP}}^s$ by introducing a smooth approximation of rank$^<$ and $\mathcal{H}$-rank$^>(k)$. An important sought property of $\mathcal{L}_{\mathcal{H}\text{-AP}}^s$ is that it is an upper bound of $\mathcal{L}_{\mathcal{H}\text{-AP}}$. To this end, we approximate $\mathcal{H}$-rank$^>(k)$ with a piece-wise linear function that is a lower bound of the Heaviside function. rank$^<$ is approximated with a smooth upper bound of the Heaviside that combines a piece-wise sigmoid function and an affine function, which has been shown to make the training more robust thanks to the induced implicit margins between positives and negatives [2, 27, 30]. More details are given in supplementary A.3 on those surrogates.

Clustering Constraint in HAPPIER. Positives only need to have a greater cosine similarity with the query than negatives in order to be correctly ranked. Yet, we cannot optimize the ranking on the entire datasets – and thus the true $\mathcal{L}_{\mathcal{H}\text{-AP}}$ – because of the batch-wise estimation performed in stochastic gradient descent. To mitigate this issue, we take inspiration from clustering methods [34, 41] to define the following objective in order to group closely the embeddings of instances that share the same fine-grained label:

$$\mathcal{L}_{\text{clust.}}(\theta) = -\log\left(\frac{\exp(\frac{v_y^T p_y}{\sigma})}{\sum_{p_z \in \mathcal{Z}}\exp(\frac{v_y^T p_z}{\sigma})}\right), \quad (6)$$

where p_y is the normalized proxy corresponding to the fine-grained class of the embedding v_y, $\mathcal{Z}$ is the set of proxies, and σ is a temperature scaling parameter. In Fig. 2, $\nabla\mathcal{L}_{\text{clust.}}$ further clusters "Lada #2" instances. $\mathcal{L}_{\text{clust.}}$ induces a reference shared across batches and thus enforces that the partial ordering in-between batches is consistent with the global ordering over the entire retrieval set.

Our resulting final objective is a linear combination of both our losses, with a weight factor $\lambda \in [0,1]$ that balances the two terms:

$$\mathcal{L}_{\text{HAPPIER}}(\theta) = (1-\lambda)\cdot\mathcal{L}_{\mathcal{H}\text{-AP}}^s(\theta) + \lambda\cdot\mathcal{L}_{\text{clust.}}(\theta).$$

4   Experiments

4.1   Experimental Setup

Datasets. We use the standard benchmark Stanford Online Products [22] (SOP) with two levels of hierarchy ($L = 2$), and iNaturalist-2018 [36] with the standard splits from [2] in two settings: i) iNat-base with two levels of hierarchy ($L = 2$) ii) iNat-full with the full biological taxonomy composed of 7 levels ($L = 7$). We also evaluate on the recent dynamic metric learning (DyML) datasets (DyML-V, DyML-A, DyML-P) introduced in [32] for the task of hierarchical image retrieval, each with 3 semantic levels ($L = 3$).

Implementation Details. Our base model is a ResNet-50 pretrained on ImageNet for SOP and iNat-base/full, and a ResNet-34 randomly initialized on DyML-V&A and pretrained on ImageNet on DyML-P, following [32]. Unless specified otherwise, all reported results are obtained with $\alpha = 1$ in Eq. (4) and $\lambda = 0.1$ for $\mathcal{L}_{\text{HAPPIER}}$. We study the impact of these parameters in Sect. 4.3.

Metrics. For SOP and iNat, we evaluate the models based on three hierarchical metrics: $\mathcal{H}$-AP – which we introduced in Eq. (3) – the Average Set Intersection (ASI) and the Normalized Discounted Cumulative Gain (NDCG), defined in supplementary B.3. We also report the AP for each semantic level. For DyML, we follow the evaluation protocols of [32] and compute AP, ASI and R@1 on each semantic scale before averaging them. We cannot compute $\mathcal{H}$-AP or NDCG on those datasets as the hierarchical tree is not available on the test set.

Baselines. We compare HAPPIER to several recent image retrieval methods optimized at the fine-grained level, which represent strong baselines for IR when training with binary labels: Triplet SH (TL_{SH}) [40], NormSoftMax (NSM) [41], ProxyNCA++ (NCA++) [34] and ROADMAP [27]. We also benchmark against hierarchical methods obtained by summing these fine-grained losses at different levels (denoted by Σ), and with respect to the recent hierarchical CSL loss [32]. Details on the experimental setup are given in supplementary B.

4.2   Main Results

Hierachical Results. We first evaluate HAPPIER on global hierarchical metrics. On Table 1, we notice that HAPPIER significantly outperforms methods trained on the fine-grained level only, with a gain on $\mathcal{H}$-AP over the best performing methods of +16.1pt on SOP, +13pt on iNat-base and 12.7pt on iNat-full. HAPPIER also exhibits significant gains compared to hierarchical methods. On $\mathcal{H}$-AP, HAPPIER has important gains on all datasets (*e.g.* +6.3pt on SOP, +4.2pt on iNat-base over the best competitor), but also on ASI and NDCG. This shows the strong generalization of the method on standard metrics. Compared to the recent CSL loss [32], we observe a consistent gain over all metrics and datasets, *e.g.* +6pt on $\mathcal{H}$-AP, +8pt on ASI and +2.6pts on NDCG on SOP. This shows the benefits of optimizing a well-behaved hierarchical metric compared to an ad-hoc proxy method.

Table 1. Comparison of HAPPIER on SOP and iNat-base/full when using hierarchical metrics. Best results in **bold**, second best underlined.

	Method	SOP			iNat-base			iNat-full		
		$\mathcal{H}$-AP	ASI	NDCG	$\mathcal{H}$-AP	ASI	NDCG	$\mathcal{H}$-AP	ASI	NDCG
Fine	Triplet SH [40]	42.2	22.4	78.8	39.5	63.7	91.5	36.1	59.2	89.8
	NSM [41]	42.8	21.1	78.3	38.0	51.6	88.9	33.3	51.7	88.2
	NCA++ [34]	43.0	21.5	78.4	39.5	57.0	90.1	35.3	55.7	89.0
	Smooth-AP [2]	42.9	20.6	78.2	41.3	64.2	91.9	37.2	60.1	90.1
	ROADMAP [27]	43.3	19.1	77.9	40.3	61.0	91.2	34.7	59.6	89.5
Hier.	ΣTL$_{SH}$ [40]	<u>53.1</u>	53.3	<u>89.2</u>	44.0	87.4	96.4	39.9	<u>85.5</u>	92.0
	ΣNSM [41]	50.4	49.7	87.0	47.9	75.8	94.4	<u>46.9</u>	74.2	**93.8**
	ΣNCA++ [34]	49.5	52.8	87.8	48.9	78.7	95.0	44.7	74.3	92.6
	CSL [32]	52.8	<u>57.9</u>	88.1	<u>50.1</u>	**89.3**	<u>96.7</u>	45.1	84.9	93.0
	HAPPIER	**59.4**	**65.9**	**91.5**	**54.3**	**89.3**	**96.9**	**47.9**	**87.2**	**93.8**

On Table 2, we evaluate HAPPIER on the recent DyML benchmarks. HAPPIER again shows significant gains in mAP and ASI compared to methods only trained on fine-grained labels, *e.g.* +9pt in mAP and +10pt in ASI on DyML-V. HAPPIER also outperforms other hierarchical baselines: +4.8pt mAP on DyML-V, +0.9 on DyML-A and +1.8 on DyML-P. In R@1, HAPPIER performs on par with other methods on DyML-V and outperforms other hierarchical baselines by a large margin on DyML-P: 63.7 *vs.* 60.8 for ΣNSM. Interestingly, HAPPIER also consistently outperforms CSL [32] on its own datasets[2].

Table 2. Performance comparison on Dynamic Metric Learning benchmarks [32].

	Method	DyML-Vehicle			DyML-Animal			DyML-Product		
		mAP	ASI	R@1	mAP	ASI	R@1	mAP	ASI	R@1
Fine	TL$_{SH}$ [40]	26.1	38.6	84.0	37.5	46.3	66.3	36.32	46.1	59.6
	NSM [41]	27.7	40.3	88.7	38.8	48.4	<u>69.6</u>	35.6	46.0	57.4
	Smooth-AP [2]	27.1	39.5	83.8	37.7	45.4	63.6	36.1	45.5	55.0
	ROADMAP [27]	27.1	39.6	84.5	34.4	42.6	62.8	34.6	44.6	<u>62.5</u>
Hier.	ΣTL$_{SH}$ [40]	25.5	38.1	81.0	38.9	47.2	65.9	<u>36.9</u>	46.3	58.5
	ΣNSM [41]	<u>32.0</u>	<u>45.7</u>	**89.4**	<u>42.6</u>	<u>50.6</u>	**70.0**	36.8	<u>46.9</u>	60.8
	CSL [32]	30.0	43.6	87.1	40.8	46.3	60.9	31.1	40.7	52.7
	HAPPIER	**37.0**	**49.8**	<u>89.1</u>	**43.8**	**50.8**	68.9	**38.0**	**47.9**	**63.7**

---

[2] CSL's score on Table 2 are above those reported in [32]; personal discussions with the authors [32] validate that our results are valid for CSL, see supplementary B.5.

Detailed Evaluation. Tables 3 and 4 shows the different methods' performances on all semantic hierarchy levels. We evaluate HAPPIER and also HAPPIER$_F$ ($\alpha > 1$ for Eq. (4) in Sect. 3.1), with $\alpha = 5$ on SOP and $\alpha = 3$ on iNat-base/full. HAPPIER optimizes the overall hierarchical performances, while HAPPIER$_F$ is meant to be optimal at the fine-grained level while still optimizing coarser levels.

Table 3. Comparison of HAPPIER *vs.* methods trained only on fine-grained labels on SOP and iNat-base. Metrics are reported for both semantic levels.

	Method	SOP			iNat-base		
		Fine		Coarse	Fine		Coarse
		R@1	AP	AP	R@1	AP	AP
Fine	TL$_{SH}$ [40]	79.8	59.6	14.5	66.3	33.3	51.5
	NSM [41]	81.3	61.3	13.4	70.2	<u>37.6</u>	38.8
	NCA++ [34]	81.4	61.7	13.6	67.3	37.0	44.5
	Smooth-AP [2]	81.3	61.7	13.4	67.3	35.2	53.1
	ROADMAP [27]	**82.2**	**62.5**	12.9	69.3	35.1	50.4
Hier.	CSL [32]	79.4	58.0	<u>45.0</u>	62.9	30.2	<u>88.5</u>
	HAPPIER	81.0	60.4	**58.4**	<u>70.7</u>	36.7	**88.6**
	HAPPIER$_F$	<u>81.8</u>	62.2	36.0	**71.0**	**37.8**	78.8

On Table 3, we observe that HAPPIER gives the best performances at the coarse level, with a significant boost compared to fine-grained methods, *e.g.* +43.9pt AP compared to the best non-hierarchical TL$_{SH}$ [40] on SOP. HAPPIER even outperforms the best fine-grained methods in R@1 on iNat-base, but is slightly below on SOP. HAPPIER$_F$ performs on par with the best methods at the finest level on SOP, while further improving performances on iNat-base, and still significantly outperforms fine-grained methods at the coarse level.

The satisfactory behaviour and the two optimal regimes of HAPPIER and HAPPIER$_F$ are confirmed and even more pronounced on iNat-full (Table 4): HAPPIER gives the best results on coarser levels (from "Order"), while being very close to the best results on finer ones. HAPPIER$_F$ gives the best results at the finest levels, even outperforming very competitive fine-grained baselines.

Again, note that HAPPIER outperforms CSL [32] on all semantic levels and datasets on Tables 3 and 4, *e.g.* +5pt on the fine-grained AP ("Species") and +3pt on the coarsest AP ("Kingdom") on Table 4.

4.3 HAPPIER Analysis

Ablation Study. In Table 5, we study the impact of our different choices regarding the direct optimization of $\mathcal{H}$-AP. The baseline method uses a sigmoid to optimize $\mathcal{H}$-AP as in [2,25]. Switching to our surrogate loss $\mathcal{L}^s_{\mathcal{H}\text{-AP}}$ Sect. 3.2 yields a +0.8pt increase in $\mathcal{H}$-AP. Finally, the combination with $\mathcal{L}_{\text{clust.}}$ in HAPPIER results in an additional 1.3pt improvement in $\mathcal{H}$-AP.

Table 4. Comparison of HAPPIER *vs.* methods trained only on fine-grained labels on iNat-Full. Metrics are reported for all 7 semantic levels.

	Method	Species		Genus	Family	Order	Class	Phylum	Kingdom
		R@1	AP	AP	AP	AP	AP	AP	AP
Fine	TL$_{\text{SH}}$ [40]	66.3	33.3	34.2	32.3	35.4	48.5	54.6	68.4
	NSM [41]	<u>70.2</u>	**37.6**	<u>38.0</u>	31.4	28.6	36.6	43.9	63.0
	NCA++ [34]	67.3	37.0	37.9	33.0	32.3	41.9	48.4	66.1
	Smooth-AP [2]	67.3	35.2	36.3	33.5	35.0	49.3	55.8	69.9
	ROADMAP [27]	69.3	35.1	35.4	29.3	29.6	46.4	54.7	69.5
Hier.	CSL [32]	59.9	30.4	32.4	36.2	50.7	<u>81.0</u>	<u>87.4</u>	<u>91.3</u>
	HAPPIER	<u>70.2</u>	36.0	37.0	<u>38.0</u>	**51.9**	**81.3**	**89.1**	**94.4**
	HAPPIER$_{\text{F}}$	**70.8**	**37.6**	**38.2**	**38.8**	<u>50.9</u>	76.1	82.2	83.1

Table 5. Impact of optimization choices for $\mathcal{H}$-AP (cf. Sect. 3.2) on iNat-base.

$\mathcal{L}^s_{\mathcal{H}\text{-AP}}$	$\mathcal{L}_{\text{clust.}}$	$\mathcal{H}$-AP
✗	✗	52.3
✓	✗	53.1
✓	✓	**54.3**

Table 6. Comparison of $\mathcal{H}$-AP (Eq. (4)) and ΣwAP from supplementary A.2.

test→ train↓	$\mathcal{H}$-AP	$\sum w$AP	NDCG
$\mathcal{H}$-AP	**53.1**	39.8	**97.0**
$\sum w$AP	52.0	**40.5**	96.4

Impact of the Relevance Function. Table 6 compares models that are trained with the relevance function of Eq. (4), *i.e.* $\mathcal{H}$-AP, and $\sum w$AP (relevance given in supplementary A.2). We report results for $\mathcal{H}$-AP, $\sum w$AP and NDCG. Both $\mathcal{H}$-AP, $\sum w$AP perform better when trained with their own metric: +1.1pt $\mathcal{H}$-AP for the model trained to optimize it and +0.7pt $\sum w$AP for the model trained to optimize it. Both models show similar performances in NDCG (96.4 *vs.* 97.0).

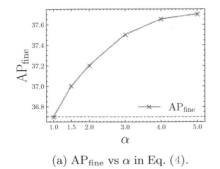

(a) AP$_{\text{fine}}$ vs α in Eq. (4).

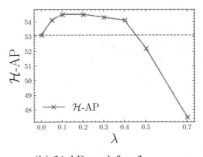

(b) $\mathcal{H}$-AP *vs.* λ for $\mathcal{L}_{\text{HAPPIER}}$.

Fig. 4. Impact on Inat-base of α in Eq. (4) for setting the relevance of $\mathcal{H}$-AP (a) and of the λ hyper-parameter on HAPPIER results (b).

Hyper-Parameters. Figure 4a studies the impact of α for setting the relevance in Eq. (4): increasing α improves the performances of the AP at the fine-grained level on iNat-base, as expected. We also show in Fig. 4b the impact of λ weighting $\mathcal{L}^s_{\mathcal{H}\text{-AP}}$ and $\mathcal{L}_{\text{clust.}}$ in HAPPIER performances: we observe a stable increase in $\mathcal{H}$-AP within $0 < \lambda < 0.5$ compared to optimizing only $\mathcal{L}^s_{\mathcal{H}\text{-AP}}$, while a drop in performance is observed for $\lambda > 0.5$. This shows the complementarity of $\mathcal{L}^s_{\mathcal{H}\text{-AP}}$ and $\mathcal{L}_{\text{clust.}}$, and how, when combined, HAPPIER reaches its best performance.

4.4   Qualitative Study

We provide here qualitative assessments of HAPPIER, including embedding space analysis and visualization of HAPPIER's retrievals.

t-SNE: Organization of the Embedding Space. In Fig. 5, we plot using t-SNE [7,35] how HAPPIER learns an embedding space on SOP ($L = 2$) that is well-organized. We plot the mean vector of each fine-grained class and we assign the color based on the coarse level. We show on Fig. 5a the t-SNE visualisation obtained using a baseline method trained on the fine-grained labels, and in Fig. 5b we plot the t-SNE of the embedding space of a model trained with HAPPIER. We cannot observe any clear clusters for the coarse level on Fig. 5a, whereas we can appreciate the the quality of the hierarchical clusters formed on Fig. 5b.

Controlled Errors. Finally, we showcase in Fig. 6 errors of HAPPIER *vs.* a fine-grained baseline. On Fig. 6a, we illustrate how a model trained with HAP-PIER makes mistakes that are less severe than a baseline model trained only on

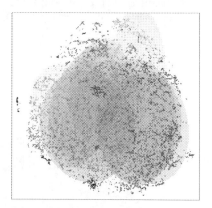 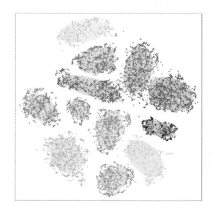

(a) t-SNE visualization of a model trained only on the fine-grained labels.

(b) t-SNE visualization of a model trained with **HAPPIER**.

Fig. 5. t-SNE visualisation of the embedding space of two models trained on SOP. Each point is the average embedding of each fine-grained label (object instance) and the colors represent coarse labels (object category, *e.g.* bike, coffee maker).

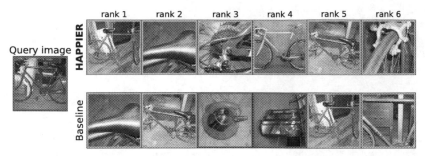

(a) HAPPIER can help make less severe mistakes. The inversion on the bottom row are with negative instances (in red), where as with HAPPIER (top row) inversions are with instances sharing the same coarse label "bike" (in orange).

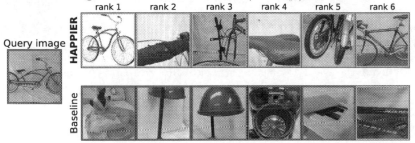

(b) In this example, the models fail to retrieve the correct fine grained images. However HAPPIER still retrieves images of very similar bikes (in orange) whereas the baseline retrieves images that are dissimilar semantically to the query (in red).

Fig. 6. Qualitative examples of failure cases from a standard fine-grained model corrected by training with HAPPIER. (Color figure online)

the fine-grained level. On Fig. 6b, we show an example where both models fail to retrieve the correct fine-grained instances, however the model trained with HAPPIER retrieves images of bikes that are visually more similar to the query.

5   Conclusion

In this work, we introduce HAPPIER, a new training method that leverages hierarchical relations between concepts to learn robust rankings. HAPPIER is based on a new metric $\mathcal{H}$-AP that evaluates hierarchical rankings and uses a combination of a smooth upper bound surrogate with theoretical guarantees and a clustering loss to directly optimize it. Extensive experiments show that HAPPIER performs on par to state-of-the-art image retrieval methods on fine-grained metrics and exhibits large improvements vs. recent hierarchical methods on hierarchical metrics. Learning more robust rankings reduces the severity of ranking errors, and is qualitatively related to a better organization of the embedding space with HAPPIER. Future works include the adaptation of HAPPIER to the unsupervised setting, e.g. for providing a relevant self-training criterion.

Acknowledgement. This work was done under a grant from the the AHEAD ANR program (ANR-20-THIA-0002). It was granted access to the HPC resources of IDRIS under the allocation 2021-AD011012645 made by GENCI.

References

1. Bertinetto, L., Mueller, R., Tertikas, K., Samangooei, S., Lord, N.A.: Making better mistakes: leveraging class hierarchies with deep networks. In: Proceedings of the IEEE/CVF Conference on Computer Vision and Pattern Recognition, pp. 12506–12515 (2020)
2. Brown, A., Xie, W., Kalogeiton, V., Zisserman, A.: Smooth-AP: smoothing the path towards large-scale image retrieval. In: Vedaldi, A., Bischof, H., Brox, T., Frahm, J.-M. (eds.) ECCV 2020. LNCS, vol. 12354, pp. 677–694. Springer, Cham (2020). https://doi.org/10.1007/978-3-030-58545-7_39
3. Bruch, S., Zoghi, M., Bendersky, M., Najork, M.: Revisiting approximate metric optimization in the age of deep neural networks. In: Proceedings of the 42nd International ACM SIGIR Conference on Research and Development in Information Retrieval, pp. 1241–1244 (2019)
4. Burges, C., et al.: Learning to rank using gradient descent. In: Proceedings of the 22nd International Conference on Machine Learning, pp. 89–96. ICML 2005, Association for Computing Machinery, New York, NY, USA (2005). https://doi.org/10.1145/1102351.1102363
5. Burges, C., Ragno, R., Le, Q.: Learning to rank with nonsmooth cost functions. In: Schölkopf, B., Platt, J., Hoffman, T. (eds.) Advances in Neural Information Processing Systems, vol. 19. MIT Press (2006). https://proceedings.neurips.cc/paper/2006/file/af44c4c56f385c43f2529f9b1b018f6a-Paper.pdf
6. Cakir, F., He, K., Xia, X., Kulis, B., Sclaroff, S.: Deep metric learning to rank. In: Proceedings of the IEEE/CVF Conference on Computer Vision and Pattern Recognition, pp. 1861–1870 (2019)
7. Chan, D.M., Rao, R., Huang, F., Canny, J.F.: GPU accelerated t-distributed stochastic neighbor embedding. J. Parallel Distrib. Comput. **131**, 1–13 (2019)
8. Chang, D., Pang, K., Zheng, Y., Ma, Z., Song, Y.Z., Guo, J.: Your "flamingo" is my "bird": fine-grained, or not. In: Proceedings of the IEEE/CVF Conference on Computer Vision and Pattern Recognition, pp. 11476–11485 (2021)
9. Chapelle, O., Chang, Y.: Yahoo! learning to rank challenge overview. In: Proceedings of the learning to rank challenge, pp. 1–24. PMLR (2011)
10. Croft, W.B., Metzler, D., Strohman, T.: Search engines: information retrieval in practice, vol. 520. Addison-Wesley Reading (2010)
11. Deng, J., Guo, J., Xue, N., Zafeiriou, S.: ArcFace: additive angular margin loss for deep face recognition. In: Proceedings of the IEEE/CVF conference on computer vision and pattern recognition. pp. 4690–4699 (2019)
12. Dhall, A., Makarova, A., Ganea, O., Pavllo, D., Greeff, M., Krause, A.: Hierarchical image classification using entailment cone embeddings. In: Proceedings of the IEEE/CVF Conference on Computer Vision and Pattern Recognition Workshops, pp. 836–837 (2020)

13. Dupret, G., Piwowarski, B.: A user behavior model for average precision and its generalization to graded judgments. In: Proceedings of the 33rd International ACM SIGIR Conference on Research and Development in Information Retrieval, pp. 531–538. SIGIR 2010, Association for Computing Machinery, New York, NY, USA (2010). https://doi.org/10.1145/1835449.1835538

14. Dupret, G., Piwowarski, B.: Model based comparison of discounted cumulative gain and average precision. J. Discrete Algorithms **18**, 49–62 (2013). https://doi.org/10.1016/j.jda.2012.10.002. https://www.sciencedirect.com/science/article/pii/S1570866712001372 Selected papers from the 18th International Symposium on String Processing and Information Retrieval (SPIRE 2011)

15. Hadsell, R., Chopra, S., LeCun, Y.: Dimensionality reduction by learning an invariant mapping. In: 2006 IEEE Computer Society Conference on Computer Vision and Pattern Recognition (CVPR'06), vol. 2, pp. 1735–1742. IEEE (2006)

16. Hjørland, B.: The foundation of the concept of relevance. J. Am. Soc. Inform. Sci. Technol. **61**(2), 217–237 (2010)

17. Järvelin, K., Kekäläinen, J.: Cumulated gain-based evaluation of IR techniques. ACM Trans. Inf. Syst. (TOIS) **20**(4), 422–446 (2002)

18. Järvelin, K., Kekäläinen, J.: IR evaluation methods for retrieving highly relevant documents. In: ACM SIGIR Forum, vol. 51, pp. 243–250. ACM New York, NY, USA (2017)

19. Kekäläinen, J., Järvelin, K.: Using graded relevance assessments in ir evaluation. J. Am. Soc. Inf. Sci. Technol. **53**(13), 1120–1129 (2002). https://doi.org/10.1002/asi.10137. https://onlinelibrary.wiley.com/doi/abs/10.1002/asi.10137

20. Law, M.T., Thome, N., Cord, M.: Learning a distance metric from relative comparisons between quadruplets of images. Int. J. Comput. Vision **121**(1), 65–94 (2017)

21. Movshovitz-Attias, Y., Toshev, A., Leung, T.K., Ioffe, S., Singh, S.: No fuss distance metric learning using proxies. In: Proceedings of the IEEE International Conference on Computer Vision, pp. 360–368 (2017)

22. Oh Song, H., Xiang, Y., Jegelka, S., Savarese, S.: Deep metric learning via lifted structured feature embedding. In: Proceedings of the IEEE Conference on Computer Vision and Pattern Recognition, pp. 4004–4012 (2016)

23. P., M.V., Paulus, A., Musil, V., Martius, G., Rolínek, M.: Differentiation of blackbox combinatorial solvers. In: ICLR (2020)

24. Qin, T., Liu, T.: Introducing LETOR 4.0 datasets. arXiv preprint arXiv:1306.2597 (2013)

25. Qin, T., Liu, T.Y., Li, H.: A general approximation framework for direct optimization of information retrieval measures. Inf. Retrieval **13**, 375–397 (2009)

26. Radenović, F., Tolias, G., Chum, O.: CNN image retrieval learns from BoW: unsupervised fine-tuning with hard examples. In: Leibe, B., Matas, J., Sebe, N., Welling, M. (eds.) ECCV 2016. LNCS, vol. 9905, pp. 3–20. Springer, Cham (2016). https://doi.org/10.1007/978-3-319-46448-0_1

27. Ramzi, E., Thome, N., Rambour, C., Audebert, N., Bitot, X.: Robust and decomposable average precision for image retrieval. Advances in Neural Information Processing Systems 34 (2021)

28. Revaud, J., Almazán, J., Rezende, R.S., Souza, C.R.D.: Learning with average precision: Training image retrieval with a listwise loss. In: Proceedings of the IEEE/CVF International Conference on Computer Vision, pp. 5107–5116 (2019)

29. Robertson, S.E., Kanoulas, E., Yilmaz, E.: Extending average precision to graded relevance judgments. In: Proceedings of the 33rd international ACM SIGIR conference on Research and development in information retrieval, pp. 603–610 (2010)

30. Rolínek, M., Musil, V., Paulus, A., Vlastelica, M., Michaelis, C., Martius, G.: Optimizing rank-based metrics with blackbox differentiation. In: Proceedings of the IEEE/CVF Conference on Computer Vision and Pattern Recognition, pp. 7620–7630 (2020)
31. Sohn, K.: Improved deep metric learning with multi-class n-pair loss objective. In: Lee, D., Sugiyama, M., Luxburg, U., Guyon, I., Garnett, R. (eds.) Advances in Neural Information Processing Systems, vol. 29. Curran Associates, Inc. (2016). https://proceedings.neurips.cc/paper/2016/file/6b180037abbebea991d8b1232f8a8ca9-Paper.pdf
32. Sun, Y., et al.: Dynamic metric learning: Towards a scalable metric space to accommodate multiple semantic scales. In: Proceedings of the IEEE/CVF Conference on Computer Vision and Pattern Recognition, pp. 5393–5402 (2021)
33. Taylor, M., Guiver, J., Robertson, S., Minka, T.: SoftRank: optimizing non-smooth rank metrics. In: Proceedings of the 2008 International Conference on Web Search and Data Mining, pp. 77–86. WSDM 2008, Association for Computing Machinery, New York, NY, USA (2008). https://doi.org/10.1145/1341531.1341544
34. Teh, E.W., DeVries, T., Taylor, G.W.: ProxyNCA++: revisiting and revitalizing proxy neighborhood component analysis. In: Vedaldi, A., Bischof, H., Brox, T., Frahm, J.-M. (eds.) ECCV 2020. LNCS, vol. 12369, pp. 448–464. Springer, Cham (2020). https://doi.org/10.1007/978-3-030-58586-0_27
35. van der Maaten, L., Hinton, G.: Visualizing high-dimensional data using t-SNE. J. Mach. Learn. Res. **9**, 2579–2605 (2008)
36. Van Horn, G., et al.: The inaturalist species classification and detection dataset. In: Proceedings of the IEEE conference on computer vision and pattern recognition, pp. 8769–8778 (2018)
37. Wang, H., et al.: CosFace: large margin cosine loss for deep face recognition. In: Proceedings of the IEEE conference on computer vision and pattern recognition, pp. 5265–5274 (2018)
38. Wang, X., Han, X., Huang, W., Dong, D., Scott, M.R.: Multi-similarity loss with general pair weighting for deep metric learning. In: Proceedings of the IEEE/CVF Conference on Computer Vision and Pattern Recognition, pp. 5022–5030 (2019)
39. Wang, X., Zhang, H., Huang, W., Scott, M.R.: Cross-batch memory for embedding learning. In: Proceedings of the IEEE/CVF Conference on Computer Vision and Pattern Recognition, pp. 6388–6397 (2020)
40. Wu, C.Y., Manmatha, R., Smola, A.J., Krahenbuhl, P.: Sampling matters in deep embedding learning. In: Proceedings of the IEEE International Conference on Computer Vision, pp. 2840–2848 (2017)
41. Zhai, A., Wu, H.Y.: Classification is a strong baseline for deep metric learning. arXiv preprint arXiv:1811.12649 (2018)

Learning Semantic Correspondence
with Sparse Annotations

Shuaiyi Huang[1]([✉]), Luyu Yang[1], Bo He[1], Songyang Zhang[2], Xuming He[3,4],
and Abhinav Shrivastava[1]

[1] University of Maryland, College Park, USA
{huangshy,loyo,bohe}@umd.edu, abhinav@cs.umd.edu
[2] Shanghai AI Laboratory, Shanghai, China
zhangsongyang@pjlab.org.cn
[3] ShanghaiTech University, Shanghai, China
hexm@shanghaitech.edu.cn
[4] Shanghai Engineering Research Center of Intelligent Vision and Imaging,
Shanghai, China

Abstract. Finding dense semantic correspondence is a fundamental
problem in computer vision, which remains challenging in complex scenes
due to background clutter, extreme intra-class variation, and a severe
lack of ground truth. In this paper, we aim to address the challenge
of label sparsity in semantic correspondence by enriching supervision
signals from sparse keypoint annotations. To this end, we first propose
a teacher-student learning paradigm for generating dense pseudo-labels
and then develop two novel strategies for denoising pseudo-labels. In particular, we use spatial priors around the sparse annotations to suppress
the noisy pseudo-labels. In addition, we introduce a loss-driven dynamic
label selection strategy for label denoising. We instantiate our paradigm
with two variants of learning strategies: a single offline teacher setting, and mutual online teachers setting. Our approach achieves notable
improvements on three challenging benchmarks for semantic correspondence and establishes the new state-of-the-art. Project page: https://
shuaiyihuang.github.io/publications/SCorrSAN.

Keywords: Semantic correspondence · Pseudo-label · Sparse
annotations

1 Introduction

Estimating pixel-wise correspondence between images is a fundamental task in
computer vision applications. Correspondences like stereo disparities [47] and
optical flow [20] are widely used for applications such as surface reconstruction

Supplementary Information The online version contains supplementary material
available at https://doi.org/10.1007/978-3-031-19781-9_16.

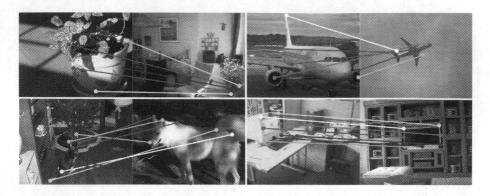

Fig. 1. Motivation. Image pairs from SPair-71k dataset [41] training split show sparse annotations for semantic correspondence.

and video analysis [3, 8]. Recently, such instance-level dense correspondence has been generalized to semantic correspondence, which, given a pair of images, aligns the object instance from the first image to the one of the same category in the second image [22, 24, 28, 40, 42, 44, 45, 52]. It has attracted growing attention due to its practical use in segmentation, style-transfer, and image editing [5, 7, 18, 26, 30, 33]. However, background clutter, intra-class variations, viewpoint changes, and particularly the severe lack of annotations make it an extremely challenging task.

Due to the high cost of dense annotation, the semantic correspondence task only provides sparse keypoint annotations in the supervised setting [31, 35, 36, 42] as shown in Fig. 1. In this paper, we are motivated by how to better utilize the limited supervision. Specifically, we explore the techniques to generate pseudo-labels. However, due to the inevitably noisy effect of pseudo-labels, filtering out noisy pseudo-labels remains a challenging problem. Our key observation is that sparse keypoint annotations and their neighborhood encode rich semantic information. By utilizing this spatial prior, one can seek reliable pseudo-labels that are more likely in the foreground region of interest.

To this end, we propose a novel teacher-student framework to cope with label sparsity. The teacher model is trained with sparse keypoint annotations to generate dense pseudo-labels. To improve pseudo-labels quality, we propose (a) using the sparse annotations as spatial prior to suppress the noisy pseudo-labels, and (b) loss-driven dynamic label selection. To train the models, we propose two variants of our strategy: (1) a single offline teacher with an online student, and (2) two online teachers that learn from each other. Both variants lead to substantial performance improvements over the state-of-the-art.

We instantiate our novel learning strategy based on our proposed simple, yet effective network architecture for semantic correspondence. The proposed network comprises three modules: (a) a feature extractor equipped with our efficient spatial context encoder, (b) a parameter-free correlation map module, and (c) a flow estimator with our designed high-resolution loss.

The contributions are summarized as follows:

- We propose a simple, yet effective model for semantic correspondence without any transformer or 4D-conv for correlation refinement. The key ingredients are an efficient spatial context encoder and a high-resolution loss.
- We introduce a novel teacher-student learning paradigm to enrich the supervision guidance when only sparse annotations are available. Two key techniques are a novel spatial-prior based label filtering and a loss-driven dynamic label selection strategy for high-quality pseudo-label generation.
- Our novel learning strategy is simple to implement, and achieves state-of-the-art results with good generalization performance on three semantic correspondence benchmarks, demonstrating the effectiveness of our method.

2 Related Work

2.1 Semantic Correspondence

Conventional approaches for semantic correspondence mostly employ hand-crafted features together with geometric models [37,49,51]. These methods establish correspondences across images via energy minimization. SIFT Flow [37] pioneers the idea of finding correspondences across similar scenes with SIFT descriptors. Ham et al. [10] utilize object proposals as the matching primitives and establish correspondence via HOG descriptors. Those methods often have difficulty dealing with background clutter, intra-class variations, and large viewpoint changes due to the lack of semantics in features.

Recently, deep CNN-based methods have been widely used in semantic correspondences due to their powerful representations. Early methods formulate semantic correspondence as a geometric alignment problem, with a major focus on developing robust geometric models [19,27,44]. Rocco et al. [44,45] propose a two-stage CNN architecture for regressing image-level transformation parameters, while other efforts regress local translation fields [26,27,29]. More recent works tend to formulate semantic correspondence as a pixel-wise matching problem and cast it as a classification problem. Among these works, there are techniques focusing on developing powerful feature representations [22,40,42], correlation map filtering with 4D/6D-conv or transformers [6,18,35,38,39], effective correspondence readout [32], and different levels of supervision [21,36,53]. However, none of these aforementioned methods have explicitly approached the task of dense semantic correspondence from the perspective of sparse annotations.

2.2 Teacher-Student Learning

Teacher-student framework has been widely used in semi-supervised learning (SSL) [13,34,48,50,54], where the predictions of the teacher model on unlabeled samples serve as pseudo-labels to guide the student model. Teacher-student

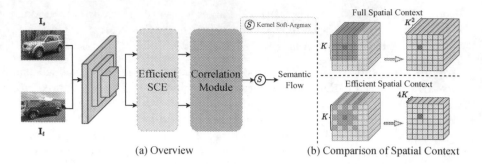

(a) Overview   (b) Comparison of Spatial Context

Fig. 2. Model overview. (a) Illustration of our network. Our network comprises three main modules, including an efficient spatial context encoder, a correlation module, and a flow estimator. (b) Comparison between our proposed efficient spatial context and the full spatial context. Please refer Sect. 3 for more details. *Best viewed in color.* (Color figure online)

framework also plays an important role in knowledge distillation [4,16,17,57,60], where knowledge from a larger teacher model can be transferred into a smaller student model without loss of validity. Recently, Xin *et al.* [36] extend teacher-student to semantic correspondence, where they distill knowledge learned from a probabilistic teacher model on synthetic data to a static student model with unlabeled real image pairs. In contrast, we directly learn from real image pairs labeled with sparse keypoints, and focus on addressing the label sparsity challenge via Siamese teacher-student network design [2]. Note that we tailor teacher-student learning specifically for the dense prediction task of semantic correspondence, where we conduct pixel-level semi-supervised learning within an image and generate pseudo-labels for unlabeled pixels, while most existing work focus on image-level semi-supervised learning.

3   Model Architecture

Semantic correspondence establishes dense correspondences between a source image $\mathbf{I}_a$ and a target image $\mathbf{I}_b$. We adopt a typical CNN-based method which computes a correlation map between the convolution features of two images, based on which a dense flow field is predicted as the final output. We additionally encode spatial context efficiently to compute high-quality correlation map and develop a novel teacher-student learning strategy to cope with label sparsity.

This section introduces our simple and powerful semantic correspondence framework. As depicted in Fig. 2, our framework comprises of three main modules: (1) a sparse spatial context feature extractor that encodes context information efficiently (Sect. 3.1), (2) a correlation operator to compute the correlation map between two convolution features (Sect. 3.2), and (3) a flow estimation operator with high-resolution loss (Sect. 3.3).

3.1   Efficient Spatial Context Encoder

Taking the conv features of the image pairs as the input, the first component of our network is an efficient spatial context encoder that incorporates spatial context into conv features. Recent methods adopt the self-similarity based descriptor to encode spatial context [22,35]. However, the time complexity of the self-similarity grows quadratically with respect to the kernel size of the self-similarity descriptor due to dense sampling patterns they used [22,35]. Inspired by the recent success of sparse attention on reducing the computational cost of non-local operation [14,23,58,59], we propose a spatial context encoder based on sparse sampling patterns, which efficiently encodes context information and reduces the time complexity from quadratic to linear.

As shown in Fig. 2(b), at location (i, j), its spatial context descriptor $\mathbf{s}_{(i,j)} \in \mathbb{R}^{4K}$ is a self-similarity vector, where K is the self-similarity operator kernel size. It is computed between its own feature vector $\mathbf{z}_{(i,j)} \in \mathbb{R}^{d_z}$ and its $4K$ neighboring feature vectors, where the neighbors are in its criss-cross and diagonal directions in an fixed ordered. In contrast to dense sampling patterns [22,35], our sparse sampling patterns reduce the time and space complexity for computing spatial descriptors from $O(K^2)$ to $O(K)$.

To combine spatial context and conv features, we employ a simple fusion step to generate the final context-aware semantic feature map $\mathbf{G}$ following [22]. Concretely, we concatenate $\mathbf{z}_{(i,j)}$ and $\mathbf{s}_{(i,j)}$ and feed the result into a linear transformation with parameter $\mathbf{W} \in \mathbb{R}^{(d_z+4K) \times d_g}$ followed by a ReLU operation, resulting in a context-aware semantic feature vector $\mathbf{g}_{(i,j)} \in \mathbb{R}^{d_g}$. We add subscripts to represent the context-aware semantic feature map $\mathbf{G}_a \in \mathbb{R}^{d_g \times h_a \times w_a}$ and $\mathbf{G}_b \in \mathbb{R}^{d_g \times h_b \times w_b}$ for the image $\mathbf{I}_a$ and $\mathbf{I}_b$, resp., where h_b, w_b (resp. h_a, w_a) is the spatial size of $\mathbf{G}_b$ (resp. $\mathbf{G}_a$).

3.2   Correlation Map Computation

We compute a 4D correlation map from the context-aware semantic feature maps $\mathbf{G}_a$, $\mathbf{G}_b$ and filter it with the mutual nearest neighbor module [46]. We denote the resulting 4D correlation map as $\mathbb{C} \in \mathcal{R}^{h_a \times w_a \times h_b \times w_b}$.

We propose to learn a high-resolution correlation map for high-quality dense matching in contrast to learning correspondence in stride16 [22,45]. We upsample the correlation map $\mathbb{C}$ (4 times) instead of upsampling the feature maps for memory efficiency. We denote the resulting upsampled correlation map as

$$\mathbf{C} = \mathcal{U}(\mathbb{C}), \quad \mathbf{C} \in \mathbb{R}^{H_a \times W_a \times H_b \times W_b}, \tag{1}$$

where H_a, W_a, H_b, and W_b are the upsampled spatial sizes, $\mathcal{U}$ is the upsample operation. Note that we achieve high performance with single layer feature, while DHPF [42] requires multi-layer features with higher complexity.

3.3   Flow Formation and High-resolution Loss

To obtain differentiable flow, we adopt the kernel soft-argmax operator [32] to transform the upsampled correlation map $\mathbf{C}$ into dense semantic flow $\hat{f}$ as below:

$$\hat{f} = \mathcal{S}(\mathbf{C}), \quad \hat{f} \in \mathbb{R}^{2 \times H_b \times W_b} \tag{2}$$

where $\mathcal{S}$ is the kernel soft-argmax operator without any learnable parameters, $\hat{f}$ is the predicted semantic flow in the direction of the target to source.

During training, as we only have sparse keypoint labels, the ground-truth flow $f^{gt} \in \mathbb{R}^{2 \times H_b \times W_b}$ have valid values only at labeled positions. We use a sparse binary label mask $\mathbf{M} \in \mathbb{R}^{H_b \times W_b}$ to indicate valid positions with ground-truth labels as below:

$$\mathbf{M}(\mathbf{p}) = \begin{cases} 1 & \text{if } \mathbf{p} \text{ is labeled,} \\ 0 & \text{otherwise,} \end{cases} \tag{3}$$

where $\mathbf{p}$ is the position index in f^{gt}.

Given $\mathbf{M}$, the objective is then defined as the L2 norm between the predicted flow and the ground-truth flow at labeled subpixel positions:

$$L^{gt}(\mathbf{p}) = \|\hat{f}(\mathbf{p}) - f^{gt}(\mathbf{p})\|_2 \cdot \mathbf{M}(\mathbf{p}) \tag{4}$$

where $L^{gt}(\mathbf{p})$ is the ground-truth loss at position $\mathbf{p}$. It is worth noting that our network does not involve any 4D-conv or transformer for correlation refinement [6, 22], but as shown later it achieves high performance thanks to our efficient spatial context encoder and high-resolution design.

4   Learning with Sparse Annotations

While our network design enables us to encode spatial context efficiently and utilize high-resolution correlation maps, the sparsely-annotated keypoint pairs (8 on average on PF-PASCAL [10]) greatly hinder the learning of the dense matching model. We address this with a novel teacher-student learning framework which we will elaborate below.

Our goal is to enrich the supervision when only sparse annotations are provided, as shown in Fig. 3. We first densify the sparse labels with a teacher-student paradigm (Sect. 4.1). Then we introduce two novel techniques to denoise the generated pseudo-labels (Sect. 4.2): (a) leveraging spatial-priors and (b) loss-driven dynamic label selection. Finally, we investigate two variants of the proposed learning paradigm (Sect. 4.3).

4.1   Sparse Label Densification via Teacher-Student Learning

To enrich the sparse supervision signals, we generate dense pseudo-labels for unlabeled region via a teacher-student paradigm, which consists of a student model $\mathbf{F}_s$ and a teacher model $\mathbf{F}_t$. The teacher model $\mathbf{F}_t$ trained with sparse

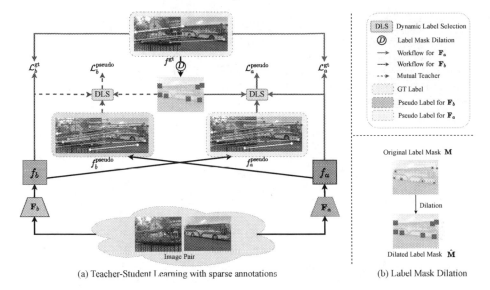

(a) Teacher-Student Learning with sparse annotations   (b) Label Mask Dilation

Fig. 3. Sparse Label Densification with Teacher-Student Learning. (a) Our Teacher-Student Learning Pipeline. Solid lines stand for Single Offline Teacher, with additional dashed lines standing for Mutual Online Teacher. (b) Illustration of Label Mask Dilation. Please refer to Sect. 4 for more details. *Best viewed in color.* (Color figure online)

annotations generates dense flows $\hat{f}_t$, providing pseudo-labels f_s^{pseudo} for the student model $\mathbf{F}_s$. Formally,

$$f_s^{\text{pseudo}} = \mathbf{F}_t(\mathbf{I}_a, \mathbf{I}_b), \quad f_s^{\text{pseudo}} \in \mathbb{R}^{2 \times H_b \times W_b}. \tag{5}$$

Then, the optimization objective $L_s(\mathbf{p})$ for the student $\mathbf{F}_s$ is a combination of the ground-truth loss $L_s^{\text{gt}}(\mathbf{p})$ and a dense pseudo-label loss $L_s^{\text{pseudo}}(\mathbf{p})$ calculated as follows:

$$L_s(\mathbf{p}) = L_s^{\text{gt}}(\mathbf{p}) + \lambda L_s^{\text{pseudo}}(\mathbf{p}) \tag{6}$$

$$L_s^{\text{gt}}(\mathbf{p}) = \|\hat{f}_s(\mathbf{p}) - f^{\text{gt}}(\mathbf{p})\|_2 \cdot \mathbf{M}(\mathbf{p}) \tag{7}$$

$$L_s^{\text{pseudo}}(\mathbf{p}) = \|\hat{f}_s(\mathbf{p}) - f_s^{\text{pseudo}}(\mathbf{p})\|_2 \tag{8}$$

where $\hat{f}_s \in \mathbb{R}^{2 \times H_t \times W_t}$ is the predicted flow of the student model $\mathbf{F}_s$ given $\mathbf{I}_a$ and $\mathbf{I}_b$, λ is the scale hyper-parameter, $\mathbf{p}$ indexes the positions in $\hat{f}_s$.

4.2   High Quality Pseudo-label Generation

The dense pseudo-labels generated by the teacher model are inevitably unreliable and inaccurate for supervision. To filter out erroneous pseudo-labels, we use: (a) label filtering based on spatial priors, and (b) loss-driven dynamic label selection.

Spatial-Prior Based Label Filtering. Our key insight is that, as the anno-tated keypoints are in the object foreground region, we are able to suppress noisy background pseudo-labels by exploiting the spatial-smoothness prior of the semantic correspondence in the neighborhood of the sparse keypoints. Moti-vated by this, we generate a densified binary label mask $\hat{\mathbf{M}}$ via dilating the sparse label mask $\mathbf{M}$ as follows, which will be used for label filtering:

$$\bar{\mathbf{M}} = \mathbf{M} * \mathcal{K} \tag{9}$$

$$\hat{\mathbf{M}}(\mathbf{p}) = \begin{cases} 1 & \text{if } \bar{\mathbf{M}}(\mathbf{p}) > 0 \\ 0 & \text{otherwise} \end{cases} \tag{10}$$

where $*$ is a convolution operator with zero padding, $\mathcal{K} \in \mathcal{R}^{k \times k}$ is a kernel filled with one with k as the kernel size. Note that dilation here refers to expanding the existing foreground region in $\mathbf{M}$. Compared with using CAM [62] or uncertainty estimation [36], our proposed label filtering technique is easy to implement and utilizes the spatial prior around the sparse annotations.

Given the dilated label mask $\hat{\mathbf{M}}$, the pseudo-loss $\hat{L}_s^{\text{pseudo}}(\mathbf{p})$ for the student model is calculated as below:

$$\hat{L}_s^{\text{pseudo}}(\mathbf{p}) = \|\hat{f}_s(\mathbf{p}) - f_s^{\text{pseudo}}(\mathbf{p})\|_2 \cdot \hat{\mathbf{M}}(\mathbf{p}). \tag{11}$$

where $\mathbf{p}$ indexes the positions. In this way, we are able to significantly suppress noisy background pseudo-labels as shown in Sect. 5.3.

Loss-Driven Dynamic Label Selection. While many background pseudo-labels can be filtered out by our dilated label mask, some noisy labels still exist due to inaccurate predictions from the teacher model. To further filter out inac-curate labels, we introduce a loss-driven label selection strategy following the small-loss principle [11]. Denoting R as the ratio of pixels being selected, we choose the pixel set $\mathcal{P}$ on the foreground region in $\hat{\mathbf{M}}$ with the smallest loss as below:

$$\mathcal{P} = \underset{\bar{\mathcal{D}}:|\bar{\mathcal{D}}| \geq R(T)N_{\hat{\mathbf{M}}} \wedge \bar{\mathcal{D}} \subseteq \hat{\mathcal{D}}}{\arg\min} \sum_{\mathbf{p} \in \bar{\mathcal{D}}} \hat{L}_s^{\text{pseudo}}(\mathbf{p}) \tag{12}$$

$$\hat{\mathcal{D}} = \{\mathbf{p} \mid \hat{\mathbf{M}}(\mathbf{p}) = 1\} \tag{13}$$

where $R(T)$ controls the selection percentage in training epoch T, $\mathbf{p}$ indexes the positions, $\hat{\mathcal{D}}$ is a collection of foreground positions in the dilated label mask $\hat{\mathbf{M}}$, $N_{\hat{\mathbf{M}}}$ is the total number of non-zero positions in $\hat{\mathbf{M}}$.

Hence the final optimization objective $\mathcal{L}_s$ for the student model $\mathbf{F}_s$ over an image pair is a combination of the sparse ground-truth loss $\mathcal{L}_s^{\text{gt}}$ at labeled positions $\mathcal{G} = \{\mathbf{p} \mid \mathbf{M}(\mathbf{p}) = 1\}$ and the dense pseudo loss $\mathcal{L}_s^{\text{pseudo}}$ at selected positions $\mathcal{P}$ as below:

$$\mathcal{L}_s = \mathcal{L}_s^{\text{gt}} + \lambda \mathcal{L}_s^{\text{pseudo}} \tag{14}$$

$$\mathcal{L}_s^{\text{pseudo}} = \frac{1}{|\mathcal{P}|} \sum_{\mathbf{p} \in \mathcal{P}} \hat{L}_s^{\text{pseudo}}(\mathbf{p}) \tag{15}$$

$$\mathcal{L}_s^{\text{gt}} = \frac{1}{|\mathcal{G}|} \sum_{\mathbf{p} \in \mathcal{G}} L_s^{\text{gt}}(\mathbf{p}) \tag{16}$$

4.3 Variants of Teacher-Student Learning

To investigate the optimization strategy of our proposed teacher-student learning, we propose two variants of our learning strategy, including (a) single offline teacher, and (b) mutual online teachers, which are detailed below.

Single Offline Teacher (ST). This variant consists of two learning stages. Specifically, we first learn a baseline network, which acts as the teacher model, given spare ground-truth annotation only. In the second stage, the pseudo-labels are generated by the fixed teacher network as described in Sect. 4.1 and Sect. 4.2. Given the enriched supervision, we then train the student model from scratch, which is used for the inference stage finally.

Mutual Online Teacher (MT). Inspired by the recent advances in multi-view learning [1,55], we additionally explore a one-stage variant with two mutual online teachers which learn from scratch. We simultaneously train two networks of the same architecture, each of which takes predictions from the other network as the pseudo-labels for optimization. These two networks can learn knowledge of correspondence with enriched pseudo-labels from each other. The one with a higher validation performance is selected for the inference stage.

Specifically, we maintain two networks $\mathbf{F}_s$ and $\mathbf{F}_t$ of the same architecture. The network $\mathbf{F}_t$ (resp. $\mathbf{F}_s$) provides its predicted flow $\hat{f}_t$ (resp. $\hat{f}_s$) as the pseudo-label f_s^{pseudo} (resp. f_t^{pseudo}) for the peer network $\mathbf{F}_s$ (resp. $\mathbf{F}_t$). Both networks use the shared dilated label mask $\hat{\mathbf{M}}$ for label filtering. For each model, the pseudo-loss filtered by dilated label masks is described as below:

$$\hat{L}_s^{\text{pseudo}}(\mathbf{p}) = \|\hat{f}_s(\mathbf{p}) - f_s^{\text{pseudo}}(\mathbf{p})\|_2 \cdot \hat{\mathbf{M}}(\mathbf{p}) \tag{17}$$

$$\hat{L}_t^{\text{pseudo}}(\mathbf{p}) = \|\hat{f}_t(\mathbf{p}) - f_t^{\text{pseudo}}(\mathbf{p})\|_2 \cdot \hat{\mathbf{M}}(\mathbf{p}), \tag{18}$$

where $\mathbf{p}$ indexes the position, f_s^{pseudo} (resp. f_t^{pseudo}) equals to $\hat{f}_t$ (resp. $\hat{f}_s$). $\hat{L}_s^{\text{pseudo}}(\mathbf{p})$ and $\hat{L}_t^{\text{pseudo}}(\mathbf{p})$ will then go through the dynamic label selection procedure as described in Sect. 4.2 to compute pseudo-label loss $\mathcal{L}_s^{\text{pseudo}}$ and $\mathcal{L}_t^{\text{pseudo}}$, respectively. The final optimization objective for each model is a combination of sparse ground-truth loss and pseudo loss as below:

$$\mathcal{L}_s = \mathcal{L}_s^{\text{gt}} + \lambda \mathcal{L}_s^{\text{pseudo}} \tag{19}$$

$$\mathcal{L}_t = \mathcal{L}_t^{\text{gt}} + \lambda \mathcal{L}_t^{\text{pseudo}} \tag{20}$$

Table 1. Comparison with SOTA methods on SPair-71k [41]. Per-class and overall PCK ($\alpha_{bbox} = 0.1$) results are shown in the table. Numbers in bold indicate the best performance and underlined ones are the second best. All models in this table use ResNet101 as the backbone. *Sup.* denotes the type of supervision. * means the backbone is finetuned. † means ground truth bbox used.

Sup.	Methods	aero	bike	bird	boat	bottle	bus	car	cat	chair	cow	dog	horse	mbike	person	plant	sheep	train	tv	all
self	CNNGeo [44]	23.4	16.7	40.2	14.3	36.4	27.7	26.0	32.7	12.7	27.4	22.8	13.7	20.9	21.0	17.5	10.2	30.8	34.1	20.6
	A2Net [19]	22.6	18.5	42.0	16.4	37.9	30.8	26.5	35.6	13.3	29.6	24.3	16.0	21.6	22.8	20.5	13.5	31.4	36.5	22.3
weak	WeakAlign [45]	22.2	17.6	41.9	15.1	38.1	27.4	27.2	31.8	12.8	26.8	22.6	14.2	20.0	22.2	17.9	10.4	32.2	35.1	20.9
	NCNet [46]	17.9	12.2	32.1	11.7	29.0	19.9	16.1	39.2	9.9	23.9	18.8	15.7	17.4	15.9	14.8	9.6	24.2	31.1	20.1
trn-none /	HPF [40]	25.2	18.9	52.1	15.7	38.0	22.8	19.1	52.9	17.9	33.0	32.8	20.6	24.4	27.9	21.1	15.9	31.5	35.6	28.2
val-strong	SCOT [38]	34.9	20.7	63.8	21.1	43.5	27.3	21.3	63.1	20.0	42.9	42.5	31.1	29.8	35.0	27.7	24.4	48.4	40.8	35.6
strong	DHPF [42]	38.4	23.8	68.3	18.9	42.6	27.9	20.1	61.6	22.0	46.9	46.1	33.5	27.6	40.1	27.6	28.1	49.5	46.5	37.3
	PMD [36]	38.5	23.7	60.3	18.1	42.7	39.3	27.6	60.6	14.0	54.0	41.8	34.6	27.0	25.2	22.1	29.9	70.1	42.8	37.4
	MMNet* [61]	43.5	27.0	62.4	27.3	40.1	50.1	37.5	60.0	21.0	56.3	50.3	41.3	30.9	19.2	30.1	33.2	64.2	43.6	40.9
	CHM [39]	49.6	29.3	68.7	29.7	45.3	48.4	39.5	64.9	20.3	60.5	56.1	46.0	33.8	44.3	38.9	31.4	72.2	55.5	46.3
	CATs†* [6]	52.0	34.7	72.2	34.3	49.9	57.5	43.6	66.5	24.4	63.2	56.5	52.0	42.6	41.7	43.0	33.6	72.6	58.0	49.9
	PMNC* [31]	54.1	35.9	74.9	36.5	42.1	48.8	40.0	72.6	21.1	67.6	58.1	50.5	40.1	54.1	43.3	35.7	74.5	59.9	50.4
	Ours (ST)*	56.9	37.0	76.2	33.9	50.1	51.7	42.4	68.2	22.4	70.7	61.0	47.7	43.6	47.8	47.8	38.6	77.0	67.1	52.4
	Ours (MT)*	57.1	40.3	78.3	38.1	51.8	57.8	47.1	67.9	25.2	71.3	63.9	49.3	45.3	49.8	48.8	40.3	77.7	69.7	55.3

5   Experiments

We evaluate our method on the supervised semantic correspondence task by conducting comprehensive experiments on three public benchmarks: PF-PASCAL [10], PF-WILLOW [9], and SPair-71k [41]. In the following sections, we first elaborate on the implementation details of our proposed method in Sect. 5.1, and follow that with the quantitative and qualitative comparison with prior state-of-the-art (SOTA) competitors in Sect. 5.2. Then, we provide ablation studies and comprehensive analysis in Sect. 5.3. For more detailed results and analysis, we refer readers to the supplementary material.

5.1   Implementation Details

Datasets. *SPair-71k* is a newly-released challenging and largest-scale benchmark [41]. There are keypoint-annotated 70,958 image pairs with large viewpoint and scale variation in diverse scenes. SPair-71k [41] is a reliable test bed for studying real problems of semantic matching. *PF-PASCAL* dataset [10] contains 1351 image pairs with limited variability and scale, which is approximately split into 700, 300, and 300 pairs for train, val, and test set, resp. *PF-WILLOW* [9] dataset consists of 900 image pairs of 4 categories, which is a widely-used benchmark for the verification of generalization ability.

Table 2. Comparison with SOTA methods on PF-PASCAL [10]. Numbers in bold indicate the best performance and underlined ones are the second best. † means ground truth bbox used.

Sup.	Methods	PCK@α_{img}			α_{bbox}
		$\alpha = 0.05$	$\alpha = 0.10$	$\alpha = 0.15$	$\alpha = 0.1$
None	PF-LOMHOG [10]	31.4	62.5	79.5	45.0
Self	CNNGeoResNet-101 [44]	41.0	69.5	89.4	68.0
weak	WeakAlignResNet-101 [45]	49.0	74.8	84.0	72.0
	NC-NetResNet-101 [46]	54.3	78.9	86.0	70.0
	DCCNetResNet-101 [22]	55.6	82.3	90.5	-
	GSFResNet-101 [25]	62.8	84.5	93.7	-
Trn-none	HPFResNet-101 [40]	60.1	84.8	92.7	78.5
Val-strong	SCOTResNet-101 [38]	63.1	85.4	92.7	-
Strong	SCNetVGG-16 [12]	36.2	72.2	82.0	48.2
	ANCNetResNet-101 [35]	-	86.1	-	-
	DHPFResNet-101 [42]	75.7	90.7	95.0	87.8
	PMDResNet-101 [36]	-	90.7	-	-
	MMNetResNet-101 [61]	77.6	89.1	94.3	-
	CHMResNet-101 [39]	80.1	91.6	-	-
	CATs†ResNet-101 [6]	75.4	92.6	<u>96.4</u>	89.2
	PMNCResNet-101 [31]	**82.4**	90.6	-	-
	Ours (ST) ResNet-101	81.4	<u>92.9</u>	96.1	<u>90.5</u>
	Ours (MT) ResNet-101	<u>81.5</u>	**93.3**	**96.6**	**91.2**

Evaluation Metric. In line with prior work, we report the percentage of correct keypoints (PCK) [56]. The predicted keypoints are considered to be correct if they lie within $\alpha \cdot \max(h, w)$ pixels from the ground-truth keypoints for $\alpha \in [0, 1]$, where h and w are the height and width of either an image (α_{img}) or an object bounding box (α_{bbox}).

Experimental Configuration. For the feature extractor, we use ResNet-101 [15] pre-trained on ImageNet with a single feature at stride 16. Learnable parameters are randomly initialized. For our base model, we set Efficient SCE kernel size $K = 7$ and $d_g = 2048$ for SPair-71k; $K = 13$ and $d_g = 1024$ for PF-PASCAL, resp. We upsample the correlation map to stride 4 for high-resolution loss. For label mask dilation, dilation kernel size $k = 7$ is set for both SPair-71k and PF-PASCAL by validation search. For dynamic label selection, we set $R(T)$ linearly increases from the ratio of 20% to 90% in a duration of 10 epochs for both SPair-71k and PF-PASCAL. λ is 10.0 for weighting pseudo-loss. We strictly follow previous work for data augmentation [6] (e.g., color jittering) except that [6] uses ground truth box for random crop while we do not. An AdamW optimizer with a learning rate of 3e-6 for the backbone and 3e-5 for the remaining parameters are used. All the implementations are in PyTorch [43].

Images of all three datasets are resized to 256×256. Our model is trained on PF-PASCAL and SPair-71k, resp. Following the previous work [6,22,42], we

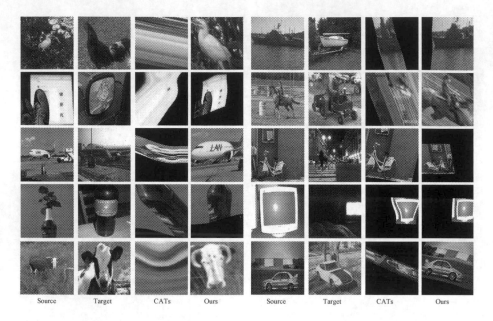

Fig. 4. **Qualitative results of our method on SPair-71k** [41]. From left to right are source image, target image, result from CATs [6], and result from ours (MT), resp.

validate the generalization ability of our method by testing on PF-WILLOW with our model trained on PF-PASCAL without any finetuning.

5.2 Comparison with State-of-the-Art Methods

SPair-71k. We compare our method with the most recent work [6,31,38,40, 42,61] on SPair-71k in Table 1. Our two variant settings (ST and MT) both achieve an overall SOTA results, with our method (MT) achieving an overall PCK ($\alpha_{\text{img}} = 0.1$) of 55.3%, outperforming the previous SOTA [31] by a large margin (4.9%). Note that our method does not involve any parameterized correlation refinement compared with [31], clearly illustrating the power of the proposed pipeline. Figure 4 shows qualitative results on SPair-71k. We observe that our method is robust to diverse variations in scale and viewpoint thanks to our enriched training signals.

PF-PASCAL. Our results on PF-PASCAL are summarized in Table 2. Our method outperforms the previous state-of-the-art [6,31] on almost all thresholds even if the performance on PF-PASCAL is near saturated, reaching a new SOTA of 93.3% PCK ($\alpha = 0.1$). Note that even if we did not use sophisticated parameterized correlation map refinement as in PMNC [31], we can still achieve comparable PCK at $\alpha = 0.05$.

Table 3. Comparison with SOTA methods on PF-WILLOW [9]. Numbers in bold indicate the best performance and underlined ones are the second best. † means ground truth bbox used.

Sup.	Methods	PCK@α_{bbox}		
		$\alpha = 0.05$	$\alpha = 0.10$	$\alpha = 0.15$
None	PF-LOMHOG [10]	28.4	56.8	68.2
Self	CNNGeoResNet-101 [44]	36.9	69.2	77.8
weak	WeakAlignResNet-101 [45]	37.0	70.2	79.9
	NC-NetResNet-101 [46]	44.0	72.7	85.4
	DCCNetResNet-101 [22]	43.6	73.8	86.5
	GSFResNet-101 [25]	47.0	75.8	88.9
Trn-none	HPFResNet-101 [40]	45.9	74.4	85.6
Val-strong	SCOTResNet-101 [38]	47.8	76.0	87.1
Strong	SCNetVGG-16 [12]	38.6	70.4	85.3
	DHPFResNet-101 [42]	49.5	77.6	89.1
	PMDResNet-101 [36]	-	75.6	-
	CHMResNet-101 [39]	52.7	<u>79.4</u>	-
	CATs†ResNet-101 [6]	50.3	79.2	**90.3**
	Ours (ST) ResNet-101	<u>53.5</u>	<u>79.4</u>	89.5
	Ours (MT) ResNet-101	**54.1**	**80.0**	<u>89.8</u>

PF-WILLOW. We test on PF-WILLOW [9] using our model trained on PF-PASCAL [10] to verify dataset generalization ability of our method. As shown in Table 3, our method (MT) outperforms the prior SOTA [6,39] in $\alpha = 0.05$, 0.1 by 1.4% and 0.6%, resp, indicating superior dataset generalization ability of our learning method. Note that our method (MT) are 0.5% behind at $\alpha = 0.15$ compared with [6], we argue that CATs [6] used ground truth bounding box during training while we did not.

5.3 Ablation Study

In this section, we conduct ablation studies to verify the effectiveness of each individual module of the proposed model. We train all the variants on the training split of SPair-71k [41] and report PCK ($\alpha_{bbox} = 0.1$) on the test split. Each ablation experiment is conducted under the same experimental setting for a fair comparison.

Effect of Individual Modules. Table 4 summarizes the ablation results of each individual module. First, we note that applying the proposed Efficient-SCE (ID A1) yields significant gain over the baseline (ID A0), showing the effectiveness of the proposed feature enhancement module. Second, enforcing high-resolution loss

Table 4. Effects of each component on SPair-71K [41] test split. HRLoss refers to high-resolution loss, FT refers to finetuning the backbone, Teacher-Student here refers to the variant with mutual online teacher.

Model ID	Efficient-SCE	HRLoss	Finetune	Teacher-Student	Label Denoise		PCK
					Dynamic Selection	Mask Dilation	
A0	-	-	-	-	-	-	14.7
A1	✓	-	-	-	-	-	33.4
A2	✓	✓	-	-	-	-	40.6
A3	✓	✓	✓	-	-	-	49.8
A4	✓	✓	✓	✓	-	-	49.7
A5	✓	✓	✓	✓	✓	-	51.5
A6	✓	✓	✓	✓	✓	✓	**55.3**

Table 5. Comparing single offline teacher and mutual online teacher setting on SPair-71K [41] test split.

Variant Setting	PCK
None	49.8
Single offline teacher	52.4
Mutual online teacher	**55.3**

Table 6. Effects of kernel size for label mask dilation on SPair-71K [41] test split.

Dilation Kernel Size	PCK
None	49.8
3	54.0
7	**55.3**
15	52.0

improves to a remarkable 49.8% after finetuning (ID A3). Our proposed network achieves competitive results without any Conv4D or transformer modules for correlation map refinement. Third, densifying labels combined with our two denoising techniques achieves 5.5% boost further and promotes the performance to 55.3%, showing the effectiveness of our proposed learning strategy. In contrast, teacher-student learning alone without any denoising provides little boost as the dense pseudo-labels might be too noisy (ID A4).

Single Offline Teacher Vs. Mutual Online Teacher. Table 5 shows the comparison between our proposed two variants of teacher-student learning. Both settings have greatly surpassed the performance of the base network (ID A3), showing the effectiveness of our proposed label densification strategy. We note that the mutual online teacher setting is 2.9% higher than the single offline teacher setting. The reason could be that performance is bounded by the fixed teacher model while the mutual online teacher setting could improve each other over the training process.

Effects of Kernel Size for Label Mask Dilation. Table 6 summarizes the results of different kernel size for label mask dilation. When increasing the kernel size, the performance rises first but then drops, with kernel size 7 being the best, which demonstrates the necessity of restricting pseudo-labels in a meaningful local neighborhood.

6 Conclusion

In this work, we propose a novel teacher-student learning paradigm in order to address the challenge of label sparsity for semantic correspondence task. In our teacher-student paradigm, we generate dense pseudo-labels by the teacher networks which are trained with sparse annotations. To improve quality of pseudo-labels, we develop two novel techniques to denoise pseudo-labels. Specifically, we first dilate the sparse label masks derived from the sparse keypoint annotations to suppress background pseudo-labels. A dynamic label selection strategy is then introduced to further filter noisy labels. We investigate two variants of the proposed learning paradigm, a single offline teacher setting, and a mutual online teacher setting. Our method achieves state-of-the-art performances on three standard datasets. The effectiveness of our method provides new insight into the problem, and is one step closer towards a more realistic application of semantic correspondence.

References

1. Blum, A., Mitchell, T.: Combining labeled and unlabeled data with co-training. In: Proceedings of the Annual Conference on Learning Theory(COLT) (1998)
2. Bromley, J., Guyon, I., LeCun, Y., Säckinger, E., Shah, R.: Signature verification using a "siamese" time delay neural network. Advances in neural information processing systems 6 (1993)
3. Chauhan, A.K., Krishan, P.: Moving object tracking using gaussian mixture model and optical flow. Int. J. Adv. Res. Comput. Sci. Softw. Eng. **3**(4) (2013)
4. Chen, T., Goodfellow, I., Shlens, J.: Net2net: accelerating learning via knowledge transfer. In: Proceedings of the International Conference on Learning Representations (ICLR) (2015)
5. Chen, Y.C., Lin, Y.Y., Yang, M.H., Huang, J.B.: Show, match and segment: joint weakly supervised learning of semantic matching and object co-segmentation. IEEE Trans. Pattern Anal. Mach. Intell. **PP**(99), 1 (2020)
6. Cho, S., Hong, S., Jeon, S., Lee, Y., Sohn, K., Kim, S.: Cats: cost aggregation transformers for visual correspondence. In: Advances in Neural Information Processing Systems (NeurIPS) (2021)
7. Dale, K., Johnson, M.K., Sunkavalli, K., Matusik, W., Pfister, H.: Image restoration using online photo collections. In: Proceedings of the International Conference on Computer Vision (ICCV) (2009)
8. Goldstein, A., Fattal, R.: Video stabilization using Epipolar geometry. ACM Trans. Graph. (TOG) **31**(5), 1–10 (2012)
9. Ham, B., Cho, M., Schmid, C., Ponce, J.: Proposal flow. In: Proceedings of the IEEE Conference on Computer Vision and Pattern Recognition (CVPR) (2016)
10. Ham, B., Cho, M., Schmid, C., Ponce, J.: Proposal flow: Semantic correspondences from object proposals. IEEE Transactions on Pattern Analysis and Machine Intelligence (2018)

11. Han, B., et al.: Co-teaching: robust training of deep neural networks with extremely noisy labels. In: Advances in Neural Information Processing Systems (NeurIPS) (2018)

12. Han, K., et al.: SCNet: learning semantic correspondence. In: Proceedings of the International Conference on Computer Vision (ICCV) (2017)

13. He, B., Yang, X., Kang, L., Cheng, Z., Zhou, X., Shrivastava, A.: ASM-Loc: action-aware segment modeling for weakly-supervised temporal action localization. In: Proceedings of the IEEE Conference on Computer Vision and Pattern Recognition (CVPR) (2022)

14. He, B., Yang, X., Wu, Z., Chen, H., Lim, S.N., Shrivastava, A.: GTA: global temporal attention for video action understanding. In: Proceedings of the British Machine Vision Conference (BMVC) (2020)

15. He, K., Zhang, X., Ren, S., Sun, J.: Deep residual learning for image recognition. In: Proceedings of the IEEE Conference on Computer Vision and Pattern Recognition (CVPR) (2016)

16. Heo, B., Kim, J., Yun, S., Park, H., Kwak, N., Choi, J.Y.: A comprehensive overhaul of feature distillation. In: Proceedings of the IEEE Conference on Computer Vision and Pattern Recognition (CVPR) (2019)

17. Hinton, G., et al.: Distilling the knowledge in a neural network. arXiv preprint arXiv:1503.02531 2(7) (2015)

18. Hong, S., Cho, S., Nam, J., Lin, S., Kim, S.: Cost aggregation with 4D convolutional Swin transformer for few-shot segmentation. arXiv preprint arXiv:2207.10866 (2022)

19. Seo, P.H., Lee, J., Jung, D., Han, B., Cho, M.: Attentive semantic alignment with offset-aware correlation kernels. In: Ferrari, V., Hebert, M., Sminchisescu, C., Weiss, Y. (eds.) ECCV 2018. LNCS, vol. 11208, pp. 367–383. Springer, Cham (2018). https://doi.org/10.1007/978-3-030-01225-0_22

20. Horn, B.K., Schunck, B.G.: Determining optical flow. Artif. Intell. **17**(1-3), 185–203 (1981)

21. Huang, S., Wang, Q., He, X.: Confidence-aware adversarial learning for self-supervised semantic matching. In: Peng, Y., et al. (eds.) PRCV 2020. LNCS, vol. 12305, pp. 91–103. Springer, Cham (2020). https://doi.org/10.1007/978-3-030-60633-6_8

22. Huang, S., Wang, Q., Zhang, S., Yan, S., He, X.: Dynamic context correspondence network for semantic alignment. In: Proceedings of the IEEE International Conference on Computer Vision (ICCV) (2019)

23. Huang, Z., Wang, X., Huang, L., Huang, C., Wei, Y., Liu, W.: CCNet: criss-cross attention for semantic segmentation. In: Proceedings of the IEEE International Conference on Computer Vision (ICCV) (2019)

24. Jeon, S., Kim, S., Min, D., Sohn, K.: PARN: pyramidal affine regression networks for dense semantic correspondence. In: Ferrari, V., Hebert, M., Sminchisescu, C., Weiss, Y. (eds.) ECCV 2018. LNCS, vol. 11210, pp. 355–371. Springer, Cham (2018). https://doi.org/10.1007/978-3-030-01231-1_22

25. Jeon, S., Min, D., Kim, S., Choe, J., Sohn, K.: Guided semantic flow. In: Vedaldi, A., Bischof, H., Brox, T., Frahm, J.-M. (eds.) ECCV 2020. LNCS, vol. 12373, pp. 631–648. Springer, Cham (2020). https://doi.org/10.1007/978-3-030-58604-1_38

26. Jeon, S., Min, D., Kim, S., Sohn, K.: Joint learning of semantic alignment and object landmark detection. In: Proceedings of the IEEE International Conference on Computer Vision (ICCV) (2019)

27. Kim, S., Lin, S., JEON, S.R., Min, D., Sohn, K.: Recurrent transformer networks for semantic correspondence. In: Advances in Neural Information Processing Systems (NeurIPS) (2018)
28. Kim, S., Min, D., Ham, B., Jeon, S., Lin, S., Sohn, K.: FCSS: fully convolutional self-similarity for dense semantic correspondence. In: Proceedings of the IEEE Conference on Computer Vision and Pattern Recognition (CVPR) (2017)
29. Kim, S., Min, D., Jeong, S., Kim, S., Jeon, S., Sohn, K.: Semantic attribute matching networks. In: Proceedings of the IEEE Conference on Computer Vision and Pattern Recognition (CVPR) (2019)
30. Lan, S., et al.: DiscoBox: weakly supervised instance segmentation and semantic correspondence from box supervision. In: Proceedings of the IEEE Conference on Computer Vision and Pattern Recognition (CVPR) (2021)
31. Lee, J.Y., DeGol, J., Fragoso, V., Sinha, S.N.: Patchmatch-based neighborhood consensus for semantic correspondence. In: Proceedings of the IEEE Conference on Computer Vision and Pattern Recognition (CVPR) (2021)
32. Lee, J., Kim, D., Ponce, J., Ham, B.: SFNet: learning object-aware semantic correspondence. In: Proceedings of the IEEE Conference on Computer Vision and Pattern Recognition (CVPR) (2019)
33. Lee, J., Kim, E., Lee, Y., Kim, D., Chang, J., Choo, J.: Reference-based sketch image colorization using augmented-self reference and dense semantic correspondence. In: Proceedings of the IEEE Conference on Computer Vision and Pattern Recognition (CVPR) (2020)
34. Li, H., Wu, Z., Shrivastava, A., Davis, L.S.: Rethinking pseudo labels for semi-supervised object detection. In: Proceedings of the AAAI Conference on Artificial Intelligence (AAAI) (2022)
35. Li, S., Han, K., Costain, T.W., Howard-Jenkins, H., Prisacariu, V.: Correspondence networks with adaptive neighbourhood consensus. In: Proceedings of the IEEE Conference on Computer Vision and Pattern Recognition (CVPR) (2020)
36. Li, X., Fan, D.P., Yang, F., Luo, A., Cheng, H., Liu, Z.: Probabilistic model distillation for semantic correspondence. In: Proceedings of the IEEE Conference on Computer Vision and Pattern Recognition (CVPR) (2021)
37. Liu, C., Yuen, J., Torralba, A.: Sift flow: dense correspondence across scenes and its applications. IEEE Trans. Pattern Anal. Mach. Intell. **33** (2011)
38. Liu, Y., Zhu, L., Yamada, M., Yang, Y.: Semantic correspondence as an optimal transport problem. In: Proceedings of the IEEE Conference on Computer Vision and Pattern Recognition (CVPR) (2020)
39. Min, J., Cho, M.: Convolutional hough matching networks. In: Proceedings of the IEEE Conference on Computer Vision and Pattern Recognition (CVPR) (2021)
40. Min, J., Lee, J., Ponce, J., Cho, M.: Hyperpixel flow: semantic correspondence with multi-layer neural features. In: Proceedings of the IEEE International Conference on Computer Vision (ICCV) (2019)
41. Min, J., Lee, J., Ponce, J., Cho, M.: Spair-71k: a large-scale benchmark for semantic correspondence. arXiv preprint arXiv:1908.10543 (2019)
42. Min, J., Lee, J., Ponce, J., Cho, M.: Learning to compose hypercolumns for visual correspondence. In: Vedaldi, A., Bischof, H., Brox, T., Frahm, J.-M. (eds.) ECCV 2020. LNCS, vol. 12360, pp. 346–363. Springer, Cham (2020). https://doi.org/10.1007/978-3-030-58555-6_21
43. Paszke, A., et al.: Automatic differentiation in pyTorch (2017)
44. Rocco, I., Arandjelović, R., Sivic, J.: Convolutional neural network architecture for geometric matching. In: Proceedings of the IEEE Conference on Computer Vision and Pattern Recognition (CVPR) (2017)

45. Rocco, I., Arandjelović, R., Sivic, J.: End-to-end weakly-supervised semantic alignment. In: Proceedings of the IEEE Conference on Computer Vision and Pattern Recognition (CVPR) (2018)
46. Rocco, I., Cimpoi, M., Arandjelović, R., Torii, A., Pajdla, T., Sivic, J.: Neighbourhood consensus networks. In: Advances in Neural Information Processing Systems (NeurIPS) (2018)
47. Scharstein, D., Szeliski, R.: A taxonomy and evaluation of dense two-frame stereo correspondence algorithms. Int. J. Comput. Vision **47**(1–3), 7–42 (2002)
48. Sohn, K., et al.: FixMatch: simplifying semi-supervised learning with consistency and confidence. In: Advances in Neural Information Processing Systems (NeurIPS) (2020)
49. Taniai, T., Sinha, S.N., Sato, Y.: Joint recovery of dense correspondence and cosegmentation in two images. In: Proceedings of the IEEE Conference on Computer Vision and Pattern Recognition (CVPR) (2016)
50. Tarvainen, A., Valpola, H.: Mean teachers are better role models: weight-averaged consistency targets improve semi-supervised deep learning results. In: Advances in Neural Information Processing Systems (NeurIPS) (2017)
51. Tola, E., Lepetit, V., Fua, P.: Daisy: an efficient dense descriptor applied to wide-baseline stereo. IEEE Trans. Pattern Anal. Mach. Intell. **32**(5), 815-830 (2010)
52. Truong, P., Danelljan, M., Timofte, R.: GLU-Net: global-local universal network for dense flow and correspondences. In: Proceedings of the IEEE Conference on Computer Vision and Pattern Recognition (CVPR) (2020)
53. Truong, P., Danelljan, M., Yu, F., Van Gool, L.: Probabilistic warp consistency for weakly-supervised semantic correspondences. In: Proceedings of the IEEE Conference on Computer Vision and Pattern Recognition (CVPR) (2022)
54. Xie, Q., Luong, M.T., Hovy, E., Le, Q.V.: Self-training with noisy student improves imageNet classification. In: Proceedings of the IEEE Conference on Computer Vision and Pattern Recognition (CVPR) (2020)
55. Yang, L., et al.: Deep co-training with task decomposition for semi-supervised domain adaptation. In: Proceedings of the International Conference on Computer Vision (ICCV) (2021)
56. Yang, Y., Ramanan, D.: Articulated human detection with flexible mixtures of parts. IEEE Trans. Pattern Anal. Mach. Intell. **35**(12), 2878-2890 (2013)
57. Yim, J., Joo, D., Bae, J., Kim, J.: A gift from knowledge distillation: fast optimization, network minimization and transfer learning. In: Proceedings of the IEEE Conference on Computer Vision and Pattern Recognition (CVPR) (2017)
58. Yue, K., Sun, M., Yuan, Y., Zhou, F., Ding, E., Xu, F.: Compact generalized non-local network. In: Advances in Neural Information Processing Systems (NeurIPS) (2018)
59. Zhang, S., He, X., Yan, S.: LatentGNN: learning efficient non-local relations for visual recognition. In: Proceedings of the International Conference on Machine Learning (ICML) (2019)
60. Zhang, Y., Xiang, T., Hospedales, T.M., Lu, H.: Deep mutual learning. In: Proceedings of the IEEE Conference on Computer Vision and Pattern Recognition (CVPR) (2018)
61. Zhao, D., Song, Z., Ji, Z., Zhao, G., Ge, W., Yu, Y.: Multi-scale matching networks for semantic correspondence. In: Proceedings of the IEEE International Conference on Computer Vision (ICCV) (2021)
62. Zhou, B., Khosla, A., Lapedriza, A., Oliva, A., Torralba, A.: Learning deep features for discriminative localization. In: Proceedings of the IEEE Conference on Computer Vision and Pattern Recognition (CVPR) (2016)

Dynamically Transformed Instance Normalization Network for Generalizable Person Re-Identification

Bingliang Jiao[1,2,3,5], Lingqiao Liu[4], Liying Gao[1,2,3], Guosheng Lin[5],
Lu Yang[1,2,3], Shizhou Zhang[1,3], Peng Wang[1,2,3(✉)], and Yanning Zhang[1,3(✉)]

[1] School of Computer Science, Northwestern Polytechnical University, Xi'an, China
bingliang.jiao@mail.nwpu.edu.cn, {peng.wang,ynzhang}@nwpu.edu.cn
[2] Ningbo Institute, Northwestern Polytechnical University, Xi'an, China
[3] National Engineering Laboratory for Integrated Aero-Space-Ground-Ocean,
Xi'an, China
[4] The University of Adelaide, Adelaide, Australia
[5] Nanyang Technological University, Singapore, Singapore

Abstract. Existing person re-identification methods often suffer significant performance degradation on unseen domains, which fuels interest in domain generalizable person re-identification (DG-PReID). As an effective technology to alleviate domain variance, the Instance Normalization (IN) has been widely employed in many existing works. However, IN also suffers from the limitation of eliminating discriminative patterns that might be useful for a particular domain or instance. In this work, we propose a new normalization scheme called Dynamically Transformed Instance Normalization (DTIN) to alleviate the drawback of IN. Our idea is to employ dynamic convolution to allow the unnormalized feature to control the transformation of the normalized features into new representations. In this way, we can ensure the network has sufficient flexibility to strike the right balance between eliminating irrelevant domain-specific features and adapting to individual domains or instances. We further utilize a multi-task learning strategy to train the model, ensuring it can adaptively produce discriminative feature representations for an arbitrary domain. Our results show a great domain generalization capability and achieve state-of-the-art performance on three mainstream DG-PReID settings.

Keywords: Person re-identification · Domain generalization · Instance Normalization · Dynamic convolution

1 Introduction

Person re-identification (PReID) aims at matching identical persons across different cameras. Many supervised PReID methods have recently achieved promising success when training and evaluating images under the same environment. However, the performance of these methods tends to significantly degrade when

© The Author(s), under exclusive license to Springer Nature Switzerland AG 2022
S. Avidan et al. (Eds.): ECCV 2022, LNCS 13674, pp. 285–301, 2022.
https://doi.org/10.1007/978-3-031-19781-9_17

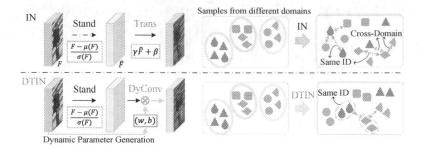

Fig. 1. The sketches of the traditional Instance Normalization (IN) and our proposed Dynamically Transformed Instance Normalization (DTIN). Stand, Trans, and DyConv represent feature standardization, affine transformation, and dynamic convolution operation, respectively. Generally, IN can alleviate domain variances between inputs by removing their statistical contrast but inevitably eliminates discriminative information of inputs. It thus leads to a relatively large distance between features of the same pedestrian (in the upper case). Therefore, in our DTIN, we employ unnormalized features to guide the transformation of normalized features into new representations to adapt individual domains and instances. In this way, features extracted by our DTIN can both generalize well and be distinguishable.

testing images from an unseen environment. It is a common belief that the change of capturing environment, e.g., change of illumination, view-angles, and seasons causes the domain shift, and existing approaches are not robust under those changes. For this reason, domain generalizable person re-identification (DG-PReID), which aims to build a ReID model that could be more robust to the domain shift and work in an unseen environment, has received increasing attention.

Recently a series of DG-PReID methods [4,5,10,12,16] have been proposed. Among these prior works, Instance Normalization (IN) [23] is widely used to produce domain-invariant features by removing the statistical contrast across feature channels. However, the removed statistics not only encode irrelevant domain-specific patterns but also contain discriminative patterns that may be useful for performing ReID for a particular domain or instance. To address this issue, Pan *et al.* [19] propose an IBN-Net which concatenates features extracted by IN and Batch Normalization (BN) together. A more sophisticated method SNR [12] tries to identify the useful information discarded by IN and then compensate it back. In this work, we aim to address the same issue as the aforementioned works but address the problem with a different principle. Rather than focusing on compensation, we shift our attention to building a module that is sufficiently flexible for learning a mapping function to combine both normalized features and unnormalized features. Our insight is that when the network is sufficiently flexible, we could use end-to-end training to discover a model that can strike a balance between eliminating irrelevant domain-specific features and adapting to individual domains or instances.

To this end, in this work, we propose a new normalization scheme named Dynamically Transformed Instance Normalization (DTIN). The main idea of our DTIN is to employ unnormalized features to guide the transformation of normalized features into a new representation that is adapted to the current domain and instance. To do so, as shown in Fig. 1, we integrate IN with a dynamic instance-aware convolution operation (DyConv). More specifically, in our DTIN, the adaptive parameters of DyConv are generated under the guidance of unnormalized features and then used to re-calibrate and transform the normalized ones. The intuition is that the unnormalized features can transform the normalized ones to dynamically capture information useful to distinguish instances in specific domains. In this way, we could achieve good generalization by adaptively creating feature representation for each individual domain or instance. In addition to utilizing unnormalized features as the control signal for the transformation, we further design a dynamic control path that makes it convenient for the network to utilize patterns from multiple layers. This design further adds the flexibility of learning a generalizable mapping function. To train the network, we adopt a multi-task learning formulation to encourage the network to generate representations that work well for an arbitrary training domain.

2   Related Works

Domain Adaptation for Person Re-Identification. Unsupervised domain adaptation (UDA) requires that the deep models trained on the source domain can adapt to the target domain and work well on it. Generally, in the UDA setting, numbers of unlabeled target domain data are allowed to be accessed during the training phase. Recently, generation-based methods [17,22] and fine-tuning methods [26,29] become two major solutions for UDA. The former type of method mainly employs style transfer algorithms like CycleGAN [34] to transfer the style of labeled source domain data to the style of the target domain. By training with these transferred data, ReID models can be adapted to target domains. In addition, fine-tuning methods try to allocate pseudo labels for unlabelled target domain data by clustering. In this way, they can fine-tune ReID models on the target domain with obtained pseudo labels. Although the UDA methods have the potential to adapt ReID models to a target domain, they highly rely on the unlabeled target domain data, which are not always available in real-world applications.

Generalizable Person Re-Identification. Domain Generalizable (DG) Person Re-Identification (PReID) aims to train a robust and generalizable person re-identification model which can perform well on unseen target domains without a further update. Recently, many relevant works [4,5,10,12,16] have been proposed to achieve this goal. Among these methods, Instance Normalization (IN) [23] has been widely employed to alleviate domain variances between input features by removing their statistical contrast. For instance, Jia *et al.* [10] simply inserts IN into the early layers of the backbone model to eliminate domain disparity. However, IN also inevitably causes discriminative information loss for input features, which limits its application. Therefore, many revisions have been given based on IN. For instance, an SNR module [12] has been proposed to disentangle identity-relevant information from features discarded by IN and reintroduce it back. In addition, Choi *et al.* [4] propose a Meta Batch-Instance Normalization (MetaBIN) model which integrates IN with Batch Normalization and balances their effort with a set of learnable trade-off parameters. However, these methods are often based on the principle of disentangling the domain-relevant information and domain-invariant information to design the model structure, which is perhaps a more challenging problem than domain generalization. Unlike the prior works, we explore dynamically transforming normalized features into appropriate representations to make them adaptive to individual domains and instances.

3   Proposed Methods

This section elaborates on our proposed Dynamically Transformed Instance Normalization (DTIN) module and multi-task training strategy. We first provide the preliminaries of Instance Normalization and Dynamic Convolution, which underpins our proposed method.

3.1   Preliminaries

Let $F \in \mathbb{R}^{C \times H \times W}$ be the convolutional feature of a given input image I, where C is the channel dimension and H, W represent the height and width of the feature map, respectively.

Instance Normalization (IN) is firstly proposed for the style transformation task [23] to remove style information from input features. Recently, it has been widely used for the DG-PReID task to extract domain-invariant representations by removing their statistical contrast across feature channels. Generally, IN consists of a standardization component and an affine transformation component, which can be respectively written as,

$$\text{Standardization: } \widehat{F} = \frac{F - \mu(F)}{\sigma(F)}$$

$$\text{Affine transformation: } \widetilde{F} = \gamma \widehat{F} + \beta, \tag{1}$$

where $\widetilde{F}$ represents the normalized features, μ and σ denote channel-wise mean and standard deviation of F; γ and β are trainable affine parameters learned from end-to-end training on the entire training dataset. By removing domain-specific factors encoded in the mean and standard deviation of the feature channels, IN can effectively enhance the robustness of ReID models by making them less sensitive to domain change.

The effectiveness of IN in extracting domain-invariant representations has been verified in many existing domain generalization studies [4,10,19]. However, the classical IN approach also faces a significant drawback. As also discussed

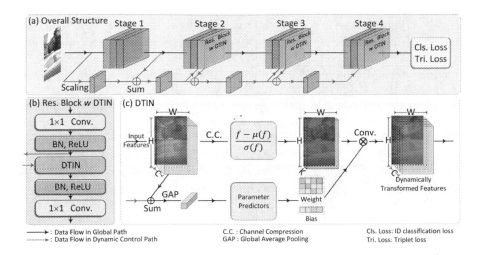

Fig. 2. The sketch of our Dynamically Transformed Instance Normalization (DTIN). The scaling modules in (a) are a set of 3×3 convolutional layers, which are responsible for matching the spatial and channel scales between multi-level features. The parameter predictors are used to generate adaptive parameters, *i.e.*, weight and bias, for the dynamic convolution operation.

in [19], the channel-wise variance contrast removed by IN also encodes discriminative patterns that might be useful for performing ReID in a particular domain or instance. Blindly removing them will have a negative impact on the DG-PReID task.

Dynamic Convolution (DyConv) can adaptively adjust the feature extraction paradigm according to the input instances. Compared to static modules, DyConv is more easily adapted to out-of-distribution inputs due to its flexiblility [1,31]. The idea of DyConv is firstly proposed in [11]. In a dynamic convolutional layer, the weight and bias of each filter are generated from the input and applied to the input. Formally, it can be written as:

$$F^{DC} = \text{conv}\left(F; w(F)\right), \tag{2}$$

where F^{DC} represents the output of DyConv; conv$(\cdot)$ indicates the convolution operation, $w(F)$ represents the filter parameters, *i.e.*, weight and bias, which are generated from input features F. So, $w(F)$ is a function of F. DyConv is an ideal structure for model adaptation since it can adjust the model parameters from the input. Motivated by that, this paper employs DyConv to modulate the normalized feature after IN.

3.2 Dynamically Transformed Instance Normalization

In this work, we propose to integrate DyConv into IN to overcome the drawback of IN. The intuition of doing so is to adaptively re-calibrate normalized features via DyConv into new representations that can adapt to individual domains and instances. To this end, in this work, we propose a Dynamically Transformed Instance Normalization (DTIN) module, as illustrated in Fig. 2c. In our DTIN, we employ a 1×1 dynamic convolution [11] to transform normalized features extracted from IN.

Intuitively, we can regard the filters generated in a DyConv as pattern detectors [3] that can adaptively adjust their sensitivity towards different visual patterns based on input image content. Also, in our design, we modify DyConv by using the unnormalized features to generate filters to process normalized (IN) features. In this way, we could make most of both normalized and unnormalized features for identifying the useful features for ReID. Please see Fig. 4 for some concrete examples that show the advantage of the proposed design. More formally, the DTIN operation can be written as:

$$
\begin{aligned}
F^{DTIN} &= \operatorname{conv}(\frac{F^c - \mu(F^c)}{\sigma(F^c)}; w(F')), \\
F^c &= \operatorname{conv}(F; \theta_0), \ w(F') = \operatorname{fc}(\operatorname{relu}(\operatorname{fc}(\operatorname{pool}((F')), \theta_1)), \theta_2),
\end{aligned}
\tag{3}
$$

where the F and the F^{DTIN} are input and output features of our DTIN; $\operatorname{conv}(F; \theta_0)$ indicates a 1×1 convolutional layer which is employed in our DTIN to reduce the channel dimension of features before IN from C to K ($K=64$); $\operatorname{pool}(\cdot)$ denotes spatially average pooling operation; θ_1 and θ_2 represent parameters of fully connected layers $\operatorname{fc}(\cdot)$. $w(F')$ denotes the produced adaptive parameters (including weight and bias); F' is the features employed for parameter prediction, which involves the unnormalized features and will be explained in the following with more details;

Note that there is no need to apply an additional static affine transformation to standardized features $\frac{F^c - \mu(F^c)}{\sigma(F^c)}$ as in IN since DyConv has already performed a dynamic transformation to them.

Details of the Control Signal F'. As mentioned above, we modify DyConv by generating the filters from F' rather than F. In our design, F' contains features before the IN layer, i.e., unnormalized features. This is motivated by our concern that IN will eliminate some domain-specific patterns that might be useful for performing ReID for an individual domain or instance. For this reason, our design is different from an architecture of using DyConv layer after IN. In the latter case, the filters are not generated from unnormalized features but from normalized features. We compare this alternative in Fig. 3 and find that it leads to much worse performance.

In addition to using unnormalized features for producing F', we also use skip-connections to bring signals from low-level features to enrich F'. We call this design the dynamic control path, and the visualization of this scheme can be found in Fig. 2a. As seen, the dynamic control path aggregates features from multiple levels, and a set of 3×3 convolutional layers are employed along the dynamic control path to ensure the consistency of feature map dimensions. After integrating these multi-level features to produce F', we fed it into DyConv to generate w as in Eq. 3.

Advantages of Our DTIN. Our DTIN integrates the IN and DyConv modules. IN plays a role in eliminating irrelevant domain-specific patterns but is with the limitation of removing discriminative domain-specific or instance-specific information. DyConv module controlled by the unnormalized features can not only re-calibrate those features to avoid over-normalization but also transform the normalized features (through convolution operation) into an appropriate representation. In this sense, our DTIN allows a flexible mapping function that could achieve a trade-off between eliminating ineffective domain-specific factors and adapting to individual domains or instances.

The Deployment of DTIN. To fairly evaluate the effectiveness of our designed DTIN, in this work, we insert it into the widely employed ResNet-50 model [8]. The empowered model is named as **DTIN-Net**, as shown in Fig. 2 (a). The ResNet-50 consists of four major stages, and each stage contains different numbers of residual blocks. To avoid introducing excess computational consumption, we only use our DTIN to replace the 3×3 convolution layer in the last residual block of the 2nd - 4th stages, to refine features extracted in these stages. Besides, we experimentally insert vanilla IN after the first residual stage to facilitate convergence [16].

3.3   Multi-task Training Strategy

Generally, we can access data from multiple source domains at the DG-PReID training stage. Here we denote all the available source domains as $\mathbf{D} = \{D_s\}_{s=1}^{S}$, where we use the S to indicate the number of available domains. In addition, $D_s - (x_i^s, y_i^s)_{i=1}^{N_s}$ represents the s-th domain, the x_i^s and y_i^s respectively represent the input image and the label of i-th sample, and the N_s denotes the number of instances in the s-th domain.

Our aim is to learn a feature extractor to adaptively generate feature representations that can achieve good ReID performance for arbitrary domain. Thus, we treat each domain as an independent task and supply those tasks with a shared feature extractor. This is equivalent to a multi-task-style training process.

Formally, for each domain, we create an ID classifier φ_s to perform ID classification. The s is the domain index. The label space of s-th classifier φ_s is the identities in the s-th domain. We also apply triplet loss to the samples randomly sampled from the s-th domain. The overall objective function of our multi-task training strategy is to minimize the average of loss in each domain:

$$L_{M.T.} = \frac{1}{S} \sum_{s=1}^{S} \frac{1}{N_s} \Big(\sum_{i=1}^{N_s} L_{ce} \left(\varphi_s \left(\psi(x_i^s) \right), y_i^s \right) + L_{tri.}(\psi(x_i^s), \psi(x_{p^i}^s), \psi(x_{n^i}^s)) \Big),$$

(4)

where the i and s are indexes of image and domain; N_s represents the number of instances in the s-th domain; $x_{p^i}^s$ and $x_{n^i}^s$ are sampled positive instance and negative instance for x_i^s; $\psi(\cdot)$ denotes the feature extraction model, *i.e.*, our DTIN-Net; the L_{ce} and $L_{tri.}$ represent cross-entropy loss and triplet loss.

Note that in contrast to the aforementioned multi-task training, one could also stack the training samples from all domains together. Then it is possible to apply ID classification loss with a classifier with the label space corresponding to all identities and also apply the triplet loss to triplets sampled from all samples. However, we empirically find this scheme leads to worse performance. The reason is that by doing so, we have the risk of encouraging the network to use features that distinguish domains but not instances. For example, if the negative image pairs in the triplet loss are from two different domains, the network could partially rely on such features to pull those negative image pairs apart.

4   Experiments

4.1   Implementation Details and Evaluation Setting

Dataset. In this paper, we employ the mainstream person ReID datasets to evaluate the generalization capability of our method, including 6 larger datasets Market1501 (M) [27], DukeMTMC-ReID (D) [30], Cuhk02 (C2) [13], Cuhk03 (C3) [14], CuhkSYSU (CS) [25], MSMT17 (MS) [24], and 4 smaller ones termed VIPeR (V) [7], PRID (P) [9], GRID (G) [18], and QMUL i-LIDS (Q) [28].

Settings. In this paper, we first adopt two widely used multiple source domain generalization protocols as in [5] to evaluate the generalization capability of our model. In Protocol-1, we employ the M, D, C2, C3, and CS to construct the training set and evaluate our model on the V, P, G, and Q, respectively. In Protocol-2, the training set comprises M, D, C3, and MS, and the test set is the same as Protocol-1. Besides, we also follow a cross-domain setting [4], in which we train our model on M (D) and test it on D (M).

Implementation Details. Before the domain generalization training, we firstly pre-train our DTIN-Net on the ImageNet [6] dataset. Under all protocols mentioned above, we train our model for 120 epochs. Particularly, for the cross-domain setting, we do not use the multi-task training strategy since it only contains one source domain. Besides, in our DTIN-Net, the dynamic control path receives and aggregates features before the first residual stage, features after the first residual stage, and features before our DTIN modules. The learning rate is initialized as 3.5×10^{-4} and divided by 10 at the 40-th and 70-th epochs, respectively. In all experiments, each image is resized to 256×128 for training and test. During the training phase, each image is flipped horizontally with a probability of 0.5. All results reported in this section are the mean of two repetitive experiments. In addition, random erasing is employed for data augmentation. The widely used metrics CMC and mAP are employed to evaluate our model.

4.2   Comparison with State-of-the-Art Methods

To clarify, in the remainder of this paper, the "Baseline" model indicates a ResNet-50 model trained with ID classification loss and triplet loss.

Multiple Source Domain Generalization. We compare our DTIN-Net with other state-of-the-art methods under the aforementioned Protocol-1 and Protocol-2 settings, and the results are shown in Table 1. As we can find, on almost all datasets, our DTIN-Net model achieves comparable or better performance than compared methods. Only on the i-LIDs dataset our DTIN-Net is worse than RaMoE [5]. Nevertheless, our DTIN-Net model outperforms all other compared algorithms in the average performance, which demonstrates the superiority of our proposed method.

Table 1. The re-identification performance comparison between our DTIN-Net and other DG-PeReID algorithms under the Protocol-1 (P-1) and Protocol-2 (P-2) settings. It can be found that our DTIN-Net outperforms the compared algorithms under both settings.

Setting	Method	GRID		VIPeR		PRID		i-LIDS		Average	
		mAP	CMC-1	mAP	CMC-1	mAP	CMC-1	mAP	CMC-1	mAP	CMC-1
P-1	DIMN [21]	41.1	29.3	60.1	51.2	52.0	39.2	78.4	70.2	57.9	47.5
	DMG-Net [2]	56.6	51.0	60.4	53.9	68.4	60.6	83.9	79.3	67.3	61.2
	RaMoE [5]	54.2	46.8	64.6	56.6	67.3	57.7	90.2	85.0	69.1	61.5
	MetaBIN [4]	57.9	48.4	68.6	59.9	81.0	74.2	87.0	81.3	73.6	66.0
	DTIN-Net	**60.6**	**51.8**	**70.7**	**62.9**	79.7	71.0	87.2	81.8	**74.6**	**66.9**
P-2	SNR [12]	41.3	30.4	65.0	55.1	60.0	49.0	91.9	87.0	64.6	55.4
	DMG-Net [2]	47.2	37.3	70.9	62.3	69.7	59.7	88.2	83.0	69.0	60.6
	RaMoE [5]	53.9	43.4	**72.2**	63.4	66.8	56.9	**92.3**	**88.4**	71.3	63.0
	DTIN-Net	**58.4**	**49.4**	71.9	**64.0**	**77.4**	**67.8**	89.2	85.3	**74.2**	**66.6**

Table 2. The performance of single domain generalization for person re-identification. Compared with other state-of-the-art algorithms, our DTIN-Net achieves a promising performance, which indicates our algorithm is also effective when training with limited data.

Method	M → D		D→ M	
	mAP	CMC-1	mAP	CMC-1
IBN-Net [19]	24.3	43.7	23.5	50.7
OSNet [33]	25.9	44.7	24.0	52.2
CrossGrad [20]	27.1	48.5	26.3	56.7
QAConv [15]	28.7	48.8	27.2	58.6
L2A-OT [33]	29.2	50.1	30.2	63.8
OSNet-AIN [32]	30.5	52.4	30.6	61.0
SNR [12]	33.6	55.1	33.9	66.7
MetaBIN [4]	33.1	55.2	35.9	69.2
DTIN-Net	**36.1**	**57.0**	**37.4**	**69.8**

Cross-Domain Generalization. To further evaluate the generalization capability of our DTIN-Net, we additionally compare it with other state-of-the-art algorithms under the cross-domain generalization setting. The experiential

Table 3. The effectiveness of our designed components, namely dynamically transformed instance normalization module (DTIN) and multi-task training strategy (M.T.). It can be found that our design is effective in enhancing the generalization capability of ReID models.

	DTIN	M.T.	GRID		VIPeR	
			mAP	CMC-1	mAP	CMC-1
Baseline	×	×	46.5	37.2	65.3	56.4
Base+IN	×	×	52.7	42.9	67.5	58.7
DTIN-Net	✓	×	$56.4^{\uparrow 3.7}$	$46.2^{\uparrow 3.3}$	$69.1^{\uparrow 1.6}$	$60.4^{\uparrow 1.7}$
DTIN-Net	✓	✓	$\mathbf{60.6}^{\uparrow 7.9}$	$\mathbf{51.8}^{\uparrow 8.9}$	$\mathbf{70.7}^{\uparrow 3.2}$	$\mathbf{62.9}^{\uparrow 4.2}$

results are shown in Table 2. Here, we do not compare our DTIN-Net with other multi-source DG-PReID methods [2,5] since these methods design their structure and training strategy based on the premise that more than one source domain can be accessed. Compared to the best competitors, i.e., SNR [12] and MetaBIN [4], our DTIN-Net achieves significantly better performance, 2.8% mAP on average (M → D). The superiority of our DTIN-Net could come from two aspects. Firstly, inheriting the advantages of dynamic convolution operation [11] and instance normalization [23], our DTIN module naturally adapts to out-of-distribution inputs and is robust to domain variance. Secondly, our DTIN integrates these two effective modules in a judicious manner. By enhancing the flexibility of ReID model, our DTIN-Net can adaptively balance normalizing domain interference and adapting to individual domains or instances. In this way, our DTIN can extract features that generalize well on unseen domains.

4.3   Ablation Studies

All experiments in this subsection are based on the protocol-1 setting. To clarify, in this subsection, "Base+IN" indicates the model (based on ResNet-50) inserted with classical IN layers before the convolutional layers we replace our DTIN with. "Base+IN+DyConv" is the two-module setup that simply concatenates IN with a 1×1 dynamic convolution module [11].

Table 4. The comparison between versions of DTIN-Net with or without our designed dynamic control path (DyCtrl). M.T. represents the multi-task training strategy. As we can find, the DyCtrl consistently enhances the generalization capability of our model.

	M.T.	DyCtrl	GRID		VIPeR	
			mAP	CMC-1	mAP	CMC-1
DTIN-Net	×	×	54.7	44.1	68.6	59.6
DTIN-Net	×	✓	$56.4^{\uparrow 1.7}$	$46.2^{\uparrow 2.1}$	$69.1^{\uparrow 0.5}$	$60.4^{\uparrow 0.8}$
DTIN-Net	✓	×	58.8	49.3	70.0	62.3
DTIN-Net	✓	✓	$60.6^{\uparrow 1.8}$	$51.8^{\uparrow 2.5}$	$70.7^{\uparrow 0.7}$	$62.9^{\uparrow 0.6}$

Effectiveness of Designed Components. To evaluate the effectiveness of our designed DTIN and multi-task training strategy (M.T.), we gradually add them to the "Baseline" model and compare the performance. For a fair comparison, in Table 3, we also give the performance of the "Base+IN" model. It can be found that the robustness of the "Baseline" model to domain variance can be relatively improved by simply applying IN to it. However, the inherent drawback of IN, *i.e.*, the loss of discriminative information, limits the capability of the "Base+IN" model. For our DTIN-Net, it achieves a significantly better recognition accuracy than the "Base+IN" model on unseen domains, about 3.7% mAP and 1.6% mAP on GRID and VIPeR datasets, respectively. The interpretation could be that the delicate parameter prediction strategy employed by our DTIN provides sufficient semantic information, based on which our DTIN can adaptively calibrate normalized features to adapt to individual domains and instances. Besides, the generalization capability of our DTIN-Net can be further improved (4.2% mAP on the GRID dataset) if we additionally utilize M.T. to train our DTIN-Net. The interpretation of the improvement could be that our training strategy can guide our DTIN to learn appropriate transformations to adapt normalized features to arbitrary domains.

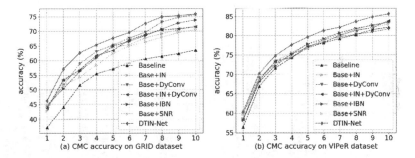

Fig. 3. The CMC-1 to CMC-10 accuracy comparison of variations on the DTIN-Net architecture. The ("Base+DyConv", "Base+SNR", and "Base+IBN") are the versions replacing our DTIN module with a 1×1 Dynamic Filter module [11], SNR module [12], and IBN module [19]. The "Base+IN+DyConv" is the two-module setup which simply concatenates IN with a 1×1 dynamic convolution module [11]. For a fair comparison, all these models above are trained without our multi-task training strategy.

Effectiveness of Dynamic Control Path. To ensure that the features to predict adaptive parameters contain sufficient semantic information, in our DTIN-Net, a dynamic control path (DyCtrl) is given to integrating low-level features as the semantic supplement to high-level features. As shown in Table 4, this straightforward operation brings 2.5% and 0.6% CMC-1 improvement to our DTIN-Net on GRID and VIPeR dataset (trained with M.T.), respectively. It indicates that additional semantic information provided by our designed DyCtrl indeed benefits to construct instance-adaptive calibration for each input and improves its distinguishability on unseen domains.

Comparison between Our DTIN and Other Relevant Modules. In Fig. 3, we give the performance of models inserting traditional IN ("Base+IN"), or dynamic convolution module [11] ("Base+DyConv") at the positions we set our DTIN to. As we can find, all these single-module setups improve the "Baseline" model. However, simply concatenating these two effective modules ("Base+IN+DyConv") together does not bring an additional improvement. The interpretation could be that the information loss caused by IN may seriously limit the flexible nature of the dynamic module. As shown in Fig. 4, the "Base+IN+DyConv" even seems to highlight the background incorrectly due to the insufficient semantic perception capability of its generated parameters. On the contrary, thanks to the delicate dynamic control path, our DTIN ensures features to parameter prediction contain sufficient semantic information. In this way, our DTIN can effectively calibrate normalized features to adapt to individual domains and instances. As shown in Fig. 4, our DTIN can effectively capture discriminative clues of each instance (like the cartoon pattern in the first case in Market1501 dataset) and thus is able to re-calibrate normalized features to be distinguishable. In addition, in Fig. 3 we also give the performance of models replacing our DTIN with SNR [12] ("Base+SNR") and IBN [19] ("Base+IBN") module. Thanks to the flexible nature of our DTIN, it can adaptively strike the

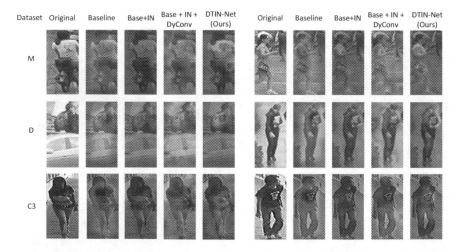

Fig. 4. The activation maps of features extracted by "Baseline" model, "Base+IN" model, two-module setup "Base+IN+DyConv" and our DTIN-Net. The M, D, C3 indicate cases in those rows are sampled from Market1501, DukeMTMC-ReID, Cuhk03 datasets. It can be found that the IN modifies the original activation map. While correctly reducing the response values for irrelevant regions, such as the "car windows" region, it also lowers the contrast between discriminative regions and background regions. Directly combining IN with DyConv, i.e., "Base+IN+DyConv", fails to overcome the limitation of IN and seems to highlight the background incorrectly. In contrast, our DTIN can correctly locate the identity-relevant patterns and remove the activations from the irrelevant regions.

right balance between normalizing irrelevant domain features and adapting to individual domains or instances and thus achieves a better performance than these compensation-based methods.

Table 5. The comparison of model capacity and computation consumption between our DTIN-Net and other state-of-the-art methods. Our DTIN-Net achieves a significantly better generalization performance than compared methods with comparable parameters (Params) and floating-point operations (FLOPs).

Method	GRID		VIPeR		Params	FLOPs
	mAP	CMC-1	mAP	CMC-1	(M)	(GMac)
Base+IN	52.3	42.5	67.3	58.4	**23.5**	4.1
RaMoE [5]	54.2	46.8	64.6	56.6	39.3	4.1
MetaBIN [4]	57.9	48.4	68.6	59.9	23.6	4.1
DTIN-Net	**60.6**	**51.8**	**70.7**	**62.9**	25.5	**3.7**

Analysis about Model Capacity. In this work, we employ a dynamic convolution module to instance-adaptively transform normalized features to make them adapt to individual domains and instances. However, it also raises the suspicion, namely, whether the superiority of our DTIN-Net is caused by increasing the capacity of the backbone model with the computation-expensive dynamic convolution modules? To explore this suspicion, we compare the generalization capability and model capacity of our DTIN-Net with other state-of-the-art models. In addition, the capacity of the "Base+IN" model is also given for comparison. The results are given in Table 5, from which we can summarize two important findings. Firstly, thanks to the light-weight design of our DTIN-Net, our DTIN-Net achieves a significantly better performance than the "Base+IN" model, about 8.3% mAP on GRID dataset, by increasing only a few additional parameters (2 M) and even saving 9.8% calculation (0.4 GMac). Secondly, compared with other state-of-the-art methods, i.e., RaMoE and Meta-BIN, our DTIN-Net achieves a better generalization performance (averagely, 4.6% mAP on GRID dataset) with comparable parameters and computations. It indicates that the superiority of our DTIN-Net on DG-PReID task is not caused by improving model capacity.

t-SNE Visualization Results. To intuitively show how our DTIN enhances the generalization capability of ReID models, in Fig. 5 we exhibit the t-SNE visualization of features extracted by the "Baseline" model, the "Base+IN" model, and our DTIN-Net. Here, we randomly sample 60 pairs of query and gallery images from each of the 4 unseen domains (GRID, PRID, VIPeR, and i-LIDS). In this experiment, the perplexity and iteration of t-SNE is set to 6 and 50. respectively. The features used for visualization are extracted after the last DTIN module for our model and the corresponding position for comparing models. As shown in Fig. 5a, we can find that the "Baseline" model is seriously influenced

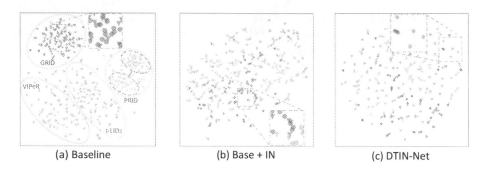

Fig. 5. The t-SNE results of features on four unseen target datasets (VIPeR, PRID, GRID, and i-LIDS). For comparison, we simultaneously give the visualization of features processed by the "Baseline" model, the "Base+IN" model, and our DTIN-Net. Dots with identical colors are from the same domain. Besides, the triangular and circular dots represent query and gallery, respectively. (Color figure online)

by domain factors, which causes clear domain boundaries. Particularly, in the selected gray section, since samples from query set and gallery set are captured under different cameras, there even exists a significant gap between the features of query samples and gallery samples. For the "Base+IN" in Fig. 5b, it effectively erases the domain interference of each input and thus breaks the domain boundaries. However, due to the loss of discriminative information, large intra-class distances exist between query instances and gallery instances (as shown in the selected section) which may limit the recognition accuracy of ReID models. For our DTIN module, by applying instance-level adaptive calibration for each normalized feature, it significantly improves the distinguishability of normalized features by IN. Specifically, we can find that our DTIN-Net forms a relatively independent sub-cluster for each person instance with clear boundaries, while the cluster boundaries in "Base-IN" are blurred.

5   Conclusion

In this work, we propose a new normalization scheme named Dynamically Transformed Instance Normalization (DTIN), which effectively alleviates the inherent drawback of Instance Normalization that inevitably impacts discriminative information of input features. By transforming normalized features into appropriate representation in a learnable and adaptive manner, our DTIN empowers ReID models to strike the right balance between normalizing irrelevant domain features and adapting to individual domains or instances. In addition, we further propose a multi-task formulation on multiple training domains to train our model. Extensive experiments demonstrate the superiority of our proposed method.

Acknowledgments. This work is supported by National Key R&D Program of China (No.2020AAA0106900), the National Natural Science Foundation of China (No. U19B2037), Shaanxi Provincial Key R&D Program (No.2021KWZ–03), Natural Science Basic Research Program of Shaanxi (No.2021JCW–03). This work is also supported by the National Research Foundation, Singapore under its AI Singapore Programme (AISG Award No: AISG–RP–2018–003), and the MOE AcRF Tier-1 research grant: RG95/20.

References

1. Akula, A., Jampani, V., Changpinyo, S., Zhu, S.C.: Robust visual reasoning via language guided neural module networks. In: Proceedings of the Advances in Neural Information Processing Systems (2021)
2. Bai, Y., et al.: Person30k: a dual-meta generalization network for person re-identification. In: Proceedings of the IEEE conference on computer vision and pattern Recognition, pp. 2123–2132 (2021)
3. Chen, L., et al.: SCA-CNN: spatial and channel-wise attention in convolutional networks for image captioning. In: Proceedings of the IEEE conference on computer vision and pattern Recognition, pp. 5659–5667 (2017)
4. Choi, S., Kim, T., Jeong, M., Park, H., Kim, C.: Meta batch-instance normalization for generalizable person re-identification. In: Proceedings of the IEEE conference on computer vision and pattern Recognition, pp. 3425–3435 (2021)
5. Dai, Y., Li, X., Liu, J., Tong, Z., Duan, L.Y.: Generalizable person re-identification with relevance-aware mixture of experts. In: Proceedings of the IEEE conference on computer vision and pattern Recognition, pp. 16145–16154 (2021)
6. Deng, J., Dong, W., Socher, R., Li, L., Li, K., FeiFei, L.: ImageNet: a large-scale hierarchical image database. In: Proceedings of the IEEE conference on computer vision and pattern Recognition, pp. 248–255 (2009)
7. Gray, D., Tao, H.: Viewpoint invariant pedestrian recognition with an ensemble of localized features. In: Forsyth, D., Torr, P., Zisserman, A. (eds.) ECCV 2008. LNCS, vol. 5302, pp. 262–275. Springer, Heidelberg (2008). https://doi.org/10.1007/978-3-540-88682-2_21
8. He, K., Zhang, X., Ren, S., Sun, J.: Deep residual learning for image recognition. In: Proceedings of the IEEE conference on computer vision and pattern Recognition, pp. 770–778 (2016)
9. Hirzer, M., Beleznai, C., Roth, P.M., Bischof, H.: Person re-identification by descriptive and discriminative classification. In: Scandinavian Conference on Image Analysis, pp. 91–102 (2011)
10. Jia, J., Ruan, Q., Hospedales, T.M.: Frustratingly easy person re-identification: generalizing person re-id in practice. arXiv preprint arXiv:1905.03422 (2019)
11. Jia, X., De Brabandere, B., Tuytelaars, T., Gool, L.V.: Dynamic filter networks. Proc. Adv. Neural Inf. Process. Syst. **29**, 667–675 (2016)
12. Jin, X., Lan, C., Zeng, W., Chen, Z., Zhang, L.: Style normalization and restitution for generalizable person re-identification. In: Proceedings of the IEEE conference on computer vision and pattern Recognition, pp. 3143–3152 (2020)
13. Li, W., Wang, X.: Locally aligned feature transforms across views. In: Proceedings of the IEEE conference on computer vision and pattern Recognition, pp. 3594–3601 (2013)

14. Li, W., Zhao, R., Xiao, T., Wang, X.: DeepReID: deep filter pairing neural network for person re-identification. In: Proceedings of the IEEE conference on computer vision and pattern recognition, pp. 152–159 (2014)
15. Liao, S., Shao, L.: Interpretable and generalizable person re-identification with query-adaptive convolution and temporal lifting. In: Vedaldi, A., Bischof, H., Brox, T., Frahm, J.-M. (eds.) ECCV 2020. LNCS, vol. 12356, pp. 456–474. Springer, Cham (2020). https://doi.org/10.1007/978-3-030-58621-8_27
16. Lin, S., Li, C.T., Kot, A.C.: Multi-domain adversarial feature generalization for person re-identification. IEEE Trans. Image Process. **30**, 1596–1607 (2020)
17. Liu, J., Zha, Z.J., Chen, D., Hong, R., Wang, M.: Adaptive transfer network for cross-domain person re-identification. In: Proceedings of the IEEE conference on computer vision and pattern Recognition, pp. 7202–7211 (2019)
18. Loy, C.C., Xiang, T., Gong, S.: Multi-camera activity correlation analysis. In: Proceedings of the IEEE conference on computer vision and pattern Recognition, pp. 1988–1995. IEEE (2009)
19. Pan, X., Luo, P., Shi, J., Tang, X.: Two at once: enhancing learning and generalization capacities via IBN-Net. In: Ferrari, V., Hebert, M., Sminchisescu, C., Weiss, Y. (eds.) ECCV 2018. LNCS, vol. 11208, pp. 484–500. Springer, Cham (2018). https://doi.org/10.1007/978-3-030-01225-0_29
20. Shankar, S., Piratla, V., Chakrabarti, S., Chaudhuri, S., Jyothi, P., Sarawagi, S.: Generalizing across domains via cross-gradient training. arXiv preprint arXiv:1804.10745 (2018)
21. Song, J., Yang, Y., Song, Y.Z., Xiang, T., Hospedales, T.M.: Generalizable person re-identification by domain-invariant mapping network. In: Proceedings of the IEEE conference on computer vision and pattern Recognition (2019)
22. Tang, Y., Yang, X., Wang, N., Song, B., Gao, X.: CGAN-TM: a novel domain-to-domain transferring method for person re-identification. IEEE Trans. Image Process. **29**, 5641–5651 (2020)
23. Ulyanov, D., Vedaldi, A., Lempitsky, V.: Instance normalization: the missing ingredient for fast stylization. arXiv preprint arXiv:1607.08022 (2016)
24. Wei, L., Zhang, S., Gao, W., Tian, Q.: Person transfer GAN to bridge domain gap for person re-identification. In: Proceedings of the IEEE conference on computer vision and pattern Recognition, pp. 79–88 (2018)
25. Xiao, T., Li, S., Wang, B., Lin, L., Wang, X.: End-to-end deep learning for person search. arXiv preprint arXiv:1604.01850 (2016)
26. Zhai, Y., et al.: AD-Cluster: augmented discriminative clustering for domain adaptive person re-identification. In: Proceedings of the IEEE conference on computer vision and pattern Recognition (2020)
27. Zheng, L., Shen, L., Tian, L., Wang, S., Wang, J., Tian, Q.: Scalable person re-identification: a benchmark. In: Proceedings of the IEEE conference on computer vision and pattern Recognition, pp. 1116–1124 (2015)
28. Zheng, W.S., Gong, S., Xiang, T.: Associating groups of people. In: Proceedings of the British Machine Vision Conference, No. 6, pp. 1–11 (2009)
29. Zheng, Y., et al.: Online pseudo label generation by hierarchical cluster dynamics for adaptive person re-identification. In: Proceedings of the IEEE conference on computer vision and pattern Recognition, pp. 8371–8381 (2021)
30. Zheng, Z., Zheng, L., Yang, Y.: Unlabeled samples generated by GAN improve the person re-identification baseline in vitro. In: Proceedings of the IEEE International Conference on Computer Vision, pp. 3754–3762 (2017)
31. Zhou, K., Liu, Z., Qiao, Y., Xiang, T., Loy, C.C.: Domain generalization in vision: a survey. arXiv preprint arXiv:2103.02503 (2021)

32. Zhou, K., Yang, Y., Cavallaro, A., Xiang, T.: Learning generalisable omni-scale representations for person re-identification. In: IEEE Transactions on Pattern Analysis and Machine Intelligence (2021)
33. Zhou, K., Yang, Y., Hospedales, T., Xiang, T.: Learning to generate novel domains for domain generalization. In: Vedaldi, A., Bischof, H., Brox, T., Frahm, J.-M. (eds.) ECCV 2020. LNCS, vol. 12361, pp. 561–578. Springer, Cham (2020). https://doi.org/10.1007/978-3-030-58517-4_33
34. Zhu, J.Y., Park, T., Isola, P., Efros, A.A.: Unpaired image-to-image translation using cycle-consistent adversarial networks. In: Proceedings of the IEEE International Conference on Computer Vision (2017)

Domain Adaptive Person Search

Junjie Li[1,2] , Yichao Yan[1(✉)] , Guanshuo Wang[2] , Fufu Yu[2] ,
Qiong Jia[2] , and Shouhong Ding[2]

[1] MoE Key Lab of Artificial Intelligence, AI Institute, Shanghai Jiao Tong
University, Shanghai, China
{junjieli00,yanyichao}@sjtu.edu.cn
[2] Tencent Youtu Lab, Shanghai, China
{mediswang,fufuyu,boajia,ericshding}@tencent.com

Abstract. Person search is a challenging task which aims to achieve
joint pedestrian detection and person re-identification (ReID). Previous
works have made significant advances under fully and weakly supervised
settings. However, existing methods ignore the generalization ability of
the person search models. In this paper, we take a further step and
present Domain Adaptive Person Search (DAPS), which aims to generalize the model from a labeled source domain to the unlabeled target
domain. Two major challenges arises under this new setting: one is how
to simultaneously solve the domain misalignment issue for both detection and ReID tasks, and the other is how to train the ReID subtask
without reliable detection results on the target domain. To address these
challenges, we propose a strong baseline framework with two dedicated
designs. 1) We design a domain alignment module including image-level
and task-sensitive instance-level alignments, to minimize the domain discrepancy. 2) We take full advantage of the unlabeled data with a dynamic
clustering strategy, and employ pseudo bounding boxes to support ReID
and detection training on the target domain. With the above designs, our
framework achieves 34.7% in mAP and 80.6% in top-1 on PRW dataset,
surpassing the direct transferring baseline by a large margin. Surprisingly, the performance of our unsupervised DAPS model even surpasses
some of the fully and weakly supervised methods. The code is available
at https://github.com/caposerenity/DAPS.

Keywords: Person search · Domain adaptation

1 Introduction

Person search [39,44] aims to detect and identify the query person from natural
images. The mainstream approaches to tacking this task is to simultaneously

---

J. Li—This work was done during Li's internship at Tencent Youtu Lab.

Supplementary Information The online version contains supplementary material
available at https://doi.org/10.1007/978-3-031-19781-9_18.

address both tasks in an end-to-end manner, where supervised learning [6,26,34, 44] that rely on both pedestrian bounding boxes annotation and identity labels have been actively investigated. However, these supervised methods may suffer from significant performance degradation on unseen domains due to domain gaps.

To address this problem, several recent works [20,40] propose the weakly supervised person search (WSPS) setting without accessible ID annotations, shown in Fig. 1. Nevertheless, several limitations are still waiting to be addressed. First, these works still require manual annotation of the ground-truth bounding boxes for the detection task, which obviously is not an economical option for real-world applications. Second, there exist several large-scale annotated person search datasets, e.g., CUHK-SYSU [39] and PRW [44], which can serve as supervised source domains and help improve the performance on the unlabeled target data. Unfortunately, the weakly supervised setting does not fully unleash the potential of the available training data. Third, these methods adopt an inconsistent training strategy with supervised detection and unsupervised ReID, which ignores the essential correlation between the two sub-tasks.

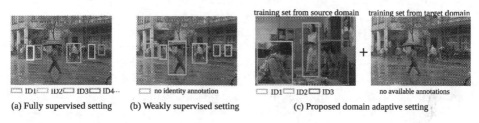

(a) Fully supervised setting   (b) Weakly supervised setting   (c) Proposed domain adaptive setting

Fig. 1. Comparison of three person search settings. (a) Fully supervised setting: bounding boxes and identity annotations are available. (b) Weakly supervised setting: only bounding boxes annotations are available. (c) Domain adaptive setting: neither bounding boxes nor identity annotations on the target domain is accessible, and there exists obvious domain gaps between different domains, e.g., the size of human crops. The network is trained with both the labeled source domain and the unlabeled target domain images.

Inspired by the unsupervised domain adaptation (UDA) [16,23,36], as shown in Fig. 1, we present the Domain Adaptive Person Search (DAPS) framework, where person search models trained on labeled source domain are transferred to unlabeled target domains. Compared to weakly supervised person search, neither the identity labels nor the bounding boxes are accessible in DAPS. Our framework faces two major challenges: (1) Both the detection and the ReID subtasks suffer from domain gap. However, detection focuses on the commonness of people regardless of the identities, while ReID needs to learn the uniqueness of different persons. This conflict can be more serious in domain adaptation. (2) Since the ground-truth detection boxes are not available, it will be extremely challenging to accurately localize the pedestrians in the target domain, which further increases the difficulty for the ReID sub-task. Therefore, directly extending WSPS methods to take advantage of target domain data is infeasible.

To address the first challenge, we explore domain alignment for robust domain invariant feature learning. In the context of pedestrian detection, this is typically achieved by domain adversarial training [8] on both image-level and instance-level features.

To tackle the second challenge, we generate pseudo bounding boxes on the target domain images iteratively, and perform the training process with GT and pseudo boxes for domain adaptation. Furthermore, we present a dynamic clustering strategy to generate pseudo identity labels on the target domain. To fully release the potential of the target domain training data, the proposed framework refines the detection task with selected proposals, and enhances the interaction between the two sub-tasks with hybrid hard case mining. Experimental results demonstrate that this design surprisingly achieves comparative performance with directly adopting ground-truth bounding boxes.

Our contributions are summarized as three-fold:

- We introduce a novel unsupervised domain adaptation paradigm for person search. This setting requires neither bounding boxes nor identity annotations on the target domain, making it more practical for real-world applications.
- We present the DAPS framework to overcome the challenges caused by cross-domain discrepancy and cross-task dependency. We propose domain alignment for person search to enhance domain-invariant feature learning. Meanwhile, a dynamic clustering and a hybrid hard case mining strategy are introduced to facilitate unsupervised target domain learning.
- Without any auxiliary label in the target domain, our framework achieves promising performance on two target person search benchmarks, surprisingly outperforming several weakly and fully supervised models.

2   Related Work

2.1   Person Search

With the development of deep learning and large scale benchmarks [39,44], person search [4] has recently become a popular research topic. Existing fully supervised person search models can be divided into two-step and one-step frameworks. Two-step frameworks typically consist of separately trained detection and ReID models [21,34]. Zheng *et al.* [44] make a systematic evaluation on different combination of detection and ReID models. Wang *et al.* [34] solve the inconsistency between detection and person ReID tasks. One-step frameworks [6,26,41] design a unified model to jointly solve detection and ReID tasks in an end-to-end manner, making the pipeline more efficient. Yan *et al.* [43] introduce a graph model to explore the impact of contextual information for identity matching. Chen *et al.* [6] disentangle the person representation into norm and angle to eliminate the cross-task conflict. Li *et al.* [26] develop a sequential structure to reduce the low-quality proposals. Several recently studies [20,40] adopt the weakly supervised setting without no accessible person ID labels. In this work, we explore a novel person search setting to generalize labeled source to unlabeled target domain without any bounding boxes and ID labels annotation.

2.2 Domain Adaptation for Person ReID

Unsupervised domain adaptation (UDA) ReID [7,10,15,17,28,32,42] typically trains a model with labelled source domain and transfers to the target domain under the unsupervised setting. Mainstream UDA ReID methods can be divided into two categories. The first category employs generative adversarial networks [19] to mitigate the style discrepancy and translate the labelled source domain data into the target domain [7,10,28]. For the second category, they generate pseudo labels by clustering [15,17,32] or assigning soft labels [35] on target domain, and use these pseudo labels to further supervise target domain training. Recently, pseudo label-based methods raise more attention due to their superior performance. However, UDA ReID requires the cropped images, which cannot be directly extended to adaptive person search due to the lack of bounding boxes on target domain. To address this, we propose a dynamic clustering strategy to generate high-quality pseudo boxes to facilitate target domain training.

2.3 Domain Adaptive Object Detection

Existing Domain Adaptive approaches object detection can be categorized into three main branches, including adversarial-based methods [8,31,33,37,45], discrepancy-based methods [2,3,24] and reconstruction based methods [1,11,27]. Adversarial based methods utilize a domain discriminator to distinguish the domain of input data, the adversarial training is performed to encourage domain confusion between the source domain and the target domain. The discrepancy-based strategy utilizes the unlabeled target domain images to fine-tune the detector, and further followed by mean-teacher learning [2] or auto-annotation [3]. The reconstruction-based approaches bridge the domain gap by reconstructing the source or target samples, which is usually realized by image-to-image translation [1,27]. In this work, we consider the conflicts between sub-tasks of person search, and develop a task-sensitive alignment module to alleviate such conflicts.

3   Methodology

3.1   Framework Overview

The general pipeline of the proposed DAPS framework is illustrated in Fig. 2. Given the input images from both the source and the target domain, the image-level feature maps are extracted with a backbone network. Then, these features are input into the Region Proposal Network (RPN) to generate candidate bounding boxes, which are subsequently fed into the ROI-Align layer to represent instance-level feature maps. To close the domain gaps for the downstream

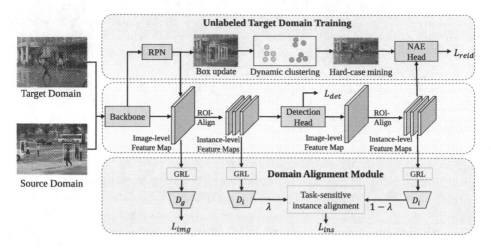

Fig. 2. Architecture of the DAPS framework. "GRL" denotes the gradient reverse layer [16]. The backbone follows SeqNet [26], and we employ a domain alignment module to minimize domain discrepancy on both image-level and instance-level. We further impose dynamic clustering, hybrid hard case mining and target detection training to take full advantage of the unlabeled target domain data

detection and ReID tasks, we design a domain alignment module (DAM) to align both image-level and instance-level features from different domains.

Subsequently, the domain-aligned instance-level feature maps are input into both the detection and the ReID branch. Since the ground-truth bounding boxes are not available in the target domain, the model will generate different pedestrian detection results for each training epoch. Therefore, it is infeasible to follow the traditional UDA ReID methods, which generally perform clustering on a fixed size of instances to generate pseudo labels. To address this issue, we design a novel dynamic clustering strategy, which continuously associates the bounding boxes generated from consecutive epochs, to guarantee the stability of instance-level ReID features. Based on dynamic clustering strategy, we further introduce the hybrid hard case mining and the target domain detection refinement to sufficiently take advantage of the unlabeled training data.

3.2 Domain Alignment Module

Image-Level Alignment. As discussed in [7,8,10,36], minimizing domain discrepancy is beneficial for both sub-tasks of person search, and an effective way is to guide the model to learn domain-invariant representation. Motivated by the recent progress in domain adaptive detectors [8,31,37,45], where intermediate features are imposed with image-level alignment constraints, we introduce a domain alignment module into our DAPS framework. As shown in Fig. 2, DAM employs a patch-based domain classifier to predict the domain where the input feature comes from. A min-max formulation is adopted to misdirect the domain classifier and encourage domain-invariant representation learning.

Suppose we have N training images $\{I_1, ..., I_N\}$ with corresponding domain labels $\{d_1, ..., d_N\}$. Particularly, $d_i = 0$ indicates that image I_i comes from the source domain, while $d_i = 1$ denotes the target domain. We denote the backbone of DAPS as Φ and the image-level domain classifier as D_g, and further represent the domain prediction result of input I_i as p_i. We apply a cross entropy loss to perform domain alignment in an adversarial training manner:

$$\mathcal{L}_{img} = - \sum_i [d_i \log p_i + (1 - d_i) \log (1 - p_i)]. \qquad (1)$$

We have tried to conduct image-level alignment on different intermediate features and multi-scale alignment, but achieve no better results.

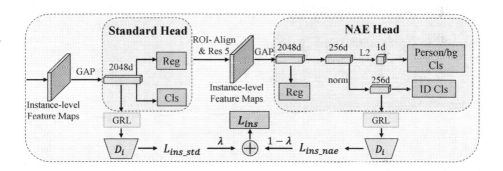

Fig. 3. Details of the two heads and the task-sensitive instance-level alignment.

Task-Sensitive Instance-Level Alignment. As illustrated in Fig. 3, our framework consists of two head networks, where the detection performance mainly depends on the first standard Faster R-CNN [18] head, while the NAE [6] head is highly relevant to ReID. When the scale of the source domain is much smaller than the unlabeled target, the target pseudo bounding boxes predicted by detector trained on the source can be severely overfitted to the smaller domain, but no reliable target detection guidance can relieve this issue. When the target is much smaller, pseudo target ID labels can be easily obtained by clustering, but these might provide insufficient generalizing for the ReID sub-task.

According to the characteristics of the up- and down-stream tasks, we propose the task-sensitive instance-level alignment module by balancing the alignment weight on instance-level features for both sub-tasks. Suppose we have K_1 instances in the standard head and K_2 instances for the NAE head, two domain classifiers $\{D_i^d, D_i^r\}$ are built in the same way with image-level alignment, and the domain predictions of the two local classifiers are denoted as $\{p_{i,1}^d, ..., p_{i,K_1}^d\}$,

$\{p_{i,1}^r, ..., p_{i,K_2}^r\}$, respectively. The instance-level loss can be formulated as:

$$\mathcal{L}_{ins} = -\lambda \sum_{i,j} \left[d_i \log p_{i,j}^d + (1 - d_i) \log \left(1 - p_{i,j}^d\right) \right]$$
$$- (1 - \lambda) \sum_{i,k} \left[d_i \log p_{i,k}^r + (1 - d_i) \log \left(1 - p_{i,k}^r\right) \right]. \tag{2}$$

where $j \in \{1, ..., K_1\}$, and $k \in \{1, ..., K_2\}$. The source and target domain contains N_s and N_t images respectively, and the balancing factor λ is obtained by

$$\lambda = \sigma \left(4 \cdot sign\left(N_t - N_s\right) \left(\frac{\max(N_s, N_t)}{\min(N_s, N_t)} - 1 \right) \right). \tag{3}$$

where the $\sigma\left(\cdot\right)$ is Sigmoid function to normalize the domain scale ratio. Moreover, we impose a L2-norm regularizer to ensure the consistency between image-level and instance-level classifiers.

3.3   Training on Unlabeled Target Domain

Dynamic Clustering. UDA ReID models typically employ the clustering strategy (e.g., DBSCAN) to generate pseudo labels for the target domain instances, and employ memory-based losses [17] for metric learning. However, without ground-truth bounding boxes on target domain, the instances can be only generated from the detection results, which varies with the training process. This makes it infeasible to directly apply typical clustering approach to DAPS. To address this issue, we propose a novel dynamic clustering strategy to make full use of the detection results for continuous ReID training.

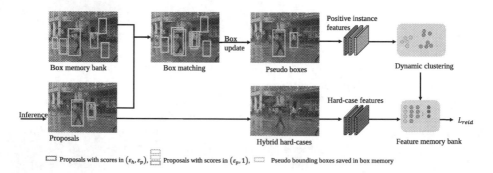

Fig. 4. Illustration of the dynamic clustering and hard case mining. At start of each epoch, we employ generated proposals, including both qualified ones and hard cases, to update the memory bank. Qualified proposals are adopted for matching pseudo boxes memory, and hard cases will directly be added.

As illustrated in Fig. 4, an asynchronized training strategy is introduced to progressively update pseudo bounding boxes with the selected proposals as ground-truth boxes on the target domain. Specifically, for the beginning α epochs, DAPS is trained only on the source dataset labeled with both bounding boxes and ID labels. After that, we maintain a bounding box memory $\mathbf{M_B} = \{B_1, ..., B_{N_t}\}$ and a feature vector memory $\mathbf{M_V} = \{V_1, ..., V_{N_t}\}$, corresponding to each of N_t target domain images. At the start of each subsequent epochs, DAPS filters out high-confidence candidate proposals $\{c_1, ..., c_m\}$ from x_i^t, and employ them to match pseudo bounding boxes in box memory $B_i = \{b_1, ..., b_n\}$ according to IOU scores. Every proposal is assigned to the most relevant box in memory if their IOU score is above the threshold, and the boxes which fail to match any qualified proposal will be removed from memory B_i. The remaining boxes in the memory are continuously updated in the Exponential Moving Average (EMA) method.

For example, suppose the proposals c_{j1}, c_{j2}, c_{j3} are mapped to the box b_k, then b_k is updated by:

$$b_k \leftarrow \gamma b_k + (1 - \gamma)\mathrm{avg}\,(c_{j1}, c_{j2}, c_{j3})\,, \tag{4}$$

where $\gamma \in [0, 1]$ controls the update rate. Eventually, the proposals without any matched box will also be fed into the memory B_i, and further, the feature memory $\mathbf{M_V}$ is updated in the same way. Afterwards, we perform clustering upon $\mathbf{M_V}$ to obtain N_t^c clusters $\{C_1, ..., C_{N_t^c}\}$ with centroids $\mathbf{W} = \{w_1, ..., w_{N_t^c}\}$, and N_t^o instances $\mathbf{F} = \{f_1, ... f_{N_t^o}\}$ not belonging to any cluster. By extracting the identity features $\mathbf{V}$ in the source domain, we eventually build a unified memory $\mathbf{M} = \{\mathbf{V}, \mathbf{W}, \mathbf{F}\}$ for ReID training. The loss function can be expressed:

$$\mathcal{L} = -\log \frac{\exp\,(x \cdot z^+/\tau)}{\sum_{k=1}^{N_t^c} \exp\,(x \cdot w_k/\tau) + \sum_{k=1}^{N_t^o} \exp\,(x \cdot f_k/\tau) + \sum_{k=1}^{N_s^c} \exp\,(x \cdot v_k/\tau)}, \tag{5}$$

where w, f, and v represents the target domain clusters, the independent instances and the source domain classes, respectively. z^+ is the corresponding class prototype of the input feature x, and $\cdot$ denotes the inner product to measure the feature similarity. The features in the memory will be updated in a momentum way during backward stage:

$$z_t \leftarrow \gamma z_t + (1 - \gamma)x, \tag{6}$$

where z_t is the t-th prototype in the memory bank $\mathbf{M}$.

Hybrid Hard Case Mining. A significant challenge for dynamic clustering is to generate reliable bounding boxes. We treat those boxes with lower confidence than a threshold as negative samples. In order to sufficiently exploit target domain information, we explore the potential of adding these "negative" samples to the ReID training. Proposals with relatively low confidence scores can be divided into highly overlapped with high-confidence boxes, the undetected persons and the background clutters. It is undesirable to enhance the ReID subtask by treating all these proposals as negative samples. As a result, we design a hierarchical scheme to categorize the candidate proposals, and employ both of the low-confidence person proposals and non-trivial background clutters to enhance the discrimination of the ReID branch.

Specifically, proposals with confidence score in the range of (ϵ_h, ϵ_p) defined by upper and lower bound thresholds are regarded as non-trivial cases. We exclude highly overlapped duplicates by further screening IOUs with positive proposals, while the hybrid of undetected persons and the negative clutters are reserved for training. The features of these hard cases will be added to $\mathbf{M}$, and be used for the contrastive learning process. The memory loss in Eq. 5 is modified as:

$$\mathcal{L} = -\log \frac{\exp\left(x \cdot z^+/\tau\right)}{\sum_{z\in\mathbf{M}} \exp\left(x \cdot z/\tau\right)},$$

$$\sum_{z\in\mathbf{M}} \exp\left(x \cdot z/\tau\right) = \sum_{k=1}^{N_t^c} \exp\left(x \cdot w_k/\tau\right) + \sum_{k=1}^{N_t^o} \exp\left(x \cdot f_k/\tau\right) \qquad (7)$$

$$+ \sum_{k=1}^{N_s^c} \exp\left(x \cdot v_k/\tau\right) + \sum_{k=1}^{N_t^n} \exp\left(x \cdot h_k/\tau\right),$$

where h denotes the hybrid hard cases. It is noteworthy that the hybrid hard cases will be involved into the dynamic clustering before the next epoch. Once a hard case is matched with new qualified proposals, it will be treated as a positive sample and updated in a momentum way.

Target Detection Training. Although DAM can minimize the domain discrepancy, the over-fitting towards the source domain is still likely to take place, especially when the source domain data is extremely less complex and comprehensive than the target domain images. To this end, simultaneously training detection with both of the source and the target domain data is beneficial for the generalization ability of model. DAM and dynamic clustering provide relatively reliable pseudo bounding boxes, and specifically, we employ such pseudo bounding boxes after the α epoch to supervise detection on the target domain. In this way, the potential of unlabeled target domain images is released for both ReID and detection training.

4   Experiment

4.1   Datasets and Evaluation Protocols

Datasets. We employ two large-scale benchmark datasets, CUHK-SYSU [39] and PRW [44] in our experiments. CUHK-SYSU is one of the largest public datasets for person search, composed of 18,184 images and 96,143 bounding boxes from 8,432 different identities. It is divided into a training set of 11,206 images with 5,532 identities, and a test set with 6,978 gallery images and 2,900 query images. The widely used PRW dataset contains 11,816 images, 43,110 annotated bounding boxes from 932 identities. The training set includes 5,704 images and 482 labelled persons, while the other 6,112 images and 2,057 probe persons from 450 identities are adopted as test set.

Evaluation Protocols. Our experiments employ the default splits for both datasets. For domain adaptation settings, the annotations of dataset used as the source domain is accessible, while neither bounding boxes nor identity labels of datasets as the target domain are available. All evaluations are performed on the test set of target domain. We adopt the widely used mean average precision (mAP) and cumulative matching characteristic (CMC) top-1 accuracy as evaluation metrics for ReID sub-task, while average precision (AP) and recall rate are adopted as the metrics for detection.

Table 1. Comparative results when combining different components. DAM: Domain Alignment Module. DC: Dynamic Clustering. HM: Hybrid hard case Mining. DTD: Detection on Target Domain.

DAM	DC	HM	DTD	Target: PRW				Target: CUHK-SYSU			
				mAP	top-1	recall	AP	mAP	top-1	recall	AP
✗	✗	✗	✗	30.3	77.7	94.0	88.3	52.5	54.8	55.2	55.1
✓	✗	✗	✗	30.9	79.3	96.3	90.7	62.2	63.6	70.8	63.1
✗	✓	✗	✗	32.2	79.4	96.8	90.3	70.9	72.3	67.8	62.2
✓	✓	✗	✗	32.7	79.6	95.9	90.4	72.6	74.3	68.3	63.2
✓	✓	✓	✗	34.5	**80.7**	97.0	91.0	73.2	74.8	70.4	64.1
✓	✓	✗	✓	33.1	79.9	96.6	**91.2**	76.8	78.7	**79.4**	**71.1**
✓	✓	✓	✓	**34.7**	80.6	**97.2**	90.9	**77.6**	**79.6**	77.7	69.9

Table 2. Comparative results of task-sensitive instance-level alignment.

Instance da	Target: PRW				Target: CUHK-SYSU			
	mAP	top-1	recall	AP	mAP	top-1	recall	AP
Normal	21.7	76.0	**96.7**	**91.1**	58.2	60.5	66.3	56.3
Task-sensitive	**30.9**	**79.3**	96.3	90.7	**62.2**	**63.6**	**70.8**	**63.1**

4.2 Implementation Details

We adopt ResNet50 [22] pretrained on ImageNet-1k [9] as our default backbone network. DBSCAN [14] with self-paced learning strategy [25] is employed as the basic clustering method, we set default hyper-parameters $\epsilon_p = 0.95$, $\epsilon_h = 0.8$ and $\lambda_t = 0.1$. During training, the input images are resized to 1500×900, and random horizontal flip is applied for data augmentation. Our model is optimized by Stochastic Gradient Descent (SGD) for 20 epochs. We set a mini-batch size of 4, and an initial learning rate of 0.0024, which is reduced by a factor of 0.1 at epoch 16 with warmed up in the first epoch. The momentum and weight decay are set to 0.9 and 5×10^{-4}, respectively. We set the momentum factor γ for memory updating to 0.2. The starting epoch of α is set to 8 when PRW is chosen as target domain, and 0 for CUHK-SYSU. All experiments are implemented with one NVIDIA Tesla A100 GPU. We also plan to support this project with MindSpore in our future work.

4.3 Ablation Study

We perform analytical experiments to verify the effectiveness of each detailed component in our proposed framework. In Table 1, we compare the baseline method with different combinations of proposed components, and report the results on both CUHK-SYSU and PRW datasets. For example, when we use CUHK-SYSU as the target domain dataset, the directly transferring baseline model achieves 52.5% mAP and 54.8% top-1. After individually adding the domain adaptive module (DAM) and dynamic clustering (DC), the performance improves 9.3% and 18.4% in terms of mAP. When combining DAM and DC, the mAP is further promoted to 72.6%, surpassing the 52.5% of baseline by a large margin. Furthermore, to make full use of the unlabeled target data, we implement the hybrid hard case mining (HM) and detection on target domain (DTD). HM improves the ReID performance by 0.6% in mAP, and DTD prominently enhances the detection branch with a 7.0% gain for AP. Eventually, DAPS achieves 77.6% mAP and 79.6%top-1 with all designed modules, outperforming the baseline by 25.1% in mAP, 24.8% in top-1, 22.5% in recall, and 14.8% in AP.

Table 3. Comparative results when employing different strategies to handle the lack of bounding boxes. 'GT' refers to using the ground truth bounding boxes for all the training process of ReID, and 'GT for init' only employs these boxes to initialize the memory bank. 'static' means directly employing the qualified proposals before each epoch.

Strategy	Target: PRW				Target: CUHK-SYSU			
	mAP	top-1	recall	AP	mAP	top-1	recall	AP
GT	34.9	79.9	94.9	89.5	73.6	76.0	74.6	68.2
GT for init	33.5	79.6	92.9	88.5	73.5	75.4	64.4	60.8
Static	25.3	77.3	96.6	90.8	64.0	66.1	67.6	62.5
Dynamic update	32.7	79.6	95.9	90.4	72.6	74.3	68.3	63.2

Table 4. Comparative results of when to start asynchronized training.

Starting epoch	Target: PRW				Target: CUHK-SYSU			
	mAP	top-1	recall	AP	mAP	top-1	recall	AP
0	31.5	79.7	95.8	89.4	**77.6**	**79.6**	**77.7**	**69.9**
4	31.4	79.4	95.8	89.1	73.6	75.3	76.6	67.7
8	**34.7**	**80.6**	97.2	**90.9**	73.2	74.7	76.7	69.0
10	33.4	80.6	**97.5**	90.7	71.4	73.3	74.8	65.8

Effectiveness of Task-Sensitive Instance-Level Alignment. To validate the effectiveness of our task-sensitive instance-level alignment design, we compare it with normal instance-level alignment, which conducts instance alignment on both head networks without balancing between them. As observed in Table 2, the task-sensitive design successfully alleviates the inner task conflicts and outperforms normal strategy by a large margin.

Effectiveness of Dynamic Clustering. As aforementioned, the key to utilizing unlabeled target domain data is generating reliable pseudo bounding boxes. To validate the quality of the pseudo bounding boxes we use, we compare different strategies of obtaining bounding boxes, and the results are reported in Table 3. We first measure the performance achieved by using ground-truth bounding boxes for training the ReID task. Furthermore, we report the performance achieved by directly employing the qualified proposals before each epoch, which is denoted as 'static' in Table 3. The results reveal that our proposed dynamic clustering strategy can generate trustworthy pseudo bounding boxes to achieve comparable performance with using ground-truth boxes.

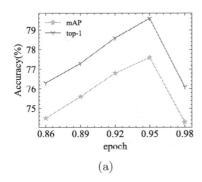 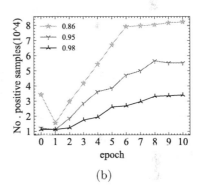

| (a) | (b) |

Fig. 5. Target domain performance with different ϵ_p on CUHK-SYSU dataset. (a): ReID accuracy results; (b): Numbers of generated positive proposals. The ground truth instances number is 55,260.

Effectiveness of Asynchronized Training. We conduct experiments for influences by the training stage hyper-parameter α on final performance. As shown in Table 4, when PRW is adopted as the target domain, the best performance is achieved with $\alpha = 8$, while with $\alpha = 0$ for CUHK-SYSU. The results might be counterintuitive but indeed validate our task-sensitive motivation. For smaller source dataset, even limited additional target information might be helpful for cross-domain generalization. In contrast, for larger source dataset, unreliable target proposals can be harmful for domain gap bridging.

Analysis on Hyper-Parameter ϵ_p. We visualize the influence of hyper-parameter ϵ_p in Fig. 5. We observe that the value of ϵ_p influences the ReID performance to a large extend, and the best performance is achieved with $\epsilon_p = 0.95$. From Fig. 5b, it can be observed that the selection of ϵ_p is a trade-off between recall rate and proposal quality. Setting it to a extremely high value leads to discarding useful proposals, while a low threshold will induct clutters to undermine the quality of clustering.

4.4   Comparison with State-of-the-Art Methods

Since no existing person search methods with such domain adaptation settings can be directly compared, we further compare DAPS with fully supervised methods in Table 5, including both of the two-step methods and one-step ones. It is surprising that our framework even surpasses some supervised methods. For example, DAPS outperforms MGTS [5], OIM [39], IAN [38], NPSM [29] and CTXGraph [43] on PRW. The comparison with the state-of-the-art fully supervised methods indicate that there exists a large performance gap, and we hope our work will encourage more explorations for this setting. Moreover, to make measure the theoretical upper limit of DAPS setting, we train some state-of-the-art method with both datasets in a supervised manner, and more details are described in the supplementary material.

Table 5. Comparison with fully supervised person search models

Method	PRW		CUHK-SYSU	
	mAP	top-1	mAP	top-1
DPM [18]	20.5	48.3	-	-
MGTS [5]	32.6	72.1	83.0	83.7
RDLR [21]	42.9	70.2	93.0	94.2
IGPN [13]	47.2	87.0	90.3	91.4
TCTS [34]	46.8	87.5	93.9	95.1
OIM [39]	21.3	49.9	75.5	78.7
IAN [38]	23.0	61.9	76.3	80.1
NPSM [29]	24.2	53.1	77.9	81.2
CTXGraph [43]	33.4	73.6	84.1	86.5
QEEPS [30]	37.1	76.7	88.9	89.1
HOIM [4]	39.8	80.4	89.7	90.8
BINet [12]	45.3	81.7	90.0	90.7
NAE [6]	44.0	81.1	92.1	92.9
AlignPS [41]	45.9	81.9	93.1	93.4
SeqNet [26]	46.7	83.4	93.8	94.6
DAPS (ours)	**34.7**	**80.6**	**77.6**	**79.6**

The comparisons with existing weakly supervised methods are shown in Table 6, and we also present the results of training R-SiamNet with both datasets in the weakly supervised manner. When evaluated on the PRW dataset, DAPS outperforms all existing weakly supervised methods by a significant margin. For the CUHK-SYSU dataset, DAPS still underperforms the state-of-the-art weakly supervised models, which is mainly caused by the limitation brought by detection capabilities. As mentioned in Sect. 4.1, the images and identities in PRW are prominently fewer than those in CUHK-SYSU, and this further leads to the poor detection performance of adopting CUHK-SYSU as target domain.

4.5 Qualitative Results

To better illustrate the distributions of our hybrid hard cases, we visualize some qualitative results from both datasets in Fig. 6. As is observed, the hybrid hard cases consist of undetected persons (column a), highly overlapped human crops (column b) and background clutters(column c, d). These qualitative results demonstrate the diversity of our hybrid hard cases, and validate the rationality of adding such cases to the memory bank.

Table 6. Comparison with weakly supervised person search models. * denotes training R-SiamNet together with both of CUHK-SYSU and PRW.

Method	PRW		CUHK-SYSU	
	mAP	top-1	mAP	top-1
CGPS [40]	16.2	68.0	80.0	82.3
R-SiamNet [20]	21.4	75.2	86.0	87.1
R-SiamNet* [20]	23.5	76.0	86.2	87.6
DAPS (ours)	**34.7**	**80.6**	77.6	79.6

(a)   (b)   (c)   (d)   (a)   (b)   (c)   (d)

Fig. 6. Visualization of some hard cases, the green bounding boxes denote the qualified proposals, while the red ones denote the undetected persons. The crops of the hybrid hard cases are presented on the right of the images. (Color figure online)

5 Conclusions

In this paper, we introduce a novel Domain Adaptive Person Search setting, where neither bounding boxes nor identity labels for target domain are required.

Based on this new setting, we propose a strong baseline framework by investigating domain alignment and taking advantage of unlabeled target domain data. Extensive results on two large-scale benchmarks demonstrate the promising performance our framework achieves and the effectiveness of designed modules. We hope this work will encourage more exploration in this direction.

Acknowledgment. This work was supported by Shanghai Municipal Science and Technology Major Project (2021SHZDZX0102), CAAI-Huawei MindSpore Open Fund.

References

1. Arruda, V.F., et al.: Cross-domain car detection using unsupervised image-to-image translation: from day to night. In: IJCNN, pp. 1–8 (2019)
2. Cai, Q., Pan, Y., Ngo, C., Tian, X., Duan, L., Yao, T.: Exploring object relation in mean teacher for cross-domain detection. In: CVPR, pp. 11457–11466 (2019)
3. Cao, Y., Guan, D., Huang, W., Yang, J., Cao, Y., Qiao, Y.: Pedestrian detection with unsupervised multispectral feature learning using deep neural networks. Inf. Fusion **46**, 206–217 (2019)
4. Chen, D., Zhang, S., Ouyang, W., Yang, J., Schiele, B.: Hierarchical online instance matching for person search. In: AAAI, pp. 10518–10525 (2020)
5. Chen, D., Zhang, S., Ouyang, W., Yang, J., Tai, Y.: Person search via a mask-guided two-stream CNN model. In: Ferrari, V., Hebert, M., Sminchisescu, C., Weiss, Y. (eds.) ECCV 2018. LNCS, vol. 11211, pp. 764–781. Springer, Cham (2018). https://doi.org/10.1007/978-3-030-01234-2_45
6. Chen, D., Zhang, S., Yang, J., Schiele, B.: Norm-aware embedding for efficient person search. In: CVPR, pp. 12612–12621 (2020)
7. Chen, Y., Zhu, X., Gong, S.: Instance-guided context rendering for cross-domain person re-identification. In: ICCV, pp. 232–242 (2019)
8. Chen, Y., Li, W., Sakaridis, C., Dai, D., Gool, L.V.: Domain adaptive faster R-CNN for object detection in the wild. In: CVPR, pp. 3339–3348 (2018)
9. Deng, J., Dong, W., Socher, R., Li, L., Li, K., Fei-Fei, L.: ImageNet: a large-scale hierarchical image database. In: CVPR, pp. 248–255 (2009)
10. Deng, W., Zheng, L., Ye, Q., Kang, G., Yang, Y., Jiao, J.: Image-image domain adaptation with preserved self-similarity and domain-dissimilarity for person re-identification. In: CVPR, pp. 994–1003 (2018)
11. Devaguptapu, C., Akolekar, N., Sharma, M.M., Balasubramanian, V.N.: Borrow from anywhere: pseudo multi-modal object detection in thermal imagery. In: CVPR Workshops, pp. 1029–1038 (2019)
12. Dong, W., Zhang, Z., Song, C., Tan, T.: Bi-directional interaction network for person search. In: CVPR, pp. 2836–2845 (2020)
13. Dong, W., Zhang, Z., Song, C., Tan, T.: Instance guided proposal network for person search. In: CVPR, pp. 2582–2591 (2020)
14. Ester, M., Kriegel, H., Sander, J., Xu, X.: A density-based algorithm for discovering clusters in large spatial databases with noise. In: KDD, pp. 226–231 (1996)
15. Fu, Y., Wei, Y., Wang, G., Zhou, Y., Shi, H., Huang, T.S.: Self-similarity grouping: a simple unsupervised cross domain adaptation approach for person re-identification. In: ICCV, pp. 6111–6120 (2019)
16. Ganin, Y., Lempitsky, V.S.: Unsupervised domain adaptation by backpropagation. In: ICML. JMLR Workshop and Conference Proceedings, vol. 37, pp. 1180–1189 (2015)

17. Ge, Y., Zhu, F., Chen, D., Zhao, R., Li, H.: Self-paced contrastive learning with hybrid memory for domain adaptive object Re-ID. In: NeurIPS (2020)
18. Girshick, R.B., Iandola, F.N., Darrell, T., Malik, J.: Deformable part models are convolutional neural networks. In: CVPR, pp. 437–446 (2015)
19. Goodfellow, I.J., et al.: Generative adversarial nets. In: NIPS, pp. 2672–2680 (2014)
20. Han, C., et al.: Weakly supervised person search with region Siamese networks. In: ICCV, pp. 12006–12015 (2021)
21. Han, C., et al.: Re-ID driven localization refinement for person search. In: ICCV, pp. 9813–9822 (2019)
22. He, K., Zhang, X., Ren, S., Sun, J.: Deep residual learning for image recognition. In: CVPR, pp. 770–778 (2016)
23. Kang, G., Jiang, L., Yang, Y., Hauptmann, A.G.: Contrastive adaptation network for unsupervised domain adaptation. In: CVPR, pp. 4893–4902 (2019)
24. Khodabandeh, M., Vahdat, A., Ranjbar, M., Macready, W.G.: A robust learning approach to domain adaptive object detection. In: ICCV, pp. 480–490 (2019)
25. Kumar, M.P., Packer, B., Koller, D.: Self-paced learning for latent variable models. In: NIPS, pp. 1189–1197 (2010)
26. Li, Z., Miao, D.: Sequential end-to-end network for efficient person search. In: AAAI, pp. 2011 2019 (2021)
27. Lin, C.: Cross domain adaptation for on-road object detection using multimodal structure-consistent image-to-image translation. In: ICIP, pp. 3029–3030 (2019)
28. Liu, C., Chang, X., Shen, Y.: Unity style transfer for person re-identification. In: CVPR, pp. 6886–6895 (2020)
29. Liu, H., et al.: Neural person search machines. In: ICCV, pp. 493–501 (2017)
30. Munjal, B., Amin, S., Tombari, F., Galasso, F.: Query-guided end-to-end person search. In: CVPR, pp. 811 820 (2019)
31. Saito, K., Ushiku, Y., Harada, T., Saenko, K.: Strong-weak distribution alignment for adaptive object detection. In: CVPR, pp. 6956–6965 (2019)
32. Song, L., et al.: Unsupervised domain adaptive re-identification: theory and practice. Pattern Recognit. **102**, 107173 (2020)
33. Tzeng, E., Hoffman, J., Saenko, K., Darrell, T.: Adversarial discriminative domain adaptation. In: CVPR, pp. 2962–2971 (2017)
34. Wang, C., Ma, B., Chang, H., Shan, S., Chen, X.: TCTS: a task-consistent two-stage framework for person search. In: CVPR, pp. 11949–11958 (2020)
35. Wang, D., Zhang, S.: Unsupervised person re-identification via multi-label classification. In: CVPR, pp. 10978–10987 (2020)
36. Wang, M., Deng, W.: Deep visual domain adaptation: a survey. Neurocomputing **312**, 135–153 (2018)
37. Wang, T., Zhang, X., Yuan, L., Feng, J.: Few-shot adaptive faster R-CNN. In: CVPR, pp. 7173–7182 (2019)
38. Xiao, J., Xie, Y., Tillo, T., Huang, K., Wei, Y., Feng, J.: IAN: the individual aggregation network for person search. Pattern Recogn. **87**, 332–340 (2019)
39. Xiao, T., Li, S., Wang, B., Lin, L., Wang, X.: Joint detection and identification feature learning for person search. In: CVPR, pp. 3376–3385 (2017)
40. Yan, Y., et al.: Exploring visual context for weakly supervised person search. In: AAAI, vol. 36, pp. 3027–3035 (2022)
41. Yan, Y., et al.: Anchor-free person search. In: CVPR, pp. 7690–7699 (2021)
42. Yan, Y., Li, J., Liao, S., Qin, J., Ni, B., Yang, X.: TAL: two-stream adaptive learning for generalizable person re-identification. CoRR abs/2111.14290 (2021)
43. Yan, Y., Zhang, Q., Ni, B., Zhang, W., Xu, M., Yang, X.: Learning context graph for person search. In: CVPR, pp. 2158–2167 (2019)

44. Zheng, L., Zhang, H., Sun, S., Chandraker, M., Yang, Y., Tian, Q.: Person re-identification in the wild. In: CVPR, pp. 3346–3355 (2017)
45. Zhu, X., Pang, J., Yang, C., Shi, J., Lin, D.: Adapting object detectors via selective cross-domain alignment. In: CVPR, pp. 687–696 (2019)

TS2-Net: Token Shift and Selection Transformer for Text-Video Retrieval

Yuqi Liu[1,2] (ID), Pengfei Xiong[2], Luhui Xu[2], Shengming Cao[2] (ID),
and Qin Jin[1(✉)] (ID)

[1] School of Information, Renmin University of China, Beijing, China
{yuqi657,qjin}@ruc.edu.cn
[2] Tencent, Shenzhen, China
devancao@tencent.com,xiongpengfei2019@gmail.com,luhuixu.cn@gmail.com

Abstract. Text-Video retrieval is a task of great practical value and has received increasing attention, among which learning spatial-temporal video representation is one of the research hotspots. The video encoders in the state-of-the-art video retrieval models usually directly adopt the pre-trained vision backbones with the network structure fixed, they therefore can not be further improved to produce the fine-grained spatial-temporal video representation. In this paper, we propose Token Shift and Selection Network (TS2-Net), a novel token shift and selection transformer architecture, which dynamically adjusts the token sequence and selects informative tokens in both temporal and spatial dimensions from input video samples. The token shift module temporally shifts the whole token features back-and-forth across adjacent frames, to preserve the complete token representation and capture subtle movements. Then the token selection module selects tokens that contribute most to local spatial semantics. Based on thorough experiments, the proposed TS2-Net achieves state-of-the-art performance on major text-video retrieval benchmarks, including new records on MSRVTT, VATEX, LSMDC, ActivityNet, and DiDeMo. Code is available at https://github.com/yuqi657/ts2_net.

Keywords: Text-video retrieval · Token shift · Token selection

1 Introduction

With advanced digital technologies, massive amount of videos are generated and uploaded online everyday. Searching for target videos based on users' text queries is a task of great practical value and has attracted increasing research attention. Over the past years, different text-video benchmarks have been established [2, 10,

---

Y. Liu—This work is done when Yuqi is an intern at Tencent.

Supplementary Information The online version contains supplementary material available at https://doi.org/10.1007/978-3-031-19781-9_19.

Two people playing basketball and the one with a *hat* makes every shot.

A guy wearing a red shirt drives a car while *talking*.

Fig. 1. The text-video retrieval examples that require fine-grained video representation. Left: the small object 'hat' is important for correctly retrieving the target video. Right: the subtle movement of 'talking' is crucial for the correct retrieval of the target video. Green boxes depict the positive video result, while red boxes are negative candidates (Color Figure Online)

25,40,44,46] and various text-video retrieval approaches have been proposed [11, 17,21,30,31,33], which usually formulate the task as a learning and matching task based on a similarity function between the text query and candidate videos in the corpus. With the success of deep neural networks [9,20,45], deep learned features have replaced manually-designed features. A text-video retrieval engine is generally composed of a text encoder and a video encoder, which maps the text query and the video candidate to the same embedding space, where the similarity can be easily computed using a distance metric.

Building a powerful video encoder to produce spatial-temporal feature encoding for videos, that can simultaneously capture motion between video frames, as well as entities in video frames, has been one of the research focuses for text-video retrieval in recent years [3,29,32]. Lately, Transformer has become the dominant visual encoder architecture, and it enables the training of video-language models with raw video and text data [4,12,19,34]. Various video transformers [3,5,8,32], considering both spatial and temporal representations, have achieved superior performance on major benchmarks. However, these models still lack fine-grained representation capacity in either spatial or temporal dimension. For example, the video encoder in models [12,19,34] normally consists of a single-frame feature extraction module followed by a global feature aggregation module, which lacks fine-grained interaction between adjacent frames and only aggregates the frame-level semantic information. Although the video encoder in Frozen [4] employs divided space-time attention, it uses only one [CLS] token as the video representation, failing to capture the find-grained spatial-temporal details. In general, all these models can effectively represent obvious motions and categorical spatial semantics in the video, but still lack the capacity for subtle movement and small objects. They will fail in cases such as illustrated in Fig. 1, where the video encoder needs to capture the small object ('hat') and subtle movement ('talking') in order to retrieve the correct target videos.

Based on the structure of video transformer, video sequence is spatially and temporally divided into consecutive patches. To enhance modeling of small objects and subtle movements, patch enhancement is an intuitive and straightforward solution. This motivates us to find a feasible way to incorporate spatial-temporal patch contexts into encoded features. The shift operation is introduced in TSM [29], which shifts parts of the channel along temporal dimension. Shift Transformer [50] applies shift in visual transformer to enhance temporal modeling. However, the architecture of transformer is different from CNN, such partial shift operation damages the completeness of each token representation.

Therefore, in this paper, we propose TS2-Net, a novel token shift and selection transformer network, to realize local patch feature enhancement. Specifically, we first adopt the token shift module in TS2-Net, which shifts the whole spatial token features back-and-forth across adjacent frames, in order to capture local movement between frames. We then design a token selection module to select top-K informative tokens to enhance the salient semantic feature modeling capability. Our token shift module treats the features of each token as a whole, and iteratively swaps token features at the same location with neighbor frames, to preserve the complete local token representation and capture local temporal semantics at the same time. The token selection module estimates the importance of each token feature of patches with a selection network, which relies on the correlation between all spatial-temporal patch features and [CLS] tokens. It then selects tokens which contributes most to local spatial semantics. Finally, we align cross-modal representation in a fine-grained manner, where we calculate the similarity between text and each frame-wise video embedding and aggregate them together. TS2-Net is optimized with video-language contrastive learning.

We conduct extensive experiments on several text-video retrieval benchmarks to evaluate our model, including MSRVTT, VATEX, LSMDC, ActivityNet, and DiDeMo. Our proposed TS2-Net achieves the state-of-the-art performance on most of the benchmarks. The ablation experiments demonstrate that the proposed token shift and token selection modules both improve the fine-grained text-video retrieval accuracy. The main contributions of this work are as follows:

- We propose a new perspective of video-language learning with local patch enhancements to improve the text-video retrieval.
- We introduce two modules, token shift transformer and token selection transformer, to better model video representation temporally and spatially.
- We report new records of retrieval accuracy on several text-video retrieval benchmarks. Thorough ablation studies demonstrate the merits of our patch enhancement concept.

2 Related Work

2.1 Video Retrieval

Various approaches have been proposed to deal with text-video retrieval task, which usually consist of off-line feature extractors and feature fusion module [11,14,17,21,30,31,43,48]. MMT [21] uses a cross-modal encoder to aggregate feature extracted by different experts. MDMMT [17] further utilizes knowledge learned from multi-domain datasets. Recent works [4,12,19,26,34] attempt to train text-video model in an end-to-end manner. ClipBERT [26] is the pioneering end-to-end text-video pretrain model. Its promising results show that jointly train high-level semantic alignment network with low-level feature extractor is beneficial. CLIP4Clip [34] and CLIP2Video [19] transfer knowledge from pretrained CLIP [37] to video retrieval task. However, these models still lack fine-grained representation capacity in either spatial or temporal dimension. Different from previous works, we aim to model fine-grained spatial and temporal information to enhance text-video retrieval.

2.2   Visual-Language Pre-training

Viusal-language pre-training models has shown promising results in visual-and-language tasks such as image retrieval, image caption and video retrieval. In works such as Unicoder-VL [27], VL-BERT [41] and VLP [51], text and visual sequence are input into a shared transformer encoder. In Hero [28], ClipBERT [26] and Univl [33], text and visual sequence are encoded independently, then a cross-encoder is used to fuse different modality. While in Frozen [4], CLIP [37], text and visual sequence are encoded independently and a contrastive loss is used to align text and visual embedding. Our work use the two-stream structure, where text feature and video feature are encoded independently, then a cross-modal contrastive loss is used to align them.

2.3   Video Representation Learning

Early works use 2D or 3D-CNN to encode video feature [9,20,29]. Recently, Visual Transformer(ViT) [16] has shown great potential in image modeling. Many works attempt to transfer ViT into video domain [3,5,8,32]. TimeSformer [5] and ViViT [3] propose variants of spatial-temporal video transformer. There are several works exploring shift operation to enable 2D network learn temporal information, including TSM [29] and Shift Transformer [50]. They shift parts of the channel along the temporal dimension. Different from previous work, we consider token shift operation, which we shift all channels of selected visual tokens to the temporal dimension rather than partial shift (i.e. shift some channels). Token selection has been used to reduce redundancy problem in transformer based visual model. Dynamic ViT [39] and STTS [42] use token selection for efficiency. Perturbed maximum is proposed in [6] to make top-K differentiable. Based on differential top-K [13], our work designs a light-weight token selection module to select informative tokens for effective temporal-spatial modeling.

3   Method

The goal of text-video retrieval is to find the best matching videos based on the text query. Figure 2 illustrates the overall structure of the proposed TS2-Net model for the text-video retrieval task, which consists of three key components:

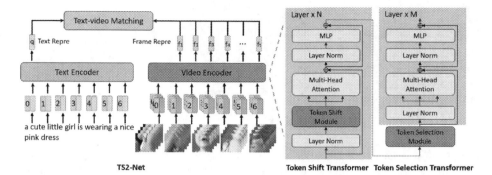

Fig. 2. Overview of the proposed TS2-Net model for text-video retrieval, which consists of three key components: the text encoder, the video encoder, and the text-video matching. The video encoder is composed of the Token Shift Transformer and Token Selection Transformer. ('Repre' is short for 'Representation')

the text encoder, the video encoder, and the text-video matching. The text encoder encodes the sequence of query words into a query representation q. In this paper, we use GPT [38] model as the text encoder. By adding a special token [EOS] at the end of query word sequence, we employ the encoding of [EOS] by the GPT encoder as the query representation q. The video encoder encodes the sequence of video frames into a sequence of frame-wise video representation $v = \{f_1, f_2, \ldots, f_t\}$. Based on the query and video representation, q and v, the text-video matching computes the cross-modal similarity between the query and video candidate. In following sections, we first elaborate the core ingredients of our video encoder, namely the token shift transformer (Sect.3.1) and the token selection transformer (Sect.3.2), and finally present our text-video matching strategy in details (Sect.3.3).

3.1 Token Shift Transformer

Token shift transformer is based on Vision Transformer (ViT) [16]. It inserts a token shift module in the transformer block. Let's review ViT model first, and then describe our modification to ViT. Given an image I, ViT first splits I into N patches $\{p_0, p_1, \ldots, p_{n-1}\}$. To eliminate ambiguity, we use *token* to represent *patch* below. After adding a [CLS] token p_{cls}, the token sequence $\{p_{cls}, p_0, p_1, \ldots, p_{n-1}\}$ is fed into a stack of transformer blocks. Then the image embedding is generated by either averaging all the visual tokens or using the [CLS] token p_{cls}. In this work, we use p_{cls} as the image embedding. Token shift transformer aims to effectively model subtle movements in a video. The proposed token shift operation is a parameter-free operation, as illustrated in Fig. 3. Suppose we have a video $V \in \mathbb{R}^{T \times N \times C}$, where T represents the number of frames, N refers to the number of tokens per frame, and C represents the feature dimension. We feed T frames into ViT to encode frame feature. In certain ViT layer, we shift some tokens from adjacent frames to the current frame to exchange

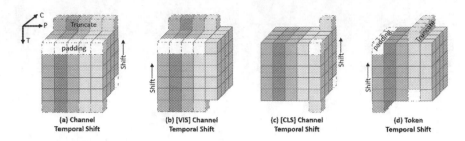

Fig. 3. Illustration of different types of Shift operation and our proposed Token Temporal Shift. 'T, P, C' refer to video temporal dimension, video token, and feature channel respectively. Each vertical cube group represents a spatial-temporal video token. Cubes with dash line represent tensor truncated, and white cubes represent tensor padding. In Shift-Transformer [50], tokens are shifted along the channel dimension, while our proposed Token Shift Module does not compromise the integrity of a video token

information of adjacent frames. Note that we use a bi-directional token shift in our implementation. By token shift operation across adjacent frames, our model is able to capture subtle movements in the local temporal interval.

Shift-Transformer [50] has also explored several shift variants on the visual transformer architecture. Figure 3 visualizes the difference between these shift variants and our proposed token shift. A naive channel temporal shift swaps part of channels of a frame tensor along temporal dimension, as shown in Fig. 3(a). Shift-Transformer [50] also presents [VIS] channel temporal shift and [CLS] channel temporal shift, as shown in Fig. 3(b)(c). They fix tensor in token dimension and shift parts of channels for chosen token along the temporal dimension. Different from these works, our token shift transformer emphasizes the token dimension, where we shift whole channels of a token back-and-forth across adjacent frames, as shown in Fig. 3(d). We believe our token shift is better for ViT architecture, because different from the CNN architecture, each token in ViT is independent and contains unique spatial information with respect to its location. Thus shifting parts of channels destroys the integrity of the information contained in a token. On the contrast, shifting a whole token with all channels can preserve complete information contained in a token and enable cross-frame interaction.

However, if we shift most of the tokens in every ViT layer, it damages the spatial modeling ability, and the information contained in these tokens is no longer accessible in the current frame. We therefore use a residual connection between original feature and token shift feature, as illustrated in Fig. 2. In addition, we assume that shallow layers are more important to model spatial features, so shifting in shallow layers could harm spatial modeling. We thus choose to apply token shift operation only in deeper layers in our implementation.

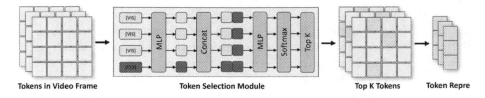

Fig. 4. Illustration of token selection module. Top-K informative tokens are selected per frame from original spatial-temporal tokens for following feature aggregation

3.2   Token Selection Transformer

Aggregating information from each frame is a necessary step in building the video representation. A naive solution to aggregate per-frame information is by adding some temporal transformer layers, or by mean pooling as CLIP4Clip [34]. We argue that aggregation with only the [CLS] token leads to missing important spatial information (i.e. some objects). An alternative way is using all tokens from all frames to aggregate information, but this introduces redundancy problem, leading to the pitfall of some background tokens with irrelevant information dominating the final video representation.

In this work, we propose the token selection transformer by inserting a token selection module, which aims to select informative tokens per frame, especially those tokens containing salient semantics of objects, for video feature aggregation. As shown in Fig. 4, top-K informative tokens are selected via the trainable token selection module every frame.

The input of the token selection module is a sequence of tokens of each frame $I = \{p_{cls}, p_0, p_1, \ldots, p_{n-1}\} \in \mathbb{R}^{(N+1) \times C}$. We first apply an MLP over I for channel dimension reduction and output $I' = \{p'_{cls}, p'_0, p'_1, \ldots, p'_{n-1}\} \in \mathbb{R}^{(N+1) \times \frac{C}{2}}$. We then use p'_{cls} as a global frame feature and concatenate it with each local token p'_i, $\hat{p}_i = [p'_{cls}, p'_i], 0 \leq i < N$. We finally feed all the concatenated token features to another MLP followed by a Softmax layer to predict the importance scores, which can be formulated as:

$$S = \mathrm{Softmax}(\mathrm{MLP}(\hat{p})) \in \mathbb{R}^{(N+1)}. \tag{1}$$

We select indices of K most informative tokens based on S, denoting as $M \in \{0, 1\}^{(N+1) \times K}$, where each column in M is a one-hot $(N+1)$ dimensional indicator. We extract top-K most informative tokens by:

$$\hat{I} = M^T I, \tag{2}$$

After top-K token select on every frame, we input the selected tokens from all frames to a joint spatial-temporal transformer, to learn global spatial-temporal video representation. We also pick the most informative token from each frame as the frame-wise video encoding.

Differentiable TopK. Until now, both top-K operation and one-hot operation are non-differentiable. To make token selection module differentiable, we employ the perturbed maximum method proposed in [6]. Specifically, a discrete optimization problem with input $\boldsymbol{S} \in \mathbb{R}^{(N+1)}$ ($\boldsymbol{S}$ is the importance score matrix in Eq. 1) and optimization variable $\mathbf{M} \in \mathbb{R}^{(N+1) \times K}$ ($\mathbf{M}$ is the index indicator matrix in Eq. 2) can be formulated as:

$$F(\boldsymbol{S}) = \max_{\mathbf{M} \in \mathcal{C}} \langle \mathbf{M}, \boldsymbol{S} \rangle, \mathbf{M}^*(\boldsymbol{S}) = \arg\max_{\mathbf{M} \in \mathcal{C}} \langle \mathbf{M}, \boldsymbol{S} \rangle, \tag{3}$$

where $F(\boldsymbol{S})$ represents the top-K selection operation, $\mathbf{M}^*(\boldsymbol{S})$ represents the optimal value. Based on Eq. 3, we can select top-K informative tokens by $F(\boldsymbol{S})$. We calculate forward and backward pass following [1,13].

3.3   Text-Video Matching

The similarity between the text query and video candidate is computed by integrating the similarity between the query and each video frame. To be specific, given the query representation q and a sequence of frame-wise video representation $v = \{f_1, f_2, ..., f_t\}$, we compute the frame-level similarity as follows:

$$s_i^* = \frac{q \cdot f_i}{\|q\| \, \|f_i\|}. \tag{4}$$

The final text-video matching similarity is defined as the weighted combination of frame-level similarities:

$$s = \sum_{i=1}^{n} \alpha_i s_i, \tag{5}$$

where $\alpha_i = \frac{\exp(\lambda s_i)}{\sum_{i=1}^{n} \exp(\lambda s_i)}$ and λ is a temperature parameter. We set λ as 4 empirically in our experiments.

Symmetric cross-entropy loss is adopted as our training objective function. For each training step with B text-video pairs, we calculate symmetric cross-entropy loss as follows:

$$\mathcal{L}_t^{t2v} = -\frac{1}{B} \sum_{i}^{B} \log \frac{\exp\left(\tau \cdot \text{sim}\left(q_i, v_i\right)\right)}{\sum_{j=1}^{B} \exp\left(\tau \cdot \text{sim}\left(q_i, v_j\right)\right)}, \tag{6}$$

$$\mathcal{L}_t^{v2t} = -\frac{1}{B} \sum_{i}^{B} \log \frac{\exp\left(\tau \cdot \text{sim}\left(q_i, v_i\right)\right)}{\sum_{j=1}^{B} \exp\left(\tau \cdot \text{sim}\left(q_j, v_i\right)\right)}, \tag{7}$$

$$\mathcal{L} = \frac{1}{2} \left(\mathcal{L}_{t2v} + \mathcal{L}_{v2t}\right), \tag{8}$$

where τ is a trainable scaling parameter and $\text{sim}\,(q, v)$ is calculated using Eq. 5. During inference, we calculate the matching score between each text and video based on Eq. 5, and return videos with the highest ranking.

4 Experiment

In this section, we carry out text-video retrieval evaluations on multiple benchmark datasets to validate our proposed model TS2-Net. We first ablate the core ingredients of our video encoder, the token shift transformer and the token selection transformer, on the dominant MSR-VTT dataset. We then compare our model with other state-of-the-art models on multiple benchmark datasets quantitatively and qualitatively.

4.1 Experimental Settings

Datasets. To demonstrate the effectiveness and generalization ability of our model, we conduct evaluations on five popular text-video benchmarks, including MSR-VTT [46], VATEX [44], LSMDC [40], ActivityNet-Caption [18,25], DiDeMo [2]. All these datasets are collected from different scenarios with various amounts of captions. Videos in different datasets also have different content styles and different lengths.

- **MSR-VTT** [46] contains 10,000 video clips with 20 captions per video. Our experiments follow 1k-A split protocol used in [21,31,35], where the training set has 9,000 videos with its corresponding captions and test set has 1,000 text-video pairs.
- **VATEX** [44] contains 34,991 video clips with several captions per video. We follow HGR [11] split protocol. There are 25,991 videos in the training set, 1,500 videos in the validation set and 1,500 videos in the test set.
- **LSMDC** [40] contains 118,081 video clips, which are extracted from 202 movies. Each video clip has one caption. There are about 100k videos in the training set, 7,408 videos in the validation set and 1,000 videos in the test set. Especially, videos in the test set are from movies disjoint with the training and validation set.
- **ActivityNet-Caption** [18,25] contains 20,000 YouTube videos. Following the same setting as in [21,34,49], we regard it as a paragraph-video retrieval by concatenate all descriptions of a video. We train our model on *train* split and test our model on *val1* split.
- **DiDeMo** [2] contains over 10k videos. There are 8,395 videos in the training set, 1,065 videos in the validation set and 1,004 videos in the test set. Following the same setting as in [26,31,34], we concatenate all descriptions of a video to retrieval videos with paragraphs.

Evaluation Metrics. We measure the retrieval performance using standard text-video retrieval metrics: Recall at K (R@K, higher is better), Median Rank (MdR, lower is better) and Mean Rank (MnR, lower is better). R@K calculates the fraction of correct videos among the top K retrieved videos. Similar to previous works [12,31,34], we use K=1,5,10 for different datasets. We also sum up all the R@K results as rsum to reflect the overall retrieval performance. MedR calculates the median rank of correct results in the retrieval ranking list and MeanR calculates the mean rank of correct results in the retrieval ranking list.

Table 1. Performance comparison with different parameter settings of the Token Shift Transformer on MSR-VTT-1k-A test split

Method	Layers	Ratio	Text ⟹ Video				Video ⟹ Text				rsum
			R@1	R@5	R@10	MnR	R@1	R@5	R@10	MnR	
Baseline	–	–	45.4	74.3	82.7	13.6	44.5	72.3	82.3	9.8	401.5
w/ Token Shift	1-12	25%	42.8	71.2	80.9	14.4	43.2	70.3	80.4	11.3	388.8
w/ Token Shift	3-12	25%	44.1	71.0	81.8	14.5	43.5	71.2	81.8	10.8	393.4
w/ Token Shift	5-12	25%	44.4	71.9	81.6	14.6	44.8	72.0	80.6	11.3	395.3
w/ Token Shift	7-12	25%	44.1	72.3	82.9	13.6	43.8	72.3	82.1	10.3	397.5
w/ Token Shift	9-12	25%	45.2	73.8	83.1	13.4	45.3	72.1	82.5	9.5	402
w/ Token Shift	11-12	12.5%	46.0	73.3	82.2	13.8	**45.8**	72.9	83.0	9.5	403.2
w/ Token Shift	11-12	50%	46.1	**74.5**	83.3	13.3	45.6	72.9	82.2	9.5	404.6
w/ Token Shift	**11-12**	**25%**	**46.2**	73.9	**83.8**	**13.0**	45.6	**73.5**	**83.2**	**9.3**	**406.2**

Table 2. Performance comparison between other shift operation variants and our proposed token shift module on MSR-VTT-1k-A test split

Method	Text ⟹ Video				Video ⟹ Text				rsum
	R@1	R@5	R@10	MnR	R@1	R@5	R@10	MnR	
Baseline	45.4	74.3	82.7	13.6	44.5	72.3	82.3	9.8	401.5
Channel Shift [50]	45.6	73.6	83.1	13.7	45.0	73.2	82.7	9.7	403.2
[VIS] Channel Shift [50]	45.1	73.8	83.5	13.9	44.7	73.3	82.2	9.8	402.6
[CLS] Channel Shift [50]	45.8	**74.3**	83.0	13.6	44.7	72.9	82.5	9.8	403.2
Token Shift	**46.2**	73.9	**83.8**	**13.0**	**45.6**	**73.5**	**83.2**	**9.3**	**406.2**

Implementation Details. The layer of GPT, token shift transformer and token selection transformer is 12, 12 and 4, respectively. The dimension of text embedding and frame embedding is 512. We initialize transformer layers in GPT, token shift transformer and token selection transformer with pre-trained weight from CLIP(ViT-B/32) [37], using parameters with similar dimension, while other modules are initialized randomly. We choose 4 most informative tokens in MSR-VTT, VATEX, ActivityNet-Caption, DiDeMo, and 1 in LSMDC. We set the max query text length as 32 and max video frame length as 12 in MSR-VTT, VATEX, LSMDC. For ActivityNet-Caption and DiDeMo, we set the max query text length and max video frame length as 64. We train our model with Adam [24] optimizer and adopt a warmup [23] setting. We choose a batch size of 128. The learning rate of GPT and token shift transformer is 1e-7 and the learning rate of token selection transformer is 1e-4.

4.2   Ablation Experiments

In this section, we evaluate the proposed token shift transformer and token selection transformer under different settings to validate their effectiveness. We conduct ablation experiments with the 1k-A test split on MSR-VTT [46]. We set our baseline model as the degraded TS2-Net model which removes the token shift and token selection modules from TS2-Net.

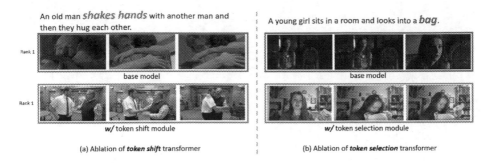

Fig. 5. The text-video retrieval results of different network architecture. Left: with *token shift transformer*, our model is able to distinguish 'shake hands', while the baseline model retrieves an incorrect video. Right: with *token selection transformer*, our model retrieves the correct video, although 'bag' is only shown in small part of video frames. Green boxes: correct target video; red boxes: incorrect target video. (Color figure online)

Ablation of Token Shift Transformer. We first analyze the impact of some factors on the token shift module in Table 1, including shift layer and shift ratio. Shift layer (in which layers should we insert token shift) and shift ratio (how many tokens should we shift) are two main factors that affect the final retrieval performance. The backbone of our token shift transformer is the 12-layer ViT. We thus experiment to insert the token shift module in different layers. As shown in Table 1, shift operation in deeper layers (i.e. 11–12 layers) brings retrieval performance improvement. But if we shift more layers (i.e. 9–12 layers), it hurts the retrieval performance, and it hurts more if we operate shift in shallower layers (i.e. 1–12 layers). We think that shallow layers in ViT are more important in modeling spatial information, so shift in shallow layers damages spatial modeling ability. We thus choose to insert the token shift module in the 11–12 layers in the following experiments. In terms of shift ratio, we find that shifting 25% tokens back-and-forth across frames achieves the best retrieval performance. Despite some slight fluctuations, token shift with different ratios achieves better results than the baseline model. The improvement is more obvious especially for R@1.

We further conduct experiments to compare our proposed token shift module with other shift operation variants proposed in Shift-ViT [50]. As shown in Table 2, our proposed token shift module outperforms all other shift operation variants. This is because our token shift operation can preserve the integrity of the token feature, posing minor impact on the spatial modeling ability. We visualize the retrieval results from the baseline model and the model with token shift transformer in Fig. 5(a). With token shift transformer, the model is able to capture subtle movement such as 'shake hand'.

Ablation of Token Selection Transformer. The token selection transformer follows the token shift transformer to select the most informative tokens for the next transformer propagation. We conduct experiment to verify what proportion of tokens is beneficial to the final retrieval in Table 3. As can be observed, selecting fewer tokens per frame tends to achieve better performance than select-

Table 3. Comparison results with different settings of Token Selection Transformer

Method	top-K	Text ⟹ Video				Video ⟹ Text				rsum
		R@1	R@5	R@10	MnR	R@1	R@5	R@10	MnR	
Token Shift	1	46.2	73.9	83.8	13.0	45.6	73.5	83.2	9.3	406.2
w/ all token	50	45.8	73.5	83.4	13.5	44.7	73.1	82.4	9.4	402.9
w/ Random select	4	46.4	73.9	83.5	13.1	45.1	73.5	82.1	9.5	404.5
w/ Select token	2	47.0	74.2	83.6	13.1	**45.6**	74.0	83.5	9.3	407.9
w/ Select token	6	46.6	74.4	**84.3**	13.2	44.5	73.8	83.2	9.2	406.8
w/ Select token	8	46.4	73.9	83.5	13.2	45.0	74.1	**83.9**	9.2	406.8
TS2-Net	4	**47.0**	**74.5**	83.8	**13.0**	45.3	**74.1**	83.7	**9.2**	**408.4**

Table 4. Retrieval results on MSR-VTT-1kA. Other SOTA methods are adopted as comparisons. Note that CLIP2TV uses patch size of 16×16, so we use TS2-Net(ViT16) for fair comparison. All results in this table do not use inverted softmax

Method	Text ⟹ Video					Video ⟹ Text				
	R@1	R@5	R@10	MdR	MnR	R@1	R@5	R@10	MdR	MnR
CE [31]	20.9	48.8	62.4	6.0	28.2	20.6	50.3	64.0	5.3	25.1
TACo citech19yang2021taco	26.7	54.5	68.2	4.0	–	–	–	–	–	–
MMT [21]	26.6	57.1	69.6	4.0	24.0	27.0	57.5	69.7	3.7	21.3
SUPPORT-SET [36]	27.4	56.3	67.7	3.0	–	26.6	55.1	67.5	3.0	–
TT-CE [14]	29.6	61.6	74.2	3.0	–	–	–	–	–	–
T2VLAD [43]	29.5	59.0	70.1	4.0	–	31.8	60.0	71.1	3.0	–
HIT-pretrained [30]	30.7	60.9	73.2	2.6	–	32.1	62.7	74.1	3.0	–
Frozen [4]	31.0	59.5	70.5	3.0	–	–	–	–	–	–
MDMMT [17]	38.9	69.0	79.7	2.0	16.5	–	–	–	–	–
CLIP [37]	39.7	72.3	82.2	2.0	12.8	11.3	22.7	29.2	5.0	–
CLIP4Clip [34]	44.5	71.4	81.6	2.0	15.3	42.7	70.9	80.6	2.0	11.6
CAMoE [12]	44.6	72.6	81.8	2.0	13.3	45.1	72.4	83.1	2.0	10.0
CLIP2Video [19]	45.6	72.6	81.7	2.0	14.6	43.5	72.3	82.1	2.0	10.2
TS2-Net	**47.0**	**74.5**	**83.8**	2.0	**13.0**	45.3	74.1	83.7	2.0	**9.2**
CLIP2TV [22]	48.3	74.6	82.8	2.0	14.9	46.5	75.4	84.9	2.0	10.2
TS2-Net(ViT16)	**49.4**	**75.6**	**85.3**	2.0	13.5	**46.6**	**75.9**	84.9	2.0	**8.9**

ing more. For example, the R@1 performance decreases from 47.0 to 45.8 while the number of selected tokens increases from 2 to 50. We consider that fewer informative tokens are sufficient to preserve the salient spatial information, while adding more tokens may bring redundancy problem. Although random selection also improves the performance slightly, it can not beat the proposed learnable token selection module. In Fig. 5(b), we show a retrieval case from the baseline model and the model with token selection transformer. With token selection transformer, the model is able to capture the small object 'bag' in video frames.

4.3   Comparisons with State-of-the-art Models

MSR-VTT-1kA. We compare our proposed TS2-Net with other state-of-the-art methods on five benchmarks. Table 4 presents the results on MSR-VTT-1kA

Table 5. Text-to-Video retrieval results on VATEX, LSMDC, ActivityNet and DiDeMo. QB-Norm uses dynamic inverted softmax during inference, while other methods report results without inverted softmax

VATEX						LSMDC					
Method	R@1	R@5	R@10	MdR	MeanR	Method	R@1	R@5	R@10	MdR	MeanR
Dual Enc. [15]	31.1	67.5	78.9	3.0	–	JSFusion [48]	9.1	21.2	34.1	36.0	–
HGR [11]	35.1	73.5	83.5	2.0	–	CE [31]	11.2	26.9	34.9	25.3	–
CLIP [37]	39.7	72.3	82.2	2.0	12.8	Frozen [4]	15.0	30.8	39.8	20.0	–
CLIP4Clip [34]	55.9	89.2	95.0	1.0	3.9	CLIP4Clip [34]	22.6	41.0	49.1	11.0	61.0
QB-Norm* [7]	58.8	88.3	93.8	1.0		QB-Norm* [7]	22.4	40.1	49.5	11.0	–
CLIP2Video [19]	57.3	90.0	**95.5**	1.0	3.6	CAMoE[12]	22.5	**42.6**	50.9	–	**56.5**
TS2-Net	**59.1**	**90.0**	95.2	**1.0**	**3.5**	**TS2-Net**	**23.4**	42.3	**50.9**	**9.0**	56.9

ActivityNet						DiDeMo					
Method	R@1	R@5	R@10	MdR	MeanR	Method	R@1	R@5	R@10	MdR	MeanR
CE [31]	20.5	47.7	63.9	6.0	23.1	ClipBERT [26]	20.4	48.0	60.8	6.0	–
ClipBERT [26]	21.3	49.0	63.5	6.0	–	TT-CE [14]	21.1	47.3	61.1	6.3	–
MMT-Pretrained [21]	28.7	61.4	–	3.3	16.0	Frozen [4]	31.0	59.8	72.4	3.0	–
CLIP4Clip [34]	40.5	73.4	–	2.0	**7.5**	CLIP4Clip [34]	**42.5**	70.2	80.6	2.0	17.5
TS2-Net	**41.0**	**73.6**	**84.5**	**2.0**	8.4	**TS2-Net**	41.8	**71.6**	**82.0**	**2.0**	**14.8**

test set. Our model outperforms previous methods across different evaluation metrics. With token shift transformer and token selection transformer, our model is able to capture subtle motion and salient objects, and thus our final video representation contains rich semantics. Compared with video-to-text retrieval, the gain on text-to-video retrieval is more significant. We consider it is because the proposed token shift and token selection modules enhance the video encoder, while a relative simple text encoder is adopted.

Other Benchmarks. Table 5 presents text-to-video retrieval results on VATEX, LSMDC, ActivityNet-Caption and DiDeMo. Results on these datasets demonstrate the generalization and robustness of our proposed model. Our model achieves consistent improvements across different datasets, which demonstrates that it is beneficial to encode spatial and temporal features simultaneously by our token shift and token selection. Note that our performance surpasses QB-Norm [7] on LSMDC and VATEX even without inverted softmax, as shown in Table 5. More detailed analysis can be found in supplementary materials.

4.4 Qualitative Results

We visualize some retrieval examples from the MSR-VTT testing set for text-to-video retrieval in Fig. 6. In the top left example, our model is able to distinguish 'hand rubbing' (in the middle picture) during a guitar-playing scene. The bottom right example shows our model can distinguish 'computer battery' from 'computer'. In the bottom left example, our model retrieves the correct video which contains all actions and objects expressed in the text query, especially

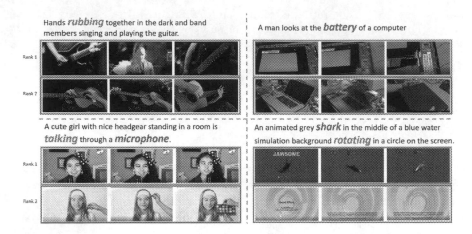

Fig. 6. Visualization of text-video retrieval examples. We sorted results based on its similarity scores. Green: ground truth; Red: incorrect (Color figure online)

the small object 'microphone' and tiny movement 'talking'. In the bottom right example, our model retrievals the correct result although 'rotating' is a periodic movement and is hard to spot.

We also select a subset from the MSR-VTT-1kA test set. Queries in this subset are selected based on their corresponding video's visual appearance, where objects mentioned in query are shown in a small part of video and movements mentioned in query is slight. Such as '*little pet shop cat* getting a bath and washed with *little brush*', 'a golf player is trying to hit the ball into the *pit*'. Since such cases account for a small proportion, so the total number of this subset is 103. During inference, we calculate similarity between queries in subset with videos in whole test set. We compare our model with another strong baseline on this subset. Our model achieves 79.6 on R@1 metric, while CLIP4Clip [34] only achieves 39.8. There is a significant margin and this verifies the effectiveness of TS2-Net in handling local subtle movements and local small entities.

5   Conclusion

In this work, we propose Token Shift and Selection Network (TS2-Net), a noval transformer architecture with token shift and selection modules, which aims to further improve the video encoder for better video representation. A token shift transformer is used to capture subtle movements, followed by a token selection transformer to enhance salient object modeling ability. Superior experimental results show our proposed TS2-Net outperforms start-of-the-art methods on five text-video retrieval benchmarks, including MSR-VTT, VATEX, LSMDC, ActivityNet-Caption and DiDeMo.

Acknowledgement. This work was partially supported by National Key R&D Program of China (No. 2020AAA0108600) and National Natural Science Foundation of China (No. 62072462).

References

1. Abernethy, J., Lee, C., Tewari, A.: Perturbation techniques in online learning and optimization. Perturbations, Optimization, and Statistics, p. 223 (2016)
2. Anne Hendricks, L., Wang, O., Shechtman, E., Sivic, J., Darrell, T., Russell, B.: Localizing moments in video with natural language. In: Proceedings of the IEEE international conference on computer vision, pp. 5803–5812 (2017)
3. Arnab, A., Dehghani, M., Heigold, G., Sun, C., Lučić, M., Schmid, C.: Vivit: A video vision transformer. In: Proceedings of the IEEE/CVF International Conference on Computer Vision, pp. 6836–6846 (2021)
4. Bain, M., Nagrani, A., Varol, G., Zisserman, A.: Frozen in time: A joint video and image encoder for end-to-end retrieval. In: Proceedings of the IEEE/CVF International Conference on Computer Vision, pp. 1728–1738 (2021)
5. Bertasius, G., Wang, H., Torresani, L.: Is space-time attention all you need for video understanding. arXiv preprint arXiv:2102.05095 (2021)
6. Berthet, Q., Blondel, M., Teboul, O., Cuturi, M., Vert, J.P., Bach, F.: Learning with differentiable pertubed optimizers. Advances in neural information processing systems, pp. 9508–9519 (2020)
7. Bogolin, S.V., Croitoru, I., Jin, H., Liu, Y., Albanie, S.: Cross modal retrieval with querybank normalisation. arXiv preprint arXiv:2112.12777 (2021)
8. Bulat, A., Perez Rua, J.M., Sudhakaran, S., Martinez, B., Tzimiropoulos, G.: Space-time mixing attention for video transformer. In: Advances in Neural Information Processing Systems (2021)
9. Carreira, J., Zisserman, A.: Quo vadis, action recognition? a new model and the kinetics dataset. In: proceedings of the IEEE Conference on Computer Vision and Pattern Recognition, pp. 6299–6308 (2017)
10. Chen, D., Dolan, W.B.: Collecting highly parallel data for paraphrase evaluation. In: Proceedings of the 49th annual meeting of the association for computational linguistics: human language technologies, pp. 190–200 (2011)
11. Chen, S., Zhao, Y., Jin, Q., Wu, Q.: Fine-grained video-text retrieval with hierarchical graph reasoning. In: Proceedings of the IEEE/CVF Conference on Computer Vision and Pattern Recognition, pp. 10638–10647 (2020)
12. Cheng, X., Lin, H., Wu, X., Yang, F., Shen, D.: Improving video-text retrieval by multi-stream corpus alignment and dual softmax loss. arXiv preprint arXiv:2109.04290 (2021)
13. Cordonnier, J.B., Mahendran, A., Dosovitskiy, A., Weissenborn, D., Uszkoreit, J., Unterthiner, T.: Differentiable patch selection for image recognition. In: Proceedings of the IEEE/CVF Conference on Computer Vision and Pattern Recognition, pp. 2351–2360 (2021)
14. Croitoru, I., et al.: Teachtext: Crossmodal generalized distillation for text-video retrieval. In: Proceedings of the IEEE/CVF International Conference on Computer Vision, pp. 11583–11593 (2021)
15. Dong, J., et al.: Dual encoding for zero-example video retrieval. In: Proceedings of the IEEE/CVF Conference on Computer Vision and Pattern Recognition, pp. 9346–9355 (2019)
16. Dosovitskiy, A., et al.: An image is worth 16x16 words: Transformers for image recognition at scale. ICLR (2021)
17. Dzabraev, M., Kalashnikov, M., Komkov, S., Petiushko, A.: Mdmmt: Multidomain multimodal transformer for video retrieval. In: Proceedings of the IEEE/CVF Conference on Computer Vision and Pattern Recognition, pp. 3354–3363 (2021)

18. Heilbron, F.C., Victor Escorcia, B.G., Niebles, J.C.: Activitynet: A large-scale video benchmark for human activity understanding. In: Proceedings of the IEEE Conference on Computer Vision and Pattern Recognition, pp. 961–970 (2015)
19. Fang, H., Xiong, P., Xu, L., Chen, Y.: Clip2video: Mastering video-text retrieval via image clip. arXiv preprint arXiv:2106.11097 (2021)
20. Feichtenhofer, C., Fan, H., Malik, J., He, K.: Slowfast networks for video recognition. In: Proceedings of the IEEE/CVF international conference on computer vision, pp. 6202–6211 (2019)
21. Gabeur, V., Sun, C., Alahari, K., Schmid, C.: Multi-modal transformer for video retrieval. In: European Conference on Computer Vision, pp. 214–229 (2020)
22. Gao, Z., Liu, J., Chen, S., Chang, D., Zhang, H., Yuan, J.: Clip2tv: An empirical study on transformer-based methods for video-text retrieval. arXiv preprint arXiv:2111.05610 (2021)
23. Goyal, P., et al.: Accurate, large minibatch sgd: Training imagenet in 1 hour. arXiv preprint arXiv:1706.02677 (2017)
24. Kingma, D.P., Ba, J.: Adam: a method for stochastic optimization. In: 3rd international Conferance for Learning Representations, San (2014)
25. Krishna, R., Hata, K., Ren, F., Fei-Fei, L., Carlos Niebles, J.: Dense-captioning events in videos. In: Proceedings of the IEEE international conference on computer vision, pp. 706–715 (2017)
26. Lei, J., et al.: Less is more: Clipbert for video-and-language learning via sparse sampling. In: Proceedings of the IEEE/CVF Conference on Computer Vision and Pattern Recognition, pp. 7331–7341 (2021)
27. Li, G., Duan, N., Fang, Y., Gong, M., Jiang, D.: Unicoder-vl: A universal encoder for vision and language by cross-modal pre-training. In: Proceedings of the AAAI Conference on Artificial Intelligence, pp. 11336–11344. No. 07 (2020)
28. Li, L., Chen, Y.C., Cheng, Y., Gan, Z., Yu, L., Liu, J.: Hero: Hierarchical encoder for video+ language omni-representation pre-training. arXiv preprint arXiv:2005.00200 (2020)
29. Lin, J., Gan, C., Han, S.: Tsm: Temporal shift module for efficient video understanding. In: Proceedings of the IEEE International Conference on Computer Vision (2019)
30. Liu, S., Fan, H., Qian, S., Chen, Y., Ding, W., Wang, Z.: Hit: Hierarchical transformer with momentum contrast for video-text retrieval. In: Proceedings of the IEEE/CVF International Conference on Computer Vision, pp. 11915–11925 (2021)
31. Liu, Y., Albanie, S., Nagrani, A., Zisserman, A.: Use what you have: Video retrieval using representations from collaborative experts. arXiv preprint arXiv:1907.13487 (2019)
32. Liu, Z., et al.: Video swin transformer. arXiv preprint arXiv:2106.13230 (2021)
33. Luo, H., et al.: Univl: A unified video and language pre-training model for multimodal understanding and generation. arXiv preprint arXiv:2002.06353 (2020)
34. Luo, H., et al.: Clip4clip: An empirical study of clip for end to end video clip retrieval. arXiv preprint arXiv:2104.08860 (2021)
35. Miech, A., Zhukov, D., Alayrac, J.B., Tapaswi, M., Laptev, I., Sivic, J.: Howto100m: Learning a text-video embedding by watching hundred million narrated video clips. In: Proceedings of the IEEE/CVF International Conference on Computer Vision, pp. 2630–2640 (2019)
36. Patrick, M., et al.: Support-set bottlenecks for video-text representation learning. arXiv preprint arXiv:2010.02824 (2020)

37. Radford, A., et al.: Learning transferable visual models from natural language supervision. In: International Conference on Machine Learning. pp. 8748–8763 (2021)
38. Radford, A., Wu, J., Child, R., Luan, D., Amodei, D., Sutskever, I., et al.: Language models are unsupervised multitask learners. OpenAI blog, p. 9 (2019)
39. Rao, Y., Zhao, W., Liu, B., Lu, J., Zhou, J., Hsieh, C.J.: Dynamicvit: Efficient vision transformers with dynamic token sparsification. Advances in neural information processing systems (2021)
40. Rohrbach, A., et al.: Movie description. International Journal of Computer Vision, pp. 94–120 (2017)
41. Su, W., et al.: Vl-bert: Pre-training of generic visual-linguistic representations. ICLR (2020)
42. Wang, J., Yang, X., Li, H., Wu, Z., Jiang, Y.G.: Efficient video transformers with spatial-temporal token selection. arXiv preprint arXiv:2111.11591 (2021)
43. Wang, X., Zhu, L., Yang, Y.: T2vlad: global-local sequence alignment for text-video retrieval. In: Proceedings of the IEEE/CVF Conference on Computer Vision and Pattern Recognition, pp. 5079–5088 (2021)
44. Wang, X., Wu, J., Chen, J., Li, L., Wang, Y.F., Wang, W.Y.: Vatex: A large-scale, high-quality multilingual dataset for video-and-language research. In: Proceedings of the IEEE/CVF International Conference on Computer Vision, pp. 4581–4591 (2019)
45. Xie, S., Sun, C., Huang, J., Tu, Z., Murphy, K.: Rethinking spatiotemporal feature learning: Speed-accuracy trade-offs in video classification. In: Proceedings of the European conference on computer vision (ECCV), pp. 305–321 (2018)
46. Xu, J., Mei, T., Yao, T., Rui, Y.: Msr-vtt: A large video description dataset for bridging video and language. In: Proceedings of the IEEE conference on computer vision and pattern recognition, pp. 5288–5296 (2016)
47. Yang, J., Bisk, Y., Gao, J.: Taco: Token-aware cascade contrastive learning for video-text alignment. In: Proceedings of the IEEE/CVF International Conference on Computer Vision, pp. 11562–11572 (2021)
48. Yu, Y., Kim, J., Kim, G.: A joint sequence fusion model for video question answering and retrieval. In: Proceedings of the European Conference on Computer Vision (ECCV), pp. 471–487 (2018)
49. Zhang, B., Hu, H., Sha, F.: Cross-modal and hierarchical modeling of video and text. In: Proceedings of the European Conference on Computer Vision (ECCV), pp. 374–390 (2018)
50. Zhang, H., Hao, Y., Ngo, C.W.: Token shift transformer for video classification. In: Proceedings of the 29th ACM International Conference on Multimedia, pp. 917–925 (2021)
51. Zhou, L., Palangi, H., Zhang, L., Hu, H., Corso, J., Gao, J.: Unified vision-language pre-training for image captioning and vqa. In: Proceedings of the AAAI Conference on Artificial Intelligence (2020)

Unstructured Feature Decoupling
for Vehicle Re-identification

Wen Qian[1,2] , Hao Luo[3] , Silong Peng[1,2] , Fan Wang[3], Chen Chen[1(✉)] ,
and Hao Li[3]

[1] Institute of Automation, Chinese Academy of Sciences, Beijing, China
{qianwen2018,silong.peng,chen.chen}@ia.ac.cn
[2] School of Artificial Intelligence, University of Chinese Academy of Sciences,
Beijing, China
[3] Alibaba group, Hangzhou, China
{michuan.lh,fan.w,lihao.lh}@alibaba-inc.com

Abstract. The misalignment of features caused by pose and viewpoint
variances is a crucial problem in Vehicle Re-Identification (ReID). Previous methods align the features by structuring the vehicles from pre-
defined vehicle parts (such as logos, windows, etc.) or attributes, which
are inefficient because of additional manual annotation. To align the features without requirements of additional annotation, this paper proposes
a **Unstructured Feature Decoupling Network** (UFDN), which consists of a transformer-based feature decomposing head (TDH) and a novel
cluster-based decoupling constraint (CDC). Different from the structured knowledge used in previous decoupling methods, we aim to achieve
more flexible unstructured decoupled features with diverse discriminative information as shown in Fig. 1. The self-attention mechanism in the
decomposing head helps the model preliminarily learn the discriminative decomposed features in a global scope. To further learn diverse but
aligned decoupled features, we introduce a cluster-based decoupling constraint consisting of a diversity constraint and an alignment constraint.
Furthermore, we improve the alignment constraint into a modulated
one to eliminate the negative impact of the outlier features that cannot
align the clusters in semantics. Extensive experiments show the proposed
UFDN achieves state-of-the-art performance on three popular Vehicle
ReID benchmarks with both CNN and Transformer backbones. Our code
is released at: https://github.com/damo-cv/UFDN-Reid.

Keywords: Unstructured feature decoupling network · Vehicle reid ·
Transformer-based decoupling head · Cluster-based decoupling
constraint

The work was supervised by Hao Luo and Chen Chen.

Supplementary Information The online version contains supplementary material
available at https://doi.org/10.1007/978-3-031-19781-9_20.

1 Introduction

Given a query vehicle image, Vehicle ReID aims to retrieve images of the same vehicle from the gallery that contains images captured by disjoint cameras. With the development of large Vehicle ReID benchmarks [13,14,18,20,31,40] and deep learning methods, Vehicle ReID achieves a great promotion in performance and is widely applied in intelligence city system [5,20,27]. However, it is challenging to deal with the misalignment of features caused by poses.

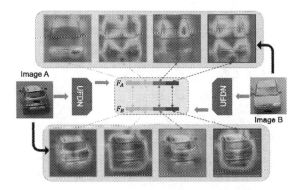

Fig. 1. Illustration of the decoupling and alignment process for vehicle features.

Existing methods [1,3,4,17,32,43] mainly tackle the misalignment of features by structuring the vehicles in two directions. 1) Some methods [25,26,32,36] spatially decompose the vehicle into several stripes or grids. Specifically, PCB [32] and its variants divide the feature maps into multi-level stripes/grids straightforwardly to integrate fine-grained information. However, these methods only decompose the feature from the spatial dimension, which is coarse and lacks semantic representations. 2) Recent methods [1,3,4,17,43] utilize prior semantic information such as pre-defined vehicle parts (e.g. lights, windows, etc.) and vehicle attributes (colors, viewpoints, etc.) to guide the feature decoupling. For example, PVEN [24] decouples the vehicle features in the view-aware feature space based on a parsing network and the pre-defined viewpoint labels. Nevertheless, such explicit alignment based on the pre-defined knowledge cannot flexibly handle missing components caused by viewpoint variances or occlusions. Structured analysis of vehicle images is usually time-consuming and inefficient since they rely heavily on manual annotation and extra modules [8,19,28]. Moreover, the local feature explored in the above methods will be only discriminative for the corresponding region due to the hand-crafted rules.

To address the aforementioned limitations, this paper studies the implicit alignment of features by first decomposing them into unstructured parts and then aligning them without using extra annotation. However, there mainly exist two challenges: 1) how to decompose the feature without using extra structured cues; 2) how to learn diverse but aligned decoupled features?

We propose a transformer-based decomposing head (TDH) to decompose the vehicle feature into unstructured parts. Different from those methods [25, 26, 32, 36] that factorize the images on the spatial dimension, TDH keeps a global receptive field of each decomposed feature through decomposing the feature map from the channel dimension. We feed each decomposed group of feature maps into a modified transformer block, and then the self-attention mechanism can automatically encode one discriminative feature in a global scope. Since the feature map is not simply divided into fixed stripes/grids, the decomposed features can implicitly learn discriminative semantic information without extra cues.

Apart from the implicit decomposing module, we propose a novel cluster-based decouple constraint (CDC) to improve the diversity and alignment of decomposed features in an annotation-free way. CDC aims to cluster the decomposed features into groups, which consists: 1) the diversity constraint: the decomposed features should be orthogonal to each other, which motivates them to focus on diverse regions of interest; 2) the alignment constraint: the decomposed features should be close to the relevant cluster centers to align with each other. However, some outlier features cannot align with the cluster centers in semantics, which will lead to useless or even inaccurate supervision. To tackle such an issue, we filter out the outlier features to mitigate their negative effects and term the final output as the decoupled features. Different from that methods [2, 16] such as ABDNet (see the details in Sect. 5.2), which only conduct the diversity constraint on the 2D features maps to keep the diversity of features, our UFDN aims to maintain the diversity and alignment of final 1D features.

Moreover, we visualize samples from different viewpoints as shown in Fig. 1 and find that the corresponding decoupled features tend to focus on similar salient regions (e.g., the lights information in the first part, the counter, and front information for the second part, etc.) We term our method as unstructured feature decoupling network (UFDN), and experiment on three popular benchmarks with two different backbones (ResNet and Swin-transformer) to evaluate the effectiveness of UFDN. **The contributions of this paper are:**

i) We propose UFDN which aims to alleviate the feature misalignment in Vehicle ReID by decoupling them into unstructured, diverse, and aligned parts without human annotation.
ii) The transformer-based unstructured feature decomposing head can decompose features into several groups from the channel dimension in a global scope which is more robust than the local specified methods.
iii) We propose a cluster-based decoupling constraint to keep the diversity and alignment of the decoupled features without human annotation and further external the outliers to eliminate the negative influence on them.
iv) Without the requirement of extra manual annotation, our UFDN outperforms other methods on three popular benchmarks consistently.

2   Related Work

Previous methods [1, 4, 5, 7, 11, 17, 18, 43] decouple vehicle features for better alignment and can be categorized into two kinds:

The Spatial Decomposing Methods. [25,26,32,36]. These methods spatially decompose the feature map into several stripes or grids to capture fine-grained information. Sun et al. [32] propose to decompose the feature map into six parts and each local feature is followed by the ReID supervision. Mo et al. [25] employ a cascaded hierarchical context-Aware Vehicle ReID network to decompose the vehicle features at multi-scale hierarchically. The coarse decomposed features provided by the spatial decomposing methods have achieved progress in ReID performance. However, the improvement brought by the spatial-wise decomposing is limited since it tends to focus on coarse local regions and neglect the important semantic information.

The Pre-defined Semantic Decoupling Methods. Different from the coarse decomposed information in previous spatial decomposing methods, the predefined semantic decoupling methods [1,4,6,7,11,17,24,39,43] decouple the vehicle feature with the assistance of the pre-defined information, e.g., pre-defined vehicle parts (wheels, logos, and windows) or vehicle attributes (color, type, and viewpoints). Zhang et al. [37] introduce a part-guided attention network to enhance the feature representation by decoupling the features under the guidance of the prominent parts. Apart from the decoupling methods guided by vehicle parts, Wang et al. [35] propose an attribute-guided module to assist the decoupling of features. Guo et al. [6] bridge the gap between the vehicle features from different models by a coarse-to-fine structured feature embedding. However, the above methods [1,17,24–26,36] rely heavily on human annotation and only focus on the fixed pre-defined regions while ignoring other potential crucial clues.

Most existing methods decouple features in a structured way, we target to study an annotation-free method that decouples the vehicle features into unstructured, diverse, and aligned ones.

3 Methodology

Figure 2 shows the illustration of UFDN, which mainly consists of a transformer-based feature decomposing head and a cluster-based decoupling constraint.

3.1 Backbone and Symbol Definition

Given an input image X, the backbone outputs a feature map which is reshaped to a base feature $F_{base} \in R^{n \times c}$, where $n = H \times W$ and c represent the spatial dimension and the channel dimension, respectively. Then we split F_{base} into k groups from the channel dimension to obtain a new feature set $F_p \in R^{k \times n \times m}$, where $c = k \times m$. The backbone can be both CNN-based or Transformer-based.

3.2 Transformer-Based Feature Decomposing Head

The transformer-based feature decomposing head (TDH) encode the unstructured information of each decomposed feature $F_p^i \in R^{n \times m}, i = 1, 2, 3, ..., k$. As

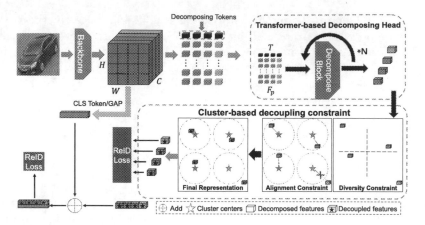

Fig. 2. Illustrative the framework of UFDN consisting of two modules: 1) the transformer-based feature decomposing head that aims to learn unstructured decomposed features of the original base feature from the attention mechanism; 2) the cluster-based decouple constraint that aims to keep the diversity and alignment of the decoupled features. Moreover, the base-feature extraction module can be either a CNN-based network (ResNet) or a Transformer-based network (Swin transformer).

shown in Fig. 2, k decomposing tokens $T^i \in R^m$ are pre-pended to the relative channel-wise feature $F_p^i \in R^{n \times m}$, respectively. The input sequence fed to transformer-based feature decomposing head (TDH) is denoted as $z_0^i = [T^i, F_p^i]$.

As illustrated in Fig. 3, TDH contains a total of L transformer blocks each of which consists of a multi-head decomposing attention (DA) module and a MLP module. Since F_p^i is a deep-layer feature that is good enough for encoding discriminative information, we follow the solution [34] to update only the decomposing token T^i to re-aggregate F_p^i, i.e. F_p^i is frozen during training to reduce computational cost. Given the input sequence $z_0^i = [T^i, F_p^i]$, we term the output decomposing token of the $l-1$ block in TDH as T_{l-1}^i and the input sequence to the l-th block as $z_{l-1}^i = [T_{l-1}^i, F_p^i]$. Then we will feed the l-th input sequence z_{l-1}^i to the DA module of the l-th block in TDH, which can be expressed as:

$$
\begin{aligned}
Q &= W_q T_{l-1}^i, \quad K = W_k z_{l-1}^i, \quad V = W_v z_{l-1}^i, \\
A &= Softmax(QK^T), \quad h_l^i = A \cdot V + T_{l-1}^i,
\end{aligned}
\tag{1}
$$

where $W_q, W_k, W_v \in R^{m \times m}$ are the projection matrices. After getting hidden variable h_l^i from the multi-head decomposing attention module, we feed it to the MLP module in the l-th TDH block as follows:

$$
T_l^i = LN(MLP(LN(h_l^i)) + h_l^i),
\tag{2}
$$

where MLP and LN denote the MLP module and the layer normalization layer, respectively. Then, the decomposing token T_l^i will be concatenated with F_p^i as the input sequence z_{l+1}^i for the next block.

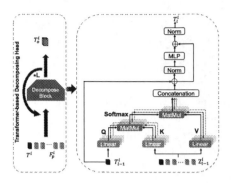

Fig. 3. The transformer-based feature decomposing head: the divided feature F_p^i and decomposing token T^i are fed into the block to calculate the composed feature T_o^i.

We integrate the decomposing token $T_o^i = T_L^i$ outputted by the last block in TDH and the relevant global feature as the decomposed feature $F^i \in R^m$:

$$F^i = T_o^i + GAP(F_p^i), \tag{3}$$

where GAP represents the global average pooling. Finally, k decomposed features $[F^1, F^2, ..., F^k]$ are concatenated to the final ReID feature $F \in R^c$. We conduct the ReID loss on each decomposed feature and the ReID feature as follows:

$$L_{THD} = L_{ReID}(F) + \sum_{i=1}^{k} L_{ReID}(F^i), \tag{4}$$

$$L_{ReID} = L_{ce} + L_{tri},$$

where L_{ce} represents the cross entropy loss and L_{tri} represents the triplet loss.

3.3   The Cluster-Based Decoupling Constraint

Apart from the implicit feature decomposing module, a cluster-based decouple constraint (CDC) is proposed to obtain the decoupled features by requiring the diversity and alignment of these decomposed features as shown in Fig. 2. Firstly, we employ a diversity constraint to enforce the diversity of the decomposed features. Secondly, to align the decomposed features between different images, we propose an alignment constraint that clusters the relevant decomposed features into groups and eliminates the negative effects of outlier features.

The Diversity Constraint. Given an input image X, we obtain k decomposed features $F = [F^1, F^2, ..., F^k]$ and hope that each of them should have diverse regions of interest (ROIs) semantically. Enforcing the diversity of decomposed features can drive the model to mine more salient and discriminate information and accelerate the decoupling process of vehicle features. For abstracting this requirement into a mathematical description, we constrain these features to be orthogonal to each other. The diversity constraint restricts the Gram matrix of F to be close to an identity matrix under Frobenius norm:

$$L_{div} = ||FF^T - I||_F \tag{5}$$

The Alignment Constraint. The diversity constraint only focuses on the relationship between features from the same image, which neglects the cross-image relationship. So we propose an alignment constraint for ranking the decomposed features from different images in the same order. For example, given two input images X_1 and X_2, we obtain the corresponding decomposed features $F_1 = [F_1^1, F_1^2, ..., F_1^k]$ and $F_2 = [F_2^1, F_2^2, ..., F_2^k]$, and hope both the blocks F_1^1 and F_2^1 can focus on the same vehicle region. We propose to build a cluster center for each group of decomposed features and require the decomposed features should be close to the relevant cluster centers to align with each other.

Clustering of Decomposed Features. Given M samples of the training set, we decompose each of them into k decomposed features and build a cluster center C^i for each group of decomposed features $[F_1^i, F_2^i, ..., F_M^i]$, $i \in [1, k]$:

$$C^i = \frac{1}{M} \sum_{j=1}^{M} F_j^i, \tag{6}$$

where F_j^i is the i-th decoupled feature of the image X_j. During the early period of the training process, the decomposed features experience severe fluctuations caused by the large network adjustment from the loss backpropagation, which leads to an unstable convergence process of the cluster centers. To smooth the convergence of cluster centers, a memory bank is adopted to store the cluster centers updated with a momentum strategy.

$$C_t^i = \alpha C_{t-1}^i + (1 - \alpha)\frac{1}{M} \sum_{j=1}^{M} F_j^i, \tag{7}$$

where α is weight for controlling the update speed of the cluster centers, and t denotes different periods of the training process. We treat the cluster-center C^i as a standard feature center since it has walked through all samples for epochs and tends to learn a more general and comprehensive feature representation.

After building the cluster centers for each group of features, we constrain the distance between the positive pair (F_j^i and its corresponding cluster center C^i) should be smaller than the smallest distance of negative pairs (other cluster centers). It is termed as the alignment constraint:

$$L_{align} = [d(F_j^i, C^i) - \min(d(F_j^i, C^{q \in [1,k]), q \neq i})) + \theta]^+, \tag{8}$$

where $d(a, b)$ measures the distance between a and b, $[\cdot]^+ = max(\cdot, 0)$, and θ is the distance threshold to control the distance between the positive and negative pairs. After constraining the decomposed features by the diversity constraint and alignment constraint, we term them as decoupled features.

Exclusion of Outlier Features. Some decoupled features may lack the semantic information of the cluster-center due to different viewpoints or poses, and we term them as outlier features. The outlier features cannot align the cluster centers in semantics, e.g., the front window information is absent in the images captured from the backside, which will result in performance degradation. So it is necessary to eliminate the negative impact from the outlier features by excluding the corresponding alignment loss back-propagation of them.

Firstly, we calculate the average distance D_i of each center C^i:

$$D_i = \frac{1}{M} \sum_{j=1}^{M} d(F_j^i, C^i).$$

(9)

Then we will exclude the loss from the feature block F_j^i if the distance between it and the corresponding cluster center is greater than the average distance D_i. After excluding the outlier features, the modulated alignment loss is:

$$L_{mod} = \begin{cases} L_{align}, & if \ d(F_j^i, C^i) < D_i \\ 0, & if \ d(F_j^i, C^i) \geq D_i. \end{cases}$$

(10)

The cluster-based decouple constraint is computed as:

$$L_{CDC} = \frac{1}{M} \sum_{j=1}^{M} [L_{div}(F_j) + \sum_{i=1}^{k} L_{mod}(F_j^i, C^i, D_i)].$$

(11)

Finally, the loss of the UFDN is: $L = L_{THD} + L_{CDC}$

4   Experiments

4.1   Datasets and Evaluation Metrics

Dataset. We experiment on three Vehicle ReID benchmarks: VeRi776 [18], VehicleID [20] and VERI-WILD [22]. VeRi776 [18] is a classic Vehicle ReID benchmark and contains 776 identities collected by 20 cameras in a real-world environment. VehicleID [20] is a large-scale dataset collected by multiple cameras during the daytime on the open road, which contains 26,267 vehicles and 221,763 images in total. VERI-WILD [22] is another large-scale dataset, and it consists of 40,671 vehicles and 416,314 images. Moreover, VERI-WILD [22] is collected by 174 cameras during a month which is a long period.

Evaluation Metrics. We use the same evaluation protocols used in previous methods [20,38,41] for evaluation: Mean Average Precision (mAP) and the cumulative matching characteristics at Rank1 (CMC@1). Moreover, VehicleID [20] reports only CMC@1 because mAP is unavailable due to its unique test sets.

Table 1. Comparison with state-of-the-art methods. It includes mAP and CMC@1 on VeRi-776; CMC@1 and mAP on three test sets of small (S), medium (M), and large (L) on VehicleID and VERI-WILD, respectively. Baseline* and UFDN* are CNN-based. Baseline and UFDN are Transformer-based. Finally, "/" indicates numbers were not reported. For a fair comparison, all methods report the results of the same-scale backbones (*i.e.* ResNet50, ViT-Small, and Swin-Tiny) pre-trained on ImageNet-1K.

Method	Backbone	VeRi-776		VehicleID			VERI-WILD		
		mAP	CMC@1	CMC@1 (S)	CMC@1 (M)	CMC@1 (L)	mAP (S)	mAP (M)	mAP (L)
PGAN [37]	Res50	79.3	96.5	77.8	/	/	74.1	/	/
PRN [7]	Res50	74.3	94.3	78.4	75.0	74.2	/	/	/
PVEN [24]	Res50	79.5	95.6	84.7	80.6	77.8	82.5	77.0	69.7
GLAMOR [33]	Res50	80.3	96.5	78.6	/	/	77.2	/	/
AGNet-ASL [35]	Res50	71.6	95.6	71.2	69.2	65.7	/	/	/
SAN [26]	Res50	72.5	93.3	79.7	78.4	75.6	/	/	/
AAVER [12]	Res50	61.2	89.0	74.7	68.6	63.5	/	/	/
SEVER [13]	Res50	79.6	96.4	79.9	77.6	75.3	83.4	78.7	71.3
VAMI [42]	Res50	61.3	89.5	63.1	52.9	47.3	/	/	/
DCDLearn [44]	Res50	70.4	92.8	82.9	78.7	75.9	/	/	/
CAL [29]	Res50	74.3	95.4	82.5	/	/	/	/	/
GB+GFB+SLB [15]	Res50	81.0	96.7	86.8	/	/	/	/	/
TransReID [10]	Vit-base	78.0	96.1	82.9	/	/	/	/	/
TransReID [10]	Swin-tiny	77.2	95.6	80.5	/	/	/	/	/
Baseline	Swin-Tiny	78.2	95.6	84.6	80.9	77.5	80.7	76.3	69.1
Baseline*	Res50	79.6	95.6	85.7	82.2	78.8	81.8	77.1	69.9
UFDN	Swin-Tiny	80.9	96.3	85.9	82.4	79.3	82.0	77.5	70.4
UFDN*	Res50	81.5	96.4	**88.4**	**84.8**	**80.6**	84.6	**79.4**	**72.0**

4.2   Implementation Details

We perform experiments in PyTorch on a machine with 8 NVIDIA V100 GPU. The images are resized to 224×224 for both training and testing, and the augmentation includes random erasing and flipping. We train the modules of UFDN together with a warmup strategy [9] and adapt different optimizers for different backbones: 1) Adam optimizer with the weight decay factor of 1e-4 for UFDN with the CNN backbone; 2) AdamW optimizer [21] with the weight decay factor of 1e-4 for UFDN with the Transformer backbone. For both optimizers, we initialize the learning rate as 3e-4. The hyper-parameter α in the memory bank mechanism is set to 0.9. Moreover, we set the hyper-parameter θ as 0.1 to control the distance between the positive distance and the negative distance.

4.3   Comparisons to State-of-the-Art Methods

We compare our UFDN with a wide range of state-of-the-art methods as shown in Table 1, including (1) part-based approaches: PGAN [37], PRN [7], PVEN [24], and GLAMOR [33]; (2) attribute-based approaches: AGNet-ASL [35] and SAN [26]; (3) attention-based approaches: AAVER [12] and SEVER [13]; (4) other interesting approaches: VAMI [42], DCDLearn [44], GB+GFB+SLB [15], CAL [29] and TransReID [10], and we get the following conclusions:

1) The CNN-based UFDN* has already achieved state-of-the-art performance on all three benchmarks by aligning the unstructured decoupled vehicle features.

Table 2. Ablation study of the components in UFDN (mAP on VeRi-776), where TDH represents the feature decomposing head and CDC represents the cluster-based decoupling constraint.

TDH	CDC	UFDN (Res50)	UFDN (Swin-Tiny)
✗	✗	79.6	78.2
✓	✗	81.0	80.0
✓	✓	81.5	80.9

Table 3. Ablation study of the depth of THD, where we report mAP for VeRi-776 and CMC@1 for VehicleID.

	UFDN (Res50)		UFDN (Swin-Tiny)	
Depth	VeRi-776	VehicleID	VeRi-776	VehicleID
1	80.2	86.5	79.2	84.3
2	**81.5**	**88.4**	**80.9**	**85.9**
3	80.9	87.5	79.7	84.9
4	79.1	85.2	78.2	83.0

The CNN-based baseline* reaches a 79.6% mAP on the VeRi-776 benchmark, and the UFDN* achieves an 81.5% mAP. Moreover, UFDN also achieves the best performance on VehicleID benchmarks that outperforms the second-best competitor GB+GFB+SLB [15] by 1.6% CMC@1 even though it borrows self-supervised representation learning to facilitate geometric features discovery.

2) We also compare UFDN with TransReID [10] (Swin-tiny). The baseline achieves 78.2% on VeRi-776 and the UFDN achieves a 2.7% mAP improvement. When comparing UFDN with the TransReID [10], we achieve a 3.7% improvement of mAP on VeRi-776 benchmark and a 5.4% improvement of CMC@1 on VehicleID benchmark. Please note that TransReID employs the camera and viewpoint labels as prior knowledge while our UFDN is annotation-free.

4.4 Ablation Study and Evaluation

Ablation Study on Components in UFDN. To evaluate the effectiveness of the two proposed modules in UFDN, we perform an ablation study by adding the components step-by-step in Table 2 on the VeRi-776 benchmark with either CNN-based or Transformer-based backbone. We keep all the hyper-parameters same to ensure a fair comparison, and get the conclusions:

1) After adding the TDH on the baseline, we observe the mAP on VeRi-776 increased by 1.4% and 1.8% on ResNet50 and Swin-Tiny backbones, respectively. It indicates that just decomposing the features from the channel dimension and then enhancing the feature representation by the transformer-based feature decomposing head can already bring a performance improvement.

2) Although TDH has already reached performance progress, it can't guarantee the diversity and alignment of the decomposed features which is realized by the CDC. We do ablation studies on CDC and find that the diversity and alignment constraint can drive an extra performance improvement over TDH, e.g., 0.5% and 0.9% mAP on the two backbones.

Ablation Study on Feature Decomposing Head. We use the self-attention mechanism in the feature decomposing head to preliminarily learn the decoupled features, and we do ablation studies on the depth of the decomposing head and the number of decoupled parts in this part.

Table 4. Ablation study of the number of decoupled parts in THD, where D_{Num} represents the number of the decoupled features. Moreover, we report CMC@1 for VehicleID and mAP for VeRi-776.

	UFDN (Swin-Tiny)		UFDN (Res50)	
D_{Num}	VeRi-776	VehicleID	VeRi-776	VehicleID
1	79.5	84.1	80.0	86.5
2	80.2	85.2	80.5	87.3
4	**80.9**	**85.9**	**81.5**	**88.4**
8	80.1	85.3	80.6	88.1
16	79.6	85.0	80.3	87.8

Table 5. Ablation study of the components in CDC (mAP on VeRi-776), where DC denotes the diversity constraint, AC denotes the alignment constraint, and MAC denotes the modulated alignment constraint.

DC	AC	MAC	UFDN (Res50)	UFDN (Swin-Tiny)
×	×	×	81.0	80.0
✓	×	×	81.2	80.5
×	✓	×	80.8	79.8
×	×	✓	81.1	80.4
✓	✓	×	81.2	80.3
✓	×	✓	**81.5**	**80.9**

1) The Depth of Feature Decomposing Head. In Table 3, we investigate the influence of the depth of the decomposing head on VeRi-776 and VehicleID benchmarks. Take UFDN (Res50) as an example: we achieve the best performance with the depth as 2, which achieves 88.4% CMC@1 on the VehicleID benchmark. But the models with a shallower depth (depth as 1) or a deeper one (depth as 4) show poor performance, which reaches 86.5% and 85.2% CMC@1 on the VehicleID benchmark, respectively. A similar phenomenon exists when experimenting with UFDN (Swin-Tiny) on both VeRi-776 and VehicleID benchmarks. Reasons for the above experiments can be concluded as: 1) TDH with shallow depth mines insufficient information from the input features and thus provides a poor performance; 2) TDH with a deeper depth is hard to train since the transformer-based networks rely heavily on large scale pre-training and the TDH added in our work is not pre-trained.

2) The Number of Decoupled Parts. We set the dimension of the output feature as 2048 in all our experiments for a fair comparison, and thus the number of decoupled parts needs to be divisible by 2048. We do ablation studies on the number of the decoupled parts in THD as shown in Table 4, and achieve the best performance when decoupling the features into four parts. Just changing the number of decoupled parts, there can be a great performance improvement, e.g., the performance improved from 86.5% CMC@1 in 1-part decoupled to 88.4% CMC@1 in 4-part decoupled on the VehicleID benchmark. We achieve two conclusions: 1) the decoupled features are not discriminative enough if decoupling it into 2 parts, and thus it's difficult to align them. 2) the ReID loss on each decoupled feature is hard to converge when we decouple a vehicle feature into too many parts and each of them contains little useful information.

Ablation Study on Cluster-Based Decoupling Constraint. We propose the cluster-based decoupling constraint to keep the diversity and alignment of the decoupled features, which consists of the diversity constraint, the alignment constraint, and the upgraded modulated alignment constraint. For figuring out

Table 6. Ablation study of the alignment constraint, and $\theta = 0$ denotes soft margin.

Scheme	θ	α	VeRi-776		VehicleID
			mAP	CMC@1	CMC@1
Scheme 1	0	0.5	80.7	95.8	87.0
Scheme 2	0.1	0.5	80.9	96.4	87.5
Scheme 3	0.3	0.5	80.7	96.1	87.2
Scheme 4	0.5	0.5	80.2	95.5	86.5
Scheme 5	0.1	0.9	**81.5**	**96.4**	**88.4**
Scheme 6	0.1	0.1	80.5	96.0	96.9

Table 7. Comparison of the computation cost and model size for different models. The mAP performance is evaluated on VeRi-776.

Method	Backbone	Speed (FPS)	Paras (M)	mAP
Baseline	Res50	234	27 M	79.6%
Baseline	Swin-T	262	32 M	78.2%
CAL	Res50	131	63 M	74.3%
TransReID	ViT-B	208	87 M	78.0%
UFDN	Res50	161	172 M	81.5%
UFDN	Swin-T	150	75 M	80.9%

how they impact the final performance, we do ablation studies on the three components and the hyperparameters in them.

1) Two steps in CDC. To explore the effect of the diversity constraint and the modulated alignment constraint, we do ablation studies as shown in Table 5: DC drives a consistent improvement on both UFDN (Res50) and UFDN (Swin-Tiny), but we should notice the different results after adding AC which shows a slight performance decrease on UFDN (Swin-Tiny). It is because the transformer-based network is more sensitive than the CNN-based network and can be easily affected by the negative impact from the outlier features.

To alleviate the problem, we propose the modulated alignment constraint (MAC) and compare the performance of AC and MAC as shown in Table 5. We speculate that MAC can eliminate the negative impact from the outlier features, and thus experiment baseline+DC+MAC outperforms experiment baseline+DC+AC on two kinds of backbones.

2) Hyper-parameters in CDC. During the design of the alignment constraint L_{align}, we validate the influence of different hyper-parameters on the final performance as shown in Table 6. Firstly, we use a margin θ in alignment constraint to control the distance between the positive distance and the negative distance, and the margin ranges between 0 and 1 since that we use Cosine distance. We experiment in Scheme 1–4 in Table 6 and find that when when $\theta = 0.1$.

Moreover, we also experiment on the memory bank parameter α which controls the smooth process of the cluster centers in Table 6 (Scheme 2,5,6). We find that a large $\alpha = 0.9$ drives the cluster centers to converge more stable and the best ReID performance.

Results of Computation Cost and Model Size. As mentioned before, we decouple the features by adding an extra transformer-based decomposing head and a decouple constraint and then examine the computation cost and model size in Table 7. Although the THD and CDC modules bring in some extra computation and time cost, UFDN has better trade-off between speed and accuracy when compared with SoTA methods like CAL or TransReID. The model size of UFDN can be further compressed by reducing feature dimension.

5   Visualizations and Discussion

5.1   Visualization of UFDN.

Figure 4 shows how the proposed UFDN decouples the Vehicle feature into four parts. The first column of images are the raw images for Vehicle ReID and the right four columns of images represent the gradient-based class activation [30] of different decoupled features. We find the information contained in the decoupled parts are diverse and unstructured, where column (b) pays more attention to the lights of vehicles, column (c) pays more attention to the background and counter information, column (d) focus more attention on the upper information and window information of vehicles, and column (e) focuses more attention on the global information such as colors, types.

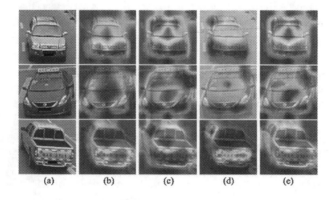

Fig. 4. The Visualization of attention map in reid branch.

Moreover, we also utilize T-SNE [23] to visualize different groups of decoupled features in Fig 5. The raw features (w/o CDC) and the decoupled features (w/ CDC) are visualized in Fig. 5(a) and Fig. 5(b), respectively. The results show that the decoupled features within the same group are clustered together and the decomposed features without CDC are indistinguishable. The above phenomenon validates the effectiveness of our cluster-based decouple constraint.

5.2   Comparison with the CNN-Based Person ReID Methods

Considering that Vehicle ReID comes under the larger object ReID, we compare with some person ReID methods [2,16] which focus on a similar problem with us. Although these methods [2,16] try to enrich the diversity of features by the diversity constraint, there still exist several differences: 1) we perform the

diverse constraint based on the one-dimensional decoupled features that contain more semantic information, but the methods [2, 16] are all operated on the
two-dimensional feature maps that are local specified and low-level. 2) UFDN
employs the self-attention (transformer) mechanism, which uses the extra decomposing token to learn the decoupled features, and thus the learning process of different decoupled parts are independent. However, the CNN-based methods [2, 16]
are limited by the reception field from the CNN, and different regions or channels
tend to have a shared region of interest. 3) UFDN further proposes the alignment
constraint for aligning the diverse features which can bring further performance
improvement as shown in Table 5. 4) We also compare UFDN with ABDNet [2]
(which is open-sourced) on VeRi-776, e.g., 81.5% (UFDN) VS 80.8% (ABDNet)
with the same backbone (ResNet50), which shows the priority of our method.

(a)   (b)

Fig. 5. Tsne visualization of a) the raw features (without decoupling constraint), b) the
decoupled features (with decoupling constraint). We decouple the vehicle feature into
four parts, where different color represents four groups of decoupled features. (Color
figure online)

Table 8. Comparison with ABDNet on VeRi-776, where 'W-norm', 'Div' and 'Align'
denote weight orthogonality regularizers, the diversity constraint and the alignment
constraint, respectively.

Method	Backbone	Diverse-method	mAP	CMC@1
ABDNet [2]	ResNet50	Div+W-norm	80.8%	96.1%
UFDN	ResNet50	Div	81.2%	96.2%
UFDN	ResNet50	Div+Align	81.5%	96.4%

6   Conclusions

In this paper, we introduce the unstructured feature decoupling network (UFDN)
that aims to decouple the vehicle features into unstructured parts and align
them without extra annotation. The UFDN consists of a transformer-based feature decomposing head (TDH) and a cluster-based decouple constraint (CDC).
We evaluate our UFDN on three popular benchmarks (VeRi-776, VehicleID, and

VERI-WILD) and two backbones (ResNet50 and Swin-Tiny) and achieve competitive results when comparing with other works. Firstly, the improvements from the TDH demonstrate that the self-attention mechanism can be used to encode the discriminative information of channel-wise decomposed features into the final decomposing tokens. Secondly, the CDC can force the relative decomposed features from different images to have a similar region of interest. Finally, the improvement achieved by UFDN proves that the decompose and decouple processes in UFDN are effective and can lead to performance progress.

Acknowledgements. This work was supported by the National Science Foundation of China under Grant NSFC 61906194 and the National Key R&D Program of China under Grant 2021YFF0602101. This work was supported by Alibaba Group through Alibaba Research Intern Program.

References

1. Chen, H., Lagadec, B., Bremond, F.: Partition and reunion: A two-branch neural network for vehicle re-identification. In: CVPR Workshops, pp. 184–192 (2019)
2. Chen, T., et al.: Abd-net: Attentive but diverse person re-identification. In: Proceedings of the IEEE/CVF International Conference on Computer Vision, pp. 8351–8361 (2019)
3. Chen, T.-S., Liu, C.-T., Wu, C.-W., Chien, S.-Y.: Orientation-aware vehicle re-identification with semantics-guided part attention network. In: Vedaldi, A., Bischof, H., Brox, T., Frahm, J.-M. (eds.) ECCV 2020. LNCS, vol. 12347, pp. 330–346. Springer, Cham (2020). https://doi.org/10.1007/978-3-030-58536-5_20
4. Chen, Y., Jing, L., Vahdani, E., Zhang, L., He, M., Tian, Y.: Multi-camera vehicle tracking and re-identification on ai city challenge 2019. In: CVPR Workshops, vol. 2 (2019)
5. Guo, H., Zhao, C., Liu, Z., Wang, J., Lu, H.: Learning coarse-to-fine structured feature embedding for vehicle re-identification. In: Thirty-Second AAAI Conference on Artificial Intelligence (2018)
6. Guo, H., Zhao, C., Liu, Z., Wang, J., Lu, H.: Learning coarse-to-fine structured feature embedding for vehicle re-identification. In: McIlraith, S.A., Weinberger, K.Q. (eds.) AAAI, pp. 6853–6860. AAAI Press (2018). www.aaai.org/ocs/index.php/AAAI/AAAI18/paper/view/16206
7. He, B., Li, J., Zhao, Y., Tian, Y.: Part-regularized near-duplicate vehicle re-identification. In: Proceedings of the IEEE Conference on Computer Vision and Pattern Recognition, pp. 3997–4005 (2019)
8. He, K., Gkioxari, G., Dollár, P., Girshick, R.: Mask r-cnn. In: Proceedings of the IEEE international conference on computer vision, pp. 2961–2969 (2017)
9. He, K., Zhang, X., Ren, S., Sun, J.: Deep residual learning for image recognition. In: 2016 IEEE Conference on Computer Vision and Pattern Recognition, CVPR 2016, Las Vegas, NV, USA, June 27–30, 2016, pp. 770–778. IEEE Computer Society (2016). https://doi.org/10.1109/CVPR.2016.90
10. He, S., Luo, H., Wang, P., Wang, F., Li, H., Jiang, W.: Transreid: Transformer-based object re-identification. In: Proceedings of the IEEE/CVF International Conference on Computer Vision (2021)

11. Khamis, S., Kuo, C.-H., Singh, V.K., Shet, V.D., Davis, L.S.: Joint learning for attribute-consistent Person re-identification. In: Agapito, L., Bronstein, M.M., Rother, C. (eds.) ECCV 2014. LNCS, vol. 8927, pp. 134–146. Springer, Cham (2015). https://doi.org/10.1007/978-3-319-16199-0_10
12. Khorramshahi, P., Kumar, A., Peri, N., Rambhatla, S.S., Chen, J.C., Chellappa, R.: A dual-path model with adaptive attention for vehicle re-identification. In: Proceedings of the IEEE International Conference on Computer Vision, pp. 6132–6141 (2019)
13. Khorramshahi, P., Peri, N., Chen, J., Chellappa, R.: The devil is in the details: self-supervised attention for vehicle re-identification. In: Vedaldi, A., Bischof, H., Brox, T., Frahm, J.-M. (eds.) ECCV 2020. LNCS, vol. 12359, pp. 369–386. Springer, Cham (2020). https://doi.org/10.1007/978-3-030-58568-6_22
14. Khorramshahi, P., Rambhatla, S.S., Chellappa, R.: Towards accurate visual and natural language-based vehicle retrieval systems. In: Proceedings of the IEEE/CVF Conference on Computer Vision and Pattern Recognition (CVPR) Workshops, pp. 4183–4192 (2021)
15. Li, M., Huang, X., Zhang, Z.: Self-supervised geometric features discovery via interpretable attention for vehicle re-identification and beyond. In: Proceedings of the IEEE/CVF International Conference on Computer Vision, pp. 194–204 (2021)
16. Li, S., Bak, S., Carr, P., Wang, X.: Diversity regularized spatiotemporal attention for video-based person re-identification. In: Proceedings of the IEEE Conference on Computer Vision and Pattern Recognition, pp. 369–378 (2018)
17. Lin, Y., et al.: Improving person re-identification by attribute and identity learning. Pattern Recogn. **95**, 151–161 (2019)
18. Liu, H., Tian, Y., Yang, Y., Pang, L., Huang, T.: Deep relative distance learning: Tell the difference between similar vehicles. In: Proceedings of the IEEE Conference on Computer Vision and Pattern Recognition, pp. 2167–2175 (2016)
19. Liu, W., Anguelov, D., Erhan, D., Szegedy, C., Reed, S., Fu, C.-Y., Berg, A.C.: SSD: single shot multibox detector. In: Leibe, B., Matas, J., Sebe, N., Welling, M. (eds.) ECCV 2016. LNCS, vol. 9905, pp. 21–37. Springer, Cham (2016). https://doi.org/10.1007/978-3-319-46448-0_2
20. Liu, X., Liu, W., Mei, T., Ma, H.: Provid: progressive and multimodal vehicle reidentification for large-scale urban surveillance. IEEE Trans. Multimedia **20**(3), 645 658 (2017)
21. Loshchilov, I., Hutter, F.: Fixing weight decay regularization in adam. CoRR abs/1711.05101 (2017), arxiv.org/abs/1711.05101
22. Lou, Y., Bai, Y., Liu, J., Wang, S., Duan, L.: Veri-wild: A large dataset and a new method for vehicle re-identification in the wild. In: Proceedings of the IEEE Conference on Computer Vision and Pattern Recognition, pp. 3235–3243 (2019)
23. Van der Maaten, L., Hinton, G.: Visualizing data using t-sne. J. Mach. Learn. Res. **9**(2605), 2579–2605 (2008)
24. Meng, D., et al.: Parsing-based view-aware embedding network for vehicle re-identification. In: Proceedings of the IEEE/CVF Conference on Computer Vision and Pattern Recognition, pp. 7103–7112 (2020)
25. Mo, W., Lv, J.: Cascaded hierarchical context-aware vehicle re-identification. In: 2021 International Joint Conference on Neural Networks (IJCNN), pp. 1–8. IEEE (2021)
26. Qian, J., Jiang, W., Luo, H., Yu, H.: Stripe-based and attribute-aware network: A two-branch deep model for vehicle re-identification. Measurement Science and Technology (2020)

27. Qian, W., He, Z., Peng, S., Chen, C., Wu, W.: Pseudo graph convolutional network for vehicle reid. In: Proceedings of the 29th ACM International Conference on Multimedia. p. 3162–3171. MM '21, Association for Computing Machinery, New York, NY, USA (2021). https://doi.org/10.1145/3474085.3475462
28. Qian, W., Yang, X., Peng, S., Yan, J., Guo, Y.: Learning modulated loss for rotated object detection. In: Proceedings of the AAAI Conference on Artificial Intelligence, vol. 35, pp. 2458–2466 (2021)
29. Rao, Y., Chen, G., Lu, J., Zhou, J.: Counterfactual attention learning for fine-grained visual categorization and re-identification. In: Proceedings of the IEEE/CVF International Conference on Computer Vision, pp. 1025–1034 (2021)
30. Selvaraju, R.R., Cogswell, M., Das, A., Vedantam, R., Parikh, D., Batra, D.: Grad-cam: Visual explanations from deep networks via gradient-based localization. In: Proceedings of the IEEE international conference on computer vision, pp. 618–626 (2017)
31. Shen, F., Xie, Y., Zhu, J., Zhu, X., Zeng, H.: Git: Graph interactive transformer for vehicle re-identification. arXiv preprint arXiv:2107.05475 (2021)
32. Sun, Y., Zheng, L., Yang, Y., Tian, Q., Wang, S.: Beyond part models: Person retrieval with refined part pooling (and a strong convolutional baseline). In: Proceedings of the European Conference on Computer Vision (ECCV), pp. 480–496 (2018)
33. Suprem, A., Pu, C.: Looking glamorous: Vehicle re-id in heterogeneous cameras networks with global and local attention. arXiv preprint arXiv:2002.02256 (2020)
34. Touvron, H., Cord, M., Sablayrolles, A., Synnaeve, G., Jégou, H.: Going deeper with image transformers. In: Proceedings of the IEEE/CVF International Conference on Computer Vision (2021)
35. Wang, H., Peng, J., Chen, D., Jiang, G., Zhao, T., Fu, X.: Attribute-guided feature learning network for vehicle re-identification. arXiv preprint arXiv:2001.03872 (2020)
36. Wang, H., Peng, J., Jiang, G., Xu, F., Fu, X.: Discriminative feature and dictionary learning with part-aware model for vehicle re-identification. Neurocomputing **438**, 55–62 (2021)
37. Zhang, X., Zhang, R., Cao, J., Gong, D., You, M., Shen, C.: Part-guided attention learning for vehicle re-identification. CoRR abs/1909.06023 (2019), arxiv.org/abs/1909.06023
38. Zhang, Z., Lan, C., Zeng, W., Jin, X., Chen, Z.: Relation-aware global attention for person re-identification. In: Proceedings of the IEEE/CVF Conference on Computer Vision and Pattern Recognition, pp. 3186–3195 (2020)
39. Zhao, Y., Shen, C., Wang, H., Chen, S.: Structural analysis of attributes for vehicle re-identification and retrieval. IEEE Trans. Intell. Transp. Syst. **21**(2), 723–734 (2019)
40. Zheng, A., Lin, X., Li, C., He, R., Tang, J.: Attributes guided feature learning for vehicle re-identification. arXiv preprint arXiv:1905.08997 (2019)
41. Zhou, J., Su, B., Wu, Y.: Online joint multi-metric adaptation from frequent sharing-subset mining for person re-identification. In: Proceedings of the IEEE/CVF Conference on Computer Vision and Pattern Recognition, pp. 2909–2918 (2020)

42. Zhou, Y., Shao, L.: Viewpoint-aware attentive multi-view inference for vehicle re-identification. In: 2018 IEEE Conference on Computer Vision and Pattern Recognition, CVPR 2018, Salt Lake City, UT, USA, June 18–22, 2018, pp. 6489–6498. IEEE Computer Society (2018). https://doi.org/10.1109/CVPR.2018. 00679.   openaccess.thecvf.com/content_cvpr_2018/html/Zhou_Viewpoint-Aware_Attentive_Multi-View_CVPR_2018_paper.html
43. Zhu, J., et al.: Vehicle re-identification using quadruple directional deep learning features. IEEE Trans. Intell. Transp. Syst. **21**(1), 410–420 (2019)
44. Zhu, R., Fang, J., Xu, H., Yu, H., Xue, J.: Dcdlearn: Multi-order deep cross-distance learning for vehicle re-identification. arXiv preprint arXiv:2003.11315 (2020)

Deep Hash Distillation for Image Retrieval

Young Kyun Jang[1], Geonmo Gu[2], Byungsoo Ko[2], Isaac Kang[1],
and Nam Ik Cho[1,3(✉)]

[1] ECE and INMC, Seoul National University, Seoul, Korea
{isaackang,nicho}@snu.ac.kr
[2] NAVER Vision, Seongnam, South Korea
[3] IPAI, Seoul National University, Seoul, Korea

Abstract. In hash-based image retrieval systems, degraded or transformed inputs usually generate different codes from the original, deteriorating the retrieval accuracy. To mitigate this issue, data augmentation can be applied during training. However, even if augmented samples of an image are similar in real feature space, the quantization can scatter them far away in Hamming space. This results in representation discrepancies that can impede training and degrade performance. In this work, we propose a novel self-distilled hashing scheme to minimize the discrepancy while exploiting the potential of augmented data. By transferring the hash knowledge of the weakly-transformed samples to the strong ones, we make the hash code insensitive to various transformations. We also introduce hash proxy-based similarity learning and binary cross entropy-based quantization loss to provide fine quality hash codes. Ultimately, we construct a deep hashing framework that not only improves the existing deep hashing approaches, but also achieves the state-of-the-art retrieval results. Extensive experiments are conducted and confirm the effectiveness of our work. Code is at https://github.com/youngkyunJang/Deep-Hash-Distillation.

Keywords: Large-scale image retrieval · Learning to hash · Self-distillation

1 Introduction

Especially for retrieval from large-scale databases, *hashing* is essential due to its practicality, *i.e.*, high search speed and low storage cost. By converting high-dimensional data points into compact binary codes with a hash function, the retrieval system can utilize a simple bit-wise XOR operation to define a distance between the images. A wide variety of works have been studied for learning to hash [15,33,38,41,43], and are still being actively pursued to build fast and accurate retrieval systems.

Supplementary Information The online version contains supplementary material available at https://doi.org/10.1007/978-3-031-19781-9_21.

Recently, techniques for hash learning have been significantly advanced by deep learning, which is called *deep hashing*, and its corresponding works are in the spotlight [1,11,20,29,39,42,46,50]. By integrating the hash function into the deep learning framework, the image encoder and hash function are simultaneously learned to generate image hash codes. Regarding the training of deep hashing, the leading techniques are pairwise similarity approaches that use sets of similar or dissimilar image pairs [3,4,21,22,44,51], and global similarity in company with classification approaches that use class labels assigned to images [23,24,47].

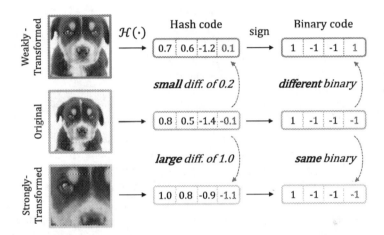

Fig. 1. Visualization of possible problems in deep hashing due to transformations. The continuous hash code generated by a deep hashing model $\mathcal{H}(\cdot)$ is changed when the input is transformed. In consequence, the binary code quantized with sign operation can also be shifted. However, the degree of transformation cannot be properly reflected in the quantized representation.

Since hash-based retrieval systems compute the distance between images with binary codes, corresponding codes need to be quantized with *sign* operation, from the continuous real space to the discrete Hamming space of $\{-1, 1\}$. In this process, the continuously optimized image representation is altered, and quantization error occurs, which in turn degrades the discriminative capability of the hash code. This becomes even more problematic when an input image is transformed and deviated from the original distribution.

To avoid performance degradation due to transformations, the most common solution is to generalize the deep model by training it with augmented data having various transformations. However, it is challenging to apply this augmentation strategy to deep hash training since discrepancy in the representation may occur. Figure 1 shows an example case that may appear: 1) *The sign of the hash code can be shifted with a slight change.* Specifically, the last element of the weakly-transformed image's hash code differs by 0.2 ($-0.1 \rightarrow 0.1$) from the original, but it results in $-1 \rightarrow 1$ shift in the Hamming space. 2) *The sign of the quantized hash code does not shift even with the big change in the hash code.* The last element of a strongly transformed image's hash code differs by 1.0 ($-0.1 \rightarrow -1.1$) from the original, resulting in no shifts in the Hamming space. Namely, the use of strong augmentation in deep hashing increases the discrepancy between Hamming and real space, which hinders finding the optimal binary code.

To resolve this issue, we introduce a novel concept dubbed *Self-distilled Hashing*, which customizes self-distillation [5,32,40,45,49] to prevent severe discrepancy in deep hash training. Specifically, based on the understanding of the relation between cosine distance and Hamming agreement [19,33,48], we minimize the cosine distance between the hash codes of two different views (transformed results) of an image to maximize the Hamming agreement between their binary outcomes. Further for stable learning, we separate the difficulties of transformations as easy and difficult, and transfer the hash knowledge from easy to difficult, inspired by [5,8,40].

Moreover, we propose two additional training objectives that optimize hash codes to enhance the self-distilled hashing: 1) a hash proxy-based similarity learning, and 2) a binary cross entropy-based quantization loss. The first term allows the deep hashing model to learn global (inter-class) discriminative hash codes with temperature-scaled cosine similarity. The second term contributes to making the hash code naturally move away from the binary threshold in a classification manner with likelihood estimators.

By combining all of our proposals, we construct a **D**eep **H**ash **D**istillation framework (DHD), which yields discriminative and transformation resilient hash codes for fast image retrieval. We conduct extensive experiments on single and multi-labeled benchmark datasets for image hashing evaluation. In addition, we validate the effectiveness of self-distilled hashing using data augmentation on the existing methods [3,4,14,47] and show the performance improvements. Furthermore, we establish that DHD is applicable with a variety of deep backbone architectures including vision Transformers [12,34,40]. Experimental results verify that self-distilled hashing strategy improves the existing works, and entire DHD framework shows the best retrieval performance.

We can summarize our contributions as follows:

- To the best of our knowledge, this is the first work to address the discrepancy between real and Hamming space provoked by data augmentation in deep hashing.
- With the introduction of self-distilled hashing scheme and training loss functions, we successfully embed the power of augmentations into the hash codes.
- Extensive experiments demonstrate the benefits of our work, improving previous deep hashing methods and achieving the state-of-the-art performances.

2   Related Works

For a better understanding, we present a brief introduction to the deep hashing methods and the research that inspired our proposal. Refer to a survey [41] to see details of the early works in non-deep hashing approaches (ITQ [15], SH [43], KSH [33], SDH [38]).

Deep Hashing Methods. Hashing algorithms using deep learning techniques such as Convolutional Neural Network (CNN) are leading the mainstream with striking results. For example, CNNH [44] utilizes a CNN to generate compact hash codes by training a network with given pairwise label information. DHN [51] learns hash codes by approximating discrete values with relaxation and trains them with supervised signals. HashNet [4] adopts the inner product to measure pairwise similarity between hash codes and tackles the data imbalance problem by employing weighted maximum likelihood estimation. DCH [3] employs Cauchy distribution to minimize the Hamming distance of the images with the same class label.

Hash Center-Based Methods. There have been several methods to find out class-wise hash representatives (centers), which can provide global similarity to hash codes by including the process of predicting image class labels with hash codes during training [21,23,24,47]. CSQ [47] uses pre-defined orthogonal binary hash targets to guarantee a certain Hamming distance between classes and makes hash codes follow the targets. DPN [14] employs randomly assigned target vectors with maximal inter-class similarity and utilizes bit-wise hinge-like loss. Unlike DPN and CSQ, which use a hash target that is not trainable, in

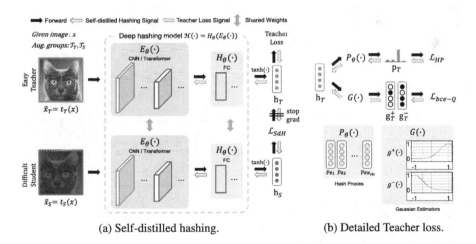

(a) Self-distilled hashing.   (b) Detailed Teacher loss.

Fig. 2. The overall training process of Deep Hash Distillation (DHD) framework. (a) From two different augmentation groups, namely Teacher $\mathcal{T}_T$ and Student $\mathcal{T}_S$, randomly sampled transformations ($t_T \sim \mathcal{T}_T$, $t_S \sim \mathcal{T}_S$) are individually applied on the input image x to produce $\tilde{x}_T$ and $\tilde{x}_S$. The deep hashing model $\mathcal{H}(\cdot)$ constructed with the deep encoder $E_\theta(\cdot)$ and the hash function $H_\theta(\cdot)$ of Fully-Connected (FC) layers yields two hash codes $\mathbf{h}_T$ and $\mathbf{h}_S$ which are learned with $\mathcal{L}_{SdH}$. We apply stop gradient operation on $\mathbf{h}_T$ for stable training. (b) Additionally, we employ trainable hash proxies $P_\theta(\cdot)$ which are used to calculate the class-wise prediction $\mathbf{p}_T$ with $\mathbf{h}_T$ to optimize with $\mathcal{L}_{HP}$, and pre-defined Gaussian estimator $G(\cdot)$ to regularize $\mathbf{h}_T$ with $\mathcal{L}_{bce\text{-}Q}$.

our DHD, the hash center is set as a trainable proxy which jointly learns the similarity with the hash codes during training.

Self-Distillation. Inspired by knowledge distillation [18], self-distillation emerged as a concept that employs a single network to generalize itself in a self-taught fashion, and plenty of works demonstrated its benefits in improving deep model performance [5,40,45,49]. Many of them utilize a simple Siamese architecture [8] to explore and learn the visual representation with data augmentation, by contrasting two different augmented results of one image. Similarly, we conduct the self-distillation with augmentations in deep hashing to see the hash codes of two different views of an image simultaneously. Additionally, in accordance with the characteristics of hashing, we consider a method of minimizing the cosine distance that behaves similarly to the distance in the Hamming space to reduce the representation discrepancy during model training.

3   Method

The goal of a deep hashing model $\mathcal{H}$ of Deep Hash Distillation (DHD) is to map an input image x to a K-dimensional binary code $b \in \{-1,1\}^K$ in Hamming space. For this purpose, $\mathcal{H}$ is optimized to find a high quality real-valued hash code h, and then sign operation is utilized to quantize h as b. Instead of including non-differentiable quantization process in model training, we learn $\mathcal{H}$ in the real space to estimate optimal b with continuously relaxed h while fully exploiting the power of data augmentation. We notate trainable components with θ as a subscript. In the following, h becomes robust to transformations in 3.1, and becomes discriminative and binary-like in 3.2.

3.1   Self-distilled Hashing

In general, $\mathcal{H}$ is trained in the real space to obtain discriminative h, which should maintain its property in the Hamming space even if quantized to b. Therefore, it is important to align h and b to carry a similar representation during training. However, when data augmentation is applied to the training input and the following change occurs in h, there can be misalignment between h and b, as shown in Figure. 1. Thus, direct use of augmentations can cause discrepancies in the representation between h and b, which degrades retrieval performance as we observed in Sec 4.3.

Hamming Distance as Cosine Distance. It is noteworthy that the Hamming distance between the binary codes can be interpreted as cosine distance (1-cosine similarity)[1] [19,33,48]. That is, the cosine similarity between hash codes h_i and h_j can be utilized to approximate the Hamming distance between the binary codes b_i and b_j as:

$$\mathcal{D}_H(b_i, b_j) \simeq \frac{K}{2}(1 - \mathcal{S}(h_i, h_j)) \tag{1}$$

---

[1] Refer supplementary material for proof.

where $b_i = \text{sign}(h_i)$, $b_j = \text{sign}(h_j)$, $\mathcal{D}_H(\cdot, \cdot)$ denotes Hamming distance, $\mathcal{S}(\cdot, \cdot)$ denotes cosine similarity. That is, the minimized cosine distance between the hash codes minimizes the Hamming distance between the binary codes.

Easy-Teacher and Difficult-Student. As shown in Fig. 2a, we propose a self-distilled hashing scheme, which supports the training of deep hashing models with augmentations. We employ weight-sharing Siamese structure [26] to contrast hash codes of two different views (augmentation results) of an image at once. According to the observations in self-distillation works [5,8], keeping the output representation of one branch steady has a significant impact on performance gain. Therefore, we configure two separate augmentation groups to provide input views with different difficulties of transformation: one is weakly-transformed easy teacher $\mathcal{T}_T$, and the other is strongly-transformed difficult student $\mathcal{T}_S$. Here, we control the difficulty in a stochastic sampling manner as: employing the same hyper-parameter s_T to all transformations in the group, and make them occur less (weakly) or more (strongly) by scaling their own probability of occurrence. While this manner makes the teacher representation stable, it has the advantage that few extreme examples that produce unstable results are not completely ruled out and contribute to learning. Besides, we stop the gradient of the teacher view's corresponding hash codes to avoid collapsing into trivial solutions [5,8].

Loss Computation. For a given image x, self-distillation is conducted with image views as: $\tilde{x}_T = t_T(x)$ and $\tilde{x}_S = t_S(x)$, where t_T, t_S are randomly sampled transformations from $\mathcal{T}_T, \mathcal{T}_S$, respectively. The deep encoder E_θ and the hash function H_θ take $\tilde{x}_T$ and $\tilde{x}_S$ as inputs and produce corresponding hash code h_T and h_S. Then, the proposed Self-distilled Hashing (SdH) loss is computed as:

$$\mathcal{L}_{SdH}(h_T, h_S) = 1 - \mathcal{S}(h_T, h_S) \qquad (2)$$

Optimizing $\mathcal{H}$ with $\mathcal{L}_{SdH}$ results in the alignment of h_T and h_S, and thus b_T and b_S as follows Eq. 1, which in turn reduces the discrepancy in representation between two differently transformed output binary codes.[2]

Flexibility. Note that self-distilled hashing is applicable to the other common deep hashing models [3,4,14,47] with regard to exploiting data augmentation during training, as shown in Sect. 4.3. Furthermore, various backbones [12,17,27, 34,40] can be utilized as deep encoder, and any hash function H_θ configuration is compatible. For simplicity, we employ a single FC layer to obtain a hash code of the desired bits, and apply *tanh* operation at the end to be bound in $[-1, 1]$.

3.2   For Better Teacher

Besides self-distilled hashing, additional training signals such as supervised learning loss, and quantization loss are required to obtain the discriminative hash codes. We only employ teacher hash codes to compute the losses, in order to transfer the learned hash knowledge to the student's codes.

---

[2] We provide a pseudo-code implementation in supplementary material.

Proxy-Based Similarity Learning. Supervised hash similarity learning with pre-defined orthogonal binary hash targets has shown great performance [14,19,47]. However, the hash target has limitations in that 1) it requires a complex initialization process, and 2) it allocates the same Hamming distance between centers so detailed distances according to semantic similarity cannot be learned. Therefore, as shown in Fig. 2b, we introduce a proxy-based representation learning [7,16,36] in deep hashing by using a collection of trainable hash proxies P_θ. It has the advantage that the proxies are simply initialized with randomness, and being able to learn semantic similarity into the proxies. In terms of training, we first use P_θ to compute class-wise prediction p_T with h_T as:

$$p_T = [\mathcal{S}(p_{\theta 1}, h_T), \mathcal{S}(p_{\theta 2}, h_T), ..., \mathcal{S}(p_{\theta N_{cls}}, h_T)] \tag{3}$$

where p_{θ_i} is a hash proxy assigned to each of the i-th class and N_{cls} denotes the number of classes to be distinguished. Then, we use p_T to learn the similarity with class label y by computing Hash Proxy (HP) loss as:

$$\mathcal{L}_{HP}(y, p_T, \tau) = H(y, \mathrm{Softmax}(p_T/\tau)) \tag{4}$$

where τ is a temperature scaling hyper-parameter, $H(u, v) = -\sum_k u_k \log v_k$ is a cross entropy, and Softmax operation is applied along the dimension of p_T. Note that, similar to Eq. 1, $\mathcal{L}_{HP}$ is designed to learn Hamming agreement with temperature scaling.

Reducing Quantization Error. To make continuous hash code elements act like binary bits, the deep hashing methods [3,4,23,47] aim to reduce the quantization error by minimizing the distance (e.g. Euclidean) between the hash code bit and its closest binary goal ($+1$ or -1) in a regression manner. However, since the purpose of hashing is to classify the sign of each bit, it is a more natural choice to view it as a binary classification: maximum likelihood problem. Hence, we adopt a pre-defined Gaussian distribution estimator $g(h)$ of mean m and standard deviation σ as:

$$g(h) = \exp\left(-\frac{(h-m)^2}{2\sigma^2}\right) \tag{5}$$

to evaluate the binary likelihood of hash code element h. By employing two estimators: g^+ of $m = 1$, and g^- of $m = -1$ with the same σ, we compute the likelihoods and a Binary Cross Entropy-based (BCE) quantization loss as:

$$\mathcal{L}_{bce\text{-}Q}(h_T) = \frac{1}{K}\sum_{k=1}^{K}\left(H_b\left(b_k^+, g_k^+\right) + H_b\left(b_k^-, g_k^-\right)\right) \tag{6}$$

where $H_b(u, v) = -(u \log v + (1-u)\log(1-v))$ is a binary cross entropy, g_k^+, g_k^- denotes k-th hash code element's estimated likelihood: $g_k^+ = g^+(h_k)$, $g_k^- = g^-(h_k)$, and b_k^+, b_k^- denotes binary likelihood labels which are obtained (refer Fig. 2b) as:

$$b_k^+ = \frac{1}{2}\left(\mathrm{sign}(h_k) + 1\right), b_k^- = 1 - b_k^+ \tag{7}$$

As a result, quantization error is reduced by a binary classification loss with the given estimators, allowing to use the merits of cross entropy presented in [2]. Note that, $\mathcal{L}_{bce\text{-}Q}$ is also applied to hash proxies to make them act as continuously relaxed binary codes.

3.3   Training

Total Training Loss. Suppose we are given a training mini-batch of N_B data points: $\mathcal{X}_B = \{(x_1, y_1), ..., (x_{N_B}, y_{N_B})\}$ where each image x_i is assigned a label $y_i \in \{0, 1\}^{N_{cls}}$. Training views are obtained as $\tilde{x}_{Ti} - t_{Ti}(x_i)$ and $\tilde{x}_{Si} = t_{Si}(x_i)$ for all data points, where $t_{Ti} \sim \mathcal{T}_T$ and $t_{Si} \sim \mathcal{T}_S$. Total loss $\mathcal{L}_T$ for DHD is computed with $\mathcal{X}_B$ as:

$$\mathcal{L}_T(\mathcal{X}_B) = \frac{1}{N_B} \sum_{n=1}^{N_B} (\mathcal{L}_{HP} + \lambda_1 \mathcal{L}_{SdH} + \lambda_2 \mathcal{L}_{bce\text{-}Q}) \qquad (8)$$

where λ_1 and λ_2 are hyper-parameters that balance the influence of the training objectives. The entire DHD framework is trained in an end-to-end fashion.

Multi-label Case. In the case of determining semantic similarity between multi-hot labeled images, the previous works [44,47,51] simply checked whether the images share at least one positive label or not. However, learning with the above similarity has limitations in that the label dependency [9] is ignored. Thus, we aim to capture the intelligence that appears in label dependency by utilizing the Softmax cross entropy with the normalized multi-hot label y. Specifically, y is converted as $y = y/\|y\|_1$ to balance the contribution of each label, and the same $\mathcal{L}_{HP}$ is computed to optimize the deep hashing model for multi-label image retrieval.

4   Experiments

4.1   Setup

Datasets. To evaluate our DHD, we conduct experiments against several conventional and modern methods. Three most popular hashing based retrieval benchmark datasets are explored[3], and we explain the composition of each dataset in Table 1.

Table 1. Description of the image retrieval datasets.

Dataset	# Database	# Train	# Query	N_c
ImageNet [37]	128,503	13,000	5,000	100
NUS-WIDE [10]	149,736	10,500	2,100	21
MS COCO [31]	117,218	10,000	5,000	80

[3] The details of each dataset are described in the supplementary material.

Evaluation Metrics. We follow the protocol utilized in deep hashing [3,4, 47] to evaluate our approach on both single-labeled and multi-labeled datasets. Specifically, we employ three metrics: 1) mean average precision (**mAP**), 2) precision-recall curves (**PR curves**), and 3) precision with respect to top-M returned image (**P@Top-M**). Regarding mAP score computation, we select the top-M images from the retrieval ranked-list results. The returned images and the query image are considered relevant whether one or more class labels are the same. We set binary code length: hash code dimensionality K as 16, 32, and 64, to examine the performance according to the code size.

4.2   Implementation Details

Data Augmentation. Following the works presented in [6], we choose family $\mathcal{T}$ of five image transformations: 1) resized crop, 2) horizontal flip, 3) color jitter, 4) grayscale, and 5) blur, where all of each are sampled uniformly with a given probability and sequentially applied to the inputs. We keep the internal parameters of each transformation equal to [6]. For self-distilled hashing, we configure two groups with $\mathcal{T}$, where the difficult student group is $\mathcal{T}_S = \mathcal{T}$, and the easy teacher group $\mathcal{T}_T$ is configured by scaling all transform occurrence with s_T, which is in the range of $(0,1]$. We set $\mathcal{T}_T$ as the default for the methods trained without SdH.

Table 2. Mean Average Precision (mAP) scores for different bits on three benchmarks.

Method	Backbone	ImageNet			NUS-WIDE			MS COCO		
		16-bit	32-bit	64-bit	16-bit	32-bit	64-bit	16-bit	32-bit	64-bit
ITQ [15]	Non-deep	0.266	0.436	0.576	0.435	0.396	0.365	**0.566**	0.562	0.502
SH [43]		0.210	0.329	0.418	0.401	0.421	0.423	0.495	0.507	0.510
KSH [43]		0.160	0.298	0.394	0.394	0.407	0.399	0.521	0.534	0.536
SDH [38]		**0.299**	**0.455**	**0.585**	**0.575**	**0.590**	**0.613**	0.554	**0.564**	**0.580**
CNNH [44]	AlexNet [27]	0.315	0.473	0.596	0.655	0.659	0.647	0.599	0.617	0.620
DNNH [28]		0.353	0.522	0.610	0.703	0.738	0.754	0.644	0.651	0.647
DHN [51]		0.367	0.522	0.627	0.712	0.739	0.751	0.701	0.710	0.735
HashNet [4]		0.425	0.559	0.649	0.720	0.745	0.758	0.685	0.714	0742
DCH [3]		0.636	0.645	0.656	0.740	0.752	0.763	0.695	0.721	0.748
DHD(Ours)		**0.657**	**0.701**	**0.721**	**0.780**	**0.805**	**0.820**	**0.749**	**0.781**	**0.792**
DPN [14]	ResNet [17]	0.828	0.863	0.872	0.783	0.816	0.838	0.796	0.838	0.861
CSQ [47]		0.851	0.865	0.873	0.810	0.825	0.839	0.750	0.824	0.852
DHD(Ours)		**0.864**	**0.891**	**0.901**	**0.820**	**0.839**	**0.850**	**0.839**	**0.873**	**0.889**
	ViT [12]	0.927	0.938	0.944	0.837	0.862	0.870	0.886	0.919	0.939
	DeiT [40]	0.932	0.943	0.948	0.839	0.861	0.867	0.883	0.913	0.925
DHD(Ours)	SwinT [34]	**0.944**	**0.955**	**0.956**	**0.848**	**0.867**	**0.875**	**0.894**	**0.930**	**0.945**

Experiments. Retrieval experiments are conducted by dividing backbones as: Non-deep, AlexNet [27], ResNet (ResNet50) [17] , and vision Transformers [12, 34,40]. For non-deep hashing approaches: ITQ [15], SH [43], KSH [33] and SDH [38], we report the results directly from the latest works [3,4,47] for comparison. We set up the same training environment by leveraging PyTorch framework and the image transformation functions of kornia [13] library for augmentation. We employ Adam optimizer [25] and decay the learning rate with cosine scheduling [35] for training deep hashing methods. Especially for DHD hyper-parameters[4], s_T is set to 0.2 for AlexNet, and 0.5 for other backbones. τ is set by considering N_{cls} as $\{0.2, 0.6, 0.4\}$ for {ImageNet, NUS-WIDE, MS COCO}, respectively. λ_1 and λ_2 are set equal to 0.1 for a balanced contributions each training objective, and σ in $\mathcal{L}_{bce\text{-}Q}$ is set to 0.5 as default.

4.3   Results and Analysis

Comparison with Others. The mAP scores are calculated by varying the top-M for each dataset as: ImageNet@1000, NUS-WIDE@5000 and MS COCO@5000 to make a fair comparison with previous works [3,4,47]. The results are listed in Table 2, where the highest score for each backbone is shown in bold, and we highlight our DHD method. Among the non-deep hashing methods, SDH

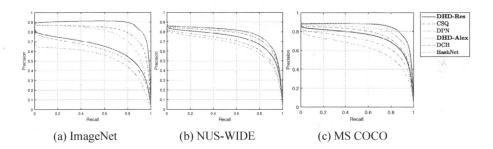

(a) ImageNet   (b) NUS-WIDE   (c) MS COCO

Fig. 3. Precision-Recall curves on three image datasets with binary codes @ 64-bits.

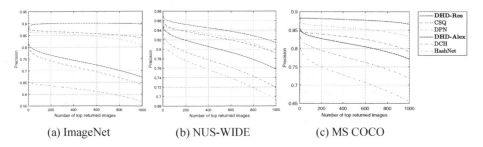

(a) ImageNet   (b) NUS-WIDE   (c) MS COCO

Fig. 4. Precision@top-1000 curves on three image datasets with binary codes @ 64-bits.

---

[4] More details can be found in the supplementary material.

shows the best retrieval results by employing supervised label signals in hash function learning. Deep hashing methods generally outperform non-deep hashing ones, since elaborately labeled annotations are fully utilized during training. For ImageNet, NUS-WIDE, and MS COCO, averaging the mAP scores of all bit lengths yields 36.3%p, 33.7%p, and 25.0%p differences between the non-deep and deep methods, respectively.

Notably, our DHD shows the best mAP scores for all datasets in every bit length with every deep backbone architecture. In particular for AlexNet backbone hashing approaches, DHD shows performance improvement of 16.3%p, 7.9%p, and 9.2%p by averaging the mAP scores of all bit lengths in three dataset results orderly, compared to others. To make a comparison with ResNet backbone methods, DHD also achieves 2.7%p, 1.8%p, and 4.7%p higher retrieval scores on average. In line with the recent trend of other computer vision tasks, we *first* introduce Transformer-based image representation learning architectures: ViT [12], DeiT [40], and SwinT [34] to the hashing community and perform retrieval experiments. As reported, when the Transformer is integrated into the DHD framework, it delivers outstanding results for the benchmark image datasets with the increase of 5.8%p, 2.2%p, and 4.8%p, in the same as above, compared to ResNet backbone DHD.

To further investigate the retrieval quality of DHD, we deploy the graph of PR curve and precision for the top 1,000 retrieved images at 64 bits. As shown in Figs. 3 and 4, DHD significantly outperforms all the comparison hashing approaches by large margins under these two evaluation metrics. Especially, DHD shows desirable retrieval results in that much higher precision are achieved at lower recall levels, and larger number of top samples are retrieved than all compared methods. These demonstrate the practicality of DHD in real world retrieval cases.

Table 3. mAP scores without ($-$) or with ($+$) Self-distilled Hashing (SdH).

Method	ImageNet				NUS-WIDE				MS COCO			
	$-$ SdH		$+$ SdH		$-$ SdH		$+$ SdH		$-$ SdH		$+$ SdH	
	16-bit	64-bit	16-bit	64-bit	16-bit	64-bit	16-bit	64-bit	16-bit	64-bit	16-bit	64-bit
HashNet [4]	0.337	0.502	0.501	0.661	0.705	0.762	0.745	0.769	0.655	0.727	0.695	0.753
DCH [3]	0.571	0.597	0.640	0.673	0.748	0.767	0.754	0.771	0.669	0.697	0.703	0.746
DPN [14]	0.562	0.656	0.630	0.708	0.753	0.787	0.757	0.801	0.672	0.760	0.710	0.772
CSQ [47]	0.569	0.658	0.634	0.711	0.757	0.793	0.759	0.804	0.670	0.752	0.707	0.765
$\mathcal{L}_{HP}$	0.574	0.660	0.642	0.715	0.759	0.798	0.766	0.812	0.706	0.759	0.725	0.775
$\mathcal{L}_{HP} + \mathcal{L}_{bce\text{-}Q}$	0.583	0.671	**0.657**	**0.721**	0.775	0.806	**0.780**	**0.820**	0.731	0.766	**0.749**	**0.792**

Self-distilled Hashing with Other Methods and Ablations. In order to prove that SdH can be applied to other deep hashing baselines [3,4,14,47], we perform retrieval experiments with AlexNet backbone and show the results in Table 3. With SdH setup, we employ $\mathcal{T}_T$ and $\mathcal{T}_S$ groups to produce input views, and for without SdH setup, we only use $\mathcal{T}_S$ to generate input views. By comparing Table 2 and the results without SdH in Table 3, we can see that the deep hashing model learned with $\mathcal{T}_S$ is inferior to the model learned with $\mathcal{T}_T$. This is because the use of $\mathcal{T}_S$ increases the chances of emerging discrepancy in representation between Hamming and real space. Otherwise, if the model adopts SdH training to utilize both $\mathcal{T}_T$ and $\mathcal{T}_S$, the retrieval performance can be improved since SdH mitigates discrepancy and properly exploits the power of data augmentation. Intending to see the ablation results of our proposals, we compare the retrieval results between $\mathcal{L}_{HP}$ with the others and find that the trainable setting for the hash centers improve the search quality. Moreover, by combining $\mathcal{L}_{HP}$ with $\mathcal{L}_{bce\text{-}Q}$, the performance gain is obtained for all the bit lengths, showing the power of binary cross entropy-based quantization. Finally, the best mAP scores are achieved when both $\mathcal{L}_{HP}$ and $\mathcal{L}_{bce\text{-}Q}$ are integrated with SdH training, confirming the effectiveness of DHD.

Trainable Hash Proxies. In DHD, we employ trainable hash proxies opposed to using predefined orthogonal hash targets [14,19,47], intending to embed detailed class-wise semantic similarity into the hash representation. We visualize[5] the pairwise cosine similarities in Fig. 5 using ResNet backbone and 64 bit codes to observe the actual alignment between hash proxies. Since hash targets are generated from a Hadamard matrix, they are orthogonal as shown in Fig. 5a. Therefore, the cosine similarity (Hamming distance) between different hash targets are equal, neglecting the semantic relevance between the hash representations of different classes. Moreover for multi-label cases, label dependencies [9] are also ignored whether the contents appear simultaneously in an image or not.

On the other hand, trainable hash proxies are designed to embed semantic similarity by themselves during training. Hence, class-wise relevance can be displayed when we compute pairwise cosine similarities as in Figs. 5b to 5d. Specifically for ImageNet, hash proxies of semantically relevant classes have higher similarity, such as *Norfolk terrier-Australian terrier* and *purse-wallet*. Moreover, for multi-label datasets hash proxies of classes that frequently appear together in an image have higher similarity, such as *sky-clouds* and *buildings-window-vehicle* in NUS-WIDE, and *bowl-cup-dining table* in MS COCO. In a nutshell overall, we can confirm that the supervised semantic signals are well guided to represent detailed similarity between the hash proxies, which in turn yields better quality search outcomes.

---

[5] Visualized results with all classes for each dataset are shown in the supplementary material.

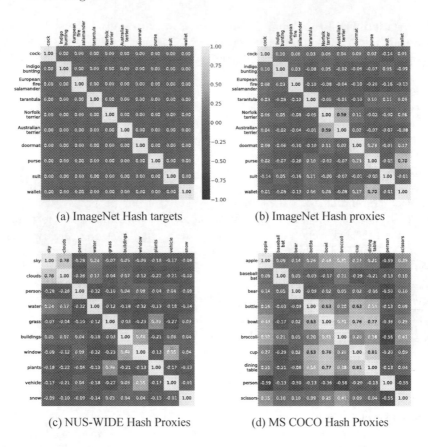

(a) ImageNet Hash targets

(b) ImageNet Hash proxies

(c) NUS-WIDE Hash Proxies

(d) MS COCO Hash Proxies

Fig. 5. Pairwise cosine similarities to verify the impact of proxy-based hash representation learning. We utilize (a) pre-defined non-trainable hash targets, and (b-d) proposed trainable hash proxies. For simplicity, we show the results of 10 classes selected.

Insensitivity to Transformations. To investigate the sensitivity to transformations, we examine how the binary code shifts when transformed images are fed to ResNet backbone methods, by using ImageNet query set. We measure the average Hamming distance between the untransformed ($s_T = 0$) images' output binary codes and the transformed (s_T in $(0, 1]$) images' output binary codes, as observed in Fig. 6. Here, CSQ [47], DPN [14], and a model learned with $\mathcal{L}_{HP}$ are trained with weakly-transformed $\mathcal{T}_T$, which in result show sensitivity to transformations due to barely used augmentation. When the augmentation is applied ($\mathcal{L}_{HP} + \mathcal{T}_S$), model is improved to be more robust to transformations, however, the mAP score decreases due to the discrepancy in representation between Hamming and Real space during training. Otherwise, the combined $\mathcal{L}_{HP} + \mathcal{L}_{SdH}$ (blue line) exhibits the highest robustness while achieving the best mAP score, by minimizing discrepancy in representation during training and successfully exploring the potential of strong augmentation.

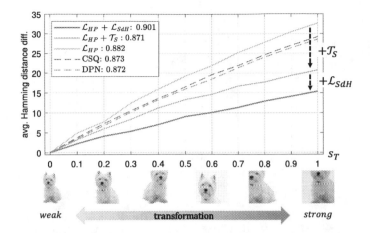

Fig. 6. Average Hamming distance difference between the original and transformed images of ImageNet query set. By varying the s_T, we measure the sensitivity to transformation of ResNet backbone methods, where the numbers in legend indicate mAP. Solid lines present DHD variants, and dotted lines present others. $+\mathcal{T}_S$ denotes strong student augmentation is applied during training. A low slope indicates insensitivity to various transformations, where the blue line (DHD) is the lowest. (Color figure online)

Table 4. mAP scores on unseen deformations.

Deformation	with SdH	without SdH
None	**0.891** (2.3% ↑)	0.871
Cutout	**0.862** (3.7% ↑)	0.827
Dropout	**0.810** (7.9% ↑)	0.765
Zoom in	**0.658** (19.0% ↑)	0.552
Zoom out	**0.816** (1.4% ↑)	0.805
Rotation	**0.856** (2.4% ↑)	0.836
Shearing	**0.842** (2.7% ↑)	0.815
Gaussian noise	**0.768** (10.5% ↑)	0.673

Robustness to Unseen Deformations. To further examine the generalization capacity of DHD, we conduct experiments with unseen (not seen during training) transformations[6] to inputs following the evaluation protocol utilized in [16]. As reported in Table 4, deep hashing model with SdH significantly outperforms the model without SdH at all deformations, showing a performance difference of up to 19% (zoom in). In particular, SdH makes deep hashing model robust to per-pixel deformations such as dropout and Gaussian noise, even though SdH has not included any pixel-level transformations.

---

[6] Detailed deformation setup is listed in the supplementary material.

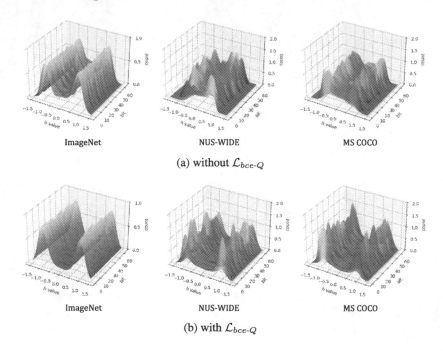

(a) without $\mathcal{L}_{bce\text{-}Q}$

(b) with $\mathcal{L}_{bce\text{-}Q}$

Fig. 7. 3D visualized histograms to verify the impact of $\mathcal{L}_{bce\text{-}Q}$. x-axis presents value of hash element h, y-axis presents bit position, and z-axis presents frequency counts.

Quantization. The effect of $\mathcal{L}_{bce\text{-}Q}$ is plotted in Fig. 7. We can see that the binary bits are distributed more evenly and binary-like in 7b. This implies that the entropy of bit distribution is much higher when $\mathcal{L}_{bce\text{-}Q}$ is applied, which can show better retrieval accuracy by representing diverse binary codes, as observed and investigated in [30].

5   Conclusion

In this paper, we proposed a novel Self-distilled Hashing (SdH) scheme which is applicable to deep hashing models during training. By maximizing the cosine similarity between hash codes of different views of an image, SdH minimizes the discrepancy in the representation due to augmentation and leads to increase the robustness of retrieval systems. Additionally, we aimed to embed elaborate semantic similarity into the hash codes with a proxy-based learning, and further impose cross entropy-based quantization loss. With all these proposals, we configured Deep Hash Distillation (DHD) framework that yields the state-of-the-art performance on popular deep hashing benchmarks.

Acknowledgement. This research was supported in part by NAVER Corporation, the National Research Foundation of Korea (NRF) grant funded by the Korean government (MSIT) (2021R1A2C2007220), and the Institute of Information & Communications Technology Planning & Evaluation (IITP) grant funded by the Korean government (MSIT) [NO.2021-0-01343, Artificial Intelligence Graduate School Program (Seoul National University)].

References

1. Bai, J., et al.: Targeted attack for deep hashing based retrieval. In: Vedaldi, A., Bischof, H., Brox, T., Frahm, J.-M. (eds.) ECCV 2020. LNCS, vol. 12346, pp. 618–634. Springer, Cham (2020). https://doi.org/10.1007/978-3-030-58452-8_36
2. Boudiaf, M., et al.: A unifying mutual information view of metric learning: cross-entropy vs. pairwise losses. In: Vedaldi, A., Bischof, H., Brox, T., Frahm, J.-M. (eds.) ECCV 2020. LNCS, vol. 12351, pp. 548–564. Springer, Cham (2020). https://doi.org/10.1007/978-3-030-58539-6_33
3. Cao, Y., Long, M., Liu, B., Wang, J.: Deep cauchy hashing for hamming space retrieval. In: CVPR, pp. 1229–1237 (2018)
4. Cao, Z., Long, M., Wang, J., Yu, P.S.: HashNet: Deep learning to hash by continuation. In: CVPR, pp. 5608–5617 (2017)
5. Caron, M., et al.: Emerging properties in self-supervised vision transformers. In: ICCV, pp. 9650–9660 (2021)
6. Chen, T., Kornblith, S., Norouzi, M., Hinton, G.: A simple framework for contrastive learning of visual representations. arXiv preprint arXiv:2002.05709 (2020)
7. Chen, W.Y., Liu, Y.C., Kira, Z., Wang, Y.C.F., Huang, J.B.: A closer look at few-shot classification. ICLR (2019)
8. Chen, X., He, K.: Exploring simple Siamese representation learning. In: CVPR, pp. 15750–15758 (2021)
9. Chen, Z.M., Wei, X.S., Wang, P., Guo, Y.: Multi-label image recognition with graph convolutional networks. In: CVPR, pp. 5177–5186 (2019)
10. Chua, T.S., Tang, J., Hong, R., Li, H., Luo, Z., Zheng, Y.: Nus-wide: a real-world web image database from national university of Singapore. In: ACM ICMR, pp. 1–9 (2009)
11. Cui, Q., Jiang, Q.-Y., Wei, X.-S., Li, W.-J., Yoshie, O.: ExchNet: a unified hashing network for large-scale fine-grained image retrieval. In: Vedaldi, A., Bischof, H., Brox, T., Frahm, J.-M. (eds.) ECCV 2020. LNCS, vol. 12348, pp. 189–205. Springer, Cham (2020). https://doi.org/10.1007/978-3-030-58580-8_12
12. Dosovitskiy, A., et al.: An image is worth 16×16 words: transformers for image recognition at scale. In: ICLR (2020)
13. Riba, E.: Kornia: an open source differentiable computer vision library for Pytorch. In: WACV (2020). arxiv.org/pdf/1910.02190pdf
14. Fan, L., Ng, K., Ju, C., Zhang, T., Chan, C.S.: Deep polarized network for supervised learning of accurate binary hashing codes. In: IJCAI, pp. 825–831 (2020)
15. Gong, Y., Lazebnik, S., Gordo, A., Perronnin, F.: Iterative quantization: a procrustean approach to learning binary codes for large-scale image retrieval. PAMI 35(12), 2916–2929 (2012)
16. Gu, G., Ko, B., Kim, H.G.: Proxy synthesis: Learning with synthetic classes for deep metric learning. In: AAAI (2021)
17. He, K., Zhang, X., Ren, S., Sun, J.: Deep residual learning for image recognition. In: CVPR, pp. 770–778 (2016)
18. Hinton, G., Vinyals, O., Dean, J.: Distilling the knowledge in a neural network. arXiv preprint arXiv:1503.02531 (2015)

19. Hoe, J.T., et al.: One loss for all: deep hashing with a single cosine similarity based learning objective. NeurIPS 34 (2021)
20. Jang, Y.K., Cho, N.I.: Deep face image retrieval for cancelable biometric authentication. In: AVSS, pp. 1–8. IEEE (2019)
21. Jang, Y.K., Cho, N.I.: Generalized product quantization network for semi-supervised image retrieval. In: CVPR, pp. 3420–3429 (2020)
22. Jang, Y.K., Cho, N.I.: Self-supervised product quantization for deep unsupervised image retrieval. In: ICCV, pp. 12085–12094 (2021)
23. Jang, Y.K., Jeong, D., Lee, S.H., Cho, N.I.: Deep clustering and block hashing network for face image retrieval. In: Jawahar, C.V., Li, H., Mori, G., Schindler, K. (eds.) ACCV 2018. LNCS, vol. 11366, pp. 325–339. Springer, Cham (2019). https://doi.org/10.1007/978-3-030-20876-9_21
24. Jeong, D.j., Choo, S.K., Seo, W., Cho, N.I.: Classification-based supervised hashing with complementary networks for image search. In: BMVC, p. 74 (2018)
25. Kingma, D.P., Ba, J.: Adam: a method for stochastic optimization. In: ICLR (2015)
26. Koch, G., Zemel, R., Salakhutdinov, R., et al.: Siamese neural networks for one-shot image recognition. In: ICML deep learning workshop, vol. 2. Lille (2015)
27. Krizhevsky, A., Sutskever, I., Hinton, G.E.: ImageNet classification with deep convolutional neural networks. In: NeurIPS, pp. 1097–1105 (2012)
28. Lai, H., Pan, Y., Liu, Y., Yan, S.: Simultaneous feature learning and hash coding with deep neural networks. In: CVPR, pp. 3270–3278 (2015)
29. Li, C., Deng, C., Li, N., Liu, W., Gao, X., Tao, D.: Self-supervised adversarial hashing networks for cross-modal retrieval. In: CVPR, pp. 4242–4251 (2018)
30. Li, Y., van Gemert, J.: Deep unsupervised image hashing by maximizing bit entropy. In: AAAI (2020)
31. Lin, T.-Y., Maire, M., Belongie, S., Hays, J., Perona, P., Ramanan, D., Dollár, P., Zitnick, C.L.: Microsoft COCO: common objects in context. In: Fleet, D., Pajdla, T., Schiele, B., Tuytelaars, T. (eds.) ECCV 2014. LNCS, vol. 8693, pp. 740–755. Springer, Cham (2014). https://doi.org/10.1007/978-3-319-10602-1_48
32. Liu, S., Wang, Y.: Few-shot learning with online self-distillation. In: ICCV, pp. 1067–1070 (2021)
33. Liu, W., Wang, J., Ji, R., Jiang, Y.G., Chang, S.F.: Supervised hashing with kernels. In: CVPR, pp. 2074–2081. IEEE (2012)
34. Liu, Z., et al.: Swin transformer: hierarchical vision transformer using shifted windows. In: ICCV (2021)
35. Loshchilov, I., Hutter, F.: SGDR: stochastic gradient descent with warm restarts. arXiv preprint arXiv:1608.03983 (2016)
36. Movshovitz-Attias, Y., Toshev, A., Leung, T.K., Ioffe, S., Singh, S.: No fuss distance metric learning using proxies. In: ICCV, pp. 360–368 (2017)
37. Russakovsky, O., et al.: ImageNet large scale visual recognition challenge. IJCV 115(3), 211–252 (2015)
38. Shen, F., Shen, C., Liu, W., Tao Shen, H.: Supervised discrete hashing. In: CVPR, pp. 37–45 (2015)
39. Shen, Y., et al.: Auto-encoding twin-bottleneck hashing. In: CVPR, pp. 2818–2827 (2020)
40. Touvron, H., Cord, M., Douze, M., Massa, F., Sablayrolles, A., Jégou, H.: Training data-efficient image transformers & distillation through attention. In: ICML, pp. 10347–10357. PMLR (2021)
41. Wang, J., Zhang, T., Sebe, N., Shen, H.T., et al.: A survey on learning to hash. PAMI 40(4), 769–790 (2017)

42. Wang, Z., Zheng, Q., Lu, J., Zhou, J.: Deep hashing with active pairwise super-vision. In: Vedaldi, A., Bischof, H., Brox, T., Frahm, J.-M. (eds.) ECCV 2020. LNCS, vol. 12364, pp. 522–538. Springer, Cham (2020). https://doi.org/10.1007/978-3-030-58529-7_31

43. Weiss, Y., Torralba, A., Fergus, R.: Spectral hashing. In: NeurIPS, pp. 1753–1760 (2009)

44. Xia, R., Pan, Y., Lai, H., Liu, C., Yan, S.: Supervised hashing for image retrieval via image representation learning. In: AAAI (2014)

45. Xu, T.B., Liu, C.L.: Data-distortion guided self-distillation for deep neural networks. In: AAAI, vol. 33, pp. 5565–5572 (2019)

46. Yang, E., Liu, T., Deng, C., Liu, W., Tao, D.: DistillHash: unsupervised deep hashing by distilling data pairs. In: CVPR, pp. 2946–2955 (2019)

47. Yuan, L., et al.: Central similarity quantization for efficient image and video retrieval. In: CVPR, pp. 3083–3092 (2020)

48. Cao, Y.: Deep quantization network for efficient image retrieval. In: AAAI (2016)

49. Yun, S., Park, J., Lee, K., Shin, J.: Regularizing class-wise predictions via self-knowledge distillation. In: CVPR, pp. 13876–13885 (2020)

50. Zhou, X., et al.: Graph convolutional network hashing. IEEE Trans. Cybernet. **50**(4), 1460–1472 (2018)

51. Zhu, H., Long, M., Wang, J., Cao, Y.: Deep hashing network for efficient similarity retrieval. In: AAAI (2016)

Mimic Embedding via Adaptive Aggregation: Learning Generalizable Person Re-identification

Boqiang Xu[1,2] , Jian Liang[1,2], Lingxiao He[3] , and Zhenan Sun[1,2(✉)]

[1] School of Artificial Intelligence, University of Chinese Academy of Sciences,
Beijing, China
boqiang.xu@cripac.ia.ac.cn

[2] Center for Research on Intelligent Perception and Computing, National Laboratory
of Pattern Recognition, Institute of Automation, Chinese Academy of Sciences,
Beijing, China
znsun@nlpr.ia.ac.cn

[3] Longfor Inc., Beijing, China
xiaomingzhidao1@gmail.com

Abstract. Domain generalizable (DG) person re-identification (ReID) aims to test across unseen domains without access to the target domain data at training time, which is a realistic but challenging problem. In contrast to methods assuming an identical model for different domains, Mixture of Experts (MoE) exploits multiple domain-specific networks for leveraging complementary information between domains, obtaining impressive results. However, prior MoE-based DG ReID methods suffer from a large model size with the increase of the number of source domains, and most of them overlook the exploitation of domain-invariant characteristics. To handle the two issues above, this paper presents a new approach called Mimic Embedding via adapTive Aggregation (META) for DG person ReID. To avoid the large model size, experts in META do not adopt a branch network for each source domain but share all the parameters except for the batch normalization layers. Besides multiple experts, META leverages Instance Normalization (IN) and introduces it into a global branch to pursue invariant features across domains. Meanwhile, META considers the relevance of an unseen target sample and source domains via normalization statistics and develops an aggregation module to adaptively integrate multiple experts for mimicking unseen target domain. Benefiting from a proposed consistency loss and an episodic training algorithm, META is expected to mimic embedding for a truly unseen target domain. Extensive experiments verify that META surpasses state-of-the-art DG person ReID methods by a large margin. Our code is available at https://github.com/xbq1994/META.

Keywords: Domain generalization · Person re-identification

S. Avidan et al. (Eds.): ECCV 2022, LNCS 13674, pp. 372–388, 2022.
https://doi.org/10.1007/978-3-031-19781-9_22

1   Introduction

Person re-identification (ReID) aims at retrieving persons of the same identity across non-overlapping cameras. Many prior works [10,25,29,39,42,43] have been devoted to the fully-supervised ReID task. Despite the promising performance when training and testing on the same domain, the performance always drops significantly when testing on an unseen domain because of the domain shift [40]. To avoid this, recent efforts are devoted to domain adaptive (DA) ReID [7,44,49] and domain generalizable (DG) ReID [4,5,17,45]. In contrast to DA ReID, DG ReID is more practical and challenging as it utilizes training data from multiple source domains and directly tests across different and unseen domains, without any target data for training or fine-tuning. In this paper, we mainly focus on the challenging DG person ReID problem.

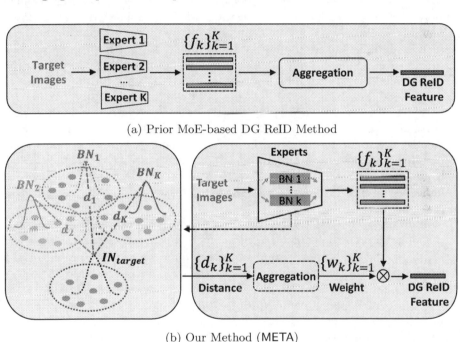

(a) Prior MoE-based DG ReID Method

(b) Our Method (META)

Fig. 1. Differences between prior MoE-based DG ReID method and our method. (a) Prior MoE-based DG ReID methods add an individual network (expert) for each source domain, suffering from a large model size with the increase of the number of source domains. (b) Experts in our method share all the parameters except for the batch normalization layers. In the testing stage, we calculate the distance between IN statistics of test samples and the BN statistics of source domains for measuring the relevance of target samples w.r.t. source domains. Such distances $\{d_k\}_{k=1}^{K}$ are exploited by an aggregation module to adaptively integrate multiple experts.

Most of the prior DG ReID methods [1,4,17,35,45] assume an identical model for different domains. However, such an assumption learns a common feature space for different source domains, which may neglect the individual domains' discriminative information and ignore the relevance of the target domain w.r.t source domains. To handle the issues above, mixture of experts (MoE) [16] has been studied for DG ReID, as shown in Fig. 1a. MoE can improve the generalization of models by integrating multiple domain-specific expert networks with the target domain's inherent relevance w.r.t. diverse source domains. Generally, prior MoE-based DG ReID methods have two potential problems: 1) As each source domain contains an individual branch network, the model size becomes fairly large with the increase of the number of source domains, limiting the practical deployment. 2) Most prior MoE-based DG ReID methods merely focus on learning domain-specific representations but overlook the domain-invariant characteristics.

To tackle the two issues above, we propose a novel DG ReID approach called Mimic Embedding via adapTive Aggregation (META), as shown in Fig. 1a. Batch Normalization (BN) statistics are computed on-the-fly during training and can be seen as statistics of the characteristics of individual domain [32]. Inspired by this, instead of adding a branch network for each source domain, we train the META as a lightweight ensemble of multiple experts sharing all the parameters except for the domain-specific BN layers (i.e., one for each source domain for collecting domain-specific BN statistics). By doing so, META is able to exploit the diversified characteristics of each source domain and meanwhile, keeping the model size from increasing as the source domain increases. To extract the domain-invariant features, we design a global branch and leverage Instance Normalization (IN) [6], which works as a style normalization layer for filtering out domain-specific contrast information, to explicitly extract domain-invariant features.

Specifically, in our META method, we exploit individual domains' discriminative information by domain-specific BN layers. Then, during testing, the characteristics of the test samples from the unseen domain can be indicated by the means of their IN statistics. By measuring the distance between the IN statistics of the test samples and the BN statistics of source domains, we can infer the relevance of the target samples w.r.t. source domains. Taking the relevance as input, we further devise a small aggregation module to integrate multiple experts for obtaining the accurate representation of the target person from an unknown domain. By doing so, those relevant source domains are able to contribute more valuable information than those less relevant domains. Moreover, we adopt episodic training [19] which simulates the test process at training time for updating the aggregation module. For each training batch, we collect training samples from the same source domain (e.g., D_k) to simulate the 'unseen target data' for other domain experts. We propose a consistency loss to push the aggregated features of other domain experts as discriminative as the features extracted by the expert of D_k. In this way, the aggregation module is learned to be able to adaptively integrate diverse domain experts for explicitly mimicking any unseen target domain.

Our major contributions can be summarized as follows:

- We propose META, a novel method to handle the DG ReID problem. Specifically, META leverages the domain-specific BN layers and designs a global branch to respectively tackle the two issues (*i.e.*, model scalability and oversight in domain invariance) in prior MoE-based DG ReID methods.
- We develop a learnable aggregation module, updated by a proposed consistency loss and an episodic training algorithm, to adaptively integrate diverse domain experts via normalization statistics for mimicking any unseen target domain.

Extensive experiments demonstrate that META surpasses state-of-the-art DG ReID methods by a large margin under various protocols.

2   Related Work

Domain Generalizable Person Re-identification. Person ReID has made great progress in recent years. Many methods [11,23,36,38,39] have been proposed to improve the ReID performance. Despite the promising performance brought by these methods when training and testing on the same domain, the performance always drops significantly when testing on an unseen domain because of the domain shift [40]. To tackle this problem, some researchers start to study the unsupervised domain adaption (UDA) methods [7,44,49]. However, UDA requires unlabeled data from the target source, which is sometimes difficult to be collected in practical applications. As a result, domain generalizable (DG) ReID [4,5,17,45] have captivated researchers recently. Generally, DG ReID utilizes training data from multiple source domains and directly tests across different and unseen domains, without any target data for training or fine-tuning.

We briefly classify prior DG ReID methods into three categories. The first category is Meta-Learning [1,4,35,45]. Meta-learning is a training strategy, which adopts the concept of 'learning to learn' by exposing the model to domain shift during training for learning more generalizable models. Zhao *et al.* [45] proposed a Memory-based Multi-Source Meta-Learning (M^3L) framework, which overcomes the unstable meta-optimization by a memory-based and non-parametric identification loss.

The second category is Domain Alignment [17], which attempts to minimize the differences between source domains for pursuing the invariant features across domains. Jin *et al.* [17] propose a Style Normalization and Restitution (SNR) module to separate the identity-relevant and identity-irrelevant features by a dual causality loss constraint.

The third category is Mixture of Experts (MoE) [5]. MoE learns diverse experts for different domains and takes the target domain's inherent relevance w.r.t. diverse source domains into consideration for better generalization. Dai *et al.* [5] proposed a method called the relevance-aware mixture of experts (RaMoE), which adds a branch network (expert) for each source domain, and designs a voting network for integrating multiple experts. However, [5] suffers from a large model size with the increase of the number of source domains, which limits the application of the RaMoE. To tackle this problem, experts in our method share all the parameters except for the batch normalization layers.

Domain-Specific Batch Normalization. The statistics of BN vary in different domains. Therefore, mixing multiple source domains' statistics may be detrimental to improving generalizable performance [50]. To tackle this problem, domain-specific BN has been studied recently [24,28,32,33]. Domain-specific BN works as constructing domain-specific classifiers but shares most of the parameters except for the BN layers.

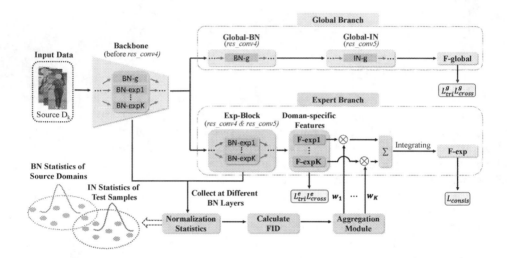

Fig. 2. Overview of the proposed META. '$\otimes$' is the operation of element-wise multiplication. '$\sum$' is a series of features' operation: element-wise division or summation. META is composed of a global branch for capturing domain-invariant features and an expert branch for exploiting complementary domain-specific information. The Exp-Block contains K domain-specific BN layers while the backbone contains K domain-specific BN layers and a global layer BN-g. We replace the BN layers in the *res_conv5* with IN layers to construct Global-IN. K domain-specific BN layers are updated by their corresponding source domain's data to capture domain-specific characteristics while BN-g and IN-g are updated by the training data from all the source domains to help extract domain-invariant features. In the expert branch, we collect the IN statistics of the test samples and BN statistics of the source domains at different BN layers and calculate the *Fréchet Inception Distance* (FID) between them to measure the relevance of target samples w.r.t. source domains. Such relevance is leveraged by an aggregation module to adaptively integrate multiple experts. Finally, we concatenate F-global and F-exp for inference.

3   Methodology

Typically, we are provided with K source domains $\{D_k\}_{k=1}^{K}$ for training a DG ReID model, which have completely disjoint label spaces. In the testing phase, we directly test on unseen target domains without additional model updating. The structure of the META is illustrated in Fig. 2.

3.1 Preliminary

In almost all the prior DG ReID methods [1,4,35,45], they share BN layers for all the source domains, which may neglect individual domains' discriminative characteristics and be detrimental to dealing with the domain gap [3,5]. To leverage the complementary information of the source domains, inspired by [2, 3,32], we adopt *domain-specific batch normalization* in META.

Let $X_k \in \mathbb{R}^{N \times C \times H \times W}$ denotes a feature map extracted from source domain D_k, where N, C, H, W respectively indicate the batch size, the number of channels, the height, and the width. BN layer normalizes features by:

$$BN(X_k) = \gamma_k^{bn} \cdot \frac{X_k - \mu_k^{bn}}{\sqrt{\sigma_k^{bn^2} + \epsilon}} + \beta_k^{bn}, \tag{1}$$

where $\gamma_k^{bn} \in \mathbb{R}^C$ and $\beta_k^{bn} \in \mathbb{R}^C$ are affine parameters, $\epsilon > 0$ is a small constant to avoid divided-by-zero. $\mu_k^{bn} \subset \mathbb{R}^C$ and $\sigma_k^{bn} \in \mathbb{R}^C$ are respectively mean value and standard deviation calculated with respect to a mini-batch and each channel:

$$\mu_k^{bn} = \frac{\sum_n \sum_{h,w} X_k}{N \cdot H \cdot W} \quad \text{and} \quad \sigma_k^{bn} = \sqrt{\frac{\sum_n \sum_{h,w} (X_k - \mu_k^{bn})^2}{N \cdot H \cdot W}}. \tag{2}$$

μ_k^{bn} and σ_k^{bn} are updated by the moving average operation [15] at training time and fixed during inference. We design individual BN layers for each source domain. Specifically, as shown in Fig. 2, Exp-Block contains K domain-specific BN layers, which are updated by the training data from the corresponding source domain to exploit domain-specific characteristics. Besides K domain-specific BN layers, another global layer BN-g is introduced in the backbone and global branch, which is updated by the training data from all the source domains to help extract domain-invariant features.

Although we have exploited the complementary information of the source domains via *domain-specific batch normalization*, it is still challenging to approximate the population statistics of the unseen target domain because target domain data cannot be accessed at training time. To do this, at testing time, we rely on IN statistics to capture the characteristics of the target samples. Given an example from target domain T_t, IN layers normalize features by:

$$IN(X_t) = \gamma_t^{in} \cdot \frac{X_t - \mu_t^{in}}{\sqrt{\sigma_t^{in^2} + \epsilon}} + \beta_t^{in}. \tag{3}$$

Different from BN, mean value μ_t^{in} and standard deviation σ_t^{in} here are calculated with respect to each sample and each channel:

$$\mu_t^{in} = \frac{\sum_{h,w} X_t}{H \cdot W} \quad \text{and} \quad \sigma_t^{in} = \sqrt{\frac{\sum_{h,w} (X_t - \mu_t^{in})^2}{H \cdot W}}. \tag{4}$$

In the next section, we explain how to measure the relevance of the target samples w.r.t. source domains via BN and IN statistics.

3.2   Expert Branch in **META**

We expect those relevant source domains to contribute more valuable information than those less relevant domains. In this section, we explain how to measure
the relevance of the target samples w.r.t. source domains via BN and IN statistics for integrating multiple experts. From Eq. (1)–Eq. (4), we can see that IN
is the degenerate case of BN with batch size N equal to 1. META is built on
such observation that BN and IN statistics are both approximations of Gaussian
distributions (*i.e.*, they are comparable) and have potential to reflect the properties of the source domains and target samples respectively. Therefore, we can
measure the relevance of the target samples w.r.t. source domains by comparing
IN and BN statistics of them.

Specifically, we collect the BN statistics of source domains at different BN
layers. Considering a source domain D_k, we denote $D_k^{(l)} = (\mu_k^{bn(l)}, \sigma_k^{bn(l)^2})$ the
BN statistics at l-th layer of k-th BN-exp. For each test sample x_t from an unseen
target domain T_t, we forward propagate x_t through the network and calculate its
IN statistics by Eq. (4) at l-th layer of k-th BN-exp as $T_t^{(l)} = (\mu_t^{in(l)}, \sigma_t^{in(l)^2})$. We
adopt *Fréchet Inception Distance* (FID) [13] to compute the distance between
the BN and IN statistics at l-th layer as:

$$
\begin{aligned}
r_{k,t}^{(l)} &= FID((\mu_k^{bn(l)}, \sigma_k^{bn(l)^2}), (\mu_t^{in(l)}, \sigma_t^{in(l)^2})) \\
&= \|\mu_k^{bn(l)} - \mu_t^{in(l)}\|_2^2 + Tr(C_k^{(l)} + C_t^{(l)} - 2(C_k^{(l)} C_t^{(l)})^{\frac{1}{2}}), \quad (5)
\end{aligned}
$$
where $C_k^{(l)} = Diag(\sigma_k^{bn(l)^2})$, $C_t^{(l)} = Diag(\sigma_t^{in(l)^2})$,

and $Diag(\cdot)$ returns a square diagonal matrix with the elements of input vector
on the main diagonal. $r_{k,t}^{(l)}$ denotes the distance between the BN statistics of
source domain D_k and IN statistics of test sample from target domain T_t at
l-th layer, $\|\cdot\|$ denotes the Euclidean norm, and $Tr(\cdot)$ denotes the trace of the
matrix. Thereafter, we concatenate $r_{k,t}^{(l)}$ at every layer as:

$$
R_k^t = [r_{k,t}^{(1)}, r_{k,t}^{(2)}, ..., r_{k,t}^{(L)}] \in \mathbb{R}^{1 \times L}. \quad (6)
$$

Then, we forward propagate R_k^t to an aggregation module $h : \mathbb{R}^L \to \mathbb{R}$ for
computing the weight of domain-specific expert:

$$
w_k = h(R_k^t), \quad (7)
$$

where h consists of two fully-connected layers. The aggregation module further
enhances the domains' relevance measure by adopting a learnable module. During testing, we get the *F-exp* as a linear combination of the multiple experts:

$$
F\text{-}exp(x) = \sum_{k=1}^{K} \frac{e^{w_k} f(x \mid k)}{\sum_j e^{w_j}}, \quad (8)
$$

where $f(x \mid k)$ is the result of a forward pass of the k-th expert in the network.
During training, we get the *F-exp* in another way, which will be introduced

in Sect. 3.4. In this way, relevant source domains are able to contribute more valuable information than those less relevant domains for better generalization performance on the target domain.

3.3 Global Branch in **META**

We design a global branch to learn the domain-invariant features, which works as a complement to the domain-specific representations extracted by the expert branch for better generalizability. IN works on normalizing features with the statistics of individual instances, by which the domain-specific information could be filtered out from the content [6]. Inspired by this, we leverage IN layers in the global branch to capture the domain-invariant features.

The global branch is designed based on the findings from [30] that adding IN layers after BN layers could significantly improve the domain generalization performance of the model. Specifically, as shown in Fig. 2, the global branch is composed of the *Global-Bn* and *Global-In* blocks. *Global-Bn* block is the same as *res_conv*4. We replace all the BN layers in the *res_conv*5 with IN layers to build the *Global-In* block. Furthermore, training samples from all the source domains are used to update the global branch.

3.4 Training Policy

At training time, each training batch is composed of the training samples collected from the same source domain. Let x denotes the current training sample collected from source domain D_i $(1 \leq i \leq K)$. As shown in Fig. 2, we freeze all the BN layers except for the BN-g and i-th BN-exp. We update the global branch by the triplet loss [12] $\mathcal{L}^g_{tri}$ and cross-entropy loss $\mathcal{L}^g_{cross}$. Meanwhile, we optimize the i-th expert by the triplet loss [12] $\mathcal{L}^e_{tri}$ and cross-entropy loss $\mathcal{L}^e_{cross}$. Combining these losses above together, we have the following overall objective:

$$\mathcal{L}_{base} = \mathcal{L}^g_{tri} + \mathcal{L}^g_{cross} + \mathcal{L}^e_{tri} + \mathcal{L}^e_{cross}. \tag{9}$$

In addition, we adopt episodic training [19] which simulates the test process at training time to update the aggregation module. When x is input to the network, domain D_i is seemed as the 'unseen target domain' to the other $K - 1$ domain-specific experts $\{f(x \mid k)\}^K_{k=1, k \neq i}$. We combine these $K - 1$ domain experts to produce the representation F-exp, which is formulated as:

$$F\text{-}exp(x) = \sum_{k=1, k \neq i}^{K} \frac{e^{w_k} f(x \mid k)}{\sum_{j, j \neq i} e^{w_j}}, \quad x \in D_i, \tag{10}$$

where w_k is the weight of k-th expert and $f(x \mid k)$ is the result of a forward pass of the k-th expert. To mimic embedding of D_i with F-exp, we propose a consistency loss to push the aggregated feature F-exp as discriminative as the feature $f(x \mid i)$ extracted by the i-th expert. The consistency loss is formulated as:

$$\mathcal{L}_{consis} = [\alpha_1 + \Gamma^+_{exp} - \Gamma^+_i]_+ + [\alpha_2 + \Gamma^-_i - \Gamma^-_{exp}]_+, \tag{11}$$

Algorithm 1: Training Procedure of META

Input: Training data x from source domain D_i; MaxIters; MaxEpochs.
Output: Feature extractor $F_\theta(\cdot)$; Domain-specific experts $\{f(x \mid k)\}_{k=1}^{K}$.

1 Initialization;
2 **for** *epoch=1* **to** *MaxEpochs* **do**
3   **for** *iter=1* **to** *MaxIters* **do**
4   **Domain-specific BN layers:**
5   Freeze all the BN layers except for the BN-g and i-th BN-exp;
6   **Global Branch:**
7   Update global branch by $\mathcal{L}_{tri}^{g}$ and $\mathcal{L}_{cross}^{g}$;
8   **Expert Branch:**
9   Update expert branch by $\mathcal{L}_{tri}^{e}$ and $\mathcal{L}_{cross}^{e}$;
10   **Aggregation Module:**
11   Combine $\{f(x \mid k)\}_{k=1, k\neq i}^{K}$ by Eq. (10) to produce *F-exp*;
12   Update aggregation module by $\mathcal{L}_{consis}$ in Eq. (11);
13   **end**
14 **end**

---

where α_1 and α_2 are margins, Γ_{exp}^{+} and Γ_{i}^{+} are hardest positive distances [12] of *F-exp* and $f(x \mid i)$ respectively, Γ_{exp}^{-} and Γ_{i}^{-} are hardest negative distances [12] of *F-exp* and $f(x \mid i)$ respectively, $[z]_{+}$ equals to $max(z, 0)$. By minimizing Eq. (11), the aggregation module is learned to explicitly mimic the target domain via multiple experts. The total loss can be formulated as:

$$\mathcal{L} = \mathcal{L}_{base} + \mathcal{L}_{consis}. \tag{12}$$

At test time, we combine K domain experts by Eq. (8) to produce *F-exp*, and concatenate it with *F-global* as the final representation. The overall training procedure is shown in Algorithm 1.

4   Experiments

4.1   Datasets and Settings

Datsets. We conduct extensive experiments on 9 public ReID or person search datasets including Market1501 [46], MSMT17 [40], CUHK02 [20], CUHK03 [21], CUHK-SYSU [41], PRID [14], GRID [26], VIPeR [8], and iLIDs [47]. The details of these datasets are illustrated in Table 1. For CUHK03, we use the 'labeled' data as [5]. For simplicity, we denote MSMT17 as MS, Market1501 as M, CUHK02 as C2, CUHK03 as C3, and CUHK-SYSU as CS. We utilize Cumulative Matching Characteristics (CMC) and mean average precision (mAP) for evaluation.

Evaluation Protocols. Because DukeMTMC-reID [48], which was widely used in previous work [1, 4, 35, 45] on DG ReID, has been taken down, we set three new

Table 1. Summary of all the datasets.

Datasets	#IDs	#Images	#Cameras
Market1501 (M) [46]	1,501	32,217	6
MSMT17 (MS) [40]	4,101	126,441	15
CUHK02 (C2) [20]	1,816	7,264	10
CUHK03 (C3) [21]	1,467	14,096	2
CUHK-SYSU (CS) [41]	11,934	34,574	1
PRID [14]	749	949	2
GRID [26]	1,025	1,275	8
VIPeR [8]	632	1,264	2
iLIDs [47]	300	4,515	2

Table 2. Evaluation protocols.

	Training Sets	Testing Sets
Protocol-1	Full-(M+C2+C3+CS)	PRID,GRID, VIPeR,iLIDs
Protocol-2	M+MS+CS	C3
	M+CS+C3	MS
	MS+CS+C3	M
Protocol-3	Full-(M+MS+CS)	C3
	Full-(M+CS+C3)	MS
	Full-(MS+CS+C3)	M

Table 3. Comparison with state-of-the-art methods under protocol-1. All the images in the source domains are used for training. The illustration of abbreviations is shown in Table 1. We report some results of other methods which leverage DukeMTMC-reID in the source domains, while we remove DukeMTMC-reID from our training sets. Although we use fewer source domains, we still get the best performance. '*' indicates that we re-implement this work based on the authors' code on Github. The best (in **bold red**), the second best (in *italic blue*).

Method	Source Domains	→PRID		→GRID		→VIPeR		→iLIDs		Average	
		mAP	Rank-1	mAP	Rank-1	mAP	Rank-1	mAP	Rank-1	mAP	Rank-1
CrossGrad [34]	M+D +C2+C3+CS	28.2	18.8	16.0	8.96	30.4	20.9	61.3	49.7	34.0	24.6
Agg_PCB [37]		45.3	31.9	38.0	26.9	54.5	45.1	72.7	64.5	52.6	42.1
MLDG [18]		35.4	24.0	23.6	15.8	33.5	23.5	65.2	53.8	39.4	29.3
PPA [31]		32.0	21.5	44.7	36.0	45.4	38.1	73.9	66.7	49.0	40.6
DIMN [35]		52.0	39.2	41.1	29.3	60.1	51.2	78.4	70.2	57.9	47.5
SNR [17]		66.5	52.1	47.7	40.2	61.3	52.9	*89.9*	*84.1*	66.4	57.3
RaMoE [5]		67.3	57.7	54.2	46.8	64.6	56.6	90.2	85.0	62.0	*61.5*
DMG-Net [1]		68.4	60.6	56.6	*51.0*	60.4	53.9	83.9	79.3	67.3	61.2
QAConv$_{50}$ [22]*	M +C2+C3+CS	62.2	52.3	57.4	48.6	66.3	57.0	81.9	75.0	67.0	58.2
M^{3}L(ResNet-50) [45]*		65.3	55.0	50.5	40.0	*68.2*	*60.8*	74.3	65.0	64.6	55.2
MetaBIN [4]*		*70.8*	*61.2*	*57.9*	50.2	64.3	55.9	82.7	74.7	*68.9*	60.5
META		**71.7**	**61.9**	**60.1**	**52.4**	**68.4**	**61.5**	83.5	79.2	**70.9**	**63.8**

protocols for DG ReID, as shown in Table 2. For protocol-1, we use all the images in the source domains (*i.e.*, including training and testing sets) for training. For PRID, GRID, VIPeR, and iLIDS, following [5], the results are evaluated on the average of 10 repeated random splits of query and gallery sets. For protocol-2, we choose one domain from M+MS+CS+C3 for testing and the remaining three domains for training. As the CS person search dataset only contains 1 camera, CS is not used for testing. The difference between protocol-2 and protocol-3 is that we use all the images in the source domains for training under protocol-3.

Implementation Details. We resize all the images to 256 × 128. ResNet50 [9] pretrained on ImageNet is used as our backbone. We set batch size to 64, including 16 identities and 4 images per identity. Similar to [5], we perform color jitter and discard random erasing for the data augmentation. We train the model for 120 epochs and adopt the warmup strategy in the first 500 iterations. The

Table 4. Comparison with state-of-the-art methods under protocol-2 and protocol-3. 'Training Sets' denotes that only the training sets in the source domains are used for training and 'Full Images' denotes that all images are leveraged at training time. The illustration of abbreviations is shown in Table 1. '*' indicates that we re-implement this work based on the authors' code on Github. The best (in bold red), the second best (in *italic blue*).

Method	Setting	M+MS+CS →C3		M+CS+C3 →MS		MS+CS+C3 →M		Average	
		mAP	Rank-1	mAP	Rank-1	mAP	Rank-1	mAP	Rank-1
SNR* [17]	Protocol-2	8.9	8.9	6.8	19.9	34.6	62.7	16.8	30.5
QAConv$_{50}$ [22]*	(Training Sets)	25.4	24.8	16.4	*45.3*	*63.1*	*83.7*	35.0	51.3
M^{3}L (ResNet-50) [45]*		20.9	31.9	15.9	36.9	58.4	79.9	31.7	49.6
M^{3}L (IBN-Net50) [45]*		*34.2*	*34.4*	16.7	37.5	61.5	82.3	*37.5*	*51.4*
MetaBIN [4]*		28.8	28.1	*17.8*	40.2	57.9	80.1	34.8	49.5
META		36.3	35.1	22.5	49.9	67.5	86.1	42.1	57.0
SNR* [17]	Protocol-3 (Full	17.5	17.1	7.7	22.0	52.4	77.8	25.9	39.0
QAConv$_{50}$* [22]	Images)	32.9	33.3	17.6	*46.6*	66.5	*85.0*	39.0	55.0
M^{3}L (ResNet-50) [45]*		32.3	33.8	16.2	36.9	61.2	81.2	36.6	50.6
M^{3}L (IBN-Net50) [45]*		35.7	36.5	17.4	38.6	62.4	82.7	38.5	52.6
MetaBIN [4]*		*43.0*	*43.1*	*18.8*	41.2	*67.2*	84.5	*43.0*	*56.3*
META		47.1	46.2	24.4	52.1	76.5	90.5	49.3	62.9

learning rate is initialized as $3e^{-4}$ and divided by 10 at the 40th and 70th epochs respectively. The margins α_1, α_2 in Eq. (11) are set to be 0.1.

4.2 Comparison with State-of-the-Art Methods

Comparison Under Protocol-1. We compare our method with other state-of-the-arts under protocol-1, as shown in Table 3. We report some results of other methods which leverage DukeMTMC-reID [48] in the source domains, while we remove it from our training sets. Although we use fewer source domains, we still get the best performance. Specifically, from the results, we can find that META achieves the best performances on the PRID, GRID and VIPeR, while RaMoE [5] gives the highest points on the iLIDs dataset. META significantly outperforms other methods by at least 2.0% and 2.3% in average mAP and Rank-1 respectively.

Comparison Under Protocol-2 and Protocol-3. We compare our method with other state-of-the-arts under protocol-2 and protocol-3, as shown in Table 4. 'Training Sets' denotes that only the training sets in the source domains are used for training and 'Full Images' denotes that all images in the source domains (*i.e.* including training and testing sets) are leveraged at training time. The results show that META outperforms other methods by a large margin on all the datasets and under both protocols. Specifically, META surpasses other methods, on average, by at least 4.6% mAP, 5.6% Rank-1 and 6.3% mAP, 6.6% Rank-1 under protocol-2 and protocol-3 respectively. The results have shown our model's superiority in domain generalization.

Table 5. Ablation study on the effectiveness of individual components and the design of global branch. The experiment is conducted under protocol-3. 'C3', 'MS', 'M' are the abbreviations of the CUHK03, MSMT17, and Market1501 respectively. The best results are highlighted in bold.

Method	Target: C3		Target: MS		Target: M		Average	
	mAP	Rank-1	mAP	Rank-1	mAP	Rank-1	mAP	Rank-1
w/o global branch	26.4	26.2	10.3	28.3	44.1	71.6	26.9	42.0
w/o expert branch	33.6	33.7	20.5	45.8	71.9	87.6	42.0	55.7
w/o aggregation module	46.0	45.5	23.3	50.9	75.1	89.6	48.1	62.0
BN-BN	43.3	43.1	21.9	48.6	71.7	88.3	45.6	60.0
BN-IBN [30]	45.2	44.0	22.7	50.2	73.2	89.8	47.0	61.3
IN-IN	41.5	40.2	18.7	46.0	68.3	86.7	42.8	57.6
META	**47.1**	**46.2**	**24.4**	**52.1**	**76.5**	**90.5**	**49.3**	**62.9**

4.3 Ablation Study

The Effectiveness of the Individual Branches. We study ablation studies on the effectiveness of individual branches, as shown in the first, second, and last rows of Table 5. The experiment is conducted under protocol-3. We train our model without the global branch or expert branch for comparison. From the results, we can find that mAP drops 20.7%, 14.1% and 32.4% on the CUHK03, MSMT17 and Market1501 respectively when the global branch is discarded. The mAP also drops 13.5%, 3.9% and 4.6% on the CUHK03, MSMT17 and Market1501 respectively when the expert branch is discarded. The results have demonstrated the effectiveness of both the global and expert branches. Furthermore, we visualize the features extracted by different branches via t-SNE [27], as shown in Fig. 3(a). Different colors denote various IDs. We find that the expert branch pushes features from different IDs away while the global branch pulls the features from same ID closer. Thus, both branches are integrated for better ReID performance.

The Effectiveness of Aggregation Module. We study ablation studies on the effectiveness of aggregation module, as shown in the third and last rows of Table 5. The experiment is conducted under protocol-3. *'w/o aggregation module'* denotes that we remove the aggregation module and directly integrate multiple experts with FID. The results show that the aggregation module gives the performance gains of 1.1%, 1.1% and 1.4% for mAP on CUHK03, MSMT17 and Market1501 respectively. The results have validated the effectiveness of the aggregation module for adaptively integrating diverse domain experts to mimic unseen target domain.

Table 6. Ablation study on the performance of the individual features under protocol-3.

Method	Target: C3		Target: MS		Target: M	
	mAP	Rank-1	mAP	Rank-1	mAP	Rank-1
F-global	46.9	46.0	24.1	52.0	76.4	90.3
F-exp	42.9	42.0	10.2	28.7	45.7	72.3
META	**47.1**	**46.2**	**24.4**	**52.1**	**76.5**	**90.5**

Table 7. Ablation study on loss functions under protocol-3.

$\mathcal{L}_{base}$	$\mathcal{L}_{cross}$	$\mathcal{L}_{tri}$	$\mathcal{L}_{consis}$	Target: MSMT17	
				mAP	Rank-1
✓				21.2	48.4
✓	✓			23.5	50.9
✓		✓		22.8	50.4
✓			✓	24.4	52.1

The Design of Global Branch. The global branch is designed based on the findings from [30] that adding IN layers after BN layers could significantly improve the domain generalization performance of the model. We compare our design with other architectures of the global branch, as shown in the last four rows of Table 5. The experiment is conducted under protocol-3. We respectively replace IN in the *Global-IN* with BN and IBN [30], and replace BN in the *Global-BN* with IN for comparison. The results show that our design achieve the best results, surpassing other architectures by 2.9%, 1.6% and 5.3% respectively in average Rank-1. The results have demonstrated the effectiveness of our design of global branch to help extract domain-invariant features.

Performance of Individual Features. We study ablation studies on the performance of individual features, as shown in Table 6. The experiment is conducted under protocol-3. We separately inference with *F-global* and *F-exp* for comparison. The results show that *F-global* has a similar performance with META which concatenates *F-global* and *F-exp* for testing. We think the reason is that the expert branch is able to help the backbone extract more generalizable features, and therefore could improve the domain generalization performance of the global branch. As a result, it is feasible to only leverage the global branch during testing for faster inference.

The Effectiveness of Loss Function Components. We study ablation studies on the effectiveness of loss function components, as shown in Table 7. The experiment is conducted under protocol-3. $\mathcal{L}_{base}$ is defined in Eq. (9) for training the global and expert branch. $\mathcal{L}_{cross}$ and $\mathcal{L}_{tri}$ indicate that we replace $\mathcal{L}_{consis}$ with cross-entropy loss and triplet loss respectively to update the aggregation module. From the first and fourth rows, we can find that $\mathcal{L}_{consis}$ gives performance gains of 3.2% and 3.7% for mAP and Rank-1 accuracy respectively. From the last three rows, we can find that $\mathcal{L}_{consis}$ achieves the best performance, which surpasses $\mathcal{L}_{cross}$ and $\mathcal{L}_{tri}$ by 1.2% and 1.7% Rank-1 accuracy respectively. The results have demonstrated the effectiveness of our proposed $\mathcal{L}_{consis}$.

The Justification for Calculating FID Between BN and IN Statistics. Both BN and IN can be seen as approximations of different Gaussian distributions, thus we can simply adopt FID to measure the difference between them. We

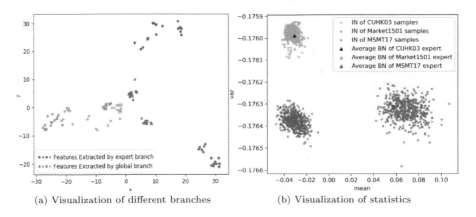

(a) Visualization of different branches   (b) Visualization of statistics

Fig. 3. (a) Visualization of the features extracted by different branches. Various colors denote different IDs. (b) Visualization of statistics, the results illustrate the justification for calculating FID between BN and IN statistics. (Color figure online)

expect through our learning scheme, BN and IN statistics could reflect the properties of the source and target domain respectively. Figure 3(b) plots the average BN of multiple experts and IN statistics of samples from different domains via t-SNE [27]. The horizontal and vertical axes represent the mean and standard deviation of the statistics respectively. The result shows that different domain clusters can be divided by their IN statistics. Additionally, IN statistics of the samples are closer to the average BN of the expert from the same domain. The result illustrates the justification for calculating FID between BN and IN statistics.

5 Conclusion

This paper presents a new approach called Mimic Embedding via adapTive Aggregation (META) for Domain generalizable (DG) person re-identification (ReID). META is a lightweight ensemble of multiple experts sharing all the parameters except for the domain-specific BN layers. Besides multiple experts, META leverages Instance Normalization (IN) and introduces it into a global branch to pursue invariant features across domains. Meanwhile, META develops an aggregation module to adaptively integrate multiple experts with the relevance of an unseen target sample w.r.t. source domains via normalization statistics. Extensive experiments demonstrate that META surpasses state-of-the-art DG ReID methods by a large margin.

Acknowledgment. The authors would like to thank reviewers for providing valuable suggestions to improve this paper. This work is supported by the National Natural Science Foundation of China (Grant No. U1836217) and the Beijing Nova Program under Grant Z211100002121108.

References

1. Bai, Y., et al.: Person30k: a dual-meta generalization network for person re-identification. In: CVPR (2021)
2. Bai, Z., Wang, Z., Wang, J., Hu, D., Ding, E.: Unsupervised multi-source domain adaptation for person re-identification. In: CVPR (2021)
3. Chang, W.G., You, T., Seo, S., Kwak, S., Han, B.: Domain-specific batch normalization for unsupervised domain adaptation. In: CVPR (2019)
4. Choi, S., Kim, T., Jeong, M., Park, H., Kim, C.: Meta batch-instance normalization for generalizable person re-identification. In: CVPR (2021)
5. Dai, Y., Li, X., Liu, J., Tong, Z., Duan, L.Y.: Generalizable person re-identification with relevance-aware mixture of experts. In: CVPR (2021)
6. Dumoulin, V., Shlens, J., Kudlur, M.: A learned representation for artistic style. arXiv (2016)
7. Fu, Y., Wei, Y., Wang, G., Zhou, Y., Shi, H., Huang, T.S.: Self-similarity grouping: a simple unsupervised cross domain adaptation approach for person re-identification. In: ICCV (2019)
8. Gray, D., Tao, H.: Viewpoint invariant pedestrian recognition with an ensemble of localized features. In: ECCV (2008)
9. He, K., Zhang, X., Ren, S., Sun, J.: Deep residual learning for image recognition. In: CVPR (2016)
10. He, L., Liang, J., Li, H., Sun, Z.: Deep spatial feature reconstruction for partial person re-identification: alignment-free approach. In: CVPR (2018)
11. He, L., et al.: Semi-supervised domain generalizable person re-identification. arXiv (2021)
12. Hermans, A., Beyer, L., Leibe, B.: In defense of the triplet loss for person re-identification. arXiv (2017)
13. Heusel, M., Ramsauer, H., Unterthiner, T., Nessler, B., Hochreiter, S.: GANs trained by a two time-scale update rule converge to a local nash equilibrium. In: NIPS (2017)
14. Hirzer, M., Beleznai, C., Roth, P.M., Bischof, H.: Person re-identification by descriptive and discriminative classification. In: Scandinavian Conference on Image Analysis (2011)
15. Ioffe, S., Szegedy, C.: Batch normalization: accelerating deep network training by reducing internal covariate shift. In: ICML (2015)
16. Jacobs, R.A., Jordan, M.I., Nowlan, S.J., Hinton, G.E.: Adaptive mixtures of local experts. Neural Computat. 3(1), 79–87 (1991)
17. Jin, X., Lan, C., Zeng, W., Chen, Z., Zhang, L.: Style normalization and restitution for generalizable person re-identification. In: CVPR (2020)
18. Li, D., Yang, Y., Song, Y.Z., Hospedales, T.M.: Learning to generalize: meta-learning for domain generalization. In: AAAI (2018)
19. Li, D., Zhang, J., Yang, Y., Liu, C., Song, Y.Z., Hospedales, T.M.: Episodic training for domain generalization. In: ICCV (2019)
20. Li, W., Wang, X.: Locally aligned feature transforms across views. In: CVPR (2013)
21. Li, W., Zhao, R., Xiao, T., Wang, X.: DeepReID: deep filter pairing neural network for person re-identification. In: CVPR (2014)
22. Liao, S., Shao, L.: Interpretable and generalizable person re-identification with query-adaptive convolution and temporal lifting. In: ECCV (2020)
23. Liu, J., Ni, B., Yan, Y., Zhou, P., Cheng, S., Hu, J.: Pose transferrable person re-identification. In: CVPR (2018)

24. Liu, Q., Dou, Q., Yu, L., Heng, P.A.: Ms-Net: multi-site network for improving prostate segmentation with heterogeneous MRI data. IEEE Trans. Med. Imaging **39**(9), 2713–2724 (2020)
25. Liu, X., Zhang, P., Yu, C., Lu, H., Yang, X.: Watching you: global-guided reciprocal learning for video-based person re-identification. In: CVPR (2021)
26. Loy, C.C., Xiang, T., Gong, S.: Time-delayed correlation analysis for multi-camera activity understanding. Int. J. Comput. Vis. **90**(1), 106–129 (2010)
27. Van der Maaten, L., Hinton, G.: Visualizing data using t-SNE. J. Mach. Learn. Res. **9**(11), 2579–2605 (2008)
28. Mancini, M., Bulo, S.R., Caputo, B., Ricci, E.: Robust place categorization with deep domain generalization. IEEE Roboti. Autom. Lett. **3**(3), 2093–2100 (2018)
29. Miao, J., Wu, Y., Liu, P., Ding, Y., Yang, Y.: Pose-guided feature alignment for occluded person re-identification. In: ICCV (2019)
30. Pan, X., Luo, P., Shi, J., Tang, X.: Two at once: enhancing learning and generalization capacities via IBN-net. In: ECCV (2018)
31. Qiao, S., Liu, C., Shen, W., Yuille, A.L.: Few-shot image recognition by predicting parameters from activations. In: CVPR (2018)
32. Segu, M., Tonioni, A., Tombari, F.: Batch normalization embeddings for deep domain generalization. arXiv (2020)
33. Seo, S., Suh, Y., Kim, D., Kim, G., Han, J., Han, B.: Learning to optimize domain specific normalization for domain generalization. In: ECCV (2020)
34. Shankar, S., Piratla, V., Chakrabarti, S., Chaudhuri, S., Jyothi, P., Sarawagi, S.: Generalizing across domains via cross-gradient training. arXiv (2018)
35. Song, J., Yang, Y., Song, Y.Z., Xiang, T., Hospedales, T.M.: Generalizable person re-identification by domain-invariant mapping network. In: CVPR (2019)
36. Su, C., Li, J., Zhang, S., Xing, J., Gao, W., Tian, Q.: Pose-driven deep convolutional model for person re-identification. In: ICCV (2017)
37. Sun, Y., Zheng, L., Li, Y., Yang, Y., Tian, Q., Wang, S.: Learning part-based convolutional features for person re-identification. IEEE Trans. Pattern Anal. Mach. Intell. **43**(3), 902–917 (2019)
38. Sun, Y., Zheng, L., Yang, Y., Tian, Q., Wang, S.: Beyond part models: person retrieval with refined part pooling (and a strong convolutional baseline). In: ECCV (2018)
39. Wang, G., Yuan, Y., Chen, X., Li, J., Zhou, X.: Learning discriminative features with multiple granularities for person re-identification. In: ACM MM (2018)
40. Wei, L., Zhang, S., Gao, W., Tian, Q.: Person transfer GAN to bridge domain gap for person re-identification. In: CVPR (2018)
41. Xiao, T., Li, S., Wang, B., Lin, L., Wang, X.: End-to-end deep learning for person search. arXiv (2016)
42. Xu, B., He, L., Liang, J., Sun, Z.: Learning feature recovery transformer for occluded person re-identification. IEEE Trans. Image Process. **31**, 4651–4662 (2022)
43. Xu, B., He, L., Liao, X., Liu, W., Sun, Z., Mei, T.: Black Re-ID: a head-shoulder descriptor for the challenging problem of person re-identification. In: ACM MM (2020)
44. Zhai, Y., et al.: Ad-cluster: augmented discriminative clustering for domain adaptive person re-identification. In: CVPR (2020)
45. Zhao, Y., et al.: Learning to generalize unseen domains via memory-based multi-source meta-learning for person re-identification. In: CVPR (2021)
46. Zheng, L., Shen, L., Tian, L., Wang, S., Wang, J., Tian, Q.: Scalable person re-identification: a benchmark. In: ICCV (2015)

47. Zheng, W.S., Gong, S., Xiang, T.: Associating groups of people. In: BMVC, pp. 1–11 (2009)
48. Zheng, Z., Zheng, L., Yang, Y.: Unlabeled samples generated by GAN improve the person re-identification baseline in vitro. In: ICCV (2017)
49. Zhong, Z., Zheng, L., Luo, Z., Li, S., Yang, Y.: Invariance matters: exemplar memory for domain adaptive person re-identification. In: CVPR (2019)
50. Zhou, K., Liu, Z., Qiao, Y., Xiang, T., Loy, C.C.: Domain generalization: a survey. arXiv (2021)

Granularity-Aware Adaptation for Image Retrieval Over Multiple Tasks

Jon Almazán[1(✉)], Byungsoo Ko[2], Geonmo Gu[2], Diane Larlus[1],
and Yannis Kalantidis[1]

[1] NAVER LABS Europe , Meylan, France
jon.almazan@naverlabs.com
[2] NAVER Corp., Seongnam-si, South Korea

Abstract. Strong image search models can be learned for a specific
domain, *i.e.* set of labels, provided that some labeled images of that domain
are available. A practical visual search model, however, should be versatile enough to solve multiple retrieval tasks simultaneously, even if those
cover very different specialized domains. Additionally, it should be able to
benefit from even unlabeled images from these various retrieval tasks. This
is the more practical scenario that we consider in this paper. We address
it with the proposed `Grappa`, an approach that starts from a strong pre-
trained model, and adapts it to tackle multiple retrieval tasks concurrently,
using only unlabeled images from the different task domains. We extend
the pretrained model with multiple independently trained sets of adaptors
that use pseudo-label sets of different sizes, effectively mimicking different pseudo-granularities. We reconcile all adaptor sets into a single unified
model suited for all retrieval tasks by learning fusion layers that we guide by
propagating pseudo-granularity attentions across neighbors in the feature
space. Results on a benchmark composed of six heterogeneous retrieval
tasks show that the unsupervised `Grappa` model improves the zero-shot
performance of a state-of-the-art self-supervised learning model, and in
some places reaches or improves over a task label-aware oracle that selects
the most fitting pseudo-granularity per task.

1 Introduction

The last few years have witnessed progress on image retrieval: successful models can be trained, provided that a set of labeled images from the domain of
interest (not necessary from the same categories) is available for training, as
in the common deep metric learning scenario. Those models are as powerful as
they are specialized: it has been shown, and we confirm in our experiments, that
one model carefully tailored for one domain (*e.g.* bird species) tend to perform
poorly to a neighboring yet different domain (*e.g.* dog breeds).

Here, we argue that a practical visual search system should be able to solve
multiple retrieval tasks simultaneously, without needing to explicitly specialize

Supplementary Information The online version contains supplementary material
available at https://doi.org/10.1007/978-3-031-19781-9_23.

for each task. Consider for example a visual search system specialized to fauna and flora. In such a system, the image database covers a broad range of fine-grained domains, *e.g.* from searching among different insect species to different kinds of mushrooms. For the system to also handle coral species, it should be as simple as providing a set of unlabeled coral images.

In parallel, the field has worked towards pretraining large and generic models for visual representations that can be used, often as a black box, to extract features for new tasks. Among those, models trained in a self-supervised way have shown to be versatile to various target tasks, including image retrieval [9,21].

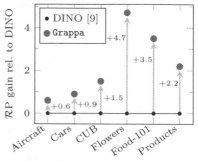

In this work, we assume access to such a large pretrained model that already provides good zero-shot performance. We also assume access to an unlabeled set of images possibly from multiple tasks. We propose to adapt the initial model so it performs

Fig. 1. Grappa is an unsupervised method that trains *a single model* with higher zero-shot performance (measured with $\mathcal{R}$-Precision or $\mathcal{R}$P) than the pretrained DINO [9] model, over several retrieval tasks.

even better on multiple image retrieval tasks *simultaneously, i.e.* when this same adapted model is used to extract features for all tasks.

This raises two questions. First, *how should we perform adaptation?* Fine-tuning is prohibitively costly especially for large pretrained models, and does not always transfer well. As an alternative to fine-tuning, and inspired by an early work on multi-task training [46] and a recent trend in natural language processing [27,41], we propose to use adaptor layers. Adaptors are embedded in between architecture blocks, and are the only weights learned, the ones from the original pretrained model remaining fixed. Our experiments show that this composite architecture allows for a versatile adaptation of a strong initial model by adjusting only a small percentage of the model parameters.

Second, *how should we reconcile various retrieval tasks in a single model?* A retrieval task focuses on a given set of visual concepts, often associated to a particular granularity. Yet, unlike in classification for which the granularity is known beforehand, the granularity of a retrieval task is context dependent, and depends on the gallery of images where visual search is performed. We therefore propose learning different sets of adaptors, each set tailored to one specific granularity. As we assume that training images are unlabeled, not even to indicate the retrieval task they correspond to, we propose to automatically define levels of granularity by partitioning the training set into more and more clusters. As a result, each partition corresponds to a different set of pseudo-labels. We then independently train one set of adaptors for each pseudo-granularity.

Next, we need to reconcile these different sets of adaptors into a single multi-purpose retrieval model. One option is to combine them with a naive fusion mechanism. The resulting model improves results on all retrieval tasks, showing the clear benefit of a multi-granularity understanding of the data. Another option is to go one step further and to achieve adaptor fusion via attention propagation. In this case, we require consistency between the adaptor attention of nearest neighbors in the feature space. We observe this fusion mechanism further improves the model.

To summarize, our contribution is threefold. First, we palliate the absence of image and task labels by creating sets of pseudo-labels, with the goal of approximating any possible granularities in a given set of retrieval tasks. Second, we propose a way to extend transformer-based architectures with adaptors, and a training framework that tailors individual sets of adaptors to different pseudo-granularities. Third, we propose a number of ways for fusing the adapter features, *e.g.* via augmentation invariance or via propagating attention from neighbors in the image features space. We validate our approach on a collection of datasets for deep metric learning and we show that `Grappa` improves over the successful DINO pretrained model, a model known to already obtain strong zero-shot performance on all these retrieval tasks (see Fig. 1).

2   Related Work

The task we tackle in this paper strongly relates to deep metric learning. It requires specific architectural changes of neural networks to extend them with adaptors. Note that our task can be seen as solving a zero-shot problem, *i.e.* it requires no labeled data from the downstream datasets and learns a single model for all tasks, something fairly uncommon in transfer learning.

Deep Metric Learning (DML). DML aims to learn a metric between data points that reflects the semantic similarity between them. It plays an important role in a wide range of tasks such as image clustering [7,26], unsupervised learning [8,10,24], and visual search [6,19,47]. Recent DML approaches typically learn visual similarity using either a pair-based loss [11,22,30,40,53] which considers pair-wise similarities, a proxy-based loss [14,23,31,57,58], which considers the similarity between samples and class representative proxies, or a contextual classification loss [5,16,50,65]. In most cases, DML approaches finetune an ImageNet pretrained model for *each* target retrieval task, and each of those finetuned models fall short when applied to other retrieval tasks. We aim at a more versatile visual search system that handles multiple retrieval tasks with a *single* model.

Neural Architectures with Adaptation Layers. Adaptation layers (or adaptors) have emerged [27,41,46,59] as a way to avoid common problems rising in sequential finetuning or multi-task learning when trying to finetune large pretrained models to solve multiple tasks, namely the issues of catastrophic forgetting [36] and task imbalance. Rebuffi *et al..* [46] were the first to introduce adaptors to visual recognition tasks, adapting a convolutional model to many classification tasks. Adaptors have also been used with transformer architectures for natural language processing [27]; bottleneck layers are added to all the blocks of a pretrained model and finetuned, keeping the underlying model fixed.

Recently, Pfeiffer *et al..* [41] introduced a way to share knowledge between adaptors using an adaptor fusion layer within a two-stage learning framework:

adaptors are trained independently in the first stage, they are kept fixed while only the fusion layer is trained in the second stage. All the methods mentioned above still result in models that specialize to a single task; e.g. [41] learns a separate fusion layer per downstream task, whereas we would like to learn a single model for all tasks.

Zero-Shot Problems. The field has recently taken an interest in pretraining large models, sometimes called zero-shot models, using large quantities of data. Those have been shown to be versatile and applicable to many target tasks. Among them, self-supervised models [8–10,24,64] are trained using self-defined pseudo-labels as supervision and typically millions of images (*e.g.* from ImageNet [13]). Recent works [20,54] exploit even larger yet uncurated, sets of unlabeled images to enhance the quality of the learned representations. Others [28,45,66] have leveraged multiple modalities, *e.g.* training visual representations so they are similar to the textual representations of their associated text. Those self-supervised or multimodal methods offer excellent initialization to be finetuned for a wide range of downstream tasks. Sometimes they are used in a zero-shot setting: a single model is used as a feature extractor, typically to solve multiple tasks. This is the regime we study here, but we further assume that a small amount of unlabeled data from the downstream tasks exists.

Relation to Other Transfer Tasks. The idea of transferring a model trained for a given task to a related one has become central to computer vision [49,51], and appears in many research fields such as task transfer [63], domain adaptation [12] or self-supervised learning [9,18,39]. Yet, in all those, the initial model is only a starting point and it is typically not only extended, but also retrained for each task of interest, leading to a multitude of specialized models. In our work, we need a *single* model to perform well across retrieval tasks. In that regard, this work is closer to zero-shot transfer of the large pretrained models discussed above. Also related are Mixtures of Experts (MoE) [44,48,52,62], an ensembling technique that decomposes a predictive problem into subtasks, training one expert for each. Although MoE architectures may look similar to ours at first glance, they typically rely on gating and pooling mechanisms that learn to predict, in a supervised way, which experts to trust, and how to combine them. Similar to typical transfer approaches, they build one specialized model for each target task. Here, we focus on a purely unsupervised task: no labels are provided to indicate image semantic content nor the retrieval task images belong to.

3   A Granularity-Aware Multi-purpose Retrieval Model

In this section we present `Grappa`, a method for adapting a pretrained model to multiple retrieval tasks simultaneously, in an unsupervised way. We first formalize our task, *i.e.* visual search over several retrieval tasks using a single model (Sect. 3.1). We then present an overview of the approach (Sect. 3.2). Next, we detail each step, *i.e.* building multiple granularities (Sect. 3.3), learning adaptors using granularity-aware pseudo-labels (Sect. 3.4), and learning to fuse them by propagating adaptor attention across feature space neighbors (Sect. 3.5).

3.1   Background

Our task of interest, **visual search on multiple retrieval tasks**, can be seen as a variant of the standard deep metric learning (DML) task. The most common protocol in DML is to a) split the *classes*[1] into disjoint train and test sets of labels; b) learn a separate model for each retrieval task on the corresponding train split; c) perform retrieval on all images of the (unseen) test split of classes.

Our setting has several key differences. First, we solve multiple retrieval tasks *simultaneously*. This means that we do not learn one model for each but a *single* model that will be used for all tasks. Second, we only assume access to a set of *unlabeled* images from each retrieval task, and do not have access to labeled training sets, unlike standard DML methods. Even more challenging, unlabeled training images are provided jointly without knowing which target retrieval task they correspond to nor the total number of retrieval tasks.

More formally, let $\mathcal{T}$ be the set of m retrieval tasks that we want to simultaneously tackle. Each task $\mathcal{T}^t$ is associated with a training and a test set. At train time, we are provided with a fused training set $\mathcal{D}$ composed of the union of all training sets of the m datasets in $\mathcal{T}$. As mentioned earlier, images are not associated to any class or task label.

With so many unknowns on the target retrieval tasks, an obvious choice is to **start with a large pretrained model**. Self-[9, 10, 24] or weakly-[28, 45] supervised learning have been shown to lead to strong models that generalize well and exhibit high zero-shot transfer performance. We assume that we are given such a model. Here, we base our work on the recently proposed Visual Transformer (ViT) [15], a popular, efficient, and highly performing architecture, pretrained in a self-supervised way with DINO [9].

We set our pretrained model $\mathcal{M}$ to be a ViT with L transformer layers and an input patch size $P \times P$ pixels. Input image $\mathbf{x} \in \mathbb{R}^{H \times W \times C}$ is reshaped into a sequence of T flattened 2D patches where $T = HW/P^2$. The transformer uses constant latent vector size D through all of its layers, so flattened patches are first mapped to D dimensions with a trainable linear projection and concatenated in $\mathbf{h}^0$, together with a prepended learnable [class] token and added position embeddings. The transformer encoder [55] consists of alternating blocks of multi-headed self-attention (MSA) and MLP (which contain two layers with a GELU non-linearity). LayerNorm (LN) is applied before every block, and residual connections after every block. Formally, each layer of $\mathcal{M}$ (shown with a gray background on Fig. 3, left) is given by:

$$\mathbf{h}^l = \text{MLP}(\text{LN}(\tilde{\mathbf{h}}^l)) + \tilde{\mathbf{h}}^l, \quad \tilde{\mathbf{h}}^l = \text{MSA}(\text{LN}(\mathbf{h}^{l-1})) + \mathbf{h}^{l-1}, \tag{1}$$

for $l = \{1 \dots L\}$. The image representation $\mathbf{z}$ is the output of the [class] token after the last layer $\mathbf{h}^L$, *i.e.* $\mathbf{z} = \text{LN}(\mathbf{h}^L)[\text{class}]$. We refer the reader to [15] for more details about the ViT architecture.

---

[1] We will use the term *classes* to refer to sets of images with the same label, whether the latter represents object instances or fine-grained classes.

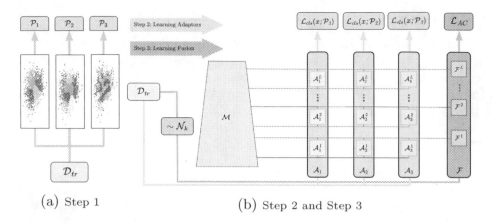

(a) Step 1   (b) Step 2 and Step 3

Fig. 2. Training of the proposed Grappa. Left: granularities correspond to pseudo-labels $\mathcal{P}_i$ obtained by multiple clusterings of the feature space (Step 1). Right: we learn the granularity-aware adaptors (Step 2, in green), and then learn how to fuse them (Step 3, in blue). For this example, $N=3$. (Color figure online)

3.2 Method Overview

Our method builds on the VIT [15] model $\mathcal{M}$ pretrained with DINO [9], that we treat as an architectural backbone. We extend and train it in an unsupervised way using $\mathcal{D}$. The training process consists of three steps (summarized in Fig. 2):

- **Step 1: Learning pseudo-labels.** We build multiple sets of pseudo-labels. Each set partitions the feature space using clustering and corresponds to a different pseudo-granularity. This process is illustrated in Fig. 2a.
- **Step 2: Learning adaptors for each pseudo-label set independently.** We learn a set of adaptors specific to each pseudo-granularity using a classification loss. This process is depicted by the green arrows in Fig. 2b.
- **Step 3: Learning to fuse adaptors.** We learn a set of fusion layers to merge the outputs of multiple adaptors using a transformation invariance or an attention propagation loss, *i.e.* neighboring images should have similar attentions over adaptors. This process is depicted by the blue arrows in Fig. 2b.

These three stages lead to a single model we denote as $\mathcal{M}^*$, that unifies the multiple granularities, and consists of: the pretrained model $\mathcal{M}$ used as a *frozen* backbone (its parameters are kept fixed during the entirety of the process), embedded adaptors $\mathcal{A}_i$, and fusion layers $\mathcal{F}$. This single model is used as a unique feature extractor for *all* retrieval tasks considered in our benchmark. We denote our method Grappa, that stands for learning **Gr**anularity-aware **A**daptors by Attention **P**ropagation. The following subsections detail the learning stages.

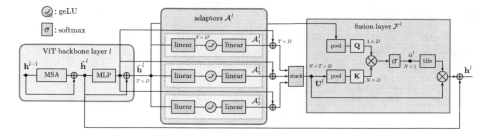

Fig. 3. Architecture of the lth layer of the Grappa model, for $N=3$ adaptors.

3.3 Step 1: Learning Pseudo-labels

We would like to build multiple sets of pseudo-labels such that they partition the feature space at different 'granularities'. We can approximate this partitioning by estimating multiple sets of clusters while varying the number of centers.

In practice, we extract features using the pretrained model $\mathcal{M}$. Let $\mathbf{z} = f(\mathbf{x}; \mathcal{M})$ be the feature of an image $\mathbf{x} \in \mathcal{D}$. Let the set of all features for training set $\mathcal{D}$ be $\mathcal{Z} = \{f(\mathbf{x}; \mathcal{M}), \forall \mathbf{x} \in \mathcal{D}\}$. To get multiple sets of pseudo-labels, we cluster the full set of features $\mathcal{Z}$ into sets of centroids $\mathcal{C}_i, i = 1..N$, of respectively k_i clusters, where k_i gets monotonically larger as i approaches N. This produces N sets of pseudo-labels $\mathcal{P}_1, \ldots, \mathcal{P}_N$. For each pseudo-label set $\mathcal{P}_i$, an image $\mathbf{x} \in \mathcal{D}$, is associated to a pseudo-label given by $\mathcal{P}_i(\mathbf{x}) = \arg\min_{\mathbf{c} \in \mathcal{C}_i} \|\mathbf{z} - \mathbf{c}\|$, for $\mathbf{z} = f(\mathbf{x}; \mathcal{M})$. We rely on the vanilla k-means clustering algorithm [34] with k-means++ [1] initialization, a common choice for the size of our benchmark. For even larger datasets, more scalable variants could be used, like hierarchical [42], approximate [2], or quantized [3] k-means. Note that other works have used k-means to define pseudolabels [7,60]. Yet, our work is the first to learn multiple sets and subsequently use all of them.

3.4 Step 2: Learning Adaptors for Each Pseudo-label Set

Given the N sets of pseudo-labels computed in the previous step, we now would like to learn adaptors tailored to each pseudo-label set, *i.e.* to each pseudo-granularity. We use the pretrained model $\mathcal{M}$ as a backbone and extend it by embedding an adaptor at every layer. We then learn the adaptor parameters while keeping the backbone ones frozen. We learn a set of L adaptors for *each* pseudo-granularity in an independent way.

Adaptor Architecture. Recent works in natural language processing [27,43,59] have embedded adaptor layers in transformer architectures. We follow a similar design and embed L adaptors, one at the end of each transformer layer of $\mathcal{M}$.

Formally, we learn a separate set of adaptors $\mathcal{A}_i$ for each pseudo-label set $\mathcal{P}_i, i = \{1, \ldots, N\}$. Each set $\mathcal{A}_i$ consists of L adaptors, denoted as $\mathcal{A}_i^1, \ldots, \mathcal{A}_i^L$. These adaptors are bottleneck layers with an intermediate dimensionality of D' (where $D' < D$), a GELU layer [25] in between, and a residual connection at the end. Since we are modifying the architecture of $\mathcal{M}$ by interleaving it with blocks, we need to revisit notations. The output of layer l in $\mathcal{M}$ (after the basic ViT block) is now defined as $\bar{\mathbf{h}}^l = \text{MLP}(\text{LN}(\tilde{\mathbf{h}}^l)) + \tilde{\mathbf{h}}^l$. The output of the new layer l (the original VIT block combined with an adaptor) can still be denoted as $\mathbf{h}^l$. Details of the overall architecture are shown in Fig. 3.

Learning the Adaptors. Given a set of pseudo-labels $\mathcal{P}_i$, we can learn the parameters of the set of adaptors $\mathcal{A}_i$ via a supervised cross entropy loss. Specifically, we use the norm-softmax loss [33,57,58,65] that, for image $\mathbf{x}$, is given by:

$$\mathcal{L}_{cls}(\mathbf{x}; y) = \log \frac{\exp(\gamma \cos \theta_y)}{\sum_{y'=1}^{k_i} \exp(\gamma \cos \theta_{y'})}, \tag{2}$$

where γ is a scale factor, $\cos \theta_y$ is the cosine similarity to the classifier of class y, and the loss is guided by the pseudo-labels, *i.e.* $y = \mathcal{P}_i(\mathbf{x})$. After learning the parameters for each set $\mathcal{P}_i$, we keep the adaptors and discard the classifiers.

3.5   Step 3: Learning to Fuse Adaptors

The process described in Sect. 3.4 leads to N separate sets of adaptors, each tailored to a different pseudo-granularity. The next step is to unify all adaptor sets into a single architecture. To that end, we append (*i.e.* stack) the N adaptors for each layer *in parallel*, as shown in Fig. 3. We then concatenate adaptor outputs in a tensor $\mathbf{U}^l \in \mathbb{R}^{N \times T \times D}$ for each layer $l = \{1, \ldots, L\}$, where each row corresponds to the output of one adaptor for this layer. Here, another residual connection is added, giving the model the opportunity to bypass the adapter if needed. Tensor $\mathbf{U}^l$ is therefore given by $\mathbf{U}^l = \{\mathcal{A}_i^l(\bar{\mathbf{h}}^l) + MLP(LN(\tilde{\mathbf{h}}^l)), i = 1..N\}$ and is then fed, together with $\bar{\mathbf{h}}^l$, to a fusion layer, as detailed below.

First Option: Fusion by Average Pooling. A straightforward way of fusing the outputs of the N adaptors is to treat them as equally important and average them. The fusion layer therefore is simply an average pooling layer that takes tensor $\mathbf{U}^l \in \mathbb{R}^{N \times T \times D}$ as input and computes the mean over its first dimension. We refer to this simpler version of our approach as `Grappa`-avg.

Second Option: Learning to Fuse. Treating all adaptors as equally important for any input image goes against our intuition that different retrieval tasks are more related to certain granularities, and hence more suited for the corresponding adaptors. We therefore design a fusion layer with trainable parameters, that can learn to weigh the different adaptor outputs. We use a simple dot-product self-attention architecture over the sequence of N adaptor outputs. Yet, we make two crucial modifications to the vanilla *query-key-value* self-attention: a) To learn an *image-level* attention, we average over the T spatial tokens; b) Given that we want to fuse the adaptors but do not want to alter the adaptor representations, we omit the linear projection of the value branch, and only learn projections for the query and key branches that affect the re-weighting of adaptor features.

Specifically, the fusion layer learns an attention vector of size N over the adaptors, given inputs $\bar{\mathbf{h}}^l$ and $\mathbf{U}^l$, by $\mathcal{F}^l(\bar{\mathbf{h}}^l, \mathbf{U}^l) = \alpha^l(\bar{\mathbf{h}}^l, \mathbf{U}^l)\mathbf{U}^l$, where vector $\alpha^l(\bar{\mathbf{h}}^l, \mathbf{U}^l) \in \mathbb{R}^N$ is given by:

$$\alpha^l(\bar{\mathbf{h}}^l, \mathbf{U}^l) = \text{softmax} \left(\frac{\left(\mathbf{Q} \sum_T \bar{\mathbf{h}}^l \right) \left(\mathbf{K} \sum_T \mathbf{U}^l \right)^T}{\sqrt{D}} \right) \tag{3}$$

where $l = \{1, \ldots, L\}$, $\mathbf{Q}$ and $\mathbf{K}$ are linear projections of size $D \times D$. A final residual connection is added after the fusion layer. The architecture details of a complete layer are shown in Fig. 3. The latter comprises the ViT block, the adaptors, and the fusion layer, all appended in a residual fashion.

Given pretrained model $\mathcal{M}$ and multiple sets of adaptors, one way to build a single model is to select one set of adaptors per image. This amounts to guessing which pseudo-granularity best fits each image. We argue that, in a generic visual search system, "picking a granularity" for a query image depends less on the image content than on the retrieval task, *i.e.* the gallery used at test time. Given a dog image query, for example, the only way to know if we are looking for any dog image or only images of the same dog breed, is by looking at the local structure of the gallery around that image. Both scenarios might favor different representations; our system reconciles them by learning a combination of adaptors. Obviously, we do *not* have access to the test images during training. Yet, we assume access to unlabeled set of images $\mathcal{D}$, representative of the target retrieval tasks, or at least of their granularity. Again, these images are provided without *task labels*, we do not know which retrieval task they correspond to.

Without any other supervisory signal, we argue that the local neighborhood in the feature space of the training set $\mathcal{D}$ can be used to approximate the "granularity" of a query. In other words, we assume that visually similar images from $\mathcal{D}$ should yield similar attention vectors over the sets of adaptors. We therefore propose to learn to fuse adaptors using a loss on neighboring image pairs in the feature space. In this step, the backbone model $\mathcal{M}$ and the adaptors remain frozen. The fusion layer only learns $\mathbf{K}$ and $\mathbf{Q}$, two linear projections that are multiplied to give the attention vectors α^l, for each ViT encoder l. This means that any loss applied to this fusion step *only re-weights* adaptor features. We denote the final model, composed of the backbone with all embedded adaptors and their fusion, as $\mathcal{M}^*$, and the corresponding feature extractor as $f^*(\mathbf{x}, \mathcal{M}^*)$.

Attention Propagation Loss. As mentioned, we propose to train the fusion layer leveraging the assumption that neighboring image pairs in the feature space should use similar attentions over adaptors. Let $\mathcal{N}_k(\mathbf{x}; \mathcal{D})$ denote the k nearest neighbors of $\mathbf{x}$ from dataset $\mathcal{D}$. We define neighbors $(\mathbf{x}_i, \mathbf{x}_j)$ as a pair of inputs such that $\mathbf{x}_j \in \mathcal{N}_k(\mathbf{x}_i; \mathcal{D})$. Although neighbors could be built using the pretrained model $\mathbf{z} = f(\mathbf{x}, \mathcal{M})$ (static k-NN), the representations $\tilde{\mathbf{z}} = f^*(\mathbf{x}, \mathcal{M}^*)$ from the learned model $\mathcal{M}^*$ provide a better estimation. This requires to periodically update neighbors during training (in practice we do it at every epoch). Given a pair of neighboring features, we bring their adaptor attentions close to each other and strive for *attention consistency*.

Attention consistency is enforced using the pairwise Barlow Twins loss [64]. Given a batch of image pairs, the loss is defined over the output representations $\tilde{\mathbf{z}}_i = f^*(\mathbf{x}_i; \mathcal{M}^*), \tilde{\mathbf{z}}_j = f^*(\mathbf{x}_j; \mathcal{M}^*)$ of our model, computed over the $D \times D$ cross-correlation matrix C and averaged over the batch, *i.e.*:

$$\mathcal{L}_{BT} = \sum_{n}(1 - C^{nn})^2 + \beta \sum_{n}\sum_{m \neq n}(C^{nm})^2, C_{nm} = \frac{\sum_b g(\hat{\mathbf{z}}_i)^{b,n} g(\hat{\mathbf{z}}_j)^{b,m}}{\sqrt{\sum_b (g(\hat{\mathbf{z}}_i)^{b,n})^2}\sqrt{\sum_b (g(\hat{\mathbf{z}}_j)^{b,m})^2}}, \quad (4)$$

where b iterates over pairs in the batch, n and m iterate over feature dimensions, β is a hyperparameter and $g(\cdot)$ is a MLP projector appended to the model and discarded after training. We refer the reader to [64] for more details.

Originally, *i.e.* in [64], this loss was defined over two transformed versions of the same image $(\mathbf{x}_i = t(\mathbf{x}), \mathbf{x}_j = t(\mathbf{x}))$. When image pairs are created using image transformations, Eq. (4) defines a transformation consistency (TC) loss or $\mathcal{L}_{TC}$. This is a variant that we consider in our benchmark, referred to as Grappa-T. However, we are interested in applying this loss on neighboring pairs in the feature space, $(\mathbf{x}_i, \mathbf{x}_j)$ such as $\mathbf{x}_j \in \mathcal{N}_k(\mathbf{x}_i; \mathcal{D})$, and using it for attention propagation. In this case, we depart from [64] and follow the recent TLDR method [29] which uses the Barlow Twins loss over neighbor pairs for learning a feature encoder for dimensionality reduction. Similarly, we use the Barlow Twins loss on image pairs defined using the k-NN graph. We denote the loss in Eq. (4) as an *attention consistency* (AC) loss, or $\mathcal{L}_{AC}$, and refer to this variant as Grappa-N.

4   Experiments

In this section we validate the proposed Grappa on several retrieval tasks. These tasks are collected in a new benchmark that we introduce, called MRT. It unifies 6 fine-grained classification datasets under a retrieval setting. We show statistics and present the evaluation protocol of this benchmark in Sect. 4.1, we present the methods we compare in Sect. 4.2 and we report all results in Sect. 4.3.

Implementation Details. We use ViT-Small [15] as a backbone architecture, with a patch size of 16 pixels and DINO [9] pre-trained weights. We generate $N=8$ sets of pseudo-labels on the training set of MRT, respectively composed of 256, 1024, 4096, 8,192, 16,384, 32,768, 65,536, and 131,072 clusters. We learn a set of adaptors for each pseudo-label set, using the norm-softmax loss from Eq. (2) and an Adam optimizer with a learning rate and weight decay of 0.001. We train the fusion layer over these adaptors using the Barlow Twins loss [64] and the LARS optimizer [61]. We used the same hyper-parameters for the scaling and β as suggested in [64], and a learning rate and weight decay of 0.5 and 0.001.

Evaluation Metrics. Recent works [17,37] in DML have questioned standard evaluation metrics (*i.e.* Recall@1) and argue they are not fair. Thus, we report the R-Precision ($\mathcal{R}$P) and MAP@R metrics recently introduced in [37].

4.1   Multiple Retrieval Tasks (MRT) Benchmark

Data. The Multiple Retrieval Tasks (MRT) benchmark combines the 6 following fine-grained datasets under a retrieval setting: Aircraft [35], Cars [32], CUB [56],

Table 1. Statistics of the Multiple Retrieval Tasks (MRT) benchmark. It is composed of 6 datasets. Classes in train and test are *disjoint*. We provide the number of classes as a reference, but labels are *never* used during training.

Dataset	Training		Testing	
	# Classes	# Images	# Classes	# Images
Aircraft [35]	50	5,000	50	5,000
Cars [32]	98	8,054	98	8,131
CUB [56]	100	5,864	100	5,924
Flowers [38]	51	3,870	51	4,319
Food-101 [4]	51	51,000	50	50,000
Products [40]	11,318	59,551	11,316	60,502
MRT	11,668	133,339	11,665	133,876

Flowers [38], Food-101 [4], and Stanford online products (Products) [40]. We follow standard practice in the DML community and, for each dataset, assign the first half of the classes (ordered alphabetically) for training and the second half for testing. We then combine images from all the training splits into a single training set $\mathcal{D}$ of 133,339 images and discard their class and dataset labels. This is the training set we use to learn the pseudo-labels, as well as the adaptor and the fusion parameters. We show statistics for all datasets in Table 1.

Evaluation Protocol. Models are trained on the combined training set $\mathcal{D}$ without task nor class labels. Evaluation is performed on the test split of each task independently, following a leave-one-out protocol: each image is used as a query once to rank all the other images in the test set. For evaluation, relevance is defined according to class labels and we report mean average precision (MAP@R or mAP) and $\mathcal{R}$-Precision ($\mathcal{R}$P) over all queries.

4.2   Compared Methods

Baselines. First and foremost, we compare with the DINO pretrained visual transformer of [9], a self-supervised model trained on ImageNet1K. It obtains impressive zero-shot performance on retrieval and constitutes a very strong baseline.[2] We denote this baseline as **DINO** or simply $\mathcal{M}$ in Table 2 and Fig. 4.

The Grappa architecture adds an extra set of parameters in the form of adaptors and fusion layers. To verify that improvements do not simply come from these extra parameters, we report results with a second baseline that has the same number of parameters as the Grappa models, but uses no pseudo-granularity-based adaptors nor the proposed attention consistency loss. Instead of following Step 2, we randomly initialize adaptors and finetune them when learning the fusion. For the latter, we use the Barlow Twins loss from Eq.(4), with transformation consistency, and train it on the training set of MRT, similar to Grappa. We denote this baseline: $\mathcal{M}^*$ **(random)**.

---

[2] We chose DINO over the CLIP [45] model as the training set of CLIP is not public. Therefore the data in MRT might be part of its 400M image-text training pairs.

Table 2. Results on the Multiple Retrieval Task (MRT) benchmark. We report MAP@R (mAP) and $\mathcal{RP}$ on the six datasets of MRT, obtained by a single model from those listed in Sect. 4.2, all unsupervised. The oracle (in gray) is not comparable as it selects the set of adaptors that performs best for each task.

| | Aircraft | | Cars | | CUB | | Flowers | | Food-101 | | Products | |
	$\mathcal{RP}$	mAP	$\mathcal{RP}$	mAP	$\mathcal{RP}$	mAP	$\mathcal{RP}$	mAP	$\mathcal{RP}$	mAP	$\mathcal{RP}$	mAP
DINO ($\mathcal{M}$)	17.5	9.2	9.0	3.5	33.6	22.9	62.5	57.0	27.0	16.1	34.1	31.5
Results without pseudo-granularity adaptors												
$\mathcal{M} + \mathcal{A}_1$	15.4	7.8	7.9	2.7	16.2	7.8	62.8	56.9	22.1	11.6	32.0	29.5
$\mathcal{M}^*$ (random)	11.8	5.0	6.2	1.7	11.0	4.5	62.7	56.2	21.9	11.2	28.9	26.5
Results using only a single set of pseudo-granularity adaptors												
$\mathcal{M} + \mathcal{P}_1$	18.0	9.5	8.9	3.4	33.3	22.7	63.3	57.6	29.0	17.9	32.2	29.6
$\mathcal{M} + \mathcal{P}_2$	18.0	9.5	9.1	3.6	34.9	24.1	66.7	61.3	29.3	18.1	32.7	30.1
$\mathcal{M} + \mathcal{P}_3$	18.3	9.7	9.6	3.8	35.3	24.5	67.9	62.7	29.5	18.4	34.5	31.9
$\mathcal{M} + \mathcal{P}_4$	18.1	9.5	9.7	3.9	35.2	24.3	67.9	62.6	29.6	18.4	35.8	33.1
$\mathcal{M} + \mathcal{P}_5$	17.6	9.1	9.8	3.9	33.5	22.6	67.5	62.2	29.5	18.4	37.7	34.9
$\mathcal{M} + \mathcal{P}_6$	17.3	8.7	9.8	3.9	31.4	20.3	66.9	61.5	29.4	18.3	39.7	36.8
$\mathcal{M} + \mathcal{P}_7$	16.3	8.0	9.4	3.7	26.5	15.8	64.9	59.1	28.8	17.8	42.4	39.6
$\mathcal{M} + \mathcal{P}_8$	12.6	5.4	7.5	2.7	14.8	6.6	56.7	49.8	22.1	11.9	46.5	43.7
Oracle ($\mathcal{O}$)	18.3	9.7	9.8	3.9	35.3	24.5	67.9	62.7	29.6	18.4	46.5	43.7
i.e. only	$\mathcal{P}_3$		$\mathcal{P}_5$-$\mathcal{P}_6$		$\mathcal{P}_3$		$\mathcal{P}_3$-$\mathcal{P}_4$		$\mathcal{P}_5$		$\mathcal{P}_8$	
Results using 8 pseudo-granularity adaptors and fusion ($\mathcal{M}^*$)												
Grappa-avg	17.9	9.3	9.8	3.9	34.1	23.1	66.5	61.2	30.4	19.2	**37.1**	**34.4**
Grappa-T	**18.1**	**9.5**	9.8	3.9	34.4	23.5	66.5	61.2	30.4	19.2	35.9	33.2
Grappa-N	**18.1**	**9.5**	9.9	4.0	**35.1**	**24.1**	67.2	61.9	30.5	19.3	36.3	33.6
vs. DINO	(↑ 0.6)	(↑ 0.3)	(↑ 0.9)	(↑ 0.5)	(↑ 1.5)	(↑ 1.2)	(↑ 4.7)	(↑ 4.9)	(↑ 3.5)	(↑ 3.2)	(↑ 2.2)	(↑ 2.1)

Proposed. We report results for models with adaptor fusion described in Sect. 3 together with results for individual adaptors. More precisely, we build N pseudo-label sets on the training set of MRT and train N sets of adaptors on these pseudo-labels independently. We report their individual performance as $\mathcal{M} + \mathcal{P}_i$. As mentioned above, we use DINO as a backbone and keep it frozen.

Then, using MRT's training set again, we combine these N adaptors into a single model and train the fusion layer using i) a Barlow Twins loss on the final representation when creating pairs from two augmented views, reported as Grappa-T, and ii) our proposed Attention Consistency framework which relies on the local neighborhood, reported as Grappa-N. We also report results for the case where the fusion is an average pooling layer, denoted as Grappa-avg.

An Adaptor Selector Oracle. What if we could choose the best performing pseudo-granularity for each retrieval task? Obviously, this requires access to the test set labels. This also results in a different representation per task, which departs from the universal representation we seek to learn. For these reasons, we only consider this variant as an oracle, and its results should not be compared with others. We still provide it as a reference, showing how much could be achieved if we set the attention as a one-hot vector that only enables the best possible pseudo-granularity adaptor. We denote the oracle as $\mathcal{O}$.

4.3   Results

We present our results in Table 2 and Fig. 4. Again, note that, unlike the common DML experimental setting, we use *a unique model for all retrieval datasets* and *no class nor task labels during training*. We make the following observations.

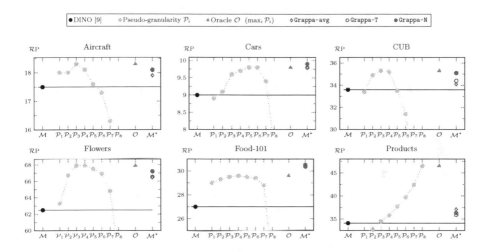

Fig. 4. Results per dataset in MRT. All 6 datasets use the same model.

Baselines. We confirm the initial observation [9] that, for common DML datasets, *DINO is a very strong baseline.* It achieves good performance even on the more challenging metrics $\mathcal{R}$P and MAP@R. Also, it turned out to be very challenging to improve over DINO by keeping its backbone frozen and embedding extra modules. We did our best to learn the additional modules from scratch, but were unsuccessful. Rows 2–3 of Table 2 report the best results after hyper-parameter tuning for a single set (no fusion) and for 8 sets of adaptors with fusion. Embedding randomly initialized modules typically deteriorates the performance. This makes separately trained pseudo-granularity adaptors all the more important.

Using a Single Set of Adaptors. In rows 4–12 of Table 2 we report results when only using adaptors from a single pseudo-granularity; these results are also visualized as lavender-colored points in Fig. 4. We observe that, for each dataset, there exists at least one pseudo-label set that improves over DINO, and that the best one (reported as Oracle $\mathcal{O}$) is different for different retrieval tasks. The oracle results use separate models, and selecting the best one for each task requires access to labels. Considering each set of adaptors as a separate model, some improve over DINO on several retrieval tasks, showing that even individual pseudo-granularities-specific sets of adaptors can be useful. Yet, as we will see next, higher gains can be achieved by fusing multiple adaptors into a single model.

Fusion and Attention Consistency. The bottom part of Table 2 shows the performance of the final model $\mathcal{M}^*$ that combines adaptors from all eight pseudo-granularities ($\mathcal{P}_1 \ldots \mathcal{P}_8$). We see that using the proposed attention consistency (AC) loss over neighborhood pairs results in an even stronger model that improves over DINO and over the Grappa-T variant that uses pairs from different augmentations of the same image, in all datasets. In fact, we see that in most cases (*i.e.* apart from Products), the proposed AC loss is able to give results *on-par with the oracle* $\mathcal{O}$ that would select the right pseudo-granularity for each dataset. Moreover, for Food-101 and Cars, Grappa outperforms the best adaptor for the dataset, showing that this oracle is not necessarily an upper bound: combining pseudo-granularities is beneficial. It is also worth noting that the simpler and parameter-free fusion of the Grappa-avg model is still improving over DINO in all cases, while it is also the best performing Grappa variant on the instance-level Products dataset.

Qualitative Results. Figure 5 presents qualitative results for two queries from the MRT benchmark. We present two cases where Grappa-N achieves significantly higher recall than DINO, *i.e.* cases where the local feature space is adapted in a way that images from the correct class are closer together.

Failure Cases. As we see from Table 2 and Fig. 4, our fusion mechanism is not able to learn that $\mathcal{P}_8$ is the best pseudo-granularity for Products. We attribute this to the different topology of that dataset and the fact that hyperparameters were chosen to optimize performance for the union of the six benchmarks.

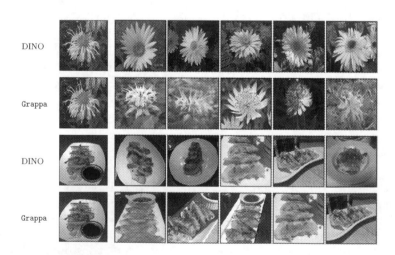

Fig. 5. Qualitative results. Queries (first column) from Flowers and Food-101 and their top 5 retrieved results (columns 2-6) by DINO [9] and Grappa-N.

5 Conclusions

We present Grappa, an unsupervised approach for adapting a large pretrained backbone to simultaneously tackle multiple retrieval tasks, given only an unlabeled set of training images associated to these retrieval tasks. We show that one can adapt a large pretrained visual transformer using a set of pseudo-granularity adaptors and simple fusion layers. Our Grappa models bring consistent gains over the strong DINO [9] baseline on all six retrieval tasks we adapt to. We envision this work as a first step towards models that dynamically adapt.

Acknowledgements. MIAI@Grenoble Alpes (ANR-19-P3IA-0003).

References

1. Arthur, D., Vassilvitskii, S.: k-means++: The advantages of careful seeding. Tech. rep, Stanford (2006)
2. Avrithis, Y., Kalantidis, Y.: Approximate gaussian mixtures for large scale vocabularies. In: Fitzgibbon, A., Lazebnik, S., Perona, P., Sato, Y., Schmid, C. (eds.) ECCV 2012. LNCS, vol. 7574, pp. 15–28. Springer, Heidelberg (2012). https://doi.org/10.1007/978-3-642-33712-3_2
3. Avrithis, Y., Kalantidis, Y., Anagnostopoulos, E., Emiris, I.Z.: Web-scale image clustering revisited. In: Proceedings of ICCV (2015)
4. Bossard, L., Guillaumin, M., Van Gool, L.: Food-101 – mining discriminative components with random forests. In: Fleet, D., Pajdla, T., Schiele, B., Tuytelaars, T. (eds.) ECCV 2014. LNCS, vol. 8694, pp. 446–461. Springer, Cham (2014). https://doi.org/10.1007/978-3-319-10599-4_29
5. Boudiaf, M., et al.: A unifying mutual information view of metric learning: crossentropy vs. pairwise losses. In: Vedaldi, A., Bischof, H., Brox, T., Frahm, J.-M. (eds.) ECCV 2020. LNCS, vol. 12351, pp. 548–564. Springer, Cham (2020). https://doi.org/10.1007/978-3-030-58539-6_33
6. Cao, B., Araujo, A., Sim, J.: Unifying deep local and global features for image search. In: Vedaldi, A., Bischof, H., Brox, T., Frahm, J.-M. (eds.) ECCV 2020. LNCS, vol. 12365, pp. 726–743. Springer, Cham (2020). https://doi.org/10.1007/978-3-030-58565-5_43
7. Caron, M., Bojanowski, P., Joulin, A., Douze, M.: Deep clustering for unsupervised learning of visual features. In: Ferrari, V., Hebert, M., Sminchisescu, C., Weiss, Y. (eds.) Computer Vision – ECCV 2018. LNCS, vol. 11218, pp. 139–156. Springer, Cham (2018). https://doi.org/10.1007/978-3-030-01264-9_9
8. Caron, M., Misra, I., Mairal, J., Goyal, P., Bojanowski, P., Joulin, A.: Unsupervised learning of visual features by contrasting cluster assignments. In: Proceedings of NeurIPS (2020)
9. Caron, M., et al.: Emerging properties in self-supervised vision transformers. In: Proceedings of ICCV (2021)
10. Chen, T., Kornblith, S., Norouzi, M., Hinton, G.: A simple framework for contrastive learning of visual representations. In: Proceedings of ICML (2020)
11. Chopra, S., Hadsell, R., LeCun, Y.: Learning a similarity metric discriminatively, with application to face verification. In: Proceedings of CVPR (2005)
12. Csurka, G. (ed.): Domain Adaptation in Computer Vision Applications. ACVPR, Springer, Cham (2017). https://doi.org/10.1007/978-3-319-58347-1

13. Deng, J., Dong, W., Socher, R., Li, L.J., Li, K., Fei-Fei, L.: Imagenet: A large-scale hierarchical image database. In: Proceedings of CVPR (2009)
14. Deng, J., Guo, J., Xue, N., Zafeiriou, S.: Arcface: Additive angular margin loss for deep face recognition. In: Proceedings of CVPR (2019)
15. Dosovitskiy, A., et al.: An image is worth 16×16 words: Transformers for image recognition at scale. In: Proceedings of ICLR (2021)
16. Elezi, I., Vascon, S., Torcinovich, A., Pelillo, M., Leal-Taixé, L.: The group loss for deep metric learning. In: Vedaldi, A., Bischof, H., Brox, T., Frahm, J.-M. (eds.) ECCV 2020. LNCS, vol. 12352, pp. 277–294. Springer, Cham (2020). https://doi.org/10.1007/978-3-030-58571-6_17
17. Fehervari, I., Ravichandran, A., Appalaraju, S.: Unbiased evaluation of deep metric learning algorithms. arXiv preprint arXiv:1911.12528 (2019)
18. Gidaris, S., Singh, P., Komodakis, N.: Unsupervised representation learning by predicting image rotations. In: Proceedings of ICLR (2018)
19. Gordo, A., Almazán, J., Revaud, J., Larlus, D.: Deep image retrieval: Learning global representations for image search. In: Leibe, B., Matas, J., Sebe, N., Welling, M. (eds.) ECCV 2016. LNCS, vol. 9910, pp. 241–257. Springer, Cham (2016). https://doi.org/10.1007/978-3-319-46466-4_15
20. Goyal, P., et al.: Self-supervised pretraining of visual features in the wild. arXiv preprint arXiv:2103.01988 (2021)
21. Grill, J.B., et al.: Bootstrap your own latent: A new approach to self-supervised learning. In: Proceedings of NeurIPS (2020)
22. Gu, G., Ko, B.: Symmetrical synthesis for deep metric learning. In: Proceedings of AAAI (2020)
23. Gu, G., Ko, B., Kim, H.G.: Proxy synthesis: Learning with synthetic classes for deep metric learning. In: Proceedings of AAAI (2021)
24. He, K., Fan, H., Wu, Y., Xie, S., Girshick, R.: Momentum contrast for unsupervised visual representation learning. In: Proceedings of CVPR (2020)
25. Hendrycks, D., Gimpel, K.: Gaussian error linear units (gelus). arXiv:1606.08415 (2016)
26. Hershey, J.R., Chen, Z., Le Roux, J., Watanabe, S.: Deep clustering: Discriminative embeddings for segmentation and separation. In: Proceedings of ICASSP (2016)
27. Houlsby, N., et al.: Parameter-efficient transfer learning for NLP. In: Proceedings of ICML (2019)
28. Jia, C., et al.: Scaling up visual and vision-language representation learning with noisy text supervision. In: Proceedings of ICML (2021)
29. Kalantidis, Y., Lassance, C., Almazán, J., Larlus, D.: TLDR: Twin learning for dimensionality reduction. In: TMLR (2022)
30. Ko, B., Gu, G.: Embedding expansion: Augmentation in embedding space for deep metric learning. In: Proceedings of CVPR (2020)
31. Ko, B., Gu, G., Kim, H.G.: Learning with memory-based virtual classes for deep metric learning. In: Proceedings of ICCV (2021)
32. Krause, J., Deng, J., Stark, M., Li, F.F.: Collecting a large-scale dataset of fine-grained cars. In: Proceedings of ICCV-W (2013)
33. Liu, W., Wen, Y., Yu, Z., Li, M., Raj, B., Song, L.: Sphereface: Deep hypersphere embedding for face recognition. In: Proceedings of CVPR (2017)
34. Lloyd, S.: Least squares quantization in pcm. TIT **28**(2), 129–137 (1982)
35. Maji, S., Rahtu, E., Kannala, J., Blaschko, M., Vedaldi, A.: Fine-grained visual classification of aircraft. arXiv:1306.5151 (2013)
36. Mccloskey, M., Cohen, N.J.: Catastrophic interference in connectionist networks: the sequential learning problem. Psychol. Learn. Motiv. **24**, 104–169 (1989)

37. Musgrave, K., Belongie, S., Lim, S.-N.: A metric learning reality check. In: Vedaldi, A., Bischof, H., Brox, T., Frahm, J.-M. (eds.) ECCV 2020. LNCS, vol. 12370, pp. 681–699. Springer, Cham (2020). https://doi.org/10.1007/978-3-030-58595-2_41
38. Nilsback, M.E., Zisserman, A.: Automated flower classification over a large number of classes. In: Proceedings of ICCVGIP (2008)
39. Noroozi, M., Favaro, P.: Unsupervised Learning of Visual Representations by Solving Jigsaw Puzzles. In: Leibe, B., Matas, J., Sebe, N., Welling, M. (eds.) ECCV 2016. LNCS, vol. 9910, pp. 69–84. Springer, Cham (2016). https://doi.org/10.1007/978-3-319-46466-4_5
40. Oh Song, H., Xiang, Y., Jegelka, S., Savarese, S.: Deep metric learning via lifted structured feature embedding. In: Proceedings of CVPR (2016)
41. Pfeiffer, J., Kamath, A., Rücklé, A., Cho, K., Gurevych, I.: AdapterFusion: Non-destructive task composition for transfer learning. In: Proceedings of EACL (2021)
42. Philbin, J., Chum, O., Isard, M., Sivic, J., Zisserman, A.: Object retrieval with large vocabularies and fast spatial matching. In: Proceedings of CVPR (2007)
43. Philip, J., Berard, A., Gallé, M., Besacier, L.: Monolingual adapters for zero-shot neural machine translation. In: Proceedings of EMNLP (2020)
44. Puigcerver, J., et al.: Scalable transfer learning with expert models. In: Proceedings of ICLR (2021)
45. Radford, A., et al.: Learning transferable visual models from natural language supervision. In: Proceedings of ICML (2021)
46. Rebuffi, S.A., Bilen, H., Vedaldi, A.: Learning multiple visual domains with residual adapters. In: Proceedings of NeurIPS (2017)
47. Revaud, J., Almazán, J., Rezende, R., de Souza, C.: Learning with average precision: Training image retrieval with a listwise loss. In: Proceedings of ICCV (2019)
48. Riquelme, C., et al.: Scaling vision with sparse mixture of experts. In: Proceedings of NeurIPS (2021)
49. Sariyildiz, M.B., Kalantidis, Y., Larlus, D., Alahari, K.: Concept generalization in visual representation learning. In: Proceedings of ICCV (2021)
50. Seidenschwarz, J., Elezi, I., Leal-Taixé, L.: Learning intra-batch connections for deep metric learning. In: Proceedings of ICML (2021)
51. Sharif Razavian, A., Azizpour, H., Sullivan, J., Carlsson, S.: CNN features off-the-shelf: An astounding baseline for recognition. In: Proceedings of CVPR-W (2014)
52. Shazeer, N., et al.: Outrageously large neural networks: The sparsely-gated mixture-of-experts layer. In: Proceedings of ICLR (2017)
53. Sohn, K.: Improved deep metric learning with multi-class n-pair loss objective. In: Proceedings of NeurIPS (2016)
54. Tian, Y., Henaff, O.J., van den Oord, A.: Divide and contrast: Self-supervised learning from uncurated data. In: Proceedings of ICCV (2021)
55. Vaswani, A., et al.: Attention is all you need. In: Proceedings of NeurIPS (2017)
56. Wah, C., Branson, S., Welinder, P., Perona, P., Belongie, S.: The Caltech-UCSD Birds-200-2011 Dataset. Tech. rep, California Institute of Technology (2011)
57. Wang, F., Cheng, J., Liu, W., Liu, H.: Additive margin softmax for face verification. IEEE Signal Process. Lett. 25, 926–930 (2018)
58. Wang, F., Xiang, X., Cheng, J., Yuille, A.L.: Normface: L2 hypersphere embedding for face verification. In: Proceedings of ACM Multimedia (2017)
59. Wang, R., et al.: K-adapter: Infusing knowledge into pre-trained models with adapters. In: ACL/IJCNLP (Findings) (2021)
60. Yan, X., Misra, I., Gupta, A., Ghadiyaram, D., Mahajan, D.: Clusterfit: Improving generalization of visual representations. In: Proceedings of CVPR (2020)

61. You, Y., Gitman, I., Ginsburg, B.: Large batch training of convolutional networks. In: Proceedings of ICLR (2018)
62. Yuksel, S.E., Wilson, J.N., Gader, D.P.: Twenty years of mixture of experts. Trans. Neural Netw. Learn. Syst. **23**, 177–1193 (2012)
63. Zamir, A., Sax, A., Shen, W., Guibas, L., Malik, J., Savarese, S.: Taskonomy: Disentangling task transfer learning. In: Proceedings of CVPR (2018)
64. Zbontar, J., Jing, L., Misra, I., LeCun, Y., Deny, S.: Barlow twins: Self-supervised learning via redundancy reduction. In: Proceedings of ICML (2021)
65. Zhai, A., Wu, H.Y.: Classification is a strong baseline for deep metric learning. In: Proceedings of BMVC (2019)
66. Zhai, X., et al.: LiT: Zero-shot transfer with locked-image text tuning. In: Proceedings of CVPR (2022)

Learning Audio-Video Modalities
from Image Captions

Arsha Nagrani[(✉)] [iD], Paul Hongsuck Seo[iD], Bryan Seybold[iD], Anja Hauth,
Santiago Manen, Chen Sun, and Cordelia Schmid

Google Research, Mountain View, USA
arsha.nagarani@gmail.com
https://a-nagrani.github.io/videocc.html

Abstract. There has been a recent explosion of large-scale image-text
datasets, as images with alt-text captions can be easily obtained online.
Obtaining large-scale, high quality data for video in the form of text-
video and text-audio pairs however, is more challenging. To close this
gap we propose a new video mining pipeline which involves transferring
captions from image captioning datasets to video clips with no additional manual effort. Using this pipeline, we create a new large-scale,
weakly labelled audio-video captioning dataset consisting of millions of
paired clips and captions. We show that training a multimodal transformer based model on this data achieves competitive performance on
video retrieval and video captioning, matching or even outperforming
HowTo100M pretraining with 20x fewer clips. We also show that our
mined clips are suitable for text-audio pretraining, and achieve state of
the art results for the task of audio retrieval.

Keywords: Data mining · Video retrieval · Captioning

1 Introduction

A key facet of human intelligence is the ability to effortlessly connect the visual and
auditory world to natural language concepts. Bridging the gap between human
perception (visual, auditory and tactile) and communication (via language) is
hence becoming an increasingly important goal for artificial agents, enabling tasks
such as text-to-visual retrieval [9,62,79], image and video captioning [44,77,84],
and visual question answering [7,47]. In the image domain in particular, this has
lead to an explosion of large scale image datasets with natural language descriptions, often by crawling alt text online [12,45,50,64,69]. In the video and audio
domains, however, obtaining natural language descriptions is more challenging.
Recent research has been either directed at modelling, for example in developing
new architectures (e.g. multimodal transformers [9,29,68]), or new training objectives (e.g. those that can deal with misaligned [55] or overly specialised [63] inputs).

Supplementary Information The online version contains supplementary material
available at https://doi.org/10.1007/978-3-031-19781-9_24.

S. Avidan et al. (Eds.): ECCV 2022, LNCS 13674, pp. 407–426, 2022.
https://doi.org/10.1007/978-3-031-19781-9_24

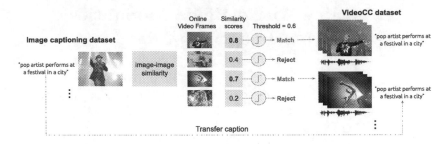

Fig. 1. Mining audio-video clips automatically. We use the images in image captioning datasets as 'seed' frames to mine related audio-visual clips. For each seed image-caption pair in a dataset, we find frames in videos with high similarity scores to the seed image. We then extract short video clips around the matching frames and transfer the caption to those clips. This gives us free captioning supervision for video and audio clips.

Annotating videos manually with clean and diverse captions is often subjective, painstaking and expensive. This means that most video-captioning datasets (e.g. MSR-VTT [82], LSMDC [66], CMD [8], ActivityNet [44] etc.) are small in size (order of magnitude 100K). Audio captioning datasets such as AudioCaps [42] and Clotho [24], are even smaller. Given the well-known benefits of pretraining, many works have proposed creative but weak forms of supervision, such as hashtags [33], titles and descriptions [72], or Automatic Speech Recognition (ASR) in instructional videos [56]. The de facto standard for video-language pretraining [5,29,48,52,62,67] has become the large HowTo100M [56] dataset, pretraining on which gives a significant boost over training from scratch. The pitfalls of using ASR however are well known; (i) there is noise in imperfect ASR transcription, (ii) continuous narration may consist of incomplete or grammatically incorrect sentences, (iii) the domain is often limited to instructional videos to increase relevance between speech and video content and finally, and (iv) ASR may not be temporally aligned with the video, or indeed may not refer to the video at all [56]. Combined, this necessitates a huge amount of training data for good performance (100s of millions of samples), and consequently, a lot of compute.

Image annotation, on the other hand, is cheaper than video and easier to obtain from web pages [12,69], and large-scale image-text pretrained models such as CLIP [64] are available online. This has led to concurrent works [10,26,54] using image-text models for video-text tasks. While this is a valuable idea, using such models beyond weight initialization requires some additional complexity. If we treat videos as a bag of sparse frames [46], we lose all the benefits of video (modalities like audio and the chance to model low-level temporal information directly from the frames) or require complicated distillation procedures from image to video models [34]. Hence we believe there is still a necessity for large-scale *video*-text datasets.

Is there another way to leverage all the existing effort that has gone into image-captioning datasets? We propose a solution in the form of a new video mining method based on *cross-modal transfer*, where we use images from image captioning datasets as seeds to find similar clips in videos online (Fig. 1). We then transfer the image captions directly to these clips, obtaining weak, albeit free video and audio captioning supervision in the process. This can also provide us with motion and audio supervision – for example, sometimes human-generated captions for images infer other modalities, e.g. the caption 'Person throws a pitch during a game against university' from the CC3M dataset [69] was written for a single, still image, but is actually describing motion that would occur in a video. Similarly, the caption 'A person singing a song', is also inferring a potential audio track. We note that like HowTo100M, our dataset curation is entirely automatic, and requires no manual input at all. However, as we show in Sect. 3, our mined data samples are more diverse than HowTo100M, are matched to better-formed captions compared to ASR, and are likely to contain at least one frame that is aligned with the text caption.

In doing so we make the following contributions: (i) We propose a new, scalable video-mining pipeline which transfers captioning supervision from image datasets to video and audio. (ii) We use this pipeline to mine paired video and captions, using the Conceptual Captions3M [69] image dataset as a seed dataset. Our resulting dataset VideoCC3M consists of millions of weakly paired clips with text captions and will be released publicly. (iii) We propose a new audio-visual transformer model for the task of video retrieval, which when trained on this weakly paired data performs on par with or better than models pre-trained on HowTo100M for video retrieval and captioning, with 20x fewer clips and 100x fewer text sentences. In particular, we show a large performance boost in the zero-shot setting. (iv) Finally, we also show that our audio-visual transformer model seamlessly transfers to *text-audio* retrieval [60] benchmarks as well, achieving state of the art results on the AudioCaps [42] and Clotho [24] datasets.

2   Related Work

Cross-Modal Supervision: Our key idea is to use labelled data in one modality (images) to aid learning in another modality (videos). A popular method for cross-modal transfer is knowledge *distillation* [37], which has shown great success for transferring supervision from RGB to depth [36], or faces to speech [4]. Another line of work enhances unimodal models via multimodal regularisations [2,3]. Ours is a related but tangential idea which involves mining new data and assigning labels to it (similar to video clips mined for action recognition using speech by [30,58]). This is particularly useful when there are large labelled datasets in one modality (here text-image retrieval [45,50,69]), but it is more challenging to obtain for a similar task in another modality (text-audio [60] or text-video [6,8,28,44,66,82,87] retrieval).

Text Supervision for Video: Existing manually annotated video captioning datasets [39,82,87] are orders of magnitude smaller than classification datasets [41]. This has led to a number of creative ideas for sourcing weakly paired text and video data. [74] use web images queried with sports activities to create temporal annotations for videos. WVT [72] mines videos from YouTube and their titles for action recognition starting from the Kinetics labels. Similarly [73] uses video level labels for the same task. Unlike these works where the labels are at a video level, our captions are localised to video clips, and are not limited to the domain of action recognition only. [33] and [49] use hashtags and titles for supervision respectively, but only to learn a better video encoder. In the movie domain, [8] uses YouTube descriptions for movie clips while [66] uses audio description (AD) from movies. The recently released WebVid2M dataset [9] comprises manually annotated captions, but given the monetary incentive on stock sites, they often contain added metatags appended, and most lack audio. Another valuable recent dataset is Spoken Moments in Time [57], however this was created with significant manual effort. The largest video-text dataset by far is HowTo100M [56] generated from ASR in instructional videos; however, this data is particularly noisy, as discussed in the introduction.

Text Supervision for Audio: Textual supervision for audio is even scarcer than it is for video. Early works perform text-audio retrieval using single word audio tags as queries [13], or class labels as text labels [25]. Even earlier, [71] linked text to audio but only using 215 animal sounds from the BBC Sound Effects Library. Unlike these works, we study unconstrained caption-like descriptions as queries. While small, manually annotated datasets such as Audio-Caps [42] and Clotho [24] do exist (and have been repurposed by [43,60] for audio-text retrieval), large-scale pretraining data for text-audio tasks is not available. Note that extracting audio from existing video-text datasets is difficult: WebVid2M [9] videos largely do not have audio, and HowTo100M captions are derived from the audio (training a model to predict HowTo100M captions from the audio might simply be learning how to do ASR). Hence we explore the link between audio and text transferred via image similarity to videos that all have audio, and show this improves text-audio retrieval. As far as we are aware, we are the first work to pre-train the same model for both *visual-focused* datasets such as MSR-VTT and *audio-focused* datasets such as AudioCaps and Clotho.

3   Text-Video Data

In this section we describe our automatic mining pipeline for obtaining video clips paired with captions. We then train text-video and text-audio models (described in Sect. 4) on this weakly paired data for 2 tasks, audiovideo retrieval and video captioning.

3.1   Mining Pipeline

The core idea of our mining pipeline is to start with an image captioning dataset, and for each image-caption pair in a dataset, find frames in videos similar to the image. We then extract short video clips around the matching frames and transfer the caption to those clips. In detail, the steps are as follows:

1. Identify Seed Images: We begin by selecting an image-captioning dataset. The images in this dataset are henceforth referred to as 'seed' images (x_{seed}).

Fig. 2. Examples of clips with captions that are mined automatically. For each seed image, we show 3 'matched' clips obtained using our automatic video mining method. For the first 2 clips, we show only a single frame, but for the third clip we present 2 frames to show motion, either of the subjects in the video (first 3 rows) or small camera motion (last 2 rows). Note the diversity in the mined clips, for example the different pitching poses and angles (first row) and the different types of statues (fourth row). Clips in the second row also contain audio relevant to the caption. Note frames may have been cropped and resized for ease of visualisation. More qualitative results are provided in the supplementary.

2. Feature Extraction: We then calculate a visual feature vector $f(x_{\text{seed}})$ for each seed image. Given our primary goal is to mine semantically similar images, we extract features using a deep model trained for image retrieval, the Graph-Regularized Image Semantic Embedding (Graph-RISE) model [40]. We then extract the same visual features $f(x_v)$ for the frames x_v of a large corpus of videos. Because of frame redundancy, we can extract features at a reduced rate (1fps) relative to the original video frame rate.

3. Identify Matches: Next, we calculate the dot product similarity between the feature vectors for each seed image in the caption data set and those for each video frame obtained from the video corpus. Pairs with a similarity above a threshold τ are deemed 'matches'. For each seed image, we keep the top 10 matches. For these top 10, we transfer the caption from the image to a short video clip extracted at a temporal span t around the matched image frame, and add it to our dataset. In Sect. 3.3, we provide brief ablations on the values of t and the threshold τ.

3.2   Video-Conceptual-Captions (VideoCC)

We ran our mining pipeline with the image captioning dataset - Conceptual Captions 3 M [69] (CC3M). We only use images which are still publicly available online, which gives us 1.25 image-caption pairs. We apply our pipeline to videos online. We filter videos for viewcount >1000, length <20 min, uploaded within the last 10 years, but at least 90 d ago, and filter using content-appropriateness signals to get 150 M videos. This gives us 10.3 M clip-text pairs with 6.3 M video clips (total 17.5 K hours of video) and 970 K unique captions. We call the resulting dataset VideoCC3M. We also run our pipeline on a more recently released seed dataset extension, called Conceptual Captions 12 M [12] (CC12M). Note that while CC3M consists of higher quality captions [69], CC12M was created by relaxing the data collection pipeline used in CC3M, and hence the captions are far noisier. Results on this dataset are provided in the supplementary material. Some examples of the matched video frames to captions for VideoCC3M are provided in Fig. 2. A preliminary fairness analysis on the data is provided in the supplementary material. The mined video clips have the following properties:

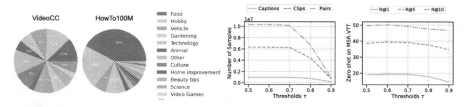

Fig. 3. Domains in VideoCC3M vs HowTo100M (left), effect of match threshold τ on mining statistics (middle) and zero-shot performance on MSR-VTT (right). VideoCC3M has a more diverse and balanced range of domains, 'Other' here includes a variety of content such as music videos, sports, politics, vlogs and so on. Note how almost half of HowTo100M videos are food-related (cooking videos). More details are provided in suppl. Effect of match threshold τ: Increasing the threshold τ beyond 0.6 decreases the size of the dataset, which leads to a corresponding performance drop on zero-shot retrieval. We use an optimal match threshold of 0.6

(i) Diversity: We compare the domains in our dataset to HowTo100M in Fig. 3 (left). Note that because VideoCC3M is mined from a general corpus of videos online (unlike HowTo100M, which is restricted to instructional videos), our dataset is more balanced. A more comprehensive bar chart is provided in the supplementary. Some of the 'Other' categories are technology, team sports, family, medicine, beauty, history, religion, gardening, music, politics–while HowTo100M videos are largely dominated by the 'Food' and 'Hobby' domains (almost half are 'cooking videos'). This is unsurprising given that HowTo100M is limited to instructional videos.

(ii) Alignment: We mine frames that have high visual similarity to the seed image. If this seed has a relevant caption (largely the case for the high quality CC3M dataset), it is likely that at least one frame in the mined clip is aligned with the caption. A manual check of a small subset of clips found this to be the case in 91% (see suppl). This is a stricter constraint than ASR based datasets, which have occasional misalignment between speech and frames. As an additional quantitative metric, we also run a commercial image classification system on the frames in both the VideoCC3M and the HowTo100M datasets. We then compute the proportion of captions for which a word in the caption exactly matches a label from the image classification system. We find the proportion to be 69.6% for VideoCC3M, whereas HowTo100M only has 19.7%.

(iii) Caption Style: The quality of the captions is transferred directly from the seed dataset. Most of the captions in CC3M are fully formed, grammatically correct sentences, unlike the distribution of sentences obtained from ASR. Each caption is matched to a mean of 10.6 clips, with some captions matched to more than 10 clips. This is possible because, while we limit the clip mining to 10 clips per seed image, the original CC3M dataset has multiple seed images with the same caption, e.g. 'an image of digital art', leading to more than 10 mined clips for these captions.[1] Having multiple pairs from the same set of captions and video clips also helps ensure that learnt video and text representations are not overly-specialised to individual samples (which can be a problem for existing datasets, as noted by [63]).

Cross-Modal Transfer from the Image Domain. Interestingly, this mining method provides us with *captioning* supervision for modalities such as video and audio that are difficult to annotate. Note that we use two existing sources of image supervision, the first is the seed image captioning dataset, and the second is the image similarity model $f(\cdot)$ which we use to mine related frames. This is not the same as simply applying a text-image model (even though that is a complementary idea) to different frames in a video for text-video retrieval. For example, our method provides some valuable supervision for new clips with motion (see the last column of retrieved clips in Fig. 2, first two rows). Many image captions in CC3M describe actions/motion, e.g. *human-human interactions* ('baby smiling down at dad while being thrown in the air'), *interactions with objects/body parts* ('person shaves hair on neck', 'rugby player fields a punt'), *movement in an environment* ('elderly couple walking on a deserted beach').[2] Our mining method, since it retrieves videos, can actually find examples of these described motions. We also obtain some free supervision for the audio stream (Fig. 2, second row and Fig. 4, right). These weakly labelled audio samples can be used for pretraining text-audio models, as we show in the results.

---

[1] Full distribution of clips per caption in VideoCC3M is provided in suppl. material.
[2] We find that interestingly, 83% of the 7.9K verbs (extracted using spacy package) in MSR-VTT (video annotated dataset), are present in CC3M.

3.3   Data Mining Ablations

In this section, we ablate the time span t and threshold τ, using zero-shot performance on the MSR-VTT test set (protocol described in Sect. 5.3).

Time Span. t: We try extracting different length clip segments t between 5 and 30 s, and found that performance increases up until 10 s, but decreases after that (results and discussion in the suppl. material). Hence we extract 10 second clips for our dataset.

Match Threshold. τ: We experiment with different match thresholds τ for the similarity in the range $\{0.5, 0.6, 0.7, 0.8.0.9\}$ and present the effect of this on mining statistics in Fig. 3 (middle) and zero-shot performance (right). The higher the match threshold, the stricter the similarity requirement on the matched frames to the caption. We note that upto a match threshold of 0.6, performance increases slightly, and there is no steep reduction in dataset size. After 0.7 however, the number of matches falls steeply as the match threshold is increased, leading to fewer videos and clips in the dataset, and a corresponding drop in downstream performance. We hence use a match threshold of 0.6 to mine clips.

4   Method

We focus on two different tasks in this paper that rely on video and text annotation-video retrieval and video captioning. We implement state of the art multimodal transformer models for each–architectures and training objectives are defined in the next two sections.

4.1   Audiovisual Video Retrieval (AVR)

For retrieval, we use a dual-stream model (one stream being an audio-video encoder and one stream being a text encoder for the caption), which when trained with a contrastive loss allows for efficient text-video retrieval. Note that the efficient dual stream approach has also been used by MIL-NCE [55] and FIT [9], but unlike these works, our video encoder is multimodal (Fig. 4, left), and utilises the audio as well. Our model is flexible, and can be used for audio-only, video-only and audio-visual retrieval.

Multimodal Video Encoder: Our encoder is inspired by the recently proposed MBT [59], which operates on RGB frames extracted at a fixed sampling rate from each video, and log-mel spectrograms used to represent audio. We first extract N non-overlapping patches from the RGB image (or the audio spectrogram), similar to the way done by ViT [23] and AST [35] respectively. The model consists of a number of transformer layers for each modality, with separate weights for each modality and fusion done via bottleneck tokens. Unlike

MBT, we use frames extracted at a larger stride (an ablation is provided in the experiments), to cover the longer videos in retrieval datasets. We implement both RGB-only, audio-only and RGB-audio fusion models.

Text Encoder: The text encoder architecture is the BERT model [21]. For the final text encoding, we use the [CLS] token output of the final layer.

Joint Embedding: For the final video encoding, we average the [CLS] tokens from both audio and RGB modalities. Both text and video encodings are then projected to a common dimension $D = 256$ via a single linear layer each. We then compute the dot product similarity between the two projected embeddings after normalisation.

Loss: We use the NCE loss [85] to learn a video and text embedding space, where matching text-video pairs in the batch are treated as positives, and all other pairwise combinations in the batch are treated as negatives. We minimise the sum of two losses, video-to-text and text-to-video [9]. At test time, inspired by FILIP [83], we sample K clips equally spaced from the video, compare each one to the text embedding, and average the similarity scores.

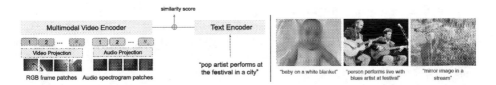

Fig. 4. (Left) Our audiovisual dual stream retrieval model (AVR), which works for both image and audio focused retrieval datasets. **(Right) Examples from VideoCC3M of automatically mined clips with relevant audio to the caption.** We show a single relevant frame from each clip as a proxy for visualising the audio. The accompanying audio contains (left to right) the sounds of a baby gurgling, music and water flowing sounds (left image intentionally blurred).

4.2 Video Captioning

For video captioning, we use an encoder-decoder style generative model. Our video encoder is the same as the one used above for retrieval.

Decoder: To generate a text caption, we adapt the autoregressive GPT-2 (117M) decoder [65], however we condition each predicted text token on video features from the video encoder as well as previously generated text tokens. More formally, given video features C as context, to generate the next token y_i in our caption Y, we first encode the previous generated tokens $Y_i = \{y_0, \ldots, y_{i-1}\}$ with a look-up table and a positional embedding to produce $H_i = \{h_0, \ldots, h_{i-1}\}$. We then encode the context C and the previous embedded tokens H_i using a single transformer. The outputs of this transformer are $\overset{C}{} \cup \overset{H}{i}$, where $\overset{H}{i} = \{\overset{h}{0}, \ldots, \overset{h}{i-1}\}$. We then predict the next token y_i from $\overset{h}{i-1}$ using a linear projection with a softmax: $y_i = \text{argmax}(\text{softmax}(\Phi \overset{h}{i-1}))$ where $\Phi \in \mathbb{R}^{\nu \times d}$ is the linear projection matrix and ν is the vocabulary size. As is standard, the first word h_0 is set using a special BOS (beginning of sentence) token, and tokens are generated until a special EOS (end of sentence) token is generated.

Loss: We minimise the negative log-likelihood of generating the ground-truth caption [17].

5   Experiments

We evaluate our text-video models on the following two tasks - (i) text-video retrieval (Sect. 5.3), which includes video retrieval on primarily *visual focused* datasets, as well as text-audio retrieval, where captions are primarily focused on *audio sounds*; and (ii) video captioning (Sect. 5.4). We use the common protocol of pretraining our models on a large dataset first, either VideoCC3M or HowTo100M, and then finetune on the target downstream dataset. Note that unlike other works, we apply the same pretrained models for both visual-focused datasets such as MSR-VTT and audio-focused datasets such as AudioCaps and Clotho. We also investigate zero-shot performance, where we apply pretrained models directly to the target task, without any finetuning at all. In this case, no supervised video-text data is used at all. We first describe datasets and metrics, then the implementation details, before finally discussing the results for each task.

5.1   Datasets and Metrics

VideoCC3M: We use the VideoCC3M dataset created using our automatic mining method described in Sect. 3.

HowTo100M. [56]: consists of 1.2 M instructional videos. Weak captions are in the form of transcribed speech, which we obtain using the YouTube ASR API [1].

MSR-VTT. [82] contains 10 K videos with 200 K descriptions. For retrieval, we follow other works [51], and train on 9 K train+val videos, reporting results on the 1K-A test set. For captioning, we use the standard splits proposed in [82].

AudioCaps. [42] contains video clips from the AudioSet dataset [32] with captions for the task of audio captioning. This dataset was then repurposed by [60] for text-audio retrieval, by taking a subset that does not overlap with the VGGSound [16] dataset. After filtering out the videos no longer available on the web, we have 47,107 training, 403 val and 778 test samples.

Clotho. [24] is an audio-only dataset of described sounds from Freesound [27]. During labelling, annotators only had access to audio (no meta tags or visual information). The data consists of a dev set and eval set of 2893 and 1045 audio samples respectively. Every audio sample is accompanied by 5 captions. We follow [60] and treat each of the 5 captions per test audio as a separate query.

Metrics. As is standard for retrieval, we report recall@K, $K \in \{1, 5, 10\}$. For captioning, we use the established metrics Bleu-4 (B-4) [61], CIDEr (C) [76], and Meteor (M) [11].

5.2 Implementation Details

In this section we describe implementation details for our models as well as certain design choices for sampling and initalisation. More details are provided in the supplementary material.

Audio-Visual Encoder: We use the ViT-Base (ViT-B, $L = 12$, $N_H = 12$, $d = 3072$), as a backbone with $B = 4$ fusion tokens and fusion layer $l_f = 8$. We sample 32 RGB frames for MSR-VTT, and 8 RGB frames for AudioCaps. For audio we extract spectrograms of size 800×128 spanning 24 seconds.

Text Encoder: We use the BERT-Base architecture ($L = 12$, $N_H = 12$, $d = 768$) with uncased wordpiece tokenization [22]. We use a total number of 32 tokens per caption during training – cropping and padding for sentences longer and shorter respectively. No text augmentation is applied.

Clip Coverage: A single segment per clip is randomly sampled at training time. We experiment with the length of this segment, controlled by the stride of the frames (32 frames at a stride of 2 frames at 25fps indicates an effective segment length of 2.5 s). We experiment with stride = 2, 6, 10, 14, 18, and find optimal performance with stride = 14 frames (effective coverage of 18s). At test time, we sample $K = 4$ clips equally spaced from the video, compare them to the text embedding, and average the similarity scores. More details are provided in the supplementary material.

Video Encoder Initialisation: Unless otherwise specified, we use Kinetics-400 [41] initialisation for both video retrieval and captioning. For audio retrieval we initialise the model with VGGSound [16] (see supplementary).

Training for Retrieval: The temperature hyperparameter σ for the NCE loss is set to 0.05, and the dimension of the common text-video projection space is set to 256. All models are trained with batch size 256, synchronous SGD with momentum 0.9, and a cosine learning rate schedule with warmup of 1.5 epochs on TPU accelerators. We pretrain for 4 epochs, and finetune for 5 epochs.

Table 1. Ablations with different initializations of the video encoder and the modalities (left) and effect of pretraining data (right) for text-video retrieval on the MSR-VTT dataset. Init. Initialisation of *video encoder only*. Modalities are **V:** RGB, **A:** Audio spectrograms. **# Caps:** Number of unique captions. (left) No VideoCC data is used in the left and we do not show audio-only results as some videos in the MSR-VTT dataset are missing audio. (right) Training on VideoCC3M provides much better performance than Howto100M, with a fraction of the dataset size (VideoCC3M has only 970K captions and 6.3M clips compared to the 130M clips in HowTo100M). The performance boost is particularly large for the zero-shot setting

Init.	Modality	R@1	R@5	R@10	PT Data	Modality	# Caps	R@1	R@5	R@10	R@1	R@5	R@10
Scratch	V	9.4	22.5	31.7			*Finetuned*				*Zero-shot*		
ImNet21K [20]	V	30.2	59.7	71.3	-	V	-	30.2	60.7	71.1	-	-	-
K400 [41]	V	30.2	60.7	71.1	HowTo100M [56]	V	130M	33.1	62.3	72.3	8.6	16.9	25.8
ImNet21k [20]	V+A	32.2	62.7	74.4	VideoCC3M	V	970K	35.0	63.1	75.1	18.9	37.5	47.1
K400 [41]	V+A	**32.3**	**64.1**	**74.6**	VideoCC3M	A+V	**970K**	**35.8**	**65.1**	**76.9**	**20.4**	**39.5**	**50.3**

Training for Captioning: We use the Adam optimizer with initial learning rate $1E-4$ and weight decay 0.01. For all models, we pretrain for 120K iterations with a batch size of 512. For finetuning, we train for 1K iterations.

5.3   Text-Audiovisual Retrieval

Video Encoder Initialisation: We first experiment with initalising the video encoder *only* (Table 1, left), and find that while ImageNet initalisation provides a significant boost over training from scratch, using Kinetics-400 (K400) only provides a very marginal further gain. This suggests that at least for retrieval, the initialisation of the video encoder is not as important as joint text-video pretraining for the entire model (as demonstrated next).

Effect of Pretraining Data: We begin by analysing the results with fine-tuning for text-video retrieval on the MSR-VTT dataset, presented in Table 1(right). We note that pretraining on VideoCC3M provides a significant boost to performance over HowTo100M, with far less data, and for an RGB-only model, yields a 5% improvement over training from scratch on R@1. This effect is even more profound in the zero-shot case, where for an RGB-only model, using VideoCC3M more than doubles the R@1 performance compared to HowTo100M pretraining. This is done with 100x fewer captions and 20x less video data. We believe that this shows the value in high-quality video-captioning pairs. Regarding audio inputs, we note that MSR-VTT is a visual benchmark (unlike Audio-Caps and Clotho), with some videos missing an audio track entirely. However we show that adding audio provides a modest performance boost. We then compare to previous works on this dataset in Table 2 (left), including recently released Frozen In Time (FIT) [9] and VideoCLIP [81]. We note that our model outperforms FIT which pretrains on 3 different datasets - CC3M, WebVid2M and COCO [18]. We were unable to train on WebVid2M due to data restrictions but believe further performance gains could be achieved by training on VideoCC3M

Table 2. Comparison to state-of-the-art results on MSR-VTT for text-to-video retrieval (left) and video captioning (right). V-T PT: Visual-text pre-training data. **#Caps:** Number of unique captions used during pretraining. † These works use numerous experts, including Object, Motion, Face, Scene, Speech, OCR and Sound classification features. ‡ Pretrained on WebVid-2M, CC3M and COCO datasets. *Numbers obtained from the authors. Modalities: **V:** RGB frames. **T:** ASR in videos.

Method	V-T PT	#Caps	R@1	R@5	R@10
Finetuned					
HERO [48]	HT100M	136M	16.8	43.4	57.7
NoiseEst. [5]	HT100M	136M	17.4	41.6	53.6
CE [51]†	-		20.9	48.8	62.4
UniVL [52]	HT100M	136M	21.2	49.6	63.1
ClipBERT [46]	Coco, VGen	5.6M	22.0	46.8	59.9
AVLnet [67]	HT100M	136M	27.1	55.6	66.6
MMT [29]†	HT100M	136M	26.6	57.1	69.6
T2VLAD [80]†	-		29.5	59.0	70.1
SupportSet [62]	HT100M	136M	30.1	58.5	69.3
VideoCLIP [81]	HT100M	136M	30.9	55.4	66.8
FIT [9]	CC3M	3M	25.5	54.5	66.1
FIT [9]	Multiple‡	6.1M	32.5	61.5	71.2
Ours	VideoCC3M	**970K**	**35.8**	**65.1**	**76.9**
Zero-shot					
MIL-NCE [56]	HT100M	136M	7.5	21.2	29.6
SupportSet [62]	HT100M	136M	8.7	23.0	31.1
EAO [70]	HT100M	136M	9.9	24.0	32.6
VideoCLIP [81]	HT100M	136M	10.4	22.2	30.0
FIT [9]	WebVid2M*	2.5M	15.4	33.6	44.1
Ours	VideoCC3M	**970K**	**20.4**	**39.5**	**50.3**

Method	V-T PT	Modality	B-4	C	M
Finetuned					
POS+CG [78]	-	V	42.00	49	28.20
POS+VCT [38]	-	V	42.30	49	29.70
SAM-SS [15]	-	V	43.80	51	28.90
ORG-TRL [86]	-	V	43.60	51	28.80
VNS-GRU [14]	-	V	45.30	53	29.90
UniVL [53]	HT100M	V+T	41.79	50	28.94
DecemBT [75]	HT100M	V	45.20	52	29.70
Ours	HT100M	V	**47.33**	55	37.11
Ours	VCC3M	V	45.47	**55**	36.96
Zero-shot					
Ours	HT100M	V	7.5	0.5	8.23
Ours	VCC3M	V	**13.23**	**8.24**	**11.34**

and WebVid jointly. We also note that by training on VideoCC3M, we outperform FIT trained only on the CC3M dataset by a big margin (R@1 25.5 to 35.3), even though the amount of manually annotated supervision is the same. This shows the benefit of mining extra video data using our data mining pipeline. On zero-shot performance, we outperform all previous works that pretrain on HowTo100M, and FIT [9] when it is trained only on video data (WebVid2M). We note that adding in various image datasets provides a huge boost to performance in FIT [9], and this complementary approach could be used with VideoCC. We could also use additional seed datasets such as COCO Captions [18] to mine more text-video clips, which we leave as future work.

Results Using CLIP. [64] Given the recent flurry of CLIP based [19,26,31,54], RGB-only works for video retrieval, in this section we show the complementarity of using CLIP [64] based models trained on the 400M pair WiT dataset such as Clip4Clip [54] finetuned on the VideoCC dataset. We reproduce Clip4Clip [54] with mean pooling in our framework (Table 3). Using CLIP (trained on 400M diverse image-caption pairs) leads to very strong zero-shot performance, however finetuning it on VideoCC *further* improves performance by over 3% R@1, showing the additional value of automatically mined *videos*. We also outperform the zero-shot SOTA from Clip4Clip which was post trained on a curated subset of HowTo100M and is the highest online number for this zero-shot benchmark (CaMoE [19] and Clip2TV [31] do not report zero-shot results). This shows the value of our automatic video mining pipeline.

Table 3. Finetuning Clip4Clip on VideoCC for zero-shot performance on MSR-VTT.

Model	Pre-training data	R@1	R@5	R@10
Clip4Clip [54]	WiT [64]	30.6	54.4	64.3
Ours	WiT [64] + VideoCC	33.7	57.9	67.9

Audio Retrieval: For text-audio retrieval we report results on two audio-centric datasets (i.e. datasets paired with natural language descriptions that focus explicitly on the content of the audio track) - AudioCaps [42] and Clotho [24]. The goal here is to retrieve the correct audio segment given a free form natural language query. While Clotho comes with only audio, AudioCaps has both audio and RGB frames. Results on the AudioCaps dataset are provided in Table 4 (left). We first show results for an audio-only encoder (we only feed spectrograms as input). We note that our model with no audio-text pretraining already outperforms the current state of the art [60] by a large margin (R@1: from 24.3 to 32.0), despite the fact that [60] uses features pretrained on VGGSound and VGG-ish features pretrained on YouTube8M. This could be because unlike their encoder, our encoder is trained end-to-end directly from spectrograms. We then show results with pretraining on the spectrograms from HowTo100M (no RGB

Table 4. Results on AudioCaps (left) and Clotho (right) for text-audio retrieval. † Higher than reported in the paper, as these are provided by authors on our test set. Inputs refers to video inputs as follows: **A:** Audio spectrograms **V:** RGB video frames. Rows highlighted in light blue show Zero-shot (ZS) performance. Note the CLOTHO dataset contains audio only (no RGB) frames.

Model	Pretraining	Modality	R@1	R@10
SOTA [60]†	-	A	24.3	72.1
Ours	-	A	32.0	82.3
Ours	HowTo100M	A	33.7	83.2
Ours	VideoCC3M	A	35.5	84.5
Ours (ZS)	HowTo100M	A	1.4	6.5
Ours (ZS)	VideoCC3M	A	8.7	37.7
SOTA [60]†	-	A+V	28.1	79.0
Ours	-	A+V	41.4	85.3
Ours	VideoCC3M	A+V	**43.2**	**88.9**
Ours (ZS)	VideoCC3M	A+V	10.6	45.2
Model	Pretraining	R@1	R@10	
SOTA [60]	-	6.7	33.3	
Ours	-	7.8	35.4	
Ours	VideoCC3M	**8.4**	**38.6**	
Ours (ZS)	VideoCC3M	3.0	17.5	
SOTA [60]	AudioCaps	9.6	40.1	
Ours	AudioCaps	11.4	43.4	
Ours	VideoCC3M+AudioCaps	**12.6**	**45.4**	

frames are used here), and find that there is some improvement. Pretraining on the audio and captions from VideoCC3M however, gives substantial performance gains to R@1 by over 3%. This improvement is particularly impressive because the captions were transferred via visual similarity to still images and no additional manual audio-text supervision was used. We also report zero-shot results, and find that unsurprisingly, pretraining on HowTo100M results in poor performance, likely because the model has learned to focus on speech. VideoCC3M provides a large improvement, however there is still a distance to finetuning performance. Finally, we also show that using an audio-visual fusion encoder and training on VideoCC3M provides a further significant improvement demonstrating the complementarity of RGB information for this task. Results on Clotho are provided in Table 4 (right). Here we show a similar trend, but as Clotho is also a much smaller dataset, we also show results with AudioCaps pre-training as is done by [60]. Combining AudioCaps supervised pretraining with VideoCC3M pretraining provides the best result.

5.4   Video Captioning

Results for video captioning are provided in Table 2 (right). For finetuning, our model pretrained on VideoCC3M outperforms previously published works. Unlike retrieval, pretraining on HowTo100M provides slight gains to the B-4 and M metrics, but VideoCC3M is still competitive with a fraction of the data size. We then compare zero-shot performance, and find that pretraining on HowTo100M performs poorly, potentially because of the large difference in style between instructional speech and human-generated captions. Training on VideoCC3M provides a substantial boost across all metrics, with a fraction of the data. Qualitative results are shown in Fig. 5.

Fig. 5. Zero-shot captioning results on MSR-VTT test set videos. We show 2 frames per clip. As expected, the style of predicted captions from HowTo100M pretraining is similar to ASR, and concepts may be tenuously related (middle). Pretraining on VideoCC3M yields captions that are closer to the ground truth.

6   Conclusion

We propose a new, automatic method for leveraging existing image datasets to mine video and audio data with captions. We apply it to the CC3M dataset [69] to mine millions of weakly labelled video-text pairs. Training a multimodal retrieval model on these clips leads to state of the art performance for video retrieval and captioning, and shows complementarity with existing image-text models such as CLIP. Future work can focus on augmenting these automatic captions with even more video related text, such as action labels, to overcome the image-centric bias in the mining pipeline. Societal impacts and fairness are discussed in supplementary.

References

1. YouTube Data API. http://developers.google.com/youtube/v3/docs/captions
2. Abavisani, M., Joze, H.R.V., Patel, V.M.: Improving the performance of unimodal dynamic hand-gesture recognition with multimodal training. In: CVPR (2019)
3. Aguilar, G., Rozgic, V., Wang, W., Wang, C.: Multimodal and multi-view models for emotion recognition. In: ACL (2019)
4. Albanie, S., Nagrani, A., Vedaldi, A., Zisserman, A.: Emotion recognition in speech using cross-modal transfer in the wild. In: Proceedings of the 26th ACM international conference on Multimedia, pp. 292–301 (2018)

5. Amrani, E., Ben-Ari, R., Rotman, D., Bronstein, A.: Noise estimation using density estimation for self-supervised multimodal learning. arXiv preprint arXiv:2003.03186 (2020)
6. Anne Hendricks, L., Wang, O., Shechtman, E., Sivic, J., Darrell, T., Russell, B.: Localizing moments in video with natural language. In: ICCV (2017)
7. Antol, S. et al.: VQA: Visual question answering. In: ICCV (2015)
8. Bain, M., Nagrani, A., Brown, A., Zisserman, A.: Condensed movies: Story based retrieval with contextual embeddings. In: ACCV (2020)
9. Bain, M., Nagrani, A., Varol, G., Zisserman, A.: Frozen in time: A joint video and image encoder for end-to-end retrieval. ICCV (2021)
10. Bain, M., Nagrani, A., Varol, G., Zisserman, A.: A clip-hitchhiker's guide to long video retrieval. arXiv preprint arXiv:2205.08508 (2022)
11. Banerjee, S., Lavie, A.: Meteor: An automatic metric for mt evaluation with improved correlation with human judgments. In: ACL Workshop on Intrinsic and Extrinsic Evaluation Measures for Machine Translation and/or Summarization (2005)
12. Changpinyo, S., Sharma, P., Ding, N., Soricut, R.: Conceptual 12m: Pushing web-scale image-text pre-training to recognize long-tail visual concepts. In: Proceedings of the IEEE/CVF Conference on Computer Vision and Pattern Recognition, pp. 3558–3568 (2021)
13. Chechik, G., Ie, E., Rehn, M., Bengio, S., Lyon, D.: Large-scale content-based audio retrieval from text queries. In: Proceedings of the 1st ACM International Conference on Multimedia Information Retrieval, pp. 105–112 (2008)
14. Chen, H., Li, J., Hu, X.: Delving deeper into the decoder for video captioning. In: ECAI (2020)
15. Chen, H., Lin, K., Maye, A., Li, J., Hu, X.: A semantics-assisted video captioning model trained with scheduled sampling. Front. Robot. AI 7 475767 (2020)
16. Chen, H., Xie, W., Vedaldi, A., Zisserman, A.: Vggsound: A large-scale audio-visual dataset. In: ICASSP 2020–2020 IEEE International Conference on Acoustics, Speech and Signal Processing (ICASSP), pp. 721–725. IEEE (2020)
17. Chen, S., Jiang, Y.G.: Motion guided spatial attention for video captioning. In: AAAI (2019)
18. Chen, X., Fang, H., Lin, T.Y., Vedantam, R., Gupta, S., Dollár, P., Zitnick, C.L.: Microsoft coco captions: Data collection and evaluation server. arXiv preprint arXiv:1504.00325 (2015)
19. Cheng, X., Lin, H., Wu, X., Yang, F., Shen, D.: Improving video-text retrieval by multi-stream corpus alignment and dual softmax loss (2021)
20. Deng, J., Dong, W., Socher, R., Li, L.J., Li, K., Fei-Fei, L.: ImageNet: A large-scale hierarchical image database. In: CVPR (2009)
21. Devlin, J., Chang, M.W., Lee, K., Toutanova, K.: BERT: Pre-training of deep bidirectional transformers for language understanding. In: NAACL-HLT (2019)
22. Devlin, J., Chang, M.W., Lee, K., Toutanova, K.: Bert: Pre-training of deep bidirectional transformers for language understanding. arXiv preprint arXiv:1810.04805 (2018)
23. Dosovitskiy, A., et al.: An image is worth 16x16 words: Transformers for image recognition at scale. In: ICLR (2021)
24. Drossos, K., Lipping, S., Virtanen, T.: Clotho: An audio captioning dataset. In: ICASSP 2020–2020 IEEE International Conference on Acoustics, Speech and Signal Processing (ICASSP), pp. 736–740. IEEE (2020)

25. Elizalde, B., Zarar, S., Raj, B.: Cross modal audio search and retrieval with joint embeddings based on text and audio. In: ICASSP 2019–2019 IEEE International Conference on Acoustics, Speech and Signal Processing (ICASSP), pp. 4095–4099. IEEE (2019)
26. Fang, H., Xiong, P., Xu, L., Chen, Y.: Clip2video: Mastering video-text retrieval via image clip. arXiv preprint arXiv:2106.11097 (2021)
27. Font, F., Roma, G., Serra, X.: Freesound technical demo. In: Proceedings of the 21st ACM International Conference on Multimedia, pp. 411–412 (2013)
28. Gabeur, V., Nagrani, A., Sun, C., Alahari, K., Schmid, C.: Masking modalities for cross-modal video retrieval. In: Proceedings of the IEEE/CVF Winter Conference on Applications of Computer Vision, pp. 1766–1775 (2022)
29. Gabeur, V., Sun, C., Alahari, K., Schmid, C.: Multi-modal transformer for video retrieval. In: ECCV (2020)
30. Gao, R., Oh, T.H., Grauman, K., Torresani, L.: Listen to look: Action recognition by previewing audio. In: Proceedings of the IEEE/CVF Conference on Computer Vision and Pattern Recognition, pp. 10457–10467 (2020)
31. Gao, Z., Liu, J., Chen, S., Chang, D., Zhang, H., Yuan, J.: Clip2tv: An empirical study on transformer-based methods for video-text retrieval. arXiv preprint arXiv:2111.05610 (2021)
32. Gemmeke, J.F., et al.: Audio set: An ontology and human-labeled dataset for audio events. In: 2017 IEEE International Conference on Acoustics, Speech and Signal Processing (ICASSP), pp. 776–780. IEEE (2017)
33. Ghadiyaram, D., Tran, D., Mahajan, D.: Large-scale weakly-supervised pre-training for video action recognition. In: Proceedings of the IEEE/CVF Conference on Computer Vision and Pattern Recognition, pp. 12046–12055 (2019)
34. Girdhar, R., Tran, D., Torresani, L., Ramanan, D.: Distinit: Learning video representations without a single labeled video. In: Proceedings of the IEEE/CVF International Conference on Computer Vision, pp. 852–861 (2019)
35. Gong, Y., Chung, Y.A., Glass, J.: Ast: Audio spectrogram transformer. arXiv preprint arXiv:2104.01778 (2021)
36. Gupta, S., Hoffman, J., Malik, J.: Cross modal distillation for supervision transfer. In: Proceedings of the IEEE Conference on Computer Vision and Pattern Recognition (CVPR) (June 2016)
37. Hinton, G., Vinyals, O., Dean, J.: Distilling the knowledge in a neural network. arXiv preprint arXiv:1503.02531 (2015)
38. Hou, J., Wu, X., Zhao, W., Luo, J., Jia, Y.: Joint syntax representation learning and visual cue translation for video captioning. In: ICCV (2019)
39. Huang, G., Pang, B., Zhu, Z., Rivera, C., Soricut, R.: Multimodal pretraining for dense video captioning. In: AACL (2020)
40. Juan, D.C., et al.: Graph-rise: Graph-regularized image semantic embedding. arXiv preprint arXiv:1902.10814 (2019)
41. Kay, W., et al.: The kinetics human action video dataset. arXiv preprint arXiv:1705.06950 (2017)
42. Kim, C.D., Kim, B., Lee, H., Kim, G.: Audiocaps: Generating captions for audios in the wild. In: Proceedings of the 2019 Conference of the North American Chapter of the Association for Computational Linguistics: Human Language Technologies, Volume 1 (Long and Short Papers), pp. 119–132 (2019)
43. Koepke, A., Oncescu, A.M., Henriques, J.F., Akata, Z., Albanie, S.: Audio retrieval with natural language queries: A benchmark study. arXiv preprint arXiv:2112.09418 (2021)

44. Krishna, R., Hata, K., Ren, F., Fei-Fei, L., Carlos Niebles, J.: Dense-captioning events in videos. In: ICCV (2017)
45. Krishan, R., et al.: Visual genome: connecting language and vision using crowd-sourced dense image annotations. Int. J. Comput. Vision **123**(1), 32–73 (2017)
46. Lei, J., et al.: Less is more: Clipbert for video-and-language learning via sparse sampling. In: CVPR (2021)
47. Lei, J., Yu, L., Bansal, M., Berg, T.L.: Tvqa: Localized, compositional video question answering. In: EMNLP (2018)
48. Li, L., Chen, Y.C., Cheng, Y., Gan, Z., Yu, L., Liu, J.: Hero: Hierarchical encoder for video+ language omni-representation pre-training. In: EMNLP (2020)
49. Li, T., Wang, L.: Learning spatiotemporal features via video and text pair discrimination. arXiv preprint arXiv:2001.05691 (2020)
50. Lin, T.Y., et al.: Microsoft COCO: Common objects in context. In: ECCV (2014)
51. Liu, Y., Albanie, S., Nagrani, A., Zisserman, A.: Use what you have: Video retrieval using representations from collaborative experts. In: BMVC (2019)
52. Luo, H., et al.: UniVL: A unified video and language pre-training model for multimodal understanding and generation. arXiv preprint arXiv:2002.06353 (2020)
53. Luo, H., et al.: UniVL: A unified video and language pre-training model for multimodal understanding and generation. arXiv e-prints (2020)
54. Luo, H., et al.: Clip4clip: An empirical study of clip for end to end video clip retrieval. arXiv preprint arXiv:2104.08860 (2021)
55. Miech, A., Alayrac, J.B., Smaira, L., Laptev, I., Sivic, J., Zisserman, A.: End-to-end learning of visual representations from uncurated instructional videos. In: CVPR (2020)
56. Miech, A., Zhukov, D., Alayrac, J.B., Tapaswi, M., Laptev, I., Sivic, J.: HowTo100M: Learning a Text-Video Embedding by Watching Hundred Million Narrated Video Clips. In: ICCV (2019)
57. Monfort, M., Jin, S., Liu, A., Harwath, D., Feris, R., Glass, J., Oliva, A.: Spoken moments: Learning joint audio-visual representations from video descriptions. In: Proceedings of the IEEE/CVF Conference on Computer Vision and Pattern Recognition, pp. 14871–14881 (2021)
58. Nagrani, A., Sun, C., Ross, D., Sukthankar, R., Schmid, C., Zisserman, A.: Speech2action: Cross-modal supervision for action recognition. In: Proceedings of the IEEE/CVF Conference on Computer Vision and Pattern Recognition, pp. 10317–10326 (2020)
59. Nagrani, A., Yang, S., Arnab, A., Jansen, A., Schmid, C., Sun, C.: Attention bottlenecks for multimodal fusion. In: NeurIPS (2021)
60. Oncescu, A.M., Koepke, A., Henriques, J.F., Akata, Z., Albanie, S.: Audio retrieval with natural language queries. arXiv preprint arXiv:2105.02192 (2021)
61. Papineni, K., Roukos, S., Ward, T., Zhu, W.J.: Bleu: a method for automatic evaluation of machine translation. In: ACL (2002)
62. Patrick, M., et al.: Support-set bottlenecks for video-text representation learning. arXiv preprint arXiv:2010.02824 (2020)
63. Patrick, M., et al.: Support-set bottlenecks for video-text representation learning. In: ICLR (2021)
64. Radford, A., et al.: Learning transferable visual models from natural language supervision (2021)
65. Radford, A., Wu, J., Child, R., Luan, D., Amodei, D., Sutskever, I.: Language models are unsupervised multitask learners. Technical Report (2019)
66. Rohrbach, A., et al.: Movie description. Int. J. Comput. Vision **123**(1), 94–120 (2017)

67. Rouditchenko, A., et al.: AVLnet: Learning audio-visual language representations from instructional videos. arXiv preprint arXiv:2006.09199 (2020)
68. Seo, P.H., Nagrani, A., Schmid, C.: Look before you speak: Visually contextualized utterances. In: CVPR (2021)
69. Sharma, P., Ding, N., Goodman, S., Soricut, R.: Conceptual captions: A cleaned, hypernymed, image alt-text dataset for automatic image captioning. In: ACL (2018)
70. Shvetsova, N., et al.: Everything at once-multi-modal fusion transformer for video retrieval. In: CVPR (2022)
71. Slaney, M.: Semantic-audio retrieval. In: 2002 IEEE International Conference on Acoustics, Speech, and Signal Processing. vol. 4, pp. IV-4108. IEEE (2002)
72. Stroud, J.C., Lu, Z., Sun, C., Deng, J., Sukthankar, R., Schmid, C., Ross, D.A.: Learning video representations from textual web supervision. arXiv preprint arXiv:2007.14937 (2020)
73. Sun, C., Shetty, S., Sukthankar, R., Nevatia, R.: Temporal localization of fine-grained actions in videos by domain transfer from web images. In: ACM Multimedia (2015)
74. Sun, C., Shetty, S., Sukthankar, R., Nevatia, R.: Temporal localization of fine-grained actions in videos by domain transfer from web images. In: ACM Multimedia (2015)
75. Tang, Z., Lei, J., Bansal, M.: Decembert: Learning from noisy instructional videos via dense captions and entropy minimization. In: NAACL (2021)
76. Vedantam, R., Lawrence Zitnick, C., Parikh, D.: Cider: Consensus-based image description evaluation. In: CVPR (2015)
77. Vinyals, O., Toshev, A., Bengio, S., Erhan, D.: Show and tell: lessons learned from the 2015 MSCOCO image captioning challenge. IEEE Trans. Pattern Anal. Mach. Intell. **39**(4), 652–663 (2016)
78. Wang, B., Ma, L., Zhang, W., Jiang, W., Wang, J., Liu, W.: Controllable video captioning with pos sequence guidance based on gated fusion network. In: ICCV (2019)
79. Wang, L., Li, Y., Lazebnik, S.: Learning deep structure-preserving image-text embeddings. In: CVPR (2016)
80. Wang, X., Zhu, L., Yang, Y.: T2vlad: Global-local sequence alignment for text-video retrieval (2021)
81. Xu, H.,et al.: Videoclip: Contrastive pre-training for zero-shot video-text understanding. arXiv preprint arXiv:2109.14084 (2021)
82. Xu, J., Mei, T., Yao, T., Rui, Y.: Msr-vtt: A large video description dataset for bridging video and language. In: CVPR (2016)
83. Yao, L., et al.: Filip: Fine-grained interactive language-image pre-training. arXiv preprint arXiv:2111.07783 (2021)
84. You, Q., Jin, H., Wang, Z., Fang, C., Luo, J.: Image captioning with semantic attention. In: CVPR (2016)
85. Zhai, A., Wu, H.Y.: Classification is a strong baseline for deep metric learning. In: BMVC (2019)
86. Zhang, Z., Shi, Y., Yuan, C., Li, B., Wang, P., Hu, W., Zha, Z.J.: Object relational graph with teacher-recommended learning for video captioning. In: CVPR (2020)
87. Zhou, L., Xu, C., Corso, J.: Towards automatic learning of procedures from web instructional videos. In: AAAI (2018)

RVSL: Robust Vehicle Similarity Learning in Real Hazy Scenes Based on Semi-supervised Learning

Wei-Ting Chen[1], I-Hsiang Chen[2], Chih-Yuan Yeh[2], Hao-Hsiang Yang[2],

Hua-En Chang[2], Jian-Jiun Ding[2], and Sy-Yen Kuo[2(✉)]

[1] Graduate Institute of Electronics Engineering, National Taiwan University, Taipei, Taiwan
`f05943089@ntu.edu.tw`
[2] Department of Electrical Engineering, National Taiwan University, Taipei, Taiwan
`{f09921058,f09921063,r10921a35,jjding,sykuo}@ntu.edu.tw,`
`https://github.com/Cihsaing/rvsl-robust-vehicle-similarity-`
`learning--ECCV22`

Abstract. Recently, vehicle similarity learning, also called re-identification (ReID), has attracted significant attention in computer vision. Several algorithms have been developed and obtained considerable success. However, most existing methods have unpleasant performance in the hazy scenario due to poor visibility. Though some strategies are possible to resolve this problem, they still have room to be improved due to the limited performance in real-world scenarios and the lack of real-world clear ground truth. Thus, to resolve this problem, inspired by CycleGAN, we construct a training paradigm called **RVSL** which integrates ReID and domain transformation techniques. The network is trained on semi-supervised fashion and does not require to employ the ID labels and the corresponding clear ground truths to learn hazy vehicle ReID mission in the real-world haze scenes. To further constrain the unsupervised learning process effectively, several losses are developed. Experimental results on synthetic and real-world datasets indicate that the proposed method can achieve state-of-the-art performance on hazy vehicle ReID problems. It is worth mentioning that although the proposed method is trained without real-world label information, it can achieve competitive performance compared to existing supervised methods trained on complete label information.

Keywords: Hazy vehicle similarity learning · Semi-supervised learning · Image dehazeing

1 Introduction

Vehicle similarity learning, also called vehicle re-identification (ReID), is a crucial technique for intelligent surveillance systems in a smart city. It is to track the vehicles with

---

W.-T. Chen and I.-H. Chen—Equal contribution.

Supplementary Information The online version contains supplementary material available at https://doi.org/10.1007/978-3-031-19781-9_25.

S. Avidan et al. (Eds.): ECCV 2022, LNCS 13674, pp. 427–443, 2022.
https://doi.org/10.1007/978-3-031-19781-9_25

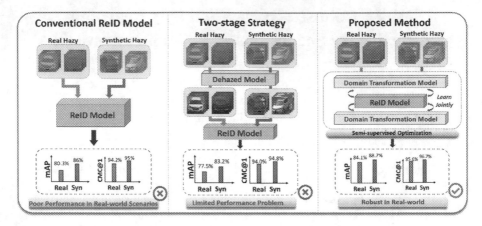

Fig. 1. Illustration of different strategies to solve hazy vehicle ReID problem. One can see that our method outperforms other existing methods in terms of the mean average precision (mAP) and CMC@1. Moreover, other strategies may have limited performance problem in real-world scenarios. We adopt CAL [33] and MPR-Net [39] as the ReID model and the dehazed model, respectively.

the same identity within a set of images captured by multiple cameras and various viewpoints. With the development of the deep convolutional neural network (DCNN), several approaches for vehicle ReID [13,18,31,41] have been proposed and achieved impressive performance. Similar to other high-level vision applications such as object detection and semantic segmentation [5], although existing methods can handle vehicle ReID effectively on normal images, they have limited performance under inclement weather, especially in hazy scenario. Haze is a common and inevitable weather phenomenon that leads to poor visual appearances and causes the loss of discriminative information by deteriorating the contents of images for vehicle ReID. Thus, this field still has room for improvement.

Inspired by previous dehazing tasks [2], we can apply an atmospheric scattering model [23] to synthesize haze images and then train the vehicle ReID models based on the rendered images and the corresponding ID labels. Though this strategy can achieve decent performance on synthetic images, they have limited performance on real haze images due to the domain gap between synthetic and real-world images [3]. While this issue can be resolved by adopting real haze images in the training stage, collecting the real haze data and labeling the correct ground truths are difficult and troublesome.

Another possible baseline strategy is to adopt the existing dehazing approaches [4, 40] or comprehensive image restoration method [39] as the pre-processing technique and then apply the ReID. Although the above strategies are shown to achieve promising results in haze removal, there is no guarantee that the selected pre-processing techniques would be able to improve ReID, since these two tasks are performed separately and existing dehazing methods are not designed for the purpose of ReID but for human perception. Moreover, most existing dehazing methods require pair data to train the model, but it is infeasible to attain the ground truths of haze images in real-world scenes. Though we can adopt synthetic data to train the network, the domain gap problem may still exist, which may generate undesired dehazed results in real haze scenes and further limit the performance of ReID.

By the above analysis, there are two reasons that hinder the development of ReID in real haze scenes: (i) the scarcity of labels of real-world data and (ii) the lack of appropriate guidance for real haze. In this paper, to mitigate these problems, we construct a novel training architecture based on the deep convolutional neural network (DCNN). Inspired by CycleGAN [21] which can transform images between any two domains, we introduce the domain transfer technique in the proposed network and combine it with the vehicle ReID. Specifically, the proposed method is trained on a semi-supervised paradigm in an end-to-end fashion and there are two parts in the training process: supervised training for synthetic data and unsupervised training for real world data. For the former part, the network can learn the knowledge of transformation between two domains and extract more discriminative features for vehicle ReID in fully supervision by paired data (i.e., synthetic hazy images and the corresponding clear ground truths). For the latter parts, we only leverage two sets of unpaired data (i.e., real hazy images and clear images) to strengthen the robustness of the domain transformation and the ReID in real-world scenes in an unsupervised manner.

The idea of our method is that, the domain transformation network transfers the input image (i.e., hazy or clear images) between two domains with the same background information and the ReID network extracts the latent features for classifying from two images (i.e., input and transferred image). Inspired by the cycle consistency [21, 37], two extracted embedding features should be identical since they are from the same vehicles. Thus, we can calculate the consistency of between two extracted features for optimizing the network.

Based on the semi-supervised training scheme, the utilization of the synthetic data can guide the unsupervised stage and prevent the network from unstable performance [21]. On the other hand, the use of real-world data can improve the generalization ability of our model to real data and further mitigate the domain gap problem when the synthetic data is applied in the training process [25]. Moreover, using the domain transformation network can assist the ReID network to learn more discriminative features for the ReID under real-world haze scenarios. By our design, the proposed method can perform vehicle ReID in hazy scenarios effectively without additional annotations in real hazy data which are usually hard to be obtained. Furthermore, our proposed training scheme can be also applied in the case that we have annotations of vehicle ID and achieve better performance.

The contribution of this paper is summarized as follows.

- A novel training paradigm based on semi-supervised learning and domain transformation is proposed to learn hazy vehicle ReID without the labels or clear ground truths of real-world data. We term it **R**obust **V**ehicle **S**imilarity **L**earning (**RVSL**). As depicted in Fig. 1, by combining domain transferring technique with the ReID network, the proposed method can achieve decent performance to learn discriminative features under real haze scenes without using ID label. Surprisingly, the proposed method achieves competitive performance compared with other existing methods trained with complete ID information.

- To constrain the unsupervised stage in the training process, we developed several loss functions such as embedding consistency loss, colinear relation constraint, and monotonously increasing dark channel loss to improve the performance. These loss functions enable the network to learn both domain transformation and ReID in an unsupervised way effectively. Experimental results prove the effectiveness of these loss functions.

2   Related Works

Vehicle Re-identification. With the great effort of data collection and annotation, several large-scale benchmarks for vehicle ReID such as VehicleID [28], VeRi-776 [29], VERI-Wild [30], and Vehicle-1M [12]) are proposed. Based on these well-developed benchmarks, several approaches [9,17,19,26,33,35] have been developed and most of them rely on DCNN. We can divide them into the following categories.

1) *Meta-information-based methods* which integrate meta-information for feature learning. For example, Zheng *et al.* [42] leveraged the additional information such as the camera view, and the vehicle type and color to guide the network. Shen *et al.* [34] integrated the visual-spatio-temporal path proposals and spatial-temporal relations to a Siamese-CNN+Path-LSTM network. Rao *et al.* [33] proposed the attention mechanism with counterfactual causality which enables the network to learn more useful attention for fine-grained features for ReID.

2) *Local information-based methods*: Meng *et al.* [31] adopted the common region information extracted by a vehicle part parser to improve the mutual representation information between different viewpoints. Khorramshahi *et al.* [22] applied Variational Auto-Encoder (VAE) to find crucial detailed information which can be regarded as the pseudo-attention map for highlighting discriminative regions. He *et al.* [13] combined local and non-local information based on a part-regularized mechanism. These strategies can preserve the variance from near-duplicate vehicles to improve the performance of vehicle ReID. Zhao *et al.* [41] applied Cross-camera Generalization Measure technique and integrated region-specific features and cross-level features together to improve the performance of ReID.

3) *Generative Adversarial Network (GAN) based methods*: Zhou *et al.* [43] applied the conditional multi-view generative network to extract global feature representation from various viewpoints and then adopted adversarial learning to facilitate feature generation. Lou *et al.* [30] designed the FDA-Net to generate hard examples in the feature space based on the GAN to improve the robustness of ReID. Yao *et al.* [38] proposed to adopt a 3D graphic engine to reduce the content gap between the existing datasets to suppress the domain gap problem.

4) *Vision Transformer (ViT) based Methods*: He *et al.* [18] leveraged the ViT to encode input images as a vector for embedding representation. To further improve representation learning, the jigsaw patch module and side information were adopted in the training scheme.

Though the above methods can achieve decent vehicle ReID performance on clear images, they are still limited in real-world hazy image scenarios.

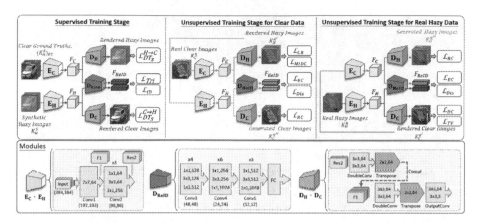

Fig. 2. The architecture of the proposed semi-supervised hazy vehicle ReID network. Our method consists of supervised and unsupervised training stages for synthetic and real-world data.

Single Image Haze Removal. Based on Koschmieder's model [23], the formation of haze can be modeled by:

$$I(x) = J(x)t(x) + A(1 - t(x)), \tag{1}$$

where $I(x)$ is the hazy image, $J(x)$ is the haze-free image and A is the global atmospheric light. $t(x) = e^{-\beta d(x)}$ is the medium transmission map where β is the scattering coefficient and $d(x)$ is the depth from the camera to the object. There are numerous haze removal methods proposed in past decades. They can be classified into prior-based and deep learning-based methods. The former class is to explore the prior knowledge between hazy and haze-free images. For example, He *et al.* [14] proposed the dark channel prior, Zhu *et al.* [44] developed the color attenuation prior, and Berman *et al.* proposed the haze-line [1] to estimate the dehazed results. The other class is to apply the DCNN. For instance, Qu *et al.* [32] proposed multi-resolution generators and discriminators for dehazing in a coarse-to-fine way. Dong *et al.* [8] used the strengthen-operate-subtract boosting strategy to improve the dehazing network. Wu *et al.* [36] proposed an auto-encoder-like framework with additive mixup operation and a dynamic feature enhancement module to improve the quality of extracted features for dehazing. Zamir *et al.* [39] proposed a multi-stage architecture that can encode a diverse set of features simultaneously to restore accurate outputs. Chen *et al.* [6] proposed a unified architecture which can learn multiple adverse weather based on a single architecture.

3   Proposed Method

3.1   Overview of the Proposed Method

As shown in Fig. 2, there are five modules in the proposed network. That is, two encoders for hazy and clear scenes ($\mathbf{E}_H$ and $\mathbf{E}_C$), and three decoders for hazy, clear, and ReID ($\mathbf{D}_H$, $\mathbf{D}_C$, and $\mathbf{D}_{ReID}$), respectively. These modules can be combined to two sub-networks called domain transformation network and re-identification network. The features extracted by $\mathbf{E}_H$ and $\mathbf{E}_C$ are termed F_H and F_C, respectively.

As mentioned in Sect. 1, due to the lack of real haze data, in this paper, inspired by semi-supervision [25], we apply both synthetic data and real data simultaneously in the learning process. At the supervised training stage, the application of the synthetic data can learn the transformation from various domains stably. At the unsupervised stage, we first take the real clear images and then take real hazy images as inputs, respectively. This operation enables our network to learn real-world information of hazy scenes, clear scenes, and ReID through the domain transformation network and ReID network simultaneously. We illustrate the details in the following subsections.

3.2 Domain Transformation Network

The goal of the domain transformation network (DT-Net) is to help the ReID network to learn haze-invariant features via transforming the domain of the input data. The detailed illustration of this network is as follows.

Architecture. The DT-Net consists of two encoders ($\mathbf{E}_H$ and $\mathbf{E}_C$) and two decoders ($\mathbf{D}_H$ and $\mathbf{D}_C$). Given a hazy input, the decoder $\mathbf{D}_C$ generates the corresponding clear image based on the features F_H extracted by encoder $\mathbf{E}_H$. On the other hand, the decoder $\mathbf{D}_H$ takes the features F_C extracted from the encoder $\mathbf{E}_C$ to produce the hazy image. The features (i.e., F_C and F_H) extracted by two encoders pass through a double convolution block and a deconvolution block for dimension matching. Then, the upsampled features are concatenated with the features extracted by the first convolution blocks in the encoders to improve the feature diversity. This operation is based on the fact that the features in the shallow layer of the network contain more fruitful spatial and contextual information which can benefit domain transformation [7,20] while the deeper layers usually consist of more high-level vision. The concatenated features are passed through a double convolution block and a deconvolution block to reconstruct the final domain transformation results. The quality of domain transformation is crucial for the ReID network since it may affect the feature extraction of the input image. Thus, for the synthetic data, we adopt the supervised loss $\mathcal{L}_{DTs}$ to optimize the networks. For real data, we adopt unsupervised losses $\mathcal{L}_{DTu}$.

Supervised Training Stage. At this stage, we can train the network in a fully supervised way since the corresponding clear ground truths and ID labels are available. First, we adopt synthetic image pairs to train the DT-Net (i.e., $\mathbf{E}_H, \mathbf{E}_C, \mathbf{D}_H$, and $\mathbf{D}_C$). Specifically, the synthetic haze image K_H^S and the corresponding clear ground truth $(K_H^S)_{GT}$ are fed into the domain transformation network to calculate the domain transformation loss $\mathcal{L}_{DT_s}$ for synthetic data. This operation aims to constrain the distance between the predicted results (i.e., the rendered hazy images and the rendered clear images) and the corresponding ground truths. The domain transformation loss $\mathcal{L}_{DTs}$ can be formulated as follows.

$$\mathcal{L}_{DT_s}^{H \to C} = \frac{1}{M} \sum_{i=1}^{M} \|\mathbf{D}_C[\mathbf{E}_H[K_H^S(i)]] - (K_H^S)_{GT}(i)\|_1 \tag{2}$$

$$\mathcal{L}_{DT_s}^{C \to H} = \frac{1}{M} \sum_{i=1}^{M} \|\mathbf{D}_H[\mathbf{E}_C[(K_H^S)_{GT}(i)]] - K_H^S(i)\|_1 \tag{3}$$

where $\| \cdot \|_1$ presents the L_1 norm and M indicates the number of images. $\mathcal{L}_{DT_s}^{C \to H}$ and $\mathcal{L}_{DT_s}^{H \to C}$ denote the domain transformation loss for 'clear to haze' and 'haze to clear', respectively. The $\mathcal{L}_{DT_s}$ loss is the summation of $\mathcal{L}_{DT_s}^{C \to H}$ and $\mathcal{L}_{DT_s}^{H \to C}$.

Unsupervised Training Stage for Clear Data. At this stage, to train the DT-Net without the hazy image ground truths, first, we adopted the cycle-consistency mechanism. The input clear image K_C^R is fed into the DT-Net to render the hazy image $K_H^{R'}$, where $K_H^{R'} = \mathbf{D}_H(\mathbf{E}_C(K_C^R))$. Then, we further take the rendered image to the DT-Net to generate the rendered clear image $K_C^{R''}$, where $K_C^{R''} = \mathbf{D}_C(\mathbf{E}_H(K_H^{R'}))$. In the same time, several loss functions are adopted to optimize the network. The loss at this stage $\mathcal{L}_{DT_{uc}}$ consists of the rendering consistency loss ($\mathcal{L}_{RC}$), the monotonously increasing dark channel loss ($\mathcal{L}_{MIDC}$), the colinear relation constraint ($\mathcal{L}_{CR}$), and the discriminative loss ($\mathcal{L}_{Dis}$). We illustrate each of them as follows.

(i) *Rendering Consistency Loss.* This loss is to constrain the learning process of the domain transformation network (i.e., $\mathbf{E}_H$, $\mathbf{E}_C$, $\mathbf{D}_H$, and $\mathbf{E}_C$). We adopt the pixel-wise difference between the clear input image K_C^R and the rendered clear image $K_C^{R''}$ to ensure that the domain transformation process can be conducted in two different domains robustly. This loss is formulated as follows.

$$\mathcal{L}_{RC} = \frac{1}{M} \sum_{i=1}^{M} \|K_C^R(i) - K_C^{R''}(i)\|_1 \tag{4}$$

(ii) *Monotonously Increasing Dark Channel Loss.* To further improve the image quality of rendered haze images, inspired by dark channel prior (DCP) [14], we propose the monotonously increasing dark channel loss $\mathcal{L}_{MIDC}$. The DCP demonstrates that for most natural clear images, the dark channel values may be close to zero. Specifically, it can be defined as:

$$DC(J)(x) = \min_{y \in \Omega(x)} \left(\min_{c \in \{r,g,b\}} J^c(y) \right) \approx 0, \tag{5}$$

where $DC(\cdot)$ is the operation of the dark channel, $J^c(y)$ is the intensity in the color channel c, and $\Omega(x)$ is a local patch with a fixed size centered at x. With this prior, it can be further extended that, given an image deteriorated by haze, its dark channel value may be higher than that of the original clear image (i.e., $DC(I)(x) \geq DC(J)(x)$). Based on this idea, we proposed $\mathcal{L}_{MIDC}$ which is determined as:

$$\mathcal{L}_{MIDC} = \frac{1}{M} \sum_{i=1}^{M} DM(i)\|DC(K_H^{R'}(i)) - DC(K_C^R)(i)\|_1 \tag{6}$$

where $DM(i)$ is a binary map that identifies the region where the dark channel values of the clear image K_C^R are higher than that of its rendered haze result $K_H^{R'}$ (i.e., $DC(K_H^{R'})(x) < DC(K_C^R)(x)$). With it, we can prevent the rendered pixels from irrational results and further improve the robustness of domain transformation.

(iii) *Colinear Relation Constraint.* Due to the inaccessibility of the ground truth of real-world hazy images, the domain transformation process may generate undesired fake image contents without appropriate training. Although we can leverage the synthetic data to guide the network at the training stage of the synthetic data, it is not applicable at the training stage of the real-world data. To further strengthen the robustness of transformation (i.e., $\mathbf{E}_C$ and $\mathbf{D}_H$), inspired by the haze-line prior [1,37], we develop colinear relation constraint $\mathcal{L}_{CR}$.

Based on the physical model of haze illustrated in (1), Berman *et al.* [1] observe that the clear image J, the hazy image I, and the atmospheric light A are colinear in the RGB space (i.e., $I(x) - A = t(x)(J(x) - A)$). We can adopt this relation to constrain the training process of the network for real-world scenarios. We define the colinear relation constraint as follows.

$$\mathcal{L}_{CR} = \frac{1}{M} \sum_{i=1}^{M} \left[1 - \phi(K_C^R(i) - A(i), K_H^{R'}(i) - A(i)) \right], \tag{7}$$

where $A(i)$ is the atmospheric light estimated by the rendered hazy image $K_H^{R'}(i)$ and $\phi(\cdot)$ means the cosine similarity. Different from [37], we adopt the atmospheric light estimation method in [14] in this loss. With this loss, the consistency of structure and color can be further constrained.

(iv) *Discriminative Loss.* To further constrain unsupervised domain transformation, we adopt the discriminative loss [10] in the training process to distinguish whether the rendered hazy image $K_H^{R'}$ is real or fake. In our method, we adopt the saturating discriminative loss [11].

Unsupervised Training Stage for Real Hazy Data. At this stage, the real hazy images are adopted to optimize the network without clean ground truths and ID labels. Like the previous stage, the hazy images are fed into the DT-Net (i.e., $\mathbf{E}_H$ and $\mathbf{D}_C$) to generate the clear images $K_C^{R'}$. Subsequently, the rendered clear images are fed into the DT-Net to obtain the rendered hazy images $K_H^{R''}$. To optimize our framework, apart from the monotonously increasing dark channel loss ($\mathcal{L}_{MIDC}$) and colinear relation constraint ($\mathcal{L}_{CR}$), the rest of losses in $\mathcal{L}_{DT_{uc}}$ are adopted. Moreover, to improve the predicted clear images by the DT-Net, we introduce two losses: the dark channel loss $\mathcal{L}_{DC}$ to curb the residual haze and total variation loss $\mathcal{L}_{TV}$ to prevent the noise generation. They can be formulated as follows.

$$\mathcal{L}_{DC} = \frac{1}{M} \sum_{i=1}^{M} \| DC(K_C^{R'}(i)) \|_1, \tag{8}$$

$$\mathcal{L}_{TV} = \frac{1}{M} \sum_{i=1}^{M} \| \nabla_x K_C^{R'}(i) \|_1 + \| \nabla_y K_C^{R'}(i) \|_1, \tag{9}$$

where ∇_x and ∇_y denote the gradient operations along the horizontal and vertical directions, respectively.

3.3   Re-identification Network

The re-identification network (ReID-Net) aims to extract discriminative features to search the images with the same identification in the gallery. The details of architecture and training are illustrated as follows.

Architecture. The ReID-Net consists two encoders (i.e., $\mathbf{E}_H$ and $\mathbf{E}_C$) and one decoder (i.e., $\mathbf{D}_{ReID}$). It adopts ResNet-50 [16] as the backbone, where we apply the first two convolution blocks as the architecture of two encoders. As shown in Fig. 2, the extracted features F_H or F_C are fed to the decoder $\mathbf{D}_{ReID}$ to generate the ReID results where the decoder consists of the rest of convolution blocks in ResNet-50 and extracted features are down-scaled by global average pooling (GAP) and batch normalization (BN) to generate 2048-d embedding features F_{ReID}. Last, we adopt the fully connected layer (FC layer) to match the number of identities for the classification. For the supervised learning of synthetic data, since we have the corresponding ID label, we adopt the triplet loss $\mathcal{L}_{Tri}$ and the ID loss $\mathcal{L}_{ID}$. For the unsupervised learning stage, due to lack of the ID label, the embedding consistency loss $\mathcal{L}_{EC}$ is adopted to constrain the network. This architecture enables our two encoders to learn domain adaptive features because the features extracted by two encoders working on different domains are fed into the same decoder.

Supervised Training Stage. At this stage, we train the re-identification network (i.e., $\mathbf{E}_C$, $\mathbf{E}_H$, and $\mathbf{D}_{ReID}$) by adopting the triplet loss [19] $\mathcal{L}_{Tri}$ and the ID loss $\mathcal{L}_{ID}$ which can be defined as follows.

$$\mathcal{L}_{Tri} = \frac{1}{M} \sum_{i-1}^{M} \sum_{k} \left[\max_{z_p \in \mathcal{P}(z_i^k)} D(z_i^k, z_p) - \min_{z_n \subset \mathcal{N}(z_i^k)} D(z_i^k, z_n) + \delta \right]_{+} \tag{10}$$

$$\mathcal{L}_{ID} = -\frac{1}{M} \sum_{i=1}^{M} \sum_{k} \log \frac{\exp(\sigma_i^{y_i^k})}{\sum_{j=1}^{C} \exp(\sigma_i^j)} \tag{11}$$

where $k \in \{(K_H^S)_{GT}, K_H^S\}$. $\mathcal{P}(z_i^k)$ and $\mathcal{N}(z_i^k)$ denote the positive and negative sample sets, respectively. z_i^k represents the extracted embedding features from the ith input sample (i.e., $((K_H^S)_{GT}(i)$ or $K_H^S(i)))$. δ is the margin of the triplet loss, $D(\cdot, \cdot)$ is the Euclidean distance, and $[\cdot]_+$ equals to $max(\cdot, 0)$. For $\mathcal{L}_{ID}$, σ_i^j is the output of the FC layer with the class j based on i^{th} input image. C presents the total number of the class, and y_i donates the ground truth class. The ReID loss $\mathcal{L}_{ReID_s}$ at this stage is the combination of $\mathcal{L}_{ID}$ and $\mathcal{L}_{Tri}$.

Unsupervised Training Stage. At this stage, we feed both real clear data and real hazy data separately. Due to the lack of labels about ID information, to train the ReID network (i.e., $\mathbf{E}_C$, $\mathbf{E}_H$, and $\mathbf{D}_{ReID}$) with real clear inputs, we develop *embedding consistency loss ($\mathcal{L}_{EC}$)* to calculate the distance of two embedding features extracted from the input clear image and the rendered haze image. Initially, given a clear image K_C^R, it is fed into the DT-Net to render the hazy image $K_H^{R'}$ and ReID-Net to extract embedding feature $(F_{ReID})_C^R$. Then, we further take the rendered image to the ReID-Net to produce the embedding feature $(F_{ReID})_H^{R'}$. We can calculate the loss between $(F_{ReID})_H^{R'}$

Fig. 3. Examples of the images in the synthetic dataset and the real-world datasets for vehicle ReID.

and $(F_{ReID})_C^R$ because they are the same vehicle. By using this loss, the haze-invariant features can be learned by the ReID-Net effectively. The mathematical expression of this loss is defined as follows.

$$\mathcal{L}_{EC} = \frac{1}{M} \sum_{i=1}^{M} ||[F_{ReID}(i)]_C^R - [F_{ReID}(i)]_H^{R'}||_1 \tag{12}$$

By contrast, when the input data is a real hazy image (i.e., K_H^R), the same mechanism is adopted.

4 Experiments

4.1 Dataset and Evaluation Protocols

Dataset Preparation. The proposed semi-supervised scheme is trained by both synthetic and real-world haze data. We select haze-free images from Vehicle-1M and VERI-Wild datasets. Subsequently, we apply the haze synthesis procedure proposed in [24] to synthesize these images. First, we adopt the method in [27] to estimate the depth map d. Then, we render the haze on these clear images by (1) with the predicted depth maps and set $\beta \in [0.4, 1.6]$ and $A \in [0.5, 1]$. Uniquely, each clear data generates a hazy image and all rendered images are divided into the training and the testing sets, respectively. For the real haze data, we survey all existing datasets and find that only Vehicle-1M and VERI-Wild datasets contain the cases in the hazy weather. Thus, we carefully select the vehicle images under hazy scenarios from two datasets. The selected images are split to the training and the testing sets. The details and examples of two types of data are presented in Table 1, Table 2 and Fig. 3, respectively.

Evaluation Protocols. Followed by the protocols of the evaluation proposed in [12, 30], we randomly select one hazy image for each vehicle and put it into the probe set. The remained images form the gallery set. We adopt the cumulative matching characteristic (CMC) curve and mean average precision (mAP) to evaluate the performance.

Table 1. Detail of the synthetic dataset.
(IDs/Images)

Set	Train	Probe	Gallery
VERI-Wild	1167/19532	389/389	389/6125
Vehicle-1M	1833/23026	611/611	611/7093
Total	3000/42558	1000/1000	1000/13218

Table 2. Detail of the real-world dataset.
(IDs/Images)

Set	Train	Probe	Gallery
VERI-Wild	156/2472	389/389	389/5985
Vehicle-1M	247/2579	611/611	611/6242
Total	403/5051	1000/1000	1000/12227

4.2 Implementation Details

Training Stage. [1]For the proposed ReID network, ResNet-50 is adopted as the backbone, whose weights are initialized from the model pre-trained on the ImageNet. In the synthetic data training stage, the dimensions of FC layers are set to 3000. The weights of the domain transformation network are initialized by Kaiming normalization [15]. The whole network is trained in an end-to-end fashion based on the training sets of synthetic and real-world datasets for learning domain transformation, vehicle ReID and ID classification simultaneously. The input image is resized to 384×384. The training batch sizes at the synthetic data and the real-world data stages are 72 $(2M)$ and 36 (M), respectively. The local patch size in the dark channel operation is 5×5. We apply the data augmentation in the training process including the random cropping and horizontal flipping techniques. The warm-up training strategy is adopted for 120 epochs. The Adam optimizer is adopted with a decay rate of 0.6. The initial learning rate is 1.09×10^{-5}, which increases to 10^{-4} after the 10th epoch. At the training stage of the synthetic data, we adopt the synthetic dataset in Table 1 and randomly select one hazy image for each vehicle and put it into the probe set. The rest of images form the gallery set. For the training stage of real-world data, we apply 5051 clear images and hazy images without ID labels, respectively. The network is trained on an Nvidia Tesla V100 GPU for 3 d and we implement it using Pytorch.

Inference Stage. At the inference stage, the encoder ($\mathbf{E}_C$) and two decoders of the domain transformation network ($\mathbf{D}_C$ and $\mathbf{D}_H$) are not involved. The computational burden caused by them can be ignored. The Euclidean distance D is computed through embedding features to evaluate the performance.

4.3 Comparison with the Existing Methods

To evaluate the performance of the proposed method, we compare the proposed algorithm with state-of-the-art ReID methods, the VRCF [9], the VOC [45], the DMT [17], the CAL [33], the VEHICLEX [38], the TransReID [18], the PVEN [31], and the HRCN [41]. For a fair and comprehensive comparison, these methods are retrained by the following training sets: (i) The ground truth clear images in the synthetic dataset; (ii) The hazy images from the training sets of both synthetic and real-world haze datasets (denoted with the '-haze'); (iii) The two-stage strategy (i.e., dehazing+ReID) which is denoted with '-dehaze'. Specifically, this strategy is the combination of the dehazing

---

[1] More details about training each stage and results are presented in the Supplementary Material.

Table 3. Quantitative evaluation on the hazy vehicle ReID scenario. The words with **bold-face** indicate the best results, and those with <u>underline</u> indicate the second-best results. The texts 'S' and 'R' indicate synthetic and real-world datasets.

Method	mAP		CMC@1		CMC@5		CMC@10	
	S	R	S	R	S	R	S	R
VRCF	25.90	36.60	61.70	63.70	76.50	78.80	81.30	83.20
VRCF-dehaze	61.50	50.80	85.40	78.00	95.10	92.00	97.20	95.40
VRCF-haze	69.00	58.00	88.60	81.10	97.60	93.80	98.40	96.80
VRCF-all	73.00	64.40	90.90	85.50	97.30	95.70	98.60	97.80
VOC	59.70	57.40	86.10	82.80	94.30	94.00	95.60	96.60
VOC-dehaze	63.40	49.20	87.00	74.10	94.80	89.90	96.50	94.30
VOC-haze	67.10	59.90	88.70	83.50	95.10	94.00	96.50	97.20
VOC-all	84.20	78.70	93.60	91.00	97.60	96.30	98.30	98.30
DMT	73.90	71.70	93.40	93.20	97.20	97.40	97.90	98.50
DMT-dehaze	75.10	71.60	93.40	92.40	96.90	97.50	98.30	98.40
DMT-haze	77.30	73.40	94.00	93.40	97.60	97.60	98.60	98.80
DMT-all	82.50	80.90	98.30	96.10	98.20	98.20	98.80	99.00
VehicleX	63.64	61.56	86.50	83.20	95.00	95.20	97.40	97.90
VehicleX-dehaze	73.06	64.82	89.70	83.90	96.70	95.10	98.20	97.60
VehicleX-haze	77.86	69.01	91.20	84.80	97.10	96.10	98.70	98.10
VehicleX-all	80.75	76.39	93.10	89.90	97.60	96.90	98.60	98.40
TransReID	62.90	64.00	82.40	77.70	92.30	88.80	98.40	94.00
TransReID-dehaze	66.80	65.30	83.00	76.60	94.10	89.90	98.10	94.60
TransReID-haze	73.90	72.10	84.80	82.60	95.20	90.70	98.70	95.60
TransReID-all	79.20	76.90	89.40	84.50	96.80	93.20	98.90	97.30
PVEN	72.83	75.36	63.73	66.48	84.39	86.53	89.65	91.20
PVEN-dehaze	81.70	78.13	73.29	69.47	92.50	89.16	96.04	93.43
PVEN-haze	84.55	81.92	76.60	74.09	95.02	92.15	97.84	95.66
PVEN-all	<u>88.63</u>	<u>84.08</u>	83.55	78.31	98.45	95.40	99.20	97.76
HRCN	81.22	71.77	92.00	85.30	97.60	95.40	99.10	97.50
HRCN-dehaze	83.44	72.78	92.20	84.60	98.00	96.10	99.00	97.80
HRCN-haze	85.40	78.64	92.80	89.40	<u>98.50</u>	96.70	99.10	98.40
HRCN-all	87.91	81.41	94.60	91.80	98.20	97.30	**99.30**	99.00
CAL	75.52	75.94	92.50	91.70	96.50	97.60	97.90	98.40
CAL-dehaze	83.21	77.49	94.80	94.00	98.30	98.00	98.90	98.80
CAL-haze	86.00	80.31	95.00	94.20	97.90	<u>98.30</u>	98.90	<u>99.10</u>
CAL-all	88.20	83.84	<u>96.30</u>	**96.00**	98.40	98.20	98.90	99.00
Ours	**88.66**	<u>84.12</u>	**96.70**	<u>95.60</u>	**98.60**	**98.60**	**99.30**	**99.30**
Ours-F	**89.14**	**87.72**	96.50	**96.90**	**98.60**	98.40	**99.40**	**99.60**

Table 4. Ablation study for each module in the real-world test set.

Method	Metric			
	mAP	CMC@1	CMC@5	CMC@10
Baseline-haze	76.17	93.40	97.50	98.50
Baseline-all	77.34	93.40	97.60	98.60
Ours w/o $\mathcal{L}_{CR}$ & $\mathcal{L}_{MIDC}$	76.02	93.50	97.20	98.60
Ours w/o $\mathcal{L}_{MIDC}$	81.19	94.20	98.20	99.00
Ours w/o $\mathcal{L}_{CR}$	82.27	95.30	98.00	99.20
Ours w/o $\mathcal{L}_{DC}$ & $\mathcal{L}_{TV}$	77.31	93.63	97.38	98.80
Ours w/o $\mathcal{L}_{DC}$	81.20	94.30	98.30	98.85
Ours w/o $\mathcal{L}_{TV}$	82.50	94.60	98.50	98.90
Ours	**84.12**	**95.60**	**98.60**	**99.30**

Table 5. Comparison of performance for using different blocks as encoders $\mathbf{E}_C$ and $\mathbf{E}_H$ in the real-world test set.

	mAP	CMC@1	CMC@5	CMC@10
Conv_2	84.12	95.60	98.60	99.30
Conv_3	83.84	94.90	98.20	98.90
Conv_4	82.56	94.40	98.20	99.00
Conv_5	80.82	94.90	98.10	99.00

Table 6. Ablation study for using different training data stages in real-world test set.

Stage	Metric			
	mAP	CMC@1	CMC@5	CMC@10
Syn	78.17	92.20	96.70	97.20
Syn+RC	80.03	94.10	98.20	98.90
Syn+RH	81.26	94.60	98.10	98.90
Ours	**84.12**	**95.60**	**98.60**	**99.30**

method for pre-processing and the ReID models trained by setting (i). For the dehazing method, we adopt one of the state-of-the-art dehazing methods called MPR-Net [39] which was retrained on hazy vehicle images. (iv) The same training images used in our method including synthetic haze, real-world clear and real-world haze datasets (denoted with '-all'). The aforementioned settings are all with complete ID labels.

The results are reported in Table 3. We can observe the following results. First, compared with other strategies, the proposed method can achieve the competitive performance on vehicle ReID in hazy weather on both synthetic and real-world datasets in terms of mAP and CMC. Second, existing methods trained on all data can obtain better performance compared to the methods trained on other training settings. Third, other methods may have limited performance in real-world scenarios, especially when they are only trained on synthetic images. Last, *surprisingly, though our method is trained without ID labels in real-world data, it can outperform most supervised methods trained with complete ID labels.*

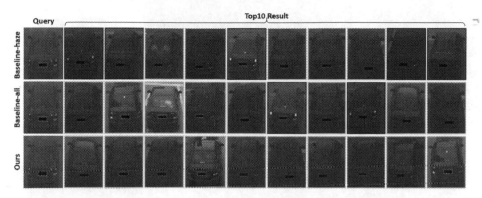

Fig. 4. Visual results of the ranking list on the real-world dataset. The query images are in the first column and the retrieved top-10 ranking results are in the rest columns. We denote the correct retrieved images with a green border while the false instances are with a red border. (Color figure online)

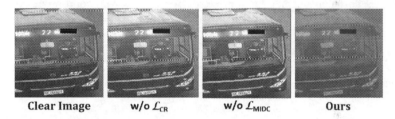

Fig. 5. Visual comparison of using the unsupervised losses $\mathcal{L}_{CR}$ **and** $\mathcal{L}_{MIDC}$ **in the domain transformation network.** These loss functions can benefit the rendering process to generate more desirable results.

We also adopt our method trained with ID labels in the real-world data training stage which is denoted with the suffix 'F'. Specifically, we introduce the triplet loss and the ID loss defined in spseqrefeq:idloss and spseqrefeq:triploss to train the real-world data stage. The result indicates that the performance can be improved if we use complete ID labels in the training stage. Our method can be also adopted in the fully supervised scenarios and obtain the decent performance.

4.4 Ablation Studies

Effectiveness of the Semi-supervised Strategy. In this paper, we proposed the semi-supervised training technique to solve hazy vehicle ReID problem. It uses the domain transformation mechanism which enables us to train the real-world data without the ID labels. We present the effectiveness of this strategy in Table 4. We adopt the our ReID network as the baseline and train with two settings for comparison, that -is, the settings (ii) and (iv) reported in Subsect. refsec4.3 with complete ID labels. One can see that our method is against the first setting favorably. Moreover, even without the ID labels of real-world data, our method can outperform the baseline trained with complete

ID labels since our methods integrate the domain transformation technique, which can improve the ReID network to learn better representation under the hazy scenes. We also show the visual results of the ranking list on the real-world dataset in Fig. 4. One can see that, the baseline retrieves wrong instances because the important features such as the window and light become ambiguous due to the degradation of haze, which may deteriorate the performance of ReID.

Moreover, in Table 5, we show the results of assigning different convolution blocks as the encoders. One can see that adopting the first two convolution blocks as the encoders can obtain the best performance. However, the proposed domain transformation architecture can assist the network to learn accurate ReID in the haze scenario.

Effectiveness of the Loss Functions. In Table 4, we verify the effectiveness of the adopted loss functions: the monotonously increasing dark channel loss $\mathcal{L}_{MICD}$ and the colinear relation constraint $\mathcal{L}_{CR}$. One can see that, with two loss functions, the performance of ReID can be improved in both mAP and CMC metrics since the DT-Net can benefit the encoders to learn more robust features with appropriate constraints which can further benefit the performance of ReID. Furthermore, using both the dark channel loss $\mathcal{L}_{DC}$ and the total variation loss $\mathcal{L}_{TV}$ can improve the performance of the network. Figure 5 presents that, with the proposed loss functions, the rendered results can be more realistic compared with other modules. The rendered results may have the color distortion problems without using $\mathcal{L}_{CR}$.

Effectiveness of Each Training Stage. We verify the effectiveness of using real clear data or real hazy data in the training process. We construct three settings for the comparison. Specifically, we adopt: (i) only the synthetic data stage (**Syn**), (ii) **Syn** with real clear data stage (**Syn+RC**), and (iii) **Syn** with real haze data stage (**Syn+RH**). The results are reported in Table 6. We can see that only adopting the synthetic data may cause limited performance in real-world scenarios due to the domain gap problem.

5   Conclusion

In this paper, to address the vehicle ReID problem under hazy scenarios, a semi-supervised training framework that integrates the domain transformation network and the ReID network is proposed. Moreover, to constrain the unsupervised training stage, several loss functions to bound the two networks are proposed. With these techniques, the proposed method can learn haze-invariant features for robust vehicle ReID. Experimental results show that, compared to existing methods trained on complete ID labels, the proposed methods can achieve decent performance even without using the ID labels in real-world data.

Acknowledgement. We thank to National Center for High-performance Computing (NCHC) for providing computational and storage resources. This research was supported by the Ministry of Science and Technology, Taiwan under Grants MOST 108–2221-E-002–072-MY3, MOST 108–2638-E-002–002-MY2, and MOST 111–2221-E-002–136-MY3.

References

1. Berman, D., Avidan, S., et al.: Non-local image dehazing. In: Proceedings of the IEEE conference on computer vision and pattern recognition, pp. 1674–1682 (2016)
2. Cai, B., Xu, X., Jia, K., Qing, C., Tao, D.: Dehazenet: An end-to-end system for single image haze removal. IEEE Trans. Image Process. **25**(11), 5187–5198 (2016)
3. Chen, W.T., Chen, I.H., Yeh, C.Y., Yang, H.H., Ding, J.J., Kuo, S.Y.: Sjdl-vehicle: Semi-supervised joint defogging learning for foggy vehicle re-identification (2022)
4. Chen, W.T., Ding, J.J., Kuo, S.Y.: Pms-net: Robust haze removal based on patch map for single images. In: Proceedings of the IEEE/CVF Conference on Computer Vision and Pattern Recognition, pp. 11681–11689 (2019)
5. Chen, W.T., et al.: All snow removed: Single image desnowing algorithm using hierarchical dual-tree complex wavelet representation and contradict channel loss. In: Proceedings of the IEEE/CVF International Conference on Computer Vision, pp. 4196–4205 (2021)
6. Chen, W.T., Huang, Z.K., Tsai, C.C., Yang, H.H., Ding, J.J., Kuo, S.Y.: Learning multiple adverse weather removal via two-stage knowledge learning and multi-contrastive regularization: Toward a unified model. In: Proceedings of the IEEE/CVF Conference on Computer Vision and Pattern Recognition, pp. 17653–17662 (2022)
7. Chen, W.T., et al.: Desmokenet: A two-stage smoke removal pipeline based on self-attentive feature consensus and multi-level contrastive regularization. In: IEEE Transactions on Circuits and Systems for Video Technology (2021)
8. Dong, H., et al.: Multi-scale boosted dehazing network with dense feature fusion. In: Proceedings of the IEEE/CVF Conference on Computer Vision and Pattern Recognition, pp. 2157–2167 (2020)
9. Gao, C., Hu, Y., Zhang, Y., Yao, R., Zhou, Y., Zhao, J.: Vehicle re-identification based on complementary features. In: Proceedings of the IEEE/CVF Conference on Computer Vision and Pattern Recognition Workshops, pp. 590–591 (2020)
10. Goodfellow, I., et al.: Generative adversarial nets. Advances in neural information processing systems, 27 (2014)
11. Gui, J., Sun, Z., Wen, Y., Tao, D., Ye, J.: A review on generative adversarial networks: Algorithms, theory, and applications. IEEE Transactions on Knowledge and Data Engineering (2021)
12. Guo, H., Zhao, C., Liu, Z., Wang, J., Lu, H.: Learning coarse-to-fine structured feature embedding for vehicle re-identification. In: Proceedings of the AAAI Conference on Artificial Intelligence, vol. 32 (2018)
13. He, B., Li, J., Zhao, Y., Tian, Y.: Part-regularized near-duplicate vehicle re-identification. In: Proceedings of the IEEE/CVF Conference on Computer Vision and Pattern Recognition, pp. 3997–4005 (2019)
14. He, K., Sun, J., Tang, X.: Single image haze removal using dark channel prior. IEEE Trans. Pattern Anal. Mach. Intell. **33**(12), 2341–2353 (2010)
15. He, K., Zhang, X., Ren, S., Sun, J.: Delving deep into rectifiers: Surpassing human-level performance on imagenet classification. In: Proceedings of the IEEE international conference on computer vision, pp. 1026–1034 (2015)
16. He, K., Zhang, X., Ren, S., Sun, J.: Deep residual learning for image recognition. In: Proceedings of the IEEE conference on computer vision and pattern recognition (2016)
17. He, S., et al.: Multi-domain learning and identity mining for vehicle re-identification. In: Proceedings of the IEEE/CVF Conference on Computer Vision and Pattern Recognition Workshops, pp. 582–583 (2020)
18. He, S., Luo, H., Wang, P., Wang, F., Li, H., Jiang, W.: Transreid: Transformer-based object re-identification. arXiv preprint arXiv:2102.04378 (2021)

19. Hermans, A., Beyer, L., Leibe, B.: In defense of the triplet loss for person re-identification. arXiv preprint arXiv:1703.07737 (2017)
20. Hui, Z., Li, J., Wang, X., Gao, X.: Image fine-grained inpainting. arXiv preprint arXiv:2002.02609 (2020)
21. Isola, P., Zhu, J.Y., Zhou, T., Efros, A.A.: Image-to-image translation with conditional adversarial networks. In: Proceedings of the IEEE conference on computer vision and pattern recognition, pp. 1125–1134 (2017)
22. Khorramshahi, P., Peri, N., Chen, J.c., Chellappa, R.: The devil is in the details: Self-supervised attention for vehicle re-identification. In: European Conference on Computer Vision, pp. 369–386. Springer (2020)
23. Koschmieder, H.: Theorie der horizontalen sichtweite. Beitrage zur Physik der freien Atmosphare pp. 33–53 (1924)
24. Li, B., et al.: Benchmarking single-image dehazing and beyond. IEEE Trans. Image Process. **28**(1), 492–505 (2018)
25. Li, L., et al.: Semi-supervised image dehazing. IEEE Trans. Image Process. **29**, 2766–2779 (2019)
26. Li, M., Huang, X., Zhang, Z.: Self-supervised geometric features discovery via interpretable attention for vehicle re-identification and beyond. In: Proceedings of the IEEE/CVF International Conference on Computer Vision, pp. 194–204 (2021)
27. Liu, F., Shen, C., Lin, G., Reid, I.: Learning depth from single monocular images using deep convolutional neural fields. IEEE Trans. Pattern Anal. Mach. Intell. **38**(10), 2024–2039 (2015)
28. Liu, H., Tian, Y., Wang, Y., Pang, L., Huang, T.: Deep relative distance learning: Tell the difference between similar vehicles. In: Proceedings of the IEEE Conference on Computer Vision and Pattern Recognition, pp. 2167–2175 (2016)
29. Liu, X., Liu, W., Mei, T., Ma, H.: Provid: Progressive and multimodal vehicle reidentification for large-scale urban surveillance. IEEE Trans. Multimedia **20**(3), 645–658 (2017)
30. Lou, Y., Bai, Y., Liu, J., Wang, S., Duan, L.: Veri-wild: A large dataset and a new method for vehicle re-identification in the wild. In: Proceedings of the IEEE/CVF Conference on Computer Vision and Pattern Recognition, pp. 3235–3243 (2019)
31. Meng, D., et al.: Parsing-based view-aware embedding network for vehicle re-identification. In: Proceedings of the IEEE/CVF Conference on Computer Vision and Pattern Recognition, pp. 7103–7112 (2020)
32. Qu, Y., Chen, Y., Huang, J., Xie, Y.: Enhanced pix2pix dehazing network. In: Proceedings of the IEEE/CVF Conference on Computer Vision and Pattern Recognition (CVPR) (June 2019)
33. Rao, Y., Chen, G., Lu, J., Zhou, J.: Counterfactual attention learning for fine-grained visual categorization and re-identification. In: Proceedings of the IEEE/CVF International Conference on Computer Vision, pp. 1025–1034 (2021)
34. Shen, Y., Xiao, T., Li, H., Yi, S., Wang, X.: Learning deep neural networks for vehicle re-id with visual-spatio-temporal path proposals. In: Proceedings of the IEEE International Conference on Computer Vision, pp. 1900–1909 (2017)
35. Wang, Z., Tang, L., Liu, X., Yao, Z., Yi, S., Shao, J., Yan, J., Wang, S., Li, H., Wang, X.: Orientation invariant feature embedding and spatial temporal regularization for vehicle re-identification. In: Proceedings of the IEEE International Conference on Computer Vision, pp. 379–387 (2017)
36. Wu, H., et al.: Contrastive learning for compact single image dehazing. In: Proceedings of the IEEE/CVF Conference on Computer Vision and Pattern Recognition, pp. 10551–10560 (2021)

37. Yan, W., Sharma, A., Tan, R.T.: Optical flow in dense foggy scenes using semi-supervised learning. In: Proceedings of the IEEE/CVF Conference on Computer Vision and Pattern Recognition, pp. 13259–13268 (2020)
38. Yao, Y., Zheng, L., Yang, X., Naphade, M., Gedeon, T.: Simulating Content Consistent Vehicle Datasets with Attribute Descent. In: Vedaldi, A., Bischof, H., Brox, T., Frahm, J.-M. (eds.) ECCV 2020. LNCS, vol. 12351, pp. 775–791. Springer, Cham (2020). https://doi.org/10.1007/978-3-030-58539-6_46
39. Zamir, S.W., et al.: Multi-stage progressive image restoration. In: Proceedings of the IEEE/CVF Conference on Computer Vision and Pattern Recognition, pp. 14821–14831 (2021)
40. Zhang, H., Patel, V.M.: Densely connected pyramid dehazing network. In: Proceedings of the IEEE conference on computer vision and pattern recognition, pp. 3194–3203 (2018)
41. Zhao, J., Zhao, Y., Li, J., Yan, K., Tian, Y.: Heterogeneous relational complement for vehicle re-identification. In: Proceedings of the IEEE/CVF International Conference on Computer Vision, pp. 205–214 (2021)
42. Zheng, A., Lin, X., Li, C., He, R., Tang, J.: Attributes guided feature learning for vehicle re-identification. arXiv preprint arXiv:1905.08997 (2019)
43. Zhou, Y., Shao, L.: Aware attentive multi-view inference for vehicle re-identification. In: Proceedings of the IEEE conference on computer vision and pattern recognition, pp. 6489–6498 (2018)
44. Zhu, Q., Mai, J., Shao, L.: A fast single image haze removal algorithm using color attenuation prior. IEEE Trans. Image Process. **24**(11), 3522–3533 (2015)
45. Zhu, X., Luo, Z., Fu, P., Ji, X.: Voc-reid: Vehicle re-identification based on vehicle-orientation-camera. In: Proceedings of the IEEE/CVF Conference on Computer Vision and Pattern Recognition Workshops, pp. 602–603 (2020)

Lightweight Attentional Feature Fusion: A New Baseline for Text-to-Video Retrieval

Fan Hu[1,2], Aozhu Chen[1,2], Ziyue Wang[1,2], Fangming Zhou[1,2], Jianfeng Dong[3], and Xirong Li[1,2(✉)]

[1] MoE Key Lab of DEKE, Renmin University of China, Beijing, China
`xirong@ruc.edu.cn`
[2] AIMC Lab, School of Information, Renmin University of China, Beijing, China
[3] College of Computer and Information Engineering, Zhejiang Gongshang University, Hangzhou, China

Abstract. In this paper we revisit *feature fusion*, an old-fashioned topic, in the new context of text-to-video retrieval. Different from previous research that considers feature fusion only at one end, let it be video or text, we aim for feature fusion for both ends within a unified framework. We hypothesize that optimizing the convex combination of the features is preferred to modeling their correlations by computationally heavy multi-head self attention. We propose Lightweight Attentional Feature Fusion (LAFF). LAFF performs feature fusion at both early and late stages and at both video and text ends, making it a powerful method for exploiting diverse (off-the-shelf) features. The interpretability of LAFF can be used for feature selection. Extensive experiments on five public benchmark sets (MSR-VTT, MSVD, TGIF, VATEX and TRECVID AVS 2016–2020) justify LAFF as a new baseline for text-to-video retrieval.

Keywords: Text-to-video retrieval · Video/text feature fusion

1 Introduction

Text-to-video retrieval is to retrieve videos *w.r.t.* to an ad-hoc textual query from many *unlabeled* videos. Both video and text have to be embedded into one or more cross-modal common spaces for text-to-video matching. The state-of-the-art tackles the task in different approaches, including novel networks for query representation learning [59,65], multi-modal Transformers for video representation learning [3,19], hybrid space learning for interpretable cross-modal matching [15,60], and more recently CLIP2Video [17] that learns text and video representations in an end-to-end manner. Differently, we look into *feature fusion*, an important yet largely underexplored topic for text-to-video retrieval.

F. Hu, A. Chen and Z. Wang.—Equal contribution.

Supplementary Information The online version contains supplementary material available at https://doi.org/10.1007/978-3-031-19781-9_26.

Given video/text samples represented by diverse features, feature fusion aims to answer a basic research question of *what is the optimal way to combine these features?* By optimal we mean the fusion shall maximize the retrieval performance. Meanwhile, the fusion process shall be explainable to interpret the importance of the individual features. As the use of each feature introduces extra computational and storage overheads, the explainability is crucial for the fusion process to be selective to balance the performance and the cost.

Feature fusion is not new by itself. In fact, the topic has been extensively studied in varied contexts such as multimedia content analysis [2,50] and multimodal or multi-view image classification [4,64]. These earlier efforts focus on combining hand-crafted features, because such kinds of features are known to be domain-specific, suffering from the semantic gap problem [49], and thus insufficient for content representation when used alone. While current deep learning features are already more powerful than their predecessors, no single feature appears to rule all. Dark knowledge about objects and scenes is better carried in pre-trained 2D convolutional neural networks (2D-CNNs) [48], while 3D-CNNs are more suited for representing actions and motions [20]. For text-to-video retrieval, there are some initial efforts on combining diverse deep video features, *e.g.* JE [41,42], CE [35] and MMT [19], whilst W2VV++ [28] and SEA [31] show the potential of combining different text features for better query representation. The recent CLIP series [17,36], due to their end-to-end learning paradigm, actually lacks the ability of exploiting existing features. Therefore, even in the era of deep learning, the need for feature fusion remains strong.

Concerning approaches to feature fusion, vector concatenation is commonly used when combining features at an early stage [15,28]. As for late fusion, multiple feature-specific common spaces are learned in parallel, with the resultant similarities combined either by averaging [31,42], empirical weighting [41] or by Mixture of Experts (MoE) ensembles [35]. As the number of features grows, vector concatenation suffers from the curse of dimensionality, while constructing common spaces per feature lacks inter-feature interactions. Moreover, the prior works focus either on the video end or on the text end. To the best of our knowledge, no attempt is made to develop a unified learning-based approach that works for both ends in the context of text-to-video retrieval, see Table 1.

One might consider feature fusion by Multi-head Self-Attention (MHSA), the cornerstone of Transformers [56]. As Fig. 1(a) shows, MHSA transforms a specific feature by blending it with information from all other features, with the blending weights produced by a self-attention mechanism termed QKV. Note that the module was initially developed for NLP tasks, for which exploiting element-wise correlations is crucial for resolving semantic ambiguity. However, as video features extracted by distinct 2D-CNNs and 3D-CNNs are meant for describing the video content from different aspects, we conjecture that optimizing their combination is preferred to modeling their correlations. Moreover, the self-attention in MHSA, computed by $\text{Softmax}((\frac{QK^T}{\sqrt{d_v}})V)$, depends largely on inter-feature correlations. It thus tends to have a group effect that features related to each other will be more attended. Consequently, the related yet relatively weak features will be over-emphasized. Hence, despite its high prevalence in varied contexts, we consider MHSA suboptimal for the current task.

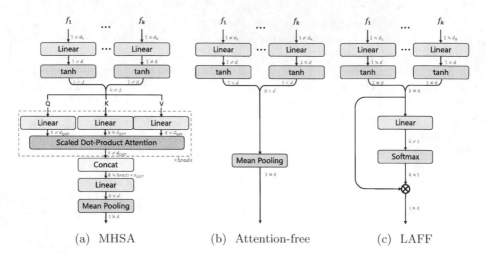

Fig. 1. Three distinct blocks for feature fusion: (a) Multi-Head Self-Attention (MHSA), (b) Attention-free, and (c) our proposed LAFF.

We propose in this paper a much simplified feature fusion block, termed Lightweight Attentional Feature Fusion (LAFF), see Fig. 1(c). LAFF is generic, working for both video and text ends. Video/text features are combined in a convex manner in a specific LAFF block, with the combination weights learned to optimize cross-modal text-to-video matching. Performing fusion at the feature level, LAFF can thus be viewed as an early fusion method. Meanwhile, with the multi-head trick as used in MHSA, multiple LAFFs can be deployed within a single network, with their resultant similarities combined in a late fusion manner. The ability to perform feature fusion at both early and late stages and at both video and text ends makes LAFF a powerful method for exploiting diverse, multi-level (off-the-shelf) features for text-to-video retrieval. In sum, our main contributions are as follows:

- We are the first to study both video-end and text-end feature fusion for text-to-video retrieval. Given the increasing availability of deep vision/language models for feature extraction, this paper presents an effective mean to harness such dark knowledge for tackling the task.
- We propose LAFF, a lightweight feature fusion block, capable of performing fusion at both early and late stages. Compared to MHSA, LAFF is much more compact yet more effective. Its attentional weights can also be used for selecting fewer features, with the retrieval performance mostly preserved.
- Experiments on five benchmarks, *i.e.* MSR-VTT, MSVD, TGIF, VATEX and TRECVID AVS 2016–2020, show that the LAFF-based video retrieval model (Fig. 2) compares favorably against the state-of-the-art, resulting in a strong baseline for text-to-video retrieval. Code is available at GitHub[1].

---

[1] https://github.com/ruc-aimc-lab/laff.

2   Related Work

Feature Fusion for Text-to-Video Retrieval. Previous methods on feature fusion focus either on the video end or on the text end, see Table 1. For video-end feature fusion, earlier works often simply use vector concatenation to merge multiple features in advance to cross-modal representation learning [14]. JE [42] and its journal extension [41] have made an initial attempt to combine diverse video features by late fusion, where multiple feature-specific common spaces are learned. Multi-space similarities are either averaged [42] or linearly combined with empirical weights [41]. CE [35] also uses late fusion, but resorts to a learning based method, *i.e.* Mixture of Experts (MoE), to determine the fusion weights on the fly. MMT [19] improves over CE by first using a multi-modal Transformer to aggregate features from the frame level to the video level, and later using MoE for combining multi-space similarities. JE, CE and MMT all use a single text feature, so these methods leave text-end feature fusion untouched.

As for text-end feature fusion, W2VV++ [28] takes an early fusion approach, combining the output of three text encoders, *i.e.* bag-of-words, word2vec and GRU, by vector concatenation. By contrast, SEA [31] opts for late fusion, first building a common space per text feature and then averaging the similarities computed within the individual spaces. The more recent TEACHTEXT [11] first trains multiple CEs with different text encoders, and later combines these models by late fusion with MoE-predicted weights.

Table 1. Taxonomy of feature fusion methods for text-to-video retrieval. Note that feature fusion is conceptually different from multi-level feature learning, *e.g.* JPoSE [59], PIE-Net [52] and Dual Encoding [15], where new features are first computed at varied levels from a single feature input and combined later. So research in that line is excluded.

Method	Feature modality	Fusion stage	Fusion block
JE, ICMR18 [42]	Video	Late fusion	Average
JE, IJMIR19 [41]	Video	Late fusion	Manual weights
CE, BMVC19 [35]	Video	Late fusion	MoE
MMT, ECCV20 [19]	Video	Hybrid fusion	Multimodal Transformer + MoE
W2VV++, MM19 [28]	Text	Early fusion	Vector concatenation
SEA, TMM21 [31]	Text	Late fusion	Average
TEACHTEXT, ICCV21 [11]	Text	Late fusion	MoE
This work	Text/Video	Hybrid fusion	LAFF

Attentional Feature Fusion in Other Contexts. LAFF is conceptually different from context gating modules [12,40], which aim to re-weight each dimension of a given feature vector, and thus produce weight per dimension. By contrast, as LAFF is to combine multiple features, it produces weight per feature vector. LAFF is technically inspired by attention-based multiple instance learning (MIL) [22], wherein there is a need of aggregating multiple instance-level features into a case-level feature. However, the text-to-video retrieval task differs from MIL in the following two aspects, making the attention-based MIL not directly applicable in the new context. First, instances in a MIL setting are of the same modality, *e.g.* patches taken from the same image [22] or images from an image array [27], so the instance-level features are homogeneous and directly comparable. By contrast, the video or text features to be fused are obtained by distinct feature extractors with varied feature dimensions and are thus incompatible. Second, MIL is typically exploited in the context of a classification task, so the case-level feature is used as input to a classification layer. By contrast, our fused features are meant for cross-modal matching. Hence, a feature fusion block at one end shall be used in pair with a fusion block at the other end. The technical novelty of LAFF lies in its overall design that effectively answers the unique challenges in video/text feature fusion for text-to-video retrieval. While combining homogeneous features have been studied in other contexts [25,34,58], our paper fills the gap between diverse feature fusion and text-to-video retrieval.

3   A New Baseline

We propose trainable feature fusion for both video and text ends. Specifically, suppose we have a specific video x represented by a set of k_1 video-level features, $\{f_{v,1}(x), \ldots, f_{v,k_1}(x)\}$, and a specific textual query q represented by a set of k_2 sentence-level features $\{f_{t,1}(q), \ldots, f_{t,k_2}(q)\}$. We shall construct two feature fusion blocks to encode respectively the video and the query into their d-dimensional cross-modal embeddings $e(x)$ and $e(q)$. Their semantic similarity $s(x,q)$ is measured in terms of the two embeddings accordingly, *i.e.*

$$\begin{cases} e(x) & := fusion_v(\{f_{v,1}(x), \ldots, f_{v,k_1}(x)\}), \\ e(q) & := fusion_t(\{f_{t,1}(q), \ldots, f_{t,k_2}(q)\}), \\ s(x,q) & := similarity(e(x), e(q)). \end{cases} \qquad (1)$$

As such, text-to-video retrieval for the given query q is achieved by sorting all videos in a test collection in light of their $s(x,q)$ in descending order. In what follows, we describe the proposed LAFF as a unified implementation of the *fusion* blocks in Eq. 1, followed by its detailed usage for text-to-video retrieval.

3.1   The LAFF Block

Without loss of generality, we are provided with a diverse set of k different features $\{f_1, \ldots, f_k\}$, sized as $d_1, \ldots, d_k$, respectively. As the features are obtained by distinct extractors and thus incompatible, we shall use a feature transformation layer to rectify the diverse features to be of the same length. To convert the i-th feature to a new d-dimensional feature, we use

$$f_i' = \sigma(Linear_{d_i \times d}(f_i)), \qquad (2)$$

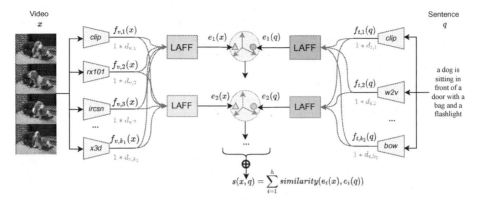

Fig. 2. Conceptual diagram of using paired LAFF blocks for text-to-video retrieval. Given a specific video x, we employ multiple (pre-trained) feature extractors to obtain a set of k_1 video-level features $\{f_{v,1}(x), \ldots, f_{v,k_1}(x)\}$. In a similar manner we extract from a query sentence q a set of k_2 sentence-level features $\{f_{t,1}(q), \ldots, f_{t,k_2}(q)\}$. Each pair of LAFFs determines a common space. In particular, a specific pair of LAFFs, indexed by i, aggregates the video (sentence) features into a d-dimensional cross-modal feature $e_i(x)$ ($e_i(q)$), and consequently computes the video-text similarity $s_i(x,q)$ per space. The sum of the similarities from all h common spaces, $i.e.$ $\sum_{i=1}^{h} s_i(x,q)$, is used for retrieval.

where σ is a nonlinear activation function. As the output of the non-linear activation in LAFF is to calculate the cosine similarity, we use *tanh* in this work[2]. The notation of $Linear_{d_i \times d}$ indicates a fully connected layer with an input of size d_i and an output of size of d. Each input feature has its own $Linear$, optional when d_i equals to d.

Although the transformed features $\{f_i'\}$ are now comparable, they are not equally important for representing the video/text content. We thus consider a weighted fusion, $i.e.$

$$\bar{f} = \sum_{i}^{k} a_i f_i', \qquad (3)$$

with weights $\{a_1, \ldots, a_k\}$ computed by a lightweight attentional layer as follow,

$$\{a_1, \ldots, a_k\} = softmax(Linear_{d \times 1}(\{f_1', \ldots, f_k'\})). \qquad (4)$$

As shown in Fig. 1(b), the Attention-free feature fusion block is a special case of LAFF when enforcing the weights in Eq. 3 to be uniform, $i.e.$ $a_i = \frac{1}{k}$. Compared to Attention-free, LAFF has d more parameters to learn, see Table 2. Such a small amount of extra parameters turn out to be important for improving

---

[2] Other non-linear activations such as ReLU and sigmoid will make each dimension non-negative, constrain the feature space be in the first quadrant, and consequently put a lower boundary of 0 on the cosine similarity. As such, the similarity will be less discriminative than the *tanh* counterpart.

the effectiveness of feature fusion, as our ablation study will show. Compared with MHSA, LAFF has much fewer trainable parameters and is thus more data-efficient. Furthermore, as the attentional weights of LAFF are directly used for a convex combination of the features, LAFF is more interpretable than MHSA.

Table 2. Complexity analysis of feature fusion blocks, with D indicating the overall dimension of the input features and d as the dimension of the output features. FLOPs are computed based on input features of shape 8×2048.

Feature fusion block	Parameters	FLOPs (M)
MHSA	$D \times d + 4 \times d^2$	94.90
Attention-free	$D \times d$	27.78
LAFF	$D \times d + d$	27.80

3.2   Paired LAFFs for Text-to-Video Retrieval

Network Architecture. We now detail the usage of LAFF for text-to-video retrieval. A straightforward solution is to substitute LAFF for the *fusion* functions in Eq. 1. As such, we have a single configuration of how the video/text features are combined. However, due to the high complexity of the video and text contents, we hypothesize that the single configuration is suboptimal for cross-modal representation and matching. Borrowing the multi-head idea of MHSA, we consider multi-head LAFF. In particular, we deploy h pairs of LAFFs, where each pair of LAFFs jointly determine a latent common space for video-text matching. In particular, a specific pair of LAFFs, denoted as $< LAFF_{v,i}, LAFF_{t,i} >$, aggregates the video/text features into a d-dimensional cross-modal embedding vector $e_i(x)/e_i(q)$, *i.e.*

$$\begin{cases} e_i(x) = LAFF_{v,i}(x) \\ e_i(q) = LAFF_{t,i}(q) \\ s_i(x,q) = similarity(e_i(x), e_i(q)) \end{cases} \tag{5}$$

where *similarity* is the widely used cosine similarity. Accordingly, we compute the final video-text similarity as the mean of the h individual similarities,

$$s(x,q) = \frac{1}{h} \sum_{i=1}^{h} s_i(x,q). \tag{6}$$

The overall architecture is illustrated in Fig. 2. In order to make the amount of trainable parameters invariant with respect to h, we set $d = \frac{d_0}{h}$, where d_0 is a constant empirically set to 2,048. As such, the multi-head version of LAFF is not an ensemble. We use $h = 8$, unless otherwise stated.

LAFF for Multi-Level Feature Fusion. So far we presume the features to be fused are already at the video level. In fact, for its high flexibility, LAFF can be extended with ease to a multi-level variant to deal with the situation wherein different frame-level and video-level features coexist. Figure 3 shows this variant, which we term *LAFF-ml.* LAFF-ml works in a bottom-up manner, where a set of specific frame-level features are aggregated via a specific LAFF block to produce a video-level feature. Suppose there are two different frame-level features, *e.g. clip* and *rx101.* Each will have its own LAFF block. The (resultant) different video features are then fused via a video-level LAFF block.

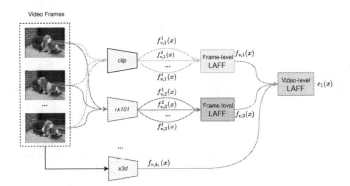

Fig. 3. LAFF-ml for multi-level feature fusion. Frame-level LAFF is applied per feature, *e.g. clip* or *rx101.* The outputs of the frame-level LAFF blocks are later combined (with other video-level features, *e.g. x3d*) by a video-level LAFF.

Network Training. Following the good practice of the previous work, we adopt as our base loss function the triplet ranking loss with hard-negative mining [16]. For a specific sentence q in a given training batch, let x_+ and x_- be videos relevant and irrelevant *w.r.t.* q, and x_-^* be the hard negative that violates the ranking constraint the most. We have

$$\begin{cases} x_-^* & = \mathrm{argmax}_{x_-} \left(s(x_-, q) - s(x_+, q) \right) \\ \mathrm{loss}(q) = \max(0, \alpha + s(x_-^*, q) - s(x_+, q)), \end{cases} \quad (7)$$

where α is a positive hyper-parameter controlling the margin of the ranking loss.

As [31] has documented, when training a cross-modal network that produces multiple similarities, combining losses per similarity gives better results than using a single loss with the combined similarity. Hence, we follow this strategy, computing $\mathrm{loss}_i(q)$, namely the loss in the i-th space by substituting s_i for s in Eq. 7. The network is trained to minimize a combined loss $\sum_{i=1}^h \mathrm{loss}_i(q)$.

4   Experiments

In order to evaluate the effectiveness of LAFF, we conduct a series of experiments. An ablation study is performed on MSR-VTT, a *de facto* benchmark, to evaluate LAFF in multiple aspects. We then compare our LAFF-based retrieval model with the state-of-the-art on MSR-VTT and three other popular benchmarks including MSVD, TGIF and VATEX. In order to assess our retrieval method on a much larger collection, a post-competition evaluation is conducted on the TRECVID AVS benchmark series.

4.1 Common Setups

Implementation Details. Eight video features and five text features are used, see Table 3. The margin α in the loss is set to 0.2 according to VSE++ [16]. We perform SGD based training, with a mini-batch size of 128 and RMSProp as the optimizer. The learning rate is initially set to 10^{-4}, decayed by a factor of 0.99 per epoch. Following [23], we half the learning rate if the validation performance does not increase in three consecutive epochs. Early stop occurs when no validation performance increase is achieved in ten consecutive epochs. The dropout rate of the *Linear* layers is set to 0.2. All experiments were done with PyTorch (1.7.1) [45] on an Nvidia GEFORCE GTX 2080Ti GPU.

Evaluation Criteria. We report three standard rank-based metrics: Recall at Rank N (R@N, N=1, 5, 10), Median rank (Med r), and mean Average Precision (mAP) for assessing the overall ranking quality.

Table 3. Video/text features used in ablation study. Video-level features are obtained by mean pooling over frames or segments, unless otherwise stated.

Feature	Dim.	Short description
Video features:		
rx101	2,048	ResNeXt-101 trained on the full set of ImageNet [39]
re152	2,048	ResNet-152 from the MXNet model zoo
wsl	2,048	ResNeXt-101 pre-trained by weakly supervised learning on 940 million public images, followed by fine-tuning on ImageNet1k [37]
clip	512	CLIP (ViT-B/32) pre-trained on web images and corpus by contrastive learning [48]
c3d	2,048	C3D trained on Kinetics400 [53]
ircsn	2,048	irCSN-152 which trained by weakly supervised learning on IG-65M [20]
tf	768	TimeSformer trained on HowTo100M [7]
x3d	2,048	X3D trained on Kinetics400 [18]
Text features:		
bow	m	m-dimensional Bag-of-words feature, with m being 7,675 (MSR-VTT), 2,916 (MSVD), 3,980 (TGIF), or 10,312 (VATEX)
w2v	500	Word2Vec trained on Flickr tags [14]
gru	1,024	Mean pooling over hidden vectors of GRU trained from scratch [14]
bert	768	The base version of BERT, pre-trained on BooksCorpus and English Wikipedia [13]
clip	512	The same CLIP as used to extract video features

4.2   Ablation Study

Our ablation study is conducted on MSR-VTT [63], which has 10k videos in total, each associated with 20 captions. We adopt the official data split: 6,513 videos for training, 497 videos for validation and the remaining 2,990 images for test. In order to distinguish this data split from other customized splits, *e.g.* JSFusion [67], we term the split **MV-test3k**.

On Combining Diverse Video/Text Features. We investigate how LAFF responds when diverse video/text features are gradually added. For the ease of lateral comparison, we include as baselines the following two models: W2VV++ [28], which simply uses vector concatenation, and SEA [31] which learns cross-modal similarities per text feature.

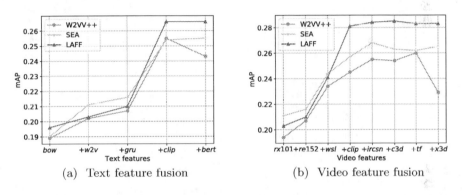

(a)   Text feature fusion   (b)   Video feature fusion

Fig. 4. Performance curves of three distinct models, *i.e.* **W2VV++, SEA and LAFF,** *w.r.t.* (a) text feature fusion, with {*rx101,re152*} as video features, and (b) video feature fusion, with {*bow,w2v,gru*} as text features. LAFF is both effective and stable for fusing diverse features. Data: MV-test3k.

Given the many video and text features investigated in this work, a complete enumeration of video-text feature combinations is impractical. We choose to reduce the computation by only varying the features at one end, with features at the other end fixed. Figure 4(a) shows the performance curves of W2VV++, SEA and LAFF *w.r.t.* text features, with {*rx101, re152*} as their common video features. The performance of all three models improves at the earlier steps when few features are fused. There is a noticeable drop in the performance curve of W2VV++ when *bert* is included. LAFF is more effective and more stable. Similar results can be observed from 4(b), which shows the performance curves of the three models *w.r.t.* video features. The above results justify the effectiveness of LAFF for combining diverse video/text features.

Comparing Feature Fusion Blocks. We compare the three feature fusion blocks by replacing LAFF in Fig. 2 with MHSA and Attention-free, respectively. For a more fair comparison, we also apply the multi-loss trick on MHSA by optimizing losses for different heads, denoted as MHSA(multi-loss). Moreover, we include as a baseline method that uses the simple feature concatenation strategy, as previously adopted in W2VV++ [28]. The performance of text-to-video retrieval with specific feature fusion blocks is reported in Table 4. LAFF performs the best, followed by Attention-free, the concatenation baseline and MHSA. Attention-free, while being extremely simple, is more effective than MHSA for combining the increasing amounts of text features, with its mAP increases from 0.264, 0.321 to 0.326. The superior performance of LAFF against Attention-free (0.358 *versus* 0.326) justifies the necessity of the attentional layer.

Table 4. Comparing feature fusion blocks. The simple feature concatenation used by W2VV++ is taken as a baseline. Numbers in parentheses are relative improvements against this baseline. Video features: all. Data: MV-test3k

Text features	Fusion block	R1	R5	R10	Med r	mAP
bow, w2v, gru	Baseline	14.0	35.9	47.7	12	0.249
	MHSA	11.7	31.9	43.4	15	0.219 (12.0%↓)
	MHSA(mulit-loss)	11.1	30.1	41.1	18	0.207 (16.8%↓)
	Attention-free	15.4	37.8	49.7	11	0.264 (6.0%↑)
	LAFF	**16.0**	**39.5**	**51.4**	**10**	**0.276** (10.8%↑)
bow, w2v, gru,clip	Baseline	19.2	43.5	55.3	8	0.310
	MHSA	18.8	43.0	54.6	8	0.305 (1.6%↓)
	MHSA(mulit-loss)	18.7	43.1	54.5	8	0.305 (1.6%↓)
	Attention-free	20.5	44.8	56.2	7	0.321 (3.5%↑)
	LAFF	**23.7**	**49.1**	**60.6**	**6**	**0.358** (15.5%↑)
bow, w2v, gru, clip, bert	Baseline	14.5	35.0	46.1	13	0.247
	MHSA	17.9	41.6	53.3	9	0.294 (19.0%↑)
	MHSA (mulit-loss)	19.0	43.4	54.9	8	0.306 (23.9%↑)
	Attention-free	20.9	45.3	56.9	7	0.326 (32.0%↑)
	LAFF	**23.8**	**49.0**	**60.3**	**6**	**0.358** (44.9%↑)

LAFF Weights for Model Interpretability and Feature Selection. Figure 5 visualizes the LAFF weights of videos and their associated captions selected from the MV-test3k test set. We observe that 3D-CNN features receive more weight when the video content contains more motions, see Fig. 5(b). For each feature, its weight averaged over samples reflects its contribution to the retrieval performance. The weights of text features in descending order are *clip* (64.3%), *bow* (15.7%), *gru* (9.5%), *w2v* (6.5%), *bert* (4.0%). For video features, the order is *clip* (38.0%), *x3d* (16.8%), *ircsn* (13.3%), *tf* (10.9%), *rx101* (7.0%), *wsl*(6.6%), *c3d* (5.1%), *re152* (1.4%). We re-train our model with the top-3 ranked video/text features. Compared to the full setup (mAP of 0.358), the reduced model obtains mAP of 0.353, meaning a relatively small performance loss of 1.4%. Hence, the LAFF weights are helpful for feature selection.

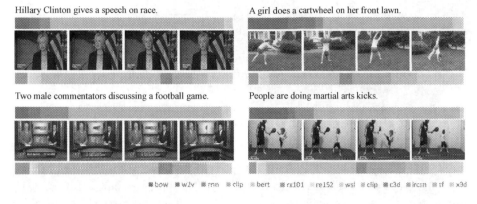

Hillary Clinton gives a speech on race.

A girl does a cartwheel on her front lawn.

Two male commentators discussing a football game.

People are doing martial arts kicks.

Fig. 5. Visualization of LAFF weights per feature, with samples from the MV-test3k test set. Green, brown, and blue mean text features, 2D video features and 3D video features, respectively. Best viewed in color. (Color figure online)

Table 5. Combined loss *versus* single loss. Video features: all.

Text features	Loss	R1	R5	R10	Med r	mAP
bow, w2v, gru	Single	14.1	36.2	47.8	12	0.250
	Combined	**16.0**	**39.5**	**51.4**	**10**	**0.276** (10.4%↑)
bow, w2v, gru, clip	Single	20.6	45.0	56.5	7	0.324
	Combined	**23.7**	**49.1**	**60.6**	**6**	**0.358** (10.5%↑)

Table 6. Effect of the number of common spaces h. Features: all.

h	R1	R5	R10	Med r	mAP
1	23.1	48.3	59.9	6	0.352
2	23.1	48.3	60.1	6	0.352
4	23.5	49.1	60.4	6	0.356
8	**23.7**	**49.1**	**60.6**	**6**	**0.358**
16	23.5	48.8	60.4	6	0.356

Combined Loss *versus* Single Loss. As Table 5 shows, LAFF trained with the combined loss produces a relative improvement of over 10% in terms of mAP, when compared to its single-loss counterpart.

The Effect of the Number of Common Spaces. Concerning the number of common spaces h, we try different values, *i.e.* {1, 2, 4, 8, 16}. As shown in Table 6, the performance improves as h increases, with the peak performance reached at $h = 8$. While using a larger h is beneficial, the relatively small gap between LAFF($h=1$) and LAFF($h=8$) suggests that the good performance of LAFF-based video retrieval is largely contributed by the LAFF block rather than the multi-space similarity.

To reveal how different are the embedding spaces to each other, we compute the Jaccard index between the top-5 video retrieval results of the individual spaces *w.r.t.* a specific query caption. The inter-space Jaccard index is lower than 0.5, suggesting sufficient divergence. Nevertheless, whether videos/captions have been separated along different axes needs further investigation.

4.3   Comparison with SOTA on Video Description Datasets

Datasets. We further include MSVD [9], TGIF [32] and VATEX [57]. For MSVD and TGIF, we follow their official data splits. For VATEX, we follow the data split as used in HGR [10]. As for MSR-VTT, in addition to the official

MV-test3k split, we also report performance on another popular data split [67], with 9k videos for training and 1k for test. We term this split *MV-test1k*.

Baselines. We compare with the SOTA that uses the same data splits as we have mentioned. In particular, the following published models are included: JE [41], W2VV++ [28], CE [35], TCE [65], HGR [10], SEA [31], MMT [19], DE [15], SSB [46], SSML [1], CLIP [47], CILP-FRL [8], and CLIP2Video (two-tower version[3]) [17]. In addition, we finetune CLIP per dataset, termed as CLIP-FT. The video/text feature extracted by CLIP-FT is denoted as *clip-ft*.

Note that the video/text features used in the above models vary. For a head-to-head comparison, we re-train LAFF, LAFF-ml, JE, W2V++, SEA and MMT using the same set of selected features[4]. Other baselines are not re-run, as they cannot handle the diverse video/text features without proper re-engineering. Since the fusion weights in JE have to be manually specified, we tried with three choices, *i.e.* JE with uniform weights, JE (0.8 for *clip-ft*) which assigns a much larger weight of 0.8 to the best *clip-ft* feature (the remains features have equal weights) and JE (0.9 for *clip-ft*). When comparing with CLIP2Video, we let LAFF include the global video/text features extracted by CLIP2Video to see whether LAFF can flexibly harness new and more powerful features.

Results. The performance of the different models on the multiple benchmarks is summarized in Table 7. Note that due to the inclusion of the better *clip-ft* feature, the performance is better than that reported in the ablation study. The baselines (JE, W2VV++ , SEA and MMT) get even worse results than using a

Table 7. Comparison with the state-of-the-art on four benchmark datasets, *i.e.* MSR-VTT (MV-test3k/MV-test1k), MSVD, TGIF and VATEX. Baselines which use video/text features different from ours and thus not directly comparable are provided in the supplement.

Model	MV-test3k			MV-test1k			MSVD			TGIF			VATEX		
	R1	R5	R10	R1	R5	R10	R1	R5	R10	R1	R5	R10	R1	R5	R10
CLIP-FT (*this paper*)	27.7	53.0	64.2	39.7	67.8	78.4	44.6	74.7	84.1	21.5	40.6	49.9	53.3	87.5	94.0
The same video and text feature as ours															
JE [41] (uniform weights)	21.2	46.5	58.4	36.0	65.9	76.4	35.9	71.0	81.8	18.7	37.5	47.1	50.2	88.7	95.4
JE (0.8 for *clip-ft*)	26.1	51.7	63.3	41.2	73.2	82.5	39.4	69.9	79.4	21.7	41.3	50.9	54.1	89.0	95.0
JE (0.9 for *clip-ft*)	25.9	51.4	63.0	40.9	72.7	82.1	38.8	69.7	78.9	21.3	40.9	50.3	53.5	88.3	94.6
W2VV++ [28]	23.0	49.0	60.7	39.4	68.1	78.1	37.8	71.0	81.6	22.0	42.8	52.7	55.8	91.2	96.0
SEA [31]	19.9	44.3	56.5	37.2	67.1	78.3	34.5	68.8	80.5	16.4	33.6	42.5	52.4	90.2	95.9
MMT [19]	24.9	50.5	62.0	39.5	68.3	78.3	40.6	72.0	81.7	22.1	42.2	51.7	54.4	89.2	95.0
LAFF	28.0	53.8	64.9	42.2	70.7	**81.2**	45.2	75.8	84.3	24.1	44.7	54.3	57.7	91.3	95.9
LAFF-ml	**29.1**	**54.9**	**65.8**	**42.6**	**71.8**	81.0	**45.4**	**76.0**	**84.6**	**24.5**	**45.0**	**54.5**	**59.1**	**91.7**	**96.3**
Comparison with arXiv SOTA															
CLIP2Video [17]	n.a	n.a	n.a	44.5	71.3	80.6	44.7	74.8	83.7	n.a	n.a	n.a	54.8	89.1	95.1
LAFF	n.a	n.a	n.a	**45.8**	**71.5**	**82.0**	**45.4**	**75.5**	**84.1**	n.a	n.a	n.a	**58.3**	**91.7**	**96.3**

---

[3] We prefer the two-tower version to the single-tower version, as the latter has to compute video and text embeddings online, making it not scalable for real applications.

[4] Video features: *clip-ft*, *x3d*, *ircsn* and *tf*. Text features: *clip-ft*, *bow*, *w2v* and *gru*.

single feature (*clip-ft*). The result suggests that one cannot take for granted that adding better features will yield better performance, and an intellectual design of feature fusion is needed. The proposed LAFF consistently performs the best on all the test sets. LAFF-ml outperforms LAFF, which shows that flexible use of LAFF in multiple levels can further improve performance.

4.4   Comparison with SOTA on TRECVID AVS 2016–2020

Setup. The test collection for TRECVID AVS 2016–2018 (TV16/TV17/TV18) is IACC.3 [44] with 335,944 video clips. The test collection for TRECVID AVS 2019–2020 (TV19/TV20) is V3C1 [6] with 1,082,659 video clips. We use the official metric, *i.e.* inferred Average Precision (infAP) [66].

Baselines. Due to the prominent performance of CLIP-FT and CLIP2Video as shown in Sect. 4.3, we again compare with the two models. Since the top-3 ranked solutions of the AVS evaluation naturally reflect the state of-the-art, we include them as well. CLIP2Video was trained on the MV-test1k split. So for a fair comparison, we train LAFF and CLIP-FT on the same split.

Table 8. State-of-the-art performance on TRECVID AVS 2016–2020.

Model	TV16	TV17	TV18	TV19	TV20	MEAN
Rank 1	0.054 [24]	0.206 [51]	0.121 [26]	0.163 [62]	**0.354** [68]	n.a
Rank 2	0.051 [38]	0.159 [54]	0.087 [21]	0.160 [29]	0.269 [30]	n.a
Rank 3	0.040 [33]	0.120 [43]	0.082 [5]	0.123 [55]	0.229 [61]	n.a
CLIP2Video	0.176	0.229	0.114	0.176	0.207	0.180
CLIP-FT	0.191	0.215	0.105	0.147	0.203	0.172
LAFF	0.211	0.285	0.137	**0.192**	0.265	**0.218**
LAFF-ml	**0.222**	**0.290**	**0.147**	0.181	0.245	0.217

Results. As shown in Table. 8, LAFF performs the best on TV16–TV19. Note that the top performer of TV20 was trained on the joint set of MSR-VTT, TGIF and VATEX. Re-training LAFF on this larger dataset results in infAP of 0.358, marginally better than the top performer. We also conduct a case study on TV20, which shows that LAFF outperforms the CLIP series for action related queries with a large margin, see supplementary materials. We attribute this result to the fact that LAFF integrates 3D-CNN features (*ircsn*, *c3d* and *tf*), which were designed to capture action and motion information in the video content.

5   Conclusions

For video retrieval by text, we propose LAFF, an extremely simple feature fusion block. LAFF is more effective than Multi-head Self-Attention, yet with much fewer parameters. Moreover, the attentional weights produced by LAFF can be used to explain the contribution of the individual video/text features for cross-modal matching. Consequently, the weights can be used for feature selection for building a more compact video retrieval model. Our LAFF-based video retrieval model surpasses the state-of-the-art on MSR-VTT, MSVD, TGIF, VATEX and TRECVID AVS 2016–2020. Given the increasing availability of (deep) video/text features, we believe our work opens up a promising avenue for further research.

Acknowledgments. This work was supported by NSFC (No. 62172420, No. 62072463), BJNSF (No. 4202033), and Public Computing Cloud, Renmin University of China.

References

1. Amrani, E., Ben-Ari, R., Rotman, D., Bronstein, A.: Noise estimation using density estimation for self-supervised multimodal learning. In: AAAI (2021)
2. Atrey, P.K., Hossain, M.A., El Saddik, A., Kankanhalli, M.S.: Multimodal fusion for multimedia analysis: a survey. Multimedia Syst. **16**, 345–379 (2010)
3. Bain, M., Nagrani, A., Varol, G., Zisserman, A.: Frozen in time: A joint video and image encoder for end-to-end retrieval. In: ICCV (2021)
4. Baltrušaitis, T., Ahuja, C., Morency, L.P.: Multimodal machine learning: a survey and taxonomy. TPAMI **41**(2), 423–443 (2018)
5. Bastan, M., et al.: NTU ROSE lab at TRECVID 2018: Ad-hoc video search and video to text. In: TRECVID (2018)
6. Berns, F., Rossetto, L., Schoeffmann, K., Beecks, C., Awad, G.: V3C1 dataset: An evaluation of content characteristics. In: ICMR (2019)
7. Bertasius, G., Wang, H., Torresani, L.: Is space-time attention all you need for video understanding? In: ICML (2021)
8. Chen, A., Hu, F., Wang, Z., Zhou, F., Li, X.: What matters for ad-hoc video search? a large-scale evaluation on TRECVID. In: ICCV Workshop on ViRal (2021)
9. Chen, D., Dolan, W.: Collecting highly parallel data for paraphrase evaluation. In: CVPR (2011)
10. Chen, S., Zhao, Y., Jin, Q., Wu, Q.: Fine-grained video-text retrieval with hierarchical graph reasoning. In: CVPR (2020)
11. Croitoru, I., et al.: TEACHTEXT: Crossmodal generalized distillation for text-video retrieval. In: ICCV (2021)
12. Dauphin, Y.N., Fan, A., Auli, M., Grangier, D.: Language modeling with gated convolutional networks. In: ICML (2017)
13. Devlin, J., Chang, M., Lee, K., Toutanova, K.: BERT: pre-training of deep bidirectional transformers for language understanding. In: NAACL-HLT (2019)
14. Dong, J., Li, X., Snoek, C.G.: Predicting visual features from text for image and video caption retrieval. TMM **20**(12), 3377–3388 (2018)
15. Dong, J., Li, X., Xu, C., Yang, X., Yang, G., Wang, X.: Dual encoding for video retrieval by text. TPAMI (2021)

16. Faghri, F., Fleet, D.J., Kiros, J.R., Fidler, S.: VSE++: improving visual-semantic embeddings with hard negatives. In: BMVC (2018)
17. Fang, H., Xiong, P., Xu, L., Chen, Y.: CLIP2Video: Mastering video-text retrieval via image clip. arXiv preprint arXiv:2106.11097 (2021)
18. Feichtenhofer, C.: X3d: Expanding architectures for efficient video recognition. In: CVPR (2020)
19. Gabeur, V., Sun, C., Alahari, K., Schmid, C.: Multi-modal transformer for video retrieval. In: ECCV (2020)
20. Ghadiyaram, D., Tran, D., Mahajan, D.: Large-scale weakly-supervised pre-training for video action recognition. In: CVPR (2019)
21. Huang, P.Y., Liang, J., Vaibhav, V., Chang, X., Hauptmann, A.: Informedia@TRECVID 2018: Ad-hoc video search with discrete and continuous representations. In: TRECVID (2018)
22. Ilse, M., Tomczak, J., Welling, M.: Attention-based deep multiple instance learning. In: ICML (2018)
23. Joulin, A., van der Maaten, L., Jabri, A., Vasilache, N.: Learning visual features from large weakly supervised data. In: ECCV (2016)
24. Le, D.D., et al.: NII-HITACHI-UIT at TRECVID 2016. In: TRECVID (2016)
25. Li, H., Chen, J., Hu, R., Yu, M., Chen, H., Xu, Z.: Action recognition using visual attention with reinforcement learning. In: MMM (2019)
26. Li, X., Dong, J., Xu, C., Cao, J., Wang, X., Yang, G.: Renmin University of China and Zhejiang Gongshang University at TRECVID 2018: Deep Cross-Modal Embeddings for Video-Text Retrieval. In: TRECVID (2018)
27. Li, X., et al.: Deep multiple instance learning with spatial attention for rop case classification, instance selection and abnormality localization. In: ICPR (2020)
28. Li, X., Xu, C., Yang, G., Chen, Z., Dong, J.: W2VV++: Fully deep learning for ad-hoc video search. In: ACMMM (2019)
29. Li, X., et al.: Renmin University of China and Zhejiang Gongshang University at TRECVID 2019: Learn to search and describe videos. In: TRECVID (2019)
30. Li, X., Zhou, F., Chen, A.: Renmin University of China at TRECVID 2020: Sentence Encoder Assembly for Ad-hoc Video Search. In: TRECVID (2020)
31. Li, X., Zhou, F., Xu, C., Ji, J., Yang, G.: SEA: sentence encoder assembly for video retrieval by textual queries. TMM **23**, 4351–4362 (2021)
32. Li, Y., et al.: TGIF: A new dataset and benchmark on animated gif description. In: CVPR (2015)
33. Liang, J., et al.: Informedia @ TRECVID 2016. In: TRECVID (2016)
34. Liu, M., Chen, X., Zhang, Y., Li, Y., Rehg, J.M.: Attention distillation for learning video representations. In: BMVC (2020)
35. Liu, Y., Albanie, S., Nagrani, A., Zisserman, A.: Use what you have: Video retrieval using representations from collaborative experts. In: BMVC (2019)
36. Luo, H., et al.: CLIP4Clip: An empirical study of clip for end to end video clip retrieval. arXiv preprint arXiv:2104.08860 (2021)
37. Mahajan, D., et al.: Exploring the limits of weakly supervised pretraining. In: ECCV (2018)
38. Markatopoulou, F., et al.: ITI-CERTH participation in TRECVID 2016. In: TRECVID (2016)
39. Mettes, P., Koelma, D.C., Snoek, C.G.M.: Shuffled ImageNet banks for video event detection and search. TOMM **16**(2), 1–21 (2020)
40. Miech, A., Laptev, I., Sivic, J.: Learning a text-video embedding from incomplete and heterogeneous data. arXiv (2018)

41. Mithun, N.C., Li, J., Metze, F., Roy-Chowdhury, A.K.: Joint embeddings with multimodal cues for video-text retrieval. Int. J. Multimedia Inf. Retrieval **8**(1), 3–18 (2019). https://doi.org/10.1007/s13735-018-00166-3
42. Mithun, N.C., Li, J., Metze, F., Roy-Chowdhury, A.K.: Learning joint embedding with multimodal cues for cross-modal video-text retrieval. In: ICMR (2018)
43. Nguyen, P.A., et al.: Vireo @ TRECVID 2017: Video-to-text, ad-hoc video search and video hyperlinking. In: TRECVID (2017)
44. Over, P., Awad, G., Smeaton, A.F., Foley, C., Lanagan, J.: Creating a web-scale video collection for research. In: The 1st Workshop on Web-scale Multimedia Corpus (2009)
45. Paszke, A.,et al.: Pytorch: An imperative style, high-performance deep learning library. In: NeurIPS (2019)
46. Patrick, M., et al.: Support-set bottlenecks for video-text representation learning. In: ICLR (2021)
47. Portillo-Quintero, J.A., Ortiz-Bayliss, J.C., Terashima-Marín, H.: A straightforward framework for video retrieval using CLIP. In: MCPR (2021)
48. Radford, A., et al.: Learning transferable visual models from natural language supervision. In: ICML (2021)
49. Smeulders, A.W.M., Worring, M., Santini, S., Gupta, A., Jain, R.C.: Content-based image retrieval at the end of the early years. TPAMI **22**(12), 1349–1380 (2000)
50. Snoek, C.G.M., Worring, M.: Multimodal video indexing: a review of the state-of-the-art. Multimedia Tools Appl. **25**(1), 5–35 (2005)
51. Snoek, C.G., Li, X., Xu, C., Koelma, D.C.: University of Amsterdam and Renmin university at TRECVID 2017: Searching video, detecting events and describing video. In: TRECVID (2017)
52. Song, Y., Soleymani, M.: Polysemous visual-semantic embedding for cross-modal retrieval. In: CVPR (2019)
53. Tran, D., Bourdev, L.D., Fergus, R., Torresani, L., Paluri, M.: Learning spatiotemporal features with 3D convolutional networks. In: ICCV (2015)
54. Ueki, K., Hirakawa, K., Kikuchi, K., Ogawa, T., Kobayashi, T.: Waseda_Meisei at TRECVID 2017: Ad-hoc video search. In: TRECVID (2017)
55. Ueki, Kazuya an Hori, T., Kobayashi, T.: Waseda_Meisei_SoftBank at TRECVID 2019: Ad-hoc video search. In: TRECVID (2019)
56. Vaswani, A., et al.: Attention is all you need. In: NeurIPS (2017)
57. Wang, X., Wu, J., Chen, J., Li, L., Wang, Y.F., Wang, W.Y.: VATEX: A large-scale, high-quality multilingual dataset for video-and-language research. In: ICCV (2019)
58. Woo, S., Park, J., Lee, J., Kweon, I.S.: CBAM: Convolutional block attention module. In: ECCV (2018)
59. Wray, M., Larlus, D., Csurka, G., Damen, D.: Fine-grained action retrieval through multiple parts-of-speech embeddings. In: ICCV (2019)
60. Wu, J., Ngo, C.W.: Interpretable embedding for ad-hoc video search. In: ACMMM (2020)
61. Wu, J., Nguyen, P.A., Ngo, C.W.: Vireo@ TRECVID 2020 ad-hoc video search. In: TRECVID (2020)
62. Wu, X., Chen, D., He, Y., Xue, H., Song, M., Mao, F.: Hybrid sequence encoder for text based video retrieval. In: TRECVID (2019)
63. Xu, J., Mei, T., Yao, T., Rui, Y.: MSR-VTT: A large video description dataset for bridging video and language. In: CVPR (2016)
64. Xue, X., Nie, F., Wang, S., Chang, X., Stantic, B., Yao, M.: Multi-view correlated feature learning by uncovering shared component. In: AAAI (2017)

65. Yang, X., Dong, J., Cao, Y., Wang, X., Wang, M., Chua, T.S.: Tree-augmented cross-modal encoding for complex-query video retrieval. In: SIGIR (2020)
66. Yilmaz, E., Aslam, J.A.: Estimating average precision with incomplete and imperfect judgments. In: CIKM (2006)
67. Yu, Y., Kim, J., Kim, G.: A joint sequence fusion model for video question answering and retrieval. In: ECCV (2018)
68. Zhao, Y., Song, Y., Chen, S., Jin, Q.: RUC_AIM3 at TRECVID 2020: Ad-hoc video search & video to text description. In: TRECVID (2020)

Modality Synergy Complement Learning with Cascaded Aggregation for Visible-Infrared Person Re-Identification

Yiyuan Zhang[1] , Sanyuan Zhao[1(✉)] , Yuhao Kang[1] ,
and Jianbing Shen[1,2]

[1] School of Computer Science, Beijing Institute of Technology, Beijing, China
zhaosanyuan@bit.edu.cn
[2] SKL-IOTSC, Department of Computer and Information Science,
University of Macau, Taipa, Macau

Abstract. Visible-Infrared Re-Identification (VI-ReID) is challenging in image retrievals. The modality discrepancy will easily make huge intra-class variations. Most existing methods either bridge different modalities through modality-invariance or generate the intermediate modality for better performance. Differently, this paper proposes a novel framework, named Modality Synergy Complement Learning Network (MSCLNet) with Cascaded Aggregation. Its basic idea is to synergize two modalities to construct diverse representations of identity-discriminative semantics and less noise. Then, we complement synergistic representations under the advantages of the two modalities. Furthermore, we propose the Cascaded Aggregation strategy for fine-grained optimization of the feature distribution, which progressively aggregates feature embeddings from the subclass, intra-class, and inter-class. Extensive experiments on SYSU-MM01 and RegDB datasets show that MSCLNet outperforms the state-of-the-art by a large margin. On the large-scale SYSU-MM01 dataset, our model can achieve 76.99% and 71.64% in terms of Rank-1 accuracy and mAP value. Our code will be available at https://github.com/bitreidgroup/VI-ReID-MSCLNet.

Keywords: VI-ReID · Modality Synergy · Cascaded Aggregation

1 Introduction

Person re-identification (ReID) is a technique that retrieves a specific person in the gallery set shot by non-overlapping cameras [5,16,40,51]. The advancement of ReID plays an important role in smart city infrastructure and public security from the perspective of intelligent surveillance systems [11,20,28]. With the increasing demands for public security, surveillance systems are expected

Supplementary Information The online version contains supplementary material available at https://doi.org/10.1007/978-3-031-19781-9_27.

  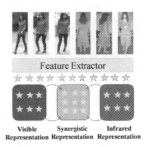

(a) **Modal Characteristics** (b) **Existing Methods**   (c) **Our Method**

Fig. 1. Idea Illustration (a) shows that, visible images usually contain discriminative fine-grained semantics, more noise and infrared images contain more similar semantics, less noise. (b) and (c) show that Synergistic features contain rich information about identities.

the ability to retrieve specific people precisely day and night. A technological requirement for Visible Infrared Person Re-Identification (VI-ReID) arises from it. In contrast to visible person ReID [4,56], VI-ReID faces huge intra-class variations mainly due to the discrepancy between visible and infrared modalities. The modality discrepancy derives from properties of lights consisting of distinct wavelengths. Yet, their images are equivalently parsed as numerical matrices. Near-infrared is smoother and loses texture details due to longer wavelengths and more scattering. It becomes much more agnostic to skin color, albedo, and illumination. Similar texture, scatter, and color can represent different semantics. Besides, it is also difficult to ensure the perspectives of camera shooting, clothing of pedestrians, occlusion, and so on. These factors all contribute to a huge challenge in VI-ReID.

To address the aforementioned difficulties, most of the existing methods chiefly pay attention to learning modality-invariance to bridge the gap between visible and infrared images [9–11,49] or generating images of intermediate or the opposite modality for person retrieval [17,41]. However, GAN-based methods usually suffer from computational complexity and noise introduction. Unfortunately, pursuing modality-invariance may cause the networks to overlook feature properties of semantic diversity, as well as loss of identity discrimination.

Differently, we consider the distinct representations and the semantic diversity between visible and infrared modalities. The success of visible person ReID validates that visible features are always discriminative enough to a large number of identities. Infrared cameras tend to capture thermal objects rather than non-thermal objects. The thermal sensitivity results in semantic loss and filtering of background noise. Infrared images represent relatively stable about the same identity and are comparatively immune to noise. Therefore, we conclude that *synergizing visible identity-discrimination and infrared noise-immunity can build noise-robust and retrieval-efficient representations for VI-ReID by learning homogeneous semantic discrimination and complementary characteristics across modalities as shown in* Fig. 1.

Furthermore, traditional approaches to hard sample mining and feature-aggregation optimize distances of feature embeddings on the instance level. This kind of coarse-grained metric learning neglects the comprehensive distribution of all instances. We target to optimize on the different levels organized in cascaded manner. The basic idea is to subdivide instances of each identity into several subclasses according to the same shooting cameras. Instances in each subclass are much easier to aggregate, whose feature embeddings have a higher intra-class similarity. In this way, we can constrain distances between feature embeddings step-by-step.

Hence, we propose a novel framework, namely, Modality Synergy Complement Learning Network (MSCLNet). It aims at reducing the intra-class variations and boosting representations of identities discrimination. Firstly, it retains the intrinsic semantic diversity and identity relevance from visible and infrared modalities by constructing a synergistic representation with the Modality Synergy module (MS). Then, it enhances the synergistic representations by the specific advantages of the two modalities as shown in Fig. 2. MC contains these two parallel complementary processes with visible and infrared representations. On one hand, it provides guidance of fine-grained and discriminative features from the visible modality. On the other, it supplies global pedestrian statistics from the infrared modality. MS and MC greatly improve the capability of the network to represent identities across modalities. In addition, we propose the Cascaded Aggregation strategy (CA) to optimize the distribution of feature embeddings. It progressively aggregates samples into sub-class, intra-class, and inter-identities. In a cascaded manner, instance belonging to the same identities are lean to aggregation, and instances belonging to different identities are mapped to dispersion.

In conclusion, the main contributions of our work can be summarized as follows: We propose a novel framework named Modality Synergy Complement Learning Network (MSCLNet) with Cascaded Aggregation for VI-ReID. To fetch more discriminative semantics, it learns enhanced feature representations by diverse semantics and specific advantages of visible and infrared modalities. And we propose a Modality Synergy module (MS) which innovatively mines the modality-specific diverse semantics and a Modality Complement module (MC) which further enhances the feature representations by two parallel guidances of modality-specific advantages. They provide a reference for further high-level identity representation. Then we design a Cascaded Aggregation strategy (CA) to optimize the distribution of feature embeddings on a fine-grained level. It progressively aggregates the overall instances in a cascaded manner and enhances the discrimination of identities. Extensive experimental results show that our proposed framework outperforms the state-of-the-art methods by a large margin on two mainstream benchmarks of VI-ReID.

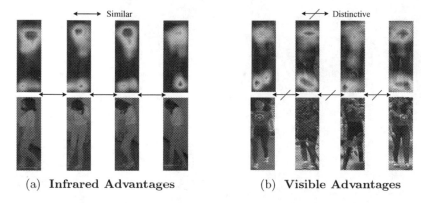

(a) **Infrared Advantages**   (b) **Visible Advantages**

Fig. 2. Demonstration of Infrared and Visible advantages. Infrared images contain similar semantics and their feature embeddings are easier to aggregate. Visible images contain distinctive semantics even they describes the same person.

2   Related Work

Single-Modality Person Re-Identification retrieves pedestrians in the set of visible images. Visible person ReID is a reliable technique which plays an important role in daily life. These methods mainly solved the single-modal ReID problem via ranking [2,29], local and global attention [38,57], camera style [3,55,59], person key-points [36], siamese network [58], similarity graph [22], network architecture searching [18], .etc.. Some works attempted domain adaptation [8,59]. And Some research dealt with the misalignment of human parts, such as cascaded convolutional module [39], refined part pooling [34], transformer [19] and so on. Beside, single-modality person re-identification contains several subdivided areas, for example, video person re-identification [26,44,60], unsupervised person re-identification which tackles pseudo labels [46,54], unsupervised domain adaption [1,31] and generalized person re-identification [16]. Due to the tremendous discrepancy between visible and infrared images, single-modal solutions are not suitable for cross-modality person re-identification, which creates a demand for the development of VI-ReID solutions.

Visible-Infrared Person Re-Identification focuses on narrowing the gap between visible and infrared modalities and learning appropriate representations for pedestrian retrieval across modalities. [43] proposed a deep zero-fill network to extract useful embedded features to reduce cross-modal variation. Dual-stream networks [21,48–51] simultaneously learned modal-shared and modal-specific features. [30] used Gaussian-based variational auto-encoder to distinguish the subspace of cross-modal features. [15] exploited samples similarity within modalities. A modality-aware learning approach [47] processed modality differences on the classifier level. Some works generated images of intermediate or the corresponding modality [7,17,35,37,40] to mitigate the effect of modality discrepancy. However, extracting modality-shared features causes the loss

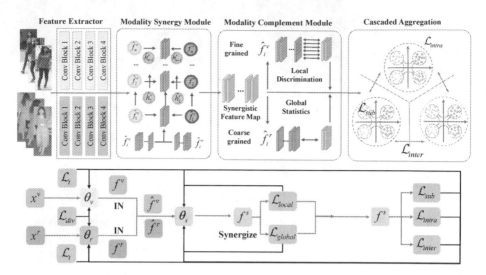

Fig. 3. Illustration of our MSCLNet. The images of visible and infrared modalities are fed into convolution blocks for visible and infrared representations. We synergize the single-modal features and complement synergistic features. Then, we design the Cascaded Aggregation strategy to fine-grained and progressively enhance feature embeddings.

of semantics related to identity discrimination, and GAN-based methods bring computational burden and non-original noise.

Differently, our work pays more attention to deep supervised knowledge synergy [32], which explores explicit information interaction between the supervised branches. We propose to make the most use of the intrinsic information of visible and infrared modalities, which learns diverse semantics and enhances feature representations by a modality synergy and complement learning scheme. To better discriminate identities, we introduce a cascaded feature aggregation strategy.

3   Modality Synergy Complement Learning

In this section, we formulate the VI-ReID problem and introduce the framework of our proposed MSCLNet (Sect. 3.1). It mainly contains three major components: Modality Synergy module (MS, Sect. 3.2), Modality Complement module (MC, Sect. 3.3), and Cascaded Aggregation strategy (CA, Sect. 3.4). We utilize MS to synergize modality-specific diverse semantics from the extractors, and then use MC to enhance feature representations under the guidance of advantages from the two modalities. To optimize the distribution of the features and aggregate instances of the same identity, we exploit CA to constrain the feature distribution in a fine-grained and progressive way. Finally, we summarize the proposed loss function (Sect. 3.5).

3.1   Problem Formulation

We take $\mathcal{V} = \{x_i^v | x_i^v \in \mathcal{V}\}$ and $\mathcal{R} = \{x_i^r | x_i^r \in \mathcal{R}\}$ to denote visible and infrared images, respectively. $\mathcal{Y}_v = \{y_i^v | x_i^v \in \mathcal{V}\}$ and $\mathcal{Y}_r = \{y_i^r | x_i^r \in \mathcal{R}\}$ indicates the corresponding identity labels. Given a query person image x_Q^v or x_Q^r, VI-ReID aims to retrieve the most precise result in the gallery set x_G^r or x_G^v. Existing methods extract modality-shared features at the cost of discarding modality-specific semantics of diversity which can well depict the person. Therefore, we take these intrinsic diverse semantics and the special advantages of each modality into consideration, to learn more precise and better discriminative representation for identities.

Figure 3 illustrates the framework of Modality Synergy Complement Learning Network (MSCLNet) with Cascaded Aggregation. It adopts a dual-stream network as the feature extractor. Firstly, based on the extracted feature representations f^v and f^r from visible and images, MSCLNet constructs synergistic representations f^s by constraining the diversity of the feature distributions between the two modalities. The synergistic feature will be further enhanced by modality complement guidance. The visible modality provides fine-grained discriminative semantics, while the infrared modality supplies with stable global pedestrian statistics. Then we aggregate feature embeddings of the same class via Cascaded Aggregation strategy which optimizes the comprehensive distribution of feature embeddings progressively on three aspects.

3.2   Modality Synergy Module

According to the differences in imaging principles and the heterogeneity of the image contents, visible and infrared images reveal quite different semantics to depict the same person. In our work, we design the network to learn and synergize the diverse semantics of the two modalities. Given a pair of visible and infrared images $x_i^v \in \mathcal{V}$, $x_i^r \in \mathcal{R}$, the dual-stream network extracts their features f_i^v and f_i^r. With the prerequisite of precise pedestrian re-identification, we concentrate on acquiring the semantic diversity to the largest extent. Features f_i^v and f_i^r are normalized by the following operations.

$$\hat{f}_i^v = \frac{f_i^v - \mathrm{E}\left[f_i^v\right]}{\sqrt{\mathrm{Var}\left[f_i^v\right] + \epsilon^v}} \times \gamma + \beta, \mathrm{E}\left[f_i^v\right] = \frac{1}{HW} \sum_{l=1}^{W} \sum_{m=1}^{H} f_{itlm}, \tag{1}$$

where $\mathrm{Var}\left[f_i^v\right] = \frac{1}{HW} \sum_{l=1}^{W} \sum_{m=1}^{H} \left(f_{itlm} - \mathrm{E}\left[f_i^v\right]\right)^2$ are calculated per-dimension separately for each instance in a mini-batch. Let $\mathcal{S}(\cdot)$ indicate the Modality Synergy module to construct synergistic feature f_i^s with label y_i on the basis of f_i^v, f_i^r:

$$f_i^s = \mathcal{S}(\hat{f}_i^v, \hat{f}_i^r, y_i, \theta_s), \tag{2}$$

where θ_s acts as parameters of the Modality Synergy module $\mathcal{S}(\cdot)$. We utilize Mogrifier LSTM [25] as a synergistic feature encoder to maximize the effect of modality synergy learning, and the synergistic feature f_i^s is encoded with visible and infrared features with their shared ground-truth label. To construct f_i^s with diverse semantics, we exploit KL-Divergence to constrain the logistic distribution of visible and infrared features f_i^v, f_i^r, which can be formulated as follows:

$$\mathcal{L}_{div} = -\text{KL}(\hat{f}^v \parallel \hat{f}^r) = -\frac{1}{N} \sum_{i=1}^{N} (\hat{f}_i^v \cdot \log \frac{\hat{f}_i^v}{\hat{f}_i^r}, \theta_v, \theta_r), \tag{3}$$

where N denotes the number of samples in a batch. θ_v and θ_r act as learned feature extractors of visible and infrared modalities respectively, which aim to maximize the diversity of semantic representation across modalities. f^v and f^r are firstly designed in the representation spaces to maximize the modality-specific discrimination among identities. Then, the synergistic feature extractor θ_s projects $\hat{f}_i^v$, $\hat{f}_i^r$ to a shared representation space and constructs synergistic features f_i^s.

Furthermore, we constrain the diverse semantics by identity-relevance, which introduces cross entropy constraining the logistic probability of visible and infrared features p_i^v and p_i^r and the ground truth label y_i.

$$\mathcal{L}_t = -\frac{1}{N} \sum_{i=1}^{N} [\hat{y}_i \cdot \log \hat{p}_i^v(\hat{f}_i^v, \theta_v)] - \frac{1}{N} \sum_{i=1}^{N} [\hat{y}_i \cdot \log \hat{p}_i^r(\hat{f}_i^r, \theta_r)] \tag{4}$$

where λ_{div} and λ_t are hype-parameters to balance the contributions of individual loss terms. The optimization processes of θ_v, θ_r separately track the gradient of $(\frac{\partial f^v}{\partial x^v}, \frac{\partial f^s}{\partial x^v})$ and $(\frac{\partial f^r}{\partial x^r}, \frac{\partial f^r}{\partial x^r})$.

$$\mathcal{L}_{Synergy} = \mathcal{L}(\theta_v, \theta_r) = \lambda_{div} \cdot \mathcal{L}_{div} + \lambda_t \cdot \mathcal{L}_t \tag{5}$$

3.3   Modality Complement Module

Although synergistic representation contains more identity-relevant diverse semantics, it is uncertain whether synergistic feature outperforms the combination of visible and infrared features $\text{Concat}(f_i^v, f_i^r)$. Due to infrared images containing global pedestrian statistics with less noise and visible images containing fine-grained discriminative semantics, we enhance the representation effectiveness of synergistic feature f_i^s from two aspects. Considering fine-grained semantics, we enhance synergistic features with advantages of visible features f_i^v in terms of local parts. And considering coarse-grained semantics, we enhance synergistic features with advantages of infrared features f_i^r about global parts.

On the fine-grained level, we split visible and synergistic features into $n = 6$ parts as MPANet [45] and get separate feature blocks as $f_i^v = [b_1^v, b_2^v \cdots, b_n^v]$, $f_i^s = [b_1^s, b_2^s \cdots, b_n^s]$. The local discrimination of synergistic features can be boosted with nuanced regions of visible modality. Cosine similarity $cos(\cdot, \cdot)$ is utilized for the optimization process.

$$\mathcal{L}_{local} = \frac{1}{N} \sum_{i=1}^{N} \sum_{j=1}^{n} (cos(b_j^v, b_j^s) + \sqrt{2 - 2cos(b_j^v, b_j^s)}) \tag{6}$$

In parallel, on the coarse-grained level, we supervise f_i^s by keeping the statistic centers of synergistic features consistent with that of the infrared feature f_i^r. The global statistics of synergistic features can get optimized with center consistency of infrared modality.

$$\mathcal{L}_{global} = \frac{1}{N} \sum_{i=1}^{N} ||C_{y_i}^s - C_{y_i}^r||_2^2, \tag{7}$$

where $C_{y_i}^s, C_{y_i}^r$ denote the center of the $y_i{}^{th}$ class for synergistic features f_i^s, f_i^r. $\mathcal{L}_{global}$ helps to coordinate semantics of the synergistic and the infrared feature and filter identity-irrelevance of the synergistic representation.

In the progress of Modality Complement module, we update the parameters of synergistic feature extractor θ_s, which aims to construct features with less noise, more diverse and more precise semantic description for each identity. θ_s is optimized as follows:

$$\mathcal{L}_{Com}(\theta_s) = \lambda_{local} \cdot \mathcal{L}_{local} + \lambda_{global} \cdot \mathcal{L}_{global}, \hat{\theta}_s = \arg\min_{\theta_s} \mathcal{L}(\theta_s), \tag{8}$$

where $\lambda_{local}, \lambda_{global}$ are hyper-parameters to balance the contributions of individual loss terms.

3.4 Cascaded Aggregation Strategy

Due to factors like shooting perspectives, clothing, and occlusion, the results of person retrieval will easily be affected [33,53]. To cope with this problem, center loss [23] and triplet loss [14] are widely adopted in ReID problems to simultaneously learn the centralized representation of feature embeddings and mine hard samples. Center loss $\mathcal{L}_c$ and Triplet loss $\mathcal{L}_{tri}$ can be formulated as:

$$\mathcal{L}_c = \frac{1}{N} \sum_{i=1}^{N} ||f_i - C_{y_i}||_2^2,$$
$$\mathcal{L}_{tri} = \sum_{i}^{N} \left[||f(x_i^a) - f(x_i^{pos})||_2^2 - ||f(x_i^a) - f(x_i^{neg})||_2^2 + \alpha \right]_+ \tag{9}$$

where x_i denotes the ith input sample, C_{y_i} is the $y_i{}^{th}$ class center, f_i is the feature embedding, x_i^a is the anchor. Center loss pays attention to aggregating feature embeddings but neglects the intrinsic differences and diverse semantics existing in the visible and the infrared modalities. Triplet loss specializes in handling hard samples separately rather than considering the comprehensive distribution across modalities, which limits the performance. Considering the diverse semantics and structural distribution across modalities, we propose Cascaded Aggregation to progressively optimize the features distribution of, as shown in Fig. 4.

1) Aggregation on Sub-class level. We utilize the identity of shooting cameras for each image as the natural sub-class, since images of the same person shot

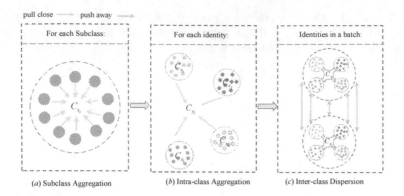

(a) Subclass Aggregation   (b) Intra-class Aggregation   (c) Inter-class Dispersion

Fig. 4. Cascaded Aggregation demonstration (a) indicates the optimization for subclass aggregation, (b) indicates the intra-class aggregation, and (c) indicates the inter-class dispersion.

by the same camera have high similarities with each other, where C_{s_i} denotes the s_i^{th} sub-class center:

$$\mathcal{L}_{sub} = \frac{1}{N} \sum_{i=1}^{N} ||f_i^s - C_{s_i}||_2^2, \tag{10}$$

2) Aggregation on the intra-class level, which keeps the structural priors of the features during the training progress. The formulation of the aggregation can be represented as follows, where N_s denotes the number of the sub-classes of each identity.

$$\mathcal{L}_{intra} = \frac{1}{N} \sum_{i=1}^{N} \sum_{j=1}^{N_s} ||C_{s_j} - C_{y_i}||_2^2, \tag{11}$$

3) Aggregation on the inter-class level. Our method of aggregation not only maximizes the similarity of intra-class instances but also maximizes the dissimilarity of inter-class instances on the whole. The dispersion between different identities and the two types of aggregation in 1) and 2) of the same identities are independent of each other. Formally, the dispersion between different identities can be represented as:

$$\mathcal{L}_{inter} = -\frac{1}{\binom{N}{2}} \sum_{i=1}^{N} \sum_{j \neq i}^{N} ||C_{y_i} - C_{y_j}||_2^2. \tag{12}$$

The loss function of CA for metric learning can be represented as:

$$\mathcal{L}_{cascade} = \mathcal{L}_{sub} + \mathcal{L}_{intra} + \mathcal{L}_{inter}$$

$$= \frac{1}{N} \sum_{i=1}^{N} ||f_i^s - C_{s_i}||_2^2 + \frac{1}{N} \sum_{i=1}^{N} \sum_{j=1}^{N_s} ||C_{s_j} - C_{y_i}||_2^2 - \frac{1}{\binom{N}{2}} \sum_{i=1}^{N} \sum_{j \neq i}^{N} ||C_{y_i} - C_{y_j}||_2^2. \tag{13}$$

Compared with Center Loss, our method begins with only a few samples of high similarity for the same shooting cameras and it will become much easier to learn sub-center representations.

Compared with Triplet Loss, our method deals with negative samples simultaneously by guiding the negative samples to the correspondent sub-class instead of easily pushing away alongside the gradient.

3.5 Objective Function

Firstly, we utilize Synergy Loss $\mathcal{L}_{Synergy}$ to enrich the representation on diverse semantics. The parameters of feature extractors θ_v and θ_r are updated as:

$$\mathcal{L}_{Synergistic} = \mathcal{L}(\theta_v, \theta_r) = \lambda_{div} \cdot \mathcal{L}_{div} + \lambda_t \cdot \mathcal{L}_t. \tag{14}$$

Then, we enhance the synergistic feature representation with the advantages of two modalities, namely, the discriminative local parts from the visible feature and global identity statistics from the infrared feature. We utilize Complementary Loss $\mathcal{L}_{Com}$ to update the modality synergy feature extractor θ_s:

$$\mathcal{L}_{com} = \mathcal{L}(\theta_s) = \lambda_{local} \cdot \mathcal{L}_{local} + \lambda_{global} \cdot \mathcal{L}_{global}. \tag{15}$$

Finally, we constrain the distribution of visible, infrared and synergistic feature f^v, f^r, f^s with cascaded aggregation strategy $\mathcal{L}_{cascaded}$:

$$\mathcal{L}_{cascaded} = \mathcal{L}(\theta_v, \theta_r, \theta_s) = \mathcal{L}_{sub} + \mathcal{L}_{intra} + \mathcal{L}_{inter}. \tag{16}$$

Overall, the objective function of our MSCLNet can be summarized as follows:

$$\mathcal{L}_{total} = \lambda_{div}\mathcal{L}_{div} + \lambda_t\mathcal{L}_t + \lambda_{local}\mathcal{L}_{local} + \lambda_{global}\mathcal{L}_{global} + \mathcal{L}_{sub} + \mathcal{L}_{intra} + \mathcal{L}_{inter} \tag{17}$$

4 Experiment

4.1 Datasets and Evaluation Protocol

SYSU-MM01 [43] is a large-scale dataset for VI-ReID which contains 491 pedestrians with total 287,628 visible images and 15,792 infrared images. It collects samples by 6 cameras, *i.e.* 4 visible and 2 infrared cameras, in the outdoor and indoor environments. It contains two different testing modes, *all-search* and *indoor-search* modes. Compared with RegDB, SYSU-MM01 is more challenging due to the large variations between samples.

RegDB [28] collects 412 identities, and each identity has 10 visible images and 10 infrared images. We randomly choose 206 identities for training and the left for testing [48]. There are two modes in testing, *visible-to-infrared* and *infrared-to-visible*. The former denotes that the model retrieves the person in the infrared gallery when given a visible image, and vice versa. We average the results for 10 trials for stable performance [40].

Evaluation Protocol. The cumulative matching characteristics (CMC) [27], and mean average precision (mAP) are used as evaluation metrics.

4.2 Implement Details

Training. We implement MSCLNet with PyTorch on a single NVIDIA RTX 2080 Ti GPU and deal with 64 images consisting of 32 visible and 32 infrared images of 8 identities in a mini-batch by randomly selecting 4 visible and 4 infrared images for each identity. Our baseline is AGW*, which means AGW [51] with Random Erasing. We adopt pre-trained ResNet-50 [13] on ImageNet as the backbone network. Then, we pre-process each image by re-scaling in to 288×144 and augment images through random cropping with zero-padding, random horizontal flipping and random erasing (80% probability, 80% max-area, 20% min-area). During the training process, we optimize the feature extractors θ_v, θ_r and modality synergy module θ_s with SGD optimizer. We set the initial learning rate $\eta = 0.1$, the momentum parameter $p = 0.9$. The learning rate is changed as $\eta = 0.05$ at 21–50 epoch, $\eta = 0.01$ at 51–100 epoch, and $\eta = 0.001$ at 101–200 epoch. The hyper-parameters $\lambda_{div}, \lambda_t, \lambda_{local}, \lambda_{global}$ are set to 0.5, 1.25, 0.8, and 1.5, respectively. We synergize visible and infrared instances to train a concise end-to-end network, which retrieves specific person across modalities.

Testing. For testing, the model works in *Single-shot* mode by extracting the query and the gallery features from a single modality by the feature extractor θ_v or θ_r. Besides, MS and MC modules **do not** participate in testing stage.

4.3 Ablation Study

In this subsection, we conduct an ablation study to evaluate the effectiveness of each component of MSCLNet, as summarized in Eq. 17. The results are demonstrated in Table 1. We evaluate how much improvement can be made by each component on the *all-search* mode of SYSU-MM01 dataset.

Table 1. Analysis of the effectiveness of MS, MC, CA on SYSU-MM01 dataset in the *all-search* mode. Rank-1 accuracy (%) and mAP (%)are reported.

Methods								Metric	
B	MS		MC			CA			
	$\mathcal{L}_{div}$	$\mathcal{L}_t$	$\mathcal{L}_{glocal}$	$\mathcal{L}_{local}$	$\mathcal{L}_{sub}$	$\mathcal{L}_{intra}$	$\mathcal{L}_{inter}$	Rank-1	mAP
✓								59.82	56.07
✓	✓							60.32	56.79
✓		✓						60.67	58.12
✓			✓					62.14	59.94
✓				✓				61.33	59.23
✓					✓			61.74	59.88
✓						✓		61.96	60.40
✓							✓	63.55	60.97
✓	✓	✓						62.82	60.25
✓			✓	✓				64.84	61.00
✓					✓	✓	✓	66.13	61.99
✓	✓	✓	✓		✓			71.16	66.30
✓	✓	✓				✓	✓	69.78	65.29
✓			✓	✓	✓	✓	✓	72.81	67.66
✓	✓	✓	✓		✓	✓	✓	76.99	71.64

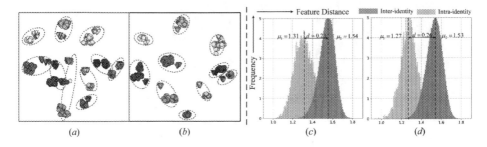

Fig. 5. Visualization Results. (a) and (b) show the feature embeddings distribution of baseline and MSCLNet via t-SNE [24], where circles and triangles in different colors denote visible and infrared modalities. (c) and (d) show the intra-and-inter distribution of feature distance.

Effectiveness of MS. Referring to the ninth row, we add the MS structure to the baseline, and the baseline obtains a rank-1 score of 62.82% and a mAP of 60.25%, improved by 3% and 19.75%. Meanwhile, we add MS to other combinations as shown in the 12th, 13th, 15th rows. MS also brings different degrees of enhancement to the model.

Effectiveness of MC. The experimental setting of Base+MC acquires 64.84% at Rank-1 and 61.00% at mAP. When baseline works with MS+MC, a further improvement is reached, where rank-1 is 71.16% and mAP is 66.3%. This illustrates that Modality Synergy Complement Learning effectively improves performance.

Effectiveness of CA. The settings of Base+MS+CA and Base+MC+CA work better than merely utilizing one of the three modules. Base+MS+MC+CA reaches the best result, in which rank-1 is 69.78%, mAP is 65.29%.

Overall. The results show that each component of MSCLNet can improve precision. At the same time, they work better when cooperating, which reveals that the three components focus on different aspects of optimization.

4.4   Visualization Analysis

To present the effectiveness of MSCLNet, we visualize the feature distribution via t-SNE [24] as shown in Fig. 5. Different colors denote different identities. For the baseline, feature embeddings of some identities entangle with each other, which indicates the baseline is confused about these identities. In comparison, MSCLNet discriminates and aggregates these feature embeddings of the same identity separately and clearly.

Meanwhile, we also visualize the feature distances analysis between baseline and MSCLNet in Fig. 5. After numerical analysis, our conclusions are as follows: 1) Distances between these distribution increase $d = 0.23 \rightarrow 0.26$ and the mean distance of intra-identity reduces $\mu = 1.31 \rightarrow 1.27$. 2) Variance of intra-identity distribution reduces prominently $\sigma = 9.5 \times 10^{-2} \rightarrow 8.8 \times 10^{-2}$ and the distribution of intra-identity aggregates better.

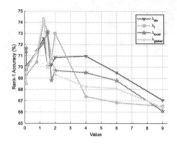

Fig. 6. Hyper-parameters sensitive graph

4.5 Parameters Analysis

We analyze the parameter sensitivity under the condition that, single hyper-parameter was selected as a variable and all other hyper-parameters are kept constant. Thus, we can obtain the curve of the effect of changing hyper-parameters on Rank-1 accuracy (%). In turn, by continuously changing the variables, we can get the sensitivity graphs of all hyper-parameters $\lambda_{div}, \lambda_t, \lambda_{global}, \lambda_{local}$ as shown in Fig. 6. It clearly shows that five hyper-parameters present different sensitivities and most of their optimal intervals of these parameters are in $[1,2]$.

4.6 Comparison with State-of-the-Art Methods

Table 2. Comparison with the state-of-the-arts on SYSU-MM01 dataset. Rank-k accuracy (%) and mAP (%) are reported.

Settings		All search				Indoor search			
Method	Venue	R1	R10	R20	mAP	R1	R10	R20	mAP
Zero-Pad [43]	ICCV17	14.80	54.12	71.33	15.95	20.58	68.38	85.79	26.92
HCML [48]	AAAI18	14.32	53.16	69.17	16.16	24.52	73.25	86.73	30.08
cmGAN [7]	IJCAI18	26.97	67.51	80.56	27.80	31.63	77.23	89.18	42.19
HSME [12]	AAAI19	20.68	32.74	77.95	23.12	–	–	–	–
AliGAN [37]	ICCV19	42.40	85.00	93.70	40.70	45.90	87.60	94.40	54.30
CMSP [42]	IJCV20	43.56	86.25	–	44.98	48.62	89.50	–	57.50
JSIA [35]	AAAI20	38.10	80.70	89.90	36.90	43.80	86.20	94.20	52.90
XIV [17]	AAAI20	49.92	89.79	95.96	50.73	–	–	–	–
MACE [47]	TIP20	51.64	87.25	94.44	50.11	57.35	93.02	97.47	64.79
MSR [9]	TIP20	37.35	83.40	93.34	38.11	39.64	89.29	97.66	50.88
Hi-CMD [6]	CVPR20	34.94	77.58	–	35.94	–	–	–	–
cm-SSFT [21]	CVPR20	47.70	–	–	54.10	–	–	–	–
AGW [51]	TPAMI21	47.50	84.39	92.14	47.65	54.17	91.14	95.98	62.97
MCLNet [11]	ICCV21	65.40	93.33	97.14	61.98	72.56	96.88	99.20	76.58
SMCL [41]	ICCV21	67.39	92.87	96.76	61.78	68.84	96.55	98.77	75.56
NFS [5]	CVPR21	56.91	91.34	96.52	55.45	62.79	96.53	99.07	69.79
CM-NAS [10]	CVPR21	61.99	92.87	97.25	60.02	67.01	97.02	99.32	72.95
MPANet [45]	CVPR21	70.58	96.21	98.80	68.24	76.74	98.21	99.57	80.95
MSCLNet	**Ours**	**76.99**	**97.63**	**99.18**	**71.64**	**78.49**	**99.32**	**99.91**	**81.17**

We compare the proposed MSCLNet with state-of-the-art methods. Table 2 and Table 3 illustrate the comparison results on the SYSU-MM01 and the RegDB datasets. MSCLNet outperforms the other methods on both of the benchmarks. On the SYSU-MM01 dataset, MSCLNet achieves the rank-1 scores of 76.99% and mAP score of 71.64% in the *all-search* mode, higher than MPANet [45] by 6.41% and 3.40%. On the RegDB dataset, MSCLNet achieves Rank-1 scores of 84.17% and 83.86% in visible-to-infrared and infrared-to-visible modes, better than NFS [10] by 3.63% and 5.91%, respectively.

Table 3. Comparison with the state-of-the-arts on RegDB dataset. Rank-1 accuracy (%) and mAP (%) are reported.

Settings		Visible to infrared		Infrared to visible	
Method	Venue	Rank-1	mAP	Rank-1	mAP
Zero-Pad [43]	ICCV'17	17.75	18.90	16.63	17.82
HCML [48]	AAAI'18	24.44	20.08	21.70	22.24
HSME [12]	AAAI'19	50.85	47.00	50.15	46.16
AliGAN [37]	ICCV'19	57.90	53.60	56.30	53.40
CMSP [42]	IJCV'20	65.07	64.50	–	–
JSIA [35]	AAAI'20	48.10	48.90	48.50	49.30
XIV [17]	AAAI'20	62.21	60.18	–	–
DG-VAE [30]	ACM MM'20	72.97	71.78	–	–
HAT [52]	TIFS'20	71.83	67.56	70.02	66.30
MSR [9]	TIP'20	48.43	48.67	–	–
MACE [47]	TIP'20	72.37	69.09	72.12	68.57
DDAG [50]	ECCV'20	69.34	63.46	68.06	61.80
Hi-CMD [6]	CVPR'20	70.93	66.04	–	–
AGW [51]	TPAMI'21	70.05	66.37	70.49	65.90
MCLNet [11]	ICCV'21	80.31	73.07	75.93	69.49
NFS [5]	CVPR'21	80.54	72.10	77.95	69.97
MPANet [45]	CVPR'21	83.70	80.90	82.80	**80.70**
MSCLNet	**Ours**	**84.17**	**80.99**	**83.86**	78.31

5   Conclusion and Discussion

In this paper, we propose a novel VI-ReID framework, which has the capability to make full use of the visible and the infrared modality semantics and learn discriminative representation of identities by synergizing and complementing instances of visible and infrared modalities. Different from existing methods pursuing modal-shared information at the risk of identity-relevant semantics loss, MSCLNet provides an innovative approach exploring high-level unity in VI-ReID task. Meanwhile, we propose Cascaded Aggregation strategy to fine-grained and progressively optimize the distribution of feature embeddings, which assists the network discriminate identities and extract more precise and more comprehensive features. Experimental results validate the merit of the framework, as well as the effectiveness of each component in this framework. In the future work, we plan to explore background scenes, gender, and appearances to construct better different sub-classes.

Acknowledgements. This work was supported in part by the National Natural Science Foundation of China under Grant 61902027, and the Start-up Research Grant (SRG) of University of Macau.

References

1. Ahmed, S.M., Lejbolle, A.R., Panda, R., Roy-Chowdhury, A.K.: Camera onboarding for person re-identification using hypothesis transfer learning. In: CVPR, pp. 12144–12153 (2020)
2. Bai, S., Tang, P., Torr, P.H., Latecki, L.J.: Re-ranking via metric fusion for object retrieval and person re-identification. In: CVPR, pp. 740–749 (2019)
3. Chen, G., Lin, C., Ren, L., Lu, J., Zhou, J.: Self-critical attention learning for person re-identification. In: ICCV, pp. 9637–9646 (2019)
4. Chen, T., et al.: ABD-net: attentive but diverse person re-identification. In: CVPR, pp. 8351–8361 (2019)
5. Chen, Y., Wan, L., Li, Z., Jing, Q., Sun, Z.: Neural feature search for RGB-infrared person re-identification. In: CVPR, pp. 587–597, June 2021
6. Choi, S., Lee, S., Kim, Y., Kim, T., Kim, C.: Hi-CMD: hierarchical cross-modality disentanglement for visible-infrared person re-identification. In: CVPR, pp. 10257–10266 (2020)
7. Dai, P., Ji, R., Wang, H., Wu, Q., Huang, Y.: Cross-modality person re-identification with generative adversarial training. In: IJCAI, pp. 677–683 (2018)
8. Deng, W., Zheng, L., Ye, Q., Kang, G., Yang, Y., Jiao, J.: Image-image domain adaptation with preserved self-similarity and domain-dissimilarity for person re-identification. In: CVPR, pp. 994–1003 (2018)
9. Feng, Z., Lai, J., Xie, X.: Learning modality-specific representations for visible-infrared person re-identification. IEEE TIP **29**, 579–590 (2019)
10. Fu, C., Hu, Y., Wu, X., Shi, H., Mei, T., He, R.: CM-NAS: cross-modality neural architecture search for visible-infrared person re-identification. In: ICCV, pp. 11823–11832, October 2021
11. Hao, X., Zhao, S., Ye, M., Shen, J.: Cross-modality person re-identification via modality confusion and center aggregation. In: ICCV, pp. 16403–16412, October 2021
12. Hao, Y., Wang, N., Li, J., Gao, X.: HSME: hypersphere manifold embedding for visible thermal person re-identification. In: AAAI, pp. 8385–8392 (2019)
13. He, K., Zhang, X., Ren, S., Sun, J.: Deep residual learning for image recognition. In: CVPR, pp. 770–778 (2016)
14. Hermans, A., Beyer, L., Leibe, B.: In defense of the triplet loss for person re-identification. arXiv preprint arXiv:1703.07737 (2017)
15. Jia, M., Zhai, Y., Lu, S., Ma, S., Zhang, J.: A similarity inference metric for RGB-infrared cross-modality person re-identification. arXiv preprint arXiv:2007.01504 (2020)

16. Jin, X., Lan, C., Zeng, W., Chen, Z., Zhang, L.: Style normalization and restitution for generalizable person re-identification. In: CVPR, pp. 3143–3152 (2020)

17. Li, D., Wei, X., Hong, X., Gong, Y.: Infrared-visible cross-modal person re-identification with an x modality. In: AAAI, pp. 4610–4617 (2020)

18. Li, H., Wu, G., Zheng, W.S.: Combined depth space based architecture search for person re-identification. In: Proceedings of the IEEE/CVF Conference on Computer Vision and Pattern Recognition, pp. 6729–6738 (2021)

19. Li, Y., He, J., Zhang, T., Liu, X., Zhang, Y., Wu, F.: Diverse part discovery: occluded person re-identification with part-aware transformer. In: Proceedings of the IEEE/CVF Conference on Computer Vision and Pattern Recognition, pp. 2898–2907 (2021)

20. Lin, Y., Xie, L., Wu, Y., Yan, C., Tian, Q.: Unsupervised person re-identification via softened similarity learning. In: CVPR, pp. 3390–3399 (2020)

21. Lu, Y., et al.: Cross-modality person re-identification with shared-specific feature transfer. In: CVPR, pp. 13379–13389 (2020)

22. Luo, C., Chen, Y., Wang, N., Zhang, Z.: Spectral feature transformation for person re-identification. In: CVPR, pp. 4976–4985 (2019)

23. Luo, H., Gu, Y., Liao, X., Lai, S., Jiang, W.: Bag of tricks and a strong baseline for deep person re-identification. In: CVPR Workshops (2019)

24. van der Maaten, L., Hinton, G.: Visualizing data using t-SNE. J. Mach. Lear. Res. 9, 2579–2605 (2008)

25. Melis, G., Kočiský, T., Blunsom, P.: Mogrifier LSTM. arXiv preprint arXiv:1909.01792 (2019)

26. Meng, J., Zheng, W.S., Lai, J.H., Wang, L.: Deep graph metric learning for weakly supervised person re-identification. IEEE Trans. Pattern Anal. Mach. Intell. 44(10), 6074–6093 (2021)

27. Moon, H., Phillips, P.J.: Computational and performance aspects of PCA-based face-recognition algorithms. Perception 30(3), 303–321 (2001)

28. Nguyen, D.T., Hong, H.G., Kim, K.W., Park, K.R.: Person recognition system based on a combination of body images from visible light and thermal cameras. Sensors 17(3), 605 (2017)

29. Paisitkriangkrai, S., Shen, C., Van Den Hengel, A.: Learning to rank in person re-identification with metric ensembles. In: CVPR, pp. 1846–1855 (2015)

30. Pu, N., Chen, W., Liu, Y., Bakker, E.M., Lew, M.S.: Dual Gaussian-based variational subspace disentanglement for visible-infrared person re-identification. In: ACMMM, pp. 2149–2158 (2020)

31. Ren, C.X., Liang, B.H., Lei, Z.: Domain adaptive person re-identification via camera style generation and label propagation. IEEE Trans. Inf. Forensics Secur. 15, 1290–1302 (2019)

32. Sun, D., Yao, A., Zhou, A., Zhao, H.: Deeply-supervised knowledge synergy. In: CVPR, pp. 6997–7006 (2019)

33. Sun, X., Zheng, L.: Dissecting person re-identification from the viewpoint of viewpoint. In: CVPR, pp. 608–617 (2019)

34. Sun, Y., Zheng, L., Yang, Y., Tian, Q., Wang, S.: Beyond part models: person retrieval with refined part pooling (and a strong convolutional baseline). In: Ferrari, V., Hebert, M., Sminchisescu, C., Weiss, Y. (eds.) ECCV 2018. LNCS, vol. 11208, pp. 501–518. Springer, Cham (2018). https://doi.org/10.1007/978-3-030-01225-0_30

35. Wang, G.A., et al.: Cross-modality paired-images generation for RGB-infrared person re-identification. In: AAAI, pp. 12144–12151 (2020)

36. Wang, G., et al.: High-order information matters: learning relation and topology for occluded person re-identification. In: CVPR, pp. 6449–6458 (2020)
37. Wang, G., Zhang, T., Cheng, J., Liu, S., Yang, Y., Hou, Z.: RGB-infrared cross-modality person re-identification via joint pixel and feature alignment. In: ICCV, pp. 3623–3632 (2019)
38. Wang, J., Zhu, X., Gong, S., Li, W.: Transferable joint attribute-identity deep learning for unsupervised person re-identification. In: CVPR, pp. 2275–2284 (2018)
39. Wang, Y., Chen, Z., Feng, W., Gang, W.: Person re-identification with cascaded pairwise convolutions. In: IEEE Conference on Computer Vision and Pattern Recognition (2018)
40. Wang, Z., Wang, Z., Zheng, Y., Chuang, Y.Y., Satoh, S.: Learning to reduce dual-level discrepancy for infrared-visible person re-identification. In: CVPR, pp. 618–626 (2019)
41. Wei, Z., Yang, X., Wang, N., Gao, X.: Syncretic modality collaborative learning for visible infrared person re-identification. In: ICCV, pp. 225–234, October 2021
42. Wu, A., Zheng, W.-S., Gong, S., Lai, J.: RGB-IR person re-identification by cross-modality similarity preservation. IJCV **128**(6), 1765–1785 (2020). https://doi.org/10.1007/s11263-019-01290-1
43. Wu, A., Zheng, W.S., Yu, H.X., Gong, S., Lai, J.: RGB-infrared cross-modality person re-identification. In: ICCV, pp. 5380–5389 (2017)
44. Wu, D., Ye, M., Lin, G., Gao, X., Shen, J.: Person re-identification by context-aware part attention and multi-head collaborative learning. IEEE Trans. Inf. Forensics Secur. **17**, 115–126 (2021)
45. Wu, Q., et al.: Discover cross-modality nuances for visible-infrared person re-identification. In: CVPR, pp. 4330–4339, June 2021
46. Xuan, S., Zhang, S.: Intra-inter camera similarity for unsupervised person re-identification. In: Proceedings of the IEEE/CVF Conference on Computer Vision and Pattern Recognition, pp. 11926–11935 (2021)
47. Ye, M., Lan, X., Leng, Q., Shen, J.: Cross-modality person re-identification via modality-aware collaborative ensemble learning. IEEE TIP **29**, 9387–9399 (2020)
48. Ye, M., Lan, X., Li, J., Yuen, P.C.: Hierarchical discriminative learning for visible thermal person re-identification. In: AAAI, pp. 7501–7508 (2018)
49. Ye, M., Lan, X., Wang, Z., Yuen, P.C.: Bi-directional center-constrained top-ranking for visible thermal person re-identification. IEEE TIFS **15**, 407–419 (2019)
50. Ye, M., Shen, J., J. Crandall, D., Shao, L., Luo, J.: Dynamic dual-attentive aggregation learning for visible-infrared person re-identification. In: Vedaldi, A., Bischof, H., Brox, T., Frahm, J.-M. (eds.) ECCV 2020. LNCS, vol. 12362, pp. 229–247. Springer, Cham (2020). https://doi.org/10.1007/978-3-030-58520-4_14
51. Ye, M., Shen, J., Lin, G., Xiang, T., Shao, L., Hoi, S.C.H.: Deep learning for person re-identification: a survey and outlook. arXiv preprint arXiv:2001.04193 (2020)
52. Ye, M., Shen, J., Shao, L.: Visible-infrared person re-identification via homogeneous augmented tri-modal learning. IEEE TIFS **16**, 728–739 (2020)
53. Yu, S., Li, S., Chen, D., Zhao, R., Yan, J., Qiao, Y.: COCAS: a large-scale clothes changing person dataset for re-identification. In: CVPR, pp. 3400–3409 (2020)
54. Zhang, X., Ge, Y., Qiao, Y., Li, H.: Refining pseudo labels with clustering consensus over generations for unsupervised object re-identification. In: Proceedings of the IEEE/CVF Conference on Computer Vision and Pattern Recognition, pp. 3436–3445 (2021)
55. Zhang, Z., Lan, C., Zeng, W., Chen, Z.: Multi-granularity reference-aided attentive feature aggregation for video-based person re-identification. In: CVPR, pp. 10407–10416 (2020)

56. Zhang, Z., Lan, C., Zeng, W., Jin, X., Chen, Z.: Relation-aware global attention for person re-identification. In: CVPR, pp. 3186–3195 (2020)
57. Zheng, F., et al.: Pyramidal person re-identification via multi-loss dynamic training. In: CVPR, pp. 8514–8522 (2019)
58. Zheng, M., Karanam, S., Wu, Z., Radke, R.J.: Re-identification with consistent attentive Siamese networks. In: CVPR, pp. 5735–5744 (2019)
59. Zhong, Z., Zheng, L., Zheng, Z., Li, S., Yang, Y.: Camera style adaptation for person re-identification. In: CVPR, pp. 5157–5166 (2018)
60. Zhu, X., Jing, X.Y., You, X., Zuo, W., Shan, S., Zheng, W.S.: Image to video person re-identification by learning heterogeneous dictionary pair with feature projection matrix. IEEE Trans. Inf. Forensics Secur. **13**, 717–732 (2017)

Cross-Modality Transformer for Visible-Infrared Person Re-Identification

Kongzhu Jiang[1], Tianzhu Zhang[1,2(✉)], Xiang Liu[3], Bingqiao Qian[1], Yongdong Zhang[1], and Feng Wu[1]

[1] University of Science and Technology of China, Hefei, China
{kzjiang,qbq}@mail.ustc.edu.cn, {tzzhang,zhyd73,fengwu}@ustc.edu.cn
[2] Deep Space Exploration Lab, Hefei, China
[3] Dongguan University of Technology of China, Dongguan, China

Abstract. Visible-infrared person re-identification (VI-ReID) is a challenging task due to the large cross-modality discrepancies and intra-class variations. Existing works mainly focus on learning modality-shared representations by embedding different modalities into the same feature space. However, these methods usually damage the modality-specific information and identification information contained in the features. To alleviate the above issues, we propose a novel Cross-Modality Transformer (CMT) to jointly explore a modality-level alignment module and an instance-level module for VI-ReID. The proposed CMT enjoys several merits. First, the modality-level alignment module is designed to compensate for the missing modality-specific information via a Transformer encoder-decoder architecture. Second, we propose an instance-level alignment module to adaptively adjust the sample features, which is achieved by a query-adaptive feature modulation. To the best of our knowledge, this is the first work to exploit a cross-modality transformer to achieve the modality compensation for VI-ReID. Extensive experimental results on two standard benchmarks demonstrate that our CMT performs favorably against the state-of-the-art methods.

Keywords: Person re-identification · Transformer · Cross-modality

1 Introduction

Person re-identification (Re-ID) aims at matching person images captured from non-overlapping camera views [40,42]. In recent years, it has gained increasing attention due to its significant practical value in video surveillance. Most of the existing methods [11,13,16,21,24,26,43,44] focus on visible (RGB) cameras and formulate the Re-ID task as a single-modality matching problem. However, the visible cameras are incapable of capturing valid appearance information of persons under poor illumination conditions (e.g., at night). To image clearly in the dark, in addition to the visible cameras, infrared (IR) cameras that are

S. Avidan et al. (Eds.): ECCV 2022, LNCS 13674, pp. 480–496, 2022.
https://doi.org/10.1007/978-3-031-19781-9_28

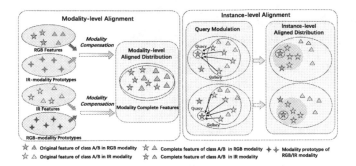

Fig. 1. Our motivation. The modality prototypes are introduced to store the global modality characteristics, which can be utilized to compensate for the missing modality features and thus contributes to the **modality-level alignment**. Then, by use of the query feature modulation, we can adaptively adjust the gallery sample features to activate query-related patterns and achieve **instance-level alignment**.

robust to illumination variants are also equipped in many surveillance scenarios. Hence, visible-infrared person re-identification (VI-ReID) [3,20,37] has recently been of great interest, which aims at retrieving IR person images of the same identity as the given RGB query and vice versa.

VI-ReID is challenging due to the cross-modal discrepancies between RGB and IR images, and the key issue is how to bridge the two modalities. To narrow the gap between two modalities, existing methods mainly focus on modality-level alignment. Some works are based on modality-shared feature learning [6,9,37,38,41], which decouple features into modality-specific and modality-shared features. Then they utilize the latter ones to align the modalities in the feature level while abandoning the modality-specific features. However, the modality-specific features also contain useful identity information that helps the final retrieval, such as colors. Therefore, with modality-shared cues only, the upper bound of the discrimination ability of the feature representation is limited. To address this limitation, modality compensation methods [20,32] have been proposed to compensate for the missing modality features. Specifically, in [20], the authors utilize the graph convolutional networks to obtain the compensated modality features based on the similarities between cross-modality samples in the current mini-batch.

By studying the previous VI-ReID methods based on modality compensation, we discover two characteristics that play an important role in achieving the robust VI-ReID. (1) **Modality-level alignment.** In previous modality compensation methods [20], the compensated features are produced solely based on the samples of the current mini-batch. This strategy suffers from a certain randomness, and thus causes the inconsistency of generated modality features when the samples are in the different mini-batches. To address this issue, an intuitive idea is to model several modality prototypes for representing global modality information. These modality prototypes can be used as the global basis for learning robust modality compensation for every sample. Therefore, it is necessary to model global modality prototypes to facilitate a better alignment between RGB and IR modalities.

(2) **Instance-level alignment.** Due to intra-class variations (e.g., viewpoint, illumination, and background clutter), the feature distribution of different samples with same ID varies greatly even under the same modality. Performing the modality-level alignment alone may lead to cases where the IR (RGB) instances are incorrectly aligned with the RGB (IR) instances of a different category. To align the cross-modal instances in the same class, most methods [9,37,38] utilize the supervised triplet loss to reduce the distances of the features of the same ID on the training set. However, in the open-set setting, because the categories of training and test sets have no overlap, the discriminative representations learned on the training set may not be optimal for the test images. Therefore, it is of vital importance to achieve dynamic instance-level alignment. In this way, the gallery instances can be adaptively refined according to the query features and be aligned to the query instances in the same class(as shown in Fig. 1).

Inspired by the above discussions, we propose a novel Cross-Modality Transformer (CMT) by jointly exploring a modality-level alignment module and an instance-level alignment module for visible-infrared person re-identification. In the **modality-level alignment module**, we introduce an encoder-decoder architecture, which can achieve the modality feature enhancement and compensation. In the encoder, we adopt a self-attention mechanism to capture the interrelationship between local human parts. Then, we introduce two sets of learnable modality prototypes to represent the RGB and IR modalities respectively, and design a decoder to compensate for the missing modality. Taking a RGB sample as the example, we take the IR modality prototypes as queries and the part features of the RGB sample as keys and values of the transformer decoder. By use of the cross-attention between the part features and the modality prototypes, we can obtain the attention scores which can be regarded as the soft correspondences between IR modality prototypes and part features. Then we can compensate for IR modality-specific features by aggregating the related part features according to the attention matrix. Besides, to guide the learning of modality prototypes, we design a modality consistency loss to constrain the compensated IR features to be aligned with the real IR modality features. Similarly, the modality compensation for IR images can be achieved in the same way. In the **instance-level alignment module**, a feature modulator is proposed to adaptively adjust the representations of instances. Concretely, given the query sample x, we can utilize its feature to generate the channel-wise modulation parameters. These parameters reveal the most discriminative patterns of sample x, and can be employed to modulate other samples of the current mini-batch in the channel dimension. Thus, the crucial x-related channels of other samples can be strengthened, and the irrelevant channels can be suppressed. In this way, we can achieve query-adaptive feature modulation during the test, which can adaptively activate coherent patterns between query instances and gallery instances and facilitate a better instance-level alignment even for the unseen test categories.

The main contributions of this paper can be summarized as follows: (1) We propose a novel Cross-Modality Transformer for VI-ReID to jointly explore modality-level alignment and instance-level alignment. To the best of our knowledge, this is the first work to exploit a cross-modality transformer to achieve the modality compensation for VI-ReID. (2) A modality-level alignment module is proposed to compensate for the missing modality-specific information via a Transformer encoder-decoder architecture. Also, we design an instance-level alignment module to adaptively adjust the sample features, which is achieved by query-adaptive feature modulation. (3) Extensive experimental results on two standard benchmarks demonstrate that the proposed model performs favorably against state-of-the-art VI-ReID methods.

2 Related Work

Single-Modality Person Re-ID. Single-modality person re-identification aims at matching pedestrian images across disjoint visible cameras. The considerable viewpoint changes and human pose variations under different visible cameras are the main challenges of single modality person Re-ID. Existing works mainly focuses on representation learning [16,21,26,30,42] and metric learning [24,36,43], and have achieved excellent performance on the widely-used datasets. However, due to the large cross-modality discrepancies, these methods may not be applicable for the VI-ReID task in the practical surveillance scenarios. Differently, our method proposes a modality-level alignment module and an instance-alignment module to learn a unified ReID framework for both RGB and IR modalities.

Visible-Infrared Person Re-ID. Visible-Infrared person Re-ID is challenging due to the cross-modal discrepancies between visible and infrared images. To address this challenge, existing methods [3,6,9,15,29,37,39,41] mainly focus on learning modality-shared feature representations to achieve modality level alignment. Some image translation-based methods [3,14,28,29] are developed to firstly achieve modality unification and then learn modality-shared representations. Wang et al. [29] propose an end-to-end alignment generative adversarial network by exploiting pixel alignment and feature alignment jointly. [28] generates cross-modality paired-images and performs both global set-level and fine-grained instance-level alignments. Another line of works [7,8,18,37] attempts to learn modality-shared features by designing various two-stream architectures. Ye et al. [37] propose a novel modality-aware collaborative ensemble learning method with the middle-level sharable two-stream network. [7] exploits the optimal two-stream architecture by neural architecture search for VI-ReID. However, the modality-specific features are generally ignored by the above methods, which limits the upper bound of the discrimination ability of the feature representation. To address this limitation, modality compensation methods [20,32] are proposed to compensate for the missing modality features. [32] generates multi-spectral images to compensate for the lacking specific information by utilizing the generative adversarial network. In [20], a cross-modality shared specific feature transfer algorithm is proposed to explore the potential of both the modality-shared information and the modality-specific features. However, the compensated features extracted by [20] only depend on the current mini-batch, which causes

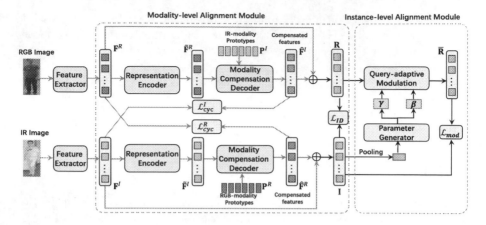

Fig. 2. Framework of our Cross-Modality Transformer. (1) The modality-level alignment module compensates for the missing modality information by the cross-attention between the modality prototypes $\mathbf{P}^I/\mathbf{P}^R$ and features $\tilde{\mathbf{F}}^R/\tilde{\mathbf{F}}^I$. A modality consistency loss $\mathcal{L}_{cyc}$ is proposed to make the modality prototypes focus on the corresponding global modality information. (2) The instance-level alignment module leverages the characteristics of the given query to automatically adapt the other instance features by the query-adaptive modulation, which helps align the query and gallery instances in the same class. For simplicity, we take an IR image $\mathbf{I}$ as an example and show the process of using $\mathbf{I}$ as the query for the modulation in the figure.

the inconsistency of generated modality features when the samples are in the different mini-batches. Differently, we introduce two sets of global modality prototypes to represent the RGB and IR modalities respectively, which can be used as the global basis to learn modality compensation for every sample.

Transformer in Person Re-ID. Transformers have recently received increasing attention for computer vision tasks, including image classification [5,19], object detection [1,45], image segmentation [19,31], and so on. Most existing Re-ID methods apply Transformer to a single modality. For example, He et al. [11] utilize a pure-transformer with a side information embedding and a jigsaw patch module to learn discriminative features. Li et al. [17] exploit a transformer architecture to discover diverse parts for occluded person Re-ID. Different from the above methods, our CMT is designed for VI-ReID to compensate for the missing modality-specific information.

3   Our Method

In this section, we introduce the details of the proposed Cross-Modality Transformer (CMT) for the VI-ReID task. As shown in Fig. 2, the proposed CMT mainly consists of two modules. (1) The modality-level alignment module aims at compensating for the missing modality-specific information via a Transformer encoder-decoder architecture. (2) The instance-level alignment module is responsible for aligning the gallery instances with the query instance in the same class by a query-adaptive feature modulation mechanism.

3.1   Modality-Level Alignment Module

In order to achieve the modality-level alignment, we follow the architecture of Transformer [27] and design a representation encoder and a modality compensation decoder, which is able to adaptively compensate for the lacking modality-special information. Different from the previous modality compensation method [20] that depends on the information in the mini-batch, we design two set of learnable modality prototypes to provide the global modality information for more robust modality compensation.

Representation Encoder. Following existing works [8,25,39,40], we adopt a two-stream network based on ResNet-50 as our feature extractor for the RGB and IR modalities, where the first two stages are parameter-independent and the latter three stages are parameter-shared. We first use the feature extractor ϕ to extract the feature maps for the given visible images and infrared images. Then, following the practice in the part-based methods [18,39], we horizontally split the feature maps into p non-overlapping parts with a region pooling strategy. In this way, the RGB and IR images can be represented by the set of the part features: $\mathbf{F}^R = \left[f_1^R; f_2^R; \ldots; f_p^R \right] \in \mathbb{R}^{p \times d}$ and $\mathbf{F}^I = \left[f_1^I; f_2^I; \ldots; f_p^I \right] \in \mathbb{R}^{p \times d}$, where $f_i^R, f_i^I \in \mathbb{R}^d$ indicate the ith part feature of two modalities. These part features are taken as the inputs of transformer encoder. For the simplicity of the description, we take the RGB image as an example.

In the representation encoder, we adopt a self-attention layer to capture the inter-relationship between the local human parts to refine the part representations. Specifically, we take part features as the query $\mathbf{Q}$, key $\mathbf{K}$ and value $\mathbf{V}$. We generate the ($\mathbf{Q}$, $\mathbf{K}$, $\mathbf{V}$) triplets by independent linear projection layers:

$$\mathbf{Q} = \mathbf{F}^R \mathbf{W}^Q, \quad \mathbf{K} = \mathbf{F}^R \mathbf{W}^K, \quad \mathbf{V} = \mathbf{F}^R \mathbf{W}^V, \tag{1}$$

where $\mathbf{W}^Q \in \mathbb{R}^{d \times d}, \mathbf{W}^K \in \mathbb{R}^{d \times d}, \mathbf{W}^V \in \mathbb{R}^{d \times d}$ are linear projections, and $\mathbf{Q}, \mathbf{K}, \mathbf{V} \in \mathbb{R}^{p \times d}$. Then, the attention weights between the query $\mathbf{Q}$ and the key $\mathbf{K}$ can be derived by the inner product with a scaling operation and a Softmax normalization. Based on the attention weights, we can obtain the refined part features as the weighted sum of values $V \in \mathbb{R}^{p \times d}$. Formally:

$$Attention\left(\mathbf{Q}, \mathbf{K}, \mathbf{V}\right) = softmax\left(\frac{\mathbf{Q}\mathbf{K}^T}{\sqrt{d}}\right)\mathbf{V}. \tag{2}$$

Equation (2) is implemented with the multi-head attention mechanism, and a feed-forward network is also applied. For more details, please refer to the work [27].

Modality Compensation Decoder. In the modality compensation decoder, we introduce two sets of learnable modality prototypes to represent the global modality information of RGB and IR modalities respectively, which are denoted as $\mathbf{P}^R = \left[p_1^R; p_2^R; \ldots; p_p^R \right], \mathbf{P}^I = \left[p_1^I; p_2^I; \ldots; p_p^I \right] \in \mathbb{R}^{p \times d}$, where p_i^I is the modality prototype for the i-th part feature in the IR modality. Following the standard architecture of the transformer [27], we first use a self-attention layer to incorporate the local context information between prototypes. The implementation is the same as the self-attention layer in the representation encoder, but the keys, queries and values arise from IR/RGB modality prototypes. Subsequently, we compensate for the missing modality features by the cross-attention between the modality prototypes and part features. Given the RGB/IR feature map of the encoder output $\tilde{\mathbf{F}}^R = \left[\tilde{f}_1^R; \tilde{f}_2^R; \ldots; \tilde{f}_p^R \right] / \tilde{\mathbf{F}}^I = \left[\tilde{f}_1^I; \tilde{f}_2^I; \ldots; \tilde{f}_p^I \right] \in \mathbb{R}^{p \times d}$ ($\tilde{f}_i^R$ and $\tilde{f}_i^I$ represent the ith part feature of RGB and IR samples, respectively), we take RGB features as an example to elaborate on the compensation process of the IR modality. Specifically, the IR modality prototypes $\mathbf{P}^I$ are taken as the queries $\mathbf{Q}^I$, and the RGB part features $\tilde{\mathbf{F}}^R$ are taken as keys $\mathbf{K}^R$ and values $\mathbf{V}^R$ of the modality compensation decoder. Formally:

$$\mathbf{Q}^I = \mathbf{P}^I \mathbf{W}^Q, \mathbf{K}^R = \tilde{\mathbf{F}}^R \mathbf{W}^K, \mathbf{V}^R = \tilde{\mathbf{F}}^R \mathbf{W}^V. \tag{3}$$

Then, we can obtain the dot-production attention scores between queries $\mathbf{Q}^I$ and keys $\mathbf{K}^R$, which can be regarded as the soft correspondences between the modality prototypes and part features. To compensate for the missing modality features, we can project the part features into the corresponding modality space according to the attention weights. Concretely, the compensated IR part features $\hat{\mathbf{F}}^I = \left[\hat{f}_1^I; \hat{f}_2^I; \ldots; \hat{f}_p^I \right]$ for the RGB sample are derived as the weighted sum over all values $\mathbf{V}^R$:

$$
\begin{aligned}
\hat{\mathbf{F}}^I &= Attention \left(\mathbf{Q}^I, \mathbf{K}^R, \mathbf{V}^R \right) \\
&= softmax \left(\frac{\mathbf{Q}^I (\mathbf{K}^R)^T}{\sqrt{d}} \right) \mathbf{V}^R,
\end{aligned} \tag{4}
$$

where $\hat{\mathbf{F}}^I \in \mathbb{R}^{p \times d}$. Similarly, the compensated RGB part features $\hat{\mathbf{F}}^R$ for the samples with the IR modality can also be derived by Eq. (3) and Eq. (4). Finally, the complete modality representations can be acquired by combining the original features and the compensated modality features:

$$\mathbf{R} = \mathbf{F}^R + \hat{\mathbf{F}}^I, \quad \mathbf{I} = \mathbf{F}^I + \hat{\mathbf{F}}^R, \tag{5}$$

where $\mathbf{R}$ and $\mathbf{I}$ are the complete RGB and IR modality representations, respectively. These complete representations are in the shared embedding space, where the samples with different modalities can be aligned well. In this way, our modality compensation decoder can achieve a robust modality-level alignment and bridge the inter-modality discrepancies, which can facilitate a better cross-modality retrieval.

Modality Consistency Loss. As we have no ground truths for the compensated modality features, the learning of the decoder is difficult. To resolve this issue, we design a modality consistency loss to guide the learning of modality prototypes, which constrains the compensated RGB/IR features to be aligned with the real RGB/IR modality features. We first compute the two centroid features of each identity for two modalities in the mini-batch:

$$\mathbf{C}_i^R = \frac{1}{K} \sum_{j=1}^{K} \mathbf{F}_{i,j}^R, \quad \mathbf{C}_i^I = \frac{1}{K} \sum_{j=1}^{K} \mathbf{F}_{i,j}^I, \tag{6}$$

where $\mathbf{F}_{i,j}^{R}, \mathbf{F}_{i,j}^{I}$ denote the jth RGB/IR image feature of the ith person in the mini-batch, and $\mathbf{C}_i^R$, $\mathbf{C}_i^I$ represent the RGB/IR centroid features of the ith person. Based on the centroids, the modality consistency loss $\mathcal{L}_{cyc}^R$ and $\mathcal{L}_{cyc}^I$ for RGB/IR modalities are defined as:

$$\mathcal{L}_{cyc}^R = \frac{1}{NK} \sum_{i=1}^{N} \sum_{j=1}^{K} \|\hat{\mathbf{F}}_{i,j}^R - \mathbf{C}_i^R\|_2, \tag{7}$$

$$\mathcal{L}_{cyc}^I = \frac{1}{NK} \sum_{i=1}^{N} \sum_{j=1}^{K} \|\hat{\mathbf{F}}_{i,j}^I - \mathbf{C}_i^I\|_2, \tag{8}$$

where $\hat{\mathbf{F}}_{i,j}^R, \hat{\mathbf{F}}_{i,j}^I$ are the compensated RGB, IR features. Constrained by $\mathcal{L}_{cyc}^R$ and $\mathcal{L}_{cyc}^I$, the modality prototypes are forced to learn the corresponding modality information to approach the real modality features, which consequently facilitates a more reliable modality compensation.

ID Loss. To guide the complete representations $\mathbf{R}$ and $\mathbf{1}$ to focus on the ID-related discriminative information, we design an ID loss consisting of an identity classification loss $\mathcal{L}_{cls}$ and a hetero-center based triplet loss $\mathcal{L}_{hc\_tri}$ following the practice in [18]. Concretely, the ID loss is formulated as:

$$\mathcal{L}_{ID} = \mathcal{L}_{cls} + \mathcal{L}_{hc\_tri} \tag{9}$$

$$\mathcal{L}_{cls} = E\left(-log\, p\left(R\right)\right) + E\left(-log\, p\left(I\right)\right) \tag{10}$$

$$\mathcal{L}_{hc\_tri} = E\left[\alpha + d_{c_a,c_p} - d_{c_a,c_n}\right]_+, \tag{11}$$

where $p()$ is the probability of correct prediction, and E represents the expectation. In Eq. (11), c_a denotes the centroid feature calculated by the RGB features $\mathbf{R}$ or IR features $\mathbf{I}$ in the current mini-batch. c_a and c_p form a positive pair of centroid features belonging to the same person but with different modalities, while c_a and c_n form a negative pair of centroid features belonging to different persons, and α is a margin parameter.

3.2   Instance-Level Alignment Module

Due to the large intra-class variations like viewpoint changes and background clutter, the feature distribution of different samples with the same ID has large differences. Therefore, we propose an instance-level alignment module, where we leverage the characteristics of the given query to automatically adapt the instance features by the query-adaptive modulator. Specifically, the modulator employs an affine transformation to excite the query-related channels by the learned modulation parameters. Next we will give the details.

Parameter Generator. The Instance-level Alignment Module is symmetry for the visible and infrared modality. Given any sample feature $X \in \mathbb{R}^{p \times d}$ in the current mini-batch from RGB or IR modality, we take it as the query and transform the query characteristics into the modulation parameters. Concretely, we propose two parameter generators g_γ and g_β to obtain the channel-wise modulation parameters, *i.e.*, the scaling parameter γ and the shifting parameter β. Each generator contains two linear layers, with the first layer followed by a ReLU activation function. Formally, the modulation parameters γ and β are generated by

$$\gamma = g_\gamma(GAP(X)), \beta = g_\beta(GAP(X)), \tag{12}$$

where $\gamma, \beta \in \mathbb{R}^d$, and GAP represents the global average pooling, which is used to aggregate the part features. After the end-to-end training, the parameter generators g_γ and g_β can extract key characteristics in the query feature, and project them into the modulation weights that indicate which channels could be useful in the instance-level alignment. Although sharing a similar network structure with SENet [12], the parameter generator is designed to modulate other samples rather than enhance the samples themselves.

Query-Adaptive Modulation. The modulation parameters reveal the most discriminative patterns of X, and can be employed to perform the query-adaptive modulation on the other sample features Y in the current mini-batch to achieve the instance-level alignment. Specifically, the query-adaptive modulation layer employs an affine transformation by the scaling parameter γ and the shifting parameterβ on Y:

$$\bar{Y}_i = Y_i \odot \gamma + \beta, \tag{13}$$

where $\odot$ denotes a point-wise vector multiplication, and Y_i represents the ith part features of the sample Y, $\bar{Y}_i$ is the modulated feature. In the modulation, the crucial query-related channels of the Y can be strengthened and the irrelevant channels can be suppressed based on the modulation weights of γ and β. In this way, the instances that have the same ID with the query can be better aligned together. During the testing, the query-adaptive feature modulation will adjust the gallery representations according to the query features, which promotes the alignment between the query and gallery with the same ID, and contributes to a better retrieval.

Modulation Discriminative Loss. Without the constraints, the modulation on the channels may cause some disturbances to the representations, which will undermine the discrimination power of each instance. To help the modulated features preserve the discriminative ability, we propose a modulation discriminative loss to restrain the modulated features, which takes the form of the triplet loss:

$$\mathcal{L}_{mod} = E\left[\alpha + d_{X,\bar{Y}_p} - d_{X,\bar{Y}_n}\right]_+, \tag{14}$$

where X and $\bar{Y}_p$ form a positive pair of feature vectors belonging to the same person, X and $\bar{Y}_n$ form a negative pair of feature vectors belonging to different persons, α is a margin parameter.

3.3   Training and Inference

For the VI-ReID task, our proposed CMT is trained by minimizing the overall objective with identity labels as defined in

$$\mathcal{L}_{CMT} = \mathcal{L}_{ID} + \mathcal{L}_{cyc}^{R} + \mathcal{L}_{cyc}^{I} + \lambda\mathcal{L}_{mod}. \tag{15}$$

During the testing stage, we first extract query features, and then generate modulation parameters according to query features to adjust the feature embedding of galleries. Finally, we reshape the feature dimension to $\mathbb{R}^{pd}$ for the feature retrieval.

4   Experiments

In this section, we first introduce datasets and implementation details. Then, we show experimental results and some visualizations.

4.1   Dataset and Evaluation Protocol

SYSU-MM01 [34] is the first large-scale benchmark dataset for VI-ReID collected by 6 cameras, including 4 visible and 2 infrared cameras. Specially, four cameras are deployed in the outdoor environments and two are deployed in the indoor environments. SYSU-MM01 contains 491 persons with a total of 287,628 visible images and 15,792 infrared images. The training set contains 395 persons, including 22258 visible images and 11909 infrared images. The test set contains 96 persons, with 3,803 IR images for query and 301/3010 (one-shot/multi-shot) randomly selected RGB images as the gallery. Meanwhile, it contains two different testing settings, all-search and indoor-search settings. Detailed descriptions of the experimental settings can be found in [34].

RegDB. [23] is collected by a dual-camera system, including one visible and one infrared camera. There are 412 identities and 8,240 images in total, with 206 identities for training and 206 identities for testing. For each person, there are 10 visible images and 10 infrared images. The testing stage also contains two evaluation settings. One is Visible to Infrared to search IR images from a RGB image. The other setting is Infrared to Visible to search RGB images from a IR image. The evaluation procedure is repeated for 10 trials to record the mean values.

Evaluation Protocol. Two evaluation metrics are used to measure the performance. The first one is the Cumulative Matching Characteristic (CMC) curves. The CMC represents the probability that a query identity appears in different sized candidate lists. We report the rank-1,10,20 accuracy in experiments. The other is the Mean Average Precision (mAP).

4.2   Implementation Details

The proposed method is implemented with the PyTorch framework on a single RTX3090Ti GPU. Following the existing methods [20,25,33], we choose ResNet-50 [10] pretrained on ImageNet as the backbone network and reduce the stride of

Table 1. Performance comparison with state-of-the-art methods on SYSU-MM01 dataset. Rank-k accuracy (%) and mAP (%) are reported.

Method	Venue	All-search								Indoor-search							
		Single-shot				Multi-shot				Single-shot				Multi-shot			
		R1	R10	R20	mAP	R1	R10	R20	mAP	R1	R10	R20	mAP	R1	R10	R20	mAP
Zero-Padding [34]	ICCV-17	14.80	54.12	71.33	15.95	19.13	61.40	78.41	10.89	20.58	68.38	85.79	26.92	24.43	75.86	91.32	18.86
cmGAN [4]	IJCAI-18	26.97	67.51	80.56	27.80	31.49	72.74	85.01	22.27	31.63	77.23	89.18	42.19	37.00	80.94	92.11	32.76
D²RL [32]	CVPR-19	28.90	70.60	82.40	29.20	–	–	–	–	–	–	–	–	–	–	–	–
Hi-CMD [3]	CVPR-20	34.94	77.58	–	35.94	–	–	–	–	–	–	–	–	–	–	–	–
JSIA-ReID [28]	AAAI-20	38.10	80.70	89.90	36.90	45.10	85.70	93.80	29.50	43.80	86.20	94.20	52.90	52.70	91.10	96.40	42.70
AlignGAN [29]	ICCV-19	42.40	85.00	93.70	40.70	51.50	89.40	95.70	33.90	45.90	87.60	94.40	54.30	57.10	92.70	97.40	45.30
cm-SSFT(sq) [20]	CVPR-20	47.70	–	–	54.10	–	–	–	–	57.40	–	–	59.10	–	–	–	–
XIV [15]	AAAI-20	49.92	89.79	95.96	50.73	–	–	–	–	–	–	–	–	–	–	–	–
DDAG [39]	ECCV-20	54.75	90.39	95.81	53.02	–	–	–	–	61.02	94.06	98.41	67.98	–	–	–	–
LbA [25]	ICCV-21	55.41	–	–	54.1	57.4	–	–	59.1	–	–	–	–	–	–	–	–
NFS [2]	CVPR-21	56.91	91.34	96.52	55.45	63.51	94.42	97.81	48.56	62.79	96.53	99.07	69.79	70.03	97.7	99.51	61.45
HCT [18]	TMM-20	61.68	93.1	97.17	57.51	–	–	–	–	63.41	91.69	95.28	68.17	–	–	–	–
CM-NAS [7]	ICCV-21	61.99	92.87	97.25	60.02	68.68	94.92	98.36	53.45	67.01	97.02	99.32	72.95	76.48	98.68	99.91	65.11
MCLNet [8]	ICCV-21	65.40	93.33	97.14	61.98	–	–	–	–	72.56	96.98	99.20	76.58	–	–	–	–
SMCL [33]	ICCV-21	67.39	92.87	96.76	61.78	72.15	90.66	94.32	54.93	68.84	96.55	98.77	75.56	79.57	95.33	98.00	66.57
MPANet [35]	CVPR-21	70.58	96.21	98.80	68.24	75.58	97.91	99.43	62.91	76.74	**98.21**	99.57	**80.95**	84.22	**99.66**	99.96	**75.11**
CMT (our)	ECCV-22	**71.88**	**96.45**	**98.87**	**68.57**	**80.23**	97.91	**99.53**	**63.13**	**76.9**	97.68	**99.64**	79.91	**84.87**	99.41	**99.97**	74.11

the last convolutional block from 2 to 1. For each mini-batch, we randomly choose 8 identities from each modality and sample 8 person images for each identity. The input images are first resized to 384×144, then we adopt random cropping with zero-padding, random horizontal flipping, and random erasing for data augmentation. In addition, we use the Adam optimizer for optimization with an initial learning rate of 3.5×10^{-4}, and the weight decay is set to 5×10^{-4}. We decay the learning rate by 0.1 and 0.01 at 60 and 90 epochs. The whole training process consists of 120 epochs. The number of part features p is set to 6. The hype-parameters λ is set to 0.2.

4.3   Comparison with the State-of-the-Art Methods

Comparisons on SYSU-MM01. We compare our CMT with various state-of-the-art methods under both all-search and single-search settings. As shown in Table 1, our CMT ranks either the first or the second among all settings, and sets the new state-of-the-art results in all-search setting, which strongly proves the effectiveness of our method. In the indoor-search setting, our method also performs comparably with the state-of-the-art methods. Based on the results, we have the following observations. (1) Compared with the methods (cmGAN [4], Hi-CMD [3], AlignGAN [29]) that only focus on learning modality-shared features by feature disentanglement, our method achieves much better performance on all settings. This is because the modality-shared features lose some useful identity information, such as colors. Therefore, with modality-shared cues only, the upper bound of the discrimination ability of the feature representation is limited. Differently, we design a modality-level alignment module to adaptively compensate for the lacking modality-special information via a transformer encoder-decoder architecture. (2) Compared with the best modality compensation method (i.e., cm-SSFT [20] in a multi-query setting), our method improves the Rank-1 accuracy and mAP by 10.28% and 5.37% in the all-search single shot setting. The reason is that the

Table 2. Comparison of the Rank-k accuracy (%) and mAP (%) performances with state-of-the-art methods on RegDB.

Method	Venue	Visible to Infrared				Infrared to Visible			
		R1	R10	R20	mAP	R1	R10	R20	mAP
Zero-Padding [34]	ICCV-17	17.75	34.21	44.35	18.90	16.63	34.68	44.25	17.82
D^{2}RL [32]	CVPR-19	43.4	66.1	76.3	44.1	–	–	–	–
JSIA-ReID [28]	AAAI-20	48.50	–	–	48.90	–	–	–	–
AlignGAN [29]	ICCV-19	57.90	–	–	53.60	56.30	–	–	53.40
XIV [15]	AAAI-20	–	–	–	–	62.21	83.13	91.72	60.18
cm-SSFT(sq) [20]	CVPR-20	65.4	–	–	65.6	63.8	–	–	64.2
DDAG [39]	ECCV-20	69.34	86.19	91.49	63.46	68.06	85.15	90.31	61.80
Hi-CMD [3]	CVPR-20	70.93	86.39	–	66.04	–	–	–	–
LbA [25]	ICCV-21	74.17	–	–	67.64	72.43	–	–	65.46
MCLNet [8]	ICCV-21	80.31	92.70	96.03	73.07	75.93	90.93	94.59	69.49
NFS [2]	CVPR-21	80.54	91.96	95.07	72.1	77.95	90.45	93.62	69.79
MPANet [35]	CVPR-21	83.7	–	–	80.9	82.8	–	–	80.7
SMCL [33]	ICCV-21	83.93	–	–	79.83	83.05	–	–	78.57
CM-NAS [7]	ICCV-21	84.54	95.18	97.85	80.32	82.57	94.51	97.37	78.31
HCT [18]	TMM-20	91.05	97.16	98.57	83.28	89.3	96.41	98.16	81.46
CMT (our)	ECCV-22	**95.17**	**98.82**	**99.51**	**87.3**	**91.97**	**97.92**	**99.07**	**84.46**

compensated features in [20] are produced solely based on the samples of the current mini-batch. This strategy suffers from a certain randomness, and does not match the default single query settings of most methods. Notably, we introduce several modality prototypes to store the global modality characteristics without relying on the current mini-batch. (3) Compared with JSIA-ReID [28] that is based on instance-level alignment between the cross-modality paired images generated by the GAN, our method acquires a better performance in all results. This is because different from JSIA-ReID, we exploit query-adaptive feature modulation to conduct more diverse and flexible instance-level alignment. In our method, the gallery instances can be adaptively refined according to the query features, while other methods do not take this into account.

Comparisons on RegDB. As shown in Table 2, it can be seen that our CMT has distinct advantages over the state-of-the-art methods on RegDB. Under the Visible to Infrared setting, compared with the state-of-the-art HCT [18], our method improves the Rank-1 accuracy and mAP by 4.12% and 4.02%. When switching to the Infrared to Visible setting, our method surpasses the HCT by 4.02% and 3% in terms of the Rank-1 accuracy and the mAP, respectively. Hence, it can be proved that our proposed method is robust against different query settings.

4.4   Ablation Study

In this section, we perform detailed ablation studies on SYSU-MM01 dataset under the all-search setting to evaluate each component of our CMT. We denote the Modality-level Alignment Module as MAM and the Instance-level Alignment Module as IAM. The results are shown in Table 3.

Table 3. Analysis of the effectiveness of different components on SYSU-MM01 dataset under the all-search setting. Rank-k accuracy (%) and mAP (%) are reported.

Base	MAM	IAM	Rank-1	Rank-10	Rank-20	mAP
✓	✗	✗	65.35	93.57	97.58	64.27
✓	✓	✗	70.55	95.27	98.21	66.50
✓	✗	✓	68.5	94.21	98.34	67.25
✓	✓	✓	**71.88**	**96.45**	**98.87**	**68.57**

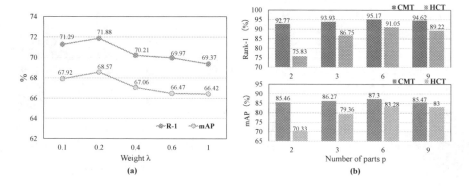

Fig. 3. The effect of weight λ in Eq. (15) on SYSU-MM01 dataset under the all-search setting and the number of parts p on RegDB dataset. Rank-1 and mAP (%) are reported.

Baseline. We adopt the HCT [18] as our baseline method, which explores the two-stream network with shared parameters and uses a hetero-center based triplet loss to improve the traditional triplet loss. In addition, we replace the optimizer with the Adam optimizer and add random erasing as extra data augmentation. The details of the implementation can be found in Sect. 4.2.

Effectiveness of the Modality-Level Alignment. Compared with the baseline model, the modality-level alignment module improves the Rank-1 accuracy and mAP by 5.2% and 2.23%. The improvements can be mainly ascribed to two reasons. For one thing, we automatically explore the modality prototypes by the modality consistency constraint, which can adaptively learn the modality-related information. The other reason is that, we conduct the modality feature compensation by the transformer, which can project the features of different modalities into a common complete space to achieve a better modality-level alignment.

Effectiveness of the Instance-Level Alignment. Compared with the baseline model, adding the instance-level alignment, the performance is greatly improved by 2.98% and up to 67.25% mAP. Besides, on top of the modality-level alignment, the instance-level alignment can still achieve 2.07% improvements in mAP. This shows that the instance-level alignment is useful to reduce the distances of the samples in the same class. The complete version of our CMT gives the best results on SYSU-MM01 dataset under all-search setting, achieving a whopping accuracy gain of 6.53% and 4.3% on in Rank-1 and mAP, which proves the effectiveness of CMT.

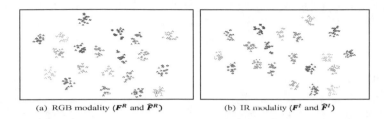

(a) RGB modality ($\boldsymbol{F}^R$ and $\hat{\boldsymbol{F}}^R$)   (b) IR modality ($\boldsymbol{F}^I$ and $\hat{\boldsymbol{F}}^I$)

Fig. 4. The t-SNE visualization of features on SYSU-MM01 dataset. The colors represent different categories. Circles represent original RGB/IR features and triangles represent compensated RGB/IR features.

4.5 Model Analysis

Parameters Analysis. We first evaluate the effect of the weight λ in Eq. (15) on SYSU-MM01 dataset under the all-search setting. The Rank-1 and the mAP results of CMT with different λ are exhibited in Fig. 3 (a). The most suitable parameter setting is to set λ as 0.2. Then, we compare the performance of CMT and our baseline model HCT [18] with different number of parts p. As shown in Fig. 3 (b), with p increasing, the performance keeps improving before p arrives 6 on RegDB dataset. This is because a bigger p allows the network to pay more attention to the details. Besides, we can observe that CMT shows surprisingly powerful results and significant improvements over the baseline under the same p setting. Compared with the HCT, CMT is more robust to p, which further verifies the effectiveness of our method.

Visualization Analysis. To further verify the effectiveness of our modality-level alignment module, we use t-SNE [22] to visualize the original modality features ($\mathbf{F}^R$ and $\mathbf{F}^I$) and the compensated modality features ($\hat{\mathbf{F}}^R$ and $\hat{\mathbf{F}}^I$). As shown in Fig. 4, compensated RGB/IR features are aligned with original RGB/IR features of the same ID in the feature space. It proves that our work can compensate for the lacking modality information to achieve a better modality-level alignment.

5 Conclusion

In this paper, we propose a novel Cross-Modality Transformer (CMT) to jointly explore a modality-level alignment module and an instance-level module for VI-REID. The proposed modality-level alignment module is able to compensate for the missing modality-specific information via a Transformer encoder-decoder architecture. We have also designed an instance-level alignment module to adaptively adjust the sample features, which is achieved by query-adaptive feature modulation. Extensive experimental results on two standard benchmarks demonstrate that our model performs favorably against state-of-the-art methods.

Acknowledgements. This work was partially supported by the National Nature Science Foundation of China (62022078, 12150007, 62021001), National Defense Basic Scientific Research Program (JCKY2020903B002), and University Synergy Innovation Program of Anhui Province No. GXXT-2019-025.

References

1. Carion, N., Massa, F., Synnaeve, G., Usunier, N., Kirillov, A., Zagoruyko, S.: End-to-end object detection with transformers. In: Vedaldi, A., Bischof, H., Brox, T., Frahm, J.-M. (eds.) ECCV 2020. LNCS, vol. 12346, pp. 213–229. Springer, Cham (2020). https://doi.org/10.1007/978-3-030-58452-8_13
2. Chen, Y., Wan, L., Li, Z., Jing, Q., Sun, Z.: Neural feature search for rgb-infrared person re-identification. In: Proceedings of the IEEE Conference on Computer Vision and Pattern Recognition, pp. 587–597 (2021)
3. Choi, S., Lee, S., Kim, Y., Kim, T., Kim, C.: Hi-cmd: Hierarchical cross-modality disentanglement for visible-infrared person re-identification. In: Proceedings of the IEEE Conference on Computer Vision and Pattern Recognition, pp. 10257–10266 (2020)
4. Dai, P., Ji, R., Wang, H., Wu, Q., Huang, Y.: Cross-modality person re-identification with generative adversarial training. In: International Joint Conference on Artificial Intelligence, vol. 1, p. 2 (2018)
5. Dosovitskiy, A., et al.: An image is worth 16×16 words: Transformers for image recognition at scale. arXiv preprint arXiv:2010.11929 (2020)
6. Feng, Z., Lai, J., Xie, X.: Learning modality-specific representations for visible-infrared person re-identification. IEEE Trans. Image Process. **29**, 579–590 (2019)
7. Fu, C., Hu, Y., Wu, X., Shi, H., Mei, T., He, R.: Cm-nas: Cross-modality neural architecture search for visible-infrared person re-identification. arXiv preprint arXiv:2101.08467 (2021)
8. Hao, X., Zhao, S., Ye, M., Shen, J.: Cross-modality person re-identification via modality confusion and center aggregation. In: Proceedings of the IEEE International Conference on Computer Vision, pp. 16403–16412 (2021)
9. Hao, Y., Wang, N., Li, J., Gao, X.: Hsme: hypersphere manifold embedding for visible thermal person re-identification. In: Proceedings of the AAAI Conference on Artificial Intelligence, vol. 33, pp. 8385–8392 (2019)
10. He, K., Zhang, X., Ren, S., Sun, J.: Deep residual learning for image recognition. In: Proceedings of the IEEE Conference on Computer Vision and Pattern Recognition, pp. 770–778 (2016)
11. He, S., Luo, H., Wang, P., Wang, F., Li, H., Jiang, W.: Transreid: Transformer-based object re-identification. arXiv preprint arXiv:2102.04378 (2021)
12. Hu, J., Shen, L., Sun, G.: Squeeze-and-excitation networks. In: Proceedings of the IEEE Conference on Computer Vision and Pattern Recognition, pp. 7132–7141 (2018)
13. Jiang, K., Zhang, T., Zhang, Y., Wu, F., Rui, Y.: Self-supervised agent learning for unsupervised cross-domain person re-identification. IEEE Trans. Image Process. **29**, 8549–8560 (2020)

14. Kniaz, V.V., Knyaz, V.A., Hladuvka, J., Kropatsch, W.G., Mizginov, V.: Thermal-gan: Multimodal color-to-thermal image translation for person re-identification in multispectral dataset. In: Proceedings of the European Conference on Computer Vision Workshops (2018)

15. Li, D., Wei, X., Hong, X., Gong, Y.: Infrared-visible cross-modal person re-identification with an x modality. In: Proceedings of the AAAI Conference on Artificial Intelligence, vol. 34, pp. 4610–4617 (2020)

16. Li, W., Zhu, X., Gong, S.: Harmonious attention network for person re-identification. In: Proceedings of the IEEE Conference on Computer Vision and Pattern Recognition, pp. 2285–2294 (2018)

17. Li, Y., He, J., Zhang, T., Liu, X., Zhang, Y., Wu, F.: Diverse part discovery: Occluded person re-identification with part-aware transformer. In: Proceedings of the IEEE/CVF Conference on Computer Vision and Pattern Recognition, pp. 2898–2907 (2021)

18. Liu, H., Tan, X., Zhou, X.: Parameter sharing exploration and hetero-center triplet loss for visible-thermal person re-identification. IEEE Trans. Multimedia **23**, 4414–4425 (2020)

19. Liu, Z., et al.: Swin transformer: Hierarchical vision transformer using shifted windows. In: Proceedings of the IEEE/CVF International Conference on Computer Vision, pp. 10012–10022 (2021)

20. Lu, Y., et al.: Cross-modality person re-identification with shared-specific feature transfer. In: Proceedings of the IEEE Conference on Computer Vision and Pattern Recognition, pp. 13379–13389 (2020)

21. Luo, H., Gu, Y., Liao, X., Lai, S., Jiang, W.: Bag of tricks and a strong baseline for deep person re-identification. In: Proceedings of the IEEE Conference on Computer Vision and Pattern Recognition Workshops (2019)

22. Van der Maaten, L., Hinton, G.: Visualizing data using t-sne. J. Mach. Learn. Res. **9**(11), 2579–2605 (2008)

23. Nguyen, D.T., Hong, H.G., Kim, K.W., Park, K.R.: Person recognition system based on a combination of body images from visible light and thermal cameras. Sensors **17**, 605 (2017)

24. Oh Song, H., Xiang, Y., Jegelka, S., Savarese, S.: Deep metric learning via lifted structured feature embedding. In: Proceedings of the IEEE Conference on Computer Vision and Pattern Recognition, pp. 4004–4012 (2016)

25. Park, H., Lee, S., Lee, J., Ham, B.: Learning by aligning: Visible-infrared person re-identification using cross-modal correspondences. In: Proceedings of the IEEE International Conference on Computer Vision, pp. 12046–12055 (2021)

26. Sun, Y., Zheng, L., Yang, Y., Tian, Q., Wang, S.: Beyond part models: Person retrieval with refined part pooling (and a strong convolutional baseline). In: Proceedings of the European Conference on Computer Vision, pp. 480–496 (2018)

27. Vaswani, A., et al.: Attention is all you need. In: Advances in Neural Information Processing Systems, pp. 5998–6008 (2017)

28. Wang, G.A., et al.: Cross-modality paired-images generation for rgb-infrared person re-identification. In: Proceedings of the AAAI Conference on Artificial Intelligence, vol. 34, pp. 12144–12151 (2020)

29. Wang, G., Zhang, T., Cheng, J., Liu, S., Yang, Y., Hou, Z.: Rgb-infrared cross-modality person re-identification via joint pixel and feature alignment. In: Proceedings of the IEEE International Conference on Computer Vision, pp. 3623–3632 (2019)

30. Wang, G., Yuan, Y., Chen, X., Li, J., Zhou, X.: Learning discriminative features with multiple granularities for person re-identification. In: Proceedings of the ACM International Conference on Multimedia, pp. 274–282 (2018)

31. Wang, W., et al.: Pyramid vision transformer: A versatile backbone for dense prediction without convolutions. In: Proceedings of the IEEE/CVF International Conference on Computer Vision, pp. 568–578 (2021)

32. Wang, Z., Wang, Z., Zheng, Y., Chuang, Y.Y., Satoh, S.: Learning to reduce dual-level discrepancy for infrared-visible person re-identification. In: Proceedings of the IEEE Conference on Computer Vision and Pattern Recognition, pp. 618–626 (2019)

33. Wei, Z., Yang, X., Wang, N., Gao, X.: Syncretic modality collaborative learning for visible infrared person re-identification. In: Proceedings of the IEEE International Conference on Computer Vision, pp. 225–234 (2021)

34. Wu, A., Zheng, W.S., Yu, H.X., Gong, S., Lai, J.: Rgb-infrared cross-modality person re-identification. In: Proceedings of the IEEE International Conference on Computer Vision, pp. 5380–5389 (2017)

35. Wu, Q., et al.: Discover cross-modality nuances for visible-infrared person re-identification. In: Proceedings of the IEEE Conference on Computer Vision and Pattern Recognition, pp. 4330–4339 (2021)

36. Yang, X., Zhou, P., Wang, M.: Person reidentification via structural deep metric learning. IEEE Trans. Neural Netw. Learn. Syst. **30**(10), 2987–2998 (2018)

37. Ye, M., Lan, X., Leng, Q., Shen, J.: Cross-modality person re-identification via modality-aware collaborative ensemble learning. IEEE Trans. Image Process. **29**, 9387–9399 (2020)

38. Ye, M., Lan, X., Li, J., Yuen, P.: Hierarchical discriminative learning for visible thermal person re-identification. In: Proceedings of the AAAI Conference on Artificial Intelligence, vol. 32 (2018)

39. Ye, M., Shen, J., J. Crandall, D., Shao, L., Luo, J.: Dynamic dual-attentive aggregation learning for visible-infrared person re-identification. In: Vedaldi, A., Bischof, H., Brox, T., Frahm, J.-M. (eds.) ECCV 2020. LNCS, vol. 12362, pp. 229–247. Springer, Cham (2020). https://doi.org/10.1007/978-3-030-58520-4_14

40. Ye, M., Shen, J., Lin, G., Xiang, T., Shao, L., Hoi, S.C.: Deep learning for person re-identification: A survey and outlook. IEEE Trans. Pattern Anal. Mach. Intell. **44**, 2872–2893 (2021)

41. Ye, M., Wang, Z., Lan, X., Yuen, P.C.: Visible thermal person re-identification via dual-constrained top-ranking. In: International Joint Conference on Artificial Intelligence, vol. 1, p. 2 (2018)

42. Zheng, L., Yang, Y., Hauptmann, A.G.: Person re-identification: Past, present and future. arXiv preprint arXiv:1610.02984 (2016)

43. Zheng, W.S., Gong, S., Xiang, T.: Reidentification by relative distance comparison. IEEE Trans. Pattern Anal. Mach. Intell. **35**(3), 653–668 (2012)

44. Zhong, Z., Zheng, L., Cao, D., Li, S.: Re-ranking person re-identification with k-reciprocal encoding. In: Proceedings of the IEEE Conference on Computer Vision and Pattern Recognition, pp. 1318–1327 (2017)

45. Zhu, X., Su, W., Lu, L., Li, B., Wang, X., Dai, J.: Deformable detr: Deformable transformers for end-to-end object detection. arXiv preprint arXiv:2010.04159 (2020)

Audio-Visual Mismatch-Aware Video Retrieval via Association and Adjustment

Sangmin Lee[✉], Sungjune Park, and Yong Man Ro

Image and Video Systems Lab, KAIST, Daejeon, South Korea
{sangmin.lee,sungjune-p,ymro}@kaist.ac.kr

Abstract. Retrieving desired videos using natural language queries has attracted increasing attention in research and industry fields as a huge number of videos appear on the internet. Some existing methods attempted to address this video retrieval problem by exploiting multimodal information, especially audio-visual data of videos. However, many videos often have mismatched visual and audio cues for several reasons including background music, noise, and even missing sound. Therefore, the naive fusion of such mismatched visual and audio cues can negatively affect the semantic embedding of video scenes. Mismatch condition can be categorized into two cases: (i) Audio itself does not exist (ii) Audio exists but does not match with visual. To deal with (i), we introduce audio-visual associative memory (AVA-Memory) to associate audio cues even from videos without audio data. The associated audio cues can guide the video embedding feature to be aware of audio information even in the missing audio condition. To address (ii), we propose audio embedding adjustment by considering the degree of matching between visual and audio data. In this procedure, constructed AVA-Memory enables to figure out how well the visual and audio in the video are matched and to adjust the weighting between actual audio and associated audio. Experimental results show that the proposed method outperforms other state-of-the-art video retrieval methods. Further, we validate the effectiveness of the proposed network designs with ablation studies and analyses.

Keywords: Video retrieval · Audio-visual mismatch · Audio association · Embedding adjustment · Memory

1 Introduction

Video retrieval is to find corresponding videos from natural language queries made by humans. Given the huge number of videos on the internet, it is highly time-consuming and labor-intensive for people to find desired video scenes manually. Thus, automatic video retrieval has attracted increasing attention in research and industry fields due to its high practicality.

Video retrieval methods utilizing deep neural networks (DNNs) have been proposed to address arising issues in video retrieval. Some works focused on

Supplementary Information The online version contains supplementary material available at https://doi.org/10.1007/978-3-031-19781-9_29.

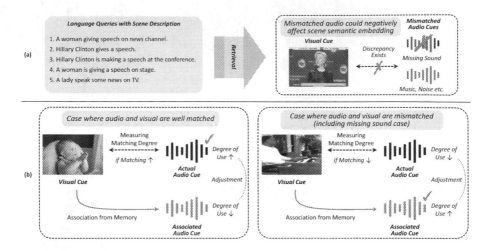

Fig. 1. (a) shows the examples of mismatching between visual and audio cues. (b) describes the concept of the proposed method for video embedding. To cope with the mismatch condition, the model adjusts the proportion of using actual audio and associated audio considering the degree of audio-visual matching.

hierarchical feature matching [46,61] for video retrieval. These methods tried to perform video-text matching in both local (*e.g.*, word-scene) and global (*e.g.*, sentence-video) levels. There exist metric learning-based video retrieval methods [45,52,53]. They formulated training metric criterion by considering similarities and relationships among samples. Several works addressed architectural aspects [4,10]. They investigated effective architectures such as multi-level embedding and pooling strategies. These works mainly focused on visual and text matching.

However, natural language queries vary greatly and often contain details related to audio cues. For example, language queries 'A woman giving speech on news channel' and 'A woman is singing while men are playing on guitars' include audio information and it is worth to give guiding the model with audio information. Some existing works tried to address such video retrieval problem using multi-modal information, especially audio-visual data of videos [17,36,50,54]. They showed that audio cues can contribute to the performance of video retrieval. Nonetheless, these works did not take into account the mismatch condition of visual and audio. In many videos, visual and audio cues are often not matched due to several reasons such as background music, noise, and even missing sound. Therefore, the naive fusion of visual and audio can negatively affect matching video with text queries. When fusing the visual and audio in the video, mismatched audio can guide the video embedding to have distracted semantics.

Our work addresses audio-visual mismatch issues for retrieving video from text, which have not been properly dealt with in previous video retrieval works. The audio-visual mismatch in video can be categorized into two cases: (*i*) Audio itself does not exist (*ii*) Audio exists but does not match with visual (*e.g.*, background music, noise). Figure 1-(a) shows that such mismatched audio cues do not help to obtain appropriate embeddings for matching language semantics.

In this paper, we introduce a novel mismatch-aware associative transformer (MA-Transformer) with association and adjustment processes to deal with the aforementioned issues. To address the issue (i), we propose audio-visual associative memory (AVA-Memory) in MA-Transformer to associate audio cues even from videos without actual audio data. AVA-Memory is composed of visual and audio memories to store information of two modalities and enables to associate different modal cues from the other modal input. The associated audio cues from visual data can guide the video embedding feature to be aware of audio information for video-text matching even in the missing audio condition.

In addition, to deal with the issue (ii), we propose audio embedding adjustment by considering the degree of matching between visual and audio data (See Fig. 1-(b)). Through AVA-Memory, the degree of audio-visual matching is determined by performing mutual associations from audio to visual and from visual to audio. It is possible to figure out how well the visual and audio of the video are matched and to adjust the weighting between actual audio and associated audio. If the visual and audio are matched well, the actual audio cue is mainly used for video embedding. Conversely, if the visual and audio are mismatched, the model adaptively lowers the use of the actual audio and performs video embedding mainly by using the associated audio. In the case of missing audio, only associated audio can be used for embedding.

The major contributions of the paper are as follows.

- We introduce a novel MA-Transformer with AVA-Memory for video retrieval. We can associate the audio cues from visual cues in the video. It enables to guide the visual embedding to be aware of audio context jointly for video-text matching even in the missing audio condition.
- We propose audio embedding adjustment which enables to address the mismatch between existing visual and audio cues. We can figure out the degree of audio-visual matching through the constructed AVA-Memory and adjust the use of audio cues based on it. To the best of our knowledge, it is the first attempt to deal with audio-visual mismatch issues in video retrieval.

2   Related Works

2.1   Video Retrieval

Video retrieval is to find corresponding videos from natural language queries made by humans. Based on the high practicality, video retrieval attracts increasing attention. Compared with image retrieval [13,24,28], video retrieval is more challenging because it necessitates thorough comprehension of temporal dynamics as well as complicated text semantics.

Early video retrieval works extended the image retrieval approach by spatiotemporal aggregation of frames for each video [8,43,47,57]. There have been methods which develop feature aggregation for video retrieval. For example, average pooling [36,42] and max pooling [41,54] were used as aggregation methods for embedding feature in retrieval. Chen et al. [4] proposed a generalized pooling operator to automatically use the best pooling strategy for each feature for retrieval. Recently, Dong *et al.* [10] proposed a dual-encoding network with multiple levels of feature capabilities for video retrieval. In [10], multi-level features are from average pooling, bi-directional GRU, and convolutional layer. Some works focused on hierarchical feature matching in terms of global and local matching. [60] estimates similarity through comparison between each word of text description and video frame. Zhang *et al.* [61] performs a paragraph-based video search using hierarchical decomposition of videos and paragraphs. Song *et al.* [46] proposed varied representations by combining global context with local features to consider polysemous videos. There were attempts to effectively pretrain the model with uncurated instruction [39] or image-level caption data [1].

These approaches do not take advantage of the further diverse information related to videos, including voice and other background sounds, which in practice can affect human-made natural language descriptions. [40,42] proposed a method using a pretrained model for audio and motion recognition. [36] investigated additional cues such as faces, OCR, speech. They proposed a more effective use of multi-modal features through a collaborative gating method. In addition to this, [12,16,17,35,38,50] introduced a video retrieval method by fusing the multi-modalities with transformer [48] structures. [50] proposed video-text matching method in terms of global and local alignment with multi-modal transformer.

However, the previous works did not take into account the visual and audio mismatch conditions. Many videos often have mismatched visual and audio cues, which can negatively affect matching video with text queries. When fusing the visual and audio in the video, mismatched audio can guide the video embedding to have distracted semantics. We introduce a novel MA-Transformer with AVA-Memory for addressing such mismatch issues.

2.2 Memory-Augmented Network

A memory-augmented network represents the neural network that includes external memory components for reading and writing historical information. Memory-augmented networks have been proposed to handle a variety of challenges in the deep learning field. There were several tasks to exploit the memory such as anomaly detection [18,44], few-shot learning [2,23,62], object tracking/detection [15,25,58], future prediction [29,37], and representation learning [19,26,30]. There exist methods that exploit the memory-augmented network for cross-modal retrieval which is image and sentence matching [22]. It utilizes memorization of shared semantic representations to address the few-shot condition.

Unlike the existing memory-augmented networks, we introduce a novel AVA-Memory for learning audio-visual correspondences with audio and visual sub-memories. Through proposed audio-visual associative learning with the memory, it is possible to recall audio cues from the videos without audio and further to measure how well visual and audio are matched. The degree of matching from AVA-Memory is utilized to adjust audio embeddings afterward.

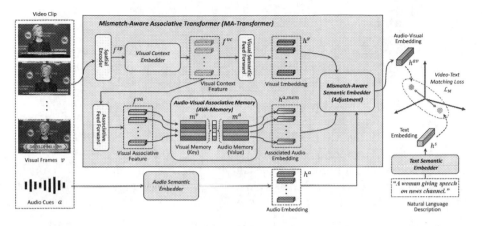

Fig. 2. Overall framework containing a proposed MA-Transformer for video retrieval. MA-Transformer mainly consists of 3 parts: a visual context embedder, an audio-visual associative memory, and a mismatch-aware semantic embedder. Each of them is for extracting spatio-temporal context of visual data, associating audio cues from visual features, and encoding video semantics jointly with visual and audio with being aware of mismatch condition, respectively.

3   Proposed Approach

Video retrieval task can be formulated as follows. Let $v = \{v_t\}_{t=1}^{n}$ denote the n frames in video clip while $a = \{a_t\}_{t=1}^{n}$ indicates audio cues with n partial audio samples. a_t is VGGish [36] feature of 1s audio and v_t is middle frame in 1s. Let s denote text sentence. A video mapping function $\mathcal{F}_v$ and a text mapping function $\mathcal{F}_s$ are optimized to make video embedding $\mathcal{F}_v(v, a)$ and text embedding $\mathcal{F}_s(s)$ be matched. As a result, it is possible to retrieve videos from text by measuring similarity between the embeddings. The goal of this work is to make video embedding to be aware of the auditory context even in the case of audio-visual mismatch conditions (*e.g.*, missing sound, background music).

To this end, we introduce a mismatch-aware associative transformer (MA-Transformer) for video embedding. MA-Transformer consists of three major parts: a visual context embedder, an audio-visual associative memory (AVA-Memory), and a mismatch-aware semantic embedder. The visual contexts are processed with self-attention mechanism to focus on the parts suitable for associating audios and embedding the semantics of visual data. AVA-Memory includes visual and audio sub-memories and is able to associate audio cues from visual cues. Finally, in the mismatch-aware semantic embedder, the degree of use between actual audio and associated audio is adjusted in consideration of the degree of matching between visual and audio. Then, it encodes audio-visual semantic embeddings that effectively match text queries. In terms of training, we propose audio-visual associative learning which enables to learn the association between visual and audio modalities in a self-supervised way.

3.1   Mismatch-Aware Associative Transformer

Figure 2 shows the overall framework of the proposed MA-Transformer. First, each frame of the input video is independently fed to a spatial encoder (CNN) to extract spatial features $f^{sp} = \{f_t^{sp}\}_{t=1}^n \in \mathbb{R}^{n \times d_{sp}}$ (d_{sp} is channel dim).

Visual Context Embedder. The extracted spatial features are received by the visual context embedder to encode the relationship between features. The visual context embedder mainly has the form of transformer [11,48] and can encode the overall spatio-temporal context of visual features through the self-attention mechanism. The overall process of the context embedder is similar to that of the transformer [48]. However, the last part of it consists of multi-head attention, not feed-forward. It can be formulated as follows.

$$
\begin{aligned}
e_0 &= [f_1^{sp}W_{vc}; f_2^{sp}W_{vc}; \ldots; f_n^{sp}W_{vc}] + E_{\text{POS}}, \\
e_l' &= \text{MHA}(\text{LN}(e_{l-1})) + e_{l-1}, & l = 1, \ldots, L_v \\
e_l &= \text{MLP}(\text{LN}(e_l')) + e_l', & l = 1, \ldots, L_v \\
f^{vc} &= \text{MHA}(\text{LN}(e_{L_v})) + e_{L_v},
\end{aligned}
\tag{1}
$$

where $W_{vc} \in \mathbb{R}^{d_{sp} \times d_v}$ indicates a fc layer and E_{POS} is positional encoding. MHA and LN represent multi-head attention and layer normalization, respectively. L_v indicates the number of layers. As a result, we obtain visual context feature $f^{vc} = \{f_t^{vc}\}_{t=1}^n \in \mathbb{R}^{n \times d_v}$. We apply multi-head attention as the last layer to aggregate f_t^{vc} differently with separate feed forward layers later. Note that the feed forward layer indicates $y = \text{MLP}(\text{LN}(x)) + x$ operation. f^{vc} separately goes through two paths. One (upper path of f^{vc} in Fig. 2) is for embedding semantic-related visual features. The other (lower path of f^{vc} in Fig. 2) is for associating audio features from the visual context.

In the upper path, visual semantic feed forward is further applied to f^{vc} to aggregate for attending the semantic-related positions of the video sequence. Through the procedure, we obtain visual embedding $h^v = \{h_t^v\}_{t=1}^n \in \mathbb{R}^{n \times d_v}$ that contains semantic characteristics of visual cues. The visual embedding is directly utilized as input of the mismatch-aware semantic embedder.

In the lower path, another feed forward layer, associative feed forward is further applied to f^{vc} to aggregate the audio-related characteristics of the video sequence. It is possible because we construct the audio-visual association based on the output of this feed forward. We get visual associative feature $f^{va} = \{f_t^{va}\}_{t=1}^n \in \mathbb{R}^{n \times d_v}$ which is used to recall audio cues from AVA-Memory.

AVA-Memory. The extracted visual associative features are utilized as memory queries for accessing a visual memory m^v and an audio memory m^a in AVA-Memory. The visual and audio memories are constructed as $m^v = \{m_r^v\}_{r=1}^k \in \mathbb{R}^{k \times d_v}$ and $m^a = \{m_r^a\}_{r=1}^k \in \mathbb{R}^{k \times d_a}$, respectively with k slots and (d_v, d_a) channels. A vector $m_r^v \in \mathbb{R}^{d_v}$ indicates the r-th memory component of m^v. AVA-Memory has key-value memory structure to map one modal space to another. In this case, the visual memory is the key and the audio memory is the value. Addressing vectors $w_t^v = \{w_{t-r}^v\}_{r=1}^k \in \mathbb{R}^k$ is individually obtained from each time component f_t^{va} of visual associative features f^{va}. Note that each addressing vector is used to access the components of audio memory m^a. The addressing procedure with input f_t^{va} can be written as

$$w_{t\_r}^v = \frac{\exp(d(f_t^{va}, m_r^v)/\tau_m)}{\sum_{r=1}^k \exp(d(f_t^{va}, m_r^v)/\tau_m)}, \tag{2}$$

$$d(f_t^{va}, m_r^v) = \frac{f_t^{va} \cdot m_r^v}{||f_t^{va}|| \, ||m_r^v||}, \tag{3}$$

where $d(\cdot, \cdot)$ indicates cosine similarity, $\exp(\cdot)/\sum \exp(\cdot)$ denotes softmax, and τ_m is a memory temperature. Each component $w_{t\_r}^v$ of w_t^v can be considered as an attention weight for audio memory slot m_r^a at time-step t. m^a outputs an associated audio feature $f_t^{a,mem} \in \mathbb{R}^{d_a}$ as follows.

$$f_t^{a,mem} = \sum_{r=1}^k w_{t\_r}^v m_r^a. \tag{4}$$

Repeating this for each time, we obtain associated audio features $f^{a,mem} = \{f_t^{a,mem}\}_{t=1}^n \in \mathbb{R}^{n \times d_a}$. It passes through a fc layer and we obtain associated audio embedding $h^{a,mem} = \{h_t^{a,mem}\}_{t=1}^n \in \mathbb{R}^{n \times d_a}$. The learning scheme of AVA-Memory is addressed in Sect. 3.2. In the meantime, we acquire the actual audio embedding $h^a = \{h_t^a\}_{t=1}^n \in \mathbb{R}^{n \times d_a}$ from audio data via audio semantic embedder which is typical transformer [48] (bottom path of Fig. 2).

Mismatch-Aware Semantic Embedder. The last part of MA-Transformer, mismatch-aware semantic embedder receives the visual embedding h^v, the actual audio embedding h^a, and the associated audio embedding $h^{a,mem}$. The positional encoding is applied separately to h^v, h^a, and $h^{a,mem}$. In addition, [CLS] token is applied for aggregating the feature afterward as [7]. The integrated input of the mismatch-aware semantic embedder is formulated as follows.

$$e_0^{av} = [[CLS]; [h_1^v; \ldots; h_n^v]; [h_1^a; \ldots; h_n^a]; [h_1^{a,mem}; \ldots; h_n^{a,mem}]] \\ + [0; E_{POS}; E_{POS}; E_{POS}]. \tag{5}$$

The mismatch-aware semantic embedder has a similar structure to the visual context embedder except for the last part (it ends with feed forward). To adjust the weighting between the associated audio and the actual audio, we define matching index α $(0\sim1)$ which indicates how well visual and audio are matched. If α is high, the degree of using actual audio h^a is increased, and vice versa, it is lowered. This matching index α is obtained by exploiting AVA-Memory and it is described in Sect. 3.2 We apply α according to each index c of e_0^{av}. In multi-head attention of the embedder, we adjust the attention with α as follows.

$$\beta_c = \begin{cases} 1 & \text{if } c \in \text{index of } [CLS] \text{ and } h^v \\ \alpha & \text{if } c \in \text{index of } h^a \\ 1-\alpha & \text{if } c \in \text{index of } h^{a,mem} \end{cases} \tag{6}$$

$$\text{Attention}(Q, K_c) = \text{Softmax}(QK_c^T/\sqrt{dim} + \log \beta_c), \tag{7}$$

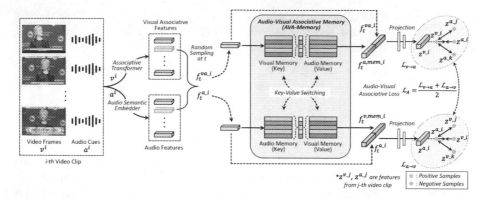

Fig. 3. Proposed audio-visual associative learning with AVA-Memory. Through the associative learning, it is possible to store audio-visual features in sub-memories and naturally associate with one another. As a result, we can obtain the corresponding audio cues from the visual features with AVA-Memory and further determine whether the visual and audio are well matched.

where c indicates the index of e_0^{av}. Q and K are query and key in multihead attention [48] while dim indicates their dimension. This attention is applied to value V from e_0^{av} to focus on parts of h^v, h^a, and $h^{a,mem}$. By applying α in log form, each attention weight is adjusted proportionally considering exponential scale in softmax function. Based on this self-attention scheme, the embedder integrally attends to information considering the reliability of audio cues. If the audio cue is not reliable (not matched with visual), it decreases the proportion of actual audio h^a while it increases the proportion of associated audio $h^{a,mem}$. Finally, we get an audio-visual embedding h^{av} by aggregating features of the semantic embedder with [CLS] output. h^{av} is used to match text embedding h^s.

3.2 Associative Learning with AVA-Memory

AVA-Memory is trained with audio-visual data as shown in Fig. 3. The goal of the associative learning is to store visual and audio features in their sub-memories and to build the link between sub-memories. As a result, we can associate different modality from the other modal input and further determine whether visual and audio are well matched or not.

With the video frames v^i and audio cues a^i from i-th video, we extract visual associative features $f^{va\text{-}i}$ and audio features $f^{a\text{-}i}$ from MA-Transformer and audio semantic embedder, respectively. Note that f^a does not pass through the last fc layer of the audio semantic embedder and thus is different from audio embedding h^a. We randomly sample paired features $(f_t^{va\text{-}i}, f_t^{a\text{-}i})$ at time t. We perform sampling a pair to construct associative learning (attract and repel) in a computationally efficient way. In addition, the combination of training data can be diversified with randomness. In terms of the visual feature $f_t^{va\text{-}i}$, the visual memory is used as a key while the audio memory is used as a value. In contrast for the audio feature $f_t^{a\text{-}i}$, those memories are switched at this time. Each feature changes its modality space through key-value memory addressing as described in the previous section. We can obtain concatenated features $[f_t^{a,mem\_i}; f_t^{va\_i}]$ and $[f_t^{a\_i}; f_t^{v,mem\_i}]$. They pass through the projection head with a fc layer to make z^{v_i} and z^{a_i}, respectively. Actually, they should be $z_t^{v\_i}$ and $z_t^{a\_i}$. However, z represents the feature for each video, so it is denoted as z^{v_i} and z^{a_i} here. If z^{v_i} and z^{a_i} are from a pair, we consider them as a positive set (e.g., z^{v_i}, z^{a_i}). Otherwise, we regard them as a negative set (e.g., z^{v_i}, z^{a_j}). The memory can associate one to another modality by making such a positive set distinctly close. However, the training data also includes videos in which visual and audio are not matched. Therefore, we introduce a new audio-visual associative loss which is the transformed version of contrastive loss [6] to reduce the influence of these mismatched samples. This is inspired by the work [21] that uses $\log(1-p)$ loss rather than the typical $-\log(p)$ (cross-entropy) to reduce the effect of noisy class labels. In case of $-\log(p)$, the gradient on the hard samples ($p \downarrow$) is higher, whereas in the case of $\log(1-p)$, the gradient on the easy samples ($p \uparrow$) is higher. Therefore, when $\log(1-p)$ is used, learning can proceed in a direction that does not force optimization on difficult mismatched samples. As a result, our audio-visual associative loss can be formulated as follows.

$$\mathcal{L}_{v \to a} = \frac{1}{N} \sum_{i=1}^{N} \log \left(1 - \frac{\exp(d(z^{v\_i}, z^{a\_i})/\tau_l)}{\sum_{j=1}^{N} \exp(d(z^{v\_i}, z^{a\_j})/\tau_l)} \right), \tag{8}$$

$$\mathcal{L}_{a \to v} = \frac{1}{N} \sum_{i=1}^{N} \log \left(1 - \frac{\exp(d(z^{a\_i}, z^{v\_i})/\tau_l)}{\sum_{j=1}^{N} \exp(d(z^{a\_i}, z^{v\_j})/\tau_l)} \right), \tag{9}$$

$$\mathcal{L}_A = \frac{\mathcal{L}_{a \to v} + \mathcal{L}_{v \to a}}{2}, \tag{10}$$

where N and τ_l indicate a batch size and a loss temperature parameter, respectively. By minimizing $\mathcal{L}_A$, we can attract one another within the positive set and repel each other within the negative set to properly associate distinct audio cues from visual and vice versa. During the training phase, the weights of m^v and m^s are updated via backpropagation as [18,29].

Furthermore, we can determine whether the visual and audio are aligned or not by utilizing AVA-Memory. When a specific i-th video (v^i, a^i) is given, we formulate the matching index α as follows.

$$\alpha = max(0, \frac{1}{n} \sum_{t=1}^{n} d(z_t^{v\_i}, z_t^{a\_i})), \tag{11}$$

where $d(\cdot, \cdot)$ indicates cosine similarity function. $z_t^{v\_i}$ and $z_t^{a\_i}$ indicates the visual and audio projections of i-th video at time t. As a result, α has a value between 0 and 1. α has a high value for a high degree of matching between visual and audio. In case of missing audio condition, matching index α is set as 0.

3.3   Video-Text Matching

To encode the text embedding h^s in the side of natural language query, we adopt common language model, BERT [7] as [17,50]. The networks are trained to match h^s with h^{av} in latent space (See Fig. 2). Based on them, triplet ranking loss [13] is applied to formulate video-text matching loss $\mathcal{L}_M$ as follows.

$$
\begin{aligned}
\mathcal{L}_M = &\ max(0, \delta + d(h^{av}, h^{s-}) - d(h^{av}, h^s)) \\
&+ max(0, \delta + d(h^{av-}, h^s) - d(h^{av}, h^s)),
\end{aligned}
\tag{12}
$$

where $d(\cdot, \cdot)$ indicates cosine similarity and δ is a margin parameter. h^{av-} and h^{s-} indicate the features from negative samples which are not paired with one another. Model training is conducted with both associative loss and matching loss concurrently. The total objective loss is defined as $\mathcal{L} = \mathcal{L}_A + \mathcal{L}_M$ for training.

At inference time, we can obtain the similarity between a video v and a text s with latent space similarity as follows.

$$
sim(v, s) = d(h^{av}, h^s).
\tag{13}
$$

As a result, the videos with high similarities can be retrieved from text queries.

4 Experiments

4.1 Datasets

MSR-VTT. MSR-VTT dataset [56] consists of 10k web videos and corresponding 200k text descriptions. Each video has 20 text captions with diverse descriptions. About 10% of the videos in MSR-VTT do not contain audio data. The *MSR-VTT-Original* partition [56] of MSR-VTT employs 6,513 clips for training, 497 clips for validation, and 2,990 clips for testing. In another partition *MSR-VTT-Miech* [40], 6,656 and 1,000 clips are used for training and testing, respectively. We evaluated these partitions for comprehensive validation.

VATEX. VATEX [51] consists of YouTube videos with multilingual text descriptions. Chinese and English descriptions exist and we adopt English descriptions only. As following the data partition [5], we perform the retrieval experiments with 25,991 training videos, 1,500 validation videos, and 1,500 testing videos.

TGIF. TGIF [33] is a GIF format dataset which does not have audio data. It contains 100,000 GIF videos and corresponding 120,000 text descriptions about the GIFs' content. According to the data split [32], we perform the experiments with 78,799 training videos, 10,705 validation videos, and 11,351 testing videos.

4.2 Implementation

For MSR-VTT and TGIF, we adopt the spatial encoder as pretrained ResNeXt-101 [55] and ResNet-152 [20]. We concatenate the two features to make a 4,096-d feature as [10]. In terms of VATEX dataset, we use the I3D [3] features with 1,024-d offered by the dataset constructor [51]. Audio data is firstly processed with pretrained VGGish as [36]. The proposed model is trained by Adam optimizer [27] with an initial learning rate of 0.00005 and a batch size of 128. Memory slot size s is fixed as 500 for all experiments. Memory and loss temperature parameters (τ_m, τ_l) are both set as 0.1 for all experiments according to temperature setup [6]. The margin parameter δ is set as 0.2. The overall detailed network structures are described in the supplementary material.

Table 1. Video retrieval performance results on *MSR-VTT-Original* dataset.

Method	Video retrieval performance				
	R@1↑(%)	R@5↑(%)	R@10↑(%)	MedR↓	mAP↑(%)
W2VV [8]	1.1	4.7	8.1	236	3.7
Francis [14]	6.5	19.3	28.0	42	–
VSE++ [13]	8.7	24.3	34.1	28	16.9
W2VV++ [31]	11.1	29.6	40.5	18	20.6
TCE [59]	7.7	22.5	32.1	30	–
HGR [5]	9.2	26.2	36.5	24	–
UWML [53]	10.9	30.4	42.3	–	–
HSL [10]	11.6	30.3	41.3	17	21.2
PSM [34]	12.0	31.7	43.0	16	21.9
T2VLAD [50]	12.7	34.8	47.1	12	–
Proposed Method	**14.7**	**37.0**	**48.6**	**11**	**25.6**

Table 2. Video retrieval performance results on *MSR-VTT-Miech* dataset.

Method	Video retrieval performance				
	R@1↑(%)	R@5↑(%)	R@10↑(%)	MedR↓	mAP↑(%)
W2VV [8]	2.7	12.5	17.3	83	7.9
VSE++ [13]	17.0	40.9	52.0	10	16.9
W2VV++ [31]	21.7	48.6	60.9	6	34.4
TCE [59]	17.1	39.9	53.7	9	–
HGR [5]	22.9	50.2	63.6	5	35.9
MMT [17]	20.3	49.1	63.9	6	–
HSL [10]	23.0	50.6	62.5	5	36.1
PSM [34]	24.2	53.0	65.3	5	37.9
T2VLAD [50]	26.1	54.7	68.1	4	–
Proposed Method	**27.8**	**57.3**	**68.7**	**4**	**41.2**

4.3   Performance Evaluation

Video retrieval performance is measured by rank-based metrics such as recall at K (R@K), median rank (MedR), and mean average precision (mAP). R@K (K=1, 5, 10) indicates the percentage of queries that find correct samples among the top K results. MedR indicates the median rank of the first correct sample in the retrieved results. mAP represents the mean of the average precision scores for each query. Higher R@K, mAP and lower MedR indicate better performances.

Video Retrieval Performance Comparison. We perform performance comparisons according to the training and testing protocol [5,10], not with the pretraining-based methods using a large amount of additional visual-text data [1,12,38]. We conduct the experiments on MSR-VTT with different types of

Table 3. Video retrieval performance results on VATEX dataset.

Method	Video Retrieval Performance				
	R@1↑(%)	R@5↑(%)	R@10↑(%)	MedR↓	mAP↑(%)
W2VV [8]	14.6	36.3	46.1	–	–
VSE++ [13]	31.3	65.8	76.4	–	–
CE [36]	31.1	68.7	80.2	–	–
W2VV++ [31]	32.0	68.2	78.8	–	–
Dual Encoding [9]	31.1	67.4	78.9	3	–
HGR [5]	35.1	73.5	83.5	2	–
HSL [10]	36.8	73.6	83.7	2	52.0
HGR (+GPO) [4]	37.3	73.4	82.4	–	–
Proposed Method	**39.0**	**75.6**	**84.1**	**2**	**55.3**

Table 4. Video retrieval performance results on TGIF dataset.

Method	Video retrieval performance				
	R@1↑(%)	R@5↑(%)	R@10↑(%)	MedR↓	mAP↑(%)
W2VV++ [31]	9.4	22.3	29.8	48	16.2
Dual Encoding [9]	9.1	21.3	28.6	50	15.7
HGR [5]	4.5	12.4	17.8	160	–
CF-GNN [49]	10.2	23.0	30.7	44	–
SEA (BERT) [32]	10.7	24.4	31.9	37	17.9
SEA (BERT+biGRU) [32]	11.1	25.2	32.8	35	18.5
Proposed Method	**11.5**	**26.3**	**34.9**	**31**	**19.2**

data partitions *MSR-VTT-Original* [56] and *MSR-VTT-Miech* [40]. Table 1 and 2 show the text-to-video retrieval performances on MSR-VTT dataset. In case of MSR-VTT, we use (ResNeXt101+ResNet152) for visual and (VGGish) for audio. [8,10,13,31,34] in tables use the same visual feature as us. While, multimodal models [17,50] use (VGGish+Speech) audio features in addition to many visual features together (DenseNet161+SENet154 +ResNet50+S3D+OCR). Ours outperforms these even with simple and fewer features. We perform the video retrieval experiments on VATEX according to the data split [5]. Table 3 shows the comparison results on VATEX dataset. In case of VATEX, we use (I3D) for visual and (VGGish) for audio. All the other methods use the same visual feature as us. The proposed method surpasses the other state-of-the-art methods in terms of all evaluation metrics. These MSR-VTT and VATEX include both mismatching cases (*i.e.*, missing audio or existing audio but not matched). To validate our method on fully missing audio condition, we conduct the experiments on TGIF which does not have audio data at all. We utilize MSR-VTT for training AVA-Memory concurrently. In terms of TGIF experiment, MSR-VTT is additional data but not supervision data because we do not use any text labels of MSR-VTT. Thus, we use the same video-text pairs with the other methods. At

Table 5. Effects of the network designs on the performances. Performance evaluations are conducted on *MSR-VTT-Miech*.

Method	Video retrieval performance				
	R@1↑(%)	R@5↑(%)	R@10↑(%)	MedR↓	mAP↑(%)
w/o Audio Association	26.5	56.4	68.1	4	40.3
w/o Mismatch-Aware Adjustment	25.0	55.0	66.6	4	39.1
Proposed Method	**27.8**	**57.3**	**68.7**	**4**	**41.2**

Table 6. Effects of using different dataset in associative learning procedure to validate the generalizability of audio-visual association on TGIF dataset.

Method	Associative learning	Video retrieval performance	
		R@1↑(%)	mAP↑(%)
w/o AVA-memory	✗	10.8	18.5
Proposed Method	✓ (w/ MSR-VTT)	**11.5**	**19.2**

testing time, only associated audio cues are used dominantly ($\alpha=0$). We conduct the video retrieval experiments on TGIF as follows the split [32]. We use only (ResNeXt101+ResNet152) for visual and state-of-the-art [32] uses the same one. As a result, ours outperforms the other methods on TGIF as shown in Table 4.

4.4 Ablation Study

Effects of Network Designs. We analyze the effects of the network design by conducting ablation studies as shown in Table 5. We investigate the effectiveness of audio association and mismatch-aware adjustment. In the table, the model 'w/o Audio Association' indicates the MA-Transformer without using associated audio features at mismatch-aware semantic embedder. It means the video embedding is made with only visual and actual audio cues (adjustment is still conducted for only actual audio). The second model 'w/o Mismatch-Aware Adjustment' indicates the model without adjusting weights between actual audio and associated audio. As shown in the table, the final proposed method clearly outperforms the base models 'w/o Audio Association' and 'w/o Mismatch-Aware Adjustment'. The results show the superiority of the proposed network designs in terms of both association and adjustment.

Generalizability of AVA-Memory. To validate the generalizability of AVA-memory in terms of audio-visual association, we observe the effect of exploiting different dataset in associative learning. Table 6 shows the results about the generalizability of audio-visual association on TGIF dataset. The first baseline model is trained without AVA-Memory. The second one is the model that utilizes a different dataset, MSR-VTT in associative learning. The training set of TGIF is mainly used for learning with the video-text matching loss $\mathcal{L}_M$ while MSR-VTT is used for learning with the associative loss $\mathcal{L}_A$. Then, the models are

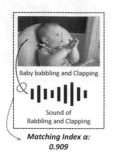 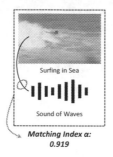  

Fig. 4. Matching index examples obtained by AVA-Memory on VATEX dataset.

evaluated on the test set of TGIF. As shown in the table, the associative learning with the different dataset (*i.e.*, MSR-VTT) also can fairly contribute to the retrieval performances. Note that TGIF does not include actual audio at all and fully exploits the associated audio cues at searching time. This result shows the generalizability of audio-visual association in terms of training data. Since the proposed audio-visual associative learning does not require any labels at all, any videos with audio-visual information are available for this learning scheme.

4.5   Qualitative Results on Matching Index

Figure 4 shows the matching index examples on VATEX test dataset. Each matching index is obtained by AVA-Memory according to equation (11). As can be seen in the figure, matching indexes are high for video samples in which visual and audio are well matched. Contrary, the matching index is low for samples where visual and audio are not matched (*e.g.*, music). Such matching index values are convincingly obtained and used in the adjustment process.

5   Conclusion

The objective of the proposed work is to address the audio-visual mismatch condition when retrieving videos from the text semantics. To this end, we propose MA-Transformer with AVA-Memory which enables to associate audio cues from visual and further to adjust the audio embeddings considering the degree of matching between visual and audio cues. As a result, the proposed method outperforms the state-of-the-art video retrieval methods on various datasets including audio-visual mismatch conditions. Further, we validate the network designs by conducting ablation studies and qualitative analyses.

Acknowledgement. This work was supported by IITP grant(No. 2020-0-00004).

References

1. Bain, M., Nagrani, A., Varol, G., Zisserman, A.: Frozen in time: A joint video and image encoder for end-to-end retrieval. In: IEEE/CVF International Conference on Computer Vision (ICCV), pp. 1728–1738 (2021)
2. Cai, Q., Pan, Y., Yao, T., Yan, C., Mei, T.: Memory matching networks for one-shot image recognition. In: IEEE/CVF Conference on Computer Vision and Pattern Recognition (CVPR), pp. 4080–4088 (2018)
3. Carreira, J., Zisserman, A.: Quo vadis, action recognition? a new model and the kinetics dataset. In: IEEE/CVF Conference on Computer Vision and Pattern Recognition (CVPR), pp. 6299–6308 (2017)
4. Chen, J., Hu, H., Wu, H., Jiang, Y., Wang, C.: Learning the best pooling strategy for visual semantic embedding. In: IEEE/CVF Conference on Computer Vision and Pattern Recognition (CVPR), pp. 15789–15798 (2021)
5. Chen, S., Zhao, Y., Jin, Q., Wu, Q.: Fine-grained video-text retrieval with hierarchical graph reasoning. In: IEEE/CVF Conference on Computer Vision and Pattern Recognition (CVPR), pp. 10638–10647 (2020)
6. Chen, T., Kornblith, S., Norouzi, M., Hinton, G.: A simple framework for contrastive learning of visual representations. In: International Conference on Machine Learning (ICML), pp. 1597–1607. PMLR (2020)
7. Devlin, J., Chang, M.W., Lee, K., Toutanova, K.: Bert: Pre-training of deep bidirectional transformers for language understanding. In: Conference of the North American Chapter of the Association for Computational Linguistics (NAACL-HLT) (2019)
8. Dong, J., Li, X., Snoek, C.G.: Predicting visual features from text for image and video caption retrieval. IEEE Trans. Multimedia 20(12), 3377–3388 (2018)
9. Dong, J., et al.: Dual encoding for zero-example video retrieval. In: IEEE/CVF Conference on Computer Vision and Pattern Recognition (CVPR), pp. 9346–9355 (2019)
10. Dong, J., et al.: Dual encoding for video retrieval by text. IEEE Trans. Pattern Anal. Mach. Intell. 44, 4065–4080 (2021)
11. Dosovitskiy, A., et al.: An image is worth 16 × 16 words: Transformers for image recognition at scale. In: International Conference on Learning Representations (ICLR) (2020)
12. Dzabraev, M., Kalashnikov, M., Komkov, S., Petiushko, A.: Mdmmt: Multidomain multimodal transformer for video retrieval. In: IEEE/CVF Conference on Computer Vision and Pattern Recognition (CVPR), pp. 3354–3363 (2021)
13. Faghri, F., Fleet, D.J., Kiros, J.R., Fidler, S.: Vse++: Improving visual-semantic embeddings with hard negatives. In: British Machine Vision Conference (BMVC) (2018)
14. Francis, D., Anh Nguyen, P., Huet, B., Ngo, C.W.: Fusion of multimodal embeddings for ad-hoc video search. In: IEEE/CVF International Conference on Computer Vision Workshops (ICCVW) (2019)
15. Fu, Z., Liu, Q., Fu, Z., Wang, Y.: Stmtrack: Template-free visual tracking with space-time memory networks. In: IEEE/CVF Conference on Computer Vision and Pattern Recognition (CVPR), pp. 13774–13783 (2021)
16. Gabeur, V., Nagrani, A., Sun, C., Alahari, K., Schmid, C.: Masking modalities for cross-modal video retrieval. In: IEEE/CVF Winter Conference on Applications of Computer Vision (WACV), pp. 1766–1775 (2022)

17. Gabeur, V., Sun, C., Alahari, K., Schmid, C.: Multi-modal transformer for video retrieval. In: Vedaldi, A., Bischof, H., Brox, T., Frahm, J.-M. (eds.) ECCV 2020. LNCS, vol. 12349, pp. 214–229. Springer, Cham (2020). https://doi.org/10.1007/978-3-030-58548-8_13

18. Gong, D., et al.: Memorizing normality to detect anomaly: Memory-augmented deep autoencoder for unsupervised anomaly detection. In: IEEE/CVF International Conference on Computer Vision (ICCV), pp. 1705–1714 (2019)

19. Han, T., Xie, W., Zisserman, A.: Memory-augmented dense predictive coding for video representation learning. In: Vedaldi, A., Bischof, H., Brox, T., Frahm, J.-M. (eds.) ECCV 2020. LNCS, vol. 12348, pp. 312–329. Springer, Cham (2020). https://doi.org/10.1007/978-3-030-58580-8_19

20. He, K., Zhang, X., Ren, S., Sun, J.: Deep residual learning for image recognition. In: IEEE/CVF Conference on Computer Vision and Pattern Recognition (CVPR), pp. 770–778 (2016)

21. Hu, P., Peng, X., Zhu, H., Zhen, L., Lin, J.: Learning cross-modal retrieval with noisy labels. In: IEEE/CVF Conference on Computer Vision and Pattern Recognition (CVPR), pp. 5403–5413 (2021)

22. Huang, Y., Wang, L.: Acmm: Aligned cross-modal memory for few-shot image and sentence matching. In: IEEE/CVF International Conference on Computer Vision (ICCV), pp. 5774–5783 (2019)

23. Kaiser, Ł., Nachum, O., Roy, A., Bengio, S.: Learning to remember rare events. In: International Conference on Learning Representations (ICLR) (2017)

24. Karpathy, A., Fei-Fei, L.: Deep visual-semantic alignments for generating image descriptions. In: IEEE/CVF Conference on Computer Vision and Pattern Recognition (CVPR), pp. 3128–3137 (2015)

25. Kim, J.U., Park, S., Ro, Y.M.: Robust small-scale pedestrian detection with cued recall via memory learning. In: IEEE/CVF International Conference on Computer Vision (ICCV), pp. 3050–3059 (2021)

26. Kim, M., Hong, J., Park, S.J., Ro, Y.M.: Multi-modality associative bridging through memory: Speech sound recollected from face video. In: IEEE/CVF International Conference on Computer Vision (ICCV), pp. 296–306 (2021)

27. Kingma, D., Ba, J.: Adam: A method for stochastic optimization. In: International Conference on Learning Representations (ICLR) (2015)

28. Kiros, R., Salakhutdinov, R., Zemel, R.S.: Unifying visual-semantic embeddings with multimodal neural language models. arXiv preprint arXiv:1411.2539 (2014)

29. Lee, S., Kim, H.G., Choi, D.H., Kim, H.I., Ro, Y.M.: Video prediction recalling long-term motion context via memory alignment learning. In: IEEE/CVF Conference on Computer Vision and Pattern Recognition (CVPR), pp. 3054–3063 (2021)

30. Lee, S., Kim, H.I., Ro, Y.M.: Weakly paired associative learning for sound and image representations via bimodal associative memory. In: IEEE/CVF Conference on Computer Vision and Pattern Recognition (CVPR), pp. 10534–10543 (2022)

31. Li, X., Xu, C., Yang, G., Chen, Z., Dong, J.: W2VV++: fully deep learning for ad-hoc video search. In: ACM International Conference on Multimedia (ACM MM), pp. 1786–1794 (2019)

32. Li, X., Zhou, F., Xu, C., Ji, J., Yang, G.: Sea: Sentence encoder assembly for video retrieval by textual queries. IEEE Trans. Multimedia **23**, 4351–4362 (2021)

33. Li, Y., et al.: Tgif: A new dataset and benchmark on animated gif description. In: IEEE/CVF Conference on Computer Vision and Pattern Recognition (CVPR), pp. 4641–4650 (2016)

34. Liu, H., Luo, R., Shang, F., Niu, M., Liu, Y.: Progressive semantic matching for video-text retrieval. In: ACM International Conference on Multimedia (ACM MM), pp. 5083–5091 (2021)
35. Liu, S., Fan, H., Qian, S., Chen, Y., Ding, W., Wang, Z.: Hit: Hierarchical transformer with momentum contrast for video-text retrieval. In: IEEE/CVF International Conference on Computer Vision (ICCV), pp. 11915–11925 (2021)
36. Liu, Y., Albanie, S., Nagrani, A., Zisserman, A.: Use what you have: Video retrieval using representations from collaborative experts. In: British Machine Vision Conference (BMVC) (2019)
37. Marchetti, F., Becattini, F., Seidenari, L., Bimbo, A.D.: Mantra: Memory augmented networks for multiple trajectory prediction. In: IEEE/CVF Conference on Computer Vision and Pattern Recognition (CVPR), pp. 7143–7152 (2020)
38. Miech, A., Alayrac, J.B., Laptev, I., Sivic, J., Zisserman, A.: Thinking fast and slow: Efficient text-to-visual retrieval with transformers. In: IEEE/CVF Conference on Computer Vision and Pattern Recognition (CVPR), pp. 9826–9836 (2021)
39. Miech, A., Alayrac, J.B., Smaira, L., Laptev, I., Sivic, J., Zisserman, A.: End-to-end learning of visual representations from uncurated instructional videos. In: IEEE/CVF Conference on Computer Vision and Pattern Recognition (CVPR), pp. 9879–9889 (2020)
40. Miech, A., Laptev, I., Sivic, J.: Learning a text-video embedding from incomplete and heterogeneous data. arXiv preprint arXiv:1804.02516 (2018)
41. Miech, A., Zhukov, D., Alayrac, J.B., Tapaswi, M., Laptev, I., Sivic, J.: Howto100m: Learning a text-video embedding by watching hundred million narrated video clips. In: IEEE/CVF International Conference on Computer Vision (ICCV), pp. 2630–2640 (2019)
42. Mithun, N.C., Li, J., Metze, F., Roy-Chowdhury, A.K.: Learning joint embedding with multimodal cues for cross-modal video-text retrieval. In: ACM International Conference on Multimedia Retrieval (ICMR), pp. 19–27 (2018)
43. Otani, M., Nakashima, Y., Rahtu, E., Heikkilä, J., Yokoya, N.: Learning joint representations of videos and sentences with web image search. In: Hua, G., Jégou, H. (eds.) ECCV 2016. LNCS, vol. 9913, pp. 651–667. Springer, Cham (2016). https://doi.org/10.1007/978-3-319-46604-0_46
44. Park, H., Noh, J., Ham, B.: Learning memory-guided normality for anomaly detection. In: IEEE/CVF Conference on Computer Vision and Pattern Recognition (CVPR), pp. 14372–14381 (2020)
45. Patrick, M., et al.: Support-set bottlenecks for video-text representation learning. In: International Conference on Learning Representations (ICLR) (2020)
46. Song, Y., Soleymani, M.: Polysemous visual-semantic embedding for cross-modal retrieval. In: IEEE/CVF Conference on Computer Vision and Pattern Recognition (CVPR), pp. 1979–1988 (2019)
47. Torabi, A., Tandon, N., Sigal, L.: Learning language-visual embedding for movie understanding with natural-language. arXiv preprint arXiv:1609.08124 (2016)
48. Vaswani, A., et al.: Attention is all you need. In: Advances in Neural Information Processing Systems (NeurIPS), pp. 5998–6008 (2017)
49. Wang, W., Gao, J., Yang, X., Xu, C.: Learning coarse-to-fine graph neural networks for video-text retrieval. IEEE Trans. Multimedia **23**, 2386–2397 (2021)
50. Wang, X., Zhu, L., Yang, Y.: T2vlad: global-local sequence alignment for text-video retrieval. In: IEEE/CVF Conference on Computer Vision and Pattern Recognition (CVPR), pp. 5079–5088 (2021)

51. Wang, X., Wu, J., Chen, J., Li, L., Wang, Y.F., Wang, W.Y.: Vatex: A large-scale, high-quality multilingual dataset for video-and-language research. In: IEEE/CVF International Conference on Computer Vision (ICCV), pp. 4581–4591 (2019)

52. Wei, J., Xu, X., Yang, Y., Ji, Y., Wang, Z., Shen, H.T.: Universal weighting metric learning for cross-modal matching. In: IEEE/CVF Conference on Computer Vision and Pattern Recognition (CVPR), pp. 13005–13014 (2020)

53. Wei, J., Yang, Y., Xu, X., Zhu, X., Shen, H.T.: Universal weighting metric learning for cross-modal retrieval. IEEE Trans. Pattern Anal. Mach. Intell. **44**, 6534–6545 (2021)

54. Wray, M., Larlus, D., Csurka, G., Damen, D.: Fine-grained action retrieval through multiple parts-of-speech embeddings. In: IEEE/CVF International Conference on Computer Vision (ICCV), pp. 450–459 (2019)

55. Xie, S., Girshick, R., Dollár, P., Tu, Z., He, K.: Aggregated residual transformations for deep neural networks. In: IEEE/CVF Conference on Computer Vision and Pattern Recognition (CVPR), pp. 1492–1500 (2017)

56. Xu, J., Mei, T., Yao, T., Rui, Y.: Msr-vtt: A large video description dataset for bridging video and language. In: IEEE/CVF Conference on Computer Vision and Pattern Recognition (CVPR), pp. 5288–5296 (2016)

57. Xu, R., Xiong, C., Chen, W., Corso, J.: Jointly modeling deep video and compositional text to bridge vision and language in a unified framework. In: AAAI Conference on Artificial Intelligence (AAAI) (2015)

58. Yang, T., Chan, A.B.: Learning dynamic memory networks for object tracking. In: Ferrari, V., Hebert, M., Sminchisescu, C., Weiss, Y. (eds.) ECCV 2018. LNCS, vol. 11213, pp. 153–169. Springer, Cham (2018). https://doi.org/10.1007/978-3-030-01240-3_10

59. Yang, X., Dong, J., Cao, Y., Wang, X., Wang, M., Chua, T.S.: Tree-augmented cross-modal encoding for complex-query video retrieval. In: International ACM SIGIR Conference on Research and Development in Information Retrieval (ACM SIGIR), pp. 1339–1348 (2020)

60. Yu, Y., Kim, J., Kim, G.: A joint sequence fusion model for video question answering and retrieval. In: Ferrari, V., Hebert, M., Sminchisescu, C., Weiss, Y. (eds.) ECCV 2018. LNCS, vol. 11211, pp. 487–503. Springer, Cham (2018). https://doi.org/10.1007/978-3-030-01234-2_29

61. Zhang, B., Hu, H., Sha, F.: Cross-modal and hierarchical modeling of video and text. In: Ferrari, V., Hebert, M., Sminchisescu, C., Weiss, Y. (eds.) ECCV 2018. LNCS, vol. 11217, pp. 385–401. Springer, Cham (2018). https://doi.org/10.1007/978-3-030-01261-8_23

62. Zhu, L., Yang, Y.: Inflated episodic memory with region self-attention for long-tailed visual recognition. In: IEEE/CVF Conference on Computer Vision and Pattern Recognition (CVPR), pp. 4344–4353 (2020)

Connecting Compression Spaces with Transformer for Approximate Nearest Neighbor Search

Haokui Zhang[1,2], Buzhou Tang[2(✉)], Wenze Hu[1], and Xiaoyu Wang[1]

[1] Intellifusion, Shenzhen, China
hkzhang1991@mail.nwpu.edu.cn
[2] Harbin Institute of Technology (Shenzhen), Shenzhen, China
tangbuzhou@hit.edu.cn

Abstract. We propose a generic feature compression method for Approximate Nearest Neighbor Search (ANNS) problems, which speeds up existing ANNS methods in a plug-and-play manner. Specifically, based on transformer, we propose a new network structure to compress the feature into a low dimensional space, and an inhomogeneous neighborhood relationship preserving (INRP) loss that aims to maintain high search accuracy. Specifically, we use multiple compression projections to cast the feature into many low dimensional spaces, and then use transformer to globally optimize these projections such that the features are well compressed following the guidance from our loss function. The loss function is designed to assign high weights on point pairs that are close in original feature space, and keep their distances in projected space. Keeping these distances helps maintain the eventual top-k retrieval accuracy, and down weighting others creates room for feature compression. In experiments, we run our compression method on public datasets, and use the compressed features in graph based, product quantization and scalar quantization based ANNS solutions. Experimental results show that our compression method can significantly improve the efficiency of these methods while preserves or even improves search accuracy, suggesting its broad potential impact on real world applications. Source code is available at https://github.com/hkzhang91/CCST.

Keywords: Approximate nearest neighbor search · Transformer · Neighborhood relationship preserving · Compression projections retrieval

1 Introduction

Approximate nearest neighbor search (ANNS) methods focus on searching for k approximate nearest neighbors from a given database to a given query node q. It is a fundamental technology in information retrieval and is widely used in applications such as search engines and recommendation systems. The common goal of ANNS approaches is to minimize the search latency while maintain a low search accuracy loss on a fixed hardware constraint.

© The Author(s), under exclusive license to Springer Nature Switzerland AG 2022
S. Avidan et al. (Eds.): ECCV 2022, LNCS 13674, pp. 515–530, 2022.
https://doi.org/10.1007/978-3-031-19781-9_30

As stated in [1], currently, the two most popular ANNS methods are graph based approaches and quantization based approaches. Many of these methods, such as product quantization (PQ) [2], HNSW [3] and NSG [4] are widely used in real-world applications. Although popular, there are rooms to further improve them. In PQ, the sub vectors are quantized into codewords using a clustering objective, which is a poor proxy to search accuracy. In graph based methods, there are a large number of distance computations in index building time (for instance, indexing complexity NSG graph is $O(kn^{\frac{1+d}{d}}\log n^{\frac{1}{d}} + n\log n)$ [4], where d is data dimension), which in real world applications can last weeks on billion scale datasets. For PQ related methods, introducing the feature compression as an intermediate step avoids direct feature quantization from high dimensional space. Such a two stage quantization strategy usually results in higher accuracy. For graph based methods, computing distance in low dimensional space reduces computational cost linearly, which significantly speeds up indexing.

Existing feature compression methods such as principal components analysis (PCA) [5], Variational Auto-Encoders(VAE) [6] focus on keeping the information in the input features instead of their neighborhood relations, but in ANNS applications the neighborhood structure affects the eventual accuracy more than the locations of the individual data points themselves. As is shown in experiments, directly applying these methods lead to significantly reduced search accuracy, which is not suitable for applications requiring both high speed and high recall.

In this paper, we propose a feature compression method for the approximate nearest neighbor search problem, which aims to retain the local neighborhood relations instead of the fidelity of the reconstructed features. Our method is composed of a compression network which connect compression spaces with transformer (CCST), and an inhomogeneous neighborhood relationship preserving (INRP) loss.

Our CCST is a combination of projection units that projects original features to low dimensions, and transformer units that compose these projections to generate the output feature. The projection unit is initialized as sparse random projection (SRP) [7], which is proved by JL (Johnson-Lindenstrauss) lemma that this projection reserves distances of an increasing number of data points with decreasing dimension reduction ratios. We design our network with multiple projections units, so that each can be used to compress a local neighborhood, and the entire space can be covered in a piece-wise manner by the ensemble of these units. We then use transformers to adaptively combine the output of these random projection units, because the transformer unit has the well known property of globally attending its inputs, which could lead to better global alignment of these projections. In other words, multiple compression spaces are connected to one via transformer. Different from standard transformer based networks, we design an input dependent, non-trivial output token named compression token, which is itself a compressed feature derived from the input feature, and is updated by the transformer layers with skip connections. The compression token is designed to provide an anchor vector to the multi-head attention units so that transformers properly mix the output of random projection units to generate the output compressed feature.

Our training loss is designed for the purpose of preserving local neighborhood structures. In most ANNS applications, user is mostly interested in the accuracy of the top few nearest neighbors. So, changes in the distances between a query and its far away points will not affect the search result, as long as their distances are still large. Inspired by this property in ANNS problems, we design the INRP loss, which assigns high losses on point pairs whose distances are within a threshold. This loss design allows certain information of the original feature to be discarded while still maintaining high retrieval accuracy.

Our experiments show that our feature compression model can be seamlessly used with popular ANNS methods. While saving 1/2 to 3/4 indexing time, graphs built using our compressed features can improve the recall slightly for HNSW and NSG methods. It also significantly improves recalls of 1@1 and recalls of 1@5 metrics by more than 10.0% points for PQ based methods, and about 1.0 or 2.0% points for a quantization methods tailored to HNSW methods [8].

Our main contributions are summarized as follows.

1. We propose a novel feature vector compression model CCST for ANNS problem. In CCST, traditional projection based compression units and emerging transformer units are jointly used to compress features for ANNS problems. To our knowledge, this is the first attempt to apply transformer to ANNS.
2. We propose an INRP loss that mainly keeps the distances of a point and its close local neighbors. This maintains top k retrieval accuracy and creates space for lossy feature compression.
3. The experiment results show that our proposed method can be used in most ANNS methods to improve efficiency. It speeds up indexing speed to $2\times$ to $4\times$ for graph based methods and it improves recalls significantly for PQ related methods.

2   Related Work

2.1   ANNS Methods

Existing ANNS approaches can be divided into four main types: 1) tree-structure based strategies [9], which partitions indexed datapoints into different subspaces based on specific conditions; 2) Locality sensitive hashing (LSH) related methods, which map similar items to the same symbol with a high probability [10]; 3) product quantization (PQ) related approaches decompose the space into a Cartesian product of low dimensional subspaces and quantize each subspace separately [2]; 4) graph based frameworks search on pre-built relative neighborhood graphs (RNG) to find the closest data points to query [3]. Compared with PQ and graph based methods, tree based and LSH based methods need ensembles of trees or hash tables to achieve similar accuracy, which consume considerably more memory and are less frequently used in large scale ANNS problems.

Product quantization based approach reduces the index memory cost by holding quantized codes and speeds up distance computation using online or offline computed distance look up tables. When combined with inverted index to further reduce the number of distance computations, PQ becomes a strong baseline for modern ANNS systems. Compared with quantization based approach, which partitions original feature space to generate shorter features, our method directly maps the original features into a new space. These two method are orthogonal. As is shown in experiments, the new space can be further quantized by PQ, which generates even better speed and accuracy trade-offs than using PQ method alone.

Graph based algorithms usually construct a navigable graph over database, and searches along the edges of this built graph to find the closest points to query. The most recent works in this category are NSG [4], HNSW [3], Disk-ANN [1] and HM-ANN [11]. NSG builds a relatively sparse indexing graph to reduce memory usage and improve search speed. HNSW employs hierarchical graph structure to reduce query latency. Disk-ANN and HM-ANN focus on reducing memory overhead via adopting quantization and heterogeneous memory. Besides, Douze et al. proposed Link and Code (L&C), which takes HNSW as basic ANNS framework and replaces full precision vectors with refined quantiztion codes [8]. Compared with brute force, these graph based methods visit 1/1000 or even less of the indexed points in each search, significantly reduce the number of needed distance computations. However, most of these methods suffer from the problem of high indexing time, as constructing a navigable graph over database has high time complexity. For instance, $O(n \log n)$ for HNSW and $O(kn^{\frac{1+d}{d}} \log n^{\frac{1}{d}} + n \log n)$ for NSG. In real world application, building a navigable graph over a billion-scale database with 30+ threads costs several days or even weeks.

2.2 Dimension Reduction/Compression

In early stage, almost all dimension reduction methods are based on mathematical theory. PCA [5] and independent components analysis (ICA) [12] are proposed to transform vectors into lower dimensional space, where the most information of high dimension vectors are retained. Random projection [13][7] project data into low dimensional space, while approximately preserving structure information in original feature space. The efficiency of random projection methods is guaranteed by Johnson-Lindenstrauss (JL) lemma [14]. This lemma states that data points in sufficiently high dimensional space may be projected into suitable low dimensional space while approximately preserving the distances between the points in original space.

In 2006, Hinton and Salakhutdiov proposed to compress high dimensional data via neural network [15], and several recent works on dimension reduction have shifted their methods to neural networks. For example, VAE maps data to low dimensional space and forces the encoded features to follow a multivariate Gaussian distribution [6].

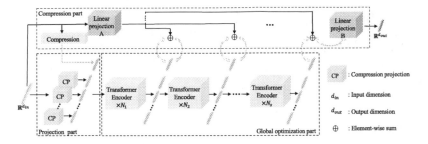

Fig. 1. Overview of the CCST model. The proposed model consists of three major parts, including compression part, projection part and global optimization part. The projection part projects input vectors into multiple subspaces and outputs a sequence of sub vectors. Global optimization part optimizes the sequence of sub vectors from global perspective and summarizes all optimized information into compression token. Compression part provides initial the compression token, and interacts with global optimization part to generate the final compression result.

Very recently, Sablayrolles et al. [16] proposed an end-to-end quantization model for similarity search, where a compression network named catalyst is designed and trained to compress vectors from d_{in}-dimensional input space to the hypersphere of a d_{out}-dimensional space, where $d_{in} > d_{out}$. Our proposed CCST is close to this work, as it involves an ANNS oriented loss and a learnable network to compress feature vectors. Compared with catalyst, our training objective is easier to implement as it does not involve offline exhaustive search for positive and negative pairs. Also, we design a deep compression model for the ANNS problem, where [16] uses multi-layer perceptrons to demonstrate their entropy maximization idea.

2.3   Transformer

Vaswani et al. first proposed the transformer architecture for the task of language modelling in [17]. Inspired by the promising performance of transformer, several works started to introduce transformer into vision tasks and proposed a new type of models. Dosovitskiy et al. proposed ViT, in which the input image is cropped into a sequence of patches to meet the input format requirement of transformer [18]. Later, Touvron et al. proposed to use knowledge distillation to overcome the difficulties of training ViT models [19]. Liu et al. proposed a hierarchical transformer, where representation is computed within shifted windows [20] to reduce computation cost. Very recently, Graham et al. mixed CNN and transformer in their LeVit model, which significantly outperforms previous CNNs and ViT models with respect to the speed/accuracy tradeoff [21].

The transformer part of our CCST model is inspired by the works above. We have made several modifications to better fit transformer with our applications, which are detailed in section *global optimization part*.

3   The Proposed Approach

In this section, we present our proposed method in detail. We first introduce the architecture of CCST. Then we elaborate on our INRP loss function.

3.1   CCST

As illustrated in Fig. 1, the proposed CCST consists of three major parts, including projection part, global optimization part and compression part.

The projection part casts the input feature vector $x \in \mathbb{R}^{d_{in}}$ to n different low dimensional spaces. The output of this part is a sequence of low dimensional vectors $[p^1(x), p^2(x), \cdots p^n(x)]$, where $p^i(x) \in \mathbb{R}^{d_{out}}, i = 1, \cdots, n$. d_{in} and d_{out} denote feature dimensions before and after compression, respectively. In addition, note that $[p^1(x), p^2(x), \cdots, p^n(x)]$ constructs a n by d_{out} matrix instead of a $n \cdot d_{out}$ dimensional vector. It is a sequence of tokens as the input to transformers instead of a concatenated vector. $p^i()$ is a compression projection function. $p^i(x) = W^i x$, where $W^i \in \mathbb{R}^{d_{in} \times d_{out}}$. The global optimization part takes $[cp(x), p^1(x), p^2(x), \cdots p^n(x)]$ as input and globally optimizes all sub vectors step by step. The global optimization part then summarizes all optimized sub vectors into a compression token $cp(x)$. The compression part is responsible for initializing the compression token, interacting with global optimization part in each step and further processes the information in compression token to generate the final compression result $f(x) \in \mathbb{R}^{d_{out}}$.

Projection Part. Our goal is compressing feature vectors into a low dimensional space, where the data neighborhood relation in original space is preserved. It seems that random projections satisfy our requirement and there are already some classical random projection methods. Unfortunately, according to JL lemma [14], the projection error of using single random projection function satisfies:

$$(1 - \epsilon)||x_i - x_j||_2^2 \leq ||p(x_i) - p(x_j)||_2^2 \leq (1 + \epsilon)||x_i - x_j||_2^2 \qquad (1)$$

where x_i and x_j represent two data points in high dimensional space and $p(x_i)$ and $p(x_j)$ are corresponding projection data points in low dimensional space. $p()$ is the projection function. ϵ satisfies equation $d_{out} > \frac{4 \ln(m)}{\epsilon^2/2 - \epsilon^3/3}$, where $0 < \epsilon < 1$ and m represents data size. Random projections are inapplicable in our case. For example, if compressing 960 dimensional vectors in GIST1M dataset to 480 dimensions, we have $0.63 < \epsilon < 1$. In GIST1M, the distance of a query to its nearest neighbor and to its hundredth neighbor are probably 1.177 and 1.5615, respectively. After projection, with the minimum ϵ, the distance of a query to its nearest neighbor and to its hundredth neighbor fall into the scope of [0.7160, 1.5027] and the scope of [0.9498, 1.9936], respectively. These two scopes have a considerable overlap, which may disturb neighborhood relationship and lead to a lower search accuracy. We can draw a consistent conclusion from our experimental results.

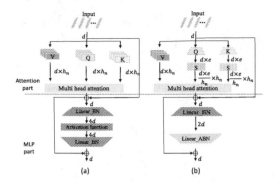

Fig. 2. Different modules. (a) Transformer module of ViT [18]. (b) Transformer module proposed in this paper. By adopting lightweight designs, transformer module in (b) has fewer parameters than that in (a).

Existing random projection methods do not address our problem directly, but they give us some inspirations. Here, we propose to use n different projections to boost compression accuracy. Specifically, we initialize the n projection matrices in $p^i(x) = W_i x$ $(i = 1, 2, \cdots, n)$ on input $x \in \mathbb{R}^{d_{in}}$ to generate a sequence of features $[p^1(x), p^2(x), \cdots p^n(x)]$. Following [7], the elements in W_i are randomly drawn from:

$$
\begin{cases}
-\sqrt{\dfrac{s}{d_{out}}} & \text{with probability} \quad 1/2s, \\
0 & \text{with probability } 1 - 1/s \\
\sqrt{\dfrac{s}{d_{out}}} & \text{with probability} \quad 1/2s
\end{cases}
\tag{2}
$$

where $s = \sqrt{d_i n}$. Different from traditional random projections, projection matrices used in here will be further optimized by our INRP loss.

Global Optimization Part. The projection part outputs a sequence of low dimensional vectors, which brings in the core problem of this global optimization part, that is how to generate the result compressed feature using these features from different sub spaces.

Here, as the problem that we need to solve is exactly what transformer model is good at, we tailor a transformer model to overcome the core problem. One biggest advantage of transformer is it good at capturing and taking use of relationships between different tokens to get a global optimization result. When we treat features from different sub spaces as different tokens, our core problem becomes optimizing a sequence of tokens from the global perspective. So, based on ViT [18] structure, we propose a new transformer model for feature compression.

Based on the characteristics of our compression problem, we have made four major modifications to the original ViT structure, including two structural modifications and two lightweight designs.

Structural modifications ensure the new proposed transformer structure is aligned with our problem. These two modifications are:

1. Discard position embedding. In transformer models for processing sentences or images, position embedding and relative position embedding are widely used, as the order of words and the positions or relative positions of images patches contain important information. However, in our case, the order of random projections is fixed. Naturally, position embedding is discarded in our transformer model. Interestingly, when we add position embedding into our transformer, the search accuracy drops. We conjecture this is because randomly initialized position coding disturbs our model.
2. Add compression token. In our transformer model, an extra token named compression token is used. Structurally, this design is similar to the transformer model in ViT, where an extra token named classification token is also employed. Compression token and classification token are added for different purposes and they have different functions. Compression token provides a base reference for optimizing and summarizing information from different projection spaces and it also works as a bridge between compression part and global optimization part. Classification token is used as a placeholder to perform classification. Different from classification token, which is a learned constant, our compression token is derived from input feature vector. Initializing compression token with random numbers is not in accordance with its role of working as base reference.

Lightweight designs are adopted to reduce the number of learnable parameters. Figure 2 shows lightweight designs for MLP and attention parts are:

1. In MLP part, we decrease the expansion ratio (changes among widths of the adjacent linear mapping layers) from 4 to 2 to reduce the number of parameters by half, as shown in the bottom half of Fig. 2 (b). In addition, the basic computing unit in our model is Linear_ABN, which, in data processing order, consists of a linear mapping layer, activation function and batch normalization. As pointed out in [22], placing batch normalization layer before ReLU leads to the conv layer updated in a suboptimal way due to the nonnegative responses of activation function. So we design our unit following the order of conv→activation→bn. This order is also recommended in [23].
2. Attention part has three intermediate variables, which are represented as Value (V), Query (Q) and Key (K). The output of head i is calculated as:

$$\text{head}_i(Q, K, V) = \text{softmax}(\frac{QW_i^Q \cdot KW_i^{K^T}}{\sqrt{d}})VW_i^V \tag{3}$$

where W_i^Q, W_i^K and W_i^V are three projection matrices. As in Eq. 3, V stores main information, Q and K are used to generate attention weights. In ViT, Q, K and V are all the same in each head. Inspired by [17], we decrease the dimension of Q and K for reducing parameters of W_i^K and W_i^V.

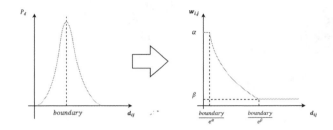

Fig. 3. The weight curve of INRP loss. d_{ij} is Euclidean distance between x_i and x_j.

Structurally, as shown in the middle part of Fig. 1, global optimization part consists of s stages, each of which has N_i transformer encoders. From stage 1 to $s-1$, projected vector from compression part is added to compression token at the end of each stage. Finally, $cp(x)$ is from the output of last stage is fed back to compression part to generate the final compression result.

Compression Part. The compression part is responsible for providing initial compression token, interacting with global optimization part in each step and generating the final compression result. Corresponding to three responsibilities three learnable modules are constructed. As shown in the top part of Fig. 1, three modules are a compression module and two linear projection modules.

The compression module is made up of a single Linear_ABN module. Each linear projection module has a linear mapping function. Compression module takes $x \in \mathbb{R}^{d_{in}}$ as input and outputs $cp(x) \in \mathbb{R}^{d_{out}}$. Linear projection A maps input $x \in \mathbb{R}^{d_{in}}$ to d_{out}-dimensional vector, which is added to compression tokens of global optimization part. Linear projection B takes compression token from the last stage of global optimization part as input and output the final compression result.

3.2 INRP Loss

To preserve neighborhood structures, a straightforward way is training the network to keep distances between all possible point pairs. However, the goal of ANNS is searching for top k approximate nearest neighbors to a query point, which generally are points that are close to query point in the original space. In other words, focusing on keeping distances between close pairs is more important, as these pairs affect search result the most. In [24], Guo et al. presented a similar idea, which is proposed for quantization instead of compression. Inspired by this, we design our INRP loss, the formula of which is:

$$loss = \frac{1}{m^2} \sum_{i=1}^{m} \sum_{j=1}^{m} w_{ij} \cdot \|\, \|f(x_i) - f(x_j)\|_2 - \|x_i - x_j\|_2 \,\|_2 \qquad (4)$$

where x_i and x_j are a pair of any two nodes from original space. m denotes data size. $f(x_i)$ and $f(x_j)$ are compression results. w_{ij} is weight for this pair and it is calculated as:

$$w_{ij} = \min(\alpha, \max(\beta, -\ln(\frac{d_{ij}}{boundary}))) \tag{5}$$

where $d_{ij} = ||x_i - x_j||_2$. α and β are two hyper-parameters, which are set to 2.0 and 0.01 respectively. $boundary$ denotes the average distance between any two nodes in original space. The curve is shown in Fig. 3. In implementation, we use all pairs inside a mini-batch to approximate Eq.4, which significantly simplifies computation.

4   Experiments

4.1   Datasets and Implementation Details

Datasest We carry out experiments on two million-scale benchmark datasets GIST1M and Deep1M[1]. GIST1M consists of 1 million of 960-dimensional hand-crafted feature vectors, which are extracted from images with GIST descriptors [25]. Deep1M contains 1 million of 256-dimensional deep feature vectors, which are extracted by using GoogleNet [26].

Training Setting. Following [27,28], we take database as training set. We train our model for 2400 epochs with the Adamw optimzier [29], where the initial learning rate and batch size are set to 1e-4 and 1024 respectively. We use poly learning rate policy with power of 0.9 to adjust the learning rate for every epoch. Just like other deep learning models, training CCST is time consuming. With a single RTX2080Ti GPU, training costs about half a day. However in practice, the trained model will be used to process huge amount of data. In Bigann-1B, indexing the 1 billion data saves 54 h, which already pays off.

Platforms. We implement our CCST using Pytorch and conduct ANNS experiments based on Faiss[2].

4.2   Speeding Up Indexing for Graph-Based Methods

Here, we focus on using our proposed CCST to speed up indexing of graph-based methods. Two most popular graph-based methods HNSW [3] and NSG [4] are employed as baselines. We construct all HNSW indices with M=48, efConstruction=512, and perform search with efSearch = 100 on GIST1M and efSearch=200 on Deep1M. Following [1,4], we build all NSG indices using R = 60, L = 70 and C = 500 and initialize K-NN graphs with NN-descent. On GIST1M, we initialize K-NN graph using GK = 400, L = 400, iter=12, S = 15 and R = 100. On Deep1M, we initialize K-NN graph using GK = 200, L = 200, iter=10, S = 12 and R = 100.

---

[1] Downloaded from https://www.cse.cuhk.edu.hk/systems/hash/gqr/datasets.html.
[2] https://github.com/facebookresearch/faiss/releases/tag/v1.7.1.

Table 1. Experiments of speeding up indexing for graph based methods. C_F represents the feature compression factor. IND and SS denote indexing time and search speed. q/s is query per second. Experiments are performed on a server with two Xeon Gold 5218 CPUs. We build indices with 32 threads and perform search with a single thread. All the speed numbers are averaged from 20 cold runs to rule out random factors. Comp_t is encoding time. Here, a single RTX2080Ti GPU is used for accelerating compressing process and the GPU memory usage is restricted to about 1GB.

C_F	Methods										Comp_t	GPU
	HNSW					NSG					(s)	(G)
	IND	SS	Recall			IND	SS	Recall				
	(s)	(q/s)	1@1	1@10	100@100	(s)	(q/s)	1@1	1@10	100@100		
GIST1M												
1	733	182	97.40	100.00	94.39	1118	179	97.40	100.00	93.97	–	–
2	454	184	97.70	100.00	94.55	778	180	98.00	99.90	94.16	21.43	1.19
4	**245**	186	**98.00**	100.00	**95.27**	**645**	182	**98.30**	100.00	**94.85**	10.00	1.17
Deep1M												
1	300	861	99.50	100.00	95.23	505	820	99.7	100.00	94.53	–	–
2	144	870	99.50	100.00	95.46	246	822	99.6	100.00	94.69	5.15	1.09
4	**100**	874	**99.80**	100.00	**95.60**	**209**	842	**99.6**	100.00	**94.92**	4.64	1.06

Table 2. Experiments of speeding up indexing for real world database and billion-scale database. Deepfeat25M is a subset of our internal 1 billion 512-d dataset.

Datasets	C_factor	indexing	Recall	Compressing	GPU
		Time (h)	1@1	Time (h)	Usage (G)
Deepfeat25M	1	15.1	95.4	–	–
	2	9.5	95.1	0.18	1.32
Bigann-1B	1	106	92.6	–	–
	2	51	92.8	1.01	1.24

We conduct three groups of experiments with different compression ratios. In all three groups, full-dimensional vectors are used to search nearest neighbors. Full-dimensional vectors, feature vectors compressed with a factor of 2 and 4 are respectively used in indexing. Experiment results listed in Table 1 show that using compressed features triples indexing speed for HNSW and doubles indexing speed for NSG. Taking experiments based on HNSW as an example, using compressed features reduce indexing time from 733 s to 454 s and 245 s for GIST1M dataset and decreases indexing time from 300 s to 144 and 100 s for Deep1M. Results on NSG show similar trends.

Interestingly, using compression feature vectors slightly improves search accuracy. For instance, on GIST1M, using 4×-compressed vectors improves recall of 1@1 to 98.00% and improves recall of 100@100 to 95.60, 0.5 and 0.37% points higher that that of baseline respectively. We conjecture this is due to that using compressed feature vectors to build index introduces some extra links and these

Table 3. Fusion experiments. Bytes denotes the size of quantized feacture vectors. Comp. and Quant. represent compression and quantization methods, respectively. Following default settings in their source code, we conduct experiments on L&C [8], except that the M of coder is set to 8 (This gets highest accuracy.). For PQ experiments, we adopt IVFADC (nlist=8) to avoid exhaustive search.

Datasets	Bytes	Comp.	ANNS	Speed	Recall		
				(q/s)	1@1	1@5	1@50
GIST1M	60	–	PQ	115	23.0	45.2	79.7
	60	CCST	PQ	117	39.5(16.5↑)	72.8(27.6↑)	97.3(17.6↑)
	60	–	L&C	1536	31.4	48.9	54.2
	60	CCST	L&C	1527	30.7(0.7↓)	50.2(1.3↑)	55.1(0.9↑)
	30	–	PQ	246	15.2	31.0	62.4
	30	CCST	PQ	251	20.9(5.7↑)	43.4(12.4↑)	79.7(17.3↑)
	30	–	L&C	2000	23.6	43.0	53.9
	30	CCST	L&C	1992	23.9(0.3↑)	43.9(0.9↑)	54.3(0.4↑)
Deep1M	32	–	PQ	236	32.3	67.5	95.9
	32	CCST	PQ	240	53.3(21.0↑)	88.5(21.0↑)	100.0(4.1↑)
	32	–	L&C	2000	47.9	72.3	76.4
	32	CCST	L&C	2020	48.7(0.9↑)	74.1(1.8↑)	78.3(1.9↑)
	16	–	PQ	520	16.8	38.6	73.6
	16	Catalyst	PQ	545	19.8(3.0↑)	47.9(9.3↑)	82.5(8.9↑)
	16	CCST	PQ	537	32.5(15.7↑)	65.6(27.0↑)	94.4(20.8↑)
	16	–	L&C	2677	28.8	55.6	71.6
	16	CCST	L&C	2676	29.2(0.4↑)	56.9(1.3↑)	73.7(2.1↑)

links improves search accuracy, just like extra links selected by select neighbors heuristic improves the accuracy of HNSW [3].

Note that the speedup is scalable to larger datasets. As shown in Table 2, for Bigann-1B[3] (1 billion 128-d hand drafted features), using our features reduces indexing time from 106 h to 51 h; On our internal dataset Deepfeat25M, comparison results shows similar trend.

4.3 Improving Accuracy and Speed for PQ Related Methods

Besides graph-based methods, PQ related methods are also very popular in real world applications. As quantization and dimension compression are orthogonal at the method level, we conjecture that our proposed CCST can be applied on PQ related methods to achieve higher accuracy. In this section, we conduct fusion experiments to verify this point. Specifically, we fuse our proposed CCST with the classical PQ [2] and most recent proposed L&C [8]. In addition, as the inspirational method catalyst [16] is close to our method in training networks to compress feature dimension, we compare our proposed model with catalyst on Deep1M, the dataset which is used in both this paper and their paper. Experiment results are listed in Table 3.

---

[3] https://dl.fbaipublicfiles.com/billion-scale-ann-benchmarks/bigann/base.1B.u8bin.

Table 4. Ablation study on lightweight designs on GIST1M (d_{in}=960). C_F, k/s and GPU denote compression factor, thousand feature vectors per second and the peak GPU memory usage during encoding. Queries are batch-processed, where batch size is set to 512 or 1024 according to GPU memory usage. # param. is the number of learnable parameters in model. GPU adopted here is RTX2080Ti.

C_F	d_{out}	Light Designs	# param (M)	HNSW		Encoding info		
				R1@1	R100@100	Time (s)	Speed (k/s)	GPU(G)
2	480	Y	10.4	97.70	94.55	21.43	46.66	1.19
2	480	N	15.9	97.70	94.57	35.21	28.40	1.37
4	240	Y	4.7	98.00	95.27	10.00	100.00	1.17
4	240	N	6.1	97.98	95.27	14.12	70.82	1.28

Table 5. Ablation study on random projection initialization (RP init) and the proposed INPR loss, using GIST1M dataset.

Settings		None	RP init	INPR loss	RP init and INPR loss
Recall	1@1	76.1	77.9	78.5	80.1
	1@5	94.3	96.1	97.2	98.4

Experiment results show that using compressed feature learned by our proposed CCST improves both search accuracy and speed for PQ related methods. On Deep1M, in experiments of quantizing feature vectors to 32 bytes, combining CCST with PQ improves recalls 1@1, 1@5 and 1@50 by 21.0, 21.0 and 4.1% points respectively. Our proposed CCST also brings more improvement than Catalyst. In experiments of coding input vectors with 16 bytes, using Catalyst improves recalls 1@1 and 1@5 by 3.0 and 9.3% points and our proposed CCST improves recalls by 11.7 and 27% points.

Using CCST also improves search accuracy for L&C, but the improvements CCST brings in here are less than that in PQ. This may result from that 2-level residual codec already optimized PQ quantized codes once and there is less room for improvement. Overall, CCST still improves search accuracy for L&C, while keeping high search speed.

4.4 Ablation Study

In this section, we conduct ablation study experiments on lightweight designs, random projection initialization and the proposed INPR loss. Experimental results are listed in Tables 4 and 5 . From results in Table 4, we can see that lightweight designs improve encoding speed and save GPU usage without sacrificing accuracy. Results in Table 5 show that using random projection units to initialize projection matrices and adopting the proposed INPR loss are beneficial for final accuracy. The best accuracy is achieved by employing both.

Table 6. Comparison experiments. Compression factor = 4.

Methods	GIST1M						Deep1M					
	HNSW			Brute force			HNSW			Brute force		
	1@1	1@5	1@10	1@1	1@5	1@10	1@1	1@5	1@10	1@1	1@5	1@10
SRP	24.9	49.4	60.7	24.8	49.7	60.9	17.3	35.6	44.4	17.3	35.6	44.4
MLP	47.9	73.8	84.2	47.9	73.8	84.1	48.3	77.6	89.1	48.4	77.6	88.9
VAE	49.2	77.0	86.0	49.3	76.8	85.8	50.0	81.9	90.2	50.0	81.9	90.2
Catalyst	–	–	–	–	–	–	57.9	89.6	92.1	57.9	89.6	92.1
CCST	**80.1**	**98.4**	**99.8**	**80.9**	**98.3**	**99.7**	**67.3**	**94.9**	**98.9**	**67.3**	**94.9**	**98.9**

4.5   Comparison with Other Compression Methods

Previous experiments have shown that using compression feature vectors learned by our proposed CCST speeds up indexing without sacrificing accuracy. Besides our CCST, there are other compression methods. In this section, we compare the proposed models with other methods to evaluate its efficiency.

Here, we employed five comparison methods, including one traditional methods and four network based learning methods. Table 6 presents comparison results. Compared with other four compression methods, our proposed CCST achieves the highest accuracy. For GIST1M, using CCST and HNSW achieves 80.1% recall 1@1, 30.9% points higher that of VAE. As is analysed in the section *the proposed method*, using single sparse random projection harms search accuracy seriously. Both VAE is classical and powerful compression method, but it focus on keeping information of the input feature instead of keeping the neighborhood structure, which is not aligned with the requirement of ANNS. They are outperformed by our proposed CCST. On Deep1M, the proposed CCST also achieves better performance, even compared with the most recent proposed method Catalyst.

5   Discussion

In this work, we have proposed a generic feature compression method for ANNS problem. The proposed method consists of a compression network (CCST) which combines traditional projection function and transformer model, and an inhomogeneous neighborhood relationship preserving (INRP) loss which is aligned with the characteristic of ANNS. The proposed method can be generalized to most ANNS methods. It speeds up indexing speed to 2× to 4 × its original speed without hurting accuracy for graph based methods. It improves recalls by several or even a dozen percentage points for PQ related methods.

Acknowledgement. B. Tang's participation was in part supported by the National Natural Science Foundations of China (U1813215)

References

1. Subramanya, S.J., Kadekodi, R., Krishaswamy, R., Simhadri, H.V.: Diskann: Fast accurate billion-point nearest neighbor search on a single node. In: Proceedings of the 33rd International Conference on Neural Information Processing Systems, pp. 13766–13776 (2019)
2. Jegou, H., Douze, M., Schmid, C.: Product quantization for nearest neighbor search. IEEE Trans. Pattern Anal. Mach. Intell. **33**(1), 117–128 (2010)
3. Malkov, Y.A., Yashunin, D.A.: Efficient and robust approximate nearest neighbor search using hierarchical navigable small world graphs. IEEE Trans. Pattern Anal. Mach. Intell. **42**(4), 824–836 (2018)
4. Fu, C., Xiang, C., Wang, C., Cai, D.: Fast approximate nearest neighbor search with the navigating spreading-out graph. arXiv preprint arXiv:1707.00143 (2017)
5. Wold, S., Esbensen, K., Geladi, P.: Principal component analysis. Chemom. Intell. Lab. Syst. **2**(1–3), 37–52 (1987)
6. Pu, Y., et al.: Variational autoencoder for deep learning of images, labels and captions. Adv. Neural. Inf. Process. Syst. **29**, 2352–2360 (2016)
7. Li, P., Hastie, T.J., Church, K.W.: Very sparse random projections. In: Proceedings of the 12th ACM SIGKDD International Conference on Knowledge Discovery and Data Mining, pp. 287–296 (2006)
8. Douze, M., Sablayrolles, A., Jégou, H.: Link and code: Fast indexing with graphs and compact regression codes. In: Proceedings of the IEEE Conference on Computer Vision and Pattern Recognition, pp. 3646–3654 (2018)
9. Silpa-Anan, C., Hartley, R.: Optimised kd-trees for fast image descriptor matching. In: 2008 IEEE Conference on Computer Vision and Pattern Recognition, pp. 1–8. IEEE (2008)
10. Andoni, A., Razenshteyn, I.: Optimal data-dependent hashing for approximate near neighbors. In: Proceedings of the Forty-Seventh Annual ACM Symposium on Theory of Computing, pp. 793–801 (2015)
11. Ren, J., Zhang, M., Li, D.: Hm-ann: Efficient billion-point nearest neighbor search on heterogeneous memory. In: Advances in Neural Information Processing Systems (2020)
12. Hyvärinen, A., Oja, E.: Independent component analysis: algorithms and applications. Neural Netw. **13**(4–5), 411–430 (2000)
13. Achlioptas, D.: Database-friendly random projections: Johnson-lindenstrauss with binary coins. J. Comput. Syst. Sci. **66**(4), 671–687 (2003)
14. Johnson, W.B., Lindenstrauss, J.: Extensions of lipschitz mappings into a hilbert space 26. Contemporary mathematics 26 (1984)
15. Hinton, G.E., Salakhutdinov, R.R.: Reducing the dimensionality of data with neural networks. Science **313**(5786), 504–507 (2006)
16. Sablayrolles, A., Douze, M., Schmid, C., Jégou, H.: Spreading vectors for similarity search. In: ICLR (2019)
17. Vaswani, A., et al.: Attention is all you need. In: Advances in neural information processing systems, pp. 5998–6008 (2017)
18. Dosovitskiy, A., et al.: An image is worth 16x16 words: Transformers for image recognition at scale. arXiv preprint arXiv:2010.11929 (2020)
19. Touvron, H., Cord, M., Douze, M., Massa, F., Sablayrolles, A., Jégou, H.: Training data-efficient image transformers & distillation through attention. In: International Conference on Machine Learning, pp. 10347–10357. PMLR (2021)

20. Liu, Z., et al.: Swin transformer: Hierarchical vision transformer using shifted windows. arXiv preprint arXiv:2103.14030 (2021)
21. Graham, B., et al.: Levit: a vision transformer in convnet's clothing for faster inference. arXiv preprint arXiv:2104.01136 (2021)
22. Chen, G., Chen, P., Shi, Y., Hsieh, C.Y., Liao, B., Zhang, S.: Rethinking the usage of batch normalization and dropout in the training of deep neural networks. arXiv preprint arXiv:1905.05928 (2019)
23. Zhuang, B., Shen, C., Tan, M., Liu, L., Reid, I.: Structured binary neural networks for accurate image classification and semantic segmentation. In: Proceedings of the IEEE/CVF Conference on Computer Vision and Pattern Recognition, pp. 413–422 (2019)
24. Guo, R., et al.: Accelerating large-scale inference with anisotropic vector quantization. In: International Conference on Machine Learning, pp. 3887–3896. PMLR (2020)
25. Oliva, A., Torralba, A.: Modeling the shape of the scene: A holistic representation of the spatial envelope. Int. J. Comput. Vision **42**(3), 145–175 (2001)
26. Krizhevsky, A., Sutskever, I., Hinton, G.E.: Imagenet classification with deep convolutional neural networks. Adv. Neural. Inf. Process. Syst. **25**, 1097–1105 (2012)
27. Muja, M., Lowe, D.G.: Scalable nearest neighbor algorithms for high dimensional data. IEEE Trans. Pattern Anal. Mach. Intell. **36**(11), 2227–2240 (2014)
28. Gong, Y., Lazebnik, S., Gordo, A., Perronnin, F.: Iterative quantization: A procrustean approach to learning binary codes for large-scale image retrieval. IEEE Trans. Pattern Anal. Mach. Intell. **35**(12), 2916–2929 (2012)
29. Loshchilov, I., Hutter, F.: Decoupled weight decay regularization. arXiv preprint arXiv:1711.05101 (2017)

SEMICON: A Learning-to-Hash Solution for Large-Scale Fine-Grained Image Retrieval

Yang Shen[1,2] , Xuhao Sun[1] , Xiu-Shen Wei[1,2,3(✉)] ,
Qing-Yuan Jiang[4] , and Jian Yang[1]

[1] School of Computer Science and Engineering, Nanjing University of Science
and Technology, Nanjing, China
{shenyang_98,sunxh,weixs,csjyang}@njust.edu.cn
[2] State Key Laboratory of Integrated Services Networks, Xidian University,
Xi'an, China
[3] State Key Laboratory for Novel Software Technology, Nanjing University,
Nanjing, China
[4] Nanjing, China

Abstract. In this paper, we propose <u>S</u>uppression-<u>E</u>nhancing <u>M</u>ask based attention and <u>I</u>nteractive <u>C</u>hannel transformati<u>ON</u> (SEMICON) to learn binary hash codes for dealing with large-scale fine-grained image retrieval tasks. In SEMICON, we first develop a suppression-enhancing mask (SEM) based attention to dynamically localize discriminative image regions. More importantly, different from existing attention mechanism simply erasing previous discriminative regions, our SEM is developed to restrain such regions and then discover other complementary regions by considering the relation between activated regions in a stage-by-stage fashion. In each stage, the interactive channel transformation (ICON) module is afterwards designed to exploit correlations across channels of attended activation tensors. Since channels could generally correspond to the parts of fine-grained objects, the part correlation can be also modeled accordingly, which further improves fine-grained retrieval accuracy. Moreover, to be computational economy, ICON is realized by an efficient two-step process. Finally, the hash learning of our SEMICON consists of both global- and local-level branches for better

Y. Shen, X. Sun, X.-S. Wei and J. Yang—Are also with Key Lab of Intelligent Perception and Systems for High-Dimensional Information of Ministry of Education, and Jiangsu Key Lab of Image and Video Understanding for Social Security, Nanjing University of Science and Technology, China. This work is supported by National Key R&D Program of China (2021YFA1001100), Natural Science Foundation of Jiangsu Province of China under Grant (BK20210340), the Fundamental Research Funds for the Central Universities (No. 30920041111, No. NJ2022028), CAAI-Huawei MindSpore Open Fund, Beijing Academy of Artificial Intelligence (BAAI), and Postgraduate Research & Practice Innovation Program of Jiangsu Province (KYCX22_0463).

Supplementary Information The online version contains supplementary material available at https://doi.org/10.1007/978-3-031-19781-9_31.

representing fine-grained objects and then generating binary hash codes explicitly corresponding to multiple levels. Experiments on five benchmark fine-grained datasets show our superiority over competing methods. (Codes are available at https://github.com/NJUST-VIPGroup/ SEMICON).

Keywords: Fine-grained image retrieval · Learning to hash · Attention mechanism · Large-scale image search

1 Introduction

The explosive growth of images on the web makes learning-to-hash methods become a promising solution for large-scale image retrieval tasks [44]. The objective of image-based hash learning aims to represent the content of an image by generating a binary code for both efficient storage and accurate retrieval [15]. Most existing deep hashing methods [4,15,17,23] merely support image retrieval for generic concepts, *e.g.*, cars or planes, which might fall short of practical demand with the rapidly growing amount of real applications associated with *fine-grained* image retrieval [16,28,32,40]. Thus, recent works on deep hashing [8,18,31,47] have begun to focus on fine-grained retrieval which is required to retrieve images accurately belonging to subordinate categories of a meta-category, *e.g.*, different species of animals or plants [40], rather than a generic (coarse-grained) category.

In the literature, existing generic hashing methods always utilized the outputs of the last CNN feature layer to generate binary hash codes [4,15]. Then, these generated hash codes naturally correspond to the holistic representations of the retrieved visual objects. On the other side, recent fine-grained hashing methods, some of which had achieved good retrieval accuracy, were proposed to be equipped with additional modules for locating fine-grained objects' parts (*e.g.*, birds tails or dogs heads) by region localization [18,31] or local feature alignment [8]. It is important to know that these located object parts are crucial for fine-grained vision tasks [19,42]. Eventually, similar to generic hashing methods, existing fine-grained hashing still fuses object- and part-level features as a unified feature, and then generates hash codes based on such unified features.

Therefore we ask: *What is the explicit meaning of these hash codes?* In order to make the learnt hash codes explicitly meaningful and interpretable, we propose **S**uppression-**E**nhancing **M**ask based attention and **I**nteractive **C**hannel transformati**ON** (SEMICON), cf. Figure 1. Our SEMICON is designed by having two branches: The one is a global feature learning branch with a single global hashing unit for representing the object-level meanings, while the other one is a local pattern learning branch with multiple local hashing units for representing the multiple (different) part-level meanings in a stage-by-stage fashion. As presented in Fig. 1, our final generated hash bits consists of a single object-level hash code and multiple part-level hash codes. Each hash code could explicitly correspond to its own semantic meaning.

In SEMICON, it has two crucial modules, including the suppression-enhancing mask based attention (SEM) module and the interactive channel transformation (ICON) module. More specifically, SEM is applied in each learning stage of the local pattern learning branch for dynamically localizing discriminative image regions one-by-one. However, different from other attention-based methods, our SEM is developed to restrain such regions and then discover other complementary regions by considering the relation between activated regions. Therefore, the image regions located in two adjacent stages will be correlated, which will be beneficial to the fine-grained tailored representations. For ICON, this module is employed upon each feature tensor (*e.g.*, $\hat{T}$ in Fig. 1) by adopting its channels as token embeddings to make interactions across different channels, cf. Section 3.3. Since channels can generally correspond to visual object parts [6, 26, 37], ICON can also model the part correlation accordingly. It could further improve fine-grained retrieval accuracy by considering the internal semantic interactions/correlations among discriminative parts [2, 31]. However, as directly calculating the correlations across all channels is computationally complex, we implement this module as a two-step process in order to be efficient and scalable. Extensive experimental results on five benchmark fine-grained retrieval datasets suggest that our method achieves the new state-of-the-art performance.

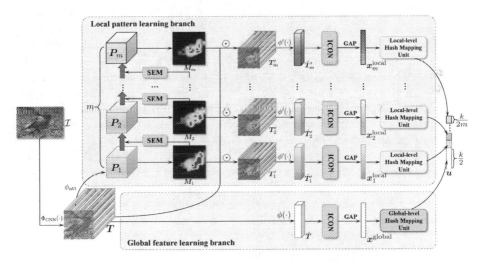

Fig. 1. Overall framework of the proposed SEMICON, which consists of two branches, *i.e.*, the global feature learning branch and the local pattern learning branch. In SEMICON, the SEM module is designed to generate m attention maps (*i.e.*, M_i) stage-by-stage and the ICON module takes each channel as token embeddings to make interactions among different channels. The whole network is end-to-end trainable.

The main contributions of our work are three-fold. (1) We propose a novel method, *i.e.*, the suppression-enhancing mask based attention and interactive channel transformation, for dealing with the fine-grained hash learning task. (2)

We design a suppression-enhancing mask based attention operation to maintain relations between different activated regions and propose a two-step interactive channel transformation module to build correlations between different channels. (3) Experimental results on five benchmark datasets show that our SEMICON achieves significant improvements over competing methods.

2 Related Work

2.1 Fine-Grained Image Retrieval

Fine-grained retrieval is a fundamental topic of fine-grained image analysis [47] which has gained more and more traction in recent years [8,34,38,46,55]. Compared with generic image retrieval, which focuses on retrieving similar images based on similarities in their content (*e.g.*, texture, color, and shape), fine-grained retrieval aims to retrieve the images of the same category type (*e.g.*, the same subordinate species of animals [40]) with only subtle differences (*e.g.*, different beak colors or claw shapes of birds).

Depending on the types of query images, fine-grained image retrieval tasks can be separated into two groups, *i.e.*, fine-grained content based image retrieval (FG-CBIR) and fine-grained sketch-based image retrieval (FG-SBIR). In concretely, SCDA [45] is one of the earliest work of FG-CBIR that used deep pre-trained networks without using explicit localization supervisions. Supervised metric learning based approaches were then proposed to overcome the retrieval accuracy limitations of unsupervised retrieval [2]. In the other research line, FG-SBIR is another interesting task related to both fine-grained image retrieval and cross-modal retrieval of which goal is to match specific photo instances using a free-hand sketch as the query modality. Existing FG-SBIR approaches generally aim to train embedding space where sketches and photos can be compared in a nearest neighbor fashion [38,49].

As all these fine-grained retrieval methods utilize the outputs of the last feature layer of deep networks to deal with retrieval tasks, they still have limitations in the face of large-scale data even if they have achieved good results. To be specific, the searching time for exact nearest neighbor is typically expensive or even impossible for the given queries in large-scale retrieval tasks. To alleviate this issue, fine-grained hashing, which aims to generate compact binary codes to represent fine-grained objects, as a promising direction has attracted the attention in the fine-grained community very recently [8,18,46,52].

2.2 Learning to Hash

Hashing has been widely-studied to transform the data item to a short code consisting of a sequence of bits (*i.e.*, hash codes). Compared to data-independent hashing [9,30,36], data-dependent hashing (*aka* learning to hash) aims to learn hash codes that are more compact yet more data-specific. Due to the discrete of hash codes and non-differentiability of binary hash functions, the optimization of learning to hash is NP-hard [15].

Specifically, data-independent hashing methods attempted to adjust hash generating from different perspectives, *e.g.*, the theory or machine learning views, to name a few: proposing random hash functions satisfying local sensitive property [9], developing better search schemes [30], providing faster computation of hash functions [36], etc. In contrast with data-independent hashing methods, since data-dependent hashing methods learn hash functions from a specific dataset to achieve similarity preserving, they can generally obtain superior retrieval accuracy. Especially for capitalizing on advances in deep learning, many well-performing methods were proposed to integrate feature learning and hash code learning into an end-to-end framework based on deep networks, e.g., [3,17,51].

Fine-grained hashing, as a more challenging and practical hashing task in the vision community, has achieved great attention in very recent years [8,18, 29,46,52]. In the literature, ExchNet [8] and DSaH [18] defined the fine-grained hashing task almost at the same time. While, they added additional modules to extract local-level features for representing objects' parts, and then aggregated both global-level features and local-level features together to generate the unified binary hash codes. SwinFGHash [29] did not add extra modules but took transformer-based architecture to model the feature interactions. The learnt hash bits of these methods seem incomprehensible and lack semantics which are meaningful to fine-grained objects as we do not know what these hash bits explicitly indicate. Although A^2-NET [46] tried to equip those learnt unified hash codes with correspondence to object attributes [11], the hash mapping component still mixes up multiple levels of features together which made the hash codes ambiguous w.r.t. clear visual semantics. In this paper, we do not aggregate all the features from different levels together to generate the unified hash codes, but generate the final hash codes corresponding to the features from different levels in a stage-by-stage fashion.

2.3 Attention Mechanism

Attention mechanisms are those methods for diverting attention to the most important regions of an image and disregarding irrelevant regions [13]. In the past years, attention mechanism has played an increasingly important role and has provided benefits in many vision tasks, *e.g.*, image classification [48], image retrieval [33] and object detection [5]. In a vision system and Deep Neural Networks (DNNs), an attention mechanism can be viewed as a step of dynamically selecting and adaptively weighting features according to the importance of inputs.

Attention mechanisms can be categorised according to data domain [13]. Besides temporal attention [22] and branch attention [24], most of the existing attention mechanisms are related to channel information. In DNNs, different channels in different feature maps usually represent different objects' parts [6,26,37]. Channel attention adaptively recalibrates the weight of each channel in DNNs which can be viewed as an object selection process [13]. In fine-grained tasks, researchers often adopt the erasing operation [25,53] on the most discriminative regions, which can also be described as the most activated channels, to mine discerning information from the rest of the channels. However, these erasing based attention methods seem less informative that the relations across different regions are completely lost.

Recently, self-attention, which has achieved great success in Natural Language Processing [41], has also shown the potential to become a dominant tool in vision tasks [10,27]. Typically, self-attention is used as a spatial attention mechanism to capture global information. Nowadays, the standard Vision Transformer usually split input images into equal-sized blocks and utilize these blocks as the token embeddings [10]. To capture fine-grained parts' correlations, we propose the interactive channel transformation (ICON) module in our SEMICON and utilize different channels as token embeddings. We further implement this module as a two-step computation process in order to reduce the computational complexity.

3   Methodology

3.1   Overall Framework and Notations

Generally, both object-level (global-level) and part-level (local-level) features are crucial in fine-grained visual tasks [47]. Therefore, the overall framework of our SEMICON maintains a global feature learning branch and a local pattern learning branch, cf. Figure 1. Correspondingly, our hash code learning component consists of two units, $i.e.$, the global-level hash mapping unit and the local-level hash mapping unit. In particular, the global-level hash mapping unit is designed to capture object-level binary codes while the local-level hash mapping unit is additionally divided into m sub linear encoder paradigms, which is beneficial to obtaining part-level binary hash codes explicitly in a stage-by-stage fashion. Thus, the final learnt hash codes contain both object-level and part-level meanings. Furthermore, our proposed suppression-enhancing mask based attention (SEM) module and interactive channel transformation (ICON) module are developed to generate both discriminative global-level features and correlated local-level features.

In concretely, for each input image $\mathcal{I}$, a backbone CNN model $\Phi_{\mathrm{CNN}}(\cdot)$ is used to extract its deep activation tensor $\boldsymbol{T}$:

$$\boldsymbol{T} = \Phi_{\mathrm{CNN}}(\mathcal{I}) \in \mathbb{R}^{C \times H \times W} . \tag{1}$$

Then, based on $\boldsymbol{T}$, a global-level transforming network $\phi(\cdot)$, which is equipped with a stack of convolution layers, is performed within the global feature learning branch as:

$$\hat{\boldsymbol{T}} = \phi(\boldsymbol{T}; \theta_{\mathrm{global}}) \in \mathbb{R}^{C' \times H' \times W'} , \tag{2}$$

where θ_{global} presents the parameters of $\phi(\cdot)$. The local pattern learning branch contains an attention guidance $\boldsymbol{P}_1 \in \mathbb{R}^{c \times H \times W}$, which is utilized to generate the attention map $\boldsymbol{M}_1$ in the first stage, cf. Section 3.2. With the help of the attention map, we can evaluate the attended deep descriptors in these $H \times W$ cells by conducting element-wise Hadamard product by:

$$\boldsymbol{T}'_1 = \boldsymbol{M}_1 \odot \boldsymbol{T}. \tag{3}$$

Then, the proposed SEM module is adopted to generate other attention maps $\boldsymbol{M}_i$ in the following $m - 1$ stages, as well as the corresponding deep activation tensors $\boldsymbol{T}'_i$. Besides, to obtain semantic-specific representations, a local-level transforming network $\phi'(\cdot)$, which has the same structure as $\phi(\cdot)$, is used to transform $\boldsymbol{T}'_i$ as

$$\hat{\boldsymbol{T}}'_i = \phi'(\boldsymbol{T}'_i; \theta_{\text{local}}) \in \mathbb{R}^{C' \times H' \times W'}, \tag{4}$$

where θ_{local} presents the parameters of $\phi'(\cdot)$. Then, the proposed ICON module is conducted over $\hat{\boldsymbol{T}}$ and $\hat{\boldsymbol{T}}'_i$ for making interactions across different channels.

Finally, by performing global average-pooling on $\hat{\boldsymbol{T}}$ and $\hat{\boldsymbol{T}}'_i$, we can obtain the object-level feature $\boldsymbol{x}^{\text{global}}$ and m part-level features $\boldsymbol{x}_i^{\text{local}}$. In order to generate the binary-like codes, a binary-like code mapping module consists of $m + 1$ linear encoder paradigms $\boldsymbol{W} = \{\boldsymbol{W}^{\text{global}}; \boldsymbol{W}_1^{\text{local}}; \boldsymbol{W}_2^{\text{local}}; \ldots; \boldsymbol{W}_m^{\text{local}}\}$ is built to project $\boldsymbol{x}^{\text{global}}/\boldsymbol{x}_i^{\text{local}}$ as $\boldsymbol{v}^{\text{global}}/\boldsymbol{v}_i^{\text{local}}$. Eventually, the hash code learning module is performed upon $\boldsymbol{v}^{\text{global}}$ and $\boldsymbol{v}_i^{\text{local}}$ to obtain the final binary hash codes $\boldsymbol{u} = [\boldsymbol{u}^{\text{global}}; \boldsymbol{u}_1^{\text{local}}; \boldsymbol{u}_2^{\text{local}}; \ldots; \boldsymbol{u}_m^{\text{local}}]$.

3.2   Suppression-Enhancing Mask Based Attention

Attention in human perception renders that humans selectively focus on several salient parts of an object, which may help better capture visual structure [20]. Inspired by this, we incorporate the attention mechanism into the local pattern learning branch to capture the patterns of fine-grained objects' parts.

In previous fine-grained vision tasks, some works adopt the mask based attention for erasing the most discriminative regions to mine the rest of the object-specific regions in different branches [25,53]. However, the simple erasing of the most discriminative regions seems trivial and will overlook the relations between the erased regions and other significant regions. To overcome such an issue, we propose the suppression-enhancing mask based attention (SEM) module to maintain relations among different activated regions. It is worth mentioning that the proposed SEM can be realized by convolutional layers sharing parameters, which could bring computational economy.

In concretely, for the given deep activation tensor $\boldsymbol{T}$ related to the input image $\mathcal{I}$, m attention maps $\mathcal{M} = \{\boldsymbol{M}_1, \boldsymbol{M}_2, \ldots, \boldsymbol{M}_m\}$ whose $\boldsymbol{M}_i \in \mathbb{R}^{H \times W}$ will be extracted. While the m attention guidances $\boldsymbol{P}_i$ which are utilized to calculate the attention maps can be expressed as:

$$\boldsymbol{P}_i = \begin{cases} \phi_{\text{att}}(\boldsymbol{T}; \theta_{\text{att}}), & i = 1 \\ f_{SEM}(\text{softmax}(\boldsymbol{M}_{i-1})) \odot \boldsymbol{P}_{i-1}, & i = \{2, 3, \ldots, m\} \end{cases}, \tag{5}$$

where ϕ_{att} is a transformation network which can be optimized in an end-to-end manner driven by the overall loss function described in Sect. 3.4 and f_{SEM} is the suppression-enhancing mask based attention operation which will be described later in this section.

More specifically, the initial attention map $\boldsymbol{M}_1$ is generated according to the attention guidance $\boldsymbol{P}_1$ w.r.t. $\boldsymbol{T}$ in the first stage while in the following $m-1$ stages, attention maps are generated by the suppression-enhancing mask based attention operation. To obtain $\boldsymbol{M}_1$, a transformation network ϕ_{att} is primarily used to obtain what to pay attention to, which can be formulated as:

$$\boldsymbol{P}_1 = \phi_{\mathrm{att}}(\boldsymbol{T}; \theta_{\mathrm{att}}), \tag{6}$$

where $\boldsymbol{P}_1 \in \mathbb{R}^{c \times H \times W}$ presents the attention guidance within the first stage and θ_{att} presents the parameters of the corresponding network w.r.t. $\boldsymbol{T}$. Then, a 1×1 convolution layer φ_1 followed by ϕ_{att} is designed to gain $\boldsymbol{M}_1$.

For the remaining attention maps $\boldsymbol{M}_i$, $i = \{2, 3, \ldots, m\}$ in the following $m-1$ stages, we perform the suppression-enhancing mask based attention operation f_{SEM} which not only helps suppress (rather than simply erasing) the previous most discriminative region but also enhance the other activated regions.

In details, we first calculate the weight of each cell in the attention map $\boldsymbol{M}_{i-1}$ of the previous stage by conducting a softmax function:

$$\boldsymbol{M}'_{i-1} = \mathrm{softmax}(\boldsymbol{M}_{i-1}) \in \mathbb{R}^{H \times W}. \tag{7}$$

Then, we record μ_{i-1}^{std} and $\mu_{i-1}^{\mathrm{mean}}$ as the standard deviation value and the mean value of all the elements in $\boldsymbol{M}'_{i-1}$. For each element $\mu_{i-1}^k \in \{\mu_{i-1}^1, \mu_{i-1}^2, \ldots, \mu_{i-1}^{H \times W}\}$ in $\boldsymbol{M}'_{i-1}$, the f_{SEM} operation is defined as follows:

$$\mu_{i-1}^k = 1 - \frac{\mu_{i-1}^k - \mu_{i-1}^{\mathrm{mean}}}{(\mu_{i-1}^{\mathrm{std}})^{\alpha}}, \tag{8}$$

where α is a hyper-parameter used to regularize the degree of suppression ratio of discriminative regions and the enhance ratio of other activated regions. Additionally, the attention guidance $\boldsymbol{P}_{i-1}$ of the previous stage is then changing to $\boldsymbol{P}_i$ by performing element-wise Hadamard product. The ith attention map is afterwards generated by the ith 1×1 convolution layer φ_i. Therefore, the representations of m attention maps $\boldsymbol{M}_i$ can be written as:

$$\boldsymbol{M}_i = \varphi_i(\boldsymbol{P}_i), i = \{1, 2, \ldots, m\}. \tag{9}$$

Thus, the final m deep activation tensors $\boldsymbol{T}'_i$ can be obtained via

$$\boldsymbol{T}'_i = \boldsymbol{M}_i \odot \boldsymbol{T}, i = \{1, 2, \ldots, m\}. \tag{10}$$

By performing this suppression-enhancing mask based attention operation, the most discriminative region in the attention guidance of the previous stage will be partially restrained. Meanwhile, those unactivated regions will be further inhibited while other activated regions will be enhanced with attention. Therefore, relations between the activated regions of the previous stage and the activated regions generated afterwards could be maintained.

121-th   222-th   323-th   424-th   525-th   626-th   727-th   828-th

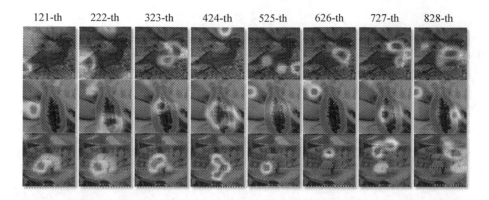

Fig. 2. Visualization of channels extracted from DNNs by highlighting their weights.

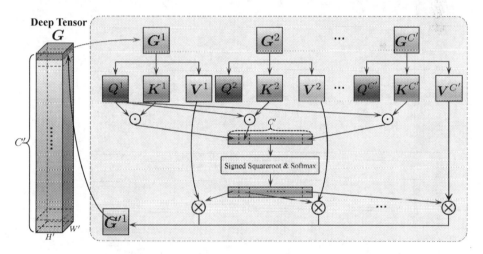

Fig. 3. The Interactive Channel transformatiON (ICON) module. It utilizes each channel as token embeddings and makes interactions across different channels.

3.3   Interactive Channel Transformation

In Deep Neural Networks (DNNs), channels are usually exploited as objects' part detectors [6,26,37]. As can be seen from Fig. 2, the activated regions of the sampled channels (highlighted in warm colors) are semantically meaningful. Therefore, we incorporate the self-attention mechanism into our model and utilize each channel as token embeddings to make interactions across different channels for capturing the correlations of fine-grained "parts", which has been proved can be greatly improved the fine-grained recognition accuracy [7,47,50,54].

In Fig. 3, an overview of the proposed interactive channel transformation (ICON) module is depicted. The computational complexity of directly performing the interactive channel transformation over all channels is considerable. Therefore, for the given deep tensor $G \in \{\hat{T}, \hat{T}'_1, \hat{T}'_2, \ldots, \hat{T}'_m\}$ of each input image $\mathcal{I}$, we split it into several portions and design a two-step interactive channel transformation module (cf. Fig. 4) which can be directly adopted in traditional deep hashing frameworks to reduce the computational consumption.

Specifically, the first step is composed of a stack of N identical parts. For each given $\boldsymbol{G}$, we split the deep tensor into N equal length portions $[\boldsymbol{G}_1; \boldsymbol{G}_2; \boldsymbol{G}_3; \ldots; \boldsymbol{G}_N]$, where $\boldsymbol{G}_i \in \mathbb{R}^{d \times H' \times W'}$ and $d = C'/N$. (H', W') is the resolution of each channel while C' is the number of channels. For each $\boldsymbol{G}_i$, the interactive channel transform operation is used to generate the transformed portion $\boldsymbol{G}'_i$ in order to make interactions over different channels within itself. The interactive channel transform operation during the first step can be described as mapping a unique query $(\boldsymbol{Q}_i)$ and key-value $(\boldsymbol{K}_i - \boldsymbol{V}_i)$ pair to an output $(\hat{\boldsymbol{G}}_i)$, where $\boldsymbol{Q}_i$, $\boldsymbol{K}_i$, $\boldsymbol{V}_i$ are generated form $\boldsymbol{G}_i$ via a 1×1 convolution layer. By following [41], we first compute the dot products $\hat{\boldsymbol{G}}_i$ of the query $\boldsymbol{Q}_i$ with the key $\boldsymbol{K}_i$ and divide by $\sqrt{d}$:

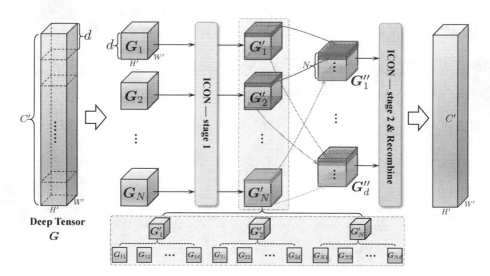

Fig. 4. Our interactive channel transformation module is implemented by a two-step process for reducing the computational consumption.

$$\hat{\boldsymbol{G}}_i = \frac{\boldsymbol{Q}_i \boldsymbol{K}_i^\top}{\sqrt{d}}. \tag{11}$$

Then, a signed squareroot step and a softmax function is applied to generate each output $\boldsymbol{G}'_i$ as:

$$\boldsymbol{G}'_i = \text{softmax}\left(\text{sign}(\hat{\boldsymbol{G}}_i) \cdot \sqrt{|\hat{\boldsymbol{G}}_i| + \delta}\right) \boldsymbol{V}_i, \tag{12}$$

where δ is a fixed positive bias.

In order to make interaction among different portions, in the second step, the tokens at the same position in $\boldsymbol{G}'_i$ are recombined into $\boldsymbol{G}''_i$. In simpler terms, for each portion $\boldsymbol{G}_i' = \{\boldsymbol{G}_{i1}; \boldsymbol{G}_{i2}; \ldots; \boldsymbol{G}_{id}\}$, where $\boldsymbol{G}_i' \in \mathbb{R}^{d \times H' \times W'}$ and $\boldsymbol{G}_{ij} \in \mathbb{R}^{H' \times W'}$ obtained from the first step, we recombine these portions by integrating those channels with the same index in preparation for the second step interactive channel transformation. To be specific, the recombined portion $\boldsymbol{G}_i''$ is consisted of N channels from the previous N portions $\boldsymbol{G}_i'$:

$$\boldsymbol{G}_i'' = \{\boldsymbol{G}_{1i}; \boldsymbol{G}_{2i}; \ldots; \boldsymbol{G}_{Ni}\}, i = \{1, 2, \ldots, d\}. \tag{13}$$

The second step ICON is then performed on $\boldsymbol{G}_i''$ with the same processes as the first step. Finally, channels which have changed their original index will be reset after performing the two-step ICON process.

Between these two steps, we employ a batch normalization and a ReLU activation. A residual connection [14] is adopted after each step. Instead of performing a single interactive channel transform operation associated with keys, values and queries, inspired by [41], we perform the two-step interactive channel transform operation in parallel. For traditional deep hashing frameworks generally use CNNs as vanilla backbones, we perform group convolutions as a substitute for multi-head linear projections. The group number across the first step is N while it will be reset as d within the second step. This two-step multi-group interactive channel transformation allows the model to jointly process information within different indexes over different channels.

3.4   Hash Code Learning

In the following, we conduct the hash code learning based on the obtained object-level features and part-level features. Assuming that we have q query data points which are denoted as $\{\boldsymbol{q}_i\}_{i=1}^q$, as well as p database points which are donated as $\{\boldsymbol{p}_j\}_{j=1}^p$. For each $\boldsymbol{q}_i$ and $\boldsymbol{p}_j$, it consists of a global feature $\boldsymbol{v}^{\text{global}}$ and m local features $\boldsymbol{v}_i^{\text{local}}$. The corresponding hash codes can be carried out by

$$\boldsymbol{u}_i = \text{sign}(\boldsymbol{q}_i), \quad \boldsymbol{z}_j = \text{sign}(\boldsymbol{p}_j), \tag{14}$$

where $\boldsymbol{u}_i, \boldsymbol{z}_j \in \{-1, +1\}^k$, and k presents the length of the final binary hash codes. The goal of hashing is to learn binary hash codes for both query points and database points and preserving their similarity simultaneously. Following [17], the formulation of the hash code learning can be written as:

$$\min_{\boldsymbol{W}, \Theta} \mathcal{L}(\mathcal{I}) = \sum_{i \in \Omega} \sum_{j \in \Gamma} \left[\text{sign}(\boldsymbol{W} \cdot F(\mathcal{I}_i; \Theta))^\top \boldsymbol{z}_j - k S_{ij} \right]^2, \quad \boldsymbol{z}_j \in \{-1, +1\}^k, \tag{15}$$

where Γ presents the indices of all the database points while $\Omega \subseteq \Gamma$ presents the indices of the query set points for we can only gain access to a set of database points $\{\boldsymbol{p}_j\}_{j=1}^p$ without query points during the training stage. $S \in \{-1, +1\}^{q \times p}$ denotes the pairwise supervised information. $\boldsymbol{W}$ presents the matrix of $m + 1$ linear projection and Θ denotes the parameters of DNNs to be learned.

By relaxation, we get the final formulation of SEMICON:

$$\min_{\boldsymbol{W},\Theta} \mathcal{L}(\mathcal{I}) = \beta \sum_{i\in\Omega} \sum_{j\in\Gamma} \left[\tanh(\boldsymbol{W} \cdot F(\mathcal{I}_i; \Theta))^\top \boldsymbol{z}_j - kS_{ij} \right]^2$$
$$+ \gamma \sum_{i\in\Omega} [\boldsymbol{z}_i - \tanh(\boldsymbol{W} \cdot F(\mathcal{I}_i; \Theta))]^2 , \tag{16}$$

where β and γ are hyper-parameters as the trade-off. The proposed SEMICON is an end-to-end deep hashing method which is able to simultaneously perform feature learning and hash code learning in such a unified framework.

4   Experiments

4.1   Datasets

By following A^2-NET [46] and ExchNet [8], our experiments are conducted on two widely used fine-grained datasets, *i.e.*, *CUB200-2011* [43] and *Aircraft* [32], as well as three popular large-scale fine-grained datasets, *i.e.*, *Food101* [1], *NABirds* [39] and *VegFru* [16]. Specifically, *CUB200-2011* contains 11,788 bird images from 200 bird species and is officially split into 5,994 images for training and 5,794 images for test. *Aircraft* contains 10,000 images of 100 aircraft variants, among which 6,667 images for training and 3,333 images for test. For large-scale datasets, *Food101* contains 101 kinds of foods with 101,000 images, where for each class, 250 test images are checked manually for correctness while 750 training images still contain a certain amount of noises. *NABirds* contains 48,562 images of North American birds with 555 sub-categories, 23,929 images for training while 24,633 images for test. *VegFru* is another large-scale fine-grained dataset covering 200 kinds of vegetables and 92 kinds of fruits with 29,200 for training, 14,600 for validation and 116,931 for test.

4.2   Baselines and Implementation Details

Baselines. In experiments, we compare our proposed method to the following competitive generic hashing methods, *i.e.*, ITQ [12], SDH [35], DPSH [23], Hash-Net [4], and ADSH [17]. Among them, DPSH, HashNet and ADSH are also deep learning based methods, while ITQ and SDH are not. Furthermore, we also compare the results of our SEMICON with state-of-the-arts of fine-grained hashing methods, including ExchNet [8] and A^2-NET [46].

Implementation Details. For fair comparisons, we follow the training setting in A^2-NET [46] and ExchNet [8]. In concretely, for *CUB200-2011*, *Aircraft* and *Food101*, we only sample 2,000 images per epoch for training, while 4,000 samples are randomly selected per epoch for *NABirds* and *VegFru*. For the training details, regarding the backbone model, we can choose any network structures as the base network for fine-grained representation learning. While, by follow-ing ExchNet [8] and A^2-NET [46], ResNet-50 [14] is employed in experiments for fair comparisons. The attention generation network ϕ_{att} is the fourth stage of ResNet-50 without downsample convolutions. The global-level transforming network $\phi(\cdot)$ and the local-level transforming network $\phi'(\cdot)$ are independent networks, sharing the same architecture with the fourth stage of ResNet-50. The total number of training epochs is 30. The iteration time is 40 for those datasets containing less than 20,000 training images while for other datasets, the iteration time is 50. For all datasets, we preprocess all images to 224×224, and the learning rate is set to 2.5×10^{-4} for all iterations. SGD with mini-batch set as 16 is used for training. We set the weight decay as 10^{-4} and momentum as 0.91. The hyper-parameters, *i.e.*, α in Eq. (8) and β, γ in Eq. (16), are set as 0.3, 1 and 200, respectively. By following ADSH [17], we adopt soft-constraints strategy [21] to avoid the similarity imbalance problem. The number of m is set as 3 which means there exists 3 attention maps $\boldsymbol{M}_i$. The length of the final hash code $\boldsymbol{u}^{\text{global}}$ and $\boldsymbol{u}_i^{\text{local}}$ is set as $\lceil \frac{k}{2} \rceil$ and $\lfloor \frac{k}{6} \rfloor$. The fixed positive bias δ is set as 10^{-5}. All experiments are conducted with one GeForce RTX 2080 Ti GPU.

Table 1. Comparisons of retrieval accuracy (% mAP) on five fine-grained datasets.

Datasets	# bits	ITQ	SDH	DPSH	HashNet	ADSH	ExchNet	A²-NET	Ours
CUB200-2011	12	6.80	10.52	8.68	12.03	20.03	25.14	33.83	**37.76**
	24	9.42	16.95	12.51	17.77	50.33	58.98	61.01	**65.41**
	32	11.19	20.43	12.74	19.93	61.68	67.74	71.61	**72.61**
	48	12.45	22.23	15.58	22.13	65.43	71.05	77.33	**79.67**
Aircraft	12	4.38	4.89	8.74	14.91	15.54	33.27	42.72	**49.87**
	24	5.28	6.36	10.87	17.75	23.09	45.83	63.66	**75.08**
	32	5.82	6.90	13.54	19.42	30.37	51.83	72.51	**80.45**
	48	6.05	7.65	13.94	20.32	50.65	59.05	81.37	**84.23**
Food101	12	6.46	10.21	11.82	24.42	35.64	45.63	46.44	**50.00**
	24	8.20	11.44	13.05	34.48	40.93	55.48	66.87	**76.57**
	32	9.70	13.36	16.41	35.90	42.89	56.39	74.27	**80.19**
	48	10.07	15.55	20.06	39.65	48.81	64.19	82.13	**82.44**
NABirds	12	2.53	3.10	2.17	2.34	2.53	5.22	**8.20**	8.12
	24	4.22	6.72	4.08	3.29	8.23	15.69	19.15	**19.44**
	32	5.38	8.86	3.61	4.52	14.71	21.94	24.41	**28.26**
	48	6.10	10.38	3.20	4.97	25.34	34.81	35.64	**41.15**
VegFru	12	3.05	5.92	6.33	3.70	8.24	23.55	25.52	**30.32**
	24	5.51	11.55	9.05	6.24	24.90	35.93	44.73	**58.45**
	32	7.48	14.55	10.28	7.83	36.53	48.27	52.75	**69.92**
	48	8.74	16.45	9.11	10.29	55.15	69.30	69.77	**79.77**

4.3   Main Results

Table 1 presents the mean average precision (mAP) results of fine-grained retrieval for comparisons with state-of-the-art hashing methods on these five aforementioned benchmark fine-grained datasets. For each dataset, we report the results of four lengths of hash bits, *i.e.*, 12, 24, 32, and 48, for evaluations. From Table 1, we can observe that the proposed SEMICON significantly outperforms the other baseline methods on these datasets.

In particular, compared with the state-of-the-art method A^2-NET [46], our SEMICON achieves 11.42% and 17.17% improvements over A^2-NET of 24-bit and 32-bit experiments on *Aircraft* and *VegFru*. Moreover, SEMICON obtains superior results on both medium-scale fine-grained datasets, *e.g.*, *CUB200-2011* and *Aircraft*, and large-scale fine-grained datasets, *e.g.*, *NABirds* and *VegFru*. These observations validate the effectiveness of the proposed SEMICON, as well as its promising practicality in real applications of fine-grained retrieval.

4.4   Ablation Studies

We demonstrate the effectiveness of these crucial modules of the proposed SEMICON, *i.e.*, the novel hash learning framework (cf. Sect. 3.1), the suppression-enhancing mask based attention (SEM) module (cf. Sect. 3.2) and the interactive channel transformation (ICON) module (cf. Sect. 3.3). In the ablation studies, we apply these modules incrementally on a vanilla backbone (*i.e.*, ResNet-50) as the baseline. As evaluated in Table 2, by stacking these modules one by one, the retrieval results are steadily improved, which justifies the effectiveness of our proposals in SEMICON.

Table 2. Retrieval accuracy (% mAP) with incremental modules of the proposed SEMICON.

Configurations	CUB200-2011				Aircraft				Food101			
	12	24	32	48	12	24	32	48	12	24	32	48
Vanilla backbone	20.03	50.33	61.68	65.43	15.54	23.09	30.37	50.65	35.64	40.93	42.89	48.81
+ SEMICON⁻*	34.93	58.73	64.71	75.66	34.18	70.14	76.50	80.23	40.59	72.75	78.98	80.15
+ SEM	36.58	64.19	71.58	79.17	43.36	73.39	**80.64**	83.99	44.95	75.44	80.07	82.40
+ ICON	**37.76**	**65.41**	**72.33**	**79.62**	**49.87**	**75.08**	80.45	**84.23**	**50.00**	**76.57**	**80.19**	**82.44**

* SEMICON⁻ represents the model generates m attention maps without performing SEM and the proposed ICON is not performed before obtaining the final hash codes.

5   Conclusion

In this paper, we proposed the Suppression-Enhancing Mask based Attention and Interactive Channel Transformation (SEMICON) for dealing with the large-scale fine-grained image retrieval task. In concretely, the SEM module was developed to restrain (rather than simply erasing) the most discriminative region under the attention guidance of the previous stage, which benefited maintaining relations between different activated regions in a stage-by-stage fashion. Moreover, as channels in DNNs could often correspond to object parts, our ICON module treated each channel as token embeddings for capturing fine-grained parts' correlations. With the hash mapping component containing two units of both global-level and local-level, the final learnt binary hash codes can be generated from different features with different levels (*i.e.*, global-level and local-level) respectively. Experiments on five fine-grained datasets demonstrated the effectiveness of our SEMICON, as well as its proposals. In the future, we would like to improve the robustness of hashing methods and conduct experiments under a more generalized retrieval setting where training classes and test classes have no overlap.

Acknowledgement. The authors would like to thank the anonymous reviewers for their critical and constructive comments and suggestions. We gratefully acknowledge the support of MindSpore, CANN (Compute Architecture for Neural Networks) and Ascend AI Processor used for this research.

References

1. Bossard, L., Guillaumin, M., Van Gool, L.: Food-101 - mining discriminative components with random forests. In: Fleet, D., Pajdla, T., Schiele, B., Tuytelaars, T. (eds.) ECCV 2014. LNCS, vol. 8694, pp. 446–461. Springer, Cham (2014). https://doi.org/10.1007/978-3-319-10599-4_29

2. Cai, S., Zuo, W., Zhang, L.: Higher-order integration of hierarchical convolutional activations for fine-grained visual categorization. In: Proceedings of the IEEE International Conference on Computer Vision, pp. 511–520 (2017)

3. Cakir, F., He, K., Sclaroff, S.: Hashing with binary matrix pursuit. In: Ferrari, V., Hebert, M., Sminchisescu, C., Weiss, Y. (eds.) ECCV 2018. LNCS, vol. 11209, pp. 344–361. Springer, Cham (2018). https://doi.org/10.1007/978-3-030-01228-1_21

4. Cao, Z., Long, M., Wang, J., Yu, P.S.: HashNet: Deep learning to hash by continuation. In: Proceedings of the IEEE International Conference on Computer Vision, pp. 5608–5617 (2017)

5. Carion, N., Massa, F., Synnaeve, G., Usunier, N., Kirillov, A., Zagoruyko, S.: End-to-end object detection with transformers. In: Proceedings of the IEEE International Conference on Computer Vision, pp. 213–229 (2020)

6. Chen, L., et al.: SCA-CNN: Spatial and channel-wise attention in convolutional networks for image captioning. In: Proceedings of the IEEE Conference on Computer Vision and Pattern Recognition, pp. 5659–5667 (2017)

7. Chen, Y., Bai, Y., Zhang, W., Mei, T.: Destruction and construction learning for fine-grained image recognition. In: Proceedings of the IEEE Conference on Computer Vision and Pattern Recognition, pp. 5157–5166 (2019)

8. Cui, Q., Jiang, Q.Y., Wei, X.S., Li, W.J., Yoshie, O.: ExchNet: A unified hashing network for large-scale fine-grained image retrieval. In: Proceedings of European Conference on Computer Vision, pp. 189–205 (2020)

9. Dasgupta, A., Kumar, R., Sarlos, T.: Fast locality-sensitive hashing. In: Proceedings of ACM SIGKDD International Conference on Knowledge Discovery & Data Mining, pp. 1073–1081 (2011)

10. Dosovitskiy, A., et al.: An image is worth 16×16 words: Transformers for image recognition at scale. arXiv preprint arXiv:2010.11929 (2020)

11. Ferrari, V., Zisserman, A.: Learning visual attributes. In: Proceedings of Advances in Neural Information Processing Systems, pp. 433–440 (2007)
12. Gong, Y., Lazebnik, S., Gordo, A., Perronnin, F.: Iterative quantization: A procrustean approach to learning binary codes for large-scale image retrieval. IEEE Trans. Pattern Anal. Mach. Intell. **35**(12), 2916–2929 (2012)
13. Guo, M.H., et al.: Attention mechanisms in computer vision: A survey. arXiv preprint arXiv:2111.07624 (2021)
14. He, K., Zhang, X., Ren, S., Sun, J.: Deep residual learning for image recognition. In: Proceedings of the IEEE Conference on Computer Vision and Pattern Recognition, pp. 770–778 (2016)
15. Hoe, J.T., Ng, K.W., Zhang, T., Chan, C.S., Song, Y.Z., Xiang, T.: One loss for all: Deep hashing with a single cosine similarity based learning objective. In: Proceedings of Advances in Neural Information Processing Systems (2021)
16. Hou, S., Feng, Y., Wang, Z.: VegFru: A domain-specific dataset for fine-grained visual categorization. In: Proceedings of the IEEE International Conference on Computer Vision, pp. 541–549 (2017)
17. Jiang, Q.Y., Li, W.J.: Asymmetric deep supervised hashing. In: Proceedings of Conference on AAAI, pp. 3342–3349 (2018)
18. Jin, S., Yao, H., Sun, X., Zhou, S., Zhang, L., Hua, X.: Deep saliency hashing for fine-grained retrieval. IEEE Trans. Image Process. **29**, 5336–5351 (2020)
19. Krause, J., Gebru, T., Deng, J., Li, L.J., Fei-Fei, L.: Learning features and parts for fine-grained recognition. In: Proceedings of International Conference on Pattern Recognition, pp. 26–33 (2014)
20. Larochelle, H., Hinton, G.: Learning to combine foveal glimpses with a third-order boltzmann machine. In: Proceedings of Advances in Neural Information Processing Systems, pp. 1243–1251 (2010)
21. Leng, C., Cheng, J., Wu, J., Zhang, X., Lu, H.: Supervised hashing with soft constraints. In: Proceedings of ACM International Conference on Information & Knowledge Management, pp. 1851–1854 (2014)
22. Li, J., Wang, J., Tian, Q., Gao, W., Zhang, S.: Global-local temporal representations for video person re-identification. In: Proceedings of IEEE International Conference on Computer Vision, pp. 3958–3967 (2019)
23. Li, W.J., Wang, S., Kang, W.C.: Feature learning based deep supervised hashing with pairwise labels. In: Proceedings of International Joint Conferences on Artificial Intelligence, pp. 1711–1717 (2015)
24. Li, X., Wang, W., Hu, X., Yang, J.: Selective kernel networks. In: Proceedings of IEEE conference on Computer Vision and Pattern Recognition, pp. 510–519 (2019)
25. Liu, C., Xie, H., Zha, Z., Yu, L., Chen, Z., Zhang, Y.: Bidirectional attention-recognition model for fine-grained object classification. IEEE Trans. Multimedia **22**(7), 1785–1795 (2019)
26. Liu, L., Shen, C., Van den Hengel, A.: The treasure beneath convolutional layers: Cross-convolutional-layer pooling for image classification. In: Proceedings of IEEE Conference on Computer Vision and Pattern Recognition, pp. 4749–4757 (2015)
27. Liu, Z., et al.: Swin transformer: Hierarchical vision transformer using shifted windows. In: Proceedings of IEEE Conference on Computer Vision and Pattern Recognition, pp. 10012–10022 (2021)
28. Liu, Z., Luo, P., Qiu, S., Wang, X., Tang, X.: DeepFashion: Powering robust clothes recognition and retrieval with rich annotations. In: Proceedings of IEEE Conference on Computer Vision and Pattern Recognition, pp. 1096–1104 (2016)

29. Lu, D., Wang, J., Zeng, Z., Chen, B., Wu, S., Xia, S.T.: SwinFGHash: Fine-grained image retrieval via transformer-based hashing network. In: Proceedings of British Machine Vision Conference, pp. 1–13 (2021)

30. Lv, Q., Josephson, W., Wang, Z., Charikar, M., Li, K.: Multi-probe LSH: Efficient indexing for high-dimensional similarity search. In: Proceedings of International Conference on Very Large Data Bases, pp. 950–961 (2007)

31. Ma, L., Li, X., Shi, Y., Wu, J., Zhang, Y.: Correlation filtering-based hashing for fine-grained image retrieval. IEEE Signal Process. Lett. **27**, 2129–2133 (2020)

32. Maji, S., Rahtu, E., Kannala, J., Blaschko, M., Vedaldi, A.: Fine-grained visual classification of aircraft. arXiv preprint arXiv:1306.5151 (2013)

33. Ng, T., Balntas, V., Tian, Y., Mikolajczyk, K.: SOLAR: second-order loss and attention for image retrieval. In: Vedaldi, A., Bischof, H., Brox, T., Frahm, J.-M. (eds.) ECCV 2020. LNCS, vol. 12370, pp. 253–270. Springer, Cham (2020). https://doi.org/10.1007/978-3-030-58595-2_16

34. Pang, K., Yang, Y., Hospedales, T.M., Xiang, T., Song, Y.Z.: Solving mixed-modal jigsaw puzzle for fine-grained sketch-based image retrieval. In: Proceedings of the IEEE Conference on Computer Vision and Pattern Recognition, pp. 10347–10355 (2020)

35. Shen, F., Shen, C., Liu, W., Shen, H.T.: Supervised discrete hashing. In: Proceedings of the IEEE Conference on Computer Vision and Pattern Recognition, pp. 37–45 (2015)

36. Shrivastava, A., Li, P.: Densifying one permutation hashing via rotation for fast near neighbor search. In: Proceedings of International Conference on Machine Learning, pp. 557–565 (2014)

37. Simon, M., Rodner, E.: Neural activation constellations: Unsupervised part model discovery with convolutional networks. In: Proceedings of the IEEE Conference on Computer Vision, pp. 1143–1151 (2015)

38. Song, J., Yu, Q., Song, Y.Z., Xiang, T., Hospedales, T.M.: Deep spatial-semantic attention for fine-grained sketch-based image retrieval. In: Proceedings of the IEEE Conference on Computer Vision, pp. 5551–5560 (2017)

39. Van Horn, G., et al.: Building a bird recognition app and large scale dataset with citizen scientists: The fine print in fine-grained dataset collection. In: Proceedings of the IEEE Conference on Computer Vision and Pattern Recognition, pp. 595–604 (2015)

40. Van Horn, G., et al.: The iNaturalist species classification and detection dataset. In: Proceedings of the IEEE Conference on Computer Vision and Pattern Recognition, pp. 8769–8778 (2018)

41. Vaswani, A., et al.: Attention is all you need. In: Proceedings of Advances in Neural Information Processing System, pp. 5998–6008 (2017)

42. Vedaldi, A., et al.: Understanding objects in detail with fine-grained attributes. In: Proceedings of the IEEE Conference on Computer Vision and Pattern Recognition, pp. 3622–3629 (2014)

43. Wah, C., Branson, S., Welinder, P., Perona, P., Belongie, S.: The Caltech-UCSD birds-200-2011 dataset. Tech. Report CNS-TR-2011-001 (2011)

44. Wang, J., Zhang, T., Sebe, N., Tao, S.H.: A survey on learning to hash. IEEE Trans. Pattern Anal. Mach. Intell. **40**(4), 769–790 (2017)

45. Wei, X.S., Luo, J.H., Wu, J., Zhou, Z.H.: Selective convolutional descriptor aggregation for fine-grained image retrieval. IEEE Trans. Image Process. **26**(6), 2868–2881 (2017)

46. Wei, X.S., Shen, Y., Sun, X., Ye, H.J., Yang, J.: A^2-Net: Learning attribute-aware hash codes for large-scale fine-grained image retrieval. In: Proceedings of Advances in Neural Information Processing System, pp. 5720–5730 (2021)

47. Wei, X.S., et al.: Fine-grained image analysis with deep learning: A survey. IEEE Trans. Pattern Anal. Mach. Intell. (2021). https://doi.org/10.1109/TPAMI.2021.3126648

48. Woo, S., Park, J., Lee, J.-Y., Kweon, I.S.: CBAM: Convolutional block attention module. In: Ferrari, V., Hebert, M., Sminchisescu, C., Weiss, Y. (eds.) ECCV 2018. LNCS, vol. 11211, pp. 3–19. Springer, Cham (2018). https://doi.org/10.1007/978-3-030-01234-2_1

49. Yu, Q., Liu, F., Song, Y.Z., Xiang, T., Hospedales, T.M., Loy, C.C.: Sketch me that shoe. In: Proceedings of the IEEE Conference on Computer Vision and Pattern Recognition, pp. 799–807 (2016)

50. Yu, Y., Tang, S., Aizawa, K., Aizawa, A.: Category-based deep cca for fine-grained venue discovery from multimodal data. IEEE Trans. Neural Netw. Learn. Syst. **30**(4), 1250–1258 (2018)

51. Yuan, X., Ren, L., Lu, J., Zhou, J.: Relaxation-free deep hashing via policy gradient. In: Ferrari, V., Hebert, M., Sminchisescu, C., Weiss, Y. (eds.) ECCV 2018. LNCS, vol. 11208, pp. 141–157. Springer, Cham (2018). https://doi.org/10.1007/978-3-030-01225-0_9

52. Zeng, Z., Wang, J., Chen, B., Dai, T., Xia, S.T.: Pyramid hybrid pooling quantization for efficient fine-grained image retrieval. arXiv preprint arXiv:2109.05206 (2021)

53. Zhang, X., Wei, Y., Feng, J., Yang, Y., Huang, T.S.: Adversarial complementary learning for weakly supervised object localization. In: Proceedings of the IEEE Conference on Computer Vision and Pattern Recognition, pp. 1325–1334 (2018)

54. Zheng, H., Fu, J., Mei, T., Luo, J.: Learning multi-attention convolutional neural network for fine-grained image recognition. In: Proceedings of the IEEE Conference on Computer Vision, pp. 5209–5217 (2017)

55. Zheng, X., Ji, R., Sun, X., Zhang, B., Wu, Y., Huang, F.: Towards optimal fine grained retrieval via decorrelated centralized loss with normalize-scale layer. In: Proceedings of Conference of AAAI, pp. 9291–9298 (2019)

CAViT: Contextual Alignment Vision Transformer for Video Object Re-identification

Jinlin Wu[1,2,3] , Lingxiao He[5], Wu Liu[4] , Yang Yang[1,2] , Zhen Lei[1,2,3(✉)] ,
Tao Mei[4] , and Stan Z. Li[6]

[1] CBSR & NLPR, Institute of Automation, Chinese Academy of Sciences,
Beijing, China
{jinlin.wu,yang.yang,zlei}@nlpr.ia.ac.cn
[2] School of Artificial Intelligence, University of Chinese Academy of Sciences,
Beijing, China
[3] Centre for Artificial Intelligence and Robotics, HKISI, CAS, Hong Kong, China
[4] JD Explore Academy, Beijing, China
{liuwu1,tmei}@jd.com
[5] Longfor Inc., Beijing, China
[6] School of Engineering, Westlake University, Hangzhou, China
Stan.ZQ.Li@westlake.edu.cn

Abstract. Video object re-identification (reID) aims at re-identifying the same object under non-overlapping cameras by matching the video tracklets with cropped video frames. The key point is how to make full use of spatio-temporal interactions to extract more accurate representation. However, there are dilemmas within existing approaches: (1) 3D solutions model the spatio-temporal interaction but are often troubled with the misalignment of adjacent frames, and (2) 2D solutions adopt a divide-and-conquer strategy against the misalignment but cannot take advantage of the spatio-temporal interactions. To address the above problems, we propose a Contextual Alignment Vision Transformer (**CAViT**) to the spatio-temporal interaction with a 2D solution. It contains a Multi-shape Patch Embedding (**MPE**) module and a Temporal Shift Attention (**TSA**) module. MPE is designed to retain spatial semantic information against the misalignment caused by pose, occlusion, or misdetection. TSA is designed to achieve contextual spatial semantic feature alignment and jointly model spatio-temporal clues. We further propose a Residual Position Embedding (**RPE**) to guide TSA in focusing on the temporal saliency clues. Experimental results on five video person reID datasets demonstrate the superiority of the proposed CAViT. Additionally, the experiment conducted on VVeRI-901-trial also shows the effectiveness of CAViT for the video vehicle reID. Our code is available on https://github.com/KimWu1994/CAViT.

Keywords: Video object reID · Vision transformer · Temporal shift attention · Residual position embedding

© The Author(s), under exclusive license to Springer Nature Switzerland AG 2022
S. Avidan et al. (Eds.): ECCV 2022, LNCS 13674, pp. 549–566, 2022.
https://doi.org/10.1007/978-3-031-19781-9_32

1   Introduction

Video object re-identification (reID) is a challenging task which matches video tracks of objects across non-overlapping cameras. The spatio-temporal relation information of video tracklets often contains diverse viewpoints and pose variations. Thus, how to learn accurate and robust spatio-temporal representations in a video track is a crucial component for video object reID.

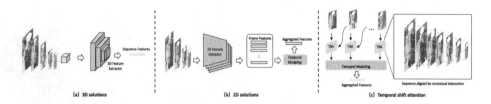

Fig. 1. Illustration of various solutions applied in spatio-temporal learning. (a) 3D solutions. (b) 2D solutions. (c) The proposed temporal shift attention jointly models spatio-temporal clues against the misalignment of adjacent frames.

Many existing methods as shown in Fig. 1(a) apply 3D convolutional neural networks to learn spatio-temporal features in a sequence of video frames. Although it can integrate feature extraction and temporal modeling in one step, it is inevitably affected with spatial misalignment caused by the movement of objects. To this end, some 2D solutions in Fig. 1(b) attempt to adopt a divide-and-conquer strategy that tackles feature representation and feature aggregation separately. However, the divide-and-conquer strategy cannot take full advantage of spatio-temporal interactions.

In this paper, we propose **C**ontextual **A**lignment **Vi**sion **T**ransformer (**CAViT**) which learns accurate and robust spatial-temporal features. Firstly, we replace the self-attention of ViT [9] with a Temporal-Shift Attention (TSA) to align the objects of adjacent frames. It naturally transfers the spatio-temporal modeling task from a 3D representation learning problem to a 2D contextual alignment problem, as shown in Fig. 1(c). To further guide TSA in focusing on the temporal saliency region, we propose a novel yet effective residual position embedding module (RPE) which utilizes the relative variation of the adjacent frames denoting the temporal position. We also design a multi-shape patch embedding (MPE) that provides rich semantic information to improve the ability of feature representation. Experiments on video person reID and vehicle reID show that CAViT achieves relatively high performance even in the presence of heavy occlusion and misdetection. Moreover, CAViT significantly outperforms the state-of-the-arts on video person / vehicle reID benchmarks. Especially, on LSVID and PRID2011, CAViT respectively achieves 89.3% rank1 and 97.5% rank1 performance.

Generally speaking, the main contributions of this paper are as follows:

– We propose a novel video representation learning framework CAViT for video object reID, which jointly learns accurate and robust spatio-temporal features with a 2D vision transformer model.
– We propose a new temporal shift attention module to replace the self-attention mechanism of the vision transformer. It aligns the adjacent frames to extract accurate pedestrian representations from an entire sequence.
– We develop a multi-shape patch embedding module to improve the scalability of the vision transformer. A novel residual position embedding is also introduced to guide our model in focusing the temporal saliency information among consecutive frames.

2   Related Work

2.1   Video Re-identification

The research of video reID has made great progress. As shown in Fig. 2, the rank1 accuracy improves from 30.7% to 91.5% in recent years on MARS dataset. We mainly review highly related video-reID methods in this subsection and give comparisons in Sec. 4 to show the superiority of our method on multiple datasets.

3D Solutions. Some approaches consider video reID as a spatio-temporal representation learning task. To make full use of the temporal clues, some 3D solutions (*e.g.*, C3D [41], P3D [40], SlowFast [11], I3D [44]) are introduced to video reID. However, due to the misalignment of adjacent frames, 3D CNNs are troubled with the background and occlusion. In order to solve this, some 3D alignment convolutional layers are proposed. For example, Li *et al.* [26] design a two-branch 3D CNN network, where one branch is used to capture optical flow clues and the other is used for spatial clues. Another approach is to develop a 3D non-local module (*e.g.*, AP3D [13], Bicnet-tks [19], RFCne [22]) and insert this module to 3D CNNs for the alignment of adjacent frames. However, limited by the locality of the convolution, these methods only align the local region of adjacent frames and cannot solve the misalignment of the whole frame.

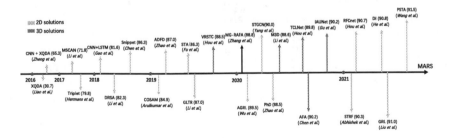

Fig. 2. Development of video reID methods on MARS. The number in parentheses for each method represents the corresponding rank1 performance.

2D Solutions. Other approaches treat video reID as a set representation learning task. To obtain the set representation, some divide-and-conquer based strategies are proposed. They firstly apply 2D CNNs as the feature extractor to obtain features of each frame and then use a feature post process module (*e.g.*, average/maximum temporal pooling, recurrent neural networks (RNNs), or attention mechanisms) to obtain the set average feature. Zhou *et al.* [54] apply a RNN to aggregate multiple frame features. Yang *et al.* [47] propose a spatial-temporal graph convolution network to model the temporal relations of different frames and spatial relations within a frame. Abandoning temporal interaction can free models from the misalignment of adjacent frames [1] but it may disregard some important temporal clues. So PSTA [43] uses aggregation module for ID switch problem. As discussed in above, all those methods cannot tackle temporal dependency, attention, and spatial misalignment simultaneously.

2.2   Transformer Based reID

Transformer breaks the locality limitation of the convolution model, and shows its superiority over convolutional architectures in many vision tasks like image classification and object detection, *e.g.*, DETR [4]. These methods are designed based on the encoder-decoder architecture of the transformer, which applies queries to read the target information from the encoding representations. However, the decoder may be not the necessary component for the visual representation learning task. The decoder-free methods are then proposed, named vision transformer, *e.g.*, ViT [9], Cross ViT [6] and Swin transformer [36]. These methods mainly adopt a patch embedding module and a self-attention mechanism for visual representation learning.

Benefiting from the development of the transformer, the object reID task also makes great progress. For the image-based object reID task, Li *et al.* [30] introduce the vanilla transformer into the partial person reID task, in which the decoder applies K queries for robust representations against the misalignment caused by occluded and partial situations. Liao *et al.* [31] propose a pair-based cross-attention strategy. They use the transformer decoder as a feature post-processing module to re-fine the similarity score of the probe-gallery pair in the unseen scene. Several vision transformer-based methods are also applied to the image-based reID task. He *et al.* [16] propose a ViT based object reID model. To learn representations suitable for cross-camera retrieval, it proposes several strategies including camera position embedding, overlapping patch embedding, jigsaw patch module, etc. Zhu *et al.* [55] propose an auto-aligned strategy in vision transformer to alleviate the misalignment of the feature matching. For the video reID task, He *et al.* [17] design a dense interaction method for transformer to obtain robust embedding. However, it is difficult for training, *i.e.*, dense interaction needs 4 GPUs and 800 epochs for convergence.

3 Methodology

3.1 Problem Formulation

Video object reID aims to retrieve the same object with a query sequence from a gallery set. Let denote $\mathcal{P}$ as the query sequence and $\mathcal{G} = \{\mathcal{G}_1, \mathcal{G}_2, \ldots, \mathcal{G}_K\}$ as the gallery set, where it contains K sequences and each sequence has multiple images. Corresponding features $f_{\mathcal{G}_k}$ for a gallery sequence and $f_\mathcal{Q}$ for the query sequence can be extracted by a video feature learning network. Video object reID retrieves the target gallery video $\mathcal{G}$ that is the most similar to the query in the video representation space, $i.e.$,

$$\mathcal{G} = \max_{\mathcal{K}} \mathcal{S}(f_{\mathcal{G}_k}, f_\mathcal{Q}), \tag{1}$$

where $\mathcal{S}$ is the similarity score of the gallery and query sequences. The key of this task is how to extract discriminative representations from the given sequence $\mathcal{T} = \{I^1, \cdots, I^\mathcal{N}\}$.

$$f^\mathcal{T} = \phi(I^1, \cdots, I^\mathcal{N}), \tag{2}$$

where ϕ is a model extracting discriminative representation from spatio-temporal clues of the video.

3.2 Contextual Alignment Feature Learning

There are two existing frameworks for designing ϕ: 3D solutions and 2D solutions. 3D solutions often apply the 3D CNN as the backbone to jointly learn representations from the whole sequence, as follow:

$$f^\mathcal{T} = \phi_{3D}(I^1, \cdots, I^\mathcal{N}), \tag{3}$$

where ϕ_{3D} denotes 3D CNN backbones. However, ϕ_{3D} is affected by the misalignment of adjacent frames and fails to extract precise representations.

To alleviate this problem, 2D solutions abandon contextual interaction in spatial clues modeling and adopt a divide-and-conquer strategy:

$$f^\mathcal{T} = \psi(\phi_{2D}(I^1), \cdots, \phi_{2D}(I^\mathcal{N})), \tag{4}$$

where ϕ_{2D} denotes 2D CNNs to extract representation for each frame. The 3D representation learning in Eq. 3 is divided into a spatial modeling module ϕ_{2D} and a temporal modeling module ψ. But the performance of ψ is limited, since there is no temporal interaction in ϕ_{2D}.

Considering the aforementioned dilemmas of 2D & 3D solutions, we model the sequence representation problem as a contextual alignment task, $i.e.$,

$$f((I^t|(I^1, \cdots, I^\mathcal{N})) = f(I^t|I^{t-1}). \tag{5}$$

Inspired by Markov chains [28], we focus on the dependencies between the current frame and the previous frame and propose an contextual alignment module $\mathcal{A}$. It models contextual interaction between x^t and x^{t-1}. The spatio-temporal joint modeling task can be formulated as follows:

$$
\begin{aligned}
f^{\mathcal{T}} &= \phi_{3D}(I^1, \cdots, I^{\mathcal{N}}) \\
&= \phi_{2D}(I^1) + \cdots + \phi_{2D}(\mathcal{A}(I^{\mathcal{N}}|I^{\mathcal{N}-1})) \\
&= \phi_{2D}(I^1) + \sum_{t=2}^{\mathcal{N}} \phi_{2D}(\mathcal{A}(I^t|I^{t-1})).
\end{aligned}
\tag{6}
$$

According to Eq. 6, CAViT transfers the spatio-temporal joint modeling task to a contextual alignment problem. The 3D representation learning task of Eq. 3 can be reduced to a 2D representation learning task. The dilemmas between the contextual interaction modeling and the misalignment robustness are also be alleviated.

3.3   Contextual Alignment Vision Transformer

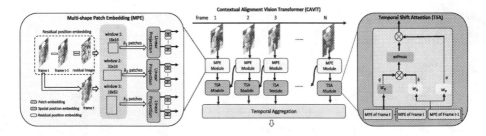

Fig. 3. Framework of **CAViT**. In the **multi-shape patch embedding module**, the input pedestrian sequence is divided into multi-shape patches with multi-shape windows and linear-projected to embeddings. The **residual position embedding** and the learnable 1D spatial position embedding are added to the patch embedding. The **temporal shift attention module** is applied to align the adjacent frames for joint spatio-temporal modeling.

Contextual Alignment Vision Transformer (CAViT) provides a feasible solution for spatio-temporal joint learning. An overview of the CAViT is presented in Fig. 3. The main pipeline of CAViT can be formulated as:

$$
\begin{aligned}
x_0^t &= \mathbf{MPE}(I^t) + \mathcal{R} + \mathcal{P} \\
\hat{x}_l^t &= [x_l^t; x_l^{t-1}] \\
y_l^t &= x_l^t + \mathbf{TSA}(\mathbf{LN}(x_l^t), \mathbf{LN}(\hat{x}_l^t)) \\
x_{l+1}^t &= y_l^t + \mathbf{FFN}(\mathbf{LN}(y_l^t)).
\end{aligned}
\tag{7}
$$

Given a pedestrian sequence$\{I^1, \cdots, I^N\}$, the multi-shape patch embedding module MPE embeds the frame to multi-shape embedding vectors with different shaped windows. Then, a learnable 1D vector $\mathcal{P}$ and the residual position embedding $\mathcal{R}$ are added to the patch embeddings. The former position embedding denotes the spatial position in the current frame, while the latter indicates the temporal variation of the current frame. After these steps, we get the patch embedding x_0^t, which is the input of temporal shift attention layers TSA in CAViT. TSA is the alignment module $\mathcal{A}$ in Eq. 6 used to align current frame I^t and the previous frame I^{t-1}. The attention mechanism of CAViT is built by stacking TSA. FFN and LN are the feed-forward network and Layer normalization of transformer attention block, respectively.

Residual Position Embedding (RPE). ViT is insensitive to the input order and treats each frame of the input sequence equally. Thus, attention power is wasted by redundant information in consecutive frames. To address this problem, we propose a residual position embedding to guide the model in focusing the temporal saliency information, as follows:

$$\mathcal{R}_{s_i}(I^t) = \mathcal{F}_{s_i}(\text{SoftMax}(I^t - I^{t-1})), \tag{8}$$

where $\mathcal{F}_{s_i}$ is the linear projection of the shape s_i. It encodes the residual of the i-th frame and the previous $(i-1)$-th frame as the position embedding. Softmax is used to normalize the residuals signal, suppress signals with small variations and amplify those with large variations caused by viewpoint changing, scale changing, and occlusions. Benefiting from MPE, CAViT extracts diversity information and learns robust representations.

Multi-shape Patch Embedding (MPE). The 16×16 patch is not scaleable enough in the origin ViT model. To perceive objects at different scales, we propose a multi-shape patch embedding module as follows:

$$x_0^t = \mathcal{F}_{s_i}(I^t) + \mathcal{P}_{s_i} + \mathcal{R}_{s_i}(I^t), \tag{9}$$

where $\mathcal{F}^{s_i}$ is the linear projection module of the i-th shape. $\mathcal{P}_{s_i}$ is the spatial position embedding. We adopt the learnable position embedding method as [9], allotting a learnable 1D vector P_{s_i} for each patch at the s_i shape. $\mathcal{R}_{s_i}(I^t)$ is the temporal position embedding as Eq. 8.

Temporal Shift Attention. For a sequence, p_i^t is the i-th patch of I^t, the t-th frame in the pedestrian sequence.

$$q_i^t = p_i^t * W_q$$
$$k_i^t = p_i^t * W_k$$
$$q_i^t = p_i^t * W_v, \tag{10}$$

where W_q, W_k, and W_v are the linear function. q_i^t, k_i^t and v_i^t are the inputs of the attention machine, respectively. The temporal shift attention (TSA) can be modeled as:

$$TSA(p_i^t) = \text{Softmax}(q_i^t \times \mathcal{K}) \times \mathcal{V}$$
$$\mathcal{K} = [k_1^t, \dots, k_N^t, k_1^{t-1} \dots, k_N^{t-1}]^T$$
$$\mathcal{V} = [v_1^t, \dots, v_N^t, v_1^{t-1}, \dots, v_N^{t-1}]. \tag{11}$$

Suppose the normalization factor of SoftMax is γ, TSA can be formulated as:

$$TSA(p_i^t) = \frac{1}{\gamma}[q_i^t * {k_1^t}^T, \dots, q_i^t * {k_N^{t-1}}^T)] \times [v_1^t, \dots, v_N^{t-1}]$$

$$= \frac{q_i^t * {k_1^t}^T}{\gamma} * v_1^t + \cdots + \frac{q_i^t * {k_N^{t-1}}^T}{\gamma} * v_N^{t-1}$$

$$= \sum_{k=k_1^{t-1}, v=v_1^{t-1}}^{k=k_N^t, v=v_N^t} \frac{q_i^t * k^T}{\gamma} * v, \tag{12}$$

where K^T concatenate all the patches of I^{t-1} and I^t. $q_i^t * k^T$ computes the similarity the patch p_i^t and all patches of adjacent frames. The similarity of patch p_i^t and I^t is the intra-frame self-attention, while the similarity of patch p_i^t and I^{t-1} is the inter-frame interaction. Specifically, if p_i^t belongs to an occluder which appears suddenly at I^t and cannot align to I^{t-1}, the response of the occluder p_i^t will be weakened. On the contrary, if p_i^t aligns to it's previous frame, the response will be enhanced. For this reason, TSA is more robust to the ID switch noise in the video object reID.

4   Experiments

4.1   Experiment Implement

Datasets. We evaluate the proposed method on five video person reID datasets and a video vehicle reID dataset, *i.e.,* MARS [53], MARS_DL [37], LSVID [25], PRID-2011 [18], iLIDS-VID [42] and VVeRI-901-trial [23]. The details of these datasets are summarized in Table 1. The bounding boxes are detected with DPM detector [12], and tracked using the GMMCP tracker [8]. The misalignment caused by the DPM detector and ID switch by the GMMCP tracker leads to confusion of video reID models. Liu *et al.* [37] clean MARS as MARS_DL. They re-detect the pedestrian bounding boxes with YOLOV4 [3] and correct the ID switch with IDE [53] model. VVeRI-901 [23] only releases a trial version VVeRI-901-trial. We validate and compare the video object reID approaches on this trial version.

Evaluation Metric. We adopt the mean Average Precision (mAP) and the Cumulative Matching Characteristics (CMC) to evaluate the performance. The evaluation protocol is followed to BiCnet-TKS [19].

Table 1. The statistics of video object reID datasets.

Dataset	# ID	# Boxes	# Tracks	# Cams	# Frames
MARS	1,261	10,675,516	20,715	6	2~920
MARS_DL	1,266	1,019,880	16,360	6	2~920
PRID2011	178	38,466	354	2	5~675
LSVID	3,772	2,982,685	14,943	15	60~2533
iLIDS-VID	300	43,800	600	2	23~192
VVeRI-901-trial	95	52,951	257	11	51~462

Training Details. For our implementation, we randomly choose 16 identities, and sample 4 sequences for each identity. For each sequence, we follow the restricted random sampling strategy [27], which divides each sequence into 8 chunks and randomly chooses one frame from each chunk. All video frames are resized to 256×128 after random data-augmentation (*i.e.,* random horizontal flipping, padding, random cropping and random erasing [15]). As for the optimizer, the SGD optimizer is employed and the learning rate is initialized as 0.01 with cosine learning rate decay. The total training epoch is set to 30. We set 3 shapes for multi-shape patch embedding to obtain diversity semantic representation: (1) 16×16, (2) 16×32, (3) 32×16.

4.2 Results on Video Person reID

In Table 2, we compare CAViT with state-of-the-arts on MARS and LSVID. CAViT achieves the best performance on all evaluation criteria. Table 2 shows the comparison on the two largest datasets (MARS and LS-VID) and Table 3 shows the comparison on the two small datasets (PRID-2011 and iLIDS-VID). In order to make a comparison with temporal shift based methods, we reproduce TSM with a ResNet50 backbone in video reID datasets. The Token shift module is reproduced by ourselves in video reID datasets. For a fair comparison, the token shift method is reproduced with the same pre-trained model (ViT_Base with 16 × 16 patch shape) and the same hyperparameters as CAViT.

CAViT *vs*. ViT Baseline. We implement a strong video reID baseline model, which adapts ViT_Base [9] as the backbone, extracting features of all frames and compute the average feature for pedestrian retrieval. (1) CAViT improves ViT baseline over all six datasets. Particularly on LS-VID, CAViT obviously outperforms ViT baseline by 2.8%/3.9% mAP/rank-1. This is because that the misaligned problem in LS-VID is more seriousness than other datasets. (2) We also note that CAViT only achieves a 0.4% rank1 improvement on MARS. This is because that MARS has a lot of ID switch noise, which is caused by the tracking and detection algorithms. As shown in Table 4, after re-detection, CAViT achieves a 1.0% improvement on MARS_DL, even though ViT baseline has achieved high performance (94.6% rank1).

Table 2. Comparison with state-of-the-arts on MARS [53], LS-VID [25] datasets. The methods are separated into two groups, the 2D neural network solutions (2D), and 3D neural network based solutions (3D).

Methods		Proc.	MARS		LS-VID	
			mAP	R-1	mAP	R-1
2D	MG-RAFA [50]	CVPR 20	85.9	88.8	–	–
	PhD [51]	CVPR 20	86.2	88.9	–	–
	AGRL [45]	TIP 20	81.9	89.5	–	–
	STGCN [47]	CVPR 20	83.7	90.0	–	–
	MGH [46]	CVPR 20	85.8	90.0	–	–
	RGTR [29]	AAA 21	84.0	89.4	–	–
	CTL [34]	CVPR 21	86.7	91.4	–	–
	GRL [35]	CVPR 21	84.8	91.0	–	–
	STRF [1]	ICCV 21	86.1	90.3	–	–
	PSTA [43]	ICCV 21	85.8	**91.5**	–	–
	DI [17]	ICCV 21	87.0	90.8	–	–
	STMN [10]	ICCV 21	84.5	90.5	69.2	82.1
	RFCnet [22]	PAMI 21	86.3	90.7	–	–
3D	I3D [5]	CVPR 17	83.0	88.6	33.9	51.0
	P3D [40]	ICCV 17	83.2	88.9	35.0	53.4
	IAUNet [21]	TNNLS 20	85.0	90.2	–	–
	M3D [26]	TPMAI 20	79.5	88.6	–	–
	TCLNet [20]	ECCV 20	85.1	89.8	–	–
	AP3D [13]	ECCV 20	85.1	90.1	–	–
	AFA [7]	ECCV 20	82.9	90.2	–	–
	STRF [1]	ICCV 21	86.1	90.3	–	–
	BiCnet-TKS [19]	CVPR 21	86.0	90.2	75.1	84.6
2D	TSM(R50) [32]	ICCV 19	81.8	88.6	66.0	78.3
	Token shift [7]	MM 21	86.6	90.2	68.7	80.4
	ViT baseline [9]	ARXIV 20	86.4	89.7	76.4	85.3
2D	**CAViT**	**Our work**	**87.2**	90.8	**79.2**	**89.2**

CAViT *vs.* **3D Solutions.** Existing joint learning solutions use 3D CNNs to jointly model the spatio-temporal clues. Compared with pure 3D CNN based methods, CAViT outperforms P3D [40] with 4.2%/2.2% mAP/rank-1 on MARS. Compared with temporal feature alignment method BiCnet-TKS, CAViT outperforms it by 1.2%/0.6% mAP/rank-1 on MARS, 4.1%/4.6% mAP/rank-1 on LSVID. Compared with temporal feature reconstructing method AP3D [13], CAViT outperforms it by 2.5%/1.2% mAP/rank-1 on MARS and 4.6% rank1 on iLIDS-VID. We argue that this is because 3D CNN is limited by the local

Table 3. Comparison with state-of-the-arts on PRID2011 [18], and iLIDS-VID [42] datasets. The methods are separated into two groups, the 2D neural network solutions (2D), and 3D neural network based solutions (3D).

Methods		Proc.	PRID-2011		iLIDS-VID	
			R-1	R-5	R-1	R-5
2D	MG-RAFA [50]	CVPR 20	95.9	99.7	88.6	98.0
	PhD [51]	CVPR 20	96.6	97.8	–	–
	AGRL [45]	TIP 20	94.6	99.1	84.5	96.7
	ADFD [52]	CVPR 19	93.9	99.5	86.3	97.4
	GLTR [25]	ICCV 19	95.5	100.0	86.0	98.0
	MGH [46]	CVPR 20	94.8	99.3	85.6	97.1
	RGTR [29]	AAAI 21	93.7	99.0	86.0	98.0
	GRL [35]	CVPR 21	96.2	99.7	90.4	98.3
	PSTA [43]	ICCV 21	95.6	98.9	91.5	98.1
	DI [17]	ICCV 21	–	–	92.0	98.0
3D	STRF [1]	ICCV 21	–	–	89.3	–
	M3D [26]	TPMAI 20	**96.6**	**100.0**	86.7	98.0
	TCLNet [20]	ECCV 20	–	–	86.6	–
	AP3D [13]	ECCV 20	–	–	88.7	–
	AFA [7]	ECCV 20	–	–	88.5	96.8
	TSM [32]	ICCV 19	87.6	93.5	69.3	81.3
2D	Token shift [7]	MM 21	91.1	95.5	86.0	98.0
	ViT baseline [9]	ARXIV 20	92.4	96.8	90.2	93.7
2D	**CAViT**	**Our work**	95.5	98.9	**93.3**	**98.0**

receptor field of the convolutional network. Neither temporal alignment methods nor temporal reconstruction methods can solve the case of misalignment between frames well. Different with them, CAViT implements alignment of the entire frames, thus solving the misalignment well.

CAViT *vs.* **2D Solutions.** 2D solutions often apply 2D CNNs to model spatial clues and then use a temporal aggregation module(*i.e.*, LSTM, RNN, GCN, transformer) to merge the spatial representations. As we can see, DI is lower than CAViT by 0.2% on mAP in MARS and 1.3% rank1 in iLIDS-VID, while PSTA is lower than CAViT by 1.8% mAP on MARS and 1.8% rank1 on iLDS-VID. This is because lacking consideration of spatio-temporal interactions, they cannot take full advantage of the complementarity of adjacent frames.

Table 4. Comparison with state-of-the-arts on the MARS_DL dataset.

Methods		Proc.	MARS_DL	
			mAP	R-1
3D	TCLNet [20]	ECCV 20	85.4	91.0
	AP3D [13]	ECCV 20	86.5	91.3
	P3D-C [40]	ICCV 17	85.0	91.0
	C2D [24]	CVPR 19	86.2	91.4
2D	Non-Local [33]	ARXIV 19	85.8	90.8
	FT-WFT [39]	AAAI20	83.8	91.0
	DL+CF-AAN [37]	ARXIV 21	86.5	91.3
	TSM+ResNet [32]	ICCV 19	86.0	93.6
	Token Shift [49]	ACMM 21	90.1	94.9
	ViT baseline [9]	ARXIV 20	89.4	94.6
2D	**CAViT**	**Our work**	**90.5**	**95.6**

CAViT *vs.* **Temporal Shift Methods.** TSM [32], token shift [49] and our CAViT use the temporal shift strategy for jointly modeling spatio-temporal clues with a 2D model. In Tab. 2, the performance gap is significant among these two methods and our CAViT. Specifically, on iLIDS-VID, CAViT outperforms TSM by 30.0% rank1 and outperforms the token shift method by 7.3% rank1. We argue that TSM directly shifts the feature channel, which may aggravate the spatial misalignment among pedestrians. The performance of Token shift is close to CAViT on almost all datasets, except LSVID, where CAViT outperforms Token shift by 10.5% mAP and 8.8% rank1. This is because the misalignment is much more serious and the CLS token worsens this misalignment, making the performance of Token shift even worse than origin ViT model.

CAViT *vs.* **Transformer-Based Methods.** Both of DI (Dense Interaction) [17] and our CAViT belong to the transformer based video reID methods. The difference is that DI applies the ResNet50 to extract spatial features and uses the transformer for temporal modeling, which is essentially a divide-and-conquer method instead of our joint modeling strategy. The proposed joint modeling method CAViT outperforms DI by 0.2% mAP on MARS and 1.3% rank1 on iLIDS-VID.

4.3   Results on Video Vehicle reID

We validate the proposed CAViT on the VVeRI-901-trial dataset. For a fair comparison, we also reproduce some widely used video representation learning methods(*e.g.*, AP3D, 3D Non-local and strong baseline of object reID) on the VVeRI-901-trial dataset in Table 5. Compared with 3D solutions, CAViT outperforms AP3D by 4.4% mAP. Compared with similar temporal shift based methods, CAViT outperforms token shift by 4.8% rank-5 and outperforms TSM 15.0% on rank1.

Table 5. Comparison with state-of-the-arts on the VVeRI-901-trial dataset.

Method		Proc.	VVeRI-901		
			mAP	R-1	R-5
3D	C2D [24]	CVPR 19	57.3	50.2	72.5
	NL3D [33]	ARXIV 19	60.5	55.0	77.5
	AP3D [13]	ECCV 20	61.2	52.5	75.0
	AP3D+NL3D [13]	ECCV 20	60.2	50.0	80.0
	BiCnet-TKS [19]	CVPR 21	50.8	41.3	70.4
2D	TSM [32]	ICCV 19	55.1	45.0	72.5
	BOT [38]	CVPRW 19	61.6	55.3	77.5
	SBS [15]	ARXIV 21	62.4	57.5	75.1
	ViT baseline [9]	ARXIV 20	62.7	52.5	84.0
	Token shift [49]	ICCV 19	**67.4**	57.5	80.0
2D	**CAViT**	**Our work**	65.6	**60.0**	**84.8**

Table 6. The ablation study of Backbones & Multi-shape Patch Embedding.

Backbone	Patch shape			MARS	
	16×16	16×32	32×16	mAP	R-1
ResNet50 [14]				83.7	87.6
ResNet101 [14]				84.0	89.9
ResNeSt101 [48]				84.3	90.1
ResNeSt200 [48]				83.2	89.1
Swin_base [36]				83.6	88.4
Swin_base 3D [37]				68.3	81.4
ViT_Base 3D [2]				81.9	87.5
ViT_Base [9]	✓			86.4	89.7
ViT_Base [9]		✓		82.9	88.6
ViT_Base [9]			✓	83.0	87.8
ViT_Base [9]	✓	✓	✓	**86.8**	**90.6**

4.4   Ablations Studies

Ablation of the Backbones and Multi-shape Patch Embedding (MPE).
In Table 6, to compare the performance of several popular backbones, we use
different backbones of extracting spatial representations and apply an average
pooling module for temporal aggregation. We can observe that, 2D backbones
perform better than 3D backbones (*i.e.,* Timeformer, Swin_base 3D), since 3D
backbones are troubled with misalignment. In addition, according to rank1 in

Table 6, we can observe that: ViT_Base + MPE > ResNeSt101 = ViT_Base > ResNet101 > ResNeSt200 > Swin_base > ResNet50. Although the 32×16 patch and the 16×32 patch are worse than 16×16 patch, the MPE which ensembles these three shapes achieves the best performance. This is because patches of different shapes focus on information of different granularity and directions.

Table 7. The ablation study of the different module in CAViT.

	MARS_DL		PRID-2011	
	mAP	R-1	R-1	R-5
ViT baseline	89.4	94.6	92.4	96.8
+ MPE	90.0	94.8	93.8	97.7
+ TSA	90.2	95.3	94.6	98.0
+ RPE	**90.5**	**95.6**	**95.5**	**98.9**

Ablation of TSA & RPE. To denote the effectiveness of Temporal Shift Attention (TSA) moudle and Residual Position Embedding (RPE) moudle, we implement ablation experiments on MARS-DL and PRID-2011 in Table 7. Compared with the ViT model with multi-shape patch embedding, TSA achieves 0.2%/0.5% mAP/rank1 increment on MARS_DL. With RPE, TSA improves 0.5%/0.8% mAP/rank1 on MARS_DL and improves 1.7%/1.2% rank1/rank5 on PRID-2011. This indicates that TSA notices temporal saliency clues, under the guidance of RPE.

Attention Map Visualization. The normalized attention maps of MPE are visualized in Fig. 4. (1) For spatial clues, according to this figure, different shape of MPE has different attention regions. MPE helps CAViT pay attention to a variety of granularity and directions and obtain more diverse spatial representations. (2) For temporal clues, the deeper the network layer, the more attention pays on adjacent frames. It also indicates that CAViT learns spatial clues in shallow layers and implements temporal alignment in deep layers.

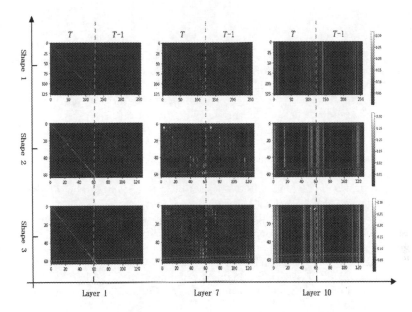

Fig. 4. Attention map visualization for MPE. (1) the first row belongs to the 16×16 patch, while the second and the third rows belong to the 16×32 patch and the 32×16 patch, respectively. (2) For each sub-figure, the left half part is the attention weight of the intra-frame, while the right part is the attention weight of the adjacent frame. (3) Different columns represent the attention map at different layers.

5 Conclusion

This paper proposes a contextual alignment vision transformer (CAViT) for the video object re-identification, which contains a multi-shape patch embedding module (MPE) and a Temporal Shift Attention (TSA) module. The former obtains diversity semantic embedding for spatial alignment in the pedestrian matching process, while the latter applies a 2D solution for jointly modeling spatio-temporal clues. We also introduce a residual position embedding (RPE) to guide the temporal shift attention in focusing on temporal saliency clues. Experimental results on five video pedestrian reID datasets and one video vehicle reID dataset demonstrate the superiority of the proposed CAViT over state-of-the-art methods.

Acknowledgments. This research was supported by the National Key R&D Program of China under Grant No.2020YFC2003901, Chinese National Natural Science Foundation Projects #61876178, #61872367, #61976229, #62176256, #62106264 and the InnoHK program.

References

1. Aich, A., Zheng, M., Karanam, S., Chen, T., Roy-Chowdhury, A.K., Wu, Z.: Spatio-temporal representation factorization for video-based person re-identification, In: ICCV (2021)
2. Bertasius, G., Wang, H., Torresani, L.: Is space-time attention all you need for video understanding? arXiv preprint arXiv:2102.05095 (2021)
3. Bochkovskiy, A., Wang, C.Y., Liao, H.Y.M.: Yolov4: Optimal speed and accuracy of object detection. arXiv preprint arXiv:2004.10934 (2020)
4. Carion, N., Massa, F., Synnaeve, G., Usunier, N., Kirillov, A., Zagoruyko, S.: End-to-end object detection with transformers. In: ECCV (2020)
5. Carreira, J., Zisserman, A.: Quo vadis, action recognition? a new model and the kinetics dataset. In: CVPR (2017)
6. Chen, C.F., Fan, Q., Panda, R.: Crossvit: Cross-attention multi-scale vision transformer for image classification. arXiv preprint arXiv:2103.14899 (2021)
7. Chen, G., Rao, Y., Lu, J., Zhou, J.: Temporal coherence or temporal motion: Which is more critical for video-based person re-identification? In: ECCV (2020)
8. Dehghan, A., Modiri Assari, S., Shah, M.: Gmmcp tracker: Globally optimal generalized maximum multi clique problem for multiple object tracking. In: CVPR (2015)
9. Dosovitskiy, A., et al.: An image is worth 16x16 words: Transformers for image recognition at scale. arXiv preprint arXiv:2010.11929 (2020)
10. Eom, C., Lee, G., Lee, J., Ham, B.: Video-based person re-identification with spatial and temporal memory networks. In: ICCV (2021)
11. Feichtenhofer, C., Fan, H., Malik, J., He, K.: Slowfast networks for video recognition. In: ICCV (2019)
12. Felzenszwalb, P.F., Girshick, R.B., McAllester, D., Ramanan, D.: Object detection with discriminatively trained part-based models. IEEE TPAMI (2009)
13. Gu, X., Chang, H., Ma, B., Zhang, H., Chen, X.: Appearance-preserving 3d convolution for video-based person re-identification. In: ECCV (2020)
14. He, K., Zhang, X., Ren, S., Sun, J.: Deep residual learning for image recognition. In: CVPR (2016)
15. He, L., Liao, X., Liu, W., Liu, X., Cheng, P., Mei, T.: Fastreid: a pytorch toolbox for real-world person re-identification. arXiv preprint arXiv:2006.02631 (2020)
16. He, S., Luo, H., Wang, P., Wang, F., Li, H., Jiang, W.: Transreid: Transformer-based object re-identification. arXiv preprint arXiv:2102.04378 (2021)
17. He, T., Jin, X., Shen, X., Huang, J., Chen, Z., Hua, X.S.: Dense interaction learning for video-based person re-identification supplementary materials. Identities (2021)
18. Hirzer, M., Beleznai, C., Roth, P.M., Bischof, H.: Person re-identification by descriptive and discriminative classification. In: Heyden, A., Kahl, F. (eds.) SCIA 2011. LNCS, vol. 6688, pp. 91–102. Springer, Heidelberg (2011). https://doi.org/10.1007/978-3-642-21227-7_9
19. Hou, R., Chang, H., Ma, B., Huang, R., Shan, S.: Bicnet-tks: Learning efficient spatial-temporal representation for video person re-identification. In: CVPR (2021)
20. Hou, R., Chang, H., Ma, B., Shan, S., Chen, X.: Temporal complementary learning for video person re-identification. In: Vedaldi, A., Bischof, H., Brox, T., Frahm, J.-M. (eds.) ECCV 2020. LNCS, vol. 12370, pp. 388–405. Springer, Cham (2020). https://doi.org/10.1007/978-3-030-58595-2_24
21. Hou, R., Ma, B., Chang, H., Gu, X., Shan, S., Chen, X.: Iaunet: Global context-aware feature learning for person reidentification. IEEE TNNLS (2020)

22. Hou, R., Ma, B., Chang, H., Gu, X., Shan, S., Chen, X.: Feature completion for occluded person re-identification. IEEE TPAMI (2021)
23. Zhao, J., Qi, F., G.R., Xu, L.: Vveri-901: Video vehicle re-identification dataset (2020). https://www.graviti.cn/open-datasets/VVeRI901'
24. Li, C., Zhong, Q., Xie, D., Pu, S.: Collaborative spatiotemporal feature learning for video action recognition. In: CVPR (2019)
25. Li, J., Wang, J., Tian, Q., Gao, W., Zhang, S.: Global-local temporal representations for video person re-identification. In: ICCV (2019)
26. Li, J., Zhang, S., Huang, T.: Multi-scale 3D convolution network for video based person re-identification. In: AAAI (2019)
27. Li, S., Bak, S., Carr, P., Wang, X.: Diversity regularized spatiotemporal attention for video-based person re-identification. In: CVPR (2018)
28. Li, S.Z.: Markov random field modeling in image analysis. Springer Science & Business Media (2009)
29. Li, X., Zhou, W., Zhou, Y., Li, H.: Relation-guided spatial attention and temporal refinement for video-based person re-identification. In: AAAI (2020)
30. Li, Y., He, J., Zhang, T., Liu, X., Zhang, Y., Wu, F.: Diverse part discovery: Occluded person re-identification with part-aware transformer. In: CVPR (2021)
31. Liao, S., Shao, L.: Transformer-based deep image matching for generalizable person re-identification. NeurIPS Workshops (2021)
32. Lin, J., Gan, C., Han, S.: Tsm: Temporal shift module for efficient video understanding. In: ICCV (2019)
33. Liu, C.T., Wu, C.W., Wang, Y.C.F., Chien, S.Y.: Spatially and temporally efficient non-local attention network for video-based person re-identification. arXiv preprint arXiv:1908.01683 (2019)
34. Liu, J., Zha, Z.J., Wu, W., Zheng, K., Sun, Q.: Spatial-temporal correlation and topology learning for person re-identification in videos. In: CVPR (2021)
35. Liu, X., Zhang, P., Yu, C., Lu, H., Yang, X.: Watching you: Global-guided reciprocal learning for video-based person re-identification. In: CVPR (2021)
36. Liu, Z., et al.: Swin transformer: Hierarchical vision transformer using shifted windows. ICCV (2021)
37. Liu, Z., et al.: Video swin transformer. arXiv preprint arXiv:2106.13230 (2021)
38. Luo, H., Gu, Y., Liao, X., Lai, S., Jiang, W.: Bag of tricks and a strong baseline for deep person re-identification. In: CVPR Workshops (2019)
39. Pathak, P., Eshratifar, A.E., Gormish, M.: Video person re-id: Fantastic techniques and where to find them. arXiv preprint arXiv:1912.05295 (2019)
40. Qiu, Z., Yao, T., Mei, T.: Learning spatio-temporal representation with pseudo-3d residual networks. In: ICCV (2017)
41. Tran, D., Bourdev, L., Fergus, R., Torresani, L., Paluri, M.: Learning spatiotemporal features with 3d convolutional networks. In: ICCV (2015)
42. Wang, T., Gong, S., Zhu, X., Wang, S.: Person re-identification by video ranking. In: Fleet, D., Pajdla, T., Schiele, B., Tuytelaars, T. (eds.) ECCV 2014. LNCS, vol. 8692, pp. 688–703. Springer, Cham (2014). https://doi.org/10.1007/978-3-319-10593-2_45
43. Wang, Y., Zhang, P., Gao, S., Geng, X., Lu, H., Wang, D.: Pyramid spatial-temporal aggregation for video-based person re-identification. In: ICCV (2021)
44. Weng, X., Kitani, K.: Learning spatio-temporal features with two-stream deep 3d cnns for lipreading. arXiv preprint arXiv:1905.02540 (2019)
45. Wu, Y., et al.: Adaptive graph representation learning for video person re-identification. IEEE TIP (2020)

46. Yan, Y., et al.: Learning multi-granular hypergraphs for video-based person re-identification. In: CVPR (2020)
47. Yang, J., Zheng, W.S., Yang, Q., Chen, Y.C., Tian, Q.: Spatial-temporal graph convolutional network for video-based person re-identification. In: CVPR (2020)
48. Zhang, H., et al.: Resnest: Split-attention networks. arXiv preprint arXiv:2004.08955 (2020)
49. Zhang, H., Hao, Y., Ngo, C.W.: Token shift transformer for video classification. In: ACM MM (2021)
50. Zhang, Z., Lan, C., Zeng, W., Chen, Z.: Multi-granularity reference-aided attentive feature aggregation for video-based person re-identification. In: CVPR (2020)
51. Zhao, J., Qi, F., Ren, G., Xu, L.: Phd learning: Learning with pompeiu-hausdorff distances for video-based vehicle re-identification. In: CVPR (2021)
52. Zhao, Y., Shen, X., Jin, Z., Lu, H., Hua, X.s.: Attribute-driven feature disentangling and temporal aggregation for video person re-identification. In: CVPR (2019)
53. Zheng, L., et al.: MARS: a video benchmark for large-scale Person re-identification. In: Leibe, B., Matas, J., Sebe, N., Welling, M. (eds.) ECCV 2016. LNCS, vol. 9910, pp. 868–884. Springer, Cham (2016). https://doi.org/10.1007/978-3-319-46466-4_52
54. Zhou, Z., Huang, Y., Wang, W., Wang, L., Tan, T.: See the forest for the trees: Joint spatial and temporal recurrent neural networks for video-based person re-identification. In: CVPR (2017)
55. Zhu, K., et al.: Aaformer: Auto-aligned transformer for person re-identification. arXiv preprint arXiv:2104.00921 (2021)

Text-Based Temporal Localization
of Novel Events

Sudipta Paul[1]([✉]) [iD], Niluthpol Chowdhury Mithun[2] [iD],
and Amit K. Roy-Chowdhury[1] [iD]

[1] University of California, Riverside, CA, USA
{spaul,amitrc}@ece.ucr.edu
[2] SRI International, Princeton, NJ, USA
niluthpol.mithun@sri.com

Abstract. Recent works on text-based localization of moments have shown high accuracy on several benchmark datasets. However, these approaches are trained and evaluated relying on the assumption that the localization system, during testing, will only encounter events that are available in the training set (i.e., *seen* events). As a result, these models are optimized for a fixed set of seen events and they are unlikely to generalize to the practical requirement of localizing a wider range of events, some of which may be *unseen*. Moreover, acquiring videos and text comprising all possible scenarios for training is not practical. In this regard, this paper introduces and tackles the problem of text-based temporal localization of novel/unseen events. Our goal is to temporally localize video moments based on text queries, where *both the video moments and text queries are not observed/available during training*. Towards solving this problem, we formulate the inference task of text-based localization of moments as a relational prediction problem, hypothesizing a conceptual relation between semantically relevant moments, e.g., a temporally relevant moment corresponding to an unseen text query and a moment corresponding to a seen text query may contain shared concepts. The likelihood of a candidate moment to be the correct one based on an unseen text query will depend on its relevance to the moment corresponding to the semantically most relevant seen query. Empirical results on two text-based moment localization datasets show that our proposed approach can reach up to 15% absolute improvement in performance compared to existing localization approaches.

Keywords: Temporal localization · Moment retrieval · Novel/Unseen events

1 Introduction

Event localization in a long and untrimmed video is an important video analysis problem. Recently, there has been a surge of works that address the task of

---

Supplementary Information The online version contains supplementary material available at https://doi.org/10.1007/978-3-031-19781-9_33.

S. Avidan et al. (Eds.): ECCV 2022, LNCS 13674, pp. 567–587, 2022.
https://doi.org/10.1007/978-3-031-19781-9_33

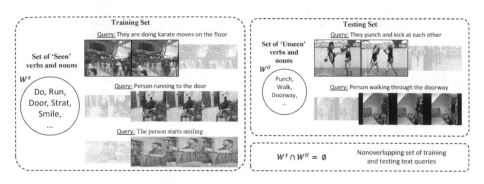

Fig. 1. Example illustration of our proposed task. We consider the task of localizing novel moments for unseen queries. The set of verbs and nouns present in the testing set is absent in the training set, e.g., training data does not have any text with verb 'walk' or noun 'doorway'. Hence, the system is required to learn transferable knowledge from the training data to perform localization for novel events based on unseen queries.

temporal grounding of text/sentence in untrimmed videos [3,12,29,36,71,74]. Most of these works utilize a set of fully supervised training data containing videos, text descriptions, and temporal boundary annotations. These works try to optimize over a fixed set of events and queries (which we call seen events and seen queries) that are available during training. However, in a real-world dynamic environment, a system is expected to encounter *previously unseen events and queries*, as shown in Fig. 1, and is required to *localize corresponding moments based on unseen text queries* in the videos. As a result, a system optimized over a fixed set of events is unlikely to generalize and perform well for unseen events. Moreover, as textual annotations are expensive and time consuming [35], it is impossible to collect videos of all possible events and textual descriptions and learn models with the collected data. Hence, the applicability of current text-based temporal localization systems are severely limited to a small set of events and the problem of localizing novel/unseen events based on unseen text queries remains unaddressed in the current literature.

In this work, our goal is to temporally localize video moments based on text queries, where both the video moments and text queries *are not observed/available during training*. Towards this goal, we learn transferable knowledge from seen events and queries and utilize it to localize novel/unseen events. We hypothesize that temporally relevant moments corresponding to unseen text queries and those corresponding to seen text queries are likely to contain shared concepts, if the unseen query and the seen query are semantically relevant. For instance, in Fig. 1, moment corresponding to the unseen text query *'They punch and kick at each other'* from the testing set has similarities to the moment corresponding to seen text query *'They are doing karate moves on the floor'* from the training set. Therefore, instead of localizing moments only based on its encoded representation, we formulate the inference task of localization as a relational prediction problem. The likelihood of a candidate moment to be the correct one based on an unseen text query depends on its relevance to the moment corresponding to the semantically most relevant seen query. We term this moment corresponding to the semantically most relevant seen query as the *support moment*. To learn a proper relational system that can localize novel events, we simulate the support moment based relational inference on the available training data during training. As a result, the system learns to localize moments based on relational reasoning, instead of directly localizing based on observed moment representations. Our motivation behind the approach is that a relational system learned on seen events/queries is transferable to the unseen events/queries [54]. We term our approach as **T**emporal **L**ocalization using **R**elational **R**easoning (**TLRR**).

Our problem is related to the zero-shot paradigm (where the objective is to adapt models to perform different tasks on the unseen or unobserved classes) as we utilize seen moment-text pairs to infer on the unseen events [26,39,64,78,87]. However, those zero-shot approaches are not directly applicable to our problem setup. For example, [79] assumes unseen classes are known in advance and uses the information to mine common semantics for seen classes and unseen classes for zero-shot temporal activity detection. However, text-based annotations of events are not limited to a fixed set of classes and the unseen queries are not known beforehand. Again, [8,27,68] perform retrieval across multiple modality data in the zero-shot setting. These works consider images with specific classes, and utilize the word embedding space to transfer knowledge between seen classes and unseen classes. However, in a video, textual descriptions refer to multiple entities, interactions of multiple entities, and different activities in a combined manner that is not expressible by a single class. As a result, directly utilizing label embeddings is not enough to transfer knowledge from seen events/queries to unseen events/queries. We will demonstrate the advantage of our proposed TLRR approach over zero-shot approaches and other recent temporal localization approaches on two benchmark datasets. The following are the main **contributions** of our work.

- We address a novel and practical problem of temporal localization of video moments based on unseen text queries.
- We hypothesize a conceptual relation between semantically relevant moments and propose a relational reasoning based temporal localization approach, TLRR, which can learn transferable knowledge from seen events and localize novel events based on unseen text queries.
- We reorganize two existing text-based temporal localization datasets (Charades-STA [12] and ActivityNet Captions [24]) for our proposed novel problem setting. Empirical results on these two text-based video moment localization datasets show that our proposed approach can reach up to 15% absolute improvement in performance compared to existing localization approaches.

2   Related Works

Temporal Localization of Moments. Temporal localization of moments in a video based on text query was introduced by [3,12]. Recently, there are many works that address the problem both in presence of strong supervision (temporal endpoints are known for each query) [5,6,9,13–18,21,22,30–34,36,38,43,47,53,57,59,60,62,65–67,71,72,74–77,80–84,86] and weak supervision (only video-text correspondence is known) [7,28,35,55,56,61,70]. Among the recent works on temporal localization of moments in the fully supervised setting, [71] performs semantic conditioned dynamic modulation, [74] relies on dense regression based approach, [36] utilizes both local and global interaction for video grounding. Recently, [37] proposed text-based temporal localization without query annotation. Unlike our setting, they have access to videos of all types of events and can optimize their model for such events in a weakly supervised manner. Hence, none of these works address the problem of localizing novel events based on unseen text queries.

Zero-Shot Learning (ZSL). ZSL aims to do inference task on classes whose instances may not have been seen during training [26,39,64,78,87]. Initial works on ZSL were attribute-based [25,41]. However, attribute-based ZSL has poor scalability and semantic embedding of labels are a good alternative for attributes [69]. Most of the works that utilize semantic embedding based learning focus on the association of visual and semantic information by linear compatibility [1,2,11,48], non-linear compatibility [52,63] or in a hybrid way [40]. To the best of our knowledge, only [79] works on activity detection in ZSL setup. However, [79] is limited to work on activity labels and can not be adapted directly for moment localization of unseen text queries.

Zero-Shot Cross Modal Retrieval (ZS-CMR). Conventional cross modal retrieval work [10] considers similar type of events are present in both training set and testing set. However, ZS-CMR aims to perform retrieval across multiple modality data in the zero-shot setting. They train the retrieval model with limited categories to support cross-modal retrieval on new categories [27]. There are few works that consider retrieval between visual and textual modality with ZS-CMR setting [8,27,68]. However, these works are limited by the use of specific class information of the images to transfer knowledge between seen classes to unseen classes.

Relational Reasoning. Relational reasoning concept has been applied to different vision applications, i.e., visual question answering [46,49], deep reinforcement learning [73], few-shot learning [54], self supervised learning [42], activity recognition [44,85]. [54] is the closest to the proposed TLRR and uses relational reasoning for zero-shot learning. However, our work differs in several ways: (i) we do not work with a fixed set of labels, (ii) our relational module learns to identify relations between visual information rather than learning to identify relations between visual and semantic information, and (iii) our proposed problem setup requires the model to identify intra-video subtle differences between moments, whereas [54] learns to differentiate classes.

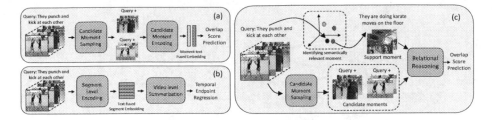

Fig. 2. A brief illustration of our novel text-based temporal localization approach. While existing works learn to encode video segments to identify the correct moment ((a) and (b)), we consider relational reasoning between two semantically relevant moment for localization purpose (c).

3   Methodology

3.1   Problem Statement

Let $\mathcal{S}^{tr} = \{(v, q, (\tau_s, \tau_e)) | v \in \mathcal{V}^{tr}, q \in \mathcal{Q}^{tr}, \tau_s, \tau_e \in [0, T]\}$ be the training set of video-sentence pairs for seen queries where $\mathcal{V}^{tr}$ is the set of all training videos with maximum duration T, $\mathcal{Q}^{tr}$ is the set of seen queries, (τ_s, τ_e) are the ground truth temporal endpoints for a query. For a given test-set $\mathcal{S}^{te} = \{(v, q) | v \in \mathcal{V}^{te}, q \in \mathcal{Q}^{te}\}$ with video-sentence pairs, our task is to predict the set of temporal endpoints $\{(\tau_s, \tau_e)\}$. We consider that $\mathcal{Q}^{tr} \cap \mathcal{Q}^{te} = \emptyset$, i.e., queries in test-set are not seen during training. As a result, $\mathcal{V}^{te}$ contains events that are not present in $\mathcal{V}^{tr}$. Additionally, we consider that S^{tr} is available during inference.

3.2   Localization Inference Schema

Existing temporal localization approaches [36,71,81] learn to encode fused moment-text representations. They either follow candidate moment sampling and encoding process to predict overlap scores (Fig. 2(a)) [71,81] or summarize the whole video based on query encoding and segment level encoding of video to regress temporal endpoints (Fig. 2(b)) [36]. In both cases, moment representations are directly optimized for available seen events. As a result, the models get tuned to the available events in the training set and do not necessarily learn to generalize for unseen events. Since, our objective is to localize events which are not available during training, we deviate from the conventional approaches and propose a novel approach on how to address the text-based temporal localization task. For our proposed TLRR, we hypothesize that the correct moment corresponding to the unseen text query and the moments corresponding to the semantically relevant seen queries will contain shared concepts or similarities. Therefore, to identify the correct moment in a video based on an unseen text query, instead of directly predicting based on the moment-text representation, we utilize semantically relevant seen events. In that regard, we formulate the localization inference as a relational reasoning problem between two semantically relevant moments.

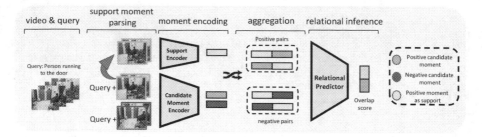

Fig. 3. Overview of the framework and the training of the relational reasoning based temporal localization approach. Candidate moment and support moment representations are aggregated to form positive pairs (positive candidate, positive support) and negative pairs (negative candidate, positive support)/(positive candidate, negative support). The relational module is trained to estimate the relational scores based on the pairs.

For a given video and an unseen text query, semantically relevant moments can be identified based on the semantics of the text query. Recent advances in Natural Language Processing (NLP) unfold many sentence encoder models which are trained on large corpus of text data in self-supervised or unsupervised manner. These models are able to capture wide range of sentence semantics and can be transferred to other NLP tasks. Our idea is to use these sentence encoders to find semantically relevant moments. In our work, we utilize universal sentence encoder [4], which is also able to capture sentence semantics, to find semantically relevant moments. Figure 2(c) clearly illustrates our localization inference scheme. Given the unseen query, instead of directly inferring overlap scores from moment-text fused representation, we first identify semantically relevant query and its corresponding moment using universal sentence encoder. We utilize this semantically relevant moment as the support moment and consider relational reasoning between the support moment and the candidate moments to identify the correct moment. Our motivation behind this approach is that this relational inference system can be learned using available training data and the learned relational model is transferable to unseen cases [54]. Our framework consists of candidate moment encoder, fusion network, support moment encoder and relational reasoning module. In the following sections, we discuss the framework and how we utilize available training data to learn a proper relational inference system.Candidate Moment Encoding and Modality Fusion

3.3 Framework

As illustrated in Fig. 3, our framework consists of a candidate moment encoder that generates a text-fused representation of candidate moments, a support moment encoder that encodes the support moment, and a relational prediction module to infer based on the relational reasoning between candidate moment and support moment. To learn the relational reasoning system utilizing available training samples, we mimic the relational inference task during training. At train-time, for seen queries in training set, we infer the overlap scores based on the relation between candidate moment and support moment, where the ground truth moment is used as the positive support moment. All the modules and the learning procedure are described in the following sections.

Visual Feature Extraction. We perform fixed interval sampling over the frames of the videos and sample l non-overlapping clips per video. For each clip, we extract 2D/3D convolutional feature, resulting in a set of l clip features $\{c_i\}_{i=1}^l$. Here, c_i is the feature representation of the ith clip.

Text Feature Extraction. We use GloVe word embedding [45] and Bi-directional LSTM network [19] for representing text queries. For each word s of the query sentence q, we use Glove word embeddings to obtain its initial embedding vectors, which are fed sequentially into a three-layer bidirectional LSTM network. The last hidden state $\hat{q}$ is used as the feature representation of the input sentence.

Candidate Moment Encoding and Modality Fusion. Clip representations $\{c_i\}_{i=1}^l$, sampled from each video is used to construct candidate moment representations. For each candidate moment, we max-pool the corresponding clip features across the specific time span. For example, moment corresponding to i^{th} to $(i+n)^{th}$ clips will be represented by $f_{i:i+n} = MaxPool(c_i, \ldots, c_{i+n})$, where $f \in \mathcal{R}^{d_f}$ (d_f is the feature dimension). Moment encodings and text encodings are projected in the same subspace and their dot product is taken as the fused moment-text representation by $e = (W^q \hat{q}).(W^f f)$. Here, W^q and W^f are the learnable parameters. We stack all moment-text representations of a video as a 2D feature map, similar to [81], and use L convolutional layers to further encode the representations. As a result, we obtain a set of candidate moment representations $\{m_i\}_{i=0}^N$, where N is the total number of candidate moments from a video and $m_i \in R^{d_m}$, where d_m is the feature dimension of the candidate moment representations.

Support Moment Encoder. We use a feed-forward network as the support moment encoder. For a support moment consisting of n consecutive clips $\{c_i\}_{i=1}^n$, where $c_i \in \mathcal{R}^{d_m}$, we first average pool the n clip representations to a single representation $s' \in \mathcal{R}^{d_m}$. If we have multiple support moments, then we average pool all the support moment representations into a single representation. Then we use a feed-forward network to obtain the final support representation s by

$$s = ReLU(W^s s' + b^s). \tag{1}$$

Here, W^s and b^s are the learnable parameters and $s \in \mathcal{R}^{d_m}$. We keep the feature dimension of support moment same as the candidate moment feature dimension d_m. The input to the support moment encoder varies in the training stage and inference stage. In the training stage, the correct candidate moment is used as the support moment. In the inference/testing stage, based on the unseen test query, most semantically relevant moments from the training set are used as the support moments. These moments work as the helper to find the correct moment from the video.

3.4   Relational Prediction

The relational module is a function $\mathcal{Z}_\theta(\cdot)$ parameterized by learnable weights θ and modeled by a feed forward neural network. Input to the relational module is a pair of two representations $\boldsymbol{x}_i$ and $\boldsymbol{x}_j$, where one element represents the selected support moment $\boldsymbol{s}$ and the other element represents a candidate moment $\boldsymbol{m}_i$ from the set of candidate moment representations $\{\boldsymbol{m}_i\}_{i=1}^N$. We use concatenation as the aggregation function to get aggregated representation of $\boldsymbol{x}_i$ and $\boldsymbol{x}_j$ as $a_{cat}(\boldsymbol{x}_i, \boldsymbol{x}_j)$. For a pair of support moment representation $\boldsymbol{s}$ and i^{th} candidate moment representation $\boldsymbol{m}_i$, the relational module outputs a overlap score ϕ_i by

$$\phi_i = \mathcal{Z}_\theta(a_{cat}(\boldsymbol{s}, \boldsymbol{m}_i)). \tag{2}$$

To confirm that the relational reasoning module $\mathcal{Z}_\theta$ predicts based on the relation between pair of representations and not based on a single representation, $\mathcal{Z}_\theta$ requires to maintain the commutative property, i.e., $\mathcal{Z}_\theta(a_{cat}(\boldsymbol{s}, \boldsymbol{m}_i)) = \mathcal{Z}_\theta(a_{cat}(\boldsymbol{m}_i, \boldsymbol{s}))$. However, the concatenation operation $a_{cat}(\cdot, \cdot)$ is not commutative. Therefore, to enforce the commutative property of the relational module, we compute the overlap score for the pair of elements $\boldsymbol{s}$ and $\boldsymbol{m}_i$ by

$$\phi_i = \frac{1}{2}\left[\mathcal{Z}_\theta(a_{cat}(\boldsymbol{m}_i, \boldsymbol{s})) + \mathcal{Z}_\theta(a_{cat}(\boldsymbol{s}, \boldsymbol{m}_i))\right]. \tag{3}$$

3.5   Learning Relational Inference

In our learning setup, a training sample consists of a video v, a text query q, and temporal ground truth information for the query (τ_s, τ_e). Instead of learning to directly predict the overlap score for each candidate moment, we learn to infer the overlap scores based on the relation with most relevant support moments. To train this relational inference system, we sample two types of support moment: i) positive support moment and ii) negative support moment. For each query in a video, we extract the ground truth segment of the video and use it as the positive support moment $\boldsymbol{s}^+$. Again, for each query in a video, we select semantically unrelated query in the trainset and use its corresponding moment as the negative support moment $\boldsymbol{s}^-$. Our objective is to distinguish intra-video candidate moments based on the support moment. To do so, we compute overlap prediction loss $\mathcal{L}^{intra}$ for a set of pairs $\mathcal{X}^1 = \{(\boldsymbol{m}_i, \boldsymbol{s}^+)\}$, which consists of pairs of all candidate moments and positive support moment in a video. To guide the learning of distinguishing intra-video candidate moments through relational inference system, we use scaled $tIoU$ (temporal Intersection-over-Union) value with ground-truth segment as the supervision signal. We compute the scaled $tIoU$ by

$$y_i = \begin{cases} 0 & g_i \leq t_{min}, \\ \frac{g_i - t_{min}}{t_{max} - t_{min}} & t_{min} < g_i < t_{max}, \\ 1 & g_i > t_{max}. \end{cases} \tag{4}$$

Here, g_i is the ground truth $tIoU$ for the ith candidate moment and t_{min}, t_{max} are two thresholds to compute y_i. For a video with N candidate moments, $\mathcal{L}^{intra}$ is realized by binary cross entropy loss as

$$\mathcal{L}^{intra} = -\frac{1}{N} \sum_{\mathcal{X}^1} \left[y_i \log(\phi_i) + (1 - y_i) \log(1 - \phi_i) \right]. \tag{5}$$

Here, ϕ_i is the overlap score computed using Eq. 3. To ensure that the model predicts the overlap score based on the relationship between the candidate moment and the support moment, we use the sampled negative support moments s^- to train the model. In each video, candidate moments with $tIoU > t_{min}$ are considered as positive candidate moment m^+. For each video with P positive candidate moments, we formulate a set of pairs $\mathcal{X}^2 = \{(m_i^+, s^-)\}$ and compute negative relational loss $\mathcal{L}^{neg}$ by

$$\mathcal{L}^{neg} = -\frac{1}{P} \sum_{\mathcal{X}^2} \log(1 - \phi_i). \tag{6}$$

The two losses are jointly considered for training our relational inference model, with λ balancing contributions as in

$$\mathcal{L}^{total} = \mathcal{L}^{intra} + \lambda \mathcal{L}^{neg}. \tag{7}$$

We compute $\mathcal{L}_{total}$ for all seen video-text query pairs in the training set and optimize the relational inference model by minimizing the total loss.

3.6 Inference for Unseen Queries

During inference, given a video and an unseen text query, we are required to localize the correct moment. We use the universal sentence encoder [4] to find semantically relevant queries from the training set. Then the corresponding moment to the relevant query is used as a support moment. Based on the video, support moments, and the unseen query, the learned relational model predicts overlap score ϕ for different temporal granularities in one forward pass. All the predicted segments are ranked and refined with non-maximum suppression (NMS) according to the predicted ϕ. Afterwards, the final temporal grounding result is obtained.

4 Experiments

4.1 Reorganized Datasets

Existing benchmark temporal moment localization dataset splits are not designed for the task of temporal localization of novel events based on unseen text queries. Instead, training set (trainset for short) and testing set (testset for short) data are sampled from the same distribution, and text queries in the testset overlap with text queries in the trainset. We reorganize two of the benchmark datasets namely Charades-STA [12] and ActivityNet Captions [24] to create splits according to our problem setting. For both datasets, we create splits based on the verbs and nouns present in the text queries. First, we combine all the annotations of the trainset and testset videos of the dataset. To create the splits, we consider a set of n_V verbs and n_N nouns present in the combined annotation. We consider it the set of seen verbs and seen nouns. Then, we identify videos that contain at least a single query that has a verb or noun not present in the mentioned set. In the selected videos, queries which do not have verbs or nouns from the mentioned set are collected as unseen testset split and, queries which have verbs or nouns from the mentioned set are collected as seen testset split. The training set is created from the rest of the videos, with queries that contain either verb or noun present in the mentioned set. We exclude queries which contains verb or noun from both seen set and unseen set. We use spaCy [20] to parse verbs and nouns from text queries. These reorganized datasets reflect a realistic setting as datasets are usually composed of recurring events of limited concepts. However, a localization system may encounter varied types of events in real-world applications. Details of the nouns and verbs selected to create the split are provided in the supplementary material. Excluding queries which contains verb or noun from both seen set and unseen set results in reduced number of moment-sentence pairs in the reorganized dataset. However, the size of the dataset doesn't have impact on the significance of our proposed problem setup, which is experimentally evaluated in the supplementary material.

Charades-STA Unseen. Charades-STA dataset contains a total of 6,670 videos where 5,336 and 1,334 are the number of training and testing videos. Textual annotations in Charades-STA has direct temporal correspondence with activity annotation of the Charades dataset [50]. We combine training and testing set annotations and consider $n_V = 20$ and $n_N = 40$ (excluding 'person' noun) for creating Charades-STA Unseen dataset. In this way, we have Charades-STA Unseen dataset with 5525, 1665, and 867 training, unseen testing, and seen testing moment-sentence pairs respectively.

ActivityNet Captions Unseen. ActivityNet Captions [24] dataset is proposed for dense video captioning task. Each video contains at least two ground truth segments and each segment is paired with one ground truth caption [66]. This dataset contains around 20k videos which are split into training, validation, and testing set with 50%, 25%, and 25% ratio respectively. Textual description for only the training and validation set is given. We combine training and validation set and consider $n_V = 70$ and $n_N = 250$ for creating ActivityNet Captions Unseen dataset. In this way, we have ActivityNet Captions Unseen dataset with 5669, 2553, and 710 training, unseen testing, and seen testing moment-sentence pairs respectively.

Table 1. This table reports **unseen** text query based temporal moment localization performance of TLRR, compared against several approaches, on Charades-STA Unseen dataset.

Method	R@1, IoU@0.5	R@1, IoU@0.7	R@5, IoU@0.5	R@5, IoU@0.7	mIoU
DeViSE [11]	29.98	11.29	71.42	39.81	-
ESZSL [48]	23.90	10.13	60.50	34.53	-
SCDM [71]	28.22	11.89	54.25	32.95	28.63
LGI [36]	29.01	12.85	-	-	29.62
2D-TAN [81]	31.05	13.33	70.75	36.94	29.88
TLRR	**33.15**	**16.22**	**77.66**	**42.40**	**31.29**

4.2 Evaluation Metric

We use "$R@k, IoU@m$", which reports the percentage of at least one of the top-k results having Intersection-over-Union (IoU) larger than m [12]. For a text query, "$R@k, IoU@m$" reflects if one of the top-k retrieved moments has IoU with the ground truth moment larger than the specified threshold m. So, "$R@k, IoU@m$" is either 1 or 0 for each text query. We compute it for all the text queries in the testing sets and report the average results for $k \in \{1, 5\}$ and $m \in \{0.50, 0.70\}$. We also compute mIoU where mIoU is the average IoU over all testing samples.

4.3 Implementation Details

We use VGG feature [51] for Charades-STA Unseen dataset. For ActivityNet Captions Unseen dataset, we use extracted C3D features [58]. The number of frames in a clip is set to 4 for Charades-STA Unseen, and 16 for ActivityNet Captions Unseen and we use non-overlapping clips for both datasets. The number of sampled clips N is set to 16 for Charades-STA Unseen, 64 for ActivityNet Captions Unseen. For the candidate moment encoder, we adopt a 4-layer convolution network with a kernel size of 5 for Charades-STA Unseen and a 4-layer convolution network with a kernel size of 9 for ActivityNet Captions Unseen. For both datasets, the support moment encoder is a single-layer feed-forward network and the relational prediction network is a two-layer feed-forward network. The proposed network is implemented in TensorFlow and trained using a single RTX 2080 GPU. We use mini-batches containing 32 video-sentence pairs and use Adam [23] optimizer with a learning rate of 0.0001. The dimension of both candidate moment representation d_m and support moment representation d_s is set to 512 for both datasets. We set $\lambda = 3$ empirically in Eq. 7 for both datasets. The scaling thresholds t_{min} and t_{max} of Eq. 4 are set to 0.5 and 1.0 respectively for both datasets. Non-maximum suppression (NMS) with a threshold of 0.5 is applied during the inference. We train TLRR for 50 epochs. We

Table 2. This table reports *unseen* text query based temporal moment localization performance of TLRR, compared against several approaches, on ActivityNet Captions Unseen dataset.

Method	R@1, IoU@0.5	R@1, IoU@0.7	R@5, IoU@0.5	R@5 IoU@0.7	mIoU
DeViSE [11]	5.07	2.00	10.46	4.05	-
ESZSL [48]	4.72	1.85	11.83	4.48	-
SCDM [71]	19.22	8.22	46.38	23.58	23.97
2D-TAN [81]	19.15	10.26	38.78	24.01	21.70
VSLNet [77]	19.23	9.99	-	-	25.32
TLRR	**23.19**	**13.24**	**53.31**	**36.66**	**26.35**

select the checkpoint which has the best average performance across metrics for seen queries.

4.4   Result Analysis

Temporal Localization Performance of Novel/Unseen Events. Since ours is the first work on temporal localization of novel events, there are no existing approaches to directly compare with. As our problem setup is closely related to zero-shot settings, we adapt two zero-shot learning approaches namely **DeViSE** [11] and **ESZSL** [48] for this problem setup. We also compare with some of the state-of-the-art temporal localization approaches with publicly available codes, e.g., **2D-TAN** [81], **SCDM** [71], **LGI** [36], and **VSLNet** [77], by training those models using our reorganized training splits.

Table 1 and Table 2 illustrate the TLRRs' performance for temporal localization of novel event based on unseen text query and compare it with other approaches for Charades-STA Unseen and ActivityNet Captions Unseen dataset respectively. For the Charades-STA Unseen dataset, the performance of different baseline approaches are comparable among them. However, TLRR provides 2%–7% absolute improvement over the best scores of compared approaches over all the reported metrics. In Table 2, baseline zero-shot approaches (DeViSE, ESZSL) are performing poorly for ActivityNet Captions Unseen dataset. This is because the text queries are complex compared to Charades-STA Unseen and it requires fine-grained analysis of longer videos in ActivityNet Caption Unseen. We observe 3%–15% absolute improvement over best scores of compared approaches in the ActivityNet Captions Unseen dataset.

Relational Reasoning Performance Analysis. Since TLRR's performance is dependent on its ability to reason on the relationship of two different moments, in Table 3, we analyze the competence of our relational reasoning module $\mathcal{Z}_\theta$ for Charades-STA Unseen dataset. We consider three scenarios: i) **Irrelevant:** based on the unseen text query, retrieve the seen query from the semantic embedding space that are furthest away or most irrelevant and use the corresponding

Table 3. This table reports *unseen* text query based novel event localization performance using different types of support moments to analyze TLRR for Charades-STA unseen dataset.

Support moment	R@1, IoU@0.5	R@1, IoU@0.7	R@5, IoU@0.5	R@5, IoU@0.7	mIoU
Irrelevant	20.30	11.05	62.58	33.93	22.48
Random	28.71	14.47	73.57	40.24	28.40
Relevant	**33.15**	**16.22**	**77.66**	**42.40**	**31.29**

moment as the support information, ii) **Random:** retrieve random seen query from the training set and use the corresponding moment as the support information, and iii) **Relevant:** retrieve the nearest/most relevant seen query from the semantic embedding space and use the corresponding moment as the support information (i.e., our proposed TLRR). We observe that when irrelevant queries are retrieved and their corresponding moment is used as the support, the performance goes down. Since the moment corresponding to a irrelevant query does not contain shared concept/similarities with the correct moment, the relational module expectedly fails to identify the correct moment. When random seen queries are selected, the performance is better compared to the irrelevant case. We obtain the best performance when the closest seen query is selected from the semantic embedding space.

Temporal Localization Performance of Seen Events. We further report the performance of different approaches when evaluated on the testing split of seen queries in both the datasets on Table 4 and Table 5. Although the main focus of this paper is temporal localization of unseen events, this experiment is presented to evaluate how the performance of different methods changes for seen events compared to localization of unseen events (Table 1 and Table 2). We expect any method to work slightly better on localizing the seen events compared to the unseen ones; however, a drastic/large change would indicate poor generalization ability of the model.

For the compared methods and baselines, we observe that there is a significant difference in performance when the same model is evaluated in the testing split of seen queries and testing split of unseen queries for both datasets comparing Table 1 and Table 2 with Table 4 and Table 5 respectively. Not surprisingly, both the conventional temporal localization approaches (i.e., SCDM and 2D-TAN) show a drastic change in performance across metrics in both datasets. The average difference in performance is reported by Δ_{avg} in Table 4 and Table 5. SCDM shows 19.80% average difference in Charades-STA and 13.24% average difference across metrics in ActivityNet in localization performance of seen queries compared to localization performance of unseen queries. Similarly, 2D-TAN shows average difference (across metrics) of 5.89% in Charades-STA and 16.18% in ActivityNet in localizing seen queries compared to unseen. Though the zero-shot based approaches (DeViSE and ESZSL) show small gap in performance between

Table 4. This table reports *seen* text query based temporal moment localization performance of TLRR on Charades-STA Unseen dataset. Here, Δ_{avg} refers to average performance difference for seen events and unseen events (Table 1) for a specific method. From the lower value of Δ_{avg}, it is evident that TLRR generalizes significantly better than other temporal localization approaches.

Method	R@1, IoU@0.5	R@1, IoU@0.7	R@5, IoU@0.5	R@5, IoU@0.7	Δ_{avg} ↓
DeViSE [11]	36.34	15.86	77.66	44.10	5.36
ESZSL [48]	37.50	18.40	72.34	42.13	10.34
SCDM [71]	50.46	28.00	73.49	54.86	19.80
2D-TAN [81]	37.95	18.45	76.70	42.56	5.89
TLRR	34.83	20.76	78.78	48.56	**3.37**

Table 5. This table reports *seen* text query based temporal moment localization performance of TLRR on ActivityNet Captions Unseen dataset. Δ_{avg} refers to average performance difference for seen events and unseen events (Table 2) for a specific method. From the lower value of Δ_{avg}, it is evident that TLRR generalizes significantly better than other temporal localization approaches.

Method	R@1, IoU@0.5	R@1, IoU@0.7	R@5, IoU@0.5	R@5, IoU@0.7	Δ_{avg} ↓
DeViSE [11]	12.07	5.40	18.18	8.52	**5.64**
ESZSL [48]	12.64	5.40	19.74	8.66	5.89
SCDM [71]	34.66	20.74	59.51	35.37	13.24
2D-TAN [81]	34.65	22.39	57.18	42.68	16.18
TLRR	27.46	17.61	60.42	49.44	7.13

seen and unseen events, which is expected due to the approaches generalization ability, they are unable to maintain a proper level of localization performance compared to other methods. However, the proposed TLRR approach shows a significantly lower change in performance, e.g., 3.37% average in Charades-STA and 7.13% average in ActivityNet Captions.

This indicates the significance of the problem setup and generalization ability of TLRR. Unlike the conventional temporal localization approaches, TLRR is not designed to specifically focus on the seen events. In Table 4 and Table 5, we observe that model optimized to do localization inference directly based on the candidate moment representation overall performs better compared to TLRR for types of events that are already seen in training. However, direct localization limits these models' capacity to a small set of events which is evident by the significant gap between performances for seen and unseen events. Instead, our proposed TLRR approach is able to retain a competitive performance for the seen queries and boost the performance for unseen queries resulting in reducing the performance gap between seen and unseen events. Also, our proposed

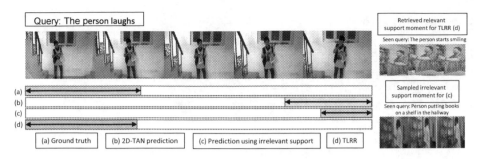

Fig. 4. Given the query 'The person laughs' and the corresponding video, this figure shows: (a) ground truth segment of the video which corresponds to the text query, (b) predicted moment by 2D-TAN, (c) predicted moment when irrelevant moment is used as support, and (d) predicted moment using retrieved relevant support moment (TLRR). While (b) and (c) result in failure, TLRR is able to detect the correct moment using relational reasoning.

TLRR is able to show comparable performance (please refer to supplementary material) on the original temporal localization dataset, even though TLRR is not optimized for seen events and have a relatively simple base architecture.

Effect of $\mathcal{L}^{neg}$ in Learning TLRR. TLRR uses $\mathcal{L}^{intra}$ and $\mathcal{L}^{neg}$ to learn relational localization system. Effectiveness of these two loss components for distinguishing intra-video moments by relational prediction is evident from Table 1, Table 2, and Table 3. We consider two setups, i) TLRR trained with $\mathcal{L}^{intra}$ and ii) TLRR trained with $\mathcal{L}^{intra} + \lambda \mathcal{L}^{neg}$. We observe that when only $\mathcal{L}^{intra}$ is used to train TLRR, there is almost no difference in performance (difference within 1%) for using relevant or irrelevant moments as input to the support encoder. However, there is 5%-5% difference in Charades-STA Unseen dataset for using relevant or irrelevant moments as input to the support encoder when $\mathcal{L}^{intra} + \lambda \mathcal{L}^{neg}$ is used to train TLRR. So, $\mathcal{L}^{neg}$ enforces the model to predict based on the relation.

Qualitative Result. In Fig. 4, we illustrate an example case of our system's success. Given the query 'The person laughs' and the corresponding video, Fig. 4 shows: (a) ground truth segment of the video which corresponds to the text query, (b) predicted moment by 2D-TAN, (c) predicted moment when irrelevant moment is used as support, and (d) predicted moment using retrieved relevant support moment. Person laughing is a difficult event to detect as it encompasses a small region of the frame and results in small temporal variation in the feature. Without any notion/previous knowledge of how the activity/event is, it becomes even harder, which is reflected by the failure case of (b) and (c). However, TLRR is able to detect the correct moment using relational reasoning.

5   Conclusion

In this paper, we address the novel problem of temporal localization of unseen/novel events based on unseen text queries. The problem of identifying novel events in video is important and practical because not every kind of event can be expected to be within the training set. This allows for generalization of temporal localization methods to novel scenarios. We propose a relational reasoning based framework hypothesizing a conceptual relation between moments corresponding to semantically relevant queries. Extensive experiments on reorganized Charades-STA and ActivityNet Captions datasets demonstrate the effectiveness of the proposed framework compared to several baselines in localizing video moments from text queries. Our code and dataset splits will be publicly available. Though support moment based relational prediction can reduce the performance gap between seen and unseen events, it is burdened with the extra computation of relevant moments, which is computationally expensive. Future work can focus on this issue.

Acknowledgment. This work was partially supported by ONR grant N00014-19-1-2264 and NSF grant 1901379.

References

1. Akata, Z., Perronnin, F., Harchaoui, Z., Schmid, C.: Label-embedding for image classification. IEEE Trans. Pattern Anal. Mach. Intell. **38**(7), 1425–1438 (2015)
2. Akata, Z., Reed, S., Walter, D., Lee, H., Schiele, B.: Evaluation of output embeddings for fine-grained image classification. In: Proceedings of the IEEE Conference on Computer Vision and Pattern Recognition, pp. 2927–2936 (2015)
3. Anne Hendricks, L., Wang, O., Shechtman, E., Sivic, J., Darrell, T., Russell, B.: Localizing moments in video with natural language. In: Proceedings of the IEEE international conference on computer vision. pp. 5803–5812 (2017)
4. Cer, D., et al.: Universal sentence encoder. arXiv preprint arXiv:1803.11175 (2018)
5. Chen, J., Chen, X., Ma, L., Jie, Z., Chua, T.S.: Temporally grounding natural sentence in video. In: Proceedings of the 2018 Conference on Empirical Methods in Natural Language Processing, pp. 162–171 (2018)
6. Chen, S., Jiang, W., Liu, W., Jiang, Y.-G.: Learning modality interaction for temporal sentence localization and event captioning in videos. In: Vedaldi, A., Bischof, H., Brox, T., Frahm, J.-M. (eds.) ECCV 2020. LNCS, vol. 12349, pp. 333–351. Springer, Cham (2020). https://doi.org/10.1007/978-3-030-58548-8_20
7. Chen, S., Jiang, Y.G.: Towards bridging event captioner and sentence localizer for weakly supervised dense event captioning. In: Proceedings of the IEEE/CVF Conference on Computer Vision and Pattern Recognition, pp. 8425–8435 (2021)
8. Chi, J., Peng, Y.: Dual adversarial networks for zero-shot cross-media retrieval. In: IJCAI, pp. 663–669 (2018)
9. Ding, X., et al.: Support-set based cross-supervision for video grounding. In: Proceedings of the IEEE/CVF International Conference on Computer Vision, pp. 11573–11582 (2021)

10. Dong, J., et al.: Dual encoding for zero-example video retrieval. In: Proceedings of the IEEE Conference on Computer Vision and Pattern Recognition, pp. 9346–9355 (2019)

11. Frome, A., et al.: Devise: a deep visual-semantic embedding model. In: Advances in Neural Information Processing Systems (NIPS), pp. 2121–2129 (2013)

12. Gao, J., Sun, C., Yang, Z., Nevatia, R.: TALL: temporal activity localization via language query. In: Proceedings of the IEEE International Conference on Computer Vision, pp. 5267–5275 (2017)

13. Gao, J., Xu, C.: Fast video moment retrieval. In: Proceedings of the IEEE/CVF International Conference on Computer Vision, pp. 1523–1532 (2021)

14. Ge, R., Gao, J., Chen, K., Nevatia, R.: MAC: mining activity concepts for language-based temporal localization. In: 2019 IEEE Winter Conference on Applications of Computer Vision (WACV), pp. 245–253. IEEE (2019)

15. Ghosh, S., Agarwal, A., Parekh, Z., Hauptmann, A.: ExCl: extractive clip localization using natural language descriptions. arXiv preprint arXiv:1904.02755 (2019)

16. Hahn, M., Kadav, A., Rehg, J.M., Graf, H.P.: Tripping through time: efficient localization of activities in videos. arXiv preprint arXiv:1904.09936 (2019)

17. He, D., Zhao, X., Huang, J., Li, F., Liu, X., Wen, S.: Read, watch, and move: reinforcement learning for temporally grounding natural language descriptions in videos. In: Proceedings of the AAAI Conference on Artificial Intelligence, vol. 33, pp. 8393–8400 (2019)

18. Hendricks, L.A., Wang, O., Shechtman, E., Sivic, J., Darrell, T., Russell, B.: Localizing moments in video with temporal language. In: Proceedings of the 2018 Conference on Empirical Methods in Natural Language Processing, pp. 1380–1390 (2018)

19. Hochreiter, S., Schmidhuber, J.: Long short-term memory. Neural Comput. $9(8)$, 1735–1780 (1997)

20. Honnibal, M., Montani, I.: spaCy 2: natural language understanding with bloom embeddings. convolutional neural networks and incremental parsing (to appear 2017)

21. Huang, J., Liu, Y., Gong, S., Jin, H.: Cross-sentence temporal and semantic relations in video activity localisation. In: Proceedings of the IEEE/CVF International Conference on Computer Vision, pp. 7199–7208 (2021)

22. Jiang, B., Huang, X., Yang, C., Yuan, J.: Cross-modal video moment retrieval with spatial and language-temporal attention. In: Proceedings of the 2019 on International Conference on Multimedia Retrieval., pp. 217–225 (2019)

23. Kingma, D., Ba, J.: Adam: a method for stochastic optimization. arXiv preprint arXiv:1412.6980 (2014)

24. Krishna, R., Hata, K., Ren, F., Fei-Fei, L., Niebles, J.C.: Dense-captioning events in videos. In: International Conference on Computer Vision (ICCV) (2017)

25. Lampert, C.H., Nickisch, H., Harmeling, S.: Attribute-based classification for zero-shot visual object categorization. IEEE Trans. Pattern Anal. Mach. Intell. $36(3)$, 453–465 (2013)

26. Li, J., Jing, M., Lu, K., Ding, Z., Zhu, L., Huang, Z.: Leveraging the invariant side of generative zero-shot learning. In: Proceedings of the IEEE/CVF Conference on Computer Vision and Pattern Recognition, pp. 7402–7411 (2019)

27. Lin, K., Xu, X., Gao, L., Wang, Z., Shen, H.T.: Learning cross-aligned latent embeddings for zero-shot cross-modal retrieval. In: Proceedings of the AAAI Conference on Artificial Intelligence, vol. 34, pp. 11515–11522 (2020)

28. Lin, Z., Zhao, Z., Zhang, Z., Wang, Q., Liu, H.: Weakly-supervised video moment retrieval via semantic completion network. In: Proceedings of the AAAI Conference on Artificial Intelligence, vol. 34, pp. 11539–11546 (2020)

29. Lin, Z., Zhao, Z., Zhang, Z., Zhang, Z., Cai, D.: Moment retrieval via cross-modal interaction networks with query reconstruction. IEEE Trans. Image Process. **29**, 3750–3762 (2020)

30. Liu, B., Yeung, S., Chou, E., Huang, D.A., Fei-Fei, L., Carlos Niebles, J.: Temporal modular networks for retrieving complex compositional activities in videos. In: Proceedings of the European Conference on Computer Vision (ECCV), pp. 552–568 (2018)

31. Liu, D., et al.: Context-aware biaffine localizing network for temporal sentence grounding. In: Proceedings of the IEEE/CVF Conference on Computer Vision and Pattern Recognition, pp. 11235–11244 (2021)

32. Liu, D., Qu, X., Liu, X.Y., Dong, J., Zhou, P., Xu, Z.: Jointly cross-and self-modal graph attention network for query-based moment localization. In: Proceedings of the 28th ACM International Conference on Multimedia, pp. 4070–4078 (2020)

33. Liu, M., Wang, X., Nie, L., He, X., Chen, B., Chua, T.S.: Attentive moment retrieval in videos. In: The 41st International ACM SIGIR Conference on Research & Development in Information Retrieval, pp. 15–24 (2018)

34. Liu, M., Wang, X., Nie, L., Tian, Q., Chen, B., Chua, T.S.: Cross-modal moment localization in videos. In: Proceedings of the 26th ACM International Conference on Multimedia, pp. 843–851 (2018)

35. Mithun, N.C., Paul, S., Roy-Chowdhury, A.K.: Weakly supervised video moment retrieval from text queries. In: The IEEE Conference on Computer Vision and Pattern Recognition (CVPR), June 2019

36. Mun, J., Cho, M., Han, B.: Local-global video-text interactions for temporal grounding. In: Proceedings of the IEEE/CVF Conference on Computer Vision and Pattern Recognition, pp. 10810–10819 (2020)

37. Nam, J., Ahn, D., Kang, D., Ha, S.J., Choi, J.: Zero-shot natural language video localization. In: Proceedings of the IEEE/CVF International Conference on Computer Vision (ICCV), pp. 1470–1479, October 2021

38. Nan, G., et al.: Interventional video grounding with dual contrastive learning. In: Proceedings of the IEEE/CVF Conference on Computer Vision and Pattern Recognition, pp. 2765–2775 (2021)

39. Niu, L., Cai, J., Veeraraghavan, A., Zhang, L.: Zero-shot learning via category-specific visual-semantic mapping and label refinement. IEEE Trans. Image Process. **28**(2), 965–979 (2018)

40. Norouzi, M., et al.: Zero-shot learning by convex combination of semantic embeddings. arXiv preprint arXiv:1312.5650 (2013)

41. Parikh, D., Grauman, K.: Relative attributes. In: 2011 International Conference on Computer Vision, pp. 503–510. IEEE (2011)

42. Patacchiola, M., Storkey, A.: Self-supervised relational reasoning for representation learning. arXiv preprint arXiv:2006.05849 (2020)

43. Paul, S., Mithun, N.C., Roy-Chowdhury, A.K.: Text-based localization of moments in a video corpus. IEEE Trans. Image Process. **30**, 8886–8899 (2021)

44. Paul, S., Torres, C., Chandrasekaran, S., Roy-Chowdhury, A.K.: Complex pairwise activity analysis via instance level evolution reasoning. In: ICASSP 2020–2020 IEEE International Conference on Acoustics, Speech and Signal Processing (ICASSP), pp. 2378–2382. IEEE (2020)

45. Pennington, J., Socher, R., Manning, C.D.: Glove: global vectors for word representation. In: Proceedings of the 2014 Conference on Empirical Methods in Natural Language Processing (EMNLP), pp. 1532–1543 (2014)

46. Raposo, D., Santoro, A., Barrett, D., Pascanu, R., Lillicrap, T., Battaglia, P.: Discovering objects and their relations from entangled scene representations. arXiv preprint arXiv:1702.05068 (2017)

47. Regneri, M., et al.: Grounding action descriptions in videos. Trans. Assoc. Comput. Linguis. **1**, 25–36 (2013)

48. Romera-Paredes, B., Torr, P.: An embarrassingly simple approach to zero-shot learning. In: International Conference on Machine Learning, pp. 2152–2161. PMLR (2015)

49. Santoro, A., et al.: A simple neural network module for relational reasoning. arXiv preprint arXiv:1706.01427 (2017)

50. Sigurdsson, G.A., Varol, G., Wang, X., Farhadi, A., Laptev, I., Gupta, A.: Hollywood in homes: crowdsourcing data collection for activity understanding. In: Leibe, B., Matas, J., Sebe, N., Welling, M. (eds.) ECCV 2016. LNCS, vol. 9905, pp. 510–526. Springer, Cham (2016). https://doi.org/10.1007/978-3-319-46448-0_31

51. Simonyan, K., Zisserman, A.: Very deep convolutional networks for large-scale image recognition. arXiv preprint arXiv:1409.1556 (2014)

52. Socher, R., Ganjoo, M., Sridhar, H., Bastani, O., Manning, C.D., Ng, A.Y.: Zero-shot learning through cross-modal transfer. arXiv preprint arXiv:1301.3666 (2013)

53. Soldan, M., Xu, M., Qu, S., Tegner, J., Ghanem, B.: VLG-Net: video-language graph matching network for video grounding. In: Proceedings of the IEEE/CVF International Conference on Computer Vision, pp. 3224–3234 (2021)

54. Sung, F., Yang, Y., Zhang, L., Xiang, T., Torr, P.H., Hospedales, T.M.: Learning to compare: relation network for few-shot learning. In: Proceedings of the IEEE Conference on Computer Vision and Pattern Recognition, pp. 1199–1208 (2018)

55. Tan, R., Xu, H., Saenko, K., Plummer, B.A.: Logan: Latent graph co-attention network for weakly-supervised video moment retrieval. In: Proceedings of the IEEE/CVF Winter Conference on Applications of Computer Vision. pp. 2083–2092 (2021)

56. Tang, H., Zhu, J., Gao, Z., Zhuo, T., Cheng, Z.: Attention feature matching for weakly-supervised video relocalization. In: Proceedings of the 2nd ACM International Conference on Multimedia in Asia, pp. 1–7 (2021)

57. Tang, H., Zhu, J., Wang, L., Zheng, Q., Zhang, T.: Multi-level query interaction for temporal language grounding. IEEE Transactions on Intelligent Transportation Systems (2021)

58. Tran, D., Bourdev, L., Fergus, R., Torresani, L., Paluri, M.: Learning spatiotemporal features with 3d convolutional networks. In: International Conference on Computer Vision (ICCV), pp. 4489–4497. IEEE (2015)

59. Wang, H., Zha, Z.J., Chen, X., Xiong, Z., Luo, J.: Dual path interaction network for video moment localization. In: Proceedings of the 28th ACM International Conference on Multimedia, pp. 4116–4124 (2020)

60. Wang, H., Zha, Z.J., Li, L., Liu, D., Luo, J.: Structured multi-level interaction network for video moment localization via language query. In: Proceedings of the IEEE/CVF Conference on Computer Vision and Pattern Recognition, pp. 7026–7035 (2021)

61. Wang, Y., Deng, J., Zhou, W., Li, H.: Weakly supervised temporal adjacent network for language grounding. IEEE Trans. Multim. **24**, 3276–3286 (2021)

62. Wu, A., Han, Y.: Multi-modal circulant fusion for video-to-language and backward. In: Proceedings of the 27th International Joint Conference on Artificial Intelligence, pp. 1029–1035 (2018)

63. Xian, Y., Akata, Z., Sharma, G., Nguyen, Q., Hein, M., Schiele, B.: Latent embeddings for zero-shot classification. In: Proceedings of the IEEE Conference on Computer Vision and Pattern Recognition, pp. 69–77 (2016)

64. Xian, Y., Schiele, B., Akata, Z.: Zero-shot learning-the good, the bad and the ugly. In: Proceedings of the IEEE Conference on Computer Vision and Pattern Recognition, pp. 4582–4591 (2017)

65. Xiao, S., et al.: Boundary proposal network for two-stage natural language video localization. In: Proceedings of the AAAI Conference on Artificial Intelligence, vol. 35, pp. 2986–2994 (2021)

66. Xu, H., He, K., Plummer, B.A., Sigal, L., Sclaroff, S., Saenko, K.: Multilevel language and vision integration for text-to-clip retrieval. In: Proceedings of the AAAI Conference on Artificial Intelligence, vol. 33, pp. 9062–9069 (2019)

67. Xu, M., et al.: Boundary-sensitive pre-training for temporal localization in videos. In: Proceedings of the IEEE/CVF International Conference on Computer Vision, pp. 7220–7230 (2021)

68. Xu, X., Song, J., Lu, H., Yang, Y., Shen, F., Huang, Z.: Modal-adversarial semantic learning network for extendable cross-modal retrieval. In: Proceedings of the 2018 ACM on International Conference on Multimedia Retrieval, pp. 46–54 (2018)

69. Xu, X., Hospedales, T., Gong, S.: Semantic embedding space for zero-shot action recognition. In: 2015 IEEE International Conference on Image Processing (ICIP), pp. 63–67. IEEE (2015)

70. Yang, W., Zhang, T., Zhang, Y., Wu, F.: Local correspondence network for weakly supervised temporal sentence grounding. IEEE Trans. Image Process. **30**, 3252–3262 (2021)

71. Yuan, Y., Ma, L., Wang, J., Liu, W., Zhu, W.: Semantic conditioned dynamic modulation for temporal sentence grounding in videos. In: Advances in Neural Information Processing Systems, pp. 534–544 (2019)

72. Yuan, Y., Mei, T., Zhu, W.: To find where you talk: Temporal sentence localization in video with attention based location regression. In: Proceedings of the AAAI Conference on Artificial Intelligence, vol. 33, pp. 9159–9166 (2019)

73. Zambaldi, V., et al.: Deep reinforcement learning with relational inductive biases. In: International Conference on Learning Representations (2018)

74. Zeng, R., Xu, H., Huang, W., Chen, P., Tan, M., Gan, C.: Dense regression network for video grounding. In: Proceedings of the IEEE/CVF Conference on Computer Vision and Pattern Recognition, pp. 10287–10296 (2020)

75. Zhang, D., Dai, X., Wang, X., Wang, Y.F., Davis, L.S.: MAN: moment alignment network for natural language moment retrieval via iterative graph adjustment. In: Proceedings of the IEEE Conference on Computer Vision and Pattern Recognition, pp. 1247–1257 (2019)

76. Zhang, H., Sun, A., Jing, W., Zhen, L., Zhou, J.T., Goh, R.S.M.: Natural language video localization: a revisit in span-based question answering framework. IEEE Trans. Pattern Anal. Mach. Intell. (2021)

77. Zhang, H., Sun, A., Jing, W., Zhou, J.T.: Span-based localizing network for natural language video localization. arXiv preprint arXiv:2004.13931 (2020)

78. Zhang, H., Long, Y., Guan, Y., Shao, L.: Triple verification network for generalized zero-shot learning. IEEE Trans. Image Process. **28**(1), 506–517 (2018)

79. Zhang, L., et al.: Zstad: zero-shot temporal activity detection. In: Proceedings of the IEEE/CVF Conference on Computer Vision and Pattern Recognition (CVPR), June 2020

80. Zhang, M., et al.: Multi-stage aggregated transformer network for temporal language localization in videos. In: Proceedings of the IEEE/CVF Conference on Computer Vision and Pattern Recognition, pp. 12669–12678 (2021)

81. Zhang, S., Peng, H., Fu, J., Luo, J.: Learning 2D temporal adjacent networks for moment localization with natural language. arXiv preprint arXiv:1912.03590 (2019)

82. Zhang, S., Su, J., Luo, J.: Exploiting temporal relationships in video moment localization with natural language. In: Proceedings of the 27th ACM International Conference on Multimedia, pp. 1230–1238 (2019)

83. Zhang, Z., Lin, Z., Zhao, Z., Xiao, Z.: Cross-modal interaction networks for query-based moment retrieval in videos. In: Proceedings of the 42nd International ACM SIGIR Conference on Research and Development in Information Retrieval, pp. 655–664 (2019)

84. Zhao, Y., Zhao, Z., Zhang, Z., Lin, Z.: Cascaded prediction network via segment tree for temporal video grounding. In: Proceedings of the IEEE/CVF Conference on Computer Vision and Pattern Recognition, pp. 4197–4206 (2021)

85. Zhou, B., Andonian, A., Oliva, A., Torralba, A.: Temporal relational reasoning in videos. In: Proceedings of the European Conference on Computer Vision (ECCV), pp. 803–818 (2018)

86. Zhou, H., Zhang, C., Luo, Y., Chen, Y., Hu, C.: Embracing uncertainty: decoupling and de-bias for robust temporal grounding. In: Proceedings of the IEEE/CVF Conference on Computer Vision and Pattern Recognition, pp. 8445–8454 (2021)

87. Zhu, Y., Long, Y., Guan, Y., Newsam, S., Shao, L.: Towards universal representation for unseen action recognition. In: Proceedings of the IEEE Conference on Computer Vision and Pattern Recognition, pp. 9436–9445 (2018)

Reliability-Aware Prediction via Uncertainty Learning for Person Image Retrieval

Zhaopeng Dou[1], Zhongdao Wang[1], Weihua Chen[2], Yali Li[1],
and Shengjin Wang[1]([✉])

[1] Department of Electronic Engineering, BNRist, Tsinghua University, Beijing, China
dcp19@mails.tsinghua.edu.cn, wgsgj@tsinghua.edu.cn
[2] Machine Intelligence Technology Lab, Alibaba Group, Hangzhou, China

Abstract. Current person image retrieval methods have achieved great improvements in accuracy metrics. However, they rarely describe the reliability of the prediction. In this paper, we propose an Uncertainty-Aware Learning (UAL) method to remedy this issue. UAL aims at providing reliability-aware predictions by considering data uncertainty and model uncertainty simultaneously. Data uncertainty captures the "noise" inherent in the sample, while model uncertainty depicts the model's confidence in the sample's prediction. Specifically, in UAL, (1) we propose a sampling-free data uncertainty learning method to adaptively assign weights to different samples during training, down-weighting the low-quality ambiguous samples. (2) we leverage the Bayesian framework to model the model uncertainty by assuming the parameters of the network follow a Bernoulli distribution. (3) the data uncertainty and the model uncertainty are jointly learned in a unified network, and they serve as two fundamental criteria for the reliability assessment: if a probe is high-quality (low data uncertainty) and the model is confident in the prediction of the probe (low model uncertainty), the final ranking will be assessed as reliable. Experiments under the risk-controlled settings and the multi-query settings show the proposed reliability assessment is effective. Our method also shows superior performance on three challenging benchmarks under the vanilla single query settings. The code is available at: https://github.com/dcp15/UAL.

Keywords: Person image retrieval · Uncertainty · Reliability assessment

1 Introduction

Person image retrieval, also known as person re-identification (ReID), aims at associating a target person across non-overlapping camera views [43, 46, 60].

Supplementary Information The online version contains supplementary material available at https://doi.org/10.1007/978-3-031-19781-9_34.

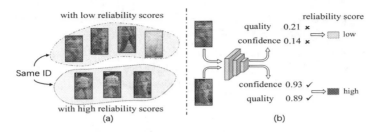

Fig. 1. Observation and Motivation. (a) In the multi-query setting, low-quality query images contain more ambiguous information. Reliability scores are required to down-weight the weights of these queries. (b) The reliability score is related to two factors: the quality of the sample and the model's confidence in the prediction of the sample.

Although current methods [4,15,17,20,21,29,31,42] have achieved promising performance on public benchmarks, they are reliability-agnostic, *i.e.*, the prediction of a probe can be generated anyway, but they rarely describe whether the prediction is reliable. However, when people are identifying pedestrians, they not only give the judgment result but also the reliability associated with it. Such a reliability assessment mechanism is important in human decision-making [9] and also essential in the ReID task. For example, in real scenarios, we often face the problem of searching for a person through his/her multiple images (*i.e.*, multi-query settings), as a pedestrian is usually captured by several cameras and one camera may capture a series of observations of the person. More generally, we can add the retrieved positive ones into the query set for further comprehensive retrieval. The quality of these query images varies, especially in complex scenes. As shown in Fig. 1(a), low-quality query images contain more ambiguous information. If we treat these query images equally, performance will degrade. At this time, reliability scores are required to down-weight the low-quality query images. However, current methods rarely consider the reliability assessment problem.

To remedy this issue, we propose a novel Uncertainty-Aware Learning (UAL) method for the ReID task. UAL aims at not only giving an accurate prediction for a sample but also providing a reliability score associated with it. The reliability score is related to two factors, *i.e.*, the quality of the sample and the confidence of the model in the prediction of the sample. These two factors are measured by considering two types of uncertainty, *i.e.*, data uncertainty and model uncertainty. Data uncertainty captures the "noise" inherent in the observation and it can describe the quality of the sample. Model uncertainty represents the model's "ignorance" and it can reflect the model's confidence in its prediction [7,24].

In this paper, we propose a unified network to learn the data uncertainty and the model uncertainty simultaneously. Specifically, first, we project a sample into a Gaussian distribution in the latent space, the mean of the distribution represents the feature, and the variance represents the data uncertainty. Different from [1,39,50] sampling feature vector from the Gaussian distribution, we propose a sampling-free method to learn the data uncertainty and adaptively down-weight low-quality ambiguous samples during training. Second, we leverage the Bayesian framework to learn the model uncertainty, in which the parameters of the network are assumed to follow the Bernoulli distribution. The model uncertainty is defined as the dispersion degree of the feature vectors caused by the distribution of the network parameter. Third, the data uncertainty and model uncertainty are jointly learned in a unified network, and they serve as two criteria to assess whether the result is reliable: as shown in Fig. 1(b), if a query image is high-quality (low data uncertainty) and the model is confident in its prediction of the query image (low model uncertainty), the final result will be assessed as reliable. Experiments under risk-controlled settings and multi-query settings show the proposed reliability assessment is effective.

The major contributions are summarized as: **(1)** We propose an uncertainty-aware learning (UAL) method that can provide reliability-aware predictions for the ReID task. **(2)** We introduce a sampling-free data uncertainty learning method, which can improve the representation by explicitly inhibiting the negative impact of low-quality samples during training without any external clues. **(3)** We propose a unified network to jointly learn data uncertainty and model uncertainty. As far as we know, this is the first work to apply data uncertainty and model uncertainty to the ReID task simultaneously. **(4)** Experiments under risk-controlled settings and multi-query settings show the reliability assessment is effective. Our method also shows superior performance in single query settings.

2   Related Work

Person ReID. Person ReID aims to associate a target person across different camera views. Existing methods can be broadly divided into two categories: hand-craft methods [32,49] and deep learning methods [4,15,18,20,29,50]. The key challenge is the large appearance variation caused by imperfect detection, different camera views, poses, and occlusions. To remedy these issues, several works [8,12–14,21,31,34,45,60] are proposed to learn local features to cope with the appearance variation. Although these methods have played a certain role, they are reliability-agnostic. That is, the model can output a prediction for a probe anyway, but it does not describe the reliability of the prediction.

Uncertainty in Person ReID. There are mainly two types of uncertainty: data uncertainty and model uncertainty [7,23,24,38]. Many tasks have considered the uncertainty to improve the robustness and interpretability of models, such as face recognition [1,25,39], semantic segmentation [19,24] and Multi-view learning [10]. In the ReID task, prior arts [22,41,50,53] consider data uncertainty to alleviate the problem of label noise or data outliers. D-Net [50] maps each person image as a Gaussian distribution in the latent space with the variance indicating the data uncertainty. PUCNN [41] extends the data uncertainty in D-Net into the part-level feature. UNRN [53] incorporates the uncertainty into a teacher-student framework to evaluate the reliability of the predicted pseudo labels for unsupervised domain adaptive (UDA) person ReID. The uncertainty is estimated as the inconsistency of these two models in terms of their predicted soft multi-labels. UMTS [22] designs an uncertainty-aware knowledge distillation loss to transfer the knowledge of the multi-shots model into the single-shot model.

Among these methods, the most relevant method to ours is D-Net [50]. Compared to D-Net, our data uncertainty learning method is sampling-free, which can explicitly suppress the ambiguous information contained in low-quality samples. We jointly learns the data uncertainty and the model uncertainty, which can utilize the complementary information provided by them during training.

3 Methodology

The reliability score is related to two factors: the quality of the sample and the confidence of the model in its prediction, which are measured by data uncertainty (Sect. 3.1) and model uncertainty (Sect. 3.2), respectively. They are incorporated into a unified network (Sect. 3.3) for joint learning. Two settings (risk-controlled and multi-query settings) are proposed to verify the effectiveness in Sect. 3.4.

3.1 Learning Data Uncertainty

Data uncertainty captures the "noise" inherent in the observation. It can reflect the quality of the sample, which is an essential factor in the reliability assessment.

Prior Method. Prior art D-Net [50] considers the data uncertainty by mapping a sample x as a Gaussian distribution in the latent space,

$$p(z|x) = \mathcal{N}(z; \mu, \sigma^2 \mathbf{I}) \tag{1}$$

where μ and σ^2 are the mean and variance vectors. μ is the feature vector and σ^2 refers to the data uncertainty of x. Then, they sample features from $p(z|x)$ by re-parameterization trick [27]: $z' = \mu + \epsilon\sigma, \epsilon \sim \mathcal{N}(\mathbf{0}, \mathbf{I})$. The sampled z' are utilized for vanilla cross-entropy loss $\mathcal{L}_{ce}$. To prevent the trivial solution of variance decreasing to zero, a regularization term $\mathcal{L}_{fu}$ is added to constrain the entropy of $\mathcal{N}(\mu, \sigma^2 \mathbf{I})$ to be larger than a constant. The final loss function is,

$$\mathcal{L} = \mathcal{L}_{ce} + \lambda\mathcal{L}_{fu} \tag{2}$$

where λ is the hyper-parameter to balance $\mathcal{L}_{ce}$ and $\mathcal{L}_{fu}$. Although this method can capture the data uncertainty, there are two limitations: (1) it is sampling-based, *i.e.*, the feature is sampled from the Gaussian distribution during training, which makes the optimization more difficult because each iteration optimizes only one point in the distribution, rather than entire distribution. (2) the objective does not explicitly distinguish samples with different data uncertainty. It is unclear how data uncertainty affects feature learning. To mitigate these two issues, we propose a sampling-free method to learn the data uncertainty and explicitly adjust the attention to the samples according to their quality.

Our Sampling-Free Data Uncertainty Learning Method. We project a sample x into a Gaussian distribution $\mathcal{N}(\mu, \sigma^2 \mathbf{I})$ in the latent space. Then the likelihood of x belonging to class i is formulated by,

$$p(x|y=i) \propto \frac{1}{(2\pi\sigma^2)^{\frac{d}{2}}} \exp\left(-\frac{\|\mu - w_i\|^2}{2\sigma^2}\right) \tag{3}$$

where $\boldsymbol{w}_i$ is the weight vector of i-th class in the classifier and d is the feature dimension. Assuming each class has the equal prior probability, the posterior of $\boldsymbol{x}$ belonging to the class i is,

$$p(y = i|\boldsymbol{x}) = \frac{\exp\left(-\frac{\|\boldsymbol{\mu}-\boldsymbol{w}_i\|^2}{2\sigma^2}\right)}{\sum_j \exp\left(-\frac{\|\boldsymbol{\mu}-\boldsymbol{w}_j\|^2}{2\sigma^2}\right)} = \frac{\exp\left(\frac{1}{\sigma^2}\boldsymbol{w}_i^T\boldsymbol{\mu}\right)}{\sum_j \exp\left(\frac{1}{\sigma^2}\boldsymbol{w}_j^T\boldsymbol{\mu}\right)} \tag{4}$$

where $\boldsymbol{\mu}$ and $\boldsymbol{w}_*$ are l_2-normalized. $p(y|\boldsymbol{x})$ can be regarded as a Boltzmann distribution. The magnitude of σ^2 controls the entropy of this distribution. The larger the σ^2, the larger the entropy. Thus σ^2 can be regarded as the data uncertainty of sample $\boldsymbol{x}$. Assuming the class label of the sample $\boldsymbol{x}$ is i, the loss function is formulated by,

$$\mathcal{L}_d(\boldsymbol{\mu}, \sigma^2) = -\log p(y = i|\boldsymbol{x}) \approx \frac{1}{\sigma^2}\mathcal{L}(\boldsymbol{\mu}) + \log \sigma^2 \tag{5}$$

where $\mathcal{L}(\boldsymbol{\mu}) = -\log \frac{\exp\left(\boldsymbol{w}_i^T\boldsymbol{\mu}\right)}{\sum_j \exp\left(\boldsymbol{w}_j^T\boldsymbol{\mu}\right)}$ is the cross entropy loss. Please see supplementary materials for derivation details.

Discussion (Difference to D-Net [50]). Compared to sampling-based method D-Net [50], our method shows several superior qualities: (1) Eq. 5 does not need to sample the representation from the Gaussian distribution. It contains entire information of the representation distribution. (2) Low-quality samples with larger data uncertainty will contribute less to learning the latent space. $\mathcal{L}(\boldsymbol{\mu})$ is weighted by $\frac{1}{\sigma^2}$, and thus it will drive the weight vector $\boldsymbol{w}$ in the classifier to be closer to high-quality samples with small σ^2. This can suppress the ambiguous information contained in low-quality samples during feature learning, which is verified in Sect. 4.4. (3) The σ^2 can not be too large or too small. If σ^2 is too small (large), the first (last) term $\frac{1}{\sigma^2}\mathcal{L}(\boldsymbol{\mu})$ ($\log \sigma^2$) becomes too large. Unlike the term $\mathcal{L}_{fu}$ in Eq. 2, we maintain the σ^2 in a unified formulation.

3.2   Learning Model Uncertainty

Model uncertainty plays an important role in the reliability assessment as it reflects the confidence of the model in its prediction of a sample. Bayesian network is often used to capture the model uncertainty [24]. Suppose we have a dataset $\mathcal{D} = (\boldsymbol{X}, \boldsymbol{Y})$, we define $f_{\boldsymbol{\theta}}$ to be a neural network such that $f_{\boldsymbol{\theta}} : \boldsymbol{X} \to \boldsymbol{Y}$ and $\boldsymbol{\theta}$ corresponds to the weight of the network. To capture model uncertainty, we assume a prior distribution on the weight, i.e., $p(\boldsymbol{\theta})$. We need to obtain the posterior $p(\boldsymbol{\theta}|\mathcal{D})$. Since $p(\boldsymbol{\theta}|\mathcal{D})$ is typically intractable, an approximate distribution $q_{\boldsymbol{\pi}}(\boldsymbol{\theta})$ parameterized by $\boldsymbol{\pi}$ is defined. $q_{\boldsymbol{\pi}}(\boldsymbol{\theta})$ is aimed to be as similar as possible to $p(\boldsymbol{\theta}|\mathcal{D})$, which is measured by the Kullback-Leibler divergence. The optimal parameters $\boldsymbol{\pi}^*$ is,

$$\boldsymbol{\pi}^* = \arg\min_{\boldsymbol{\pi}} \text{KL}[q_{\boldsymbol{\pi}}(\boldsymbol{\theta})\|p(\boldsymbol{\theta}|\mathcal{D})] = \arg\min_{\boldsymbol{\pi}} \text{KL}[q_{\boldsymbol{\pi}}(\boldsymbol{\theta})\|p(\boldsymbol{\theta})] - \mathbb{E}_{q_{\boldsymbol{\pi}}(\boldsymbol{\theta})}[\log p(\mathcal{D}|\boldsymbol{\theta})] \tag{6}$$

Please see supplementary materials for derivation details. Similar to [6], we assume the prior distribution $p(\boldsymbol{\theta})$ as the Bernoulli distribution and perform Monte Carlo integration to the second term. Then the objective loss can be written as $\mathcal{L}_m = -\frac{1}{T}\sum_{t=1}^{T}[\log p(\mathcal{D}|\boldsymbol{\theta}_t)]$, where $\boldsymbol{\theta}_t$ is sampled from $q_\pi(\boldsymbol{\theta})$. In $q_\pi(\boldsymbol{\theta})$, $\boldsymbol{\theta}_{ij} = \boldsymbol{\pi}_{ij} * z_{ij}$, where $z_{ij} \sim \text{Bernoulli}(\rho)$. ρ is the hyper-parameters and $\boldsymbol{\pi}$ is the set of parameters to be optimized. For a sample $\boldsymbol{x}$, we denote the extracted feature is $\boldsymbol{\mu}_t = f_{\boldsymbol{\theta}_t}(\boldsymbol{x})$ when the weight of network is $\boldsymbol{\theta}_t$. The model uncertainty is defined by the variance of the features over the network parameter distribution,

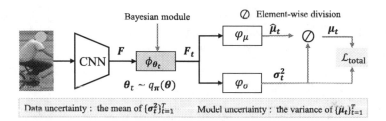

Fig. 2. Overview of the proposed method. An image is fed into the backbone (CNN) to obtain the feature maps $\boldsymbol{F}$. $\boldsymbol{F}$ is further input into the Bayesian module ϕ whose parameter is $\boldsymbol{\theta}_t$ to obtain $\boldsymbol{F}_t$. Then $\boldsymbol{F}_t$ is input to two deterministic modules φ_μ and φ_σ to predict the feature $\hat{\boldsymbol{\mu}}_t$, $\boldsymbol{\mu}_t$ and the data uncertainty σ_t^2, respectively. The model uncertainty is the variance of $\{\hat{\boldsymbol{\mu}}_t\}_{t=1}^{T}$, and the data uncertainty is the mean of $\{\sigma_t^2\}_{t=1}^{T}$.

$$\sigma_m^2 = \frac{1}{T}\sum_{t=1}^{T}(\boldsymbol{\mu}_t - \bar{\boldsymbol{\mu}})^2 \tag{7}$$

where $\bar{\boldsymbol{\mu}} = \frac{1}{T}\sum_{t=1}^{T}\boldsymbol{\mu}_t$ and squaring operations in Eq. 7 are element-wise.

3.3 Jointly Learning Data and Model Uncertainty

To avoid the additional overhead caused by separately learning the data uncertainty and the model uncertainty, and at the same time leverage the complementary information provided by them for representation learning, we integrate them into a unified network for joint learning. As shown in Fig. 2, given a sample $\boldsymbol{x}$, we first feed it into a CNN backbone to get the feature maps $\boldsymbol{F} \in \mathbb{R}^{h \times w \times c}$, where h, w, c are height, weight and channel, respectively. $\boldsymbol{F}$ is further fed into a Bayesian module ϕ to obtain $\boldsymbol{F}_t = \phi_{\boldsymbol{\theta}_t}(\boldsymbol{F})$, where $\boldsymbol{\theta}_t \sim q_\pi(\boldsymbol{\theta})$ corresponds to the parameters of ϕ. Then we input $\boldsymbol{F}_t$ into two deterministic modules φ_μ and φ_σ to obtain the embedding $\hat{\boldsymbol{\mu}}_t = \varphi_\mu(\boldsymbol{F}_t)$ and data uncertainty $\sigma_t^2 = \varphi_\sigma(\boldsymbol{F}_t)$, respectively. Here, $\boldsymbol{F}_t$, $\hat{\boldsymbol{\mu}}_t$, σ_t^2 have the same shape as $\boldsymbol{F}$. We further regularize $\hat{\boldsymbol{\mu}}_t$ by quality-aware pooling: $\boldsymbol{\mu}_t = \text{GAP}(\hat{\boldsymbol{\mu}}_t \oslash \sigma_t^2) \in \mathbb{R}^c$, where $\oslash$ represents the element-wise division and GAP refers to the global average pooling. Thus, for the sample $\boldsymbol{x}$, we get its feature distribution $\mathcal{N}(\boldsymbol{\mu}_t, \sigma_t^2\mathbf{I})$, where σ_t^2 is the mean of σ_t^2 across all elements. During training, the loss function is,

$$\mathcal{L}_{\text{total}} = \mathcal{L}_d(\boldsymbol{\mu}_t, \sigma_t^2) + \mathcal{L}_{\text{tri}}(\boldsymbol{\mu}_t) \tag{8}$$

where $\mathcal{L}_d$ is defined in Eq. 5 and $\mathcal{L}_{\text{tri}}(\boldsymbol{\mu}_t)$ is the triplet loss [16] on $\boldsymbol{\mu}_t$.

In inference, for a sample $\boldsymbol{x}$, we can sample T times from $q_{\pi}(\boldsymbol{\theta})$ and obtain $\{\hat{\boldsymbol{\mu}}_t\}_{t=1}^T$, $\{\boldsymbol{\mu}_t\}_{t=1}^T$ and $\{\sigma_t^2\}_{t=1}^T$. The data uncertainty is formulated by $\sigma_d^2 = \frac{1}{T}\sum_{t=1}^T \sigma_t^2$. Then $\boldsymbol{\sigma}_m^2$ is estimated as the variance of $\{\hat{\boldsymbol{\mu}}_t\}_{t=1}^T$ according to Eq. 7 and the model uncertainty σ_m^2 is defined as the mean of $\boldsymbol{\sigma}_m^2$ across all elements. The final representation is calculated by $\bar{\boldsymbol{\mu}} = \frac{1}{T}\sum_{t=1}^T \boldsymbol{\mu}_t$.

Note that σ_t^2 is the data uncertainty of $\boldsymbol{x}$ under the parameter $\boldsymbol{\theta}_t$. In practice, we train the network to predict $\boldsymbol{s}_t$, and $\sigma_t^2 := \log(1 + \exp(\boldsymbol{s}_t))$. This is because that it is more numerically stable than directly predicting the σ_t^2, as it ensures that each element of σ_t^2 is greater than zero. ϕ, φ_μ and φ_σ are all light weighted modules and their architectures are detailed in supplementary materials.

3.4   Reliability Assessment

Based on the risk-controlled settings and the multi-query settings, we introduce how to leverage the learned data and model uncertainty for reliability assessment.

Risk-Controlled Settings. When facing complex application scenarios, to control the cost of errors, we would expect the model to reject input images (probes) if it can not deal with them. We show the proposed uncertainty mechanism can be naturally utilized as such a "risk indicator". Given a probe $\boldsymbol{x}$, we can obtain its data uncertainty σ_d^2 and model uncertainty σ_m^2. The probe will be assessed as safety (the prediction is reliable) only if $\frac{1}{\sigma_d^2} > \gamma_d$ and $\frac{1}{\sigma_m^2} > \gamma_m$, where γ_d and γ_m are two thresholds that can be set based on the risk tolerance.

Multi-query Settings. In real scenarios, how to utilize multiple query images from the same identity to search this person is an essential issue as an interested pedestrian is usually captured by several cameras. Our method is naturally suitable for such settings because it can suppress the negative impact of ambiguous queries according to the reliability score. Considering $\mathcal{X} = \{\boldsymbol{x}_1, \ldots, \boldsymbol{x}_n\}$ are query images from the same identity and $\boldsymbol{y}$ is an image in the gallery set. The key issue is how to measure the similarity between $\mathcal{X}$ and $\boldsymbol{y}$. Let $\sigma_{d,i}^2$ ($\sigma_{m,i}^2$) be the data (model) uncertainty of $\boldsymbol{x}_i$. To combine the data uncertainty and model uncertainty without being affected by their numerical scale, we project the data (model) uncertainty into the interval $[\tau_{\min}, \tau_{\max}]$. Specifically, $\sigma_{d,i}^2$ is projected to $d_i = \beta_i \tau_{\min} + (1 - \beta_i)\tau_{\max}$, where $\beta_i = \frac{\sigma_{d,i}^2 - \sigma_{d,\min}^2}{\sigma_{d,\max}^2 - \sigma_{d,\min}^2}$ and $\sigma_{d,\max}^2$ ($\sigma_{d,\min}^2$) is the maximum (minimum) value in $\{\sigma_{d,i}^2\}_{i=1}^n$. The model uncertainty $\sigma_{m,i}^2$ is mapped to m_i in the similar way. The **reliability score** of $\boldsymbol{x}_i$ is $w_i = \frac{m_i d_i}{\sum_{i=1}^n m_j d_j}$. The similarity between $\mathcal{X}$ and $\boldsymbol{y}$ is calculated by $s = \sum_{i=1}^n w_i s_i$, where s_i is the similarity between $\boldsymbol{x}_i$ and $\boldsymbol{y}$. The similarity s considers the reliability of each element in $\mathcal{X}$. If an element has larger reliability score, it plays a more important role.

4   Experiments

4.1   Datasets and Evaluation Metrics

The datasets we use include Market-1501 [54], MSMT17 [47], CUHK03 [30, 57], Occluded-Duke [34], Occluded-REID [61] and Partial-REID [55]. **Market-1501** has 12,936 training, 3,368 query and 19,732 gallery images. **CUHK03** contains 13,164 images of 1,467 identities. We adopt the new testing protocol proposed in [57]. **MSMT17** is the largest image dataset for person ReID. It contains 126,441 images of 4,101 identities. **Occluded-Duke** is reconstructed from DukeMTMC-reID [36] by selecting occluded images as query set. It has 15,618 training images, 2,210 query and 17,661 gallery images. **Occluded-REID** contains 2,000 images belonging to 200 identities. Each identity has five occluded person images and five-full-body images. **Partial-REID** contains 600 images from 60 person, with five partial images and five full-body images per person.

Evaluation Metrics. We use the Cumulative Matching Characteristic (CMC) curve and mean average precision (mAP) as the evaluation metric.

4.2   Implementation Details

Input images are resized to 256×128. During training, images are augmented by random cropping, random horizontal flipping, and random erasing [58]. Following [43], we adopt ResNet50 [11] pre-trained on ImageNet as the backbone. For a fair comparison with methods employing more complex backbone networks, following ISP [60], we also employ HRNet-W32 [40] as our backbone. There are 64 images from 16 identities in a mini-batch. The initial learning rate is 3.5×10^{-4}. For MSMT17 [47] (other datasets), we train our model 160 (120) epochs, and the learning rate is decreased to its 0.1 and 0.01 at the 70th (40th) and 110th (70th) epochs, respectively. Each epoch has 200 iterations. The optimizer is Adam [26]. The ρ of Bernoulli distribution in the Bayesian module ϕ is empirically set as 0.7. The $\tau_{\min}$ and $\tau_{\max}$ in the multi-query settings are set as 0.5 and 1.0, respectively.

During testing, we first obtain the parameters $\{\boldsymbol{\theta}_t\}_{t=1}^T$ of the Bayesian module ϕ by sampling T times from $q_\pi(\boldsymbol{\theta})$. For each $\boldsymbol{\theta}_t$, we use it to handle all samples. This ensures that the features of different samples are extracted by the same network, which makes it possible to employ our method with a small T. As the Bayesian module ϕ is at the back of the entire network, we only need to repeatedly input it into ϕ, φ_μ and φ_σ for T times, rather than entire network, which saves a lot of expenditure (related results are shown in Table 5).

4.3   Experiments on the Reliability Assessment

Risk-Controlled Settings. Here, we design experiments to verify whether the proposed uncertainty can serves as the "risk indicator" as described in Sect. 3.4. Specifically, we allow the model to filter out some queries it is diffident to maintain higher performance. Here, we consider four criteria for filtering, i.e., (1)

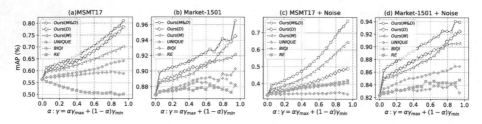

Fig. 3. Comparison of the risk-controlled settings. Filtering out a proportion of query images according to the thresholds of different criteria. D: data uncertainty; M: model uncertainty; $M\&D$: model uncertainty and data uncertainty; $BIQI$ [35]: blind image quality indices. RE: the reciprocal of the entropy of the predicted category distribution. $UNIQUE$ [51]: unified no-reference image quality and uncertainty evaluator.

Fig. 4. Visualization of different quality-degrading transformations.

$UNIQUE$ [51]: unified no-reference image quality and uncertainty evaluator; (2) $BIQI$ [35]: blind image quality indices; (3) RE: the reciprocal of the entropy of the predicted category distribution; (4) $Ours$: the reciprocal of the proposed uncertainty as described in Sect. 3.4. (M: model uncertainty alone; D: data uncertainty alone; $M\&D$: combing model uncertainty and data uncertainty). For each criterion, we first calculate its maximum and minimum on all queries, denoted as $\gamma_{\max}$ and $\gamma_{\min}$. Then we setting the threshold as $\gamma = \alpha \gamma_{\max} + (1 - \alpha)\gamma_{\min}$. For $M\&D$, a query will be kept only if $\frac{1}{\sigma_d^2} > \gamma_{d,\alpha}$ and $\frac{1}{\sigma_m^2} > \gamma_{m,\alpha}$, where σ_d^2 and σ_m^2 are the data uncertainty and the model uncertainty, and $\gamma_{d,\alpha}$ and $\gamma_{m,\alpha}$ are their thresholds at parameter α. We report the mAP score against the α under different settings. Figure 3(a-b) shows the results on MSMT17 and Market-1501. In Fig. 3(c–d), we add Gaussian noise to the images in the query set with a probability of 0.5. From the results, we can make several observations. First, for our methods (M, D, and $M\&D$), as α increasing, the mAP score of the remained queries increases. This shows that the proposed uncertainty mechanism help filter out queries whose prediction is unreliable (the model can not deal with) while retaining queries the model is capable of. Second, under different settings, model uncertainty or data uncertainty alone already outperforms $UNIQUE$, $BIQI$ and RE. When they are combined, the mAP score of retained queries is higher. This shows the proposed uncertainty mechanism is a better indicator of the reliability of the prediction, and thus helps for making risk-controlled decisions.

Multi-query Settings. We find that there are few images with the same personal identity and same camera identity in the query set of existing datasets. Thus, we reconstruct the test sets of Market-1501, Occluded-Duke and MSMT17

Table 1. Results of the multi-query settings on Market-1501 [54] and MSMT17 [47] and Occluded-Duke [34]. To simulate the complex scenes, we transform the query images via different quality-degrading transformations. MT17 → Market: the model is trained on MSMT17 and directly tested on Market-1501. "*w/* R" ("*w/o* R") means we use (don't use) the reliability score to adjust the weights for different queries.

Complex scenes	Method	Market-1501		MSMT17		Occlu-Duke		MT17 → Market	
		R1	mAP	R1	mAP	R1	mAP	R1	mAP
Gaussian noise	D-Net [50]	54.5	52.6	39.8	26.5	43.8	35.9	25.0	12.9
	Ours (*w/o* R)	55.0	53.1	49.9	34.5	51.3	43.7	32.2	18.3
	Ours (*w/* R)	**76.7**	**71.4**	**62.9**	**45.4**	**61.5**	**53.8**	**36.8**	**20.6**
Random crop	D-Net [50]	36.5	33.6	52.6	33.7	23.8	20.3	22.5	12.7
	Ours (*w/o* R)	37.4	36.0	58.4	38.6	27.0	23.4	28.6	16.4
	Ours (*w/* R)	**46.3**	**44.4**	**63.1**	**42.8**	**35.1**	**31.3**	**32.7**	**18.8**
Motion blur	D-Net [50]	66.9	59.3	55.9	38.0	60.0	50.2	25.5	14.2
	Ours (*w/o* R)	69.7	63.4	66.7	47.1	62.0	53.3	35.5	20.8
	Ours (*w/* R)	**77.4**	**71.4**	**75.3**	**54.9**	**66.4**	**58.5**	**46.3**	**26.3**
Adding fog	D-Net [50]	62.1	56.6	50.9	33.7	50.3	42.1	29.0	15.7
	Ours (*w/o* R)	65.1	60.3	58.3	40.2	55.9	47.5	37.4	21.9
	Ours (*w/* R)	**77.1**	**71.9**	**72.0**	**52.0**	**64.1**	**56.6**	**47.0**	**26.9**

Table 2. Results of multi-query settings on Occluded-REID [61], Occluded-Duke [34] and Partial-REID [55]. The query images are **not** transformed to degrade the quality.

Method	Occluded-Duke		Occluded-REID		Partial-REID	
	R1	mAP	R1	mAP	R1	mAP
D-Net [50]	75.9	67.0	82.0	72.3	80.0	74.4
Ours (*w/o* R)	76.6	68.3	80.0	72.2	85.0	77.1
Ours (*w/* R)	**78.0**	**70.5**	**83.0**	**73.5**	**88.3**	**80.0**

to evaluate the proposed reliability assessment under multi-query settings. Specifically, for each dataset, we first collect the images belonging to the same personal identity and same camera identity from the query set and gallery set. Then, we randomly select half of these images to be allocated to the reorganized query set and the other half to the reorganized gallery set. Simultaneously, to simulate the complex scenes in reality, we transform images in the reorganized query set with a certain probability. The transformation includes: (1) add Gaussian noise to mimic camera quality differences; (2) crop the image to simulate pedestrians are partially out of the camera's field of view; (3) add motion blur to mimic fast-moving pedestrians; (4) add fog to imitate complex weather. Please refer to the supplementary materials for the transformation details. Figure 4 shows some examples of these transformations. In inference, the images belonging to the same personal identity and same camera identity in the query set are regarded as a set of templates that need to be associated. We report the performance of our method against the D-Net [50] in Table 1. "*w/* R" ("*w/o* R") means we use (don't use) the reliability score to adjust the weights for different queries. From the results, we can make several observations. First, for different complex scenes and different datasets, our method ("*w/o* R") already outperforms D-Net [50]. This shows that our method learns better embedding space, in which the features are more discriminative. When we additionally use the reliability score to adjust the weights for different queries, the performance is further improved. This shows that the proposed reliability score is credible, which can help mine more valuable queries and suppress distractions from low-quality ones.

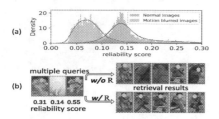

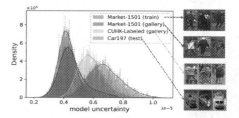

Fig. 5. (a) Distributions of the reliability scores under multi-query settings. (Normal images *v.s.* Motion-blurred images). (b) An example. "*w/o* R" ("*w/* R") means we use (don't use) the reliability score to adjust the weights for different queries.

Fig. 6. The role of model uncertainty in reliability assessment. The model is trained with the training set of the Market-1501. The learned model uncertainty is proportional to the degree of deviation between test set domain and training set domain.

We also evaluate our method on Occluded-REID [61], Occluded-Duke [34] and Partial-REID [55]. As query images in these three datasets are occluded or partial, we directly test on them without any quality degradation transformation. Results are shown in Table 2. We can draw the same conclusions as in Table 1.

4.4   Analysis of the Reliability Assessment

Is the Reliability Score Reasonable? Here, we analyze the reasonableness of the reliability score based on the multi-query settings. For each identity in the test set of Market-1501, we randomly select 10 images to form multiple queries, in which 5 images are downgraded in quality by adding motion blur. Figure 5 (a) shows the distributions of reliability scores for normal images and motion-blurred images. We can see that, on average, the reliability score of normal images is greater than that of motion-blurred images, showing the reliability score is credible. The retrieval example in Fig. 5 (b) shows that when we adjust the weights of different queries according to the reliability scores, the ambiguous information from the low-quality query is suppressed, resulting in a better performance.

The Role of Model Uncertainty in Reliability Assessment. We design experiments to verify whether the proposed model uncertainty can describe the confidence of the model in its prediction of the sample. We use the training set of Market-1501 to train the model. Then we estimate the model uncertainty of samples from the training set of Market-1501, the gallery set of Market-1501, the gallery set of CUHK03 and the test set of Car197 [28], respectively. Figure 6 shows the results. On average, with the increase of the domain deviation, the model uncertainty gradually grows. This shows that the estimated model uncertainty is related to the model's prediction confidence. For out-of-distribution inputs, the model is diffident about the predictions, and the model uncertainty is large.

The Role of Data Uncertainty in Reliability Assessment. We design experiments to verify whether the estimated data uncertainty can capture the "noise" inherent in the data. We first regard samples in the gallery set of Market-1501 [54] as clean data, and then add noise to pollute them to generate the noisy (low-quality) ones. Specifically, for an image tensor, *i.e.*, x, we generate a noise tensor ε, where $\varepsilon \sim \mathcal{N}(\mathbf{0}, \mathbf{I})$. Then we pollute the origin data by $x \leftarrow x + \eta \varepsilon$, where η control the strength of the pollution. We gradually vary the size of η to see how data uncertainty changes. The Gaussian kernel density estimation [37] of the estimated data uncertainty are shown in Fig. 7. When η increases, the data uncertainty of noisy samples grows correspondingly. This shows that the proposed data uncertainty can capture the quality of the sample.

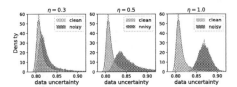

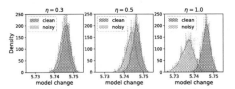

Fig. 7. The role of data uncertainty in reliability assessment. It can be seen that samples with lower quality (larger noise) have larger data uncertainty.

Fig. 8. The role of data uncertainty in the training process. It can be seen that samples with lower quality (larger noise) cause less model change in the training process.

The Role of Data Uncertainty in the Training Process. We regard the samples in the training set of Market-1501 as clean samples and then pollute them to generate the noisy ones. We investigate how the clean and noisy samples affect model learning. Specifically, for each sample, we use it to individually train the model for 10 iterations, and count the model change. The model change is defined as the mean of the absolute difference of the model parameters before and after training. This can reflect the impact of the sample on the learning process. As shown in Fig. 8, when η increases, the model change caused by noisy samples reduces, showing the proposed sampling-free data uncertainty learning method can suppress the contribution of low-quality samples during training.

4.5   Comparison with State-of-the Art Methods

We also compare our method with state-of-the-art methods under the single-query settings. As shown in Table 3, the compared methods are divided into two categories. One category employs ResNet50 [11] or slightly modifies ResNet50 without changing the main structure of the network. The other category employs more powerful backbones than ResNet50, such as HRNet-W32 [40] and Transformer [44]. For fair comparisons, we adopt ResNet50 and HRNet-W32 as our backbone, respectively. From the results, we can make several observations. (1) Whether using ResNet50 or HRNet-W32 as the backbone, our method consistently achieves comparable or superior performance on these datasets. Specifically, when HRNet-W32 is used as the backbone, our method outperforms the previous state-of-the-art methods by +4.5% and +4.0% in terms of mAP scores on MSMT17 and CUHK03-NP (detected), respectively. (2) Our method and ISP [60] have the same backbone and our method is much more effective. Especially, on CUHK03-NP (detected), our method outperforms ISP by +7.1% mAP score. (3) Compared with semantic-based methods, e.g., FPR [14] and HOReID [45], our method does not need any additional external cues. Compared with methods that use more complex network structures, such as PAT [31] using the transformer [44], our method still shows promising performance.

Table 3. Comparison with state-of-the-art methods under the single-query settings.

Methods	Backbone	Market		MSMT17		CUHK03-NP			
						Labeled		Detected	
		R1	mAP	R1	mAP	R1	mAP	R1	mAP
PCB+RPP [43]	ResNet50	93.8	81.6	–	–	63.7	57.5	–	–
MGN [46]	ResNet50	**95.7**	86.9	–	–	68.0	67.4	66.8	66.0
CAMA [48]	ResNet50	94.7	84.5	–	–	70.1	66.5	66.6	64.2
MHN-6 [2]	ResNet50	95.1	85.0	–	–	77.2	72.4	71.7	65.4
FPR [14]	ResNet50	95.4	86.6	–	–	76.1	72.3	–	–
HOReID [45]	ResNet50	94.2	84.9	–	–	–	–	–	–
DGNet [56]	ResNet50	94.8	86.0	77.2	52.3	–	–	–	–
UAL *(Ours)*	ResNet50	95.2	**87.0**	**80.0**	**56.5**	**78.2**	**75.6**	**76.1**	**72.0**
RGA-SC [52]	R50-RGA	**96.1**	88.4	80.3	57.5	81.1	77.4	79.6	74.5
ABD-Net [3]	ABD-Net	95.6	88.3	82.3	60.8	–	–	–	–
BAT-net [5]	BAT-net	95.1	87.4	79.5	56.8	78.6	76.1	76.2	73.2
OSNet [59]	OSNet	94.8	84.9	78.7	52.9	–	–	72.3	67.8
ISP [60]	HRNet	95.3	88.6	–	–	76.5	74.1	75.2	71.4
PAT [31]	Transformer	95.4	88.0	–	–	–	–	–	–
UAL *(Ours)*	HRNet	95.7	**89.5**	**84.7**	**65.3**	**83.7**	**81.0**	**81.0**	**78.5**

4.6   Ablation Study

In this part, we conduct ablation studies to show the effectiveness of each component of the proposed method.

The Effectiveness of Data Uncertainty and Model Uncertainty. Besides based on the backbones ResNet50 and HRNet-W32 implemented by ours, we also conduct the experiments based on the more in-data baseline in **fast-reid**, *i.e.*, BOT [33], with different backbones including ResNet50 (R50), a variant with IBN layers (R50-ibn) and ResNeSt (S50). Experiments are conducted on MSMT17 [47] dataset. As shown in Table 4, both learning data uncertainty and learning model uncertainty improve the performance, and learning data uncertainty provides a larger improvement. The results show these two uncertainties can provide complementary information for learning discriminative latent space.

Impact of Hyper-parameter T. During testing, for a sample x, we need to sample T times from $q_\pi(\mathbf{0})$ to obtain the $\bar{\mu}$ and σ_m^2 (as described in Sect. 3.3). Here, we show how T affects the performance on CUHK03-NP [30] and Market-1501 [54]. The results are shown in Table 5. We also report the time cost of the entire testing process, from extracting features to calculating the Rank-1 scores. As we can see, our method can work well with a small T. Further, our method outperforms HOReID [45] and PGFA [34] by a large margin while the time cost is much smaller, which shows the effectiveness and efficiency of our method.

Table 4. Effectiveness of data uncertainty and model uncertainty. Experiments are conducted on MSMT17 [47]. [†] indicates the experiments are conducted on the backbones in publicly available repository fast-reid: https://github.com/JDAI-CV/fast-reid.

Uncertainty		ResNet50		[†]BOT(R50)		[†]BOT(R50-ibn)		[†]BOT(S50)		HRNet-W32	
Data	Model	Rank-1	mAP	Rank-1	mAP	Rank-1	mAP	Rank-1	mAP	Rank-1	mAP
✗	✗	76.5	52.0	73.9	49.9	79.1	55.4	81.0	59.4	81.6	59.6
✓	✗	77.7	53.1	77.1	51.4	80.4	56.0	81.9	59.5	83.4	62.2
✗	✓	76.7	52.3	76.6	51.2	80.8	56.9	82.6	60.8	81.9	60.0
✓	✓	**80.0**	**56.5**	**78.7**	**53.6**	**82.4**	**59.1**	**84.1**	**62.1**	**84.7**	**65.3**

Table 5. Impact of the hyper-parameter T. We use the AMD EPYC 7742 CPU and GeForce RTX 3090 GPU. The backbone is ResNet50.

T	CUHK03 (Labeled)			Market-1501		
	Time	R1	mAP	Time	R1	mAP
HOReID [45]	–	–	–	236 s	94.2	84.9
PGFA [34]	–	–	–	193 s	91.2	76.8
Ours ($T=5$)	17 s	78.4	75.5	66 s	95.0	86.9
Ours ($T=10$)	24 s	**78.6**	75.6	87 s	**95.2**	87.0
Ours ($T=20$)	39 s	78.3	**75.6**	138 s	95.1	**87.0**

5   Conclusions

In this paper, we propose an Uncertainty-Aware Learning (UAL) method for the ReID task to provide reliability-aware predictions, which is achieved by considering two types of uncertainty: data uncertainty and model uncertainty. These two types of uncertainty are integrated into a unified network for joint learning without any external clues. Comprehensive experiments under the risk-controlled settings and the multi-query settings verify that the proposed reliability assessment is effective. Our method also shows superior performance under the single query settings. Meanwhile, we also provide quantitative analyses of the learned data uncertainty and model uncertainty. We expect that our method will provide new insights and attract more interest in the reliability issue in person ReID.

Acknowledgement. This work was supported by the state key development program in 14th Five-Year under Grant Nos. 2021YFF0602103, 2021YFF0602102, 2021QY1702. We also thank for the research fund under Grant No. 2019GQG0001 from the Institute for Guo Qiang, Tsinghua University.

References

1. Chang, J., Lan, Z., Cheng, C., Wei, Y.: Data uncertainty learning in face recognition. In: CVPR, pp. 5710–5719 (2020)
2. Chen, B., Deng, W., Hu, J.: Mixed high-order attention network for person re-identification. In: ICCV, pp. 371–381 (2019)
3. Chen, T., et al.: ABD-Net: attentive but diverse person re-identification. In: ICCV, pp. 8351–8361 (2019)
4. Dai, Y., Liu, J., Sun, Y., Tong, Z., Zhang, C., Duan, L.Y.: IDM: an intermediate domain module for domain adaptive person re-id. In: ICCV, pp. 11864–11874 (2021)
5. Fang, P., Zhou, J., Roy, S.K., Petersson, L., Harandi, M.: Bilinear attention networks for person retrieval. In: ICCV, pp. 8030–8039 (2019)
6. Gal, Y., Ghahramani, Z.: Bayesian convolutional neural networks with Bernoulli approximate variational inference. arXiv preprint arXiv:1506.02158 (2015)
7. Gal, Y., Ghahramani, Z.: Dropout as a Bayesian approximation: representing model uncertainty in deep learning. In: ICML, pp. 1050–1059 (2016)
8. Gao, S., Wang, J., Lu, H., Liu, Z.: Pose-guided visible part matching for occluded person ReID. In: CVPR, pp. 11744–11752 (2020)
9. Gawlikowski, J., et al.: A survey of uncertainty in deep neural networks. arXiv preprint arXiv:2107.03342 (2021)
10. Geng, Y., Han, Z., Zhang, C., Hu, Q.: Uncertainty-aware multi-view representation learning. In: AAAI, pp. 7545–7553 (2021)
11. He, K., Zhang, X., Ren, S., Sun, J.: Deep residual learning for image recognition. In: CVPR, pp. 770–778 (2016)
12. He, L., Liang, J., Li, H., Sun, Z.: Deep spatial feature reconstruction for partial person re-identification: Alignment-free approach. In: CVPR, pp. 7073–7082 (2018)
13. He, L., Sun, Z., Zhu, Y., Wang, Y.: Recognizing partial biometric patterns. arXiv preprint arXiv:1810.07399 (2018)
14. He, L., Wang, Y., Liu, W., Zhao, H., Sun, Z., Feng, J.: Foreground-aware pyramid reconstruction for alignment-free occluded person re-identification. In: ICCV, pp. 8450–8459 (2019)

15. He, S., Luo, H., Wang, P., Wang, F., Li, H., Jiang, W.: TransReID: transformer-based object re-identification. arXiv preprint arXiv:2102.04378 (2021)
16. Hermans, A., Beyer, L., Leibe, B.: In defense of the triplet loss for person re-identification. arXiv preprint arXiv:1703.07737 (2017)
17. Hou, R., Ma, B., Chang, H., Gu, X., Shan, S., Chen, X.: Feature completion for occluded person re-identification. IEEE TPAMI **44**, 4894–4912 (2021)
18. Huang, Y., Zha, Z.J., Fu, X., Hong, R., Li, L.: Real-world person re-identification via degradation invariance learning. In: CVPR, pp. 14084–14094 (2020)
19. Isobe, S., Arai, S.: Deep convolutional encoder-decoder network with model uncertainty for semantic segmentation. In: INISTA, pp. 365–370 (2017)
20. Isobe, T., Li, D., Tian, L., Chen, W., Shan, Y., Wang, S.: Towards discriminative representation learning for unsupervised person re-identification. In: ICCV, pp. 8526–8536 (2021)
21. Jia, M., et al.: Matching on sets: conquer occluded person re-identification without alignment. In: AAAI, pp. 1673–1681 (2021)
22. Jin, X., Lan, C., Zeng, W., Chen, Z.: Uncertainty-aware multi-shot knowledge distillation for image-based object re-identification. In: AAAI, pp. 11165–11172 (2020)
23. Kendall, A., Cipolla, R.: Modelling uncertainty in deep learning for camera relocalization. In: ICRA, pp. 4762–4769 (2016)
24. Kendall, A., Gal, Y.: What uncertainties do we need in Bayesian deep learning for computer vision? arXiv preprint arXiv:1703.04977 (2017)
25. Khan, S., Hayat, M., Zamir, S.W., Shen, J., Shao, L.: Striking the right balance with uncertainty. In: CVPR, pp. 103–112 (2019)
26. Kingma, D.P., Ba, J.: Adam: a method for stochastic optimization. arXiv preprint arXiv:1412.6980 (2014)
27. Kingma, D.P., Welling, M.: Auto-encoding variational Bayes. arXiv preprint arXiv:1312.6114 (2013)
28. Krause, J., Stark, M., Deng, J., Fei-Fei, L.: 3D object representations for fine-grained categorization. In: ICCVW, pp. 554–561 (2013)
29. Li, H., Wu, G., Zheng, W.S.: Combined depth space based architecture search for person re-identification. In: CVPR, pp. 6729–6738 (2021)
30. Li, W., Zhao, R., Xiao, T., Wang, X.: DeepReID: deep filter pairing neural network for person re-identification. In: CVPR, pp. 152–159 (2014)
31. Li, Y., He, J., Zhang, T., Liu, X., Zhang, Y., Wu, F.: Diverse part discovery: Occluded person re-identification with part-aware transformer. In: CVPR, pp. 2898–2907 (2021)
32. Liao, S., Hu, Y., Zhu, X., Li, S.Z.: Person re-identification by local maximal occurrence representation and metric learning. In: CVPR, pp. 2197–2206 (2015)
33. Luo, H., Gu, Y., Liao, X., Lai, S., Jiang, W.: Bag of tricks and a strong baseline for deep person re-identification. In: CVPRW (2019)
34. Miao, J., Wu, Y., Liu, P., Ding, Y., Yang, Y.: Pose-guided feature alignment for occluded person re-identification. In: ICCV, pp. 542–551 (2019)
35. Moorthy, A.K., Bovik, A.C.: A two-step framework for constructing blind image quality indices. IEEE Signal Process. Lett. **17**(5), 513–516 (2010)
36. Ristani, E., Solera, F., Zou, R., Cucchiara, R., Tomasi, C.: Performance measures and a data set for multi-target, multi-camera tracking. In: Hua, G., Jégou, H. (eds.) ECCV 2016. LNCS, vol. 9914, pp. 17–35. Springer, Cham (2016). https://doi.org/10.1007/978-3-319-48881-3_2
37. Scott, D.W.: Multivariate Density Estimation: Theory, Practice, and Visualization. Wiley, New York (2015)

38. Shen, Y., Zhang, Z., Sabuncu, M.R., Sun, L.: Real-time uncertainty estimation in computer vision via uncertainty-aware distribution distillation. In: WACV, pp. 707–716 (2021)
39. Shi, Y., Jain, A.K.: Probabilistic face embeddings. In: ICCV, pp. 6902–6911 (2019)
40. Sun, K., Xiao, B., Liu, D., Wang, J.: Deep high-resolution representation learning for human pose estimation. In: CVPR, pp. 5693–5703 (2019)
41. Sun, W., Xie, J., Qiu, J., Ma, Z.: Part uncertainty estimation convolutional neural network for person re-identification. In: ICIP, pp. 2304–2308 (2021)
42. Sun, Y., et al.: Perceive where to focus: learning visibility-aware part-level features for partial person re-identification. In: CVPR, pp. 393–402 (2019)
43. Sun, Y., Zheng, L., Yang, Y., Tian, Q., Wang, S.: Beyond part models: person retrieval with refined part pooling (and a strong convolutional baseline). In: Ferrari, V., Hebert, M., Sminchisescu, C., Weiss, Y. (eds.) ECCV 2018. LNCS, vol. 11208, pp. 501–518. Springer, Cham (2018). https://doi.org/10.1007/978-3-030-01225-0_30
44. Vaswani, A., et al.: Attention is all you need. In: NeurIPS, pp. 5998–6008 (2017)
45. Wang, G., et al.: High-order information matters: learning relation and topology for occluded person re-identification. In: CVPR, pp. 6449–6458 (2020)
46. Wang, G., Yuan, Y., Chen, X., Li, J., Zhou, X.: Learning discriminative features with multiple granularities for person re-identification. In: ACM MM, pp. 274–282 (2018)
47. Wei, L., Zhang, S., Gao, W., Tian, Q.: Person transfer GAN to bridge domain gap for person re-identification. In: CVPR, pp. 79–88 (2018)
48. Yang, W., Huang, H., Zhang, Z., Chen, X., Huang, K., Zhang, S.: Towards rich feature discovery with class activation maps augmentation for person re-identification. In: CVPR, pp. 1389–1398 (2019)
49. Yang, Y., Liao, S., Lei, Z., Li, S.Z.: Large scale similarity learning using similar pairs for person verification. In: AAAI (2016)
50. Yu, T., Li, D., Yang, Y., Hospedales, T.M., Xiang, T.: Robust person re-identification by modelling feature uncertainty. In: ICCV, pp. 552–561 (2019)
51. Zhang, W., Ma, K., Zhai, G., Yang, X.: Uncertainty-aware blind image quality assessment in the laboratory and wild. IEEE TIP **30**, 3474–3486 (2021)
52. Zhang, Z., Lan, C., Zeng, W., Jin, X., Chen, Z.: Relation-aware global attention for person re-identification. In: CVPR, pp. 3186–3195 (2020)
53. Zheng, K., Lan, C., Zeng, W., Zhang, Z., Zha, Z.J.: Exploiting sample uncertainty for domain adaptive person re-identification. arXiv preprint arXiv:2012.08733 (2020)
54. Zheng, L., Shen, L., Tian, L., Wang, S., Wang, J., Tian, Q.: Scalable person re-identification: a benchmark. In: ICCV, pp. 1116–1124 (2015)
55. Zheng, W.S., Li, X., Xiang, T., Liao, S., Lai, J., Gong, S.: Partial person re-identification. In: ICCV, pp. 4678–4686 (2015)
56. Zheng, Z., Yang, X., Yu, Z., Zheng, L., Yang, Y., Kautz, J.: Joint discriminative and generative learning for person re-identification. In: CVPR, pp. 2138–2147 (2019)
57. Zhong, Z., Zheng, L., Cao, D., Li, S.: Re-ranking person re-identification with k-reciprocal encoding. In: CVPR, pp. 1318–1327 (2017)
58. Zhong, Z., Zheng, L., Kang, G., Li, S., Yang, Y.: Random erasing data augmentation. In: AAAI, pp. 13001–13008 (2020)
59. Zhou, K., Yang, Y., Cavallaro, A., Xiang, T.: Omni-scale feature learning for person re-identification. In: ICCV, pp. 3702–3712 (2019)

60. Zhu, K., Guo, H., Liu, Z., Tang, M., Wang, J.: Identity-guided human semantic parsing for person re-identification. In: Vedaldi, A., Bischof, H., Brox, T., Frahm, J.-M. (eds.) ECCV 2020. LNCS, vol. 12348, pp. 346–363. Springer, Cham (2020). https://doi.org/10.1007/978-3-030-58580-8_21
61. Zhuo, J., Chen, Z., Lai, J., Wang, G.: Occluded person re-identification. In: ICME, pp. 1–6. IEEE (2018)

Relighting4D: Neural Relightable Human from Videos

Zhaoxi Chen⬤ and Ziwei Liu$^{(\boxtimes)}$⬤

S-Lab, Nanyang Technological University, Singapore, Singapore
{zhaoxi001,ziwei.liu}@ntu.edu.sg

Abstract. Human relighting is a highly desirable yet challenging task. Existing works either require expensive one-light-at-a-time (OLAT) captured data using light stage or cannot freely change the viewpoints of the rendered body. In this work, we propose a principled framework, **Relighting4D**, that enables free-viewpoints relighting from only human videos under unknown illuminations. Our key insight is that the space-time varying geometry and reflectance of the human body can be decomposed as a set of neural fields of normal, occlusion, diffuse, and specular maps. These neural fields are further integrated into reflectance-aware physically based rendering, where each vertex in the neural field absorbs and reflects the light from the environment. The whole framework can be learned from videos in a self-supervised manner, with physically informed priors designed for regularization. Extensive experiments on both real and synthetic datasets demonstrate that our framework is capable of relighting dynamic human actors with free-viewpoints. Codes are available at https://github.com/FrozenBurning/Relighting4D.

Keywords: Neural rendering · Dynamic scenes · Inverse rendering

1 Introduction

The emergence of metaverse has fueled the demands for photorealistic rendering of human characters, which benefits applications like digital 3D human and virtual reality. Among all factors, lighting is the most crucial one for rendering quality. Recently, remarkable success in relighting humans has been achieved [4, 11, 17, 25, 26, 31, 45, 51, 60, 62]. However, the impressive quality of these methods heavily relies on the data captured by Light Stage [8]. The complicated hardware setup makes relighting systems expensive and only applicable in the constrained environment. On the other hand, a number of recent works propose to relight human images from a perspective of inverse rendering [13, 15, 21, 39, 47, 64]. They succeed in relighting 2D images, yet fail to relight with novel views. A lack of underlying 3D representations impedes their flexibility of application.

Supplementary Information The online version contains supplementary material available at https://doi.org/10.1007/978-3-031-19781-9_35.

Videos of Dynamic Humans   Relighting with Free Viewpoints

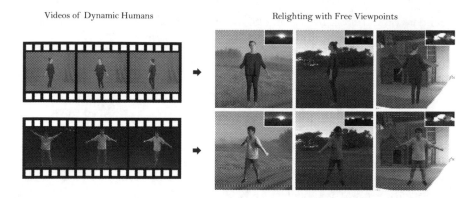

Fig. 1. Relighting of dynamic humans with free viewpoints. *Relighting4D* ;takes only videos as input, decomposing them into geometry and reflectance, which enables relighting of dynamic humans with free viewpoints by a physically based renderer.

In this paper, we focus on the problem of relighting dynamic humans from only videos, as illustrated in Fig. 1. The setting significantly reduces the cost of a flexible relighting system and broadens its scope of application. It has been proved [5,6,19,24,28,30,32,34,35,41,42,54,58,63] that a scene can be represented as neural fields to enable novel view synthesis and relighting. Among above methods, some [5,6,24,42,59,63] deal with relighting static objects but fail to model dynamic scenes. In sum, none of those methods successfully incorporate illuminations and scene dynamics simultaneously.

Different from existing methods on novel view synthesis of the human body that are either non-relightable or require expensive OLAT captured images, we seek to estimate plausible geometry and reflectance from posed human videos.

To this end, we propose **Relighting4D**, to relight dynamic humans with free viewpoints from videos given the 4D coordinates (x, y, z, t) and the desired illumination. Specifically, our method first aggregates observations from posed human videos through space and time by a neural field conditioned on a deformable human model. Then, we decompose the neural field into geometry and reflectance counterparts, namely normal, occlusion, diffuse, and specular maps, which drive a physically based renderer to perform relighting.

We evaluate our approach on both monocular and multi-view videos. Overall, *Relighting4D* outperforms other methods on perceptual quality and physical correctness. It relights dynamic humans in high fidelity, and generalizes to novel views. Furthermore, we demonstrate our capability of relighting under novel illuminations, especially the challenging OLAT setting, by creating a synthetic dataset called BlenderHuman for quantitative evaluations.

We summarize our contributions as follows: **1)** We present a principled framework, *Relighting4D*, which is the first to relight dynamic humans with free viewpoints using only videos. **2)** We propose to disentangle reflectance and geometry from input videos under unknown illuminations by leveraging multiple physically informed priors in a physically based rendering pipeline. **3)** Extensive experiments on both synthetic and real datasets demonstrate the feasibility and significant improvements of our approach over prior arts.

2   Related Work

Neural Scene Representation. [14,18,20,28,34–36,40,41,43,46,49,56] has witnessed significant progress in representing a 3D scene with deep neural networks. NeRF [28] proposes to model the scene as a 5D radiance field. To model dynamic humans, Neural Body [34] proposes to attach a set of latent codes to a deformable human body model (i.e., SMPL [23]). However, these methods implicitly incorporate all color information in the radiance field, which impedes their application towards relighting a dynamic human.

Inverse rendering aims to disentangle the appearance from observed images into geometry, material, and lighting condition. Previous works [3,16, 21,22,27,37,53,57] seek to address it by conditioning on physically based priors or synthetic data. However, they fail in novel view synthesis due to the lack of underlying 3D representations. Recently, NeRF based methods [5,6,42,59,63] propose to learn 3D reflectance fields or light transport fields from input images to enable free-viewpoint relighting. However, none of them is applicable to relight dynamic humans with space-time varying features.

Relighting of human face, avatar and body has wide-range applications [31,39,44,51,64]. As for full-body human relighting, convolutional methods [13,47] fail to relight from novel viewpoints as there is no underlying 3D representation. Other methods [11,62] heavily relies on one-light-at-a-time [8] (OLAT) images, which is neither cheap to capture nor publicly available. *Relighting4D* differentiates itself from aforementioned methods in that we achieve free-viewpoint relighting of dynamic full-body humans without the requirement on expensive capture setup.

3   Our Approach

Given a human video, *Relighting4D* can synthesize videos with free viewpoints under novel illuminations. We denote the input video as $I = \{I_1, I_2, ..., I_t\}$, where t is the time step. In general, our model learns a physically based renderer from I. During inference, it takes a 3D position $\boldsymbol{x} \in \mathbb{R}^3$, a time step t, a camera view $\boldsymbol{\omega}_o \in \mathbb{R}^3$, a desired light probes $\boldsymbol{L}_i \in \mathbb{R}^{16 \times 32 \times 3}$ as inputs, and outputs the corresponding outgoing radiance $\boldsymbol{L}_o \in \mathbb{R}^3$.

Framework Overview. We first give an overview of *Relighting4D* (Figure 2). It first derives latent features from the video, which is achieved by estimating a neural field. Based on the latent features, *Relighting4D* decomposes the human performer into geometry and reflectance information which drive our physically based renderer. The space-time varying geometry and reflectance of the full human body are parameterized by four multilayer perceptrons. Note that, *Relighting4D* enables relighting of dynamic humans with free viewpoints using **only** videos, without training on any captured data (e.g., OLAT or flash images).

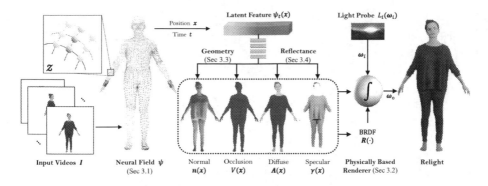

Fig. 2. Overview of _Relighting4D_. Given the input video frame at time step t, _Relighting4D_ represents the human as a neural field ψ on latent vectors $\mathcal{Z}$ anchored to a deformable human model. The value of the neural field $\psi_t(x)$ at any 3D point x and time t is taken as latent feature and fed into multilayer perceptrons to obtain geometry and reflectance, which are normal, occlusion, diffuse, and specular maps respectively. Finally, a physically based renderer is raised to render the human subject to the input light probe under novel illumination.

3.1 Neural Field as Human Representation

Extracting 4D representations of dynamic human performers is a non-trivial task. Compared to the static scenes where NeRF [28] fits well, dynamic scenes in videos have factors like motion, occlusion, non-rigid deformation, and illumination that vary through space and time which hampers an accurate estimation.

Inspired by the local implicit representations [10,34], we introduce a 4D neural field ψ conditioned on a parametric human model (SMPL [23] or SMPL-X [33]) to represent a dynamic human performer, which maps the position x and time step t to the latent feature $\psi_t(x)$. Specifically, at frame I_t, we obtain the parameters of human model (i.e. locations of vertices) using this tool [12]. Then, a set of latent vectors $\mathcal{Z} \in \mathbb{R}^{N \times 16}$ is assigned to the vertices of human model, where $N = 6890$ for SMPL [23] and $N = 10475$ for SMPL-X [33]. Then we query the neural field by the 4D coordinates (x, t), extracting the latent feature $\psi_t(x) \in \mathbb{R}^{256}$ from $\mathcal{Z}$ via trilinear interpolation of its nearby vertices.

NeuralBody [34] employs a similar strategy on human representations. But it's not relightable in the way that it fails to disentangle geometry and reflectance from the latent codes. In contrast, _Relighting4D_ learns a distinct neural field that can be decomposed into geometry (Sect. 3.3) and reflectance (Sect. 3.4), which serves the physically based renderer (Sect. 3.2) for relighting.

3.2 Physically Based Rendering

While differentiable volume rendering has been used in recent works [28,34,56], these methods focus on novel view synthesis with radiance fields. In general, to enable relighting with neural representations, instead of modeling the human body as a field of vertices that *emit* light, we represent the human as a field of vertices that *reflect* the light from the environment. Specifically, we leverage a physically based renderer, which models a reflectance-aware rendering process. Mathematically, our rendering pipeline is driven by the following equation:

$$L_o(\boldsymbol{x}, \boldsymbol{\omega}_o) = \int_{\Omega} R(\boldsymbol{x}, \boldsymbol{\omega}_i, \boldsymbol{\omega}_o, \boldsymbol{n}(\boldsymbol{x})) L_i(\boldsymbol{x}, \boldsymbol{\omega}_i)(\boldsymbol{\omega}_i \cdot \boldsymbol{n}(\boldsymbol{x})) d\boldsymbol{\omega}_i, \tag{1}$$

where $L_o(\boldsymbol{x}, \boldsymbol{\omega}_o) \in \mathbb{R}^3$ is the outgoing radiance at point $\boldsymbol{x}$ viewed from $\boldsymbol{\omega}_o$. $L_i(\boldsymbol{x}, \boldsymbol{\omega}_i) \in \mathbb{R}^3$ is the incident radiance arriving at $\boldsymbol{x}$ from direction $\boldsymbol{\omega}_i$. Ω is an unit sphere that models all possible light directions, and $\boldsymbol{n}(\boldsymbol{x}) \in \mathbb{R}^3$ is the normal. $R(\boldsymbol{x}, \boldsymbol{\omega}_i, \boldsymbol{\omega}_o, \boldsymbol{n}(\boldsymbol{x}))^2$ is the Bidirectional Reflectance Distribution Function (BRDF) which defines how the incident light is reflected at the surface, and $d\boldsymbol{\omega}_i$ is the solid angle of incident light at $\boldsymbol{\omega}_i$. We use a discrete set of light samples to approximate Eq. 1 in the following way:

$$L_o(\boldsymbol{x}, \boldsymbol{\omega}_o) \approx \sum_{\boldsymbol{\omega}_i} R(\boldsymbol{x}, \boldsymbol{\omega}_i, \boldsymbol{\omega}_o, \boldsymbol{n}(\boldsymbol{x})) L_i(\boldsymbol{x}, \boldsymbol{\omega}_i)(\boldsymbol{\omega}_i \cdot \boldsymbol{n}(\boldsymbol{x})) \Delta\boldsymbol{\omega}_i, \tag{2}$$

where $\Delta\boldsymbol{\omega}_i$ is sampled from a light probe that depicts the distribution of light sources in space. We represent the environment light $L_i(\boldsymbol{\omega}_i)$ as a light probe image in latitude-longitude format with a resolution of $16 \times 32 \times 3$, which facilitates relighting applications by replacing the estimated light probe with an external one. Figure 3 illustrates our physically based renderer at surface $\boldsymbol{x}$.

Note that previous work [28,34] implicitly encodes $R(\cdot)$ in the radiance fields without modeling the reflectance. To enable flexible relighting applications, we leverage the microfacet model [50] to approximate a differentiable reflectance function parameterized by the surface normal $\boldsymbol{n}(\boldsymbol{x})$, the diffuse map $\boldsymbol{A}(\boldsymbol{x})$ and the specular roughness $\gamma(\boldsymbol{x})$. Due to the limited space, we introduce the implementation of $R(\cdot)$ in the supplementary.

To encode harsh shadow and occlusion, we mask the incident light $L_i(\boldsymbol{x}, \boldsymbol{\omega}_i)$ by the occlusion map $V(\boldsymbol{x}, \boldsymbol{\omega}_i)$ at $\boldsymbol{x}$:

$$L_i(\boldsymbol{x}, \boldsymbol{\omega}_i) = V(\boldsymbol{x}, \boldsymbol{\omega}_i) L_i(\boldsymbol{\omega}_i). \tag{3}$$

Physical Characteristics Disentanglement. Driven by Eq. 2, the renderer requires physical characteristics, i.e., geometry, reflectance, and light, of a given human performer, which are disentangled and estimated by *Relighting4D* from input videos. The details are introduced in the following two sections.

3.3   Volumetric Geometry

In terms of geometry, our renderer requires a normal map $\boldsymbol{n}(\boldsymbol{x}) \in \mathbb{R}^3$ and an occlusion map $V(\boldsymbol{x}, \boldsymbol{\omega}_i) \in \mathbb{R}$ as inputs. Moreover, we render on the surface to keep the computing process tractable, which requires an estimation of surface position. It can be easily obtained by querying a density field.

---

^{2} For simplicity, we also use $R(\cdot)$ to denote BRDF when necessary.

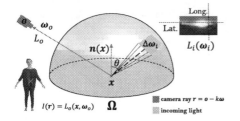 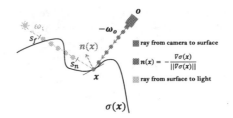

Fig. 3. Illustration of our physically based rendering pipeline. The environment light is represented as spherical coordinates in latitude-longtitude (Lat.-Long.) format. Given the surface location x, the incoming light from ω_i with the area of $\Delta\omega_i$ is scattered by the microfacet that is parameterized by BRDF $R(\cdot)$, normal $n(x)$, and $\cos\theta = \omega_i \cdot n(x)$. Then the outgoing radiance $L_o(x, \omega_o)$ along the ray $r = o - k\omega_o$ is calculated according to Eq. 2, which equals to the corresponding pixel value.

Fig. 4. The process of baking geometry. Note that we perform two different types of ray marching during training. The one is marching along the camera ray $r = o - k\omega_o$ to the expected depth of termination k to get the geometry surface x, while the other is marching from the surface x to the light coming from direction ω_i to calculate the accumulated transmittance (occlusion map).

We first reconstruct the geometry of the given scene using an auxiliary density field $f_\sigma : (x, \psi_t(x)) \rightarrow \sigma(x)$. It's derived from the latent feature $\psi_t(x)$ using an MLP. As shown in Fig. 4, *Relighting4D* leverages the auxiliary density field by baking it into surface maps, normal maps and occlusion maps.

Surface map is the 3D coordinates of points at the expected termination of depth given the camera view ω_o. We march the camera ray r from its origin o along the direction $-\omega_o$ to the expected termination of depth k to get the surface $x = o - k\omega_o$.

Normal map is computed on the surface as the normalized negative gradient of the density field: $\tilde{n}(x) = -\nabla\sigma(x)/||\nabla\sigma(x)||$.

Occlusion map denotes the transmittance of surface points from a specific direction. We compute the occlusion map by marching the ray $r(s, x, \omega_i) = x + s\omega_i$ from the surface of the human body to the corresponding light at ω_i: $\tilde{V}(x, \omega_i) = 1 - exp(-\int_{s_n}^{s_f} \sigma(r(s, x, \omega_i))ds)$, where s_n and s_f is the near and far bounds along the direction of the light. We set $s_n = 0, s_f = 0.5$ for all scenes. In other words, occlusion map considers the visibility at the given surface x by querying the density fields from s_n to s_f along the incident light direction ω_i.

Unfortunately, directly using the baked geometry causes numerous queries of f_σ(e.g., for occlusion map, we should trace $16 \times 32 = 512$ rays from all possible lighting directions for one 3D point), which is not tractable during training and rendering. Thus, we use an MLP $f_n : (x, \psi_t(x)) \rightarrow n(x)$ to reparameterize the surface and latent features to the normal map, and another MLP $f_V : (x, \omega_i, \psi_t(x)) \rightarrow V(x)$ to map the surface, light direction and features to the occlusion map V. The weights of f_V, f_n are trained with the geometry reconstruction loss, intending to recover the baked geometry:

$$\mathcal{L}_{geo} = ||V(\boldsymbol{x}) - \tilde{V}(\boldsymbol{x})||_2^2 + ||\boldsymbol{n}(\boldsymbol{x}) - \tilde{\boldsymbol{n}}(\boldsymbol{x})||_2^2. \tag{4}$$

Smoothness Regularization. We regularize f_V, f_n by L1 penalty to keep the smoothness of their outputs:

$$\tau_V = |V(\boldsymbol{x}) - V(\boldsymbol{x} + \boldsymbol{\epsilon})|_1 \quad \tau_n = |\boldsymbol{n}(\boldsymbol{x}) - \boldsymbol{n}(\boldsymbol{x} + \boldsymbol{\epsilon})|_1, \tag{5}$$

where we measure the local smoothness by adding 3D perturbation $\boldsymbol{\epsilon}$ to $\boldsymbol{x}$ which is sampled from a Gaussian distribution with zero mean and standard deviation 0.01. Several works [29,63] have validated the use of similar smoothness losses for the aim of shape reconstruction.

Temporal Coherence Regularization. It is crucial for a 4D representation to incorporate temporal coherence. Otherwise, the rendered sequence will contain jitter appearance. Moreover, an accurate geometry is also important for artifact-free physically based rendering. Therefore, we add the following regularization term to encourage a temporally smooth geometry:

$$\mathcal{L}_{temp} = \frac{1}{N} \sum_{i=1}^{N} |\sigma_t(\hat{\boldsymbol{x}}_i) - \sigma_{t+1}(\hat{\boldsymbol{x}}_i)|_1, \tag{6}$$

where $\hat{\boldsymbol{x}}_i$ is the 3D position of i-th vertex of SMPL model. Eqn. 6 explicitly constrains the temporal coherence of the geometry, and also implicitly regularize the latent feature $\psi_t(\boldsymbol{x})$ which benefits the following reflectance estimation.

3.4   Reflectance

In terms of reflectance, our physically based renderer requires the BRDF $R(\cdot)$ and the light probe $L_i(\boldsymbol{\omega}_i)$ as inputs. As presented in Sect. 3.2, our BRDF estimation consists of a Lambertian RGB diffuse component $\boldsymbol{A}(\boldsymbol{x}) \in \mathbb{R}^3$ and a specular component $\gamma(\boldsymbol{x}) \in \mathbb{R}$. We parameterize the diffuse map at $\boldsymbol{x}$ with latent features $\psi_t(\boldsymbol{x})$ as an MLP $f_A : (\boldsymbol{x}, \psi_t(\boldsymbol{x})) \rightarrow \boldsymbol{A}(\boldsymbol{x})$, and parameterize the specular map as another MLP $f_\gamma : (\boldsymbol{x}, \psi_t(\boldsymbol{x})) \rightarrow \gamma(\boldsymbol{x})$.

Local Smoothness Prior. The problem that decomposes BRDF from video frames under unknown illumination is highly ill-posed. As the color information is entangled, and there is no off-the-shelf supervision on the reflectance. Inspired by work [3,6,37,59,63] on intrinsic decomposition which leverages piece-wise smoothness prior on albedo, we regularize the optimization of f_A by L1 penalty:

$$\tau_A = |\boldsymbol{A}(\boldsymbol{x}) - \boldsymbol{A}(\boldsymbol{x} + \boldsymbol{\epsilon})|_1, \tag{7}$$

where $\boldsymbol{\epsilon}$ is the same type of perturbation as Eq. 5.

Global Sparsity Prior. However, given this under-constrained problem, the local smoothness regularization in Eq. 7 is not sufficient for a plausible estimation of the diffuse map, as shown in Fig. 7. Thus, we further leverage global minimum-entropy sparsity prior on diffuse map which has been previously explored [2,3,9,38] on shadow removal. From a perspective of physically based rendering, the diffuse map represents the base color, indicating that the palette should be sparse enough. Intuitively, the diffuse map of clothes should contains a small number of colors. Thus, we minimize the Shannon entropy of diffuse map, denoted as H_A, to impose this prior on our model. Since the diffuse map $A(x)$ is a continuous variable whose probability density function (PDF) is unknown, a naive way to estimate its entropy is using histogram to get PDF. But, it's not differentiable. Instead, it's always possible to use a soft and differentiable generalization of Shannon entropy (i.e. quadratic entropy [55]). However, it's quadratically expensive to the number of sampled camera rays.

This motivates our novel approximation of minimizing H_A in a both differentiable and efficient way. The key insight is that the PDF of $A(x)$, $p(A(x))$, can be estimated by a Gaussian KDE (Kernel Density Estimator). Given a diffuse map $A(x)$, we leverage a KDE as its PDF approximation:

$$\tilde{p}(A(x)) = \frac{1}{n} \sum_{i=1}^{n} K_G(A(x) - A_i(x)), \qquad (8)$$

where K_G is the standard normal density function, n is the number of sampled rays during training, and $A_i(x)$ is the value of diffuse map at the i-th camera ray. Then the entropy of $A(x)$ is computed as an expectation:

$$H_A = \mathbb{E}[-\log(\tilde{p}(A(x)))]. \qquad (9)$$

In addition, as the input video is captured under unknown illuminations, we randomly initialize the light probe $L_i(\omega_i)$ as a trainable parameter, optimizing it during the training phase to estimate a plausible ambient light of the scene. It can be replaced by a new HDR map for relighting after training.

3.5   Progressive End-to-End Learning

In the training phase, we randomly sample 1024 camera rays for each input frame. Besides, we employ a progressive training strategy which allows the resolution of video to gradually increase. In specific, before ray sampling, the input video is scaled to the resolution of $\alpha H \times \alpha W$ where $\alpha \in (0,1]$ is a monotonically increasing function of the number of iterations.

Furthermore, we embed the surface position x and the light direction ω_i using the positional encoding [28,48] before concatenating them with latent features $\psi_t(x)$. The maximum frequency of set to 2^{10} and 2^4, respectively. We use four fully-connected ReLU layers with 256 channels for each MLP.

Our full loss function is a summation:

$$\mathcal{L} = \lambda_{rgb}\mathcal{L}_{rgb} + \lambda_{geo}\mathcal{L}_{geo} + \lambda_{temp}\mathcal{L}_{temp} + \lambda_V \tau_V + \lambda_n \tau_n + \lambda_A \tau_A + \lambda_H H_A, \quad (10)$$

where $\mathcal{L}_{rgb}$ is the reconstruction loss against the ground-truth pixel color value. We train each model for 260k iterations with a Tesla V100 GPU. Details of training hyperparameters are deferred to the supplementary.

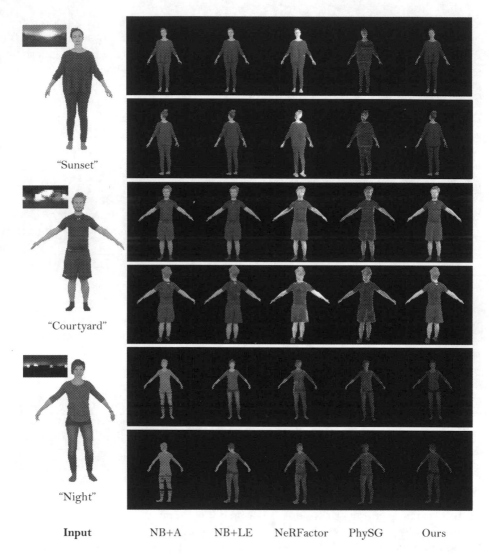

Input   NB+A   NB+LE   NeRFactor   PhySG   Ours

Fig. 5. Free-Viewpoint Relighting on the People-Snapshot Dataset. Two variants of NeuralBody [34] (NB+A and NB+LE) fail to incorporate the lighting in a physical way, thus are unable to reasonably relight the human actor. NB+A learns the wrong mapping between the target light and the appearance. And NB+LE reconstruct the input video well yet fails to generalize to novel lightings. NeRFactor [63] and PhySG [59] seem to model physically correct illuminations but gives blurry results due to the incapability of modeling dynamics. *Relighting4D* significantly outperforms comparison methods. **We show more results in both ambient lighting and OLAT setting in the supplementary videos.**

4   Experiments

Rendering Settings. We render humans in both the ambient lighting and the OLAT setting. For ambient lighting, we use publicly available[2] HDRi maps as light probes. Furthermore, for the OLAT setting, we simulate point lights by generating one-hot light probes given the incoming light directions.

Real Datasets. We validate our method on the People-Snapshot [1] dataset and ZJU-Mocap [34] dataset qualitatively. People-Snapshot [1] captures monocular videos with dynamic performers that keep rotating. And ZJU-Mocap [34] captures dynamic humans with complex motions using a multi-camera system.

Synthetic Dataset. To further demonstrate the effectiveness of *Relighting4D*, we create a dataset, **BlenderHuman**, using the Blender engine [7] for quantitative evaluation. Details will be deferred to the supplementary.

Comparison Methods. We compare *Relighting4D* with several competitive methods. **NeRFactor** [63] requires a pretrained NeRF as a geometry proxy and learns a data-driven BRDF to perform relighting, but it fails to represent dynamic scenes. **PhySG** [59] adopts a spherical Gaussians reflectance model which cannot handle high-frequency lights, and its geometry representation cannot model dynamic scenes. Moreover, to demonstrate the importance of physically based rendering, we implement two variants on top of NeuralBody (NB) [34] which succeeds in novel view synthesis of dynamic humans but fails to incorporate lighting and reflectance. **NB+Ambient Light** (NB+A) uses a flattened light probe as the latent code which contributes to the prediction of its color model, while **NB+Learnable Embedding** (NB+LE) maps the light probe into a latent code using an MLP with two layers.

Evaluation Metrics. For quantitative analysis, we use Peak Signal-to-Noise Ratio (PSNR), Structural Similarity Index Measure [52] (SSIM), and Learned Perceptual Image Patch Similarity [61] (LPIPS) as metrics. In addition, we use the error of degree($°$) to measure the normal map estimations.

4.1   Results on Real Datasets

Performance on Relighting with Novel Views. Figure 5 shows qualitative results on People-Snapshot dataset. All methods train a separate model for each human performer and re-render the human given the input light probes. Two variants of NeuralBody [34], NB+A and NB+LE, are good at reconstructing appearance but fail to incorporate novel illuminations in a perceptually salient way. They fail to learn the underlying physics of rendering. For example, NB+A maps the input light probe to artifacts of texture while NB+LE even seems to discard the features from lightings. NeRFactor [63] and PhySG [59] give blurry results, which show that they cannot aggregate space-time varying geometry and reflectance of dynamic humans, leading to degraded rendering results. In contrast, our method generates photorealistic relit novel views.

---

[2] https://polyhaven.com/.

Input

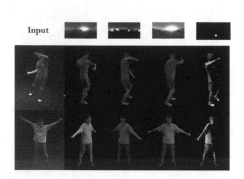

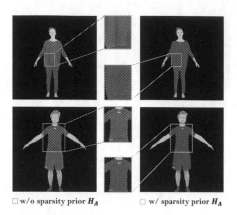

☐ w/o sparsity prior H_A ☐ w/ sparsity prior H_A

Fig. 6. Relighting of dynamic humans with complex motions on ZJU-Mocap dataset. *Relighting4D* renders high-fidelity human actors with time-varying poses under novel illuminations. **Please check the supplementary videos for more results.**

Fig. 7. Here we visualize how our model benefits from incorporating minimum-entropy sparsity prior by minimizing H_A. Without this prior, the estimation of diffuse map would suffer from shadow residuals as shown on the left side.

We also present our qualitative results in the challenging OLAT setting. Since the point light comes from only one direction, these OLAT illuminations induce hard cast shadows, effectively revealing rendering artifacts due to inaccurate geometry and materials. *Relighting4D* synthesizes shadows cast by limbs and clothes in a physically correct way. Please refer to the supplementary for details.

We demonstrate that our method is capable of relighting dynamic humans with complex motions from multi-views videos on the ZJU-Mocap dataset [34]. Figure 6 shows our qualitative results on the "Twirl" and "Swing" scenes.

Decomposition of Geometry and Reflectance. We demonstrate that our method is able to extract geometry and reflectance representations from the input videos and disentangle them into surface normals, diffuse maps and occlusion maps, which may facilitate downstream graphics tasks. The visualizations on People-Snapshot videos are presented in Fig. 10, and quantitative results on the BlenderHuman dataset are shown in Table 1. Note that, we directly take the albedo generated by NeRFactor [63] and PhySG [59] as their diffuse maps for comparisons. However, NeRFactor [63] estimates the diffuse map of the dynamic human with incorrect base color and facial details, and fails to capture the accurate geometry of dynamic humans. Though PhySG [59] captures the correct base color of clothes, due to its incapability of handling dynamic scenes, the facial details of diffuse map remains artifacts when the viewpoint changes. With the latent representation of human body, *Relighting4D* can integrate geometry information through space and time, successfully capturing the fine-grained details of the normal map and the correct color of diffuse map.

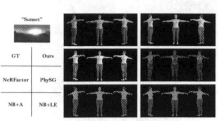
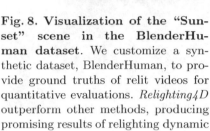

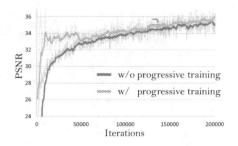

Fig. 8. Visualization of the "Sunset" scene in the BlenderHuman dataset. We customize a synthetic dataset, BlenderHuman, to provide ground truths of relit videos for quantitative evaluations. *Relighting4D* outperform other methods, producing promising results of relighting dynamic human under novel illuminations.

Fig. 9. The PSNR v.s. training iterations on People-Snapshot dataset. Progressive training helps the model reconstruct the scene faster and better. When trained with a constant spatial resolution, the reconstruction error falls into sub-optimal at the end (PSNR drops from 36.65 to 34.56).

Table 1. Results on the BlenderHuman Dataset. The top two techniques for each metric are highlighted in red and orange respectively. The reported numbers are the arithmetic averages of 16 different scenes. We relight the human actor with 8 HDR ambient light probes and 8 OLAT conditions. *Relighting4D* achieves the best overall performance across all metrics.

Method	Relighting			Normal Map	Diffuse Map		
	PSNR ↑	SSIM ↑	LPIPS ↓	Degree° ↓	PSNR ↑	SSIM ↑	LPIPS ↓
NB [34]+A	20.9348	0.8559	0.2368	-	-	-	-
NB [34]+LE	22.7957	0.8721	0.2145	-	-	-	-
NeRFactor [63]	22.8037	0.8830	0.2045	43.7012	27.0585	0.9202	0.1929
PhySG [59]	23.8810	0.8427	0.2959	50.5721	28.0852	0.9350	0.1810
Ours	26.1475	0.9118	0.1639	32.1803	28.9517	0.9279	0.1502

4.2 Results on Synthetic Dataset

We quantitatively evaluate comparison methods on the BlenderHuman dataset. We show results in this section and defer the visualizations in the supplementary.

Table 1 shows the results on the BlenderHuman dataset. Overall, our model achieves the best performance. The results indicate that *Relighting4D* better handles the dynamics across video frames while feasibly modeling the light transport to relight dynamic humans.

4.3 Ablation Studies

We conduct ablation studies on BlenderHuman dataset, as presented in Table 2.

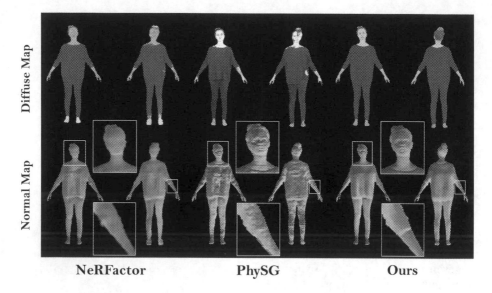

Diffuse Map

Normal Map

NeRFactor   PhySG   Ours

Fig. 10. Comparisons of geometry and reflectance decomposition. *Relighting4D* is able to estimate fine-grained details of geometry and physically correct reflectance. We defer the visualization of the occlusion map in the supplementary.

Impact of the Human Representation. We train our model without $\psi_t(x)$ that is proposed in Sect. 3.1. In other words, the geometry and reflectance MLPs take only the surface coordinates of the human body as inputs. The result indicates that incorporating our latent feature $\psi_t(x)$ is crucial for relighting dynamic human videos. The incapability of modeling scene dynamics leads to significant performance drops in terms of relighting quality and inverse rendering quality, which explains why NeRFactor [63] fails so badly.

Effectiveness of the smoothness regularization is validated in Table 2. We train our model without τ_V, τ_n, which leads to the decreased rendering quality.

Impact of the Baked Geometry. We train *Relighting4D* without the supervision of the baked geometry. The results indicate that the baked geometry improves the relighting performance. It reveals that poor renderings are caused by the inaccurate geometry as the error of normals drops by a large margin.

Effectiveness of the global sparsity prior on diffuse map is validated in Table 2. Without minimizing the entropy of the diffuse map, the relighting quality is perceptually decreased due to the degraded inverse rendering, which induces shadows in the estimation of diffuse map, as shown in Fig. 7.

Impact of Progressive Training. Without progressive spatial resolutions during training, the relighting quality decreases as shown in the last row of Table 2. We believe the progressive strategy helps the model quickly learn coarse geometry in the early training phase, which is even validated on real datasets (Fig. 9).

We plot the reconstruction error (PSNR) versus iterations on one training scene in People-Snapshot dataset, discovering that progressive training helps model reconstruct the given scene faster and better.

Table 2. Ablation Studies on the BlenderHuman Dataset. We take the average metrics on all 16 scenes. The top three techniques for each metric are highlighted in red, orange, and yellow respectively.

Method	Relighting			Normal Map	Diffuse Map		
	PSNR ↑	SSIM ↑	LPIPS ↓	Degree° ↓	PSNR ↑	SSIM ↑	LPIPS ↓
Full model	26.1475	0.9118	0.1639	32.1803	28.9517	0.9279	0.1502
w/o $\psi_t(x)$	21.1163	0.8407	0.2372	36.5699	25.6806	0.9008	0.1896
w/o τ_V, τ_n	25.3504	0.8800	0.2061	32.9243	28.3329	0.9224	0.1660
w/o $\tilde{V}, \tilde{n}$	22.4221	0.8559	0.2285	57.0452	27.6652	0.9165	0.1425
w/o H_A	27.7545	0.9042	0.1717	30.6685	24.0195	0.8950	0.1767
w/o progressive	25.5562	0.9031	0.1742	30.1662	24.2455	0.8958	0.1760

5   Discussion and Conclusion

Limitations. We have demonstrated the capability of *Relighting4D* on relighting dynamic humans with free viewpoints. Nevertheless, there are a few limitations. First, for tractable training and rendering, we consider only the one-bounce direct environment light, thus our method cannot relight furry appearances. Second, as we leverage a fully physically based renderer, the rendering quality is tied with the accuracy of geometry. Dense scenes with multiple people, which may negatively impact the estimation of geometry, will lead to poor performance. Finally, if the texture patterns are complicated or the lighting is harsh during training, the decomposition of reflectance and geometry is hard to solve due to the ambiguity of color scale, causing poor relighting quality. It can be alleviated by incorporating more information other than self-supervision from videos into the network (e.g., other supervision signals or data-driven priors).

In this paper, we present a principled rendering scheme called *Relighting4D*, a method that enables relighting with free viewpoints from only posed human videos under unknown illuminations. Our method exploits the physically based rendering pipeline and decomposes the appearance of humans into geometry and reflectance. All components are parameterized by MLPs based on the neural field conditioned on the deformable human model. Extensive experiments on synthetic and real datasets demonstrate that *Relighting4D* is capable of high-quality relighting of dynamic human performers with free viewpoints.

Acknowledgements. This work is supported by the National Research Foundation, Singapore under its AI Singapore Programme (AISG Award No: AISG2-PhD-2021-08-019), NTU NAP, MOE AcRF Tier 2 (T2EP20221-0033), and under the RIE2020 Industry Alignment Fund - Industry Collaboration Projects (IAF-ICP) Funding Initiative, as well as cash and in-kind contribution from the industry partner(s).

References

1. Alldieck, T., Magnor, M., Xu, W., Theobalt, C., Pons-Moll, G.: Video based reconstruction of 3D people models. In: 2018 IEEE/CVF Conference on Computer Vision and Pattern Recognition, pp. 8387–8397. IEEE (2018). https://doi.org/10.1109/CVPR.2018.00875, https://ieeexplore.ieee.org/document/8578973/
2. Alldrin, N.G., Mallick, S.P., Kriegman, D.J.: Resolving the generalized bas-relief ambiguity by entropy minimization. In: 2007 IEEE Conference on Computer Vision and Pattern Recognition.,pp. 1–7 (2007). https://doi.org/10.1109/CVPR.2007.383208
3. Barron, J.T., Malik, J.: Shape, albedo, and illumination from a single image of an unknown object. In: 2012 IEEE Conference on Computer Vision and Pattern Recognition, pp. 334–341. IEEE (2012). https://doi.org/10.1109/CVPR.2012.6247693, https://ieeexplore.ieee.org/document/6247693/
4. Bi, S., et al.: Deep relightable appearance models for animatable faces. ACM Trans. Graph. **40**(4), 1–15 (2021). https://doi.org/10.1145/3476576.3476647, https://dl.acm.org/doi/10.1145/3476576.3476647
5. Bi, S., et al.: Neural reflectance fields for appearance acquisition. arXiv:2008.03824 [cs] (2020)
6. Boss, M., Braun, R., Jampani, V., Barron, J.T., Liu, C., Lensch, H.P.A.: NeRD: neural reflectance decomposition from image collections. arXiv:2012.03918 [cs] (2021)
7. Community, B.O.: Blender - a 3D modelling and rendering package. Blender Foundation (2018). http://www.blender.org
8. Debevec, P., Hawkins, T., Tchou, C., Duiker, H.P., Sarokin, W.: Acquiring the Reflectance Field of a Human Face. In: SIGGRAPH (2000)
9. Finlayson, G.D., Drew, M.S., Lu, C.: Entropy minimization for shadow removal. Int. J. Comput. Vis. **85**(1), 35–57 (2009). https://doi.org/10.1007/s11263-009-0243-z
10. Genova, K., Cole, F., Sud, A., Sarna, A., Funkhouser, T.: Local deep implicit functions for 3D Shape. In: 2020 IEEE/CVF Conference on Computer Vision and Pattern Recognition (CVPR), pp. 4856–4865. IEEE (2020). https://doi.org/10.1109/CVPR42600.2020.00491, https://ieeexplore.ieee.org/document/9157823/
11. Guo, K., et al.: The relightables: volumetric performance capture of humans with realistic relighting. ACM Trans. Graph. **38**(6), 1–19 (2019) https://doi.org/10.1145/3355089.3356571, https://dl.acm.org/doi/10.1145/3355089.3356571
12. Joo, H., Simon, T., Sheikh, Y.: Total capture: a 3D deformation model for tracking faces, hands, and bodies. In: 2018 IEEE/CVF Conference on Computer Vision and Pattern Recognition, pp. 8320–8329 (2018). https://doi.org/10.1109/CVPR.2018.00868
13. Kanamori, Y., Endo, Y.: Relighting humans: occlusion-aware inverse rendering for full-body human images. ACM Trans. Graph. **37**(6), 1–11 (2019). https://doi.org/10.1145/3272127.3275104, https://arxiv.org/abs/1908.02714
14. Kwon, Y., Kim, D., Ceylan, D., Fuchs, H.: Neural Human Performer: Learning Generalizable Radiance Fields for Human Performance Rendering. arXiv:2109.07448 [cs] (2021)
15. Lagunas, M., et al.: Single-image Full-body human relighting. arXiv:2107.07259 [cs] (2021). https://doi.org/10.2312/sr.20211300

16. LeGendre, C., et al.: DeepLight: learning illumination for unconstrained mobile mixed reality. CoRR abs/1904.01175 (2019), http://arxiv.org/abs/1904.01175
17. LeGendre, C., et al.: Learning illumination from diverse portraits. In: SIGGRAPH Asia 2020 Technical Communications. SA 2020, Association for Computing Machinery (2020). https://doi.org/10.1145/3410700.3425432
18. Li, J., Feng, Z., She, Q., Ding, H., Wang, C., Lee, G.H.: Mine: towards continuous depth MPI with nerf for novel view synthesis. In: ICCV (2021)
19. Li, Z., Niklaus, S., Snavely, N., Wang, O.: Neural scene flow fields for space-time view synthesis of dynamic scenes. arXiv:2011.13084 [cs] (2021)
20. Liu, L., Habermann, M., Rudnev, V., Sarkar, K., Gu, J., Theobalt, C.: Neural actor: neural free-view synthesis of human actors with pose control. arXiv:2106.02019 [cs] (2021)
21. Liu, Y., Neophytou, A., Sengupta, S., Sommerlade, E.: Relighting images in the wild with a self-supervised siamese auto-encoder. In: 2021 IEEE Winter Conference on Applications of Computer Vision (WaACV), pp. 32–40. IEEE (2021). https://doi.org/10.1109/WACV48630.2021.00008, https://ieeexplore.ieee.org/document/9423347/
22. Liu, Y., Li, Y., You, S., Lu, F.: Unsupervised learning for intrinsic image decomposition from a single image. arXiv:1911.09930 [cs] (2020)
23. Loper, M., Mahmood, N., Romero, J., Pons-Moll, G., Black, M.J.: SMPL: a skinned multi-person linear model. ACM Trans. Graphics (Proc. SIGGRAPH Asia) 34(6), 248:1–248:16 (2015)
24. Martin-Brualla, R., Radwan, N., Sajjadi, M.S.M., Barron, J.T., Dosovitskiy, A., Duckworth, D.: NeRF in the Wild: Neural radiance fields for unconstrained photo collections. In: CVPR (2021)
25. Meka, A., et al.: Deep reflectance fields: high-quality facial reflectance field inference from color gradient illumination. ACM Trans. Graph. 38(4), 1–12 (2019). https://doi.org/10.1145/3306346.3323027, https://dl.acm.org/doi/10.1145/3306346.3323027
26. Meka, A., et al.: Deep relightable textures: volumetric performance capture with neural rendering. ACM Trans. Graph. 39(6), 1–21 (2020)
27. Meka, A., Shafiei, M., Zollhoefer, M., Richardt, C., Theobalt, C.: Real-time global illumination decomposition of videos. ACM Trans. Graph. 40(3), 1–16 (2021). https://doi.org/10.1145/3374753, http://arxiv.org/abs/1908.01961
28. Mildenhall, B., Srinivasan, P.P., Tancik, M., Barron, J.T., Ramamoorthi, R., Ng, R.: NeRF: representing Scenes as neural radiance fields for view synthesis. arXiv:2003.08934 [cs] (2020)
29. Oechsle, M., Peng, S., Geiger, A.: UNISURF: unifying neural implicit surfaces and radiance fields for multi-view reconstruction. arXiv:2104.10078 [cs] (2021)
30. Ost, J., Mannan, F., Thuerey, N., Knodt, J., Heide, F.: Neural Scene Graphs for Dynamic Scenes. arXiv:2011.10379 [cs] (2021)
31. Pandey, R., et al.: Total relighting: learning to relight portraits for background replacement. ACM Trans. Graph. 40(4), 1–21 (2021). https://doi.org/10.1145/3476576.3476588, https://dl.acm.org/doi/10.1145/3476576.3476588
32. Park, K., et al.: Nerfies: deformable neural radiance fields. arXiv:2011.12948 [cs] (2021)
33. Pavlakos, G., et al.: Expressive body capture: 3D hands, face, and body from a single image. In: Proceedings IEEE Conference on Computer Vision and Pattern Recognition (CVPR), pp. 10975–10985 (2019)
34. Peng, S., et al.: Neural body: implicit neural representations with structured latent codes for novel view synthesis of dynamic humans. arXiv:2012.15838 [cs] (2021)

35. Pumarola, A., Corona, E., Pons-Moll, G., Moreno-Noguer, F.: D-NeRF: neural radiance fields for dynamic scenes. arXiv:2011.13961 [cs] (2020)
36. Raj, A., Tanke, J., Hays, J., Vo, M., Stoll, C., Lassner, C.: ANR: articulated neural rendering for virtual avatars. In: Proceedings of the IEEE/CVF Conference on Computer Vision and Pattern Recognition (CVPR), pp. 3722–3731 (2021)
37. Sengupta, S., Gu, J., Kim, K., Liu, G., Jacobs, D., Kautz, J.: Neural Inverse rendering of an indoor scene from a single image. In: 2019 IEEE/CVF International Conference on Computer Vision (ICCV), pp. 8597–8606. IEEE (2019). https://doi. org/10.1109/ICCV.2019.00869, https://ieeexplore.ieee.org/document/9008823/
38. Shen, L., Yeo, C.: Intrinsic images decomposition using a local and global sparse representation of reflectance. In: CVPR 2011, pp. 697–704 (2011). https://doi.org/ 10.1109/CVPR.2011.5995738
39. Shu, Z., Yumer, E., Hadap, S., Sunkavalli, K., Shechtman, E., Samaras, D.: Neural face editing with intrinsic image disentangling. In: 2017 IEEE Conference on Computer Vision and Pattern Recognition (CVPR), pp. 5444–5453. IEEE (2017). https://doi.org/10.1109/CVPR.2017.578, https://ieeexplore. ieee.org/document/8100061/
40. Sitzmann, V., Martel, J., Bergman, A., Lindell, D., Wetzstein, G.: Implicit neural representations with periodic activation functions. In: Advances in Neural Information Processing Systems, vol. 33, pp. 7462–7473. Curran Associates, Inc. (2020). https://proceedings.neurips.cc/paper/2020/hash/ 53c04118df112c13a8c34b38343b9c10-Abstract.html
41. Sitzmann, V., Zollhöfer, M., Wetzstein, G.: Scene representation networks: continuous 3D-structure-aware neural scene representations. arXiv:1906.01618 [cs] (2020)
42. Srinivasan, P.P., Deng, B., Zhang, X., Tancik, M., Mildenhall, B., Barron, J.T.: NeRV: neural reflectance and visibility fields for relighting and view synthesis. arXiv:2012.03927 [cs] (2020)
43. Sun, G., et al.: Neural free-viewpoint performance rendering under complex human-object interactions. arXiv:2108.00362 [cs] (2021)
44. Sun, T., et al.: Single Image Portrait Relighting. ACM Trans. Graph. **38**(4), 1–12 (2019). —DOIurl10.1145/3306346.3323008, https://arxiv.org/abs/1905.00824
45. Sun, T., Lin, K., Bi, S., Xu, Z., Ramamoorthi, R.: Nelf: Neural light-transport field for portrait view synthesis and relighting. CoRR abs/2107.12351 (2021). https:// arxiv.org/abs/2107.12351
46. Suo, X., et al.: NeuralHumanFVV: real-time neural volumetric human performance rendering using RGB cameras. arXiv:2103.07700 [cs] (2021)
47. Tajima, D., Kanamori, Y., Endo, Y.: Relighting humans in the wild: Monocular full-body human relighting with domain adaptation (2021)
48. Tancik, M., et al.: Fourier features let networks learn high frequency functions in low dimensional domains. arXiv:2006.10739 [cs] (2020)
49. Tucker, R., Snavely, N.: Single-view view synthesis with multiplane images. In: 2020 IEEE/CVF Conference on Computer Vision and Pattern Recognition (CVPR). pp. 548–557. IEEE (2020). https://doi.org/10.1109/CVPR42600.2020.00063, https:// ieeexplore.ieee.org/document/9156372/
50. Walter, B., Marschner, S.R., Li, H., Torrance, K.E.: Microfacet models for refraction through rough surfaces. In: Proceedings of the 18th Eurographics Conference on Rendering Techniques, pp. 195–206. EGSR 2007, Eurographics Association (2007). https://doi.org/10.2312/EGWR/EGSR07/195-206
51. Wang, Z., Yu, X., Lu, M., Wang, Q., Qian, C., Xu, F.: Single image portrait relighting via explicit multiple reflectance channel modeling. ACM Trans. Graph.

39(6), 1–13 (2020). https://doi.org/10.1145/3414685.3417824, https://dl.acm.org/doi/10.1145/3414685.3417824

52. Wang, Z., Bovik, A., Sheikh, H., Simoncelli, E.: Image quality assessment: from error visibility to structural similarity. IEEE Trans. Image Process. **13**(4), 600–612 (2004). https://doi.org/10.1109/TIP.2003.819861

53. Wang, Z., Philion, J., Fidler, S., Kautz, J.: Learning indoor inverse rendering with 3D spatially-varying lighting. In: Proceedings of International Conference on Computer Vision (ICCV) (2021)

54. Xian, W., Huang, J.B., Kopf, J., Kim, C.: Space-time neural irradiance fields for free-viewpoint video. arXiv:2011.12950 [cs] (2021)

55. Xu, D., Principe, J.: Learning from examples with quadratic mutual information. In: Neural Networks for Signal Processing VIII. Proceedings of the 1998 IEEE Signal Processing Society Workshop (Cat. No.98TH8378), pp. 155–164 (1998). https://doi.org/10.1109/NNSP.1998.710645

56. Yu, A., Ye, V., Tancik, M., Kanazawa, A.: pixelNeRF: neural radiance fields from one or few images (2020)

57. Yu, Y., Smith, W.A.P.: InverseRenderNet: Learning single image inverse rendering. arXiv:1811.12328 [cs] (2018)

58. Zhang, J., et al.: Editable free-viewpoint video using a layered neural representation. ACM Trans. Graph. **40**(4), 1–18 (2021). https://doi.org/10.1145/3450626.3459756, https://arxiv.org/abs/2104.14786

59. Zhang, K., Luan, F., Wang, Q., Bala, K., Snavely, N.: PhySG: inverse rendering with spherical gaussians for physics-based material editing and relighting. arXiv:2104.00674 [cs] (2021)

60. Zhang, L., Zhang, Q., Wu, M., Yu, J., Xu, L.: Neural video portrait relighting in real-time via consistency modeling. arXiv:2104.00484 [cs] (2021)

61. Zhang, R., Isola, P., Efros, A.A., Shechtman, E., Wang, O.: The unreasonable effectiveness of deep features as a perceptual metric. In: CVPR (2018)

62. Zhang, X., et al.: Neural light transport for relighting and view synthesis. ACM Trans. Graph. **40**(1), 1–17 (2021). https://dl.acm.org/doi/10.1145/3446328

63. Zhang, X., Srinivasan, P.P., Deng, B., Debevec, P., Freeman, W.T., Barron, J.T.: Nerfactor. ACM Trans. Graph. **40**(6), 1–18 (2021). https://doi.org/10.1145/3478513.3480496, https://dx.doi.org/10.1145/3478513.3480496

64. Zhou, H., Hadap, S., Sunkavalli, K., Jacobs, D.: Deep single-image portrait relighting. In: 2019 IEEE/CVF International Conference on Computer Vision (ICCV), pp. 7193–7201. IEEE (2019). https://doi.org/10.1109/ICCV.2019.00729, https://ieeexplore.ieee.org/document/9010718/

Real-Time Intermediate Flow Estimation for Video Frame Interpolation

Zhewei Huang[1], Tianyuan Zhang[1], Wen Heng[1], Boxin Shi[2,3,4(✉)], and Shuchang Zhou[1(✉)]

[1] Megvii Research, Beijing, China
{huangzhewei,zhangtianyuan,hengwen,zsc}@megvii.com
[2] NERCVT, School of Computer Science, Peking University, Beijing, China
shiboxin@pku.edu.cn
[3] Institute for Artificial Intelligence, Peking University, Beijing, China
[4] Beijing Academy of Artificial Intelligence, Beijing, China
https://github.com/megvii-research/ECCV2022-RIFE

Abstract. Real-time video frame interpolation (VFI) is very useful in video processing, media players, and display devices. We propose RIFE, a Real-time Intermediate Flow Estimation algorithm for VFI. To realize a high-quality flow-based VFI method, RIFE uses a neural network named IFNet that can estimate the intermediate flows end-to-end with much faster speed. A privileged distillation scheme is designed for stable IFNet training and improve the overall performance. RIFE does not rely on pre-trained optical flow models and can support arbitrary-timestep frame interpolation with the temporal encoding input. Experiments demonstrate that RIFE achieves state-of-the-art performance on several public benchmarks. Compared with the popular SuperSlomo and DAIN methods, RIFE is 4–27 times faster and produces better results. Furthermore, RIFE can be extended to wider applications thanks to temporal encoding. https://github.com/megvii-research/ECCV2022-RIFE

1 Introduction

Video Frame Interpolation (VFI) aims to synthesize intermediate frames between two consecutive video frames. VFI supports various applications like slow-motion generation, video compression [56], and video frame predition [57]. Moreover, real-time VFI methods running on high-resolution videos have many potential applications, such as reducing bandwidth requirements for live video streaming, providing video editing services for users with limited computing resources, and video frame rate adaption on display devices.

Supplementary Information The online version contains supplementary material available at https://doi.org/10.1007/978-3-031-19781-9_36.

VFI is challenging due to the complex, non-linear motions and illumination changes in real-world videos. Recently, flow-based VFI algorithms have offered a framework to address these challenges and achieved impressive results [3,22, 28,30,40,60,61]. Common approaches for these methods involve two steps: 1) warping the input frames according to approximated optical flows and 2) fusing the warped frames using Convolutional Neural Networks (CNNs).

Optical flow models can not be directly used in VFI. Given the input frames I_0, I_1, flow-based methods [3,22,30] need to approximate the intermediate flows $F_{t \to 0}, F_{t \to 1}$ from the perspective of the frame I_t that we are expected to synthesize. There is a "chicken-and-egg" problem between intermediate flows and frames because I_t is not available beforehand, and its estimation is a difficult problem [22,44]. Many practices [3,22,28,60] first compute bi-directional flows from optical flow models, then reverse and refine them to generate intermediate flows. However, such flows may have flaws in motion boundaries, as the object position changes from frame to frame ("object shift" problem). Appearance Flow [65], A pioneering work in view synthesis, proposes to estimate flow starting from the target view using CNNs. DVF [30] extend it to the voxel flow of dynamic scenes to jointly model the intermediate flow and blend mask to estimate them end-to-end. AdaCoF [27] further extends intermediate flows to adaptive collaborative flows. BMBC [44] designs a bilateral cost volume operator for obtaining more accurate intermediate flows (bilateral motion). In this paper, we aim to build a lightweight pipeline that achieves state-of-the-art (SOTA) performance while maintaining the conciseness of direct intermediate flow estimation. Our pipeline has these main design concepts:

1) Not requiring additional components, like image depth model [3], flow refinement model [22] and flow reversal layer [60], which are introduced to compensate for the defects of intermediate flow estimation. We also want to eliminate reliance on pre-trained SOTA optical flow models that are not tailored for VFI tasks.
2) End-to-end learnable motion estimation: we demonstrate experimentally that instead of introducing some inaccurate motion modeling, it is better to make the CNN learn the intermediate flow end-to-end. This methodology has been proposed [30]. However, the follow-up works do not fully inherit this idea.
3) Providing direct supervision for the approximated intermediate flows: most VFI models are trained with only the final reconstruction loss. Intuitively, propagating gradients of pixel-wise loss across warping operator is not efficient for flow estimation [11,35,37]. Lacking supervision explicitly designed for flow estimation degrades the performance of VFI models.

We propose IFNet, which directly estimates intermediate flow from adjacent frames and a temporal encoding input. IFNet adopts a coarse-to-fine strategy [20] with progressively increasing resolution: it iteratively updates the intermediate flows and soft fusion mask via successive IFBlocks. Intuitively, according to the iteratively updated flow fields, we could move corresponding pixels from two input frames to the same location in a latent intermediate frame and use a fusion mask to combine pixels from two input frames. To make our model efficient, unlike most previous optical flow models [15,19,20,53,55], IFBlocks do not contain expensive operators like cost volume and only use 3×3 convolution and deconvolution as building blocks, which are suitable for resource-constrained devices. Furthermore, plain Conv is highly supported by NPU embedded in display devices and provides convenience for customized requirements. Thanks related researchers for the exploration of efficient models [14,36,47].

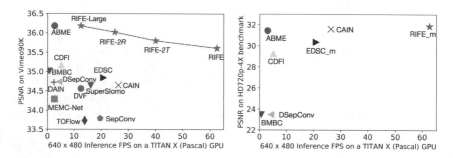

Fig. 1. Performance comparison. Results are reported for Vimeo90K [61] and HD-4× [3] benchmark. More details are in the experimental section

Employing intermediate supervision is very important. When training the IFNet end-to-end using the final reconstruction loss, our method produces worse results than SOTA methods because of the inaccurate optical flow estimation. The situation dramatically changes after we design a privileged distillation scheme that employs a teacher model with access to the intermediate frames to guide the student to learn.

Combining these designs, we propose the Real-time Intermediate Flow Estimation (**RIFE**). RIFE trained from scratch can achieve satisfactory results, without requiring pre-trained models or datasets with optical flow labels. We illustrate the RIFE's performance compared with other methods in Fig. 1.

To sum up, our main contributions include:

- We design an effective IFNet to approximate the intermediate flows and introduce a privileged distillation scheme to improve the performance.
- Our experiments demonstrate that RIFE achieves SOTA performance on several public benchmarks, especially in the scene of arbitrary-time frame interpolation.
- We show RIFE can be extended to applications such as depth map interpolation and dynamic scene stitching, thanks to its flexible temporal encoding.

2   Related Works

Optical Flow Estimation. Optical flow estimation is a long-standing vision task that aims to estimate the per-pixel motion, useful in many downstream

tasks [33,54,63,64]. Since the milestone work of FlowNet [15], flow model architectures have evolved for several years, yielding more accurate results while being
more efficient, such as FlowNet2.0 [20], PWC-Net [53] and LiteFlowNet [19].
Recently Teed et al. [55] introduce RAFT, which iteratively updates a flow
field through a recurrent unit and achieves a remarkable breakthrough in this
field. Another important research direction is unsupervised optical flow estimation [23,35,37] which tackles the difficulty of labeling.

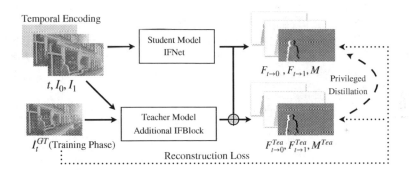

Fig. 2. Overview of RIFE pipeline. Given two input frames I_0, I_1 and temporal
encoding t (timestep encoded as an separate channel [18,39]), we directly feed them
into the IFNet to approximate intermediate flows $F_{t\rightarrow 0}, F_{t\rightarrow 1}$ and the fusion map M.
During the training phase, a privileged teacher refines student's results based on ground
truth I_t using a special IFBlock

Video Frame Interpolation. Recently, optical flow has been a prevalent component in video interpolation. In addition to the method of directly estimating
the intermediate flow [27,30,44], Jiang et al. [22] propose SuperSlomo using the
linear combination of the two bi-directional flows as an initial approximation
of the intermediate flows and then refining them using U-Net. Reda et al. [49]
and Liu et al. [29] propose to improve intermediate frames using cycle consistency. Bao et al. [3] propose DAIN to estimate the intermediate flow as a
weighted combination of bidirectional flow. Niklaus et al. [41] propose SoftSplat
to forward-warp frames and their feature map using softmax splatting. Xu et
al. [60] propose QVI to exploit four consecutive frames and flow reversal filter
to get the intermediate flows. Liu et al. [28] further extend QVI with rectified
quadratic flow prediction to EQVI.

Along with flow-based methods, flow-free methods have also achieved remarkable progress. Meyer et al. [38] utilize phase information to learn the motion
relationship for multiple video frame interpolation. Niklaus et al. [43] formulate
VFI as a spatially adaptive convolution whose convolution kernel is generated
using a CNN given the input frames. Cheng et al. propose DSepConv [8] to
extend kernel-based method using deformable separable convolution and. Choi et
al. [10] propose an efficient flow-free method named CAIN, which employs the
PixelShuffle operator and channel attention to capture the motion information

implicitly. Some work further focus on increasing the resolution and frame rate of the video together and has achieved good visual effect [58,59]. In addition, large-motion and animation frame interpolation is also fields of great interest [6,48,51]. **Knowledge Distillation.** Our privileged distillation [31] for intermediate flow conceptually belongs to the knowledge distillation [17], which originally aims to transfer knowledge from a large model to a smaller one. In privileged distillation, the teacher model gets more input than the student model, such as scene depth, images from other views, and even image annotation. Therefore, the teacher model can provide more accurate representations to guide the student model to learn. This idea is applied to some computer vision tasks, such as hand pose estimation [62], re-identification [45] and video style transfer [7]. Our work is also related to codistillation [1] where the student and teacher have the same architecture and different inputs during training.

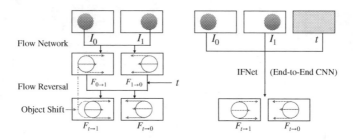

Fig. 3. Compare indirect intermediate flow estimation [3,22,60] (left) with IFNet (right). As the object shifts, flow reversal modules may have flaws in motion boundaries. Rather than hand-engineering flow reversal layers, CNNs can learn intermediate flow estimates end-to-end

3   Method

3.1   Pipeline Overview

We illustrate the overall pipeline of RIFE in Fig. 2. Given a pair of consecutive RGB frames, I_0, I_1 and target timestep t ($0 \leq t \leq 1$), our goal is to synthesize an intermediate frame $\widehat{I}_t$. We estimate the intermediate flows $F_{t \to 0}$, $F_{t \to 1}$ and fusion map M by feeding input frames and t as an additional channel into the IFNet. We can get reconstructed image $\widehat{I}_t$ using following formulation:

$$\widehat{I}_t = M \odot \widehat{I}_{t \leftarrow 0} + (1 - M) \odot \widehat{I}_{t \leftarrow 1}, \tag{1}$$

$$\widehat{I}_{t \leftarrow 0} = \overleftarrow{\mathcal{W}}(I_0, F_{t \to 0}), \quad \widehat{I}_{t \leftarrow 1} = \overleftarrow{\mathcal{W}}(I_1, F_{t \to 1}). \tag{2}$$

where $\overleftarrow{\mathcal{W}}$ is the image backward warping, $\odot$ is an element-wise multiplier, and M is the fusion map ($0 \leq M \leq 1$). We use another encoder-decoder CNNs named RefineNet following previous methods [22, 41] to refine the high-frequency area of $\widehat{I}_t$ and reduce artifacts of the student model. Its computational cost is similar to the IFNet. The RefineNet finally produce a reconstruction residual Δ ($-1 \leq \Delta \leq 1$). And we will get a refined reconstructed image $\widehat{I}_t + \Delta$. The detailed architecture of RefineNet is in the **Appendix**.

3.2   Intermediate Flow Estimation

Some previous VFI methods reverse and refine bi-directional flows [3,22,28,60] as depicted in Fig. 3. The flow reversal process is usually cumbersome due to the difficulty of handling the changes of object positions. Intuitively, the previous flow reversal method hopes to perform spatial interpolation on the optical flow field, which is not trivial because of the "object shift" problem. The role of our IFNet is to directly and efficiently predict $F_{t\rightarrow0}, F_{t\rightarrow1}$ and fusion mask M given two consecutive input frames I_0, I_1 and timestep t. When $t = 0$ or $t = 1$, IFNet is similar to the classical optical flow models.

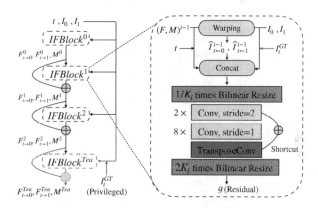

Fig. 4. Left: The IFNet is composed of several stacked IFBlocks operating at different resolution. **Right**: In an IFBlock, we first backward warp the two input frames based on current approximated flow F^{i-1}. Then the input frames I_0, I_1, warped frames $\widehat{I}_{t\leftarrow0}, \widehat{I}_{t\leftarrow1}$, the previous results F^{i-1}, M^{i-1} and timestep t are fed into the next IFBlock to approximate the residual of flow and mask. The privileged information I_t^{GT} is only provided for teacher

To handle the large motion encountered in intermediate flow estimation, we employ a coarse-to-fine strategy with gradually increasing resolution, as illustrated in Fig. 4. Specifically, we first compute a rough prediction of the flow on low resolution, which is believed to capture large motions easier, then iteratively refine the flow fields with gradually increasing resolution. Following this design, our IFNet has a stacked hourglass structure, where a flow field is iteratively refined via successive IFBlocks:

$$\begin{bmatrix} F^i \\ M^i \end{bmatrix} = \begin{bmatrix} F^{i-1} \\ M^{i-1} \end{bmatrix} + \text{IFB}^i(\begin{bmatrix} F^{i-1} \\ M^{i-1} \end{bmatrix}, t, \widehat{I}^{i-1}), \tag{3}$$

where F^{i-1} and M^{i-1} denote the current estimation of the intermediate flows and fusion map from the $(i-1)^{th}$ IFBlock, and IFB^i represents the ith IFBlock. We use a total of 3 IFBlocks, and each has a resolution parameter, $(K^0, K^1, K^2) = (4, 2, 1)$. During inference time, the final estimation is F^n and $M^n(n=2)$. Each IFBlock has a feed-forward structure consisting of serveral convolutional layers and an up-sampling operator. Except for the layer that outputs the optical flow residuals and the fusion map, we use PReLU [16] as the activation function. The cost volume [15] operator is computationally expensive and usually ties the starting point of optical flow to the input image. So it is not directly transferable.

Table 1. Average inference time on the 640×480 **frames.** Recent VFI methods [3, 22, 41] run the optical flow model twice to obtain bi-directional flows

Method	FlowNet2.0 [20]	PWC-Net [53]	LiteFlownet [19]	RAFT [55]	IFNet
Runtime	2×207 ms	2×21 ms	2×73 ms	2×52 ms	**7 ms**

Fig. 5. Results of DVF [30] **(Vimeo90K).** After feeding the edge map of intermediate frames (privileged information) into the model, the estimated flows can be significantly improved, resulting in better reconstruction on validation set

We compare the runtime of the SOTA optical flow models [19,53,55] and IFNet in Table 1. Current flow-based VFI methods [3,22,41] usually need to run their flow models twice then process the bi-directional flows. Therefore the intermediate flow estimation in RIFE runs at a faster speed. Although these optical models can estimate inter-frame motion accurately, they are not suitable for direct migration to VFI tasks.

3.3 Priveleged Distillation for Intermediate Flow

We use an experiment to show that directly approximating the intermediate flows is challenging without access to the intermediate frame. We train DVF [30] model to estimate intermediate flow on Vimeo90K [61] dataset. As a comparison, we add an additional input channel to the DVF model, containing the edge map [12] of intermediate frames (denoted as "Privileged DVF"). Figure 5 shows that the quantization result of Privileged DVF is surprisingly high, while the flows estimated by DVF are blurry. Similar conclusions are also demonstrated in deferred rendering, showing that VFI will be simpler with some intermediate information [6]. This demonstrates that estimating optical flow between two images is easier for the model than estimating intermediate flow. This inspire us to design a privileged model to teach the original model.

We design a privileged distillation loss to IFNet. We stack an additional IFBlock (teacher model IFB^{Tea}, $K^{Tea} = 1$) that refines the results of IFNet referring to the target frame I_t^{GT}:

$$\begin{bmatrix} F^{Tea} \\ M^{Tea} \end{bmatrix} = \begin{bmatrix} F^n \\ M^n \end{bmatrix} + IFB^{Tea}(\begin{bmatrix} F^n \\ M^n \end{bmatrix}, t, \widehat{I}^n, I_t^{GT}). \tag{4}$$

With the access of I_t^{GT} as privileged information, the teacher model produces more accurate flows. We define the distillation loss $\mathcal{L}_{dis}$ as follows:

$$\mathcal{L}_{dis} = \sum_{i \in \{0,1\}} ||F_{t \to i} - F_{t \to i}^{Tea}||_2. \tag{5}$$

We apply the distillation loss over the full sequence of predictions generated from the iteratively updating process in the student model. The gradient of this loss will not be backpropagated to the teacher model. The teacher block will be discarded after the training phase, hence this would incur no extra cost for inference. It makes more stable training and faster convergence.

3.4   Implementation Details

Supervisions. Our training loss $\mathcal{L}_{total}$ is a linear combination of the reconstruction losses $\mathcal{L}_{rec}, \mathcal{L}_{rec}^{Tea}$ and privileged distillation loss $\mathcal{L}_{dis}$:

$$\mathcal{L}_{total} = \mathcal{L}_{rec} + \mathcal{L}_{rec}^{Tea} + \lambda_d \mathcal{L}_{dis}, \tag{6}$$

where we set $\lambda_d = 0.01$ to balance the scale of losses.

The reconstruction loss $\mathcal{L}_{rec}$ models the reconstruction quality of the intermediate frame. The reconstruction loss has the formulation of:

$$\mathcal{L}_{rec} = d(\widehat{I}_t, I_t^{GT}), \mathcal{L}_{rec}^{Tea} = d(\widehat{I}_t^{Tea}, I_t^{GT}), \tag{7}$$

where d is often a pixel-wised loss. Following previous work [40,41], we use L_1 loss between two Laplacian pyramid representations of the reconstructed image and ground truth (denoted as L_{Lap}, the pyramidal level is 5).

Training Dataset. We use the Vimeo90K dataset [61] to train RIFE. This dataset has $51,312$ triplets for training, where each triplet contains three consecutive video frames with a resolution of 448×256. We randomly augment the training data using horizontal and vertical flipping, temporal order reversing, and rotating by $90°$.

Training Strategy. We train RIFE on the Vimeo90K training set and fix $t = 0.5$. RIFE is optimized by AdamW [32] with weight decay 10^{-4} on 224×224 patches. Our training uses a batch size of 64. We gradually reduce the learning rate from 10^{-4} to 10^{-5} using cosine annealing during the whole training process. We train RIFE on 8 TITAN X (Pascal) GPUs for 300 epochs in 10 h.

We use the Vimeo90K-Septuplet [61] dataset to extend RIFE to support arbitrary-timestep frame interpolation [9,24]. This dataset has $91,701$ sequence with a resolution of 448×256, each of which contains 7 consecutive frames. For each training sample, we randomly select 3 frames $(I_{n_0}, I_{n_1}, I_{n_2})$ and calculate the target timestep $t = (n_1 - n_0)/(n_2 - n_0)$, where $0 \le n_0 < n_1 < n_2 < 7$. So we can write RIFE's temporal encoding to extend it. We keep other training setting unchanged and denote the model trained on Vimeo90K-Septuplet as RIFE_m.

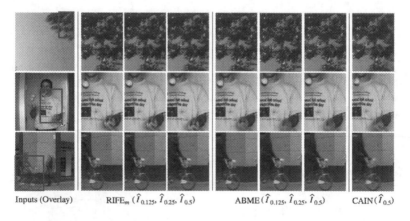

Inputs (Overlay) RIFE_m $(\hat{I}_{0.125}, \hat{I}_{0.25}, \hat{I}_{0.5})$ ABME $(\hat{I}_{0.125}, \hat{I}_{0.25}, \hat{I}_{0.5})$ $\text{CAIN}(\hat{I}_{0.5})$

Fig. 6. Interpolating multiple frames using RIFE_m. These images are from HD [3], M.B. [2], Vimeo90K [61] benchmarks, respectively. We attach the results of CAIN [10] and ABME [44]. RIFE_m provides smooth and continuous motions

4   Experiments

We first introduce the benchmarks for evaluation. Then we provide variants of our models with different computational costs. We compare these models with representative SOTA methods. In addition, we show the capability of generating arbitrary-timestep frames and other applications using RIFE. An ablation study is carried out to analyze our design. Finally, we discuss some limitations of RIFE.

4.1   Benchmarks and Evaluation Metrics

We train our models on the Vimeo90K training dataset and directly test it on the following benchmarks.

Vimeo90K. There are 3,782 triplets in the Vimeo90K testing set [61] with resolution of 448×256. This dataset is widely evaluated in recent VFI methods.

UCF101. The UCF101 dataset [52] contains videos with various human actions. There are 379 triplets with a resolution of 256×256.

HD. Bao et al. [4] collect 11 videos for evaluation. The HD benchmark consists of four 1080p, three 720p and four 1280×544 videos. Following the author of this benchmark, we use the first 100 frames of each video for evaluation.

X4K-1000FPS. A recently released high frame rate 4K dataset [50] containing 15 scenes for testing. We follow the evaluation of [44].

We measure the peak signal-to-noise ratio (PSNR), structural similarity (SSIM), and interpolation error (IE) for quantitative evaluation. All the methods are tested on a TITAN X (Pascal) GPU. To report the runtime, we test all models for processing a pair of 640×480 images using the same device. Disagreements with some of the published results are explained in the **Appendix**.

Table 2. Quantitative evaluation (PSNR) for $4\times$ interpolation on the HD [4] and $8\times$ interpolation on X4K-1000FPS [50] benchmark

Method	Arbitrary-timestep	HD544p	HD720p	HD1080p	X4K-1000FPS
DAIN [3]	✓	22.17	30.25	OOM	26.78†
CAIN [10]	-	21.81	31.59	31.08	-
BMBC [44]	✓	19.51	23.47	OOM	OOM
DSepconv [8]	-	19.28	23.48	OOM	OOM
CDFI [13]	-	21.85	29.28	OOM	OOM
EDSC$_m$ [9]	✓	21.89	30.35	30.91	-
ABME [44]	-	22.46	31.43	33.22	30.16†
RIFE$_m$	✓	**22.95**	**31.87**	**34.25**	30.58‡

†: copy from [44].
‡: estimate flows on 1/4 downsampled videos.

4.2 Comparisons with Previous Methods

We compare RIFE with previous VFI models [3,4,8–10,13,27,41,43,44,61]. These models are officially released except SoftSplat [41]. A recently unofficial reproduction [48] report SoftSplat [41] is slower than ABME [44], and we can not verify it with the available materials. In addition, we train DVF [30] model and SuperSlomo [22] using our training pipeline on Vimeo90K dataset because the released models of these methods are trained on early datasets.

Interpolating Arbitrary-Timestep Frame. Arbitrary-timestep VFI is important in frame-rate conversion. We apply $RIFE_m$ to interpolate multiple intermediate frames at different timesteps $t \in (0,1)$, as shown in Fig. 6. $RIFE_m$ can successfully handle $t = 0.125$ (8×) which is not included in the training data.

To provide a quantitative comparison of multiple frame interpolation, we further extract every fourth frame of videos from HD benchmark [4] and use them to interpolate other frames. We divide the HD benchmark into three subsets with different resolution to test these methods. We show the quantitative PSNR between generated frames and frames of the original videos in Table 2. Note that DAIN [3], BMBC [44] and $EDSC_m$ [8] can generate a frame at an arbitrary timestep. Some other methods can only interpolate the intermediate frame at $t = 0.5$. Thus we use them recursively to produce 4× results. Specifically, we firstly apply the single interpolation method once to get intermediate frame $\widehat{I}_{0.5}$. Then we feed I_0 and $\widehat{I}_{0.5}$ to get $\widehat{I}_{0.25}$ and so on. Furthermore, we test 8× interpolation in a recently released dataset, X4K-1000FPS [50]. Overall, $RIFE_m$ is very effective in the multiple frame interpolation.

Model Scaling. To scale our models that can be compared with existing methods, we introduce two modifications following: test-time augmentation and resolution multiplying. 1) We flip the input images horizontally and vertically to get augmented test data. We infer and average (with flipping) these two results finally. This model is denoted as RIFE-2T. 2) We remove the first downsample layer of IFNet and add a downsample layer before its output to match the origin pipeline. We also perform this modification on RefineNet. It enlarges the process resolution of the feature maps and produces a model named RIFE-2R. We combine these two modifications to extend RIFE to RIFE-Large (2T2R).

Middle Timestep Interpolation. We report the performance of middle timestep interpolation in Table 3. For ease of comparison, we group the models by running speed. RIFE achieve very high performance compared to other small models. Meanwhile, RIFE needs only about 3 gigabytes of GPU memory to process 1080p videos. We get a larger version of our model (RIFE-Large) by model scaling, which runs about 4× faster than ABME [44] with comparable performance. We provide a visual comparison of video clips with large motions from the Vimeo90K testing set in Fig. 7, where SepConv [43] and DAIN [3] produce ghosting artifacts, and CAIN [10] causes missing-parts artifacts. Overall, RIFE (with small computation) can produce more reliable results.

Table 3. Quantitative comparisons on several benchmarks. The images of each dataset are directly inputted to each model. Some models are unable to run on 1080p images due to exceeding the memory available on our graphics card (denoted as "OOM"). We use gray backgrounds to mark the methods that require pre-trained depth models or optical flow models

Method	# Parameters (Million)	Runtime (ms)	UCF101 [52] PSNR	UCF101 [52] SSIM	Vimeo90K [61] PSNR	Vimeo90K [61] SSIM	M.B. [2] IE	HD [3] PSNR
DVF [61]	1.6	80	34.92	0.968	34.56	0.973	2.47	31.47
TOFlow [2]	**1.1**	84	34.58	0.967	33.73	0.968	2.15	29.37
DAIN [3]	24.0	436	35.00	0.968	34.71	0.976	2.04	31.64[†]
DSepConv [8]	21.8	236	35.08	0.969	34.73	0.974	2.03	OOM
SoftSplat [41][†]	7.7	-	35.39	**0.970**	36.10	0.980	**1.81**	-
BMBC [44]	11.0	1580	35.15	0.969	35.01	0.976	2.04	OOM
CDFI [13]	5.0	198	35.21	0.969	35.17	0.977	1.98	OOM
ABME [44]	18.1	339	35.37	**0.970**	36.18	**0.981**	1.88	32.17
RIFE-Large	9.8	80	**35.41**	**0.970**	**36.19**	**0.981**	1.82	**32.31**
Relatively Fast Models								
CAIN [10]	42.8	38	34.98	**0.969**	34.65	0.973	2.28	31.77
Superslomo [22]	19.8	62	35.15	0.968	34.64	0.974	2.21	31.55
SepConv [42]	21.6	51	34.78	0.967	33.79	0.970	2.27	30.87
AdaCoF [27]	21.8	34	34.91	0.968	34.27	0.971	2.31	31.43
EDSC [9]	**8.9**	46	35.13	0.968	34.84	0.975	2.02	31.59
RIFE	9.8	**16**	**35.28**	0.969	35.61	0.978	**1.96**	32.14
RIFE$_m$[‡]	9.8	**16**	35.22	0.969	35.46	0.978	2.16	**32.31**

†: copy from the original papers. ‡: trained on Vimeo90K-Septuplet dataset.

4.3   General Temporal Encode

In the VFI task, our temporal encoding t is used to control the timestep. To show its generalization capability, we demonstrate that we can control this encoding to implement diverse applications. As shown in Fig. 8, if we input a gradient encoding t_p, the RIFE$_m$ will synthesize the two images from dynamic scenes in a "panoramic" view (use different timestamps for each column). The position relation of the vehicle in $\widehat{I}_p$ is between I_0 and I_1. In other words, if I_0, I_1 are from the binocular camera, the shooting time of I_1 is later than that of I_0. $\widehat{I}_p$ is the result of a wider FOV camera scan in columns. Similarly, this method may potentially eliminate the rolling shutter of the videos by having different timestamps for each horizontal row.

Inputs   SepConv [42]   DAIN [3]   CAIN [10]   RIFE (Ours)   GT

Fig. 7. Qualitative comparison on Vimeo90K [61] testing set

Fig. 8. Synthesize images from two views on one "panoramic" image $\widehat{I}_p$ **using** **RIFE**$_m$. $\widehat{I}_p$ has been stretched for better visualization

4.4   Image Representation Interpolation

RIFE$_m$ can interpolate other image representations using the intermediate flows and fusion map approximating from images. For instance, we interpolate the results of MiDaS [46] which is a popular monocular depth model, shown in Fig. 9. The synthesis formula is simply as follows:

$$\widehat{D}_t = M \odot \overleftarrow{W}(D_0, F_{t\to0}) + (1 - M) \odot \overleftarrow{W}(D_1, F_{t\to1}), \tag{8}$$

where D_0, D_1 are estimated by MiDas [46] and F, M are estimated by RIFE$_m$. RIFE may potentially be used to extend some models and provide visually plausible effects when we ignore z-axis motion of objects.

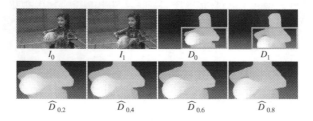

Fig. 9. Interpolation for depth map using RIFE$_m$.

4.5   Ablation Studies

We design some ablation studies on the intermediate flow estimation, distillation scheme, model design and loss function, shown in Table 4. These experiments use the same hyper-parameter setting and evaluation on Vimeo90K [61] and MiddleBury [2] benchmarks.

IFNet vs. Flow Reversal. We compare IFNet with previous intermediate flow estimation methods. Specifically, we use RAFT [55] and PWC-Net [53] with officially pre-trained parameters to estimate the bi-directional flows. Then we implement three flow reversal methods, including linear reversal [22], using a hidden convolutional layer with 128 channels, and the flow reversal layer from EQVI [28]. The optical flow models and flow reversal modules are combined together to replace the IFNet. Furthermore, we try to use the forward warping [41] operator to bypass flow reversal. These models are jointly fine-tuned with RefineNet. Because these models can not directly approximate the fusion map, the fusion map is subsequently approximated by RefineNet. As shown in Table 4, RIFE is more efficient and gets better interpolation performance. These flow models can estimate accurate bi-directional optical flow, but the flow reversal has difficulties in dealing with the object shift problem illustrated in Fig. 3.

Ablation on the Distillation Scheme. We observe that removing the distillation framework makes model training sometimes divergent. Furthermore, we show the importance of distillation design in following experiments. **a1**) Remove the privileged teacher block and use the last IFBlock's results to guide the first two IFBlocks, denoted as "self-consistency"; **a2**) Use pre-trained RAFT [55] to estimate the intermediate flows based on the ground truth image, denoted as "RAFT-KD". This guidance is inspired by the pseudo-labels method [26]. However, this implementation relies on the pre-trained optical flow model and extremely increases the training duration ($3\times$). We found **a1** and **a2** suffer in quality. These experiments demonstrate the importance of optical flow supervision. Some recent work [25,34] has also echoed the improvement using suitable optical flow distillation.

Ablation on RIFE's Architecture and Loss Function. To verify the coarse-to-fine strategy of IFNet, we removed the first IFBlock and the first two IFBlocks in two experiments, respectively. We also try some other popular techniques, such as Batch Normlization (BN) [21]. BN does stabilize the training, but degrades final performance and increases inference overhead. We provide a pair of experiments to show L_{Lap} [40,41] is quantitatively better than $\mathcal{L}_1$.

Limitations. Our work may not cover some practical application requirements. Firstly, RIFE focuses on using two input frames and multi-frame input [24,28,60] is left to future work. One straightforward approach is to extend IFNet to use more frames as input. Secondly, most experiments are done with SSIM and PSNR as quantitative indexes. If human perception quality is preferred, RIFE can readily be changed to use the perceptually related losses [5,42]. Thirdly, additional training data may be necessary for extending RIFE to various applications, such as interpolation for depth map and animation videos [51].

Table 4. Ablation study

Setting	Vimeo90K [61]	MiddleBury [2]	Runtime
	PSNR	IE	640 × 480
Intermediate fow estimation			
RAFT [55] + linear reversal [22]	34.68	2.31	60 ms
RAFT [55] + CNN reversal	34.82	2.24	65 ms
RAFT [55] + Reversal layer [28]	35.16	2.04	101 ms
PWC-Net [53] + Reversal layer [28]	35.24	2.06	83 ms
PWC-Net [53] + Forward warping [41]	35.48	2.02	52 ms
RIFE	**35.61**	1.96	16 ms
Distillation scheme			
RIFE w/self-consistency	35.37	2.02	16 ms
RIFE w/RAFT-KD	35.52	1.98	16 ms
RIFE (priviledged distillation)	**35.61**	1.96	16 ms
Model design			
RIFE w/one IFBlock	35.17	2.12	**12 ms**
RIFE w/two IFBlocks	35.46	1.97	14 ms
RIFE + BN [21]	35.49	2.02	21 ms
RIFE	**35.61**	1.96	16 ms
Loss function			
RIFE w/L_1	35.51	**1.94**	16 ms
RIFE w/L_{Lap}	**35.61**	1.96	16 ms

5   Conclusion

We develop an efficient and flexible algorithm for VFI, namely RIFE. A separate neural module IFNet directly estimates the intermediate optical flows, supervised by a privileged distillation scheme, where the teacher model can access the ground truth intermediate frames. Experiments confirm RIFE can effectively process videos of different scenes. Furthermore, an extra input with temporal encoding enables RIFE for arbitrary-timestep frame interpolation. The lightweight nature of RIFE makes it much more accessible for downstream tasks.

Acknowledgement. This work is supported by National Key R&D Program of China (2021ZD0109803) and National Natural Science Foundation of China under Grant No. 62136001, 62088102.

References

1. Anil, R., Pereyra, G., Passos, A., Ormandi, R., Dahl, G.E., Hinton, G.E.: Large scale distributed neural network training through online distillation. In: Proceedings of the International Conference on Learning Representations (ICLR) (2018)
2. Baker, S., Scharstein, D., Lewis, J., Roth, S., Black, M.J., Szeliski, R.: A database and evaluation methodology for optical flow. In: International Journal of Computer Vision (IJCV) (2011)
3. Bao, W., Lai, W.S., Ma, C., Zhang, X., Gao, Z., Yang, M.H.: Depth-aware video frame interpolation. In: Proceedings of the IEEE Conference on Computer Vision and Pattern Recognition (CVPR) (2019)
4. Bao, W., Lai, W.S., Zhang, X., Gao, Z., Yang, M.H.: MEMC-Net: motion estimation and motion compensation driven neural network for video interpolation and enhancement. In: IEEE Transactions on Pattern Analysis and Machine Intelligence (IEEE TPAMI) (2018). https://doi.org/10.1109/TPAMI.2019.2941941
5. Blau, Y., Michaeli, T.: The perception-distortion tradeoff. In: Proceedings of the IEEE Conference on Computer Vision and Pattern Recognition (CVPR) (2018)
6. Briedis, K.M., Djelouah, A., Meyer, M., McGonigal, I., Gross, M., Schroers, C.: Neural frame interpolation for rendered content. ACM Trans. Graph. **40**(6), 1–13 (2021)
7. Chen, X., Zhang, Y., Wang, Y., Shu, H., Xu, C., Xu, C.: Optical flow distillation: Towards efficient and stable video style transfer. In: Proceedings of the European Conference on Computer Vision (ECCV) (2020)
8. Cheng, X., Chen, Z.: Video frame interpolation via deformable separable convolution. In: AAAI Conference on Artificial Intelligence (2020)
9. Cheng, X., Chen, Z.: Multiple video frame interpolation via enhanced deformable separable convolution. In: IEEE Transactions on Pattern Analysis and Machine Intelligence (TPAMI) (2021). https://doi.org/10.1109/TPAMI.2021.3100714
10. Choi, M., Kim, H., Han, B., Xu, N., Lee, K.M.: Channel attention is all you need for video frame interpolation. In: AAAI Conference on Artificial Intelligence (2020)
11. Danier, D., Zhang, F., Bull, D.: Spatio-temporal multi-flow network for video frame interpolation. arXiv preprint arXiv:2111.15483 (2021)
12. Ding, L., Goshtasby, A.: On the canny edge detector. Pattern Recogn. **34**(3), 721–725 (2001)
13. Ding, T., Liang, L., Zhu, Z., Zharkov, I.: CDFI: compression-driven network design for frame interpolation. In: Proceedings of the IEEE Conference on Computer Vision and Pattern Recognition (CVPR) (2021)
14. Ding, X., Zhang, X., Ma, N., Han, J., Ding, G., Sun, J.: RepVGG: making VGG-style convnets great again. In: Proceedings of the IEEE Conference on Computer Vision and Pattern Recognition (CVPR) (2021)
15. Dosovitskiy, A., et al.: Learning optical flow with convolutional networks. In: Proceedings of the IEEE International Conference on Computer Vision (ICCV) (2015)
16. He, K., Zhang, X., Ren, S., Sun, J.: Delving deep into rectifiers: surpassing human-level performance on ImageNet classification. In: Proceedings of the IEEE International Conference on Computer Vision (ICCV) (2015)
17. Hinton, G., Vinyals, O., Dean, J.: Distilling the knowledge in a neural network. arXiv preprint arXiv:1503.02531 (2015)
18. Huang, Z., Heng, W., Zhou, S.: Learning to paint with model-based deep reinforcement learning. In: Proceedings of the IEEE International Conference on Computer Vision (ICCV) (2019)

19. Hui, T.W., Tang, X., Change Loy, C.: LiteFlowNet: a lightweight convolutional neural network for optical flow estimation. In: Proceedings of the IEEE Conference on Computer Vision and Pattern Recognition (CVPR) (2018)

20. Ilg, E., et al.: Evolution of optical flow estimation with deep networks. In: Proceedings of the IEEE Conference on Computer Vision and Pattern Recognition (CVPR) (2017)

21. Ioffe, S., Szegedy, C.: Batch normalization: Accelerating deep network training by reducing internal covariate shift. arXiv preprint arXiv:1502.03167 (2015)

22. Jiang, H., Sun, D., Jampani, V., Yang, M.H., Learned-Miller, E., Kautz, J.: Super SloMo: high quality estimation of multiple intermediate frames for video interpolation. In: Proceedings of the IEEE Conference on Computer Vision and Pattern Recognition (CVPR) (2018)

23. Jonschkowski, R., Stone, A., Barron, J.T., Gordon, A., Konolige, K., Angelova, A.: What matters in unsupervised optical flow. In: Proceedings of the European Conference on Computer Vision (ECCV) (2020)

24. Kalluri, T., Pathak, D., Chandraker, M., Tran, D.: FLAVR: Flow-agnostic video representations for fast frame interpolation. arXiv preprint arXiv:2012.08512 (2020)

25. Kong, L., et al.: IfrNet: intermediate feature refine network for efficient frame interpolation. In: Proceedings of the IEEE Conference on Computer Vision and Pattern Recognition (CVPR) (2022)

26. Lee, D.H., et al.: Pseudo-label: The simple and efficient semi-supervised learning method for deep neural networks. In: Proceedings of the IEEE International Conference on Machine Learning Workshops (ICMLW) (2013)

27. Lee, H., Kim, T., Chung, T.y., Pak, D., Ban, Y., Lee, S.: AdaCOF: adaptive collaboration of flows for video frame interpolation. In: Proceedings of the IEEE Conference on Computer Vision and Pattern Recognition (CVPR) (2020)

28. Liu, Y., Xie, L., Siyao, L., Sun, W., Qiao, Y., Dong, C.: Enhanced quadratic video interpolation. In: Proceedings of the European Conference on Computer Vision (ECCV) (2020)

29. Liu, Y.L., Liao, Y.T., Lin, Y.Y., Chuang, Y.Y.: Deep video frame interpolation using cyclic frame generation. In: Proceedings of the 33rd Conference on Artificial Intelligence (AAAI) (2019)

30. Liu, Z., Yeh, R.A., Tang, X., Liu, Y., Agarwala, A.: Video frame synthesis using deep voxel flow. In: Proceedings of the IEEE International Conference on Computer Vision (ICCV) (2017)

31. Lopez-Paz, D., Bottou, L., Schölkopf, B., Vapnik, V.: Unifying distillation and privileged information. In: Proceedings of the International Conference on Learning Representations (ICLR) (2016)

32. Loshchilov, I., Hutter, F.: Fixing weight decay regularization in Adam. arXiv preprint arXiv:1711.05101 (2017)

33. Lu, G., Ouyang, W., Xu, D., Zhang, X., Cai, C., Gao, Z.: DVC: an end-to-end deep video compression framework. In: Proceedings of the IEEE Conference on Computer Vision and Pattern Recognition (CVPR) (2019)

34. Lu, L., Wu, R., Lin, H., Lu, J., Jia, J.: Video frame interpolation with transformer. In: Proceedings of the IEEE Conference on Computer Vision and Pattern Recognition (CVPR) (2022)

35. Luo, K., Wang, C., Liu, S., Fan, H., Wang, J., Sun, J.: UPFlow: upsampling pyramid for unsupervised optical flow learning. In: Proceedings of the IEEE Conference on Computer Vision and Pattern Recognition (CVPR) (2021)

36. Ma, N., Zhang, X., Zheng, H.T., Sun, J.: ShuffleNet v2: practical guidelines for efficient CNN architecture design. In: Proceedings of the European conference on computer vision (ECCV) (2018)
37. Meister, S., Hur, J., Roth, S.: UnFlow: unsupervised learning of optical flow with a bidirectional census loss. In: AAAI Conference on Artificial Intelligence (2018)
38. Meyer, S., Wang, O., Zimmer, H., Grosse, M., Sorkine-Hornung, a.: Phase-based frame interpolation for video. In: Proceedings of the IEEE Conference on Computer Vision and Pattern Recognition (CVPR) (2015)
39. Mnih, V., et al.: Playing Atari with deep reinforcement learning. arXiv preprint arXiv:1312.5602 (2013)
40. Niklaus, S., Liu, F.: Context-aware synthesis for video frame interpolation. In: Proceedings of the IEEE Conference on Computer Vision and Pattern Recognition (CVPR) (2018)
41. Niklaus, S., Liu, F.: SoftMax splatting for video frame interpolation. In: Proceedings of the IEEE Conference on Computer Vision and Pattern Recognition (CVPR) (2020)
42. Niklaus, S., Mai, L., Liu, F.: Video frame interpolation via adaptive convolution. In: Proceedings of the IEEE Conference on Computer Vision and Pattern Recognition (CVPR) (2017)
43. Niklaus, S., Mai, L., Liu, F.: Video frame interpolation via adaptive separable convolution. In: Proceedings of the IEEE International Conference on Computer Vision (ICCV) (2017)
44. Park, J., Lee, C., Kim, C.S.: Asymmetric bilateral motion estimation for video frame interpolation. In: Proceedings of the IEEE International Conference on Computer Vision (ICCV) (2021)
45. Porrello, A., Bergamini, L., Calderara, S.: Robust re-identification by multiple views knowledge distillation. In: Proceedings of the European Conference on Computer Vision (ECCV) (2020)
46. Ranftl, R., Lasinger, K., Hafner, D., Schindler, K., Koltun, V.: Towards robust monocular depth estimation: Mixing datasets for zero-shot cross-dataset transfer. In: IEEE Transactions on Pattern Analysis and Machine Intelligence (TPAMI) (2020)
47. Ranjan, A., Black, M.J.: Optical flow estimation using a spatial pyramid network. In: Proceedings of the IEEE Conference on Computer Vision and Pattern Recognition (CVPR) (2017)
48. Reda, F., Kontkanen, J., Tabellion, E., Sun, D., Pantofaru, C., Curless, B.: Frame interpolation for large motion. arXiv (2022)
49. Reda, F.A., et al.: Unsupervised video interpolation using cycle consistency. In: Proceedings of the IEEE International Conference on Computer Vision (ICCV) (2019)
50. Sim, H., Oh, J., Kim, M.: XVFI: extreme video frame interpolation. In: Proceedings of the IEEE International Conference on Computer Vision (ICCV) (2021)
51. Siyao, L., et al.: Deep animation video interpolation in the wild. In: Proceedings of the IEEE Conference on Computer Vision and Pattern Recognition (CVPR) (2021)
52. Soomro, K., Zamir, A.R., Shah, M.: Ucf101: a dataset of 101 human actions classes from videos in the wild. arXiv preprint arXiv:1212.0402 (2012)
53. Sun, D., Yang, X., Liu, M.Y., Kautz, J.: PWC-Net: CNNs for optical flow using pyramid, warping, and cost volume. In: Proceedings of the IEEE Conference on Computer Vision and Pattern Recognition (CVPR) (2018)

54. Sun, S., Kuang, Z., Sheng, L., Ouyang, W., Zhang, W.: Optical flow guided feature: a fast and robust motion representation for video action recognition. In: IEEE/CVF Conference on Computer Vision and Pattern Recognition (CVPR) (2018)
55. Teed, Z., Deng, J.: RAFT: recurrent all-pairs field transforms for optical flow. In: Proceedings of the European Conference on Computer Vision (ECCV) (2020)
56. Wu, C.Y., Singhal, N., Krahenbuhl, P.: Video compression through image interpolation. In: Proceedings of the European Conference on Computer Vision (ECCV) (2018)
57. Wu, Y., Wen, Q., Chen, Q.: Optimizing video prediction via video frame interpolation. In: Proceedings of the IEEE Conference on Computer Vision and Pattern Recognition (CVPR) (2022)
58. Xiang, X., Tian, Y., Zhang, Y., Fu, Y., Allebach, J.P., Xu, C.: Zooming slow-MO: fast and accurate one-stage space-time video super-resolution. In: IEEE/CVF Conference on Computer Vision and Pattern Recognition (CVPR) (2020)
59. Xu, G., Xu, J., Li, Z., Wang, L., Sun, X., Cheng, M.: Temporal modulation network for controllable space-time video super-resolution. In: IEEE/CVF Conference on Computer Vision and Pattern Recognition (CVPR) (2021)
60. Xu, X., Siyao, L., Sun, W., Yin, Q., Yang, M.H.: Quadratic video interpolation. In: Advances in Neural Information Processing Systems (NIPS) (2019)
61. Xue, T., Chen, B., Wu, J., Wei, D., Freeman, W.T.: Video enhancement with task-oriented flow. In: International Journal of Computer Vision (IJCV) (2019)
62. Yuan, S., Stenger, B., Kim, T.K.: RGB-based 3d hand pose estimation via privileged learning with depth images. In: Proceedings of the IEEE International Conference on Computer Vision Workshops (ICCVW) (2019)
63. Zhao, Z., Wu, Z., Zhuang, Y., Li, B., Jia, J.: Tracking objects as pixel-wise distributions. In: Proceedings of the European conference on computer vision (ECCV) (2022)
64. Zhou, M., Bai, Y., Zhang, W., Zhao, T., Mei, T.: Responsive listening head generation: a benchmark dataset and baseline. In: Proceedings of the European Conference on Computer Vision (ECCV) (2022)
65. Zhou, T., Tulsiani, S., Sun, W., Malik, J., Efros, A.A.: View synthesis by appearance flow. In: Proceedings of the European Conference on Computer Vision (ECCV) (2016)

PixelFolder: An Efficient Progressive Pixel Synthesis Network for Image Generation

Jing He[1], Yiyi Zhou[1]([✉]), Qi Zhang[2], Jun Peng[1], Yunhang Shen[2], Xiaoshuai Sun[1,3], Chao Chen[2], and Rongrong Ji[1,3]

[1] Media Analytics and Computing Lab, Department of Artificial Intelligence, School of Informatics, Xiamen University, Xiamen, China
{blinghe,pengjun}@stu.xmu.edu.cn, {zhouyiyi,xssun,rrji}@xmu.edu.cn
[2] Youtu Lab, Tencent, Shenzhen, China
{merazhang,aaronccchen}@tencent.com
[3] Institute of Artificial Intelligence, Xiamen University, Xiamen, China

Abstract. Pixel synthesis is a promising research paradigm for image generation, which can well exploit pixel-wise prior knowledge for generation. However, existing methods still suffer from excessive memory footprint and computation overhead. In this paper, we propose a progressive pixel synthesis network towards efficient image generation, coined as *PixelFolder*. Specifically, PixelFolder formulates image generation as a progressive pixel regression problem and synthesizes images by a multi-stage paradigm, which can greatly reduce the overhead caused by large tensor transformations. In addition, we introduce novel *pixel folding* operations to further improve model efficiency while maintaining pixel-wise prior knowledge for end-to-end regression. With these innovative designs, we greatly reduce the expenditure of pixel synthesis, *e.g.*, reducing 89% computation and 53% parameters compared to the latest pixel synthesis method called *CIPS*. To validate our approach, we conduct extensive experiments on two benchmark datasets, namely FFHQ and LSUN Church. The experimental results show that with much less expenditure, PixelFolder obtains new state-of-the-art (SOTA) performance on two benchmark datasets, *i.e.*, 3.77 *FID* and 2.45 *FID* on FFHQ and LSUN Church, respectively. Meanwhile, PixelFolder is also more efficient than the SOTA methods like *StyleGAN2*, reducing about 72% computation and 31% parameters, respectively. These results greatly validate the effectiveness of the proposed PixelFolder. Our source code is available at https://github.com/BlingHe/PixelFolder.

Keywords: Pixel synthesis · Image generation · Pixel folding

Supplementary Information The online version contains supplementary material available at https://doi.org/10.1007/978-3-031-19781-9_37.

1   Introduction

As an important task of computer vision, image generation has made remarkable progress in recent years, which is supported by a flurry of generative adversarial networks [4,5,7,9,15,18–20,25,42]. One of the milestone works is the StyleGAN series [19,20], which borrows the principle of style transfer [14] to build an effective generator architecture. Due to the superior performance in image quality, this style-driven modeling has become the mainstream paradigm of image generation [19,20], which also greatly influences and promotes the development of other generative tasks, such as image manipulation [8,21,49,51,55], image-to-image translation [6,16,17,27,36,54] and text-to-image generation [26,39,41,50].

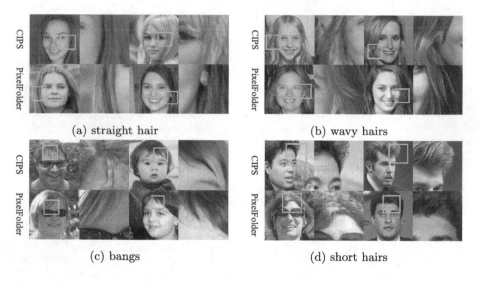

(a) straight hair

(b) wavy hairs

(c) bangs

(d) short hairs

Fig. 1. Comparison of the generated faces by CIPS [2] and PixelFolder on FFHQ. Compared with CIPS, PixelFolder synthesizes more vivid faces and can also alleviate local incongruities via its novel network structure.

In addition to the StyleGAN series, pixel synthesis [2,45] is another paradigm of great potential for image generation. Recently, Anokin *et al.* [2] propose a novel Conditionally-Independent Pixel Synthesis (CIPS) network for adversarial image generation, which directly computes each pixel value based on the random latent vector and positional embeddings. This end-to-end pixel regression strategy can well exploit pixel-wise prior knowledge to facilitate the generation of high-quality images. Meanwhile, it also simplifies the design of generator architecture, *e.g.*, only using 1×1 convolutions, and has a higher generation ability with non-trivial topologies [2]. On multiple benchmarks [19,42], this method exhibits comparable performance against the StyleGAN series, showing a great potential in image generation. In this paper, we also follow the principle of pixel synthesis to build an effective image generation network.

Despite the aforementioned merits, CIPS still has obvious shortcomings in model efficiency. Firstly, although CIPS is built with a simple network structure, it still requires excessive memory footprint and computation during inference. Specifically, this is mainly attributed to its high-resolution pixel tensors for end-to-end pixel regression, $e.g.$, $256 \times 256 \times 512$, which results in a large computational overhead and memory footprint, as shown in Fig. 2a. Meanwhile, the learnable coordinate embeddings also constitute a large number of parameters, making CIPS taking about 30% more parameters than StyleGAN2 [20]. These issues greatly limit the applications of CIPS in high-resolution image synthesis.

To address these issues, we propose a novel progressive pixel synthesis network towards efficient image generation, termed *PixelFolder*, of which structure is illustrated in Fig. 2b. Firstly, we transform the pixel synthesis problem to a progressive one and then compute pixel values via a multi-stage structure. In this way, the generator can process the pixel tensors of varying scales instead of the fixed high-resolution ones, thereby reducing memory footprint and computation greatly. Secondly, we introduce novel *pixel folding* operations to further improve model efficiency. In PixelFolder, the large pixel tensors of different stages are folded into the smaller ones, and then gradually unfolded (expanded) during feature transformations. These pixel folding (and unfolding) operations can well preserve the independence of each pixel, while saving model expenditure. These innovative designs help PixelFolder achieves high-quality image generations with superior model efficiency, which are also shown to be effective for *local imaging incongruity* found in CIPS [2], as shown in Fig. 1.

To validate the proposed PixelFolder, we conduct extensive experiments on two benchmark datasets of image generation, $i.e.$, FFHQ [19] and LSUN Church [42]. The experimental results show that PixelFolder not only outperforms CIPS in terms of image quality on both benchmarks, but also reduces parameters and computation by 53% and 89%, respectively. Compared to the state-of-the-art model, $i.e.$, StyleGAN2 [20], PixelFolder is also very competitive and obtains new SOTA performance on FFHQ and LSUN Church, $i.e.$, 3.77 FID and 2.45 FID, respectively. Meanwhile, the efficiency of PixelFolder is still superior, with 31% less parameters and 72% less computation than StyleGAN2.

To sum up, our contribution is two-fold:

1. We propose a progressive pixel synthesis network for efficient image generation, termed *PixelFolder*. With the multi-stage structure and innovative pixel folding operations, PixelFolder greatly reduces the computational and memory overhead while keeping the property of end-to-end pixel synthesis.
2. Retaining much higher efficiency, the proposed PixelFolder not only has better performance than the latest pixel synthesis method CIPS, but also achieves new SOTA performance on FFHQ and LSUN Church.

2   Related Work

Recent years have witnessed the rapid development of image generation supported by a bunch of generative adversarial network (GAN) [9] based methods

[1, 11, 28, 30, 33, 38, 40, 46, 48]. Compared with previous approaches [23, 47], GAN-based methods model the domain-specific data distributions better through the specific adversarial training paradigm, *i.e.*, a discriminator is trained to distinguish whether the images are true or false for the optimization of the generator. To further improve the quality of generations, a flurry of methods [3, 5, 7, 10, 42] have made great improvements in both GAN structures and objective functions. Recent advances also resort to a progressive structure for high-resolution image generation. PGGAN [18] proposes a progressive network to generate high-resolution images, where both generator and discriminator start their training with low-resolution images and gradually increase the model depth by adding-up the new layers during training. StyleGAN series [19, 20] further borrow the concept of *"style"* into the image generation and achieve remarkable progress. The common characteristic of these progressive methods is to increase the resolution of hidden features by up-sampling or deconvolution operations. Differing from these methods, our progressive modeling is based on the principle of pixel synthesis with pixel-wise independence for end-to-end regression.

In addition to being controlled by noise alone, some methods exploit coordinate information for image generation. CoordConv-GAN [32] introduces pixel coordinates in every convolution based on DCGAN [42], which proves that pixel coordinates can better establish geometric correlations between the generated pixels. COCO-GAN [29] divides the image into multiple patches with different coordinates, which are further synthesized independently. CIPS [2] builds a new paradigm of using coordinates for image generation, *i.e.*, pixel regression, which initializes the prior matrix based on pixel coordinates and deploys multiple 1×1 convolutions for pixel transformation. This approach not only greatly simplifies the structure of generator, but also achieves competitive performance against existing methods. In this paper, we also follow the principle of pixel regression to build the proposed PixelFolder.

Our work is also similar to a recently proposed method called INR-GAN [45], which also adopts a multi-stage structure. In addition to the obvious differences in network designs and settings, PixelFolder is also different from INR-GAN in the process of pixel synthesis. In INR-GAN, the embeddings of pixels are gradually up-sampled via *nearest neighbor interpolation*, which is more in line with the progressive models like StyleGAN2 [20] or PGGAN [18]. In contrast, PixelFolder can well maintain the independence of each pixel during multi-stage generation, and preserve the property of end-to-end pixel regression via pixel folding operations.

3   Preliminary

Conditionally-Independent Pixel Synthesis (CIPS) is a novel generative adversarial network proposed by Anokhin *et al.* [2]. Its main principle is to synthesis each pixel conditioned on a random vector $z \in Z$ and the pixel coordinates (x, y), which can be defined by

$$I = \{G(x, y; \mathbf{z})|(x, y) \in mgrid(H, W)\}, \tag{1}$$

where $mgrid(H, W) = \{(x, y) | 0 \le x \le W, 0 \le y \le H\}$ is the set of integer pixel coordinates, and $G(\cdot)$ is the generator. Similar to StyleGAN2 [20], z is turned into a style vector w via a mapping network and then shared by all pixels. Afterwards, w is injected into the generation process via ModFC layers [2].

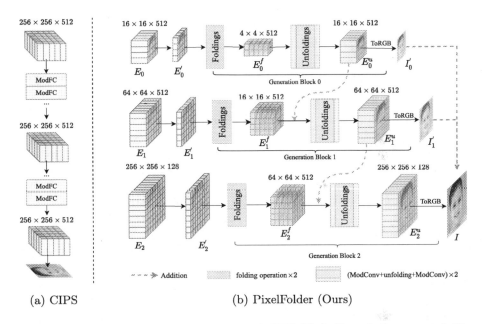

(a) CIPS   (b) PixelFolder (Ours)

Fig. 2. A comparison of the architectures of CIPS [2] (left) and the proposed PixelFolder (right). PixelFolder follows the pixel synthesis principle of CIPS, but regards image generation as a multi-stage regression problem, thereby reducing the cost of large tensor transformations. Meanwhile, novel *pixel folding* operations are also applied in PixelFodler to further improve model efficiency.

An important design in CIPS is the positional embeddings of synthesized pixels, which are consisted of Fourier features and coordinate embeddings. The Fourier feature of each pixel $e_{fo}(x, y) \in \mathbb{R}^d$ is computed based on the coordinate (x, y) and transformed by a learnable weight matrix $B_{fo} \in \mathbb{R}^{2 \times d}$ and sin activation. To improve model capacity, Anokhin *et al.* also adopt the coordinate embedding $e_{co}(x, y) \in \mathbb{R}^d$, which has $H \times W$ learnable vectors in total. Afterwards, the final pixel vector $e(x, y) \in \mathbb{R}^d$ is initialized by concatenating these two types of embeddings and then fed to the generator.

Although CIPS has a simple structure and can be processed in parallel [2], its computational cost and memory footprint are still expensive, mainly due to the high-resolution pixel tensor for end-to-end generation. In this paper, we follow the principle of CIPS defined in Eq. 1 to build our model and address the issue of model efficiency via a progressive regression paradigm.

4   PixelFolder

4.1   Overview

The structure of the proposed PixelFodler is illustrated in Fig. 2. To reduce the high expenditure caused by end-to-end regression for large pixel tensors, we first transform pixel synthesis to a multi-stage generation problem, which can be formulated as

$$I = \sum_{i=0}^{K-1} \{G_i(x_i, y_i; \mathbf{z}) | (x_i, y_i) \in mgrid(H_i, W_i)\}, \tag{2}$$

where i denotes the index of generation stages. At each stage, we initialize a pixel tensor $\mathbf{E}_i \in \mathbb{R}^{H_i \times W_i \times d}$ for generation. The RGB tensors $I'_i \in \mathbb{R}^{H_i \times W_i \times 3}$ predicted by different stages are then aggregated for the final pixel regression. This progressive paradigm can avoid the constant use of large pixel tensors to reduce excessive memory footprint. In literature [18,45,52,53], it is also shown effective to reduce the difficulty of image generation.

To further reduce the expenditure of each generation stage, we introduce novel *pixel folding* operations to PixelFolder. As shown in Fig. 2, the large pixel tensor is first projected onto a lower-dimension space, and their local pixels, *e.g.*, in 2×2 patch, are then concatenated to form a new tensor with a smaller resolution, denoted as $\mathbf{E}_i^f \in \mathbb{R}^{\frac{H_i}{k} \times \frac{W_i}{k} \times d}$, where k is the scale of folding. After passing through the convolution layers, the pixel tensor is decomposed again (truncated from the feature dimension), and combined back to the original resolution. We term these parameter-free operations as *pixel folding* (and unfolding). Folding features is not uncommon in computer vision, which is often used as an alternative to the operations like *down-sampling* or *pooling* [31,34,35,44]. But in PixelFolder, it not only acts to reduce the tensor resolution, but also serves to maintain the independence of folded pixels.

To maximize the use of pixel-wise prior knowledge at different scales, we further combine the folded tensor E_i^f with the unfolded pixel tensor E_{i-1}^u of the previous stage, as shown in Fig. 2b. With the aforementioned designs, PixelFolder can significantly reduce memory footprint and computation, while maintaining the property of pixel synthesis.

4.2   Pixel Folding

The illustration of *pixel folding* is depicted in Fig. 3a, which consists of two operations, namely *folding* and *unfolding*. The folding operation spatially decomposes the pixel tensor into multiple local patches, and straighten each of the patches to form a smaller but deeper tensor. On the contrary, the unfolding operation will truncate the folded pixel vectors from the feature dimension to recover the tensor resolution.

Particularly, pixel folding can effectively keep the independence and spatial information of each pixel regardless of the varying resolutions of the hidden

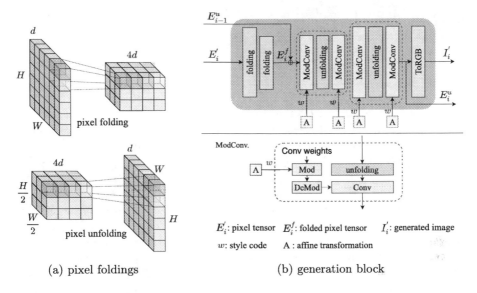

(a) pixel foldings   (b) generation block

Fig. 3. (a) The illustrations of pixel folding and unfolding operations. These parameter-free operations can maintain the pixel-wise independence when changing the tensor resolution. (b) The detailed structure of the generation block in PixelFolder. The number of parameterized layers in PixelFolder is much smaller than those of CIPS and StyleGAN2.

tensors. This also enables the pixel-wise prior knowledge to be fully exploited for image generation. In addition, when the pixels are folded, they can receive more interactions via convolutions, which is found to be effective for the issue of *local imagery incongruity* caused by insufficient local modeling [2].

4.3   Pixel Tensor Initialization

Similar to CIPS [2], we also apply Fourier features and coordinate embeddings to initialize the pixel tensors. Specifically, given the coordinate of a pixel (x, y), Fourier feature $e_{fo}(x, y)$ is obtained by

$$e_{fo}(x, y) = \sin\left[B_{fo}(x', y')^T\right], \tag{3}$$

where $x' = \frac{2x}{W_i - 1} - 1$ and $y' = \frac{2y}{H_i - 1} - 1$, and $B_{fo} \in \mathbb{R}^{2 \times d}$ is the projection weight matrix. The coordinate embedding is a parameterized vector, denoted as $e_{co}(x, y) \in \mathbb{R}^d$. Afterwards, these two types of embeddings are concatenated and projected to obtain the new pixel tensor, denoted as $\mathbf{E}_i \in \mathbb{R}^{H_i \times W_i \times d}$.

In principle, Fourier features serve to preserve the spatial information and capture the relationships between pixels [2,32]. The learnable coordinate embeddings can increase model capacity to improve image quality, *e.g.*, to avoid wave-like artifacts [2]. In PixelFolder, we only apply coordinate embeddings to the first generation stage to keep model compactness, and we found this trade-off has little detriment to image quality during experiments.

4.4   Generation Blocks

The detailed structure of generation blocks in PixelFolder is given in Fig. 3b. After folding operations, a modulated convolution (*ModConv*) layer [20] is deployed for feature transformation. Then unfolding operations are used to recover the resolution, each followed by another ModConv layer. In practice, we use two folding and unfolding operations to gradually reduce and recover the tensor resolution, respectively, which is to avoid the drastic change of tensor resolution during feature transformation. The convolution filter is set to 3×3, considering the issue of local imaging incongruity. Besides, we also carefully set the resolution and folded pixels of each generation stage to ensure that the output tensor of current stage can be integrated into the next stage. Similar to StyleGAN2 [20], the style vector w is injected into the ModConv layers via modulating their convolution filter, *i.e.*, being mapped to scale vector s with an affine network. Finally, the recovered pixel tensors are linearly projected onto RGB space as the output of each stage, which are then aggregated for the final regression. Due to our efficient modeling strategy, PixelFolder uses only 12 convolution layers in all generation stages, thus having much fewer parameters than the SOTA methods like StyleGAN2 [20] and CIPS [2].

5   Experiments

To validate the proposed PixelFolder, we conduct extensive experiments on two benchmark datasets[1], namely Flickr Faces-HQ [19] and LSUN Church [42], and compare it with a set of state-of-the-art (SOTA) methods including CIPS [2], StyleGAN2 [20] and INR-GAN [45].

5.1   Datasets

Flickr Faces-HQ (FFHQ) [19] consistes of $70,000$ high-quality human face images, which all have a resolution of 1024×1024. The images were crawled from Flickr and automatically aligned and cropped.
LSUN Church is the sub-dataset of Large-scale Scene UNderstanding(LSUN) benchmark [42]. It contains about $126,000$ images of churches in various architectural styles, which are collected from natural surroundings.

5.2   Metrics

To validate the proposed PixelFolder, we conduct evaluations from the aspects of image quality and model efficiency, respectively. The metrics used for image quality include *Fréchet Inception Distance* (FID) [12] and *Precision and Recall* (P&R) [24,43]. FID measures the distance between the real images and the generated ones from the perspective of mean and covariance matrix. P&R evaluates

---

[1] More experiments on other datasets and high-resolution are available in the supplementary material.

the ability of fitting the true data distribution. Specifically, for each method, we randomly generate 50,000 images for evaluation. In terms of model efficiency, we adopt the number of parameters (#Params), *Giga Multiply Accumulate Operations* (GMACs) [13], and generation speed (im/s) to measure model compactness, computation overhead and model inference, respectively.

Table 1. Comparison between PixelFolder, StyleGAN2, CIPS and INR-GAN in terms of parameter size (#Params), computation overhead (GMACs) and inference speed. Here, "M" denotes millions, and "im/s" is image per-second. ↑ denotes that lower is better, while ↓ is *vice verse*. PixelFolder is much superior than other methods in both model compactness and efficiency, which well validates its innovative designs.

	#Parm (M) ↓	GMACs ↓	Speed (im/s) ↑
INR-GAN [45]	107.03	38.76	**84.55**
CIPS [2]	44.32	223.36	11.005
StyleGAN2 [20]	30.03	83.77	44.133
PixelFolder (ours)	**20.84**	**23.78**	77.735

5.3   Implementation

In terms of the generation network, we deploy three generation stages for PixelFolder, and their resolutions are set to 16, 64 and 256, respectively. In these operations, the scale of folding and unfolding k is set to 2, *i.e.*, the size of local patches is 2×2. The dimensions of initialized tensors are all 512, except for the last stage which is set to 128. Then these initialized tensors are all reduced to 32 via linear projections before pixel folding. The recovered pixel tensors after pixel unfolding are also projected to RGB by linear projections. For the discriminator, we use a residual convolution network following the settings in StyleGAN2 [20] and CIPS [2], which has *FusedLeakyReLU* activation functions and minibatch standard deviation layers [18].

In terms of training, we use *non-saturating logistic GAN* loss [20] with *R1* penalty [37] to optimize PixelFolder. *Adam* optimizer [22] is used with a learning rate of 2×10^{-3}, and its hyperparameters β_0 and β_1 are set to 0 and 0.99, respectively. The batch size is set to 32, and the models are trained on 8 NVIDIA V100 32GB GPUs for about four days.

5.4   Quantitative Analysis

Comparison with the State-of-the-Arts. We first compare the efficiency of PixelFolder with CIPS [2], StyleGAN2 [20] and INR-GAN [45] in Table 1. From this table, we can find that the advantages of PixelFolder in terms of parameter size, computation complexity and inference speed are very obvious. Compared with CIPS, our method can reduce parameters by 53%, while the reduction in computation complexity (GMACs) is more distinct, about 89%. The inference

Table 2. The performance comparison of PixelFolder and the SOTA methods on FFHQ [20] and LSUN Church [42]. The proposed PixelFolder not only has better performance than existing pixel synthesis methods, *i.e.*, INR-GAN and CIPS, but also achieves new SOTA performance on both benchmarks.

Method	FFHQ, 256×256			LSUN Church, 256×256		
	FID ↓	Precision ↑	Recall ↑	FID ↓	Precision ↑	Recall ↑
INR-GAN [45]	4.95	0.631	0.465	4.04	0.590	0.465
CIPS [2]	4.38	0.670	0.407	2.92	0.603	0.474
StyleGAN2 [20]	3.83	0.661	**0.528**	3.86	-	-
PixelFolder(Ours)	**3.77**	**0.683**	0.526	**2.45**	**0.630**	**0.542**

Table 3. Ablation study on FFHQ. The models of all settings are trained with 200k steps for a quick comparison. These results show the obvious advantages of pixel folding (Fold+Unfold) over down-sampling and DeConv.

Settings	#Parm (M) ↓	GMACs ↓	FID ↓	Precision ↑	Recall ↑
Fold+Unfold (base)	**20.84**	**23.78**	**5.49**	**0.679**	**0.514**
Fold+DeConv	29.41	86.53	5.60	0.667	0.371
Down-Sampling+DeConv	29.21	89.38	5.53	0.679	0.456

speed is even improved by about 7×. These results strongly confirm the validity of our progressive modeling paradigm and pixel folding operations applied to PixelFolder. Meanwhile, compared with StyleGAN2, the efficiency of PixelFolder is also superior, which reduces 31% parameters and 72% GMACs and speed up the inference by about 76%. Also as a multi-stage method, INR-GAN is still inferior to the proposed PixelFolder in terms of parameter size and computation overhead, *i.e.*, nearly 5× more parameters and 1.6× more GMACs. In terms of inference, INR-GAN is a bit faster mainly due to its optimized implementation[2]. Conclusively, these results greatly confirm the superior efficiency of PixelFolder over the compared image generation methods.

We further benchmark these methods on FFHQ and LUSN Church, of which results are given in Table 2. From this table, we can first observe that on all metrics of two datasets, the proposed PixelFolder greatly outperforms the latest pixel synthesis network, *i.e.*, CIPS [2] and INR-GAN [45], which strongly validates the motivations of our method about efficient pixel synthesis. Meanwhile, we can observe that compared to StyleGAN2, PixelFolder is also very competitive and obtains new SOTA performance on FFHQ and LSUN Church, *i.e.*, 3.77 FID and 2.45 FID, respectively. Overall, these results suggest that PixelFolder is a method of great potential in image generation, especially considering its high efficiency and low expenditure.

---

[2] INR-GAN optimizes the CUDA kernels to speed up inference.

Ablation Studies. We further ablates pixel folding operations on FFHQ, of which results are given in Table 3. Specifically, we replace the pixel folding and unfolding with down-sampling and deconvolution (DeConv.) [20], respectively.

Table 4. Ablation study on LSUN Church. The models of all settings are trained with 200k steps for a quick comparison. "*w/o design*" is not cumulative and only represents the performance of PixelFolder without this design/setting.

Settings	#Parm (M) ↓	GMACs ↓	FID ↓	Precision ↑	Recall ↑
PixelFolder	20.84	23.78	**4.78**	**0.602**	**0.517**
w/o coordinate embeddings	**20.32**	**23.64**	4.95	0.598	0.500
w/o multi-stage connection	20.84	23.78	5.46	0.532	0.441

Fig. 4. Comparison of the image interpolations by CIPS [2] and PixelFolder. The interpolation is computed by $z = \alpha z_1 + (1 - \alpha)z_2$, where z_1 and z_2 refer to the left-most and right-most samples, respectively.

From these results, we can observe that although these operations can also serve to reduce or recover tensor resolutions, their practical effectiveness is much inferior than our pixel folding operations, *e.g.* 5.49 FID (fold+unfold) *v.s.* 8.36 FID (down-sampling+DeConv). These results greatly confirm the merit of pixel folding in preserving pixel-wise independence, which can help the model exploit pixel-wise prior knowledge. In Table 4, we examine the initialization of pixel tensor and the impact of multi-stage connection. From this table, we can see that only using Fourier features without coordinate embeddings slightly reduces model performance, but this impact is smaller than that in CIPS [2]. This result also subsequently suggests that PixelFolder do not rely on large parameterized tensors to store pixel-wise prior knowledge, leading to better model compactness. Meanwhile, we also notice that without the multi-stage connection, the performance drops significantly, suggesting the importance of joint multi-scale pixel regression, as discussed in Sect. 4.1. Overall, these ablation results well confirm the effectiveness of the designs of PixelFolder.

5.5   Qualitative Analysis

To obtain deep insight into the proposed PixelFolder, we further visualize its synthesized images as well as the ones of other SOTA methods.

Comparison with CIPS. We first compare the image interpolations of PixelFolder and CIPS on two benchmarks, *i.e.*, FFHQ and LSUN Church, as shown in Fig. 4. It can be obviously seen that the interpolations by PixelFolder are more natural and reasonable than those of CIPS, especially in terms of local imaging. We further present more images synthesized by two methods in Fig. 1 and Fig. 5. From these examples, a quick observation is that the overall image quality of PixelFolder is better than CIPS. The synthesized faces by PixelFolder look more natural and vivid, which also avoid obvious deformations. Meanwhile, the surroundings and backgrounds of the generated church images by PixelFolder are more realistic and reasonable, as shown in Fig. 5c–5d. In terms of local imaging, the merit of PixelFolder becomes more obvious. As discussed in this paper, CIPS is easy to produce local pixel incongruities due to its relatively independent pixel modeling strategy [2]. This problem is reflected in its face generations, especially the hair details. In contrast, PixelFolder well excels in local imaging, such as the synthesis of accessories and hat details, as shown in Fig. 5a–5b. Meanwhile, CIPS is also prone to wavy textures and distortions in the church images, while these issues are greatly alleviated by PixelFolder. Conclusively, these findings well validate the motivations of PixelFolder for image generation.

(a) FFHQ-eyeglasses   (b) FFHQ-headwear

(c) LSUN Church-wavy texture   (d) LSUN Church-pixel offset

Fig. 5. Comparison of the generated images by CIPS [2] and PixelFolder on FFHQ and LSUN Church. The overall quality of images generated by PixelFolder is better than that of CIPS. Meanwhile, PixelFolder can better handle the local imagery incongruity, confirming the effectiveness of its designs.

Comparison of Stage-Wise Visualizations. We also compare PixelFolder with CIPS, StyleGAN2 and INR-GAN by visualizing their stage-wise results, as shown in Fig. 6. From these examples, we can first observe that the intermediate results of other progressive methods, *i.e.*, StyleGAN2 and INR-GAN, are too blurry to recognize. In contrast, PixelFolder and CIPS can depict the outline of generated faces in the initial and intermediate stages. This case suggests that PixelFolder and CIPS can well exploit the high-frequency information provided by Fourier features [2], verifying the merits of end-to-end pixel regression. We can also see that PixelFolder can learn more details than CIPS in the intermediate features, which also suggests the superior efficiency of PixelFolder in face generation. Meanwhile, the progressive refinement (from left to right) also makes PixelFolder more efficient than CIPS in computation overhead and memory footprint. We attribute these advantages to the pixel folding operations and the multi-stage paradigm of PixelFolder, which can help the model exploit prior knowledge in different generation stages.

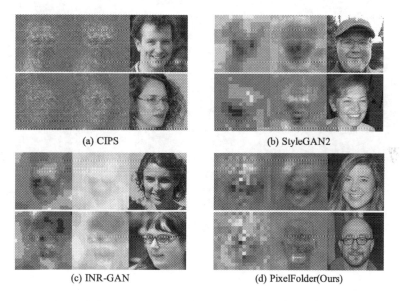

(a) CIPS   (b) StyleGAN2

(c) INR-GAN   (d) PixelFolder(Ours)

Fig. 6. Comparison of the stage-wise synthesis by the SOTA methods and PixelFolder. The color spaces of the first two hidden images are uniformly adjusted for better observation. We chose the hidden images of all methods from the same number of convolution layers. Pixel-synthesis based methods, such as CIPS [2] and PixelFolder, present more interpretable results in initial steps, where PixelFolder can also provide better outline details.

Comparison of Pixel Folding and Its Alternatives. In Fig. 7, we visualize the generations of PixelFolder with pixel folding operations and the alternatives mentioned in Table 3. From these examples, we can find that although down-sampling and DeConv. can also serve to change the resolution of hidden pixel tensors, their practical effectiveness is still much inferior than that of pixel folding. We attribute these results to the unique property of pixel folding in preserving pixel-wise prior knowledge for end-to-end pixel regression. Meanwhile, we also note that when using these alternatives, there is still the problem of local image incongruity, which however can be largely avoided by pixel foldings. These results greatly validate the motivation and effectiveness of the pixel folding operations.

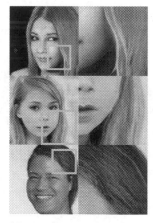

(a) folding+unfolding   (b) folding+DeConv   (c) downsample+DeConv

Fig. 7. Comparisons of PixelFolder with pixel folding operations (folding+unfolding) and the alternatives (*i.e.*, folding+DeConv. and down-sampling+DeConv). Compared with these alternatives, pixel folding operations can well preserve pixel-wise prior knowledge for generation, leading to much better image quality. Meanwhile, pixel folding can also well tackle with local imagery incongruities.

6   Conclusions

In this paper, we propose a novel pixel synthesis network towards efficient image generation, termed *PixelFolder*. Specifically, PixelFolder considers the pixel synthesis as a problem of progressive pixel regression, which can greatly reduce the excessive overhead caused by large tensor transformations. Meanwhile, we also apply novel *pixel folding* operations to further improve model efficiency while preserving the property of end-to-end pixel regression. With these novel designs, PixelFolder requires much less computational and memory overhead than the latest pixel synthesis methods, such as CIPS and INR-GAN. Meanwhile, compared with the state-of-the-art method StyleGAN2, PixelFolder is also more efficient.

With much higher efficiency, the proposed PixelFolder exhibits new SOTA performance on FFHQ and LSUN Church benchmarks, *i.e.*, 3.77 FID and 2.45 FID, respectively, yielding a great potential in image generation.

Acknowledgement. This work is supported by the National Science Fund for Distinguished Young (No. 62025603), the National Natural Science Foundation of China (No. 62025603, No. U1705262, No. 62072386, No. 62072387, No. 62072389, No. 62002305, No.61772443, No. 61802324 and No. 61702136) and Guangdong Basic and Applied Basic Research Foundation (No. 2019B1515120049).

References

1. Afifi, M., Brubaker, M.A., Brown, M.S.: HistoGAN: controlling colors of GAN-generated and real images via color histograms. In: Proceedings of the IEEE Conference on Computer Vision and Pattern Recognition, pp. 7941–7950 (2021)
2. Anokhin, I., Demochkin, K., Khakhulin, T., Sterkin, G., Lempitsky, V., Korzhenkov, D.: Image generators with conditionally-independent pixel synthesis. In: Proceedings of the IEEE Conference on Computer Vision and Pattern Recognition, pp. 14278–14287 (2021)
3. Arjovsky, M., Chintala, S., Bottou, L.: Wasserstein generative adversarial networks. In: International Conference on Machine Learning, pp. 214–223. PMLR (2017)
4. Brock, A., Donahue, J., Simonyan, K.: Large scale GAN training for high fidelity natural image synthesis. In: International Conference on Learning Representations (2018)
5. Chen, X., Duan, Y., Houthooft, R., Schulman, J., Sutskever, I., Abbeel, P.: Info-GAN: interpretable representation learning by information maximizing generative adversarial nets. In: Proceedings of the International Conference on Neural Information Processing Systems, pp. 2180–2188 (2016)
6. Choi, Y., Choi, M., Kim, M., Ha, J.W., Kim, S., Choo, J.: StarGAN: unified generative adversarial networks for multi-domain image-to-image translation. In: Proceedings of the IEEE Conference on Computer Vision and Pattern Recognition, pp. 8789–8797 (2018)
7. Denton, E.L., Chintala, S., Fergus, R., et al.: Deep generative image models using a Laplacian pyramid of adversarial networks. In: Advances in Neural Information Processing Systems, vol. 28 (2015)
8. Dolhansky, B., Ferrer, C.C.: Eye in-painting with exemplar generative adversarial networks. In: Proceedings of the IEEE Conference on Computer Vision and Pattern Recognition, pp. 7902–7911 (2018)
9. Goodfellow, I., et al.: Generative adversarial nets. In: Advances in Neural Information Processing Systems, vol. 27 (2014)
10. Gulrajani, I., Ahmed, F., Arjovsky, M., Dumoulin, V., Courville, A.C.: Improved training of Wasserstein GANs. In: Advances in Neural Information Processing Systems, vol. 30 (2017)
11. He, Z., Kan, M., Shan, S.: EigenGAN: layer-wise eigen-learning for GANs. In: Proceedings of the IEEE International Conference on Computer Vision, pp. 14408–14417 (2021)

12. Heusel, M., Ramsauer, H., Unterthiner, T., Nessler, B., Hochreiter, S.: GANs trained by a two time-scale update rule converge to a local Nash equilibrium. In: Advances in Neural Information Processing Systems, vol. 30 (2017)

13. Howard, A.G., et al.: MobileNets: efficient convolutional neural networks for mobile vision applications. arXiv preprint arXiv:1704.04861 (2017)

14. Huang, X., Belongie, S.: Arbitrary style transfer in real-time with adaptive instance normalization. In: Proceedings of the IEEE International Conference on Computer Vision, pp. 1501–1510 (2017)

15. Hudson, D.A., Zitnick, C.L.: Generative adversarial transformers. IN: Advances in Neural Information Processing Systems, vol. 139 (2021)

16. Isola, P., Zhu, J.Y., Zhou, T., Efros, A.A.: Image-to-image translation with conditional adversarial networks. In: Proceedings of the IEEE Conference on Computer Vision and Pattern Recognition, pp. 1125–1134 (2017)

17. Ji, J., Ma, Y., Sun, X., Zhou, Y., Wu, Y., Ji, R.: Knowing what to learn: a metric-oriented focal mechanism for image captioning. IEEE Trans. Image Process. **31**, 4321–4335 (2022). https://doi.org/10.1109/TIP.2022.3183434

18. Karras, T., Aila, T., Laine, S., Lehtinen, J.: Progressive growing of GANs for improved quality, stability, and variation. In: International Conference on Learning Representations (2018)

19. Karras, T., Laine, S., Aila, T.: A style-based generator architecture for generative adversarial networks. In: Proceedings of the IEEE Conference on Computer Vision and Pattern Recognition, pp. 4401–4410 (2019)

20. Karras, T., Laine, S., Aittala, M., Hellsten, J., Lehtinen, J., Aila, T.: Analyzing and improving the image quality of StyleGAN. In: Proceedings of the IEEE Conference on Computer Vision and Pattern Recognition, pp. 8110–8119 (2020)

21. Kim, H., Choi, Y., Kim, J., Yoo, S., Uh, Y.: Exploiting spatial dimensions of latent in GAN for real-time image editing. In: Proceedings of the IEEE Conference on Computer Vision and Pattern Recognition, pp. 852–861 (2021)

22. Kingma, D.P., Ba, J.: Adam: a method for stochastic optimization. arXiv preprint arXiv:1412.6980 (2014)

23. Kingma, D.P., Welling, M.: Auto-encoding variational Bayes. arXiv preprint arXiv:1312.6114 (2013)

24. Kynkäänniemi, T., Karras, T., Laine, S., Lehtinen, J., Aila, T.: Improved precision and recall metric for assessing generative models. In: Advances in Neural Information Processing Systems, vol. 32 (2019)

25. Lee, K., Chang, H., Jiang, L., Zhang, H., Tu, Z., Liu, C.: VitGAN: training GANs with vision transformers. arXiv preprint arXiv:2107.04589 (2021)

26. Li, B., Qi, X., Lukasiewicz, T., Torr, P.: Controllable text-to-image generation. In: Advances in Neural Information Processing Systems, vol. 32 (2019)

27. Li, X., et al.: Image-to-image translation via hierarchical style disentanglement. In: Proceedings of the IEEE Conference on Computer Vision and Pattern Recognition, pp. 8639–8648 (2021)

28. Liang, J., Zeng, H., Zhang, L.: High-resolution photorealistic image translation in real-time: a Laplacian pyramid translation network. In: Proceedings of the IEEE Conference on Computer Vision and Pattern Recognition, pp. 9392–9400 (2021)

29. Lin, C.H., Chang, C.C., Chen, Y.S., Juan, D.C., Wei, W., Chen, H.T.: Coco-GAN: generation by parts via conditional coordinating. In: Proceedings of the IEEE International Conference on Computer Vision, pp. 4512–4521 (2019)

30. Lin, J., Zhang, R., Ganz, F., Han, S., Zhu, J.Y.: Anycost GANs for interactive image synthesis and editing. In: Proceedings of the IEEE Conference on Computer Vision and Pattern Recognition, pp. 14986–14996 (2021)

31. Liu, H., Navarrete Michelini, P., Zhu, D.: Deep networks for image-to-image translation with mux and demux layers. In: Leal-Taixé, L., Roth, S. (eds.) ECCV 2018. LNCS, vol. 11133, pp. 150–165. Springer, Cham (2019). https://doi.org/10.1007/978-3-030-11021-5_10

32. Liu, R., et al.: An intriguing failing of convolutional neural networks and the Coord-Conv solution. In: Advances in Neural Information Processing Systems, vol. 31 (2018)

33. Liu, R., Ge, Y., Choi, C.L., Wang, X., Li, H.: DivCo: diverse conditional image synthesis via contrastive generative adversarial network. In: Proceedings of the IEEE Conference on Computer Vision and Pattern Recognition, pp. 16377–16386 (2021)

34. Luo, G., et al.: Towards language-guided visual recognition via dynamic convolutions. arXiv preprint arXiv:2110.08797 (2021)

35. Luo, G., et al.: Towards lightweight transformer via group-wise transformation for vision-and-language tasks. IEEE Trans. Image Process. **31**, 3386–3398 (2022)

36. Ma, Y., et al.: Knowing what it is: semantic-enhanced dual attention transformer. IEEE Trans. Multimedia, 1 (2022). https://doi.org/10.1109/TMM.2022.3164787

37. Mescheder, L., Geiger, A., Nowozin, S.: Which training methods for GANs do actually converge? In: International Conference on Machine Learning, pp. 3481–3490. PMLR (2018)

38. Park, T., Efros, A.A., Zhang, R., Zhu, J.-Y.: Contrastive learning for unpaired image-to-image translation. In: Vedaldi, A., Bischof, H., Brox, T., Frahm, J.-M. (eds.) ECCV 2020. LNCS, vol. 12354, pp. 319–345. Springer, Cham (2020). https://doi.org/10.1007/978-3-030-58545-7_19

39. Park, T., et al.: Swapping autoencoder for deep image manipulation. Adv. Neural. Inf. Process. Syst. **33**, 7198–7211 (2020)

40. Patashnik, O., Wu, Z., Shechtman, E., Cohen-Or, D., Lischinski, D.: StyleCLIP: text-driven manipulation of StyleGAN imagery. In: Proceedings of the IEEE International Conference on Computer Vision, pp. 2085–2094 (2021)

41. Peng, J., et al.: Knowledge-driven generative adversarial network for text-to-image synthesis. IEEE Trans. Multimedia (2021)

42. Radford, A., Metz, L., Chintala, S.: Unsupervised representation learning with deep convolutional generative adversarial networks. arXiv preprint arXiv:1511.06434 (2015)

43. Sajjadi, M.S., Bachem, O., Lucic, M., Bousquet, O., Gelly, S.: Assessing generative models via precision and recall. In: Advances in Neural Information Processing Systems, vol. 31 (2018)

44. Shi, W., et al.: Real-time single image and video super-resolution using an efficient sub-pixel convolutional neural network. In: Proceedings of the IEEE Conference on Computer Vision and Pattern Recognition, pp. 1874–1883 (2016)

45. Skorokhodov, I., Ignatyev, S., Elhoseiny, M.: Adversarial generation of continuous images. In: Proceedings of the IEEE Conference on Computer Vision and Pattern Recognition, pp. 10753–10764 (2021)

46. Tang, H., Bai, S., Zhang, L., Torr, P.H.S., Sebe, N.: XingGAN for person image generation. In: Vedaldi, A., Bischof, H., Brox, T., Frahm, J.-M. (eds.) ECCV 2020. LNCS, vol. 12370, pp. 717–734. Springer, Cham (2020). https://doi.org/10.1007/978-3-030-58595-2_43

47. Van Den Oord, A., Vinyals, O., et al.: Neural discrete representation learning. In: Advances in Neural Information Processing Systems, vol. 30 (2017)

48. Wang, Y., Qi, L., Chen, Y.C., Zhang, X., Jia, J.: Image synthesis via semantic composition. In: Proceedings of the IEEE International Conference on Computer Vision, pp. 13749–13758 (2021)
49. Wang, Y., et al.: HifiFace: 3D shape and semantic prior guided high fidelity face swapping. arXiv preprint arXiv:2106.09965 (2021)
50. Xu, T., et al.: AttnGAN: fine-grained text to image generation with attentional generative adversarial networks. In: Proceedings of the IEEE Conference on Computer Vision and Pattern Recognition, pp. 1316–1324 (2018)
51. Yu, J., Lin, Z., Yang, J., Shen, X., Lu, X., Huang, T.S.: Generative image inpainting with contextual attention. In: Proceedings of the IEEE Conference on Computer Vision and Pattern Recognition, pp. 5505–5514 (2018)
52. Zhang, H., et al.: StackGAN++: realistic image synthesis with stacked generative adversarial networks. IEEE Trans. Pattern Anal. Mach. Intell. **41**(8), 1947–1962 (2018)
53. Zhang, Z., Xie, Y., Yang, L.: Photographic text-to-image synthesis with a hierarchically-nested adversarial network. In: Proceedings of the IEEE Conference on Computer Vision and Pattern Recognition, pp. 6199–6208 (2018)
54. Zhu, J.Y., Park, T., Isola, P., Efros, A.A.: Unpaired image-to-image translation using cycle-consistent adversarial networks. In: Proceedings of the IEEE International Conference on Computer Vision, pp. 2223–2232 (2017)
55. Zhu, P., Abdal, R., Qin, Y., Wonka, P.: SEAN: image synthesis with semantic region-adaptive normalization. In: Proceedings of the IEEE Conference on Computer Vision and Pattern Recognition, pp. 5104–5113 (2020)

StyleSwap: Style-Based Generator Empowers Robust Face Swapping

Zhiliang Xu[1], Hang Zhou[1(✉)], Zhibin Hong[1], Ziwei Liu[2], Jiaming Liu[1], Zhizhi Guo[1], Junyu Han[1], Jingtuo Liu[1], Errui Ding[1], and Jingdong Wang[1]

[1] Department of Computer Vision Technology (VIS), Baidu Inc., Beijing, China
{xuzhiliang,zhouhang09,liujiaming03,dingerrui,wangjingdong}@baidu.com
[2] S-Lab, Nanyang Technological University, Singapore, Singapore
ziwei.liu@ntu.edu.sg

Abstract. Numerous attempts have been made to the task of person-agnostic face swapping given its wide applications. While existing methods mostly rely on tedious network and loss designs, they still struggle in the information balancing between the source and target faces, and tend to produce visible artifacts. In this work, we introduce a concise and effective framework named **StyleSwap**. Our core idea is to *leverage a style-based generator to empower high-fidelity and robust face swapping, thus the generator's advantage can be adopted for optimizing identity similarity.* We identify that with only minimal modifications, a StyleGAN2 architecture can successfully handle the desired information from both source and target. Additionally, inspired by the ToRGB layers, a *Swapping-Driven Mask Branch* is further devised to improve information blending. Furthermore, the advantage of StyleGAN inversion can be adopted. Particularly, a *Swapping-Guided ID Inversion* strategy is proposed to optimize identity similarity. Extensive experiments validate that our framework generates high-quality face swapping results that outperform state-of-the-art methods both qualitatively and quantitatively.

Keywords: Face swapping · Style based generator · GAN

1 Introduction

The task of face swapping has drawn great attention [4,5,22,23,27,28,32,47,49] due to its wide applications in the fields of entertainment, film making, virtual human creation, privacy protection, *etc.*. It aims at transferring the facial identity from a source person to a target frame or video while preserving attributes information including pose, expressions, lighting condition, and background [27].

Z. Xu and H. Zhou—Equal contribution.

Supplementary Information The online version contains supplementary material available at https://doi.org/10.1007/978-3-031-19781-9_38.

Fig. 1. Qualitative results on 512×512 **resolution.** Our method is robust under complicated conditions and achieves high-fidelity face swapping results

With the development of deep learning, generative models have been leveraged to boost face swapping quality [22,23,26,28,47,50]. However, the task is still challenging, particularly under the identity-agnostic setting where only a single frame is provided as the source image but targeting different sorts of scenarios. The key challenges lie in two essential parts: how to explicitly capture the identity information; and how to blend the swapped face into the target seamlessly while preserving the implicit attributes unchanged.

To tackle the above problems, previous studies take two different paths. **1)** Graphics-based methods [4,5] have involved the strong prior knowledge of intermediate structural representations such as landmarks and 3D models [11] into face swapping long ago. Recent researchers combine this information with generative adversarial networks (GANs) [13] for identity and expression extraction [28,29,47,50]. However, the inaccuracy of structural information would greatly influence the stableness and coherence of generated results, particularly in videos. **2)** Other methods explore pure learning-based pipelines [7,12,23,54]. Most of them rely on tedious loss and network structure designs [7,12,23] for balancing the information between source and target images. Such designs make training difficult and fail in expressing desired information, which leads to non-similar or non-robust results with visible artifacts.

Recently, StyleGAN architectures [17–19] and their variants have been verified effective on various facial generative tasks, including face attributes editing [1,2,36,37], face enhancement [46,51], and even face reenactment [6,52]. It is owing to style-based generator's strong expressibility and its advantages in latent space manipulation. But the exploration of such architectures in face swapping [48,54] is still insufficient. Specifically, the lighting conditions are greatly condemned in Zhu *et al.* [54] due to the limited distribution covered by the fixed generator. The structure of their feature blending procedure is also designed in a hand-crafted and layer-specific manner, which requires complicated human tuning. Concurrently, Xu *et al.* [48] aggregate the StyleGAN2 features with another designed encoder and decoder. Thus, a natural question arises: can we avoid tedious layer-by-layer structure design by adopting a versatile style-based generator [18,19] with only minimal modifications?

To this end, we propose **StyleSwap**, a concise and effective pipeline that empowers face swapping by a style-based generator. It produces results with higher fidelity, identity similarity and is more robust (*i.e.* avoids visible artifacts creation) under different scenarios compared with previous methods. Moreover, it is easy to implement and friendly for training. The key is to *adapt StyleGAN2 architecture to face swapping data flows through simple modifications, and adopt the generator's advantage for identity optimization.* Detailedly, we first achieve the restoration of the target image's attributes with a simple layer-fusion strategy. The same idea has been proven to maintain the original StyleGAN's capability [51]. Then we argue that the identity information can be injected by mapping extracted identity features to the $\mathcal{W}$ space. In this way, the identity information can be implicitly blended into the attributes in the convolution operations. Additionally, we propose a *Swapping-Driven Mask Branch* which is identical to the *ToRGB* branch. It naturally enforces the network to focus less on the target's high-level information and benefits final image blending.

We further illustrate the advantages of this architecture by involving a simple optimization strategy for improving identity similarity. As the identity feature is mapped to the $\mathcal{W}$ space, a natural inspiration from recent StyleGAN inversion studies [1,2] is to optimize a powerful $\mathcal{W}^+$ space through self-reconstruction. To avoid mode collapse, we introduce a novel *Swapping-Guided ID Inversion* strategy by iteratively performing feature optimization and face swapping. Armed with these tools, we show that our StyleSwap generates high-fidelity results with simple video training paradigms. It is particularly robust and can be supported with enhanced data for generating high-resolution results.

We summarize our contributions as follows: **(1)** We present the **StyleSwap** framework, which adopts a style-based generator into the person-agnostic face swapping task by simple modifications and the design of a *Swapping-Driven Mask Branch.* It is easy to implement and train. **(2)** By taking the advantage of the StyleGAN model, we design the novel *Swapping-Guided ID Inversion* strategy to improve the identity similarity. **(3)** Extensive experiments demonstrate that our method outperforms the state of the arts on person-agnostic face swapping. Particularly, it demonstrates great robustness and has the capability of generating high-quality results.

2   Related Work

2.1   Face Swapping

Structural Prior-Guided Face Swapping. The task of face swapping has long been a research interest for both the computer graphics and computer vision community. Structural information such as 3D models and landmarks provide strong prior knowledge. Blanz *et al.* [5] leverage 3DMM, and Bitouk *et al.* use 3D lighting basis to design adjustment-based methods. Both of them rely on manual interaction, and can hardly change the source's expressions. Nirkin *et al.* [29] involve 3DMM with learned masks, but they render unrealistic results. Recent studies [15,28,47,50] combine structural information with GANs for identity-agnostic face swapping. Xu *et al.* and Wang *et al.* both inject the parameters of 3DMM into self-designed architectures. Though high-fidelity results can be generated, the inaccuracy of 3D models and the need for inpainting greatly harm the temporal coherence and robustness of these methods under the video face swapping setting.

Reconstruction-Based Face Swapping. On the other hand, pure reconstruction based methods with GANs have also shown success. Korshunova *et al.* [22] train a network for swapping paired identities. The popular Deepfakes and DeepFaceLab [32] share the same setting. However, these methods cannot generalize to arbitrary identities, which limits their practical usage.

As for person-agnostic face swapping, Li *et al.* [23] build the Faceshifter network. SimSwap [7] improves the expression consistency. However, they generate low-quality results with visible artifacts under certain circumstances. Recently, InfoSwap [12] creates high quality results by building a pipeline that relies on careful loss designs. It involves multi-stage of finetuning on various datasets. Different from these methods, we would like to ease the network design procedure with a style-based generator.

Specifically, Wang *et al.* [45] firstly leverage a pretrained StyleGAN generator for high resolution face swapping. However, in order to adapt to the latent spaces of a pretrained StyleGAN generator, the authors design layer-specific fusion strategies, which involves a large number of hyper-parameters and ablative studies. Moreover, their method cannot keep the lighting conditions of the target frames. In our work, we retrain the style-based generator with simple modifications that preserve the attribute information better.

2.2   Facial Editing with Style-Based Generator

The strong ability of StyleGAN [17–19] has been shown in various facial editing tasks including facial attributes editing [1,2,33,36,37,41], blind face restoration [46,51], face reenactment [6,24,52], hairstyle editing [53], and so on [3,40,41,53].

StyleGAN Inversion. Most face attribute editing framework [36,37] fix the pretrained generator unchanged and perform StyleGAN inversion. Abdal *et al.* [1,2] expand the original $\mathcal{W}$ latent space to the $\mathcal{W}^+$ space during inversion and achieve better image reconstruction results. Recent studies invert images with StyleGAN specific encoders [3,33,41] for fast inversion. In our work, we take the inspiration of StyleGAN inversion and expand our identity feature to $\mathcal{W}^+$ space for boosting identity similarity. The usage of the StyleGAN specific encoders is left as a future work.

Face Reenactment with Style-Based Generator. Face reenactment is very similar to face swapping and even serves as part of the face swapping procedure in certain methods [28,32,38]. The difference is that it aims at keeping the source image's identity and background. Burkov *et al.* [6] encode identity and expression information into the $\mathcal{W}$ space and re-train the generator in a simple pipeline. Later studies [24,52] then expand this pipeline to the audio-driven setting.

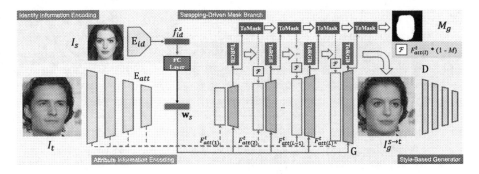

Fig. 2. Our StyleSwap framework. The building blocks in **Blue** indicate the original structure of StyleGAN2. The source I_s is encoded to f_{id}^s by E_{id} (**Red**). The attribute feature maps $\mathbf{F}_{att}^t$ are encoded from I_t by E_{att} and concatenated to the StyleGAN2 generator blocks (**Yellow**). Specifically, we devise a Swapping-Driven Mask Branch (**Green**) to predict a mask M_g by leveraging the same structure of ToRGB layers (Color figure online)

3   Our Approach

Framework Overview. The goal of person-agnostic face swapping is to swap the identity information from a source image I_s onto a target frame I_t while preserving I_t's attribute information unchanged.

In this section, we present our **StyleSwap** framework (illustrated in Fig. 2) which empowers face swapping by slightly modifying a style-based generator. In Sect. 3.1 we introduce how to adapt a style-based generator to the face swapping task. Section 3.2 illustrates the simple training paradigm. Importantly, in Sect. 3.3, we propose the *Swapping-Guided ID Inversion* which specifically takes the advantage of GAN inversion for optimizing identity similarity.

3.1   Adapting Style-Based Generator to Face Swapping

Revisiting StyleGAN2. We first recap the general setting of StyleGAN2 [19]. The original generator takes a constant feature map at the lowest resolution, then a latent vector $\mathbf{z}$ is sampled and mapped to a feature vector $\mathbf{w}$ (lies in the $\mathcal{W}$ space). Afterwards, $\mathbf{w}$ is sent into each layer of the $2L$-layer generator by affine transformations as the "*style*" to modulate the convolutional kernels' weights. At each resolution, a ToRGB layer is designed to render a three-channel RGB image progressively. The rough structures are depicted on Fig. 2 in *blue*.

The attribute disentanglement ability in StyleGAN2 is implicitly achieved in the $\mathcal{W}$ or the expanded $\mathcal{W}^+$ space (if feeding different $\mathbf{w}$ features to different layers). As a result, operations on existing faces require inverting faces to latent vectors, which harms the preservation of spatial information. In our face swapping task, the problem lies in how to modify the generator so that the information of the target frame and identity can be sufficiently used.

Infusing Attribute Information. We propose to infuse the spatial information of the target frame as feature maps, rather than feature vectors, to preserve attribute information. A recent face restoration work [51] verifies two important properties of StyleGAN2. Firstly, concatenating a noise map to each layer of the generator would not affect the network's generative ability. Secondly, such noise maps can be replaced by encoded spatial feature maps so that both the generative prior of StyleGAN2 model and the structural information of the input image can be kept.

Inspired by this observation, we apply a similar modification. Specifically, we leverage a simple encoder E_{att} that encodes I_t to different scales of spatial feature maps $\mathbf{F}_{att}^t = \{F_{att(l)}^t | l \in [1, L]\}$. They are then infused into the StyleGAN2 architecture at each $2l$-th layer of the generator by concatenation following [51].

Injecting Identity Feature. In order to adapt to different target views, it is natural to encode the identity information into feature vectors. Here we use a pretrained identity encoder provided by ArcFace [10] to encode the identity feature $f_{id}^s = E_{id}(I_s)$.

As the attributes on the face are already fused, we identify that modulated convolutions in StyleGAN2 are naturally suitable for the blending and shape-shifting of facial organs. Thus we directly map f_{id}^s to $\mathbf{w}_s$ in the $\mathcal{W}$ space ($\mathbf{w}_s = FC_{\mathbf{w}}(f_{id}^s)$) through fully connected layers $FC_{\mathbf{w}}$.

So far, a swapping result $I_g^{s \to t} = G(\mathbf{F}_{att}^t, \mathbf{w}_s)$ from the source to the target frame can already be rendered by the generator.

Swapping-Driven Mask Branch. We then identify that learning a rough one-channel facial mask would benefit the whole face swapping step from two perspectives. 1) With a mask in the image domain, the areas that do not require modifications such as the background and hair can be directly kept unchanged. 2) If the mask can be progressively and implicitly learned along with the generator, coarse masks at lower resolutions can help balance attribute and identity information in a similar way as [23]. We thus propose an additional modification by devising a *Swapping-Driven Mask Branch* which takes the advantage of the StyleGAN2 model. Its structure is directly borrowed from the "ToRGB" branch, and denoted as "ToMask". Here we leverage a soft mask with values between 0 and 1. The details of the mask branch are illustrated in Fig. 3 (a).

Let $M_{g(l)}'$ denote the one-channel output of the l-th ToMask network, which has the same resolution with the output of the l-th ToRGB layer and the $(l+1)$-th $\mathbf{F}_{att}^t$. The non-normalized mask $\tilde{M}_{g(l)}$ at the l-th layer of the mask branch is the combination of the $(l-1)$-th layer's result and $M_{g(l)}'$.

$$\tilde{M}_{g(l)} = upsample(\tilde{M}_{g(l-1)}) + M_{g(l)}', \tag{1}$$

where the bilinear *upsample* is used. The softmasks we used are the normalized results $M_{g(l)} = \text{Sigmoid}(\tilde{M}_{g(l)})$. $M_{g(1)}$ is the normalized output of the first ToMask network, and M_g is the final predicted mask. The face swapping result can be updated as:

$$\hat{I}_g^{s \to t} = M_g * I_g^{s \to t} + (1 - M_g) * I_t, \tag{2}$$

where the $*$ denotes element-wise multiplication with broadcasting and $\mathbf{1}$ is the tensor with all ones. The mask repeats itself channel-wise three times to match the RGB channels.

Masking Attribute Information. Our design above infuses facial attribute information into all layers. However, spatial information provided at mid- and low-level layers might influence facial structures. Though we expect the network to automatically perform information balancing, we identify that this procedure can be eased by blocking the attribute information relying on an implicitly learned mask. Thus we multiply our learned mask $M_{g(l)}$ at each resolution with the next layer's attribute feature map $F_{att(l+1)}^t$ to an updated version:

$$\hat{F}_{att(l+1)}^t = F_{att(l+1)}^t * (\mathbf{1} - M_{g(l)}).\tag{3}$$

Note that $M_{g(l)}$ and $F_{att(l+1)}^t$ share the same spatial resolution, and there is no masking operation on $F_{att(1)}^t$ which provides the initial facial attributes. In this way, as $M_{g(l)}$ progressively grows to reach the ground truth mask, it also implicitly prevents the attribute information from influencing the identity similarity.

3.2   Training Paradigm

With our StyleSwap architecture, person-agnostic face swapping results can be learned in a simple end-to-end training paradigm. Particularly, we propose to involve certain video data for training a more robust swapping model.

Given a source image I_s and a target video $\{I_T | T \in \{t(1) \dots t(K)\}\}$, we generate the following frames: **(1)** The face swapping results from the source to any target frame: $I_g^{s \to t} = \mathrm{G}(\mathbf{F}_{att}^t, \mathbf{w}_s)$, where $\mathbf{F}_{att}^t = \mathrm{E}_{att}(I_t)$ and $\mathbf{w}_s = \mathrm{FC}_{\mathbf{w}}(f_{id}^s) = \mathrm{FC}_{\mathbf{w}}(\mathrm{E}_{id}(I_s))$. **(2)** The self-reconstruction results on the source frame itself: $I_g^{s \to s} = \mathrm{G}(\mathbf{F}_{att}^s, \mathbf{w}_s)$. **(3)** Particularly, we sample two target frames $I_{t(a)}$, $I_{t(b)}$ from a same video of a same person, and generate $I_g^{t(a) \to t(b)} = \mathrm{G}(\mathbf{F}_{att}^{t(b)}, \mathbf{w}_{t(a)})$, $\mathbf{w}_{t(a)} = \mathrm{FC}_{\mathbf{w}}(f_{id}^{t(a)})$.

The training objectives consist of mainly four parts: the identity loss, feature matching loss, adversarial loss applied to all generated results, and the reconstruction loss applied to self- and cross-view reconstruction results.

Identity Loss. The identity loss is built upon the cosine distances $D_{\cos}(f_a, f_b) = \frac{f_a{}^\mathrm{T} \cdot f_b}{\|f_a\|_2 \|f_b\|_2}$ between extracted identity features from E_{id}. Given any sampled data I_i, I_j where $i, j \in \{s, T\}$:

$$\mathcal{L}_{id} = 1 - D_{\cos}(f_{id}^i, \mathrm{E}_{id}(I_g^{i \to j})).\tag{4}$$

Note that in order to disentangle identity information with illuminations, all images are augmented with color jittering when sent to the identity encoder.

Adversarial Loss and Feature Matching Loss. We directly adopt the original discriminator D and adversarial loss functions of StyleGAN2 [18]. For any $I_g^{i \to j} (i, j \in \{s, T\})$, I_j is provided as the real image when applying the adversarial training. We denote this loss function as $\mathcal{L}_{adv}$ and omit the details.

Similar to [7], we leverage a weak feature matching loss from the feature maps of the discriminator.

$$\mathcal{L}_{FM} = \sum_{m=n_D}^{N_D} (\|\mathrm{D}_m(I_g^{i \to j}) - \mathrm{D}_m(I_j)\|_1), \tag{5}$$

where D_m denotes the m-th layer's output of the discriminator, and n_D is the layer that starts computing the feature matching loss. This loss accounts for preserving the expression and certain low-level attribute information.

Reconstruction Loss. The reconstruction loss consists of the L_1 loss and the VGG perceptual loss [30,44] when the pixel-level supervision can be provided.

$$\mathcal{L}_{rec} = \|I_g^{i \to j} - I_j\|_1 + \sum_{m=1}^{N_{vgg}} \|\mathrm{VGG}_m(I_g^{i \to j}) - \mathrm{VGG}_m(I_j)\|_1, \tag{6}$$

where VGG_m denotes the m-th layer's output of a pre-trained VGG19 network. When applying the reconstruction loss, we set $i, j = s$ or $i, j \in \{T\}$. The reconstruction training provides supervision in the pixel space and has proven to be crucial in previous studies. Particularly, cross-view reconstruction is widely used in face reenactment training [6,20,52]. It benefits the attribute information's preservation by providing samples with different expressions on the source and target, and forces the network to learn with strong supervision.

Mask Loss. We leverage a pretrained facial mask predictor [39] to predict only a rough facial mask for each image in the training set. For each generated image $I_g^{i \to j} = \mathrm{G}(\mathbf{F}_{att}^j, \mathbf{w}_i)$, it is supervised with the binary mask of the target frame M_j. The supervision can be written as:

$$\mathcal{L}_{mask} = \mathrm{BCELoss}(M_g, M_j), \tag{7}$$

where BCELoss denotes the point-wise binary cross entropy loss.

It is worth mentioning that the mask branch is a plug-in module that requires fine-tuning. When the **Masking Attribute Information** in Sect. 3.1 is activated, all generated images are updated to $\hat{I}_g^{i \to j}$ as illustrated in Eq. 2 and all $\mathbf{F}_{att}$ are updated according to Eq. 3.

Overall Loss Function. The overall loss function for training the StyleSwap framework is the combination of the losses introduced above. It can be represented as:

$$\mathcal{L}_{total} = \mathcal{L}_{adv} + \lambda_{id}\mathcal{L}_{id} + \lambda_{FM}\mathcal{L}_{FM} + \lambda_{rec}\mathcal{L}_{rec} + \lambda_{mask}\mathcal{L}_{mask}, \tag{8}$$

where the λs are balancing coefficients.

3.3   Swapping-Guided ID Inversion

Inspired by StyleGAN inversion, we illustrate one interesting property of our $\mathcal{W}$ space design: the identity similarity can be optimized in a GAN inversion manner

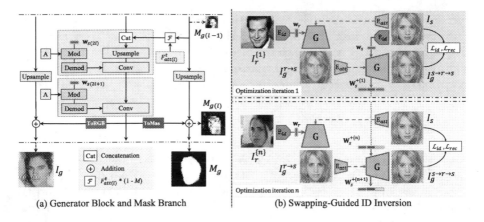

(a) Generator Block and Mask Branch   (b) Swapping-Guided ID Inversion

Fig. 3. (a) The details of the generator blocks and the *Swapping-Guided Mask Branch*. The figures show the block with visualized masks when $l = 4$ and the final outputs. $\mathbf{w}_{s(2l)} = \mathbf{w}_{s(2l+1)} = \mathbf{w}_s$ when the $\mathbf{w}_s$ is not optimized. **(b)** The procedure for *Swapping-Guided ID Inversion*. At each iteration, a different random $I_r^{\{n\}}$ is selected. The first and the n-th iterations are shown. $\mathbf{w}_s$ is taken as initialization and updated to the $\mathcal{W}^+$ space

through a *Swapping-Guided ID Inversion* strategy, where the generator is fixed. As the encoded $\mathbf{w}_s$ feature from the source lies in the $\mathcal{W}$ space, it can naturally be served as a good initialization. We expect to optimize $\mathcal{W}$ or expanded $\mathcal{W}^+$ space for a specific person's identity. The optimization procedure on $\mathcal{W}^+$ space is depicted in Fig. 3 (b). The reconstruction loss $\mathcal{L}_{rec}$ and ID loss $\mathcal{L}_{id}$ are used.

Challenges. We first take $\mathcal{W}$ space optimization as an example, and expand it to the $\mathcal{W}^+$ space. An intuitive operation is to directly optimize $\mathbf{w}_s$ (to any $\mathbf{w}_s + \Delta\mathbf{w}_s$) using gradient descent at each self-reconstruction step $I_g^{s \to s} = G(\mathbf{F}_{att}^s, \mathbf{w}_s + \Delta\mathbf{w}_s)$. However, such practice would easily lead to mode collapse, *i.e.*, direct reconstruction by ignoring the $\mathbf{w}_s$ information. To tackle this problem, an alternative way is to perform face swapping before the optimization step (as shown in the optimization iteration 1 in Fig. 3). A randomly sampled face I_r of arbitrary identity is used to build $I_g^{r \to s}$ which is later sent to the attribute encoder as the target frame. However, this way of restricting $I_g^{s \to r \to s} = G(I_g^{r \to s}, \mathbf{w}_s)$ still relies on one fixed input image, which also leads to mode collapse during implementation.

Iterative Identity Optimization and Swapping. The key to our method is to feed a different face to the source image at each iteration of the optimization procedure. Specifically, at each iteration n, we randomly sample any $I_r^{\{n\}}$ to update the identity information on the source frame, so that the network perceives different identities.

Moreover, a unique advantage of Style-based generator is to optimize the set of $2L$ different style features $\mathbf{W}_s^+ = [\mathbf{w}_{s(1)}, \ldots, \mathbf{w}_{(2L)}]$ as performed in StyleGAN inversion studies [1,2]. At each iteration, it is updated to $\mathbf{W}_s^{+\{n\}}$. After optimizing $\mathbf{W}_s^+$, it can be directly leveraged by the generator to create face swapping result: $I_g^{s\rightarrow t} = \mathbf{G}(\mathbf{F}_{att}^t, \mathbf{W}_s^+)$. Such operation can hardly be performed on traditional face swapping pipelines, and has never been explored before. More details about the optimization algorithm are shown in the Supplementary Materials.

4   Experiments

Datasets. We train our model on the VGGFace [31] and a small part of Vox-Celeb2 [8]. Due to the limitation of data quality, the original data from both datasets only supports 256×256 resolution training. To show our capability of handling higher resolutions, we enhance the datasets with GPEN [51] and train a 512×512 model. Part of the evaluation is conducted on FaceForensics++ (FF++) dataset [34] which contains 1,000 Internet face videos and 1,000 Deep-fakes [9] and 1000 official FaceShifter results. Evaluations on high-quality images leverage CelebA-HQ [16,25].

Implementation Details. Our model is trained with batch size 64 on one NVIDIA Tesla A100 GPU for 256×256 resolution and batch size 12 for 512×512 resolution. We use the ADAM optimizer [21] with learning rate fixed at 1×10^{-4}. The λ_{id} is set at 10, λ_{rec} and λ_{FM} are set at 100. The other coefficients do not affect the generation results much thus empirically set at 1.

A total of $2L = 14$ layers are used on 256×256 resolution images, and $2L = 16$ layers on 512×512 resolution. We leverage the modified ResNet50 provided by Arcface [10] as the identity encoder and fix it through out the experiments. The number of iterations required for optimizing the $\mathcal{W}$ and $\mathcal{W}^+$ for one identity is set empirically at 50. Please refer to supplementary materials for more details.

Competing Methods. We compare our methods with previous state-of-the-arts reconstruction-based face swapping methods. They are **Faceshifter** [23] and **SimSwap** [7] which leverage self-designed network structures; **InfoSwap** [12] that relies on tedious training procedure for high-resolution results; **MegaFS** [54] which borrows a fixed StyleGAN2 pretrained model; and **Deep-fakes** [9] which is a famous open-sourced tool. We use the official Deepfakes and FaceShifter results from the FF++ dataset and the officially released codes and models for producing swapped results of other methods. We refer our results without identity optimization as **StyleSwap** and depict the optimized one separately as **StyleSwap w/ $\mathcal{W}^+$**.

4.1   Quantitative Evaluation

Evaluation Metrics. The quantitative experiments are carried out on the FF++ dataset [34]. Following [23], 10 frames are uniformly sampled from the 1000 videos to get 10K faces for evaluation. We leverage five popularly used metrics. The *ID cosine similarity* and *ID retrieval scores* are measured between swapped results and the source using another pretrained identity-recognition network [43]. Particularly, the calculation process of the ID retrieval score is the

Table 1. Quantitative Results. We report the percentage of successfully retrieved images on the *ID Retrieval* metric. For the ID correlated metrics the higher the better, and it is the lower the better for other metrics

Method\Metric	ID Retrieval ↑	ID Cosine ↑	Pose error↓	Exp. error ↓	FID ↓
Deepfakes [9]	86.43%	0.438	3.96	8.98	4.07
FaceShiter [23]	90.04%	0.510	2.19	6.77	3.50
SimSwap [7]	93.07%	0.578	**1.36**	**5.07**	3.04
MegaFS [54]	89.12%	0.497	3.69	10.12	4.62
InfoSwap [12]	95.82%	0.635	2.54	6.99	4.74
StyleSwap (Ours)	**97.05%**	**0.677**	1.56	5.28	**2.72**
StyleSwap w/ $\mathcal{W}^+$	**97.87%**	**0.706**	1.51	5.27	**2.58**

same as [23]. The pose error is evaluated on the poses' L_2 distances produced by a pose estimator [35]. The expression error is the L_2 distance between the results' and targets' expression embedding extracted from a facial expression extraction model [42].

Quantitative Results. We list the results in Table 1. As can be seen that our **StyleSwap** model directly achieves state-of-the-art results on ID-related metrics and FID compared to previous ones, which shows the high similarity and image quality of our method. The *Swapping-Driven ID Inversion* strategy further improves ID similarity. Meanwhile, we achieve comparable pose and expression errors with SimSwap [7] and outperform the others. The qualitative results further prove that we preserve the attribute information at a robust level.

4.2 Qualitative Evaluation

Subjective evaluation plays an important role in face swapping. Note that for fair comparisons, we do not perform $\mathcal{W}^+$ space optimization. We provide image-based comparisons below. Demo videos and resources are available at https:// hangz-nju-cuhk.github.io/projects/StyleSwap.

Qualitative Results on 256×256 Resolution. Here we show the image-based comparisons on 256×256 resolution in Fig. 4. The source and target figures are selected from FF++ [34]. It can be seen that MegaFS [54] keeps the lighting conditions and texture badly, and InfoSwap [12] also fails to preserve accurate information. Though for some cases, SimSwap renders the best expression (the first row), it sometimes produces visible artifacts. Our method generates robust results with the highest similarity and competitive attribute preservation.

Qualitative Results on 512×512 Resolution. Particularly, we are able to generate high-resolution results with 512×512 resolution. Some results are shown in Fig. 1. We compare our results with MegaFS [54] and InfoSwap [12] which are also able to produce high-resolution results. As shown in Fig. 5, MegaFS generates a false skin tune of the target person and InfoSwap generates visible artifacts. Our method clearly outperforms theirs on high-resolution results.

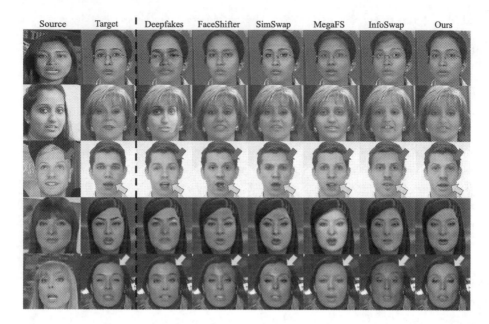

| Source | Target | Deepfakes | FaceShifter | SimSwap | MegaFS | InfoSwap | Ours |

Fig. 4. Qualitative results on FF++ with 256×256 **resolution.** Please pay attention to the identity information at the **red** arrows, and the attribute information at the **yellow** arrows (Color figure online)

Table 2. User Study's Ranking Scores. Larger is higher, with the maximum value to be 5

Perspective\Method	Faceshiter [23]	SimSwap [7]	MegaFS [54]	InfoSwap [12]	**Ours**
ID Similarity	3.05	2.59	2.80	2.89	**3.68**
Att. Preservation	3.16	2.94	2.21	3.36	**3.82**
Naturalness	2.95	2.62	2.43	3.11	**3.88**

User Study. We further conduct a user study for subjective evaluations. A total of 30 users are involved to discriminate 20 different samples from the FF++ dataset. The users are asked to rank the quality of the fake images from the following perspectives: 1) **Id similarity** with the source image; 2) **Attribute (Att) Preservation** including expressions and backgrounds referring to the target; and 3) **Naturalness.** Whether there are visible artifacts on the face and does the figure look like a real person. Detailed instructions and training are provided. As Deepfakes cannot produce plausible results, we only conduct the user study on the other 4 methods and our StyleSwap. Thus we define the highest score to be 5 and the lowest score to be 1 for each case.

The results are shown in Table 2. All ranks are given corresponding scores with 5 being the highest. It can be seen that our method performs the best on

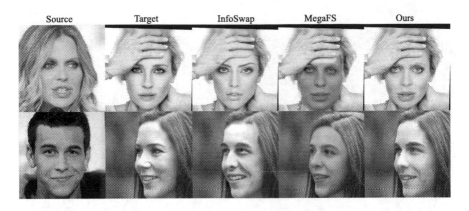

Fig. 5. **Qualitative results** on CelebA-HQ with 512 × 512 resolution

Table 3. Quantitative Ablation Study. The StyleSwap model with $\mathcal{W}^+$ space optimization achieves the best results on all metrics

Method\Metric	ID Retrieval ↑	ID Cosine ↑	Pose error↓	Exp. error ↓	FID ↓
StyleSwap (vector)	96.68%	0.668	2.81	8.44	4.54
StyleSwap w/o Mask	95.56%	0.653	1.80	5.78	3.18
StyleSwap	97.05%	0.677	1.56	5.28	2.72
StyleSwap w/ $\mathcal{W}$	97.54%	0.693	1.52	5.28	2.60
StyleSwap w/ $\mathcal{W}^+$	**97.87%**	**0.706**	**1.51**	**5.27**	**2.58**

given all metrics. Specifically, the low score of SimSwap with respect to attribute preservation is led by their artifacts.

4.3   Further Analysis

Ablation Studies. We conduct ablation studies on several important designs of our network. (1) The effectiveness of the $\mathcal{W}^+$ space optimization and its comparison with the $\mathcal{W}$ space optimization **StyleSwap w/ $\mathcal{W}$**; (2) Our StyleSwap without the Mask Branch **StyleSwap w/o Mask**. (3) Our design of concatenating the target's spatial feature maps into the style-based generator. An alternative way is to map both source and target information into the $\mathcal{W}$ as performed in [6,52]. This model is denoted as **StyleSwap (vector)**. The cross-reconstruction training paradigm does not affect numerical results and visualization in most cases, thus the comparison will be given.

We show the quantitative ablation results in Table 3 under the same evaluation protocols on FF++, and the qualitative ablation results on Fig. 6 which is generated on CelebA-HQ [16]. It can be seen that without the spatial feature maps, StyleSwap (vector) cannot preserve the attribute information correctly and thus loses details. Moreover, the identity similarity improves along with the addition of the *Mask Branch*, the $\mathcal{W}$ and $\mathcal{W}^+$ space optimization.

Source	Target	StyleSwap (vector)	StyleSwap w/o Mask	StyleSwap	StyleSwap w/ $\mathcal{W}$	StyleSwap w/ $\mathcal{W}^+$

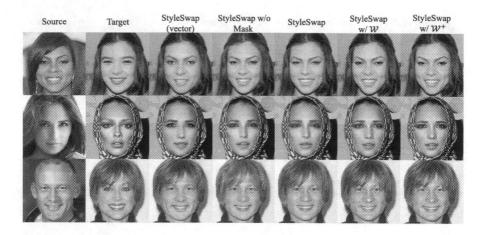

Fig. 6. Qualitative Ablation Study on CelebA-HQ.

Effectiveness on Face Forgery Detection. Importantly, our method can contribute to the face forgery detection community [14,27,34]. We conduct experiments under the standard face forgery detection pipeline and additionally provide the same amount of generated data by our method and FaceShifter [23]. The evaluation results on 4 different datasets validate that our method assists the forgery detection better. The details can be found in the Supplementary Materials.

5   Conclusion and Discussion

Conclusion. In this paper, we propose **StyleSwap**, a concise and effective framework that adapts a style-based generator for high-fidelity face swapping. We emphasize several key properties of our method: 1) With only minor modifications to the StyleGAN2 generator, our method is easy to implement and friendly to train, which saves a lot of human labor. 2) With the strong capability of the style-based generator and the simple design of the *Swapping-Guided Mask Branch*, our results are not only with high quality, similarity but enjoy high robustness. 3) Our method can take the advantage of GAN inversion and optimize the $\mathcal{W}^+$ space for improving identity similarity. 4) Our method can benefit face forgery detection by providing realistic fake results.

Broader Impact. Face swapping technique could create deepfake results for malicious purposes. We also take this issue into serious consideration and show the effectiveness of our method in the face forgery detection community. We will strictly limit the usage of this work for research purposes only.

Acknowledgement. This work is supported by NTU NAP, MOE AcRF Tier 1 (2021-T1-001-088), and under the RIE2020 Industry Alignment Fund - Industry Collaboration Projects (IAF-ICP) Funding Initiative, as well as cash and in-kind contributions from the industry partner(s).

References

1. Abdal, R., Qin, Y., Wonka, P.: Image2StyleGAN: how to embed images into the StyleGAN latent space? In: Proceedings of the IEEE/CVF International Conference on Computer Vision, pp. 4432–4441 (2019)
2. Abdal, R., Qin, Y., Wonka, P.: Image2StyleGAN++: how to edit the embedded images? In: Proceedings of the IEEE/CVF Conference on Computer Vision and Pattern Recognition, pp. 8296–8305 (2020)
3. Alaluf, Y., Patashnik, O., Cohen-Or, D.: Restyle: a residual-based StyleGAN encoder via iterative refinement. In: Proceedings of the IEEE/CVF International Conference on Computer Vision, pp. 6711–6720 (2021)
4. Bitouk, D., Kumar, N., Dhillon, S., Belhumeur, P., Nayar, S.K.: Face swapping: automatically replacing faces in photographs. In: ACM SIGGRAPH 2008 papers, pp. 1–8 (2008)
5. Blanz, V., Scherbaum, K., Vetter, T., Seidel, H.P.: Exchanging faces in images. In: Computer Graphics Forum, vol. 23, pp. 669–676. Wiley Online Library (2004)
6. Burkov, E., Pasechnik, I., Grigorev, A., Lempitsky, V.: Neural head reenactment with latent pose descriptors. In: Proceedings of the IEEE/CVF Conference on Computer Vision and Pattern Recognition, pp. 13786–13795 (2020)
7. Chen, R., Chen, X., Ni, B., Ge, Y.: SimSwap: an efficient framework for high fidelity face swapping. In: Proceedings of the 28th ACM International Conference on Multimedia, pp. 2003–2011 (2020)
8. Chung, J.S., Nagrani, A., Zisserman, A.: VoxCeleb2: deep speaker recognition. arXiv preprint arXiv:1806.05622 (2018)
9. Deepfakes: Faceswap. https://github.com/deepfakes/faceswap
10. Deng, J., Guo, J., Xue, N., Zafeiriou, S.: ArcFace: additive angular margin loss for deep face recognition. In: Proceedings of the IEEE/CVF Conference on Computer Vision and Pattern Recognition, pp. 4690–4699 (2019)
11. Egger, B., et al.: 3D morphable face models-past, present, and future. ACM Trans. Graph. (TOG) **39**(5), 1–38 (2020)
12. Gao, G., Huang, H., Fu, C., Li, Z., He, R.: Information bottleneck disentanglement for identity swapping. In: Proceedings of the IEEE/CVF Conference on Computer Vision and Pattern Recognition, pp. 3404–3413 (2021)
13. Goodfellow, I., et al.: Generative adversarial nets. In: Advances in Neural Information Processing Systems, vol. 27 (2014)
14. Guan, J., et al.: Delving into sequential patches for DeepFake detection. arXiv preprint arXiv:2207.02803 (2022)
15. Jiang, L., Li, R., Wu, W., Qian, C., Loy, C.C.: DeeperForensics-1.0: a large-scale dataset for real-world face forgery detection. In: Proceedings of the IEEE/CVF Conference on Computer Vision and Pattern Recognition, pp. 2889–2898 (2020)
16. Karras, T., Aila, T., Laine, S., Lehtinen, J.: Progressive growing of GANs for improved quality, stability, and variation. arXiv preprint arXiv:1710.10196 (2017)
17. Karras, T., et al.: Alias-free generative adversarial networks. In: Proceedings NeurIPS (2021)
18. Karras, T., Laine, S., Aila, T.: A style-based generator architecture for generative adversarial networks. In: Proceedings of the IEEE/CVF Conference on Computer Vision and Pattern Recognition, pp. 4401–4410 (2019)
19. Karras, T., Laine, S., Aittala, M., Hellsten, J., Lehtinen, J., Aila, T.: Analyzing and improving the image quality of StyleGAN. In: Proceedings of the IEEE/CVF Conference on Computer Vision and Pattern Recognition, pp. 8110–8119 (2020)

20. Kim, H., et al.: Deep video portraits. ACM Trans. Graph. (TOG) **37**, 1–14 (2018)
21. Kingma, D.P., Ba, J.: Adam: a method for stochastic optimization. arXiv preprint arXiv:1412.6980 (2014)
22. Korshunova, I., Shi, W., Dambre, J., Theis, L.: Fast face-swap using convolutional neural networks. In: Proceedings of the IEEE International Conference on Computer Vision, pp. 3677–3685 (2017)
23. Li, L., Bao, J., Yang, H., Chen, D., Wen, F.: FaceShifter: towards high fidelity and occlusion aware face swapping. In: CVPR (2020)
24. Liang, B., et al.: Expressive talking head generation with granular audio-visual control. In: Proceedings of the IEEE/CVF Conference on Computer Vision and Pattern Recognition (CVPR), pp. 3387–3396, June 2022
25. Liu, Z., Luo, P., Wang, X., Tang, X.: Deep learning face attributes in the wild. In: Proceedings of the IEEE International Conference on Computer Vision, pp. 3730–3738 (2015)
26. Natsume, R., Yatagawa, T., Morishima, S.: RSGAN: face swapping and editing using face and hair representation in latent spaces. arXiv preprint arXiv:1804.03447 (2018)
27. Nguyen, T.T., Nguyen, Q.V.H., Nguyen, C.M., Nguyen, D., Nguyen, D.T., Nahavandi, S.: Deep learning for DeepFakes creation and detection: a survey. arXiv preprint arXiv:1909.11573 (2019)
28. Nirkin, Y., Keller, Y., Hassner, T.: FSGAN: subject agnostic face swapping and reenactment. In: Proceedings of the IEEE/CVF International Conference on Computer Vision, pp. 7184–7193 (2019)
29. Nirkin, Y., Masi, I., Tuan, A.T., Hassner, T., Medioni, G.: On face segmentation, face swapping, and face perception. In: 2018 13th IEEE International Conference on Automatic Face & Gesture Recognition (FG 2018), pp. 98–105. IEEE (2018)
30. Park, T., Liu, M.Y., Wang, T.C., Zhu, J.Y.: Semantic image synthesis with spatially-adaptive normalization. In: Proceedings of the IEEE Conference on Computer Vision and Pattern Recognition (2019)
31. Parkhi, O.M., Vedaldi, A., Zisserman, A.: Deep face recognition (2015)
32. Perov, I., et al.: DeepFaceLab: integrated, flexible and extensible face-swapping framework. arXiv preprint arXiv:2005.05535 (2020)
33. Richardson, E., et al.: Encoding in style: a StyleGAN encoder for image-to-image translation. In: Proceedings of the IEEE/CVF Conference on Computer Vision and Pattern Recognition, pp. 2287–2296 (2021)
34. Rössler, A., Cozzolino, D., Verdoliva, L., Riess, C., Thies, J., Nießner, M.: FaceForensics++: learning to detect manipulated facial images. In: International Conference on Computer Vision (ICCV) (2019)
35. Ruiz, N., Chong, E., Rehg, J.M.: Fine-grained head pose estimation without keypoints. In: Proceedings of the IEEE Conference on Computer Vision and Pattern Recognition Workshops, pp. 2074–2083 (2018)
36. Shen, Y., Yang, C., Tang, X., Zhou, B.: InterFaceGAN: interpreting the disentangled face representation learned by GANs. TPAMI (2020)
37. Shen, Y., Zhou, B.: Closed-form factorization of latent semantics in GANs. In: CVPR (2021)
38. Shu, C., et al.: Few-shot head swapping in the wild. In: Proceedings of the IEEE/CVF Conference on Computer Vision and Pattern Recognition (CVPR), pp. 10789–10798, June 2022
39. Sun, K., et al.: High-resolution representations for labeling pixels and regions. arXiv preprint arXiv:1904.04514 (2019)

40. Sun, Y., Zhou, H., Liu, Z., Koike, H.: Speech2Talking-Face: inferring and driving a face with synchronized audio-visual representation. In: IJCAI, vol. 2, p. 4 (2021)

41. Tov, O., Alaluf, Y., Nitzan, Y., Patashnik, O., Cohen-Or, D.: Designing an encoder for StyleGAN image manipulation. ACM Trans. Graph. (TOG) **40**, 1–14 (2021)

42. Vemulapalli, R., Agarwala, A.: A compact embedding for facial expression similarity. In: Proceedings of the IEEE/CVF Conference on Computer Vision and Pattern Recognition, pp. 5683–5692 (2019)

43. Wang, H., et al.: CosFace: large margin cosine loss for deep face recognition. In: Proceedings of the IEEE Conference on Computer Vision and Pattern Recognition, pp. 5265–5274 (2018)

44. Wang, T.C., Liu, M.Y., Zhu, J.Y., Tao, A., Kautz, J., Catanzaro, B.: High-resolution image synthesis and semantic manipulation with conditional GANs. In: Proceedings of the IEEE Conference on Computer Vision and Pattern Recognition (2018)

45. Wang, T.C., Mallya, A., Liu, M.Y.: One-shot free-view neural talking-head synthesis for video conferencing. In: Proceedings of the IEEE/CVF Conference on Computer Vision and Pattern Recognition, pp. 10039–10049 (2021)

46. Wang, X., Li, Y., Zhang, H., Shan, Y.: Towards real-world blind face restoration with generative facial prior. In: The IEEE Conference on Computer Vision and Pattern Recognition (CVPR) (2021)

47. Wang, Y., et al.: HifiFace: 3D shape and semantic prior guided high fidelity face swapping. In: IJCAI (2021)

48. Xu, Y., Deng, B., Wang, J., Jing, Y., Pan, J., He, S.: High-resolution face swapping via latent semantics disentanglement. In: Proceedings of the IEEE/CVF Conference on Computer Vision and Pattern Recognition (CVPR), pp. 7642–7651, June 2022

49. Xu, Z., Hong, Z., Ding, C., Zhu, Z., Han, J., Liu, J., Ding, E.: MobileFaceSwap: a lightweight framework for video face swapping. In: AAAI (2022)

50. Xu, Z., et al.: FaceController: controllable attribute editing for face in the wild. In: AAAI (2021)

51. Yang, T., Ren, P., Xie, X., Zhang, L.: Gan prior embedded network for blind face restoration in the wild. In: Proceedings of the IEEE/CVF Conference on Computer Vision and Pattern Recognition, pp. 672–681 (2021)

52. Zhou, H., Sun, Y., Wu, W., Loy, C.C., Wang, X., Liu, Z.: Pose-controllable talking face generation by implicitly modularized audio-visual representation. In: Proceedings of the IEEE Conference on Computer Vision and Pattern Recognition (CVPR) (2021)

53. Zhu, P., Abdal, R., Femiani, J., Wonka, P.: Barbershop: GAN-based image compositing using segmentation masks. arXiv preprint arXiv:2106.01505 (2021)

54. Zhu, Y., Li, Q., Wang, J., Xu, C.Z., Sun, Z.: One shot face swapping on megapixels. In: Proceedings of the IEEE/CVF Conference on Computer Vision and Pattern Recognition (CVPR), pp. 4834–4844 (2021)

Paint2Pix: Interactive Painting Based Progressive Image Synthesis and Editing

Jaskirat Singh[1,2]([⊠]) [iD], Liang Zheng[1] [iD], Cameron Smith[2] [iD],
and Jose Echevarria[2] [iD]

[1] Australian National University, Canberra, Australia
{jaskirat.singh,liang.zheng}@anu.edu.au
[2] Adobe Research, San Jose, USA
{casmith,echevarr}@adobe.com

Abstract. Controllable image synthesis with user scribbles is a topic of keen interest in the computer vision community. In this paper, for the first time we study the problem of photorealistic image synthesis from incomplete and primitive human paintings. In particular, we propose a novel approach *paint2pix*, which learns to predict (and adapt) "what a user wants to draw" from rudimentary brushstroke inputs, by learning a mapping from the manifold of incomplete human paintings to their realistic renderings. When used in conjunction with recent works in autonomous painting agents, we show that *paint2pix* can be used for progressive image synthesis from scratch. During this process, *paint2pix* allows a novice user to progressively synthesize the desired image output, while requiring just few coarse user scribbles to accurately steer the trajectory of the synthesis process. Furthermore, we find that our approach also forms a surprisingly convenient approach for real image editing, and allows the user to perform a diverse range of custom fine-grained edits through the addition of only a few well-placed brushstrokes. Source code and demo is available at https://github.com/1jsingh/paint2pix.

1 Introduction

The human painting process represents a powerful mechanism for the expression of our inner visualizations. However, accurate depiction of the same is often quite time consuming and limited to those with sufficient artistic skill. Conditional image synthesis provides a popular solution to this problem, and simplifies output image synthesis based on higher-level input modalities (segmentation, sketch) which can be easily expressed using coarse user scribbles. For instance, segmentation based image generation methods [8,9,27,40] allow for control over output image attributes based on user-editable semantic segmentation maps. However, they have obvious disadvantage of requiring large-scale dense semantic

---

Supplementary Information The online version contains supplementary material available at https://doi.org/10.1007/978-3-031-19781-9_39.

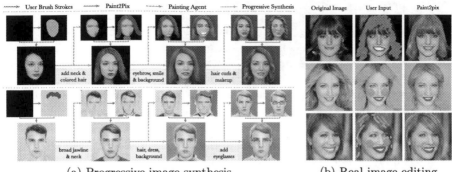

(a) Progressive image synthesis   (b) Real image editing

Fig. 1. Overview. We propose *paint2pix* which helps the user directly express his/her ideas in visual form by learning to predict user-intention from a few rudimentary brush-strokes. The proposed approach can be used for (a) synthesizing a desired image output directly from scratch wherein it allows the user to control the overall synthesis trajectory using just few coarse brushstrokes (blue arrows) at key points, or, (b) performing a diverse range of custom edits directly on real image inputs. Best viewed zoom-in. (Color figure online)

segmentation annotations for training, which makes them not easily scalable to new domains. Unsupervised sketch based image synthesis has also been explored [6,12,24], but they do not provide control over non-edge image areas.

In this paper, we explore the use of another modality in this direction, by studying the problem of photorealistic image synthesis from *incomplete and primitive human paintings*. This is motivated from the observation that when constrained to a particular domain (*e.g.*, faces), a lot of information about the final image output can be inferred from fairly rudimentary and partially drawn human paintings. We thus propose a novel approach *paint2pix*, which learns to predict (and adapt) "what the user intends to draw" from rudimentary brush-stroke inputs, by learning a mapping from the manifold of incomplete human paintings to their realistic renderings. However, learning the manifold of incomplete human paintings is challenging as it would require extensive collection of human painting trajectories for each target domain. For this challenge, we show that a fair approximation of this manifold can still be obtained by using painting trajectories from recent works on autonomous human-like painting agents [31].

While predicting photo-realistic outputs from partially drawn paintings might be helpful for capturing certain parts of a user's visualization (*e.g.*, face shape, hairstyle), fine grain control over different image attributes might be missing. In order to address this need for fine-grained control, we introduce an interactive synthesis strategy, wherein *paint2pix* when used in conjunction with an autonomous painting agent, allows a novice user to progressively synthesize and refine the desired image output using just few rudimentary brushstrokes. The overall image synthesis (refer Fig. 1a) is performed in a progressive fashion wherein *paint2pix* and the autonomous painting agent are used in successive steps. Starting with an empty canvas, the user begins by making few rudimentary strokes (*e.g.*, describing face shape, color) to obtain an initial user-intention prediction (through *paint2pix*). The painting agent then uses this prediction to paint until a user-controlled timestep, at which point, the user again provides a coarse brushstroke input (*e.g.*, describing finer details like hair color) to change the trajectory of the synthesis process. By iterating between these steps till the end of painting trajectory, the human artist is able to gain significant control over final image contents whilst requiring to input only few coarse scribbles (blue arrows in Fig. 1a) at key points of the autonomous painting process.

In addition to progressive image synthesis, the proposed approach can also be used to perform fine-grained editing on real-images (Fig. 1b). As compared with previous latent space manipulation methods [3, 28, 30], we find that our approach forms a surprisingly convenient alternative for making a diverse range of custom fine-grained modifications through the use of a few user scribbles. Furthermore, we show that once the user is satisfied with a custom edit on one image (*e.g.*, adding smile, changing makeup), the same edit can then be transferred to another image in a semantically-consistent manner (refer Sect. 6).

To summarize, the main contributions of this paper are **1)** We introduce a novel task of photorealistic image synthesis from incomplete and primitive human paintings. **2)** We propose *paint2pix* which learns to predict (and adapt) "what a user wants to ultimately draw" from rudimentary brushstroke inputs. **3)** We finally demonstrate the efficacy of our approach for (a) progressively synthesizing an output image from scratch, and, (b) performing a diverse range of custom edits directly on real image inputs.

2   Related Work

Autonomous Painting Agents. In recent years, substantial research efforts [15, 19, 25, 31, 32, 35, 41] have been focused on developing autonoumous painting agents which can learn an unsupervised stroke decomposition for the recreation of a given target image. Despite their efficacy, previous works in this area are often limited to the non-photorealistic recreation of *a provided target image*. This assumes that the user already has a fixed reference image that he/she wants to recreate. However, in practical applications the intended image output may not be available and has to be synthesized in a progressive fashion. Our work thus proposes to develop a new application for autonomous painting agents by predicting user-intention from incomplete canvas frames.

Segmentation Based Image Generation. Image to image translation frameworks have been extensively studied for controllable generation of highly realistic image outputs based on a more simplified image representation. For instance, [11, 16, 21, 26, 27, 33, 40] use conditional generative adversarial networks for controllable image synthesis using user-provided semantic segmentation maps. While effective, these works require large-scale semantic segmentation annotations for training, which limits their scalability to new domains. Furthermore,

Table 1. Related work overview. Broad positioning of our work with respect to other methods for controllable image synthesis with user scribbles. (refer Sect. 2 for details)

Method Attribute	Paint2Pix	GAN-Inversion	Seg2Photo	Sketch2Photo
From scratch	✓	✗	✓	✓
Responsiveness	✓	✓	✗	✗
No user-expertise	✓	✗	✓	✓
Data efficiency	✓	✓	✗	✓

making fine-grained changes within each semantic contour after image synthesis is non-trivial and often relies on style encoding methods [9,40], which require the user to first find a set of reference images which best describe the nature of each intended change (*e.g.* adding makeup or changing hair style for facial images). In contrast, our work allows for a range of custom fine-grained image editions through the addition of just few well placed brush strokes.

Sketch based image generation has also been explored [5,6,22–24,36,37]. For instance, Ghosh *et al.* [12] predict possible image outputs from rudimentary sketches of simple objects. While effective in controlling initial aspects of the image output, the use of sketches (compared to paintings) is less effective as it offers limited control and sensitivity to changes made in non-edge areas.

GAN Inversion. Interactive image generation and editing with user scribbles has also been explored in the context of GAN-inversion [1,2,39]. Zhu *et al.* [39] propose a hybrid optimization approach for projecting user-given strokes onto the natural image manifold. Similarly, [1,2] use GAN-inversion to perform local image edits with user scribbles. While effective for small-scale photorealistic manipulations, these methods often lack means to learn the distribution of user-inputs (manifold of rudimentary paintings in our case) and thus are limited to performing a pure color-based optimization. As shown in Sect. 5, this leads to poor performance on from-scratch synthesis and semantic edits on real images.

Positioning Our Work. Table 1 summarizes the positioning of our approach with respect to previous methods performing controllable image synthesis using user-given brushstrokes/scribbles. In particular, we posit the comparative benefits of our approach with respect to the following desirable properties.

- *Image synthesis from scratch.* While paint2pix, segmentation and sketch based methods allow for direct synthesis of the primary image from scratch, GAN-inversion methods perform a more color-based optimization and thereby show poor performance on image synthesis from scratch (Sect. 5.1).
- *Responsiveness (control) over all image areas.* Due to the one-to-many nature of learned mappings, segmentation based methods fail to provide fine-grained control over attributes within each semantic region. Similarly, sketch-based methods lack sensitivity to changes in non-edge areas.

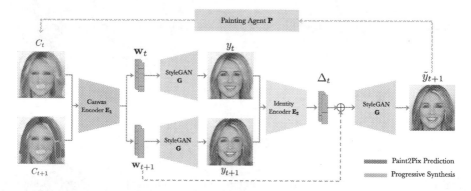

Fig. 2. Model Overview. The *paint2pix* model helps simplify the image synthesis task by predicting user-intention from rudimentary canvas state C_t, while also allowing the user to accurately steer the synthesis trajectory using coarse brushstroke inputs in C_{t+1}. This is done in two steps. First, the canvas encoder $\mathbf{E}_1$ learns a mapping between the manifold of incomplete paintings and real images to predict realistic user-intention predictions $\{y_t, y_{t+1}\}$ from $\{C_t, C_{t+1}\}$ respectively. These intermediate predictions are then fed into a second identity encoder $\mathbf{E}_2$ to predict a latent-space correctional term Δ_t, which ensures that the final prediction $\tilde{y}_{t+1}$ preserves the identity of the prediction from the original canvas C_t, while at the same time incorporating changes made by the user input brushstrokes in C_{t+1}. The progressive synthesis process can then be continued by feeding final prediction $\tilde{y}_{t+1}$ to an autonomous painting agent which paints it till a user-controlled timestep, at which point, the user can again add coarse brushstroke inputs in order to better express her inner ideas in the final image output.

– *Usability by novice artists*. A key advantage of our method is that it allows a novice artist to control the synthesis process while using fairly rudimentary brushstrokes. In contrast, GAN-inversion based methods require the user to make sufficiently detailed strokes in order to preserve closeness to the real image manifold (refer Sect. 5.2 for more details).
– *Data efficiency*. Our method is largely self-supervised and uses [31] to approximate the manifold of incomplete human paintings. In contrast, segmentation based methods require large-scale dense semantic maps for training on each target domain, which limits their scalability.

3   Our Method

The *paint2pix* model uses a two-step decoupled encoder-decoder architecture (refer Fig. 2) for predicting user intention from incomplete user paintings.

3.1   Canvas Encoding Stage

The goal of the canvas encoding stage is two-fold: 1) predict user-intention by learning a mapping between the manifold of incomplete user paintings to their realistic output renderings, while at the same time 2) allow for modification in the progressive synthesis trajectory based on coarse user-brushstrokes.

In particular, given current canvas state C_t and the updated canvas state after coarse user-brushstroke input C_{t+1}, we first use a canvas encoder $\mathbf{E}_1$ to predict a tuple of initial latent vector predictions $\{\mathbf{w}_t, \mathbf{w}_{t+1}\}$ as,

$$\{\mathbf{w}_t, \mathbf{w}_{t+1}\} = \mathbf{E}_1(C_t, C_{t+1}). \tag{1}$$

These latent predictions are then fed into a StyleGAN [18] decoder network $\mathbf{G}$, in order to get realistic user-intention predictions $\{y_t, y_{t+1}\}$ corresponding to input canvas tuple $\{C_t, C_{t+1}\}$ respectively, i.e. $\mathbf{y}_t, \mathbf{y}_{t+1} = G(\mathbf{w}_t), G(\mathbf{w}_{t+1})$.

Losses. Given realistic output ground-truth annotation $\hat{y}_t$ corresponding to canvas C_t (refer Sect. 3.3), the canvas encoder $\mathbf{E}_1$ is trained to learn to predict user-intention with the following prediction loss $\mathcal{L}_{pred}$,

$$\mathcal{L}_{pred} = \mathcal{L}_2(y_t, \hat{y}_t) + \lambda_1 \mathcal{L}_{lpips}(y_t, \hat{y}_t) + \lambda_2 \mathcal{L}_{id}(y_t, \hat{y}_t), \tag{2}$$

where $\mathcal{L}_{lpips}$ is the perceptual similarity loss [38] and $\mathcal{L}_{id}$ represents the Arcface [10] / MoCo-v2 [7] features based identity similarity loss from Tov et al. [34].

As previously mentioned, we would also like to ensure that the output predictions are modified in order to reflect the changes added by the user in C_{t+1}. This is then achieved by the following edition loss $\mathcal{L}_{edit}$,

$$\mathcal{L}_{edit} = \mathcal{L}_{lpips}(\Delta C_t, \Delta y_t) + \lambda_3 \mathcal{L}_{adv}(w_{t+1}) + \lambda_4 \|w_{t+1} - w_t\|_2, \tag{3}$$

where $\Delta C_t = C_{t+1} - C_t$ and $\Delta y_t = y_{t+1} - y_t$ represent the changes in the original canvas and output predictions respectively. $\mathcal{L}_{adv}$ refers to the latent discriminator loss from e4e [34] to ensure realism of the latent space prediction. Finally, the last term ensures that the codes $\{w_t, w_{t+1}\}$ for consecutive image outputs $\{y_t, y_{t+1}\}$ lie close in the StyleGAN [18] latent space.

3.2   Identity Embedding Stage

While enforcing closeness of consecutive latent vector codes $\{w_t, w_{t+1}\}$ (Eq. 3), helps in ensuring that the updated output prediction y_{t+1} is derived from the original prediction y_t, inconsistencies might still arise due to subtle changes in the identity of the underlying prediction (Fig. 2). Thus, the goal of the second stage is to preserve the underlying identity between consecutive image predictions and thereby ensure semantic consistency of the overall image synthesis process.

To address this, we train a second identity encoder $\mathbf{E}_2$ which ensures that the final prediction $\tilde{y}_{t+1}$ preserves identity of the original prediction y_t while still reflecting the changes made by the user in canvas C_{t+1}. In particular, given output image predictions $\{y_t, y_{t+1}\}$ from the canvas encoding stage, the identity encoder $\mathbf{E}_2$ predicts a correctional term Δ_t to update the latent codes as,

$$\tilde{w}_{t+1} = w_{t+1} + \Delta_t, \quad \text{where} \quad \Delta_t = \mathbf{E}_2(y_t, y_{t+1}). \tag{4}$$

The updated latent code $\tilde{w}_{t+1}$ is then used to predict the final output prediction $\tilde{y}_{t+1}$ using the StyleGAN [18] decoder $\mathbf{G}$ as,

$$\tilde{y}_{t+1} = \mathbf{G}(\tilde{w}_{t+1}). \tag{5}$$

Losses. The identity encoder is trained using the following loss,

$$\mathcal{L}_{embed} = \mathcal{L}_2(y_{t+1}, \tilde{y}_{t+1}) + \lambda_5 \mathcal{L}_{lpips}(y_{t+1}, \tilde{y}_{t+1}) + \lambda_6 \|\Delta_t\|_2 + \lambda_7 \mathcal{L}_{id}(y_t, \tilde{y}_{t+1}) \tag{6}$$

where the first three terms ensure the preservation of edits made by the user in C_{t+1}, while the last term enforces that the final prediction $\tilde{y}_{t+1}$ preserves the identity of the original image prediction y_t, thereby ensuring consistency of the overall progressive synthesis process.

Reason for Decoupled Encoders. While its feasible to design a model architecture wherein both $\mathcal{L}_{canvas}$ and $\mathcal{L}_{embed}$ are applied using a single encoder, the use of a decoupled identity encoder offers several practical advantages. For instance, while ensuring identity consistency is usually important (*e.g.*, making fine-grained changes), a change in underlying identity might sometimes be actually desirable, especially at the beginning of the progressive synthesis process. The decoupling of canvas encoding and identity embedding stage is therefore useful, as it allows the user to apply identity correction depending on the nature of the intended change. Furthermore, as shown in Sect. 7, decoupling the two stages allows our model to perform multi-modal synthesis without requiring any special architecture for producing multiple output predictions.

3.3 Overall Training

Total Loss. The overall *paint2pix* model is trained using a combination of canvas-encoding $\{\mathcal{L}_{pred}, \mathcal{L}_{edit}\}$ and identity embedding $\mathcal{L}_{embed}$ losses,

$$\mathcal{L}_{total} = \mathcal{L}_{pred} + \lambda_{edit}\,\mathcal{L}_{edit} + \lambda_{embed}\,\mathcal{L}_{embed}. \tag{7}$$

Ground Truth Painting Annotations. As discussed before, a key requirement of our approach is the ability to learn a mapping between the manifold of incomplete human-user paintings to their ideal realistic outputs. This requirement is challenging as it would need large-scale collection of human painting trajectories for each target domain, making our method intractable for most practical applications. To address this, we propose to instead use the recent works on autonomous painting agents for obtaining a decent approximation for the manifold of incomplete user paintings. The accuracy of such an approximation would depend highly on the domain gap between the incomplete paintings made by human users as compared to those made by a painting agent. We reduce this domain gap by using the recently proposed *Intelli-paint* [31] method, which has been shown to generate intermediate canvas frames which are more intelligible to actual human artists as opposed to previous works [15, 25, 32, 41].

In particular, for each painting trajectory trying to recreate a given target image $I_{target} \in \mathcal{D}$ ($\mathcal{D}$ is input domain, *e.g.*, FFHQ [17] for faces), we collect input canvas annotations by uniformly sampling 20 tuples of consecutive canvas frames $\{C_t, C_{t+1}\}$ observed during the painting process. The output image annotation $\hat{y}_t$ for all sampled canvas tuples (from the same trajectory) is then set to the original target image I_{target}. Furthermore, we collect painting annotations under various brushstroke counts $N_{strokes} \in [200, 500]$, as it helps capture the diverse degrees of abstraction observed in paintings made by actual human artists.

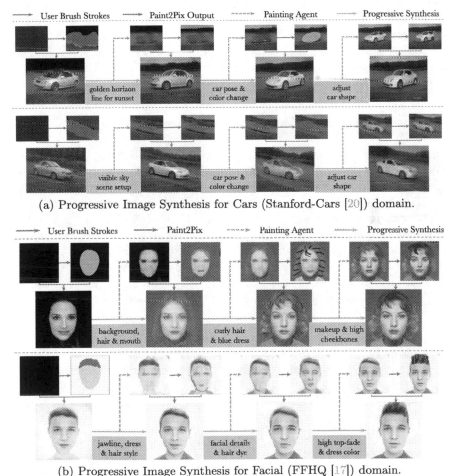

(a) Progressive Image Synthesis for Cars (Stanford-Cars [20]) domain.

(b) Progressive Image Synthesis for Facial (FFHQ [17]) domain.

Fig. 3. Paint2pix for Progressive Image Synthesis. Best viewed zoomed-in.

4   Paint2pix for Progressive Image Synthesis

Figure 3 demonstrates the use of *paint2pix* for progressive image synthesis from scratch. A potential user would start the painting process by adding a few rudimentary brushstrokes on the canvas (*e.g.*, background scene for cars or face shape, color for faces). The *paint2pix* network then outputs a set of possible realistic image renderings (refer Sect. 7 for more details on multi-modal synthesis) that the user might be interested in drawing. The user may then select the image that most closely resembles his/her idea to obtain a user-intention prediction. The progressive synthesis process can then be continued by feeding this prediction to an autonomous painting agent which paints it till a user-controlled timestep, at which point, the user can again add coarse scribbles (*e.g.*, describing finer details like sky color for cars or hairstyle for faces) in order to steer the synthesis trajectory according to his/her ideas. By continuing this iterative process till the end of the painting process, a novice user can gain significant control over the final image contents while requiring to only input few rudimentary brushstrokes at key points in the autonomous painting trajectory.

5   Comparison with Inversion Methods

Interactive image generation and editing with user brushstrokes has also been explored in the context of GAN-inversion methods [1, 2, 39], which use an encoder or optimization based inversion approach in order to project user scribbles onto the real image manifold. In this section, we present extensive quantitative and qualitative results comparing our approach with existing GAN-inversion methods for image manipulation with user-scribbles. In particular, we demonstrate the efficacy of our approach in terms of both 1) from scratch synthesis: *i.e.*, predicting user intention from fairly rudimentary paintings (Sect. 5.1), and 2) real image editing: allowing a potential user to make a range of custom fine-grain edits directly by just using a few coarse input brushstrokes (Sect. 5.2).

Baselines. We compare our results with recent state-of-the-art encoder based methods from Restyle [4], e4e [34] and pSp [29]. In addition, we report results for *optimization* based encoding approach from Karras *et al.* [18] and *hybrid* strategy from Zhu *et al.* [39]. Please note that in order to get best output quality, results for [39] are reported while using a pretrained ReStyle [4] encoder.

5.1   Predicting User-Intention from Rudimentary Paintings

Qualitative Results. Figure 4a shows qualitative comparisons while predicting photo-realistic (user-intention) outputs from rudimentary paintings. We clearly see that our approach results in much more photorealistic predictions for the user-intended final output. In contrast, the predominantly color-based optimization nature of previous GAN-inversion works leads to non-photorealistic projections when all color details are yet to be added by the user. For instance, while drawing a human face, it is quite common for an artist to first draw a coarse brushstroke for the face region without adding in the finer facial details. However, this leads to poor performance while using color-based optimization as it leads the model to instead predict an output face where the finer facial details are hardly noticeable. Adversarial loss in e4e [29] helps improve the realism of output images but it still performs worse than *paint2pix* for this task.

Quantitative Results. We also report quantitative results for this task (and image editing tasks from Sect. 5.2) in Table 2. Results are reported in terms of the Fréchet inception distance (FID) [14], which is used capture the output image quality from different methods. Furthermore, we perform a human user-study (details in supp. material) and report the percentage of human users which prefer our method as opposed to competing works. As shown in Table 2, we observe that *paint2pix* produces better quality images (lower FID scores) and is preferred by majority of human users over competing methods.

5.2   Real-Image Editing

In addition to being able to perform progressive synthesis from scratch, *paint2pix* also offers a surprisingly convenient approach for making a diverse range of custom semantic edits (*e.g.*, add smile for faces) on real images by simply initializing the canvas input C_t with a real image. We next compare our method with previous GAN-inversion works on performing real image editing with user scribbles.

Semantic Image Edits. As shown in Fig. 4b, we observe that our approach performs much better when the nature of the underlying edit is not purely color-based. For instance, consider the first example from Fig. 4b. Our method is able to correctly interpret that coarse white brushstrokes near the mouth region implies that the user is trying to add smile to the underlying facial image. In contrast, due to the predominantly color-based-optimization, gan-inversion methods fail to understand the change in semantics of face, and thus predict output faces in which the mouth region has been artificially-colored white.

Color-Based Custom Edits. Even when the custom-edits are color-based, we show that *paint2pix* leads to outputs which are 1) more photorealistic, 2) exhibit a greater level of detail at the edit locations, 3) modify non-edit locations (in addition to edit locations) in order to maintain coherence of the resulting image and 4) better preserve the identity of the original image input.

Results are shown in Fig. 4c. Consider the first example (row-1). The increased realism of *paint2pix* outputs can be clearly seen by the more photorealistic and detailed representation at edit locations (*e.g.*, hair, eyebrows). Furthermore, note that our method shows a more global understanding of image semantics and subtly modifies the skin tone and the eye shading of the face to maintain consistency with user-given edits. In contrast, the color-based optimization of GAN-inversion methods exhibits a lower level of detail at edit locations

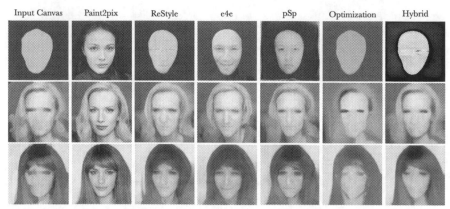

(a) Predicting user intention from rudimentary paintings.

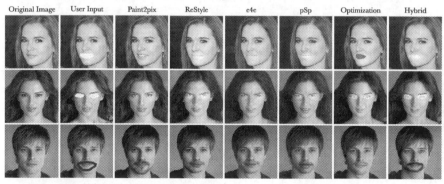

(b) Paint2pix for achieving semantic image edits.

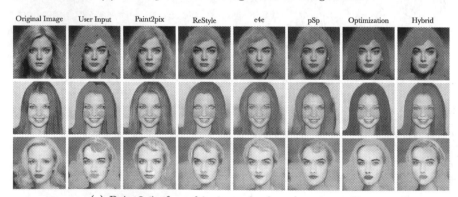

(c) Paint2pix for achieving color-based custom edits.

Fig. 4. Qualitative comparisons with GAN-inversion methods. Best viewed zoomed-in.

(*e.g.*, hair in row 1–3 and makeup in row 2,3). Furthermore, we find that our method shows better performance in preserving the identity of the original image in the final output (*e.g.*row 2,3), which is highly essential for real-image editing.

Table 2. Quantitative evaluation. Col 2–7: FID results for comparing output image quality on different image synthesis, editing tasks. Col-8: Human user-study results reporting percentage of users which prefer Paint2pix outputs over other methods.

Task	FID ($\downarrow$) comparison						User study
	Paint2pix	Restyle	e4e	pSp	Optimization	Hybrid	Paint2pix Preference
From scratch	**40.96**	85.98	79.69	89.83	107.2	91.62	97.32%
Semantic edits	**40.24**	45.27	42.32	46.08	47.16	49.29	94.04%
Color edits	**63.56**	100.2	93.11	107.3	116.4	114.2	93.85%

6   Inferring Global Edit Directions

We next show that the custom edits (*e.g.*, adding glasses, changing makeup) learned through *paint2pix* are not limited to the image on which the modifications were originally performed but instead show semantically-consistent generalization across the input domain. Put another way, once the user is satisfied with the output of a given custom edit on one image, the same edit can then be applied across different images from the input data distribution without requiring the user to repeat similar brushstrokes on each individual image.

In particular, consider $\{x_0, x_1\}$ be the original and edited image tuple with stylegan latent space vectors $\{\mathbf{w}_0, \mathbf{w}_1\}$ respectively. The custom edit $x_0 \rightarrow x_1$ can then be applied to another image x (with stylegan latent code $\mathbf{w}$) by computing a modified latent space edit direction $\delta_{edit}(x)$ as,

$$\delta_{edit}(x) = \delta_{edit}(x_0) + \mathbf{E}_2(x, \mathbf{G}(\mathbf{w} + \delta_{edit}(x_0))), \qquad (8)$$

where $\delta_{edit}(x_0) = \mathbf{w}_1 - \mathbf{w}_0$ represents the original edit direction from $x_0 \rightarrow x_1$, and the second term ensures identity preservation in the transferred edit.

The original edit can then be transferred to the input image x as,

$$x' = \mathbf{G}(\mathbf{w} + \alpha \, \delta_{edit}(x)), \qquad (9)$$

where α is the edit strength and $\mathbf{G}$ is the StyleGAN [18] decoder network.

Results are shown in Fig. 5. We are able to clearly see that custom edits learned on one image can be extended to different images in a semantic-consistent manner. Furthermore, we observe that the strength of the intended edit can be varied by simply adjusting the edit-strength parameter α. This helps us to use extrapolation in order to achieve edits which would be otherwise difficult to draw using coarse scribbles alone. For instance, while adding smile (using white brushstrokes) is easy, drawing a fully laughing face might be difficult for a novice artist. However, the same can be easily achieved by using a higher edit strength which allows us to extrapolate the original smiling edit to a laughing face edit (ow-1, Fig. 5). Similarly, different levels of facial wrinkles (or aging) can also be achieved in an analogous fashion (row-2, Fig. 5).

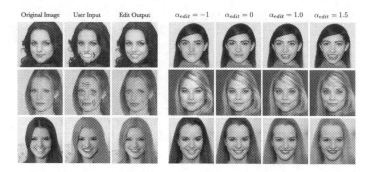

Fig. 5. Inferring of global edit directions using Paint2pix. Left: Original edit using Paint2pix. Right: Same edit transferred to another image with different edit strengths.

7 Multi-modal Synthesis

Predicting a single output for inferring user intention from an incomplete painting might not be always useful if the user's ideas are vastly different from the output prediction. The use of decoupled encoders in *paint2pix* is helpful in this regard, as it allows our approach to perform multi-modal synthesis for the final output without requiring special architecture changes.

In practice, given an incomplete canvas C_t, multi-modal synthesis is achieved by sampling a random image as the identity input (y_t) to the identity encoder network. Results are shown in Fig. 6. We observe that the above approach forms a convenient method for predicting multiple possible image completions from incomplete paintings. This provides the user with a wider range of choices to select the best direction for the synthesis process. Furthermore, we note that this idea can also be used to perform identity conditioned synthesis, by using the same identity image (*e.g.*, Chris Hemsworth) throughout the painting trajectory.

Fig. 6. Paint2pix for (left) multi-modal synthesis and (right) *id*-conditioned generation.

8   Ablation Study

In this section, we perform several ablation studies in order to study the importance of different losses $\{\mathcal{L}_{pred}, \mathcal{L}_{edit}, \mathcal{L}_{embed}\}$ in the performance of *paint2pix*. Please note that in order to still get meaningful results, the experiments without $\mathcal{L}_{pred}$ are performed while using a pretrained restyle [4] network for independently predicting intermediate outputs $\{y_{t+1}, y_t\}$ from canvas frames $\{C_{t+1}, C_t\}$, and, the ablation $\{$w/o $\mathcal{L}_{edit}\}$ is done using $\mathbf{w}_{t+1} = \mathbf{w}_t$ for the canvas encoder.

Results are shown in Fig. 7. We observe that $\{$w/o $\mathcal{L}_{pred}\}$ the model lacks an understanding of the manifold of incomplete paintings and thus produces outputs which are not fully photorealistic. In contrast, $\{$w/o $\mathcal{L}_{edit}\}$ shows high quality outputs but does not incorporate the edits made by user brushstrokes. Finally, we see that the use of $\mathcal{L}_{embed}$ helps the model produce images which preserve the identity of the original image in the final prediction.

Fig. 7. Results for ablation study for different losses in Paint2pix. Please zoom-in for better comparison.

9   Discussion and Limitations

In-distribution Predictions. A key advantage of *paint2pix* is that allows a novice user to synthesize and manipulate an output image on the real image manifold, while using fairly rudimentary and crude brushstrokes. While this is desirable in most scenarios, it also limits our method as it prevents a potential user from intentionally performing out-of-distribution (or non-realistic) facial manipulations (*e.g.*, blue eyebrows, ghost like faces *etc.*).

Invertibility for Real-Image Editing. Much like other GAN-inversion and latent space manipulation methods [3,4,13,28,29,39], accurate real-image editing with *paint2pix* is highly dependent on the ability of used encoder architecture to invert the original real image into StyleGAN [18] latent space.

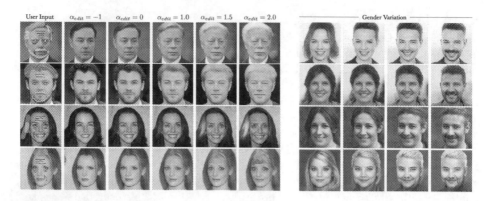

Fig. 8. Analysing paint2pix usage for age (left) and gender (right) variation edits.

Advanced Edits. Another limitation is that *paint2pix* does not provide a direct approach for achieving advanced semantic edits like age, gender manipulation. Nevertheless, as show in Fig. 8, age variation edits can still be achieved using extrapolation of edit strength α. Similarly, gender variation edits are possible by using progressive synthesis to infer the gender edit direction. Further details and analysis for gender variation edits are provided in the supp. material.

10   Conclusion

In this paper, we explore a novel task of performing photorealistic image synthesis and editing using primitive user paintings and brushstrokes. To this end, we propose *paint2pix* which can be used for 1) progressively synthesizing a desired image output from scratch using just few rudimentary brushstrokes, or, 2) real image editing: wherein it allows a human user to directly perform a range of custom edits without requiring any artistic expertise. As shown through extensive experimentation, we find that *paint2pix* forms a highly convenient and simple approach for directly expressing a potential user's inner ideas in visual form.

References

1. Abdal, R., Qin, Y., Wonka, P.: Image2styleGAN: how to embed images into the styleGAN latent space? In: Proceedings of the IEEE/CVF International Conference on Computer Vision, pp. 4432–4441 (2019)
2. Abdal, R., Qin, Y., Wonka, P.: Image2styleGAN++: how to edit the embedded images? In: Proceedings of the IEEE/CVF Conference on Computer Vision and Pattern Recognition, pp. 8296–8305 (2020)
3. Abdal, R., Zhu, P., Mitra, N.J., Wonka, P.: StyleFlow: attribute-conditioned exploration of styleGAN-generated images using conditional continuous normalizing flows. ACM Trans. Graph. (TOG) **40**(3), 1–21 (2021)

4. Alaluf, Y., Patashnik, O., Cohen-Or, D.: ReStyle: a residual-based StyleGAN encoder via iterative refinement. In: Proceedings of the IEEE/CVF International Conference on Computer Vision (ICCV), October 2021

5. Chen, T., Cheng, M.M., Tan, P., Shamir, A., Hu, S.M.: Sketch2Photo: internet image montage. ACM Trans. Graph. (TOG) **28**(5), 1–10 (2009)

6. Chen, W., Hays, J.: SketchyGAN: towards diverse and realistic sketch to image synthesis. In: Proceedings of the IEEE Conference on Computer Vision and Pattern Recognition, pp. 9416–9425 (2018)

7. Chen, X., Fan, H., Girshick, R., He, K.: Improved baselines with momentum contrastive learning. arXiv preprint arXiv:2003.04297 (2020)

8. Choi, Y., Choi, M., Kim, M., Ha, J.W., Kim, S., Choo, J.: StarGAN: unified generative adversarial networks for multi-domain image-to-image translation. In: Proceedings of the IEEE Conference on Computer Vision and Pattern Recognition, pp. 8789–8797 (2018)

9. Choi, Y., Uh, Y., Yoo, J., Ha, J.W.: StarGAN v2: diverse image synthesis for multiple domains. In: Proceedings of the IEEE/CVF Conference on Computer Vision and Pattern Recognition, pp. 8188–8197 (2020)

10. Deng, J., Guo, J., Xue, N., Zafeiriou, S.: ArcFace: additive angular margin loss for deep face recognition. In: Proceedings of the IEEE/CVF Conference on Computer Vision and Pattern Recognition, pp. 4690–4699 (2019)

11. Esser, P., Rombach, R., Ommer, B.: Taming transformers for high-resolution image synthesis. In: Proceedings of the IEEE/CVF Conference on Computer Vision and Pattern Recognition, pp. 12873–12883 (2021)

12. Ghosh, A., et al.: Interactive sketch & fill: Multiclass sketch-to-image translation. In: Proceedings of the IEEE/CVF International Conference on Computer Vision, pp. 1171–1180 (2019)

13. Härkönen, E., Hertzmann, A., Lehtinen, J., Paris, S.: GANSpace: discovering interpretable GAN controls. In: Advances in Neural Information Processing Systems, vol. 33, pp. 9841–9850 (2020)

14. Heusel, M., Ramsauer, H., Unterthiner, T., Nessler, B., Hochreiter, S.: GANs trained by a two time-scale update rule converge to a local nash equilibrium. In: Advances in Neural Information Processing Systems 30 (2017)

15. Huang, Z., Heng, W., Zhou, S.: Learning to paint with model-based deep reinforcement learning. In: Proceedings of the IEEE International Conference on Computer Vision, pp. 8709–8718 (2019)

16. Isola, P., Zhu, J.Y., Zhou, T., Efros, A.A.: Image-to-image translation with conditional adversarial networks. In: Proceedings of the IEEE Conference on Computer Vision and Pattern Recognition, pp. 1125–1134 (2017)

17. Karras, T., Laine, S., Aila, T.: A style-based generator architecture for generative adversarial networks. In: Proceedings of the IEEE/CVF Conference on Computer Vision and Pattern Recognition, pp. 4401–4410 (2019)

18. Karras, T., Laine, S., Aittala, M., Hellsten, J., Lehtinen, J., Aila, T.: Analyzing and improving the image quality of StyleGAN. In: Proceedings of the IEEE/CVF Conference on Computer Vision and Pattern Recognition, pp. 8110–8119 (2020)

19. Kotovenko, D., Wright, M., Heimbrecht, A., Ommer, B.: Rethinking style transfer: from pixels to parameterized brushstrokes. arXiv preprint arXiv:2103.17185 (2021)

20. Krause, J., Stark, M., Deng, J., Fei-Fei, L.: 3D object representations for fine-grained categorization. In: 4th International IEEE Workshop on 3D Representation and Recognition (3dRR-13), Sydney, Australia (2013)

21. Lee, C.H., Liu, Z., Wu, L., Luo, P.: MaskGAN: towards diverse and interactive facial image manipulation. In: Proceedings of the IEEE/CVF Conference on Computer Vision and Pattern Recognition, pp. 5549–5558 (2020)
22. Lee, J., Kim, E., Lee, Y., Kim, D., Chang, J., Choo, J.: Reference-based sketch image colorization using augmented-self reference and dense semantic correspondence. In: Proceedings of the IEEE/CVF Conference on Computer Vision and Pattern Recognition, pp. 5801–5810 (2020)
23. Li, X., Zhang, B., Liao, J., Sander, P.V.: Deep sketch-guided cartoon video synthesis. CoRR (2020)
24. Liu, R., Yu, Q., Yu, S.X.: Unsupervised sketch to photo synthesis. In: Vedaldi, A., Bischof, H., Brox, T., Frahm, J.-M. (eds.) ECCV 2020. LNCS, vol. 12348, pp. 36–52. Springer, Cham (2020). https://doi.org/10.1007/978-3-030-58580-8_3
25. Liu, S., et al.: Paint transformer: feed forward neural painting with stroke prediction. In: Proceedings of the IEEE/CVF International Conference on Computer Vision, pp. 6598–6607 (2021)
26. Liu, X., Yin, G., Shao, J., Wang, X., et al.: Learning to predict layout-to-image conditional convolutions for semantic image synthesis. In: Advances in Neural Information Processing Systems 32 (2019)
27. Park, T., Liu, M.Y., Wang, T.C., Zhu, J.Y.: Semantic image synthesis with spatially-adaptive normalization. In: Proceedings of the IEEE/CVF Conference on Computer Vision and Pattern Recognition, pp. 2337–2346 (2019)
28. Patashnik, O., Wu, Z., Shechtman, E., Cohen-Or, D., Lischinski, D.: Styleclip: text-driven manipulation of StyleGAN imagery. In: Proceedings of the IEEE/CVF International Conference on Computer Vision, pp. 2085–2094 (2021)
29. Richardson, E., et al.: Encoding in style: a StyleGAN encoder for image-to-image translation. In: IEEE/CVF Conference on Computer Vision and Pattern Recognition (CVPR), June 2021
30. Shen, Y., Zhou, B.: Closed-form factorization of latent semantics in GANs. In: Proceedings of the IEEE/CVF Conference on Computer Vision and Pattern Recognition, pp. 1532–1540 (2021)
31. Singh, J., Smith, C., Echevarria, J., Zheng, L.: Intelli-paint: towards developing human-like painting agents. In: European Conference on Computer Vision. Springer (2022)
32. Singh, J., Zheng, L.: Combining semantic guidance and deep reinforcement learning for generating human level paintings. In: Proceedings of the IEEE/CVF International Conference on Computer Vision (2021)
33. Sushko, V., Schönfeld, E., Zhang, D., Gall, J., Schiele, B., Khoreva, A.: You only need adversarial supervision for semantic image synthesis. arXiv preprint arXiv:2012.04781 (2020)
34. Tov, O., Alaluf, Y., Nitzan, Y., Patashnik, O., Cohen-Or, D.: Designing an encoder for stylegan image manipulation. arXiv preprint arXiv:2102.02766 (2021)
35. Wang, Q., Guo, C., Dai, H.N., Li, P.: Self-stylized neural painter. In: SIGGRAPH Asia 2021 Posters, pp. 1–2 (2021)
36. Xiang, X., Liu, D., Yang, X., Zhu, Y., Shen, X., Allebach, J.P.: Adversarial open domain adaptation for sketch-to-photo synthesis. In: Proceedings of the IEEE/CVF Winter Conference on Applications of Computer Vision, pp. 1434–1444 (2022)
37. Yang, S., Wang, Z., Liu, J., Guo, Z.: Controllable sketch-to-image translation for robust face synthesis. IEEE Trans. Image Process. **30**, 8797–8810 (2021)
38. Zhang, R., Isola, P., Efros, A.A., Shechtman, E., Wang, O.: The unreasonable effectiveness of deep features as a perceptual metric. In: CVPR (2018)

39. Zhu, J.-Y., Krähenbühl, P., Shechtman, E., Efros, A.A.: Generative visual manipulation on the natural image manifold. In: Leibe, B., Matas, J., Sebe, N., Welling, M. (eds.) ECCV 2016. LNCS, vol. 9909, pp. 597–613. Springer, Cham (2016). https:// doi.org/10.1007/978-3-319-46454-1_36
40. Zhu, P., Abdal, R., Qin, Y., Wonka, P.: Sean: Image synthesis with semantic region-adaptive normalization. In: Proceedings of the IEEE/CVF Conference on Computer Vision and Pattern Recognition, pp. 5104–5113 (2020)
41. Zou, Z., Shi, T., Qiu, S., Yuan, Y., Shi, Z.: Stylized neural painting. In: Proceedings of the IEEE/CVF Conference on Computer Vision and Pattern Recognition, pp. 15689–15698 (2021)

FurryGAN: High Quality Foreground-Aware Image Synthesis

Jeongmin Bae, Mingi Kwon, and Youngjung Uh$^{(\boxtimes)}$

Yonsei University, Seoul, Korea
{jaymin.bae,kwonmingi,yj.uh}@yonsei.ac.kr

Abstract. Foreground-aware image synthesis aims to generate images as well as their foreground masks. A common approach is to formulate an image as a masked blending of a foreground image and a background image. It is a challenging problem because it is prone to reach the trivial solution where either image overwhelms the other, i.e., the masks become completely full or empty, and the foreground and background are not meaningfully separated. We present FurryGAN with three key components: 1) imposing both the foreground image and the composite image to be realistic, 2) designing a mask as a combination of coarse and fine masks, and 3) guiding the generator by an auxiliary mask predictor in the discriminator. Our method produces realistic images with remarkably detailed alpha masks which cover hair, fur, and whiskers in a fully unsupervised manner. Project page: https://jeongminb.github.io/FurryGAN/.

1 Introduction

As the quality of images from generative adversarial networks (GANs) improves [9,13–16], discovering the semantics in their latent space is useful to control the generation process [2,27,28,40] or to edit real images through latent inversion [4,25,26,42]. Localizing the semantics in the latent space is another important research direction for understanding how GANs work. Some methods tackle local editing by separating parts in the intermediate feature maps [7,17].

Meanwhile, a few recent works tackle foreground-aware image synthesis by modeling an image as a composition of foreground and background images according to a mask. While previous methods achieve some success, they explicitly prepare a background distribution by removing the foreground with an off-the-shelf object detector [29], assume that images with shifted foreground objects should look real [5,37], or require multi-stage training with dataset-tailored hyperparameters [1]. These ingredients are obstacles that block general solutions for foreground-aware synthesis.

In this paper, we propose FurryGAN which learns to synthesize images with the explicit understanding of the foreground given only a collection of images. Intuitions in our method include the following. 1) We encourage the foreground images

---

Supplementary Information The online version contains supplementary material available at https://doi.org/10.1007/978-3-031-19781-9_40.

Fig. 1. Example images and the corresponding foreground masks. Both are simultaneously *generated* by our model. FurryGAN learns not only to generate realistic images but also to synthesize alpha masks with fine details such as hair, fur, and whiskers in a fully unsupervised manner (left). Our model also can be trained on various datasets (right)

and the composite images to resemble the training distribution. It prevents the foreground from losing the objects. 2) We introduce coarse and fine masks. The coarse mask captures the rough shape, and the fine mask captures details such as whiskers and hair. 3) We introduce an auxiliary task for the discriminator to predict the mask from the generated image so that the generator produces the foreground image aligned with the mask.

Compared to the previous works, our method does not require off-the-shelf networks, the assumption for perturbation, multi-stage training, or careful early stopping. Experiments demonstrate the superiority of our framework compared to previous methods regarding high quality alpha masks. We also provide thorough ablation studies to justify each component of our method.

Figure 1 shows example synthesized images and the corresponding alpha masks. They catch unprecedented levels of fine details, especially in hair and whiskers. Consequently, the detailed masks enable the natural composition of the foreground part and any background (Fig. 5). As a byproduct, GAN inversion on our method achieves unsupervised object segmentation with the same level of details.

2   Related Work

GANs and Semantic Interpretation. GANs [14–16] synthesize astonishingly high quality images from random latent codes. Understanding semantic interpretation of the latent codes is an important research topic so that users can control the generation process or edit real images through latent inversion [26, 27, 30, 35, 42]. Instead, we focus on teaching GANs spatial understanding of foreground objects.

Foreground-Aware GANs. Although semantic interpretation has some correlation with spatial separation, incorporating the notion of foreground objects has been tackled in the orthogonal direction, mostly by modeling an image as a combination of foreground and background according to a mask. PSeg [5] and improved layered GAN [37] rely on the assumption that the composite image with spatially transformed foreground should still be realistic. However, the parameters for the transformation should be determined for each dataset, and the assumption does not hold when the foreground region touches a border of the image. In [20,32,33], they identify the latent directions in a pretrained generator for changing the background to separate the foreground and background. Labels4Free [1] trains an alpha mask network that produces masks for combining foregrounds and backgrounds, generated by pretrained StyleGAN2 and pretrained pseudo-background StyleGAN2, respectively. Whereas it requires multi-stage training with tailored hyperparameters, our framework is trained in an end-to-end fashion and produces remarkably fine details in the masks.

Unsupervised Segmentation. Early image segmentation methods rely on the clustering of color and coordinates [3,8]. In order to cluster the regions regarding semantics, learning deep networks for maximizing mutual information within the cluster [12,24] or for contrasting different instances [31] have been successful. These objectives assume multiple classes and are not straightforward to be applied in foreground-background separation. Given a generator for foreground-aware image synthesis, inverting a real image to the latent space inherently leads to unsupervised foreground segmentation. Thus we focus on a better understanding of the foreground in GANs.

3D-Aware GANs. 3D-aware GANs based on NeRF [22] represent a scene as a neural network which receives 3D coordinates and outputs their color or feature vector with occupancy. Furthermore, recent approaches divide the scene into foreground and background by a depth threshold [10,41] or separate feature fields [23]. However, they aim to understand the 3D geometry of the scene and do not explicitly learn to generate high quality foreground alpha masks.

3   Method

In this section, we overview our framework (Sect. 3.1), describe the networks (Sect. 3.2), and explain their training techniques including loss functions (Sect. 3.3). To begin with, we briefly introduce a common formulation.

Common Formulation. We follow the common formulation [1,5,29,37] for generating images: an image is a masked combination of a foreground image x_{fg} and a background image x_{bg}, according to an alpha mask $\mathbf{m}$. Formally,

$$x_{comp} = \mathbf{m} \odot x_{fg} + (1 - \mathbf{m}) \odot x_{bg}, \tag{1}$$

where $\odot$ denotes pixel-wise multiplication.

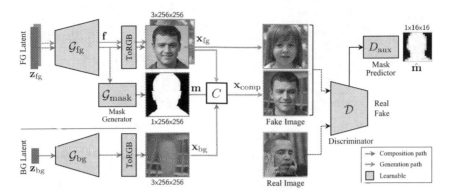

Fig. 2. Our framework. consists of a foreground generator, a mask generator, a background generator, and a discriminator with a mask predictor. The alpha mask specifies the combination of the foreground image and the background image to produce the composite image. We feed both the foreground images and the composite images to the discriminator as fake images

3.1 Framework Overview and Dual Fake Input Strategy

Figure 2 shows the framework overview. FurryGAN has three generators for the foreground, background, and mask to produce the composite images according to Eq. (1). Then the discriminator guides the generator to produce realistic images.

Dual Fake Input Strategy. Guiding the generator to produce realistic *composite* images solely does not guarantee the separation of the foreground and background. The motivation of the dual fake input is the following. The foreground images should contain salient objects (e.g., a person in FFHQ) so that there exists a solution for the masks to produce realistic composite images including the foreground images. Otherwise, the mask will favor excluding the foreground images from the composite images. Hence, we ensure the foreground images to contain salient objects by imposing a sufficient condition: being realistic by themselves. The fake mini-batch for the discriminator consists of the foreground images and the composite images (Fig. 3a). Then the discriminator tries to classify them as fakes, and the generator tries to produce realistic images in both the foreground and the composite images. We find that the dual fake input strategy helps prevent improper foreground separation.

3.2 Architecture

Generators. The foreground and background generators, $\mathcal{G}_{\text{fg}}$ and $\mathcal{G}_{\text{bg}}$, synthesize images $\mathbf{x}_{\text{fg}}$ and $\mathbf{x}_{\text{bg}}$ from latent codes $\mathbf{z}_{\text{fg}}$ and $\mathbf{z}_{\text{bg}}$, respectively. The two generators do not share any parameters. The mask generator $\mathcal{G}_{\text{mask}}$ synthesizes $\mathbf{m}$ from the penultimate feature maps of the foreground generator. Then, their simple alpha-blending produces the composite image $\mathbf{x}_{\text{comp}}$ (Eq. (1)). Note that the composite

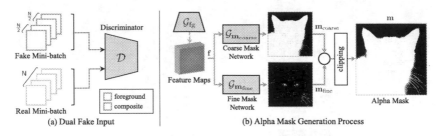

Fig. 3. Dual fake input strategy and mask generators. (a) Our discriminator receives the foreground images and the composite images as fake. They evenly share a fake mini-batch. (b) Our mask generator produces an alpha mask as a combination of a coarse mask and a fine mask

function causes unexpected additional degree of freedom: $\mathbf{x}_{\mathrm{fg}} \odot \mathbf{m} = 2 \cdot \mathbf{x}_{\mathrm{fg}} \odot 0.5 \cdot \mathbf{m}$. Thus we restrict the generators' outputs to be in the range of $[-1, 1]$ by adding a tanh function at the output of the ToRGB layer.

Coarse and Fine Mask Generator. As shown in Fig. 3b, our mask generator consists of a coarse mask network $\mathcal{G}_{\mathbf{m}_{\mathrm{coarse}}}$ and a fine mask network $\mathcal{G}_{\mathbf{m}_{\mathrm{fine}}}$. We expect the coarse mask network to cover the overall shape and the fine mask network to make up for the details missed in the coarse mask (e.g., cat whiskers, fur, and hair). Each mask is normalized to the range of [0,1] by min-max normalization and their summation becomes the final alpha mask. The final mask $\mathbf{m}$ is computed as:

$$\mathbf{m}_{\mathrm{coarse}} = \mathcal{G}_{\mathbf{m}_{\mathrm{coarse}}}(\mathbf{f}), \quad \mathbf{m}_{\mathrm{fine}} = \mathcal{G}_{\mathbf{m}_{\mathrm{fine}}}(\mathbf{f}), \tag{2}$$

$$\mathbf{m} = \mathrm{clip}(\mathbf{m}_{\mathrm{coarse}} + \gamma \mathbf{m}_{\mathrm{fine}}, 0, 1), \tag{3}$$

where $\mathbf{f}$ denotes the penultimate feature maps of the foreground generator. The design details are described in the Appendix E. For stability, we fade in the fine mask by linearly increasing γ from 0 to 1 over the first 5K iterations.

Discriminator with a Mask Predictor. We follow the discriminator architecture of StyleGAN2 [16] and add an auxiliary mask predictor. The mask predictor tries to reconstruct the mask of an input image given the 16×16 feature maps. It has minimal capacity for predicting the masks, i.e., two 1×1 convolutional layers and residual connections. How it guides the generators will be discussed in the following section (Eqs. (4) and (5)).

3.3   Training Objectives

Adversarial Loss. As described in Sect. 3.1, we impose adversarial losses on the foreground image and the composite image. We adopt non-saturating loss [9] and lazy R1-regularization [16,21] and skip defining trivial equations $L_{\mathrm{adv}}^{\mathcal{D}}$, $L_{\mathrm{R1}}^{\mathcal{D}}$, and $L_{\mathrm{adv}}^{\mathcal{G}}$ for brevity. Adversarial losses act as the primary source for driving foreground-aware image synthesis.

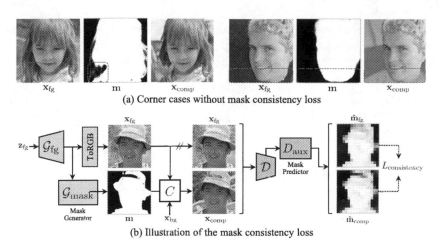

(a) Corner cases without mask consistency loss

(b) Illustration of the mask consistency loss

Fig. 4. Mask consistency loss. (a) Results without mask consistency loss show inconsistency between foreground images and composite images, e.g., cutting off long hair or adding shoulders. (b) Mask consistency loss computes discrepancy between the predicted masks of the foreground and composite images. $//$ denotes a stop gradient operator

Mask Prediction Loss. The auxiliary mask predictor $\mathcal{D}_{\mathrm{aux}}$ in the discriminator aims to regress the generated mask given the generated image:

$$L_{\mathrm{pred}} = \frac{1}{|\hat{\mathbf{m}}|} \|\texttt{Downsample}(\mathbf{m}) - \hat{\mathbf{m}}\|_2^2, \qquad (4)$$

where $\hat{\mathbf{m}}$ is the output of $\mathcal{D}_{\mathrm{aux}}$ for the generated images ($\mathbf{x}_{\mathrm{fg}}$ and $\mathbf{x}_{\mathrm{comp}}$). We use bilinear interpolation for $\texttt{Downsample}$. The $16{\times}16$ prediction will be useful for guiding the generator in conjunction with Eq. (5).

Mask Consistency Loss. We observe that object regions of foreground $\mathbf{x}_{\mathrm{fg}}$ and composite $\mathbf{x}_{\mathrm{comp}}$ can be inconsistent. For example, the mask may cut off the long hair in the foreground so that the composite image becomes a face with short hair. As another example, the missing body part of the foreground object may be supplemented from the background (both cases are shown in Fig. 4a).

Hence, we demand the mask predicted from the composite image to be consistent to the mask predicted from the foreground image (Fig. 4b):

$$L_{\mathrm{consistency}} = \frac{1}{|\hat{\mathbf{m}}_{\mathrm{comp}}|} \|\mathcal{D}_{\mathrm{aux}}(\texttt{stopgrad}(\mathbf{x}_{\mathrm{fg}})) - \hat{\mathbf{m}}_{\mathrm{comp}}\|_2^2, \qquad (5)$$

where $\hat{\mathbf{m}}_{\mathrm{comp}} = \mathcal{D}_{\mathrm{aux}}(\mathbf{x}_{\mathrm{comp}})$ and $\texttt{stopgrad}(\cdot)$ denotes a stop gradient operator.

As the mask predictor regresses the mask from the mask generator given a composite image, imposing consistency between the two masks encourages the foreground object region and the generated mask to resemble each other.

Coarse Mask Loss. We adopt binarization loss and area loss following previous methods [1,5]. The binarization loss pushes the alpha values in the masks to either 0 or 1:

$$L_{binary} = \mathbb{E}[\min(\mathbf{m}_{coarse}, 1 - \mathbf{m}_{coarse})]. \tag{6}$$

The area loss penalizes the ratio of a mask being less than ϕ_1 to promote using the foreground images more than ϕ_1, i.e., preventing the background image from taking charge of everywhere, which is a degenerate solution:

$$L_{area}^{coarse} = \max(0, \phi_1 - \frac{1}{|\mathbf{m}_{coarse}|} \sum \mathbf{m}_{coarse}), \tag{7}$$

where $|\mathbf{m}|$ denotes the number of pixels in the mask image and ϕ_1 is set to 0.35 for all experiments (unless otherwise noted). The final coarse mask loss is:

$$L_{\mathbf{m}_{coarse}} = L_{binary} + L_{area}^{coarse}. \tag{8}$$

Fine Mask Loss. The fine mask aims to capture details like hair, fur, and whiskers. Such a thin body becomes transparent due to the property of light. Hence, we do not use the binarization loss to free the masks to bear medium values between 0 and 1. Instead, we impose an inverse area loss to prevent the fine mask from taking charge of too large an area:

$$L_{\mathbf{m}_{fine}} = L_{area}^{fine} = \max(0, \phi_2 - \frac{1}{|\tilde{\mathbf{m}}_{fine}|} \sum (1 - \tilde{\mathbf{m}}_{fine})), \tag{9}$$

where $\tilde{\mathbf{m}}_{fine} = \mathbf{m} - \mathbf{m}_{coarse}$ for penalizing the area where the fine mask actually contributes after clipping. ϕ_2 is set to 0.01 in all experiments. More details about the mask are illustrated in Appendix E.

Background Participation Loss. We sometimes observe that the alpha mask tries to employ foreground images excessively. As a remedy, we penalize the difference between the composite image and the background image. It indirectly removes the excessive spread of the alpha mask.

$$L_{reg} = \frac{1}{|\mathbf{x}_{comp}|} \|\mathbf{x}_{comp} - \mathbf{x}_{bg}\|_2^2 \tag{10}$$

Intuitively, an easy way to reduce the difference between the composite image and the background is to remove unnecessary foreground areas that do not harm the realism of the composite image.

Overall Objective. Consequently, our full loss functions are:

$$L_{\text{total}}^{\mathcal{D}} = L_{\text{adv}}^{\mathcal{D}} + L_{\text{R1}}^{\mathcal{D}} + L_{\text{pred}}, \tag{11}$$

$$L_{\text{total}}^{\mathcal{G}} = L_{\text{adv}}^{\mathcal{G}} + L_{\text{consistency}} + \lambda_{\text{coarse}} L_{\mathbf{m}_{\text{coarse}}} + \lambda_{\text{fine}} L_{\mathbf{m}_{\text{fine}}} + L_{\text{reg}}. \tag{12}$$

4   Experiments

4.1   Implementation Detail

Our foreground generator and background generator are based on StyleGAN2 [16]. For simplicity, we remove output skip connections in the synthesis network and use a shallow mapping network [13]. The foreground and background generators, $\mathcal{G}_{\text{fg}}$ and $\mathcal{G}_{\text{bg}}$, use a slightly modified StyleGAN2 structure. The number of channels of the latent codes and the feature maps in $\mathcal{G}_{\text{fg}}$ and $\mathcal{G}_{\text{bg}}$ become $\frac{3}{4}$ and $\frac{1}{4}$, respectively. As a result, the total number of parameters reduces by about half. Similar to [37], background codes are shared with foreground codes. More precisely, we borrow the front part of the $\mathbf{z}_{\text{fg}} \sim \mathcal{N}(0, I)$ and use it as $\mathbf{z}_{\text{bg}}$.

We train our model on a single RTX-3090 for a period of about 100 h. In all experiments, we trained our model for 300K iterations with a batch size of 16. We follow training parameters from StyleGAN2 but do not use mixing regularization. Mask consistency loss and background participation loss update the model every other iteration. We set $\lambda_{\text{coarse}} = \lambda_{\text{fine}} = 5$. The coefficient of binarization loss is linearly reduced to 0.5 over the first 5K iterations.

4.2   Setup

Datasets. We evaluate our model on FFHQ [15] and AFHQv2-Cat [6,14]. FFHQ has 70,000 high-quality images of human faces. It has faces of various races and poses and also has good coverage of accessories such as eyeglasses, hats, etc. AFHQv2-Cat contains 5000 images of cat faces. The rebuilt (v2) dataset has higher quality due to proper resizing and compression. We also trained our model on unaligned datasets such as LSUN-Object [39], and CUB [34] (see Appendix F for details and results). All models are trained at 256×256 resolution.

Pseudo Ground Truth Masks. We evaluate the generated mask quality to show foreground-background separation performance. Because the generated images do not contain ground truth masks for evaluating the generated foreground masks, we adopt TRACER [19] to prepare pseudo ground truth masks. It provides detailed masks including hair and whiskers, which are not captured by segmentation networks used in PSeg [5] and Labels4Free [1]. Please refer to Appendix B for their comparison.

(a) Generated foreground on generated background   (b) Generated foreground on real background

Fig. 5. Composite images. The same person is placed on the vertical axis, and the same background is placed on the horizontal axis

Fig. 6. Latent space interpolation. We show the mask changes naturally as the image changes

Metrics. To quantitatively measure the quality of images, we compute Fréchet Inception Distance (FID) [11] between generated foreground images and all training images. Unless otherwise specified, all results were obtained with 50,000 generated images following [13,14]. To quantitatively measure the quality of masks, we employ intersection over union (IoU) for the foreground and background, and their mean (mIoU). IoU and mIoU measure the overlap between prediction masks and ground truth masks. Furthermore, we report standard segmentation metrics: precision, recall, F1 score, and segmentation accuracy following [1].

4.3   Experiments About Masked Foregrounds

Thanks to the high quality mask generated by our model, the masked foreground object can be naturally combined with various backgrounds, as shown in Fig. 5. We sample random background latent codes for the backgrounds used in Fig. 5 (a). Figure 6 shows that the interpolation in the foreground latent space not only changes the image but also changes the shape of masks correspondingly.

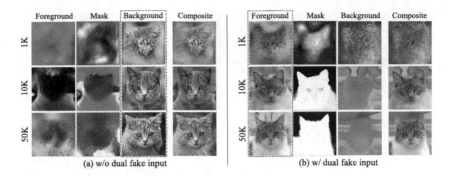

Fig. 7. Ablation of dual fake input strategy. Each row shows the early results as training proceeds. (a) without dual fake input, the foreground object has shown in the background image. (b) With dual fake input, it is separated naturally

4.4   Ablation Study

Dual Fake Input Strategy. Figure 7a shows that the foreground and mask generator fail to synthesize meaningful foreground and mask, respectively, without the dual fake input strategy. We suppose that it is easier for the generators to focus on the background to synthesize realistic composite images because the foreground generator has a more complicated task: producing foreground images and the masks with more parameters. On the other hand, with the dual fake input strategy, the foreground images have clear objects as they should resemble the training images (Fig. 7b).

Ablation of Losses. Figure 8 visually compares the results without one component at a time and our full method. Without background participation loss (Eq. (10)), the masks tend to be wider than the foreground object area. Without mask consistency loss (Eq. (5)), the mask does not align correctly with the foreground object region. Without the fine mask network, the fine details in the mask tend to be less accurate, especially on the region between hair and background. With all components combined, our method produces fine masks aligned to the foreground object region.

Table 1 provides quantitative ablation study. The decrease in the quality of masks shows the necessity of background participation loss and mask consistency loss. Ablating any of the components harms FID, implying that spatial understanding in the generators is important for the quality of images. We suppose that the influence of the fine mask generator is negligible in the metrics for the masks because the metrics are not sensitive enough to reflect changes in a small area.

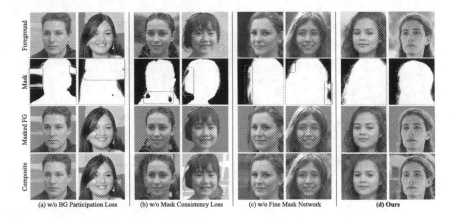

(a) w/o BG Participation Loss   (b) w/o Mask Consistency Loss   (c) w/o Fine Mask Network   (d) Ours

Fig. 8. Ablation of our methods. Three columns show the result without one component. (a) The masks have background area. (b) The masks are not align correctly with the foreground object region. (c) The masks bring the surrounding background. (d) our method produces fine-grained masks

Table 1. Quantitative comparison of ablation study on FFHQ

Setting	IoU(fg/bg)	mIoU	recall	precision	F1	Accuracy	FID
w/o Fine Mask Network	**0.93/0.83**	**0.88**	0.95	**0.98**	**0.96**	**0.93**	9.48
w/o BG Participation	0.91/0.77	0.84	**0.97**	0.94	0.95	0.91	9.79
w/o Mask Consistency	0.92/0.81	0.86	0.93	0.98	**0.96**	0.92	9.53
Full ours	**0.93**/0.82	**0.88**	0.95	**0.98**	**0.96**	**0.93**	**8.72**

4.5 Comparisons

Competitors. We choose PSeg[1] [5] and Labels4Free[2] (L4F in short, [1]) as our competitors. For a fair comparison, we trained L4F in FFHQ and AFHQ under the same conditions, i.e., batch size, data augmentations, training iterations. We pretrain StyleGAN2[3] for 256×256 resolution and then train the alpha network with its official setting. As Labels4Free does not conduct experiments on AFHQ, we train their alpha network for the same number of iterations on FFHQ (=1K), and manually find the working hyperparameters: $\lambda_2 = 3$ and $\phi_2 = 0.2$[4]. For PSeg, we add additional layers to their networks for 256×256 resolution since PSeg conducts their experiments in 128×128 resolution. When training Pseg, we followed the default setting reported in the paper. We do not include FineGAN [29] because it does not focus on foreground-background separation and it requires an external pretrained object detector for supervision.

---

[1] https://github.com/adambielski/perturbed-seg.
[2] https://github.com/RameenAbdal/Labels4Free.
[3] https://github.com/rosinality/stylegan2-pytorch.
[4] Without setting ϕ_2, all masks of Labels4Free saturate to 1.

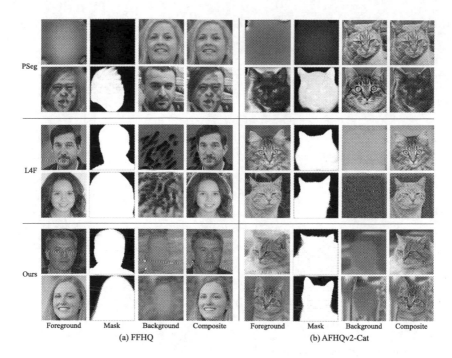

| Foreground | Mask | Background | Composite | Foreground | Mask | Background | Composite |

(a) FFHQ   (b) AFHQv2-Cat

Fig. 9. Qualitative comparison of image composition results on FFHQ and AFHQv2-Cat

Qualitative Results. Figure 9 provides a qualitative comparison between the methods. PSeg rarely succeeds in synthesizing proper foreground images and mostly draws objects on the background images. We suppose the reason to be the unmet assumption: for faces, the foreground cannot help touching the edges. Thus shifting the foreground will not be realistic. Labels4Free somewhat successes separating foregrounds. However, their masks are not accurate enough and the composition leaves artifacts on the boundaries. In contrast, our method produces masks that accurately capture the foreground object, even including hair, fur, and whiskers. Uncurated samples can be found in Appendix A

Quantitative Results. Table 2 reports how well the masks align with the foreground object region. The pseudo ground truth masks are obtained by feeding the foreground images to TRACER [19]. Our method consistently outperforms the competitors in all settings: different levels of truncation and datasets. Appendix C provides the results with other choices of pseudo ground truth.

Table 3 quantitatively compares the visual quality of the generated images. Our method achieves FIDs comparable to Labels4Free whose foreground generator equals the pretrained StyleGAN2 while drastically improving the masks.

Table 2. Quantitative comparison of alpha mask results on FFHQ and AFHQv2-Cat. We report the result with/without truncation($\psi = 1.0$, 0.7) and the threshold for the mask is 0.5 (Ours, PSeg) and 0.9 (L4F)

	ψ	method	IoU(fg/bg)	mIoU	recall	precision	F1	Accuracy
FFHQ	1.0	PSeg	0.05/0.23	0.14	0.05	0.18	0.07	0.05
		L4F	0.87/0.70	0.78	0.92	0.94	0.93	0.87
		Ours	**0.93/0.82**	**0.88**	**0.95**	**0.98**	**0.96**	**0.93**
	0.7	PSeg	0.01/0.23	0.12	0.01	0.04	0.01	0.01
		L4F	0.91/0.79	0.85	0.94	0.97	0.95	0.91
		Ours	**0.95/0.88**	**0.91**	**0.95**	**0.99**	**0.97**	**0.95**
AFHQv2-Cat	1.0	PSeg	0.06/0.23	0.15	0.06	0.16	0.07	0.06
		L4F	0.91/0.80	0.86	0.93	**0.98**	0.95	0.91
		Ours	**0.94/0.82**	**0.88**	**0.98**	0.96	**0.97**	**0.94**
	0.7	PSeg	0.01/0.19	0.10	0.01	0.12	0.01	0.01
		L4F	0.91/0.79	0.85	0.94	**0.97**	0.95	0.91
		Ours	**0.95/0.87**	**0.91**	**0.98**	**0.97**	**0.97**	**0.95**

Table 3. Quantitative comparison of generated foreground images on FFHQ and AFHQv2-Cat. Foreground generator of L4F equals to the pretrained StyleGAN2

	FID	
	FFHQ	AFHQv2-Cat
Pseg	62.44	12.71
Labels4Free (=StyleGAN2)	**6.51**	**5.19**
Ours	8.72	6.34

4.6   Segmenting Real Images

In addition, we demonstrate an extension of our method for segmenting real images. Following Labels4Free, we use 1K images and their ground truth segmentation masks from CelebAMask-HQ dataset [18] for evaluation. We employ the original inversion method from StyleGAN2. While Table 4 shows that our method achieves similar performance, Fig. 10 shows that our method produces much more accurate and finer masks.

Table 4. Quantitative comparison of alpha masks from inverted real images on CelebAMask-HQ. We report the result with the original inversion method from Style-GAN2 and the threshold for the mask is 0.5(Ours) and 0.9(Labels4Free)

	IoU(fg/bg)	mIoU	Recall	Precision	f1	Accuracy
Labels4Free	**0.93/0.81**	**0.87**	**0.97**	0.95	**0.96**	**0.93**
Ours	0.92/0.81	**0.87**	0.95	**0.97**	**0.96**	0.92

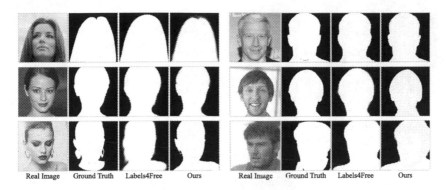

Real Image   Ground Truth   Labels4Free   Ours   Real Image   Ground Truth   Labels4Free   Ours

Fig. 10. Visual comparison on segmenting real images

Fig. 11. Various kinds of failures in our model (foreground-mask pair)

5   Conclusion and Discussion

Understanding spatial semantics in the synthesized images is an important research problem in GANs. In this paper, we proposed a GAN framework for foreground-aware image synthesis, generating images as a combination of foreground and background according to a mask. Our method achieves dramatic improvement in the fine details of the masks without any supervision or dataset-tailored assumption. Our method generalizes to a broad range of datasets including unaligned objects, e.g., CUB, LSUN-Church, and LSUN-Horse.

However, we observe exceptional cases where the mask generator struggles in Fig. 11. We suggest one of the main reasons to be the ambiguity of the task itself. In CompCars [36], the road below the vehicles is often marked as foreground. It is a reasonable choice because the road is physically close to the vehicles. Using a minimal amount of human supervision for resolving such ambiguity would be a sensible research direction, e.g., specifying foreground or background by scribbles on one or a few images. In some cases, the mask misses a small portion of the object area. This might be because the composite image is natural enough, even if the mask is inappropriate. We hope that our success in the common datasets in GAN literature sheds light on foreground-aware image synthesis.

Acknowledgement. This work was supported by the National Research Foundation of Korea(NRF) grant (No. 2022R1F1A107624111) funded by the Korea government (MSIT).

References

1. Abdal, R., Zhu, P., Mitra, N.J., Wonka, P.: Labels4Free: unsupervised segmentation using styleGAN. In: Proceedings of the IEEE/CVF International Conference on Computer Vision, pp. 13970–13979 (2021)
2. Abdal, R., Zhu, P., Mitra, N.J., Wonka, P.: Styleflow: attribute-conditioned exploration of styleGAN-generated images using conditional continuous normalizing flows. ACM Transactions on Graphics (TOG) **40**(3), 1–21 (2021)
3. Achanta, R., Shaji, A., Smith, K., Lucchi, A., Fua, P., Süsstrunk, S.: Slic superpixels compared to state-of-the-art superpixel methods. IEEE Trans. Pattern Anal. Mach. Intell. **34**(11), 2274–2282 (2012)
4. Alaluf, Y., Patashnik, O., Cohen-Or, D.: Restyle: a residual-based styleGAN encoder via iterative refinement. In: Proceedings of the IEEE/CVF International Conference on Computer Vision, pp. 6711–6720 (2021)
5. Bielski, A., Favaro, P.: Emergence of object segmentation in perturbed generative models. Advances in Neural Information Processing Systems 32 (2019)
6. Choi, Y., Uh, Y., Yoo, J., Ha, J.W.: StarGAN v2: diverse image synthesis for multiple domains. In: Proceedings of the IEEE Conference on Computer Vision and Pattern Recognition (2020)
7. Collins, E., Bala, R., Price, B., Susstrunk, S.: Editing in style: uncovering the local semantics of gans. In: Proceedings of the IEEE/CVF Conference on Computer Vision and Pattern Recognition, pp. 5771–5780 (2020)
8. Comaniciu, D., Meer, P.: Mean shift: a robust approach toward feature space analysis. IEEE Trans. Pattern Anal. Mach. Intell. **24**(5), 603–619 (2002)
9. Goodfellow, I., et al.: Generative adversarial nets. Advances in neural information processing systems 27 (2014)
10. Gu, J., Liu, L., Wang, P., Theobalt, C.: StyleNeRF: a style-based 3D-aware generator for high-resolution image synthesis. arXiv preprint arXiv:2110.08985 (2021)
11. Heusel, M., Ramsauer, H., Unterthiner, T., Nessler, B., Hochreiter, S.: GANs trained by a two time-scale update rule converge to a local nash equilibrium. Advances in neural information processing systems 30 (2017)
12. Ji, X., Henriques, J.F., Vedaldi, A.: Invariant information clustering for unsupervised image classification and segmentation. In: Proceedings of the IEEE/CVF International Conference on Computer Vision, pp. 9865–9874 (2019)
13. Karras, T., Aittala, M., Hellsten, J., Laine, S., Lehtinen, J., Aila, T.: Training generative adversarial networks with limited data. Adv. Neural. Inf. Process. Syst. **33**, 12104–12114 (2020)
14. Karras, T., et al.: Alias-free generative adversarial networks. Advances in Neural Information Processing Systems 34 (2021)
15. Karras, T., Laine, S., Aila, T.: A style-based generator architecture for generative adversarial networks. In: Proceedings of the IEEE/CVF conference on computer vision and pattern recognition, pp. 4401–4410 (2019)
16. Karras, T., Laine, S., Aittala, M., Hellsten, J., Lehtinen, J., Aila, T.: Analyzing and improving the image quality of styleGAN. In: Proceedings of the IEEE/CVF conference on computer vision and pattern recognition, pp. 8110–8119 (2020)
17. Kim, H., Choi, Y., Kim, J., Yoo, S., Uh, Y.: Exploiting spatial dimensions of latent in gan for real-time image editing. In: Proceedings of the IEEE/CVF Conference on Computer Vision and Pattern Recognition, pp. 852–861 (2021)
18. Lee, C.H., Liu, Z., Wu, L., Luo, P.: MaskGAN: towards diverse and interactive facial image manipulation. In: IEEE Conference on Computer Vision and Pattern Recognition (CVPR) (2020)

19. Lee, M.S., Shin, W., Han, S.W.: Tracer: extreme attention guided salient object tracing network. arXiv preprint arXiv:2112.07380 (2021)
20. Melas-Kyriazi, L., Rupprecht, C., Laina, I., Vedaldi, A.: Finding an unsupervised image segmenter in each of your deep generative models. arXiv preprint arXiv:2105.08127 (2021)
21. Mescheder, L., Geiger, A., Nowozin, S.: Which training methods for GANs do actually converge? In: International conference on machine learning, pp. 3481–3490. PMLR (2018)
22. Mildenhall, B., Srinivasan, P.P., Tancik, M., Barron, J.T., Ramamoorthi, R., Ng, R.: NeRF: representing scenes as neural radiance fields for view synthesis. In: Vedaldi, A., Bischof, H., Brox, T., Frahm, J.-M. (eds.) ECCV 2020. LNCS, vol. 12346, pp. 405–421. Springer, Cham (2020). https://doi.org/10.1007/978-3-030-58452-8_24
23. Niemeyer, M., Geiger, A.: Giraffe: representing scenes as compositional generative neural feature fields. In: Proceedings of the IEEE/CVF Conference on Computer Vision and Pattern Recognition, pp. 11453–11464 (2021)
24. Ouali, Y., Hudelot, C., Tami, M.: Autoregressive unsupervised image segmentation. In: Vedaldi, A., Bischof, H., Brox, T., Frahm, J.-M. (eds.) ECCV 2020. LNCS, vol. 12352, pp. 142–158. Springer, Cham (2020). https://doi.org/10.1007/978-3-030-58571-6_9
25. Patashnik, O., Wu, Z., Shechtman, E., Cohen-Or, D., Lischinski, D.: StyleCLIP: text-driven manipulation of styleGAN imagery. In: Proceedings of the IEEE/CVF International Conference on Computer Vision, pp. 2085–2094 (2021)
26. Richardson, E., et al.: Encoding in style: a styleGAN encoder for image-to-image translation. In: Proceedings of the IEEE/CVF Conference on Computer Vision and Pattern Recognition, pp. 2287–2296 (2021)
27. Shen, Y., Yang, C., Tang, X., Zhou, B.: InterFaceGAN: interpreting the disentangled face representation learned by GANs. IEEE Transactions on Pattern Analysis and Machine Intelligence (2020)
28. Shen, Y., Zhou, B.: Closed-form factorization of latent semantics in GANs. In: Proceedings of the IEEE/CVF Conference on Computer Vision and Pattern Recognition, pp. 1532–1540 (2021)
29. Singh, K.K., Ojha, U., Lee, Y.J.: FineGAN: unsupervised hierarchical disentanglement for fine-grained object generation and discovery. In: Proceedings of the IEEE/CVF Conference on Computer Vision and Pattern Recognition, pp. 6490–6499 (2019)
30. Tov, O., Alaluf, Y., Nitzan, Y., Patashnik, O., Cohen-Or, D.: Designing an encoder for styleGAN image manipulation. ACM Trans. Graph. (TOG) 40(4), 1–14 (2021)
31. Van Gansbeke, W., Vandenhende, S., Georgoulis, S., Van Gool, L.: Unsupervised semantic segmentation by contrasting object mask proposals. In: Proceedings of the IEEE/CVF International Conference on Computer Vision, pp. 10052–10062 (2021)
32. Voynov, A., Babenko, A.: Unsupervised discovery of interpretable directions in the gan latent space. In: International conference on machine learning, pp. 9786–9796. PMLR (2020)
33. Voynov, A., Morozov, S., Babenko, A.: Object segmentation without labels with large-scale generative models. In: International Conference on Machine Learning, pp. 10596–10606. PMLR (2021)
34. Wah, C., Branson, S., Welinder, P., Perona, P., Belongie, S.: The caltech-UCSD birds-200-2011 dataset (2011)
35. Wu, Z., Lischinski, D., Shechtman, E.: StyleSpace analysis: disentangled controls for stylegan image generation. In: Proceedings of the IEEE/CVF Conference on Computer Vision and Pattern Recognition, pp. 12863–12872 (2021)

36. Yang, L., Luo, P., Change Loy, C., Tang, X.: A large-scale car dataset for fine-grained categorization and verification. In: Proceedings of the IEEE Conference on Computer Vision and Pattern Recognition, pp. 3973–3981 (2015)

37. Yang, Y., Bilen, H., Zou, Q., Cheung, W.Y., Ji, X.: Learning foreground-background segmentation from improved layered GANs. In: Proceedings of the IEEE/CVF Winter Conference on Applications of Computer Vision, pp. 2524–2533 (2022)

38. Yu, C., Wang, J., Peng, C., Gao, C., Yu, G., Sang, N.: BiSeNet: bilateral segmentation network for real-time semantic segmentation. In: Ferrari, V., Hebert, M., Sminchisescu, C., Weiss, Y. (eds.) ECCV 2018. LNCS, vol. 11217, pp. 334–349. Springer, Cham (2018). https://doi.org/10.1007/978-3-030-01261-8_20

39. Yu, F., Seff, A., Zhang, Y., Song, S., Funkhouser, T., Xiao, J.: LSUN: construction of a large-scale image dataset using deep learning with humans in the loop. arXiv preprint arXiv:1506.03365 (2015)

40. Yüksel, O.K., Simsar, E., Er, E.G., Yanardag, P.: LatentCLR: a contrastive learning approach for unsupervised discovery of interpretable directions. In: Proceedings of the IEEE/CVF International Conference on Computer Vision, pp. 14263–14272 (2021)

41. Zhang, K., Riegler, G., Snavely, N., Koltun, V.: NeRF++: analyzing and improving neural radiance fields. arXiv preprint arXiv:2010.07492 (2020)

42. Zhu, J., Shen, Y., Zhao, D., Zhou, B.: In-domain GAN inversion for real image editing. In: Vedaldi, A., Bischof, H., Brox, T., Frahm, J.-M. (eds.) ECCV 2020. LNCS, vol. 12362, pp. 592–608. Springer, Cham (2020). https://doi.org/10.1007/978-3-030-58520-4_35

SCAM! Transferring Humans Between Images with Semantic Cross Attention Modulation

Nicolas Dufour[1,2]($\boxtimes$) (iD), David Picard[1] (iD), and Vicky Kalogeiton[2] (iD)

[1] LIGM, Ecole des Ponts, Univ Gustave Eiffel, CNRS, Marne-la-Vallée, France
{nicolas.dufour,david.picard}@enpc.fr
[2] LIX, CNRS, Ecole Polytechnique, IP Paris, Palaiseau, France
vicky.kalogeiton@lix.polytechnique.fr
https://imagine.enpc.fr/~dufourn/scam

Abstract. A large body of recent work targets semantically conditioned image generation. Most such methods focus on the narrower task of pose transfer and ignore the more challenging task of subject transfer that consists in not only transferring the pose but also the appearance and background. In this work, we introduce SCAM (Semantic Cross Attention Modulation), a system that encodes rich and diverse information in each semantic region of the image (including foreground and background), thus achieving precise generation with emphasis on fine details. This is enabled by the Semantic Attention Transformer Encoder that extracts multiple latent vectors for each semantic region, and the corresponding generator that exploits these multiple latents by using semantic cross attention modulation. It is trained only using a reconstruction setup, while subject transfer is performed at test time. Our analysis shows that our proposed architecture is successful at encoding the diversity of appearance in each semantic region. Extensive experiments on the iDesigner, CelebAMask-HD and ADE20K datasets show that SCAM outperforms competing approaches; moreover, it sets the new state of the art on subject transfer.

Keywords: Semantic generation · Semantic editing · Generative adversarial networks · Subject transfer

1 Introduction

Being able to perform subject transfer between two images is a key challenge for many applications, from post-processing in game or art industries to software addressing the needs of the public. For instance, in film industries, one could replace a stunt performer by the main actors, thus alleviating the need of finding a look-alike performer; hence, increasing the filmmakers freedom. Similarly, it could enable finishing a film when an actor is indisposed.

Supplementary Information The online version contains supplementary material available at https://doi.org/10.1007/978-3-031-19781-9_41.

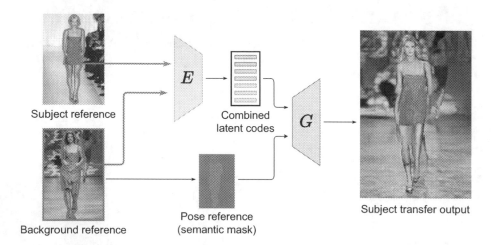

Fig. 1. Subject transfer with the proposed SCAM. We first encode the desired subject with the encoder E and get the subject latent codes. Then, we encode the background and the semantic mask for the pose and background reference. Finally, the generator G synthesizes an image, where the subject is transferred with the desired background and pose. Pictures taken from the Internet (Kate Moss picture by JB Villareal/Shoot Digital, Natalia Vodianova picture by Karl Prouse/Catwalking).

Given a source and a target subject, the idea of subject transfer is for the source subject to replace the target subject in the target image *seamlessly*. The target image should keep the same background, the same interactions between subject and objects, and the same spatial configuration, to account for possible occlusions. Figure 1 illustrates this. Note, in contrast to faces, buildings, or landscapes, human bodies are malleable with high morphological diversity, thus casting the task hard to model.

Most methods focus either on pose transfer [2,44,49], where the pose changes, or on style transfer [31,54], where the pose remains fixed but the subject's styling changes. These are limited and cannot be used out of the box for our task, as they are: (1) restrictive; they only work on uniform backgrounds, failing in complex ones (PISE [49], SEAN [54], [44]), and (2) expensive; they require hard training [44] or training one model per human (Everybody Dance Now [2]). Instead, subject transfer changes both the pose and the style/identity of the subject. Thus, a successful system is decoupled in both pose and style transfer and performs both tasks simultaneously.

Semantic editing is a related task, consisting in controlling the output of a generative network by a segmentation mask. Indeed, subject transfer can be performed with semantic editing by using the mask of the target subject with the style of the source. However, modern methods cannot handle complex layout and rich structure (like full-bodies) with in the wild backgrounds. For instance, SPADE [31] fails to control each region style independently, while SEAN [54] fails to handle complex, detailed scenes, such as multiple background objects.

To this end, we propose the **S**emantic **C**ross **A**ttention **M**odulation system (**SCAM**), a semantic editing model that accounts for all aforementioned challenges relevant to subject transfer. SCAM captures fine image details inside the semantic region by having multiple latents per semantic regions. This enables capturing unsupervised semantic information inside the semantic labels, which allows for better handling coarse semantic labels, such as background. Our model can generate more complex backgrounds than previous methods and outperforms SEAN [54] both on the subject transfer and semantic reconstruction tasks.

We propose three architectural contributions: First, we propose the Semantic Cross Attention (**SCA**) that performs attention between a set of latents (each linked to a semantic region) and an image feature map. SCA constrains the attention, such as the latents only attend the regions on the image feature map that correspond to the relevant semantic label. Secondly, we introduce the **SAT** operation and encoder (**S**emantic **A**ttention **T**ransformer) that relies on cross attention to decide which information to gather in the image and for which latent, thus allowing for richer information to be encoded. Third, we propose the **SCAM**-Generator (after which SCAM is named) that modulates the feature maps using the SCAM-Operation, which allows every pixel to attend to the semantically meaningful latents. Note, the whole architecture is trained using a reconstruction setup only, and subject transfer is performed at test time.

2   Related Work

Image to Image Synthesis with GANs. GANs [10] generate images by processing a random vector sampled from a predefined distribution with a dedicated network [3,18,32]. A major improvement is StyleGAN [19,20] that allows to modulate the feature map at each resolution according to a given style vector. Typically, unconditional GANs allow for minimal control over the generator's output. For more flexibility, Pix2Pix [14,45] trains a generator coupled to an encoder, allowing for output control with multiple modalities (sketches, keypoints). One of its drawbacks is the need for data pairs, which can be hard to collect (drawings). To tackle this, CycleGAN [53] uses unpaired data by enforcing cycle consistency across domains. However, acquiring paired data is feasible when leveraging external models, such as semantic masks from segmentation models. In our case, we do not have access to ground-truth images where the subject has been transferred as it would require both the subject and the reference to have exactly the same pose and occlusion. We circumvent this by training on a reconstruction proxy task and performing subject transfer at test time.

Semantic Image Generation. Even if Pix2Pix [14,45] manage to control the output image with a segmentation mask, it suffers from a semantic information washing-up. SPADE [31] propose to fix this problem by introducing layer-wise semantic conditioning. CLADE [38] propose a more efficient version of SPADE to reduce runtime complexity. Other approaches such as [7,11,25,27,36,38,40,46] propose improvements over spade. However, these approaches work well generating images from semantic information, but they do not focus on the case where we want to re-generate a given image to then be able to edit it. To do

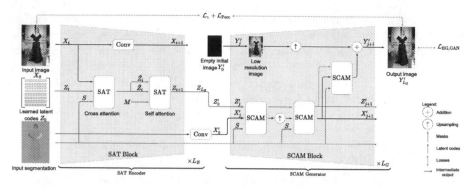

Fig. 2. Training setup of the proposed SCAM architecture. It consists of the SAT-Encoder (pink) and the SCAM-Generator (yellow). The SAT-Encoder allows the latents to retrieve information from an image, exploiting both the raw image and the convolution feature maps. Once the image is encoded, the latents are fed to the SCAM-Generator, which captures top-down and bottom-up interactions with a semantic constraint, allowing to easily alter the desired regions thanks to the latents that are dedicated to a given region. (Color figure online)

so, the style of the image must be carefully encoded and move from the single style vector used in SPADE and its variants. SEAN [54] propose to introduce a single style vector per semantic label by performing average pooling on the encoder CNN features. GroupDNet [56] propose to solve this by encoding each semantic region separately using grouped convolution. INADE [37] propose to use instance conditioned convolution to extract a single style code per instance. Although this allows better control of the output and richer representation, it still has two problems: (1) it is limited when handling coarse labels with diverse objects, (2) it creates a single vector per semantic region. In our approach, SCAM, we solve this by introducing the SAT-Encoder which can extract rich representation from images and is able to output multiple and diverse latents per semantic region. In turn, the SEAN-Generator modulates the output by both the semantic mask and the extracted style. This, however, modulates each pixel of a semantic region with the same style vector. Instead, our SCAM-Generator, uses attention to leverage different tokens to interact with, leading to different modulations per pixel, and hence enabling the emergence of unsupervised semantic structure in the semantic regions. Other approaches propose to use diffusion process approaches [30] to perform this editing process. However, these diffusion approaches are very expensive to sample from.

Attention in Computer Vision. Despite their remarkable success [1,4,33,34], transformers suffer from a quadratic complexity problem, which makes it hard to use. To tackle this, most vision methods [6,28,41] subdivide images into patches, resulting in losing information. Instead, the recent Perceiver [15,16] tackles this complexity issue by replacing self attention by cross attention. The image pixels are attended by a significatively smaller set of learned tokens.

Attention in GANS. It [8,17,22,47,48,51] has shown great progress over the past years. GANsformer [13] leverages cross attention to exploit multiple style codes between style vectors and feature maps. These architectures use attention for unconditional generation; however, they do not focus on subject transfer. Instead, our proposed SCA improves upon GANsformer's duplex attention for semantically constrained generation by assigning latents to semantic regions.

Pose Transfer. Using keypoints for pose transfer [23,29,39,55] typically results in coarse representation of bodies. To tackle this, some methods use semantic masks [5,12,49,50]. These, however, focus only on pose transfer, which does not alter the background. Instead, we aim at subject transfer, where preserving the background is crucial. Most methods are limited to simple backgrounds. [2] overfit a GAN to a video and regenerate the background; hence, it cannot be used for dynamic scenes. [44] address this by adapting the weights of the generator, but this ties the subject to the background, not allowing for subject transfer. Here, we focus on subject transfer that changes both pose and background.

3 Method

Our goal is to perform semantic editing with a focus on subject transfer. We propose the **SCAM** method (Semantic Cross Attention Modulation, Fig. 2). It relies on **SCA** (Semantic Cross Attention), i.e. a novel mechanism that masks the attention according to segmentation masks, thus encoding semantically meaningful latent variables (Sect. 3.1). SCAM consists of: (a) **SAT-Encoder** (Semantic Attention Transformer) that relies on cross attention to decide which information to gather in the image and for which latent (Sect. 3.2); and (b) **SCAM-Generator** (Semantic Cross Attention Modulation) that captures rich semantic information in an unsupervised way (Sect. 3.3).

Notation. Let $X \in \mathbb{R}^{n \times C}$ be the feature map with n the number of pixels, and C the number of channels. Let $Z \in \mathbb{R}^{m \times d}$ be a set of m latents of dimension d and s the number of semantic labels. Each semantic label is attributed k latents, such that $m = k \times s$. Each semantic label mask is assigned k copies in $S \in \{0; 1\}^{n \times m}$. $\sigma(.)$ is the softmax operation.

Motivation. Since many regions of the image have visually diverse content (e.g., the background), we propose to encode this varied information in several complementary latents. Motivated by the findings of Gansformers [13], we use attention to introduce both a constraint on which part of the image a latent can get information from and a competing mechanism between latents attending the same region so as to specialize them. Reciprocally, using duplex attention, we introduce the same strategy by limiting the latents that the feature map can attend to and introduce a competition between parts of a semantic region that can attend the same latent.

3.1 Semantic Cross Attention (SCA)

Definition. The goal of SCA is two-fold depending on what is the query and what is the key. Either it allows to give the feature map information from a semantically restricted set of latents or, respectively, it allows a set of latents to retrieve information in a semantically restricted region of the feature map. It is defined as:

$$\text{SCA}(I_1, I_2, I_3) = \sigma \left(\frac{QK^T \odot I_3 + \tau (1 - I_3)}{\sqrt{d_{in}}} \right) V \quad , \tag{1}$$

where I_1, I_2, I_3 the inputs, with I_1 attending I_2, and I_3 the mask that forces tokens from I_1 to attend only specific tokens from I_2[1], $Q = W_Q I_1$, $K = W_K I_2$ and $V = W_V I_2$ the queries, keys and values, and d_{in} the internal attention dimension.

We use three types of SCA. *(a) SCA with pixels X attending latents Z*: $\text{SCA}(X, Z, S)$, where $W_Q \in \mathbb{R}^{n \times d_{in}}$ and $W_K, W_V \in \mathbb{R}^{m \times d_{in}}$. The idea is to force the pixels from a semantic region to attend latents that are associated with the same label. *(b) SCA with latents Z attending pixels X*: $\text{SCA}(Z, X, S)$, where $W_Q \in \mathbb{R}^{m \times d_{in}}$, $W_K, W_V \in \mathbb{R}^{n \times d_{in}}$. The idea is to semantically mask attention values to enforce latents to attend semantically corresponding pixels. *(c) SCA with latents Z attending themselves*: $\text{SCA}(Z, Z, M)$, where $W_Q, W_K, W_V \in \mathbb{R}^{n \times d_{in}}$. We denote $M \in \mathbb{N}^{m \times m}$ this mask, with $M_{\text{latents}}(i, j) = 1$ if the semantic label of latent i is the same as the one of latent j; 0 otherwise. The idea is to let the latents only attend latents that share the same semantic label.

3.2   SAT-Encoder

Following [16], our SAT-Encoder relies on cross attention. It consists of L_E consecutive layers of **SAT-Blocks**, where the input of the $i+1$-th one is the output of the i-th one (Fig. 2 (left)). Given a set of learned queries Z_0 (i.e., parameters updated with gradient descent using back-propagation), it outputs latents Z_{L_E} that have encoded the input image. This allows to create multiple latents per semantic regions resulting in specialized latents for different part of each semantic region of the image. The encoder is also flexible enough to easily assign a different number of latent for each semantic region, allowing to optimize the representation power given to each semantic region. At each layer, the latent code retrieves information from the image feature maps at different scales.

The **SAT-Block** is composed of three components: two **SAT-Operations** and a strided convolution. SAT-Operations are transformer-like [43] operations, replacing self-attention by our proposed SCA. They are defined as:

$$\text{SAT}(I_1, I_2, S) = \text{LN}(f(\text{LN}(\text{SCA}(I_1, I_2, S) + I_1)) + I_1), \tag{2}$$

with LN the layer norm and f a simple 2-layer feed forward network. The first SAT-Operation, $\text{SAT}(X, Z, S)$, let the latents retrieve information from the image feature map (case *(b)* from Sect. 3.1). The second SAT-Operation, $\text{SAT}(Z, Z, M)$, is refining the latents using SCA in a self attention setup where the latents attend themselves, keeping the semantic restriction (case *(c)* from Sect. 3.1). The strided convolution encode the previous layer image feature map, reducing its spatial dimension. Implementation details and reference code about SAT-Blocks and SAT-Operation are in the supplementary material.

---

[1] The attention values requiring masking are filled with $-\infty$ before the softmax. (In practice $\tau = -10^9$).

3.3   SCAM-Generator

Definition and Architecture. SCAM-Generator takes as an input the latent codes Z'_0 $(= Z_{L_E}$ encoder's output) and the segmentation mask S and outputs the generated image Y'_{L_G}. It consists of L_G **SCAM-Blocks** (Fig. 2 (right)). The input latent of the generator is given by the encoder's output, whereas the input latents of each block within the generator are the output latents of previous blocks. Similarly, the input feature maps of each block are the feature map outputs at each resolution, while for the first features, we encode the segmentation with a convolutional layer following [31,54].

The **SCAM-Block** has a progressive growing architecture similar to the one of StyleGAN2 [20] and consists of 3 SCAM-Operations (See Fig. 2 (right)). 2 SCAM-Operations process the generator feature-map, with an upscaling operation between the two, while a parallel SCAM-operation retrieves information from the feature maps and generate the image in the RGB space. Implementation details and reference code are in the supplementary.

SCAM-Operation. It aims at exchanging information between pixels/features and latents of the same semantic label and is depicted in Fig. 3. Each SCAM-Operation has inputs: (1) the set of input latents Z'_{in},(2) the input feature map X'_{in} , and (3) the segmentation mask S. Its outputs are: (1) the output latents Z'_{out}, and (2) the output feature map X'_{out}. SCAM-Operation consists of three parts: (a) the latent SAT, (b) the feature SCA, (c) the Modulation operation.
(a) The latent SAT. It uses a SAT operation to update the current latents Z'_{in} based on the current feature map: $Z'_{out}:=SAT(Z'_{in}, X'_{in}, S)$. This allows for latent refinement while enforcing the latents semantic constraint thanks to the SCA operation inside SAT.
(b) The feature SCA performs latent to image attention: it incorporates the latent information to the pixels/features using SCA. Given X'_{in}, the out-

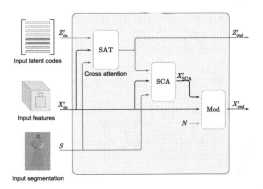

Fig. 3. SCAM-Operation. It modulates a feature map according to a segmentation map, allowing each pixel to retrieve information from a semantically restricted set of latents. It enables both top-bottom (latents retrieve information from the feature map) and bottom-top interactions (the map gets information from latents).

put latents from the SAT-Operation Z'_{out} and the mask S, it outputs X'_{SCA}:

$$X'_{\text{SCA}} = \text{SCA}(X'_{\text{in}}, Z'_{\text{out}}, S) \quad . \tag{3}$$

(c) The Modulation operation takes as input the X'_{in} maps and the X'_{SCA} from SCA, and outputs features that are passed through a convolution layer $g(.)$ to produce the final output image features X'_{out}. It is defined as:

$$X'_{out} = g\left(\gamma(X'_{\text{SCA}}) \odot \text{IN}(X'_{\text{in}}) + \mu(X'_{\text{SCA}}) + N\right) \quad . \tag{4}$$

It consists of the $\gamma(.)$ and $\mu(.)$ operators that determine the scale and bias factor for the modulation operation. We use Instance Normalization (IN) [42] to perform the feature map normalization. Following [19,20], we also add noise $N \in \mathbb{R}^{n \times C}$, with $N \sim \mathcal{N}(0, \sigma^2 I_{n \times C})$, and σ learned parameter. This encourages the modulation to account for stochastic variation.

Discussion. SCAM-Operation outputs both the modulated feature map and the updated latents. Thus, the information from the previous maps is propagated to the latents and to the feature maps. This brings several advantages: first, it preserves the semantic constraint; second, it provides finer refinement within the semantic mask by attending to multiple latents; and third it allows each pixel to choose which latent to use for modulation.

3.4   Training Losses

We train SCAM with GAN and reconstruction losses. We denote by D the discriminator, G the generator and E the encoder. For the **discriminator**, we follow [14,31,45,45] and use PatchGAN, as it discriminates for patches of the given image instead of the global image. For the **GAN loss**, we use Hinge GAN loss [26]: $\mathcal{L}_{\text{EG,GAN}}, \mathcal{L}_{\text{D,GAN}}$. For the **reconstruction loss**, we use the perceptual loss $\mathcal{L}_{\text{Perc}}$ as in [54] and the $\mathcal{L}_1$ between the input and the reconstructed input. The **final losses** are $\mathcal{L}_{\text{EG}} = \mathcal{L}_{\text{EG,GAN}} + \lambda_{\text{perc}}\mathcal{L}_{\text{Perc}} + \lambda_1\mathcal{L}_1$ and $\mathcal{L}_{\text{D}} = \mathcal{L}_{\text{D,GAN}}$ with λ_{perc} and λ_1 hyperparameters. We use $\lambda_{\text{perc}} = \lambda_1 = 10$ in our experiments. Training details are in the supplementary.

3.5   Subject Transfer

Once SCAM is trained for reconstruction, at test time we perform subject transfer. Given two images X_A, X_B with their respective segmentation masks (S_A, S_B), we retrieve their latent codes as Z_A and Z_B using the SAT-Encoder. To transfer the subject from X_B to the context of X_A, we create Z_{mix}, where the style codes related to the background come from Z_A and the remainder codes come from Z_B. Then, we retrieve $Y'_{\text{mix}} = G(Z_{\text{mix}}, S_B)$. See Fig. 1.

4   Experiments

We now present experimental results for SCAM. More results and ethical impacts are discussed in supplementary.

Table 1. Comparison on iDesigner [35] and CelebAMask-HQ [21] and ADE20K [52].

Method	iDesigner					CelebAMask-HQ			ADE20K	
	PSNR ↑	R-FID ↓	S-FID ↓	REIDSim ↑	REIDAcc ↑	PSNR ↑	R-FID ↓	S-FID ↓	PSNR ↑	R-FID ↓
SPADE [31] [CVPR19]	10.4	66.7	67.5	0.67	0.26	10.9	38.2	38.3	10.7	59.7
CLADE [38] [TPAMI21]	11.3	45.4	46.1	0.68	0.29	10.8	41.8	42.0	10.4	53.7
SEAN-CLADE [38] [TPAMI21]	15.3	48.4	56.1	0.75	0.31	16.2	19.8	24.3	14.0	38.7
INADE [37] [CVPR21]	12.0	33.0	33.9	0.72	0.34	12.24	22.7	23.4	11.3	48.6
SEAN [54] [CVPR20]	14.9	53.5	58.7	0.74	0.30	16.2	18.9	22.8	14.6	47.6
SCAM (Ours)	21.4	13.2	26.9	0.81	0.56	21.9	15.5	19.8	20.0	27.5

Implementation Details. We train all models for 50 k steps, with batch size of 32 on 4 Nvidia V100 GPUs. We set $k = 8$ latents per label of dim $d = 256$. We generate images of resolution 256px.

4.1 Datasets and Metrics

iDesigner. [35] contains images from designer fashion shows, including 50 different designers with 50k train and 10k test samples. We segment human parts with [24] and then merge the labels to end up with: face, body and background labels.

CelebAMask-HQ. [21] contains celebrity faces from CelebA-HQ [18] with 28k train and 2k test images labelled with 19 semantic labels of high quality.

ADE20K. [52] contains diverse scenes images with 20k train and 2k test images. The images are labelled with 150 semantic labels of high quality.

Metrics. We use PSNR, reconstruction FID (R-FID), and swap FID (S-FID). R-FID is computed as the FID between the train set and the reconstructed test set, while S-FID is between the test set and a set of subject transfer images computed on the test set. We introduce REIDAcc and REIDSim, computed in the latent space of a re-identification network [9]. REIDSim computes the average cosine similarity between the subject image and the subject transfer image. REIDAcc accounts for the proportion of images where the cosine similarity of the transferred subject is higher with the subject than with the background.

4.2 Comparison to the State of the Art

We first compare our proposed SCAM to INADE [37], SEAN [54], CLADE [38], SEAN-CLADE [38] and SPADE [31]. We reproduce all the methods and provide code for our implementations in the supplementary material.

Results on iDesigner are shown in Table 1 (left). Overall, SCAM outperforms competing approaches for all metrics. Specifically, for PSNR it outperforms SEAN-CLADE by approximately +5.9dB, whereas it reaches 13.2 R-FID vs 33.0 for INADE. These major boosts show that our reconstructed images better preserve the details of the initial images, meaning the representation power of SCAM is higher than that of other approaches. The difference is also notable for

Table 2. Ablation study on iDesigner [35] and CelebAMask-HQ [21]

	Encoder		Generator	Losses			iDesigner						CelebAMask-HQ					
	Conv	SA	SAT	$\mathcal{L}_1$	$\mathcal{L}_{Perc}$	$\mathcal{L}_{GAN}$	PSNR↑	+Δ	R-FID↓	-Δ	S-FID↓	-Δ	PSNR↑	+Δ	R-FID↓	-Δ	S-FID↓	-Δ
i	✓	✓	✓	✓	✓	✓	21.0		13.2		27.1		22.0		15.7		20.7	
ii	✗	✓	✓	✓	✓	✓	19.2	−1.8	26.3	−13.1	34.1	−7.0	**22.1**	+0.1	**15.5**	+0.2	20.2	+0.5
iii	✓	✗	✓	✓	✓	✓	21.3	+0.3	**12.7**	+0.5	**26.1**	+1.0	22.0	0.0	16.5	−0.8	**20.0**	+0.7
iv	✗	✗	✓	✓	✓	✓	18.5	−2.5	24.8	−11.6	29.8	−2.7	19.8	−2.2	19.0	−3.3	21.6	−0.9
v	✓	✓	✗	✓	✓	✓	21.1	+0.1	15.6	−2.4	27.7	−0.6	21.8	−0.2	15.6	+0.1	21.7	−1.0
vi	✗	✓	✗	✓	✓	✓	17.7	−3.3	49.8	−36.6	55.9	−28.8	20.2	−1.8	21.7	−6.0	25.3	−4.6
vii	✓	✗	✗	✓	✓	✓	21.0	0.0	16.7	−3.5	32.3	−5.2	21.4	−0.6	16.7	−1.0	21.8	−1.1
viii	✓	✓	✓	✗	✓	✓	19.5	−1.5	16.1	−2.9	30.7	−3.6	-	-	-	-	-	-
ix	✓	✓	✓	✓	✗	✓	**22.9**	+1.9	43.2	−30.0	91.3	−64.2	-	-	-	-	-	-
x	SEAN		SCAM	✓	✓	✓	17.5	−3.5	27.6	−14.4	32.7	−5.6	17.6	−2.4	20.4	−5.3	24.2	−3.5

S-FID, with INADE reaching 33.9 vs 26.9 for SCAM, showing that our method perform better subject transfer on datasets with coarse semantic labels than other approaches. We also observe a superiority of SCAM on REIDSim (+0.06 compared to INADE) and REIDAcc (+0.22 compared to INADE). Overall REIDAcc is a hard metric. This can be explained by the fact that the subject transfer image shares the same semantic information with the background image.

Results on CelebAMask-HQ are shown in Table 1 (center), where we observe that SCAM outperforms all methods. For instance, for PSNR, SCAM outperforms SEAN by +5.7dB. SCAM also improves over SEAN by 3.4 R-FID points (15.5 vs 19.8). For subject transfer, SCAM outperforms SEAN by a S-FID decrease of almost 3 points (19.8 vs 22.8), clearly indicating that our method is also better at transferring. We observe that even for a dataset that has precise labelling, our approach still outperforms competing approaches.

Results on ADE20K are shown in Table 1 (right), where we observe that SCAM outperforms all methods. SCAM has the best PSNR of 20.0 whereas second to best SEAN has a PSNR of 14.6. SCAM also beats SEAN-CLADE by 11.2 R-FID points (27.5 vs 38.7). We cannot evaluate S-FID on this dataset since it is hard to select what is the main subject in the image, and not all images share the same semantic labels.

4.3 Ablations

Here, we perform several ablations to validate the effectiveness of all components of SCAM. We denote by − experiments that do not converge.

Ablations of SAT-Encoder and SCAM-Generator. We examine the effectiveness of our encoder and generator by modifying either the SAT-Encoder or the SCAM-Generator with baseline versions and report the results on iDesigner in Table 2. The first row (i) corresponds to our SCAM. We benchmark some variants of our SAT-Encoder: Conv denotes whether we use convolutions in SAT or not. SA denotes whether we use self-attention SAT block or not. The results show that overall convolutions in the SAT-Encoder provide a big encoding advantage. This is especially true for the complex iDesigner dataset. Indeed (i) outperforms (ii, iv, vi) by a high margin on iDesigner e.g., by -13.1 R-FID and -7.0 S-FID for (ii), which validates our use of multiple resolutions in the encoder. We also examine a variation of the generator by removing the SAT block in the SCAM block. Having SAT in SCAM leads to better results; such as -2.4 in R-FID, and -0.6 in S-FID for (v) on iDesigner. Similar results can be observed in (vi and vii).

In (x), we use the same encoder as in SEAN and the SCAM generator. To manage to have multiple latents per semantic latents, we split the SEAN encoding in 8 smaller latents for each semantic latent. We observe here that our SAT-Encoder is better than the SEAN encoder at extracting information from images. Indeed, we obtain better R-FID and S-FID (-14,4/-5.6) with our encoder than the SEAN encoder.

Ablation of Losses. Table 2 (i,viii,ix) ablates the three $\mathcal{L}_1, \mathcal{L}_{\text{Perc}}, \mathcal{L}_{\text{GAN}}$ losses used in SCAM. The full combination (i) reaches the best results. Interestingly, removing $\mathcal{L}_{\text{Perc}}$ (ix) results in the best PSNR point (22.9db) while having among the worst R-FID and S-FID (43.2, 91.3, respectively). This is expected, as removing the perceptual loss makes the generator rely only on the L_1 loss, and may artificially increase PSNR at the cost of realism.

Number k of latents per semantic label. We've studied the impact of the number of latents k per semantic label on the performance of the model. We tested $k \in [4, 8, 16, 32]$ on iDesigner [35]. We find that the R-FID increase with the number of latents where the S-FID decreases. Indeed for $k = 4$ we have R-FID of 15.9 and a S-FID of 27.3. For $k = 32$, we have R-FID of 9.7 and a S-FID of 30.3. We settle for $k = 8$ in our experiments since it offers the best trade-off between S-FID and R-FID. We also privilege a smaller k since the time complexity of our model is $O(k)$.

Visualization of the Attention Matrix. To investigate how using multiple latents per region is handled by SCAM, we visualize in Fig. 4 the last SCAM layer attention matrix. We colour each pixel according to the corresponding latent with the highest attention value. Overall, we observe that for each semantic region, the latents attend to different subregions, capturing semantic information without supervision. The first example (a) shows that even without specialized segmentation labels, SCAM specializes some latents to reconstruct the complex face pattern (eyes, mouth, and hair) and others for the different body parts (dress and shoes). The second row displays an interesting case: SCAM is capable of assigning different latents to the humans in the background even if they are not labelled as such.

Fig. 4. (a) SCAM specializes latents on complex patterns (shirt, eyes, shoes); (b) it learns semantic information *on its own* inside the semantic labels (background people).

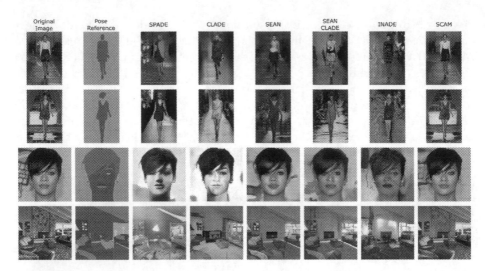

Fig. 5. Reconstructions on iDesigner [35], CelebAMask-HQ [21] and ADE20K [52].

4.4 Qualitative Results on Reconstruction

Overall, we observe that the reconstruction quality of SCAM is superior to competing approaches. Figure 5 displays the input images, their masks (first two columns) and the reconstructions by SPADE [31], CLADE [38], SEAN [54], SEAN-CLADE [38], INADE [37] and SCAM (last six columns). The first two rows are samples from iDesigner, the third is from CelebAMask-HQ and the lastrow from ADE20K.

Subject Reconstruction on iDesigner. SCAM reconstructs more structure than SEAN, both in the background and the human. INADE, CLADE and SPADE tend to generate images that doesn't match the style of the original image. For the *background*, we observe that the curtains and window frame of the second row are well-reconstructed by SCAM, in contrast to SEAN that includes colors but no other frame-cues. This highlights the rich generation capabilities of SCAM-Generator, which manage to generate complex backgrounds where competing approaches fail. For the *subject*, SCAM results in finer reconstructions compared to other approaches. For instance, in the first row SCAM reconstructs coherent clothes, while SEAN generate a blurred out version of the clothes.

Reconstruction on CelebAMask-HQ. Overall, SCAM generates crisper and more realistic results than competing approaches. For instance, in the third row, SEAN fails entirely to reconstruct the background by producing an averaged texture. Our method does a better job at this, figuring out a better positioning for the logo and capturing better colors and shapes. On the subject, we also observe that SEAN fails to capture small details such as eye colors.

Reconstruction on ADE20K. SCAM has a more reliable reconstruction than competing approaches. In the fourth row, we observe that SCAM is the only

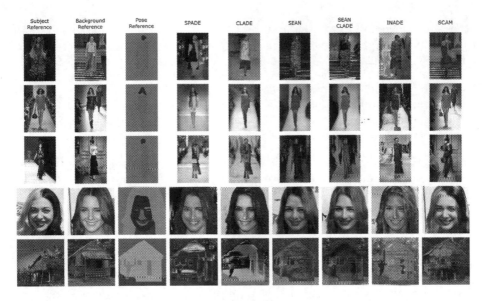

Fig. 6. Subject Transfer on the test set of iDesigner [35], CelebAMask-HQ [21] and ADE20K [52]. Note the hard case in row 3, where only SCAM rotates the subject. For ADE20K, we consider the house as the subject in the 5th row.

approach that manage to reconstruct some texture on the chimney wall. We also observe that overall, SCAM has more detail in the reconstructed object whereas approaches such as SEAN tend to have more averaged textures Fig. 6.

4.5 Qualitative Results on Subject Transfer

Figure 5 displays the subject and background images and the segmentation mask of the pose (first three columns). Then we show the subject transfer for SPADE [31], CLADE [38], SEAN [54], SEAN-CLADE [38], INADE [37] and SCAMSCAM (last six columns). The first three rows are samples from iDesigner, the fourth is from CelebAMask-HQ and the last one is on ADE20K. Overall, amongst all methods, SCAM is the one to successfully preserve all components of subject transfer: subject appearance, background appearance and pose.

Subject Transfer on iDesigner. SCAM leads to superior subjects (e.g., washed out colors in SEAN vs coherent structured clothing in SCAM in the first row) and background reconstruction (i.e., global background structure, positioning of people on the sides and colours) than SEAN. For instance, in the first row, SEAN completely fails to reconstruct the background, while SEAN-CLADE reconstructs some texture but lacks detail. Similarly, in the second row, SCAM captures the model in the background whereas other approaches miss it; note also the precise people reconstruction in the left part of the catwalk for SCAM compared to competing approaches. In the third row, we have a very hard case, where the subject has a completely different pose than the pose reference. While other approaches fail to rotate the subject, SCAM does succeed in doing, even if the quality of the generated image is low. Overall, subjects have better appearance in SCAM than in other methods, like details in clothes, shoes, or faces.

Subject Transfer on CelebAMask-HQ. The fourth row shows that SCAM recovers more details in the transferred image, such as the colour of the skin or facial expression. Notably, SCAM does capture the bicolor separation of the hair, while SEAN, SEAN-CLADE and INADE display an averaged hair color.

Subject Transfer on ADE20K. In the fifth row, we consider the house as the subject we want to transfer. We can see that SCAM does transfer the hut like appearance of the subject, whereas competing approaches fail to do so. We also observe that most approaches have difficulties with the person generation, only SCAM generates a human that is coherent with the background reference.

4.6   User Study on iDesigner

We perform an user study on iDesigner. We compare 3 models one against each other: INADE, SEAN and SCAM. We have asked 38 different people to select among 20 reconstruction images which method in their opinion had the best reconstruction. We denote the percentage of best pick for each method as R-UP (Reconstruction User Preference). Similarly, we have the same people asked to select among 20 subject transfer images which method performed the best subject transfer in their opinion. Participant were asked to take into account image quality and quality of transfer. We denote the percentage of best pick for each method as ST-UP (Subject Transfer User Preference).

We observe that our method, SCAM, outperforms competing methods by a high margin. Indeed, 98.4% of users chose SCAM as the best image reconstruction technique, 1.5% picked SEAN and 0.1% for INADE. As for subject transfer 92.8% of users preferred SCAM, 5.0% chose SCAM and 2.2% INADE.

5   Conclusion

We introduced SCAM that performs semantic editing and in particular subject transfer in images. The architecture contributions of SCAM are: first, the semantic cross attention (**SCA**) mechanism performing attention between features and a set of latents under the constraint that they only attend to semantically meaningful regions; second, the Semantic Attention Transformer Encoder (**SAT**) retrieving information based on a semantic attention mask; third, the Semantic Cross Attention Modulation Generator (**SCAM**) performing semantic-based generation. SCAM sets the new state of the art by leveraging multiple latents per semantic region and by providing a finer encoding of the latent vectors both at encoding and decoding stages.

Acknowledgments. We would like to thank Dimitrios Papadopoulos, Monika Wysoczanska, Philippe Chiberre and Thibaut Issenhuth for proofreading. This work was granted access to the HPC resources of IDRIS under the allocation 2021-AD011012630 made by GENCI and was supported by a DIM RFSI grant and ANR project TOSAI ANR-20-IADJ-0009.

References

1. Brown, T., et al.: Language models are few-shot learners. In: NeurIPS (2020)
2. Chan, C., Ginosar, S., Zhou, T., Efros, A.A.: Everybody dance now. In: ICCV (2019)
3. Denton, E., Chintala, S., Szlam, A., Fergus, R.: Deep generative image models using a Laplacian pyramid of adversarial networks. In: NeurIPS (2015)
4. Devlin, J., Chang, M.W., Lee, K., Toutanova, K.: BERT: pre-training of deep bidirectional transformers for language understanding. In: NAACL (2019)
5. Dong, H., Liang, X., Gong, K., Lai, H., Zhu, J., Yin, J.: Soft-gated warping-GAN for pose-guided person image synthesis. In: NeurIPS (2018)
6. Dosovitskiy, A., et al.: An image is worth 16x16 words: transformers for image recognition at scale. In: ICLR (2021)
7. Endo, Y., Kanamori, Y.: Diversifying semantic image synthesis and editing via class-and layer-wise VAEs. Comput. Graph. Forum **39**(7), 519–530 (2020)
8. Esser, P., Rombach, R., Ommer, B.: Taming transformers for high-resolution image synthesis. In: CVPR (2021)
9. Fu, D., et al.: Unsupervised pre-training for person re-identification. In: CVPR (2021)
10. Goodfellow, I., et al.: Generative adversarial nets. In: NeurIPS (2014)
11. Gu, S., Bao, J., Yang, H., Chen, D., Wen, F., Yuan, L.: Mask guided portrait editing with conditional GANs. In: CVPR (2019)
12. Han, X., Hu, X., Huang, W., Scott, M.R.: Clothflow: A flow-based model for clothed person generation. In: ICCV (2019)
13. Hudson, D.A., Zitnick, C.L.: Generative adversarial transformers. In: Proceedings ICML (2021)
14. Isola, P., Zhu, J.Y., Zhou, T., Efros, A.A.: Image-to-image translation with conditional adversarial networks. In: CVPR (2017)
15. Jaegle, A., et al.: Perceiver IO: a general architecture for structured inputs & outputs. arXiv preprint arXiv:2107.14795 (2021)
16. Jaegle, A., Gimeno, F., Brock, A., Zisserman, A., Vinyals, O., Carreira, J.: Perceiver: general perception with iterative attention. In: Proceedings ICML (2021)
17. Jiang, Y., Chang, S., Wang, Z.: TransGAN: two transformers can make one strong GAN, and that can scale up. arXiv preprint arXiv:2102.07074 (2021)
18. Karras, T., Aila, T., Laine, S., Lehtinen, J.: Progressive growing of GANs for improved quality, stability, and variation. In: ICLR (2018)
19. Karras, T., Laine, S., Aila, T.: A style-based generator architecture for generative adversarial networks. In: CVPR (2019)
20. Karras, T., Laine, S., Aittala, M., Hellsten, J., Lehtinen, J., Aila, T.: Analyzing and improving the image quality of styleGAN. In: CVPR (2020)
21. Lee, C.H., Liu, Z., Wu, L., Luo, P.: MaskGAN: towards diverse and interactive facial image manipulation. In: CVPR (2020)
22. Lee, K., Chang, H., Jiang, L., Zhang, H., Tu, Z., Liu, C.: ViTGAN: training GANs with vision transformers. arXiv preprint arXiv:2107.04589 (2021)

23. Li, K., Zhang, J., Liu, Y., Lai, Y.K., Dai, Q.: PoNa: pose-guided non-local attention for human pose transfer. In: IEEE Trans. Image Process. **29**, 9584–9599 (2020)
24. Li, P., Xu, Y., Wei, Y., Yang, Y.: Self-correction for human parsing. In: IEEE TPAMI (2020)
25. Li, Y., Li, Y., Lu, J., Shechtman, E., Lee, Y.J., Singh, K.K.: Collaging class-specific GANs for semantic image synthesis. In: ICCV (2021)
26. Lim, J.H., Ye, J.C.: Geometric GAN. arXiv preprint arXiv:1705.02894 (2017)
27. Liu, X., Yin, G., Shao, J., Wang, X., Li, h.: Learning to predict layout-to-image conditional convolutions for semantic image synthesis. In: NeurIPS (2019)
28. Liu, Z., et al.: Swin transformer: hierarchical vision transformer using shifted windows. In: ICCV (2021)
29. Ma, L., Jia, X., Sun, Q., Schiele, B., Tuytelaars, T., Gool, L.V.: Pose guided person image generation. In: NeurIPS (2017)
30. Meng, C., et al.: SDEdit: guided image synthesis and editing with stochastic differential equations. In: ICLR (2022)
31. Park, T., Liu, M.Y., Wang, T.C., Zhu, J.Y.: Semantic image synthesis with spatially-adaptive normalization. In: CVPR (2019)
32. Radford, A., Metz, L., Chintala, S.: Unsupervised representation learning with deep convolutional generative adversarial networks. In: ICLR (2016)
33. Radford, A., Narasimhan, K., Salimans, T., Sutskever, I.: Improving language understanding by generative pre-training. In: ArXiv (2018)
34. Radford, A., Wu, J., Child, R., Luan, D., Amodei, D., Sutskever, I.: Language models are unsupervised multitask learners. In: ArXiv (2019)
35. R.J. Lehman, H.H.: iDesigner 2019. In: FGVC6 (2019)
36. Schönfeld, E., Sushko, V., Zhang, D., Gall, J., Schiele, B., Khoreva, A.: You only need adversarial supervision for semantic image synthesis. In: ICLR (2021)
37. Tan, Z., et al.: Diverse semantic image synthesis via probability distribution modeling. In: CVPR (2021)
38. Tan, Z., et al.: Efficient semantic image synthesis via class-adaptive normalization. In: IEEE TPAMI (2021)
39. Tang, H., Bai, S., Zhang, L., Torr, P.H.S., Sebe, N.: XingGAN for person image generation. In: Vedaldi, A., Bischof, H., Brox, T., Frahm, J.-M. (eds.) ECCV 2020. LNCS, vol. 12370, pp. 717–734. Springer, Cham (2020). https://doi.org/10.1007/978-3-030-58595-2_43
40. Tang, H., Xu, D., Yan, Y., Torr, P.H., Sebe, N.: Local class-specific and global image-level generative adversarial networks for semantic-guided scene generation. In: CVPR (2020)
41. Touvron, H., Cord, M., Douze, M., Massa, F., Sablayrolles, A., Jegou, H.: Training data-efficient image transformers & distillation through attention. In: Proceedings ICML (2021)
42. Ulyanov, D., Vedaldi, A., Lempitsky, V.: Instance normalization: the missing ingredient for fast stylization. In: CVPR (2020)
43. Vaswani, A., et al.: Attention is all you need. In: NeurIPS (2017)
44. Wang, T.C., Liu, M.Y., Tao, A., Liu, G., Kautz, J., Catanzaro, B.: Few-shot video-to-video synthesis. In: NeurIPS (2019)
45. Wang, T.C., Liu, M.Y., Zhu, J.Y., Tao, A., Kautz, J., Catanzaro, B.: High-resolution image synthesis and semantic manipulation with conditional GANs. In: CVPR (2018)
46. Wang, Y., Qi, L., Chen, Y.C., Zhang, X., Jia, J.: Image synthesis via semantic composition. In: ICCV (2021)

47. Zhang, B., et al.: StyleSwin: Transformer-based GAN for high-resolution image generation. arXiv preprint arXiv:2112.10762 (2021)
48. Zhang, H., Goodfellow, I., Metaxas, D., Odena, A.: Self-attention generative adversarial networks. In: Proceedings ICML (2019)
49. Zhang, J., Li, K., Lai, Y.K., Yang, J.: PISE: person image synthesis and editing with decoupled GAN. In: CVPR (2021)
50. Zhang, J., Liu, X., Li, K.: Human pose transfer by adaptive hierarchical deformation. In: CGF (2020)
51. Zhao, L., Zhang, Z., Chen, T., Metaxas, D., Zhang, H.: Improved transformer for high-resolution GANs. In: NeurIPS, vol. 34 (2021)
52. Zhou, B., Zhao, H., Puig, X., Fidler, S., Barriuso, A., Torralba, A.: Scene parsing through ade20k dataset. In: CVPR (2017)
53. Zhu, J.Y., Park, T., Isola, P., Efros, A.A.: Unpaired image-to-image translation using cycle-consistent adversarial networks. In: ICCV (2017)
54. Zhu, P., Abdal, R., Qin, Y., Wonka, P.: SEAN: image synthesis with semantic region-adaptive normalization. In: CVPR (2020)
55. Zhu, Z., Huang, T., Shi, B., Yu, M., Wang, B., Bai, X.: Progressive pose attention transfer for person image generation. In: CVPR (2019)
56. Zhu, Z., Xu, Z., You, A., Bai, X.: Semantically multi-modal image synthesis. In: CVPR (2020)

Sem2NeRF: Converting Single-View Semantic Masks to Neural Radiance Fields

Yuedong Chen[1]($\boxtimes$) ⓘ, Qianyi Wu[1] ⓘ, Chuanxia Zheng[2] ⓘ, Tat-Jen Cham[2] ⓘ, and Jianfei Cai[1] ⓘ

[1] Monash University, Melbourne, Australia
{yuedong.chen,qianyi.wu,jianfei.cai}@monash.edu
[2] Nanyang Technological University, Singapore, Singapore
chuanxia001@e.ntu.edu.sg, astjcham@ntu.edu.sg

Abstract. Image translation and manipulation have gain increasing attention along with the rapid development of deep generative models. Although existing approaches have brought impressive results, they mainly operated in 2D space. In light of recent advances in NeRF-based 3D-aware generative models, we introduce a new task, Semantic-to-NeRF translation, that aims to reconstruct a 3D scene modelled by NeRF, conditioned on one single-view semantic mask as input. To kick-off this novel task, we propose the Sem2NeRF framework. In particular, Sem2NeRF addresses the highly challenging task by encoding the semantic mask into the latent code that controls the 3D scene representation of a pretrained decoder. To further improve the accuracy of the mapping, we integrate a new region-aware learning strategy into the design of both the encoder and the decoder. We verify the efficacy of the proposed Sem2NeRF and demonstrate that it outperforms several strong baselines on two benchmark datasets. Code and video are available at https://donydchen.github.io/sem2nerf/.

Keywords: NeRF-based generation · Conditional generative model · 3D deep learning · Neural radiance fields · Image-to-image translation

1 Introduction

Controllable image generation, translation, and manipulation have seen rapid advances in the last few years along with the emergence of Generative Adversarial Networks (GANs) [12]. Current systems are able to freely change the image appearance through referenced images [18,19,68], modify scene content via semantic masks [27,38,51], and even accurately manipulate various attributes in feature space [22,54,55]. Despite impressive performance and wide applicability, these systems are mainly focused on 2D images, without directly considering the 3D nature of the world and the objects within.

Supplementary Information The online version contains supplementary material available at https://doi.org/10.1007/978-3-031-19781-9_42.

Concurrently, significant progress has been made for 3D generation by using deep generative networks [12,24]. Methods were developed for different 3D shape representations, including voxels [52], point clouds [34], and meshes [11]. More recently, Neural Radiance Fields (NeRF) [36] has been a new paradigm for 3D representation, providing accurate 3D shape and view-dependent appearance simultaneously. Based on this new representation, seminal 3D generation approaches [2,3,13,37,44] have been proposed that aim to generate photorealistic images from a given distribution in a 3D-aware and view-consistent manner. However, these techniques are primarily developed purely for high-quality 3D generation, leaving controllable 3D manipulation and editing unsolved.

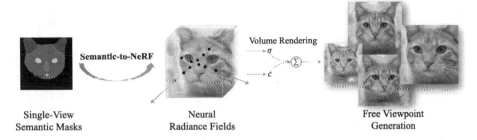

Fig. 1. Illustration of the Semantic-to-NeRF translation task, which aims to achieve free-viewpoint image generation by taking only a single-view semantic mask as input

It would be a dramatic enhancement if we can *freely manipulate and edit an object's content and appearance in 3D space, while only leveraging easily obtained 2D input information.* In this paper, we take an initial step toward this grand goal by introducing a new task, termed **Semantic-to-NeRF translation**, analogous to a 2D Semantic-to-Image translation task but operating on 3D space. Specifically, Semantic-to-NeRF translation (see Fig. 1) takes as input a single-view 2D semantic mask, yet output a NeRF-based 3D representation that can be used to render photorealistic images in a 3D-aware view-consistent manner. More importantly, it allows free editing of the object's content and appearance in 3D space, by modifying the content only via a single-view 2D semantic mask.

However, generating 3D structure from a single 2D image is already an ill-posed problem, and it will be even more so from a single 2D semantic mask. There are also two other major issues in this novel task:

1. *Large information gap between 3D structure and 2D semantics.* A single-view 2D semantic mask neither holds any 3D shape or surface information, nor provides much guidance for plausible appearances, making it tough to generate a neural radiance field with comprehensive details.
2. *Imbalanced semantic distribution.* Since semantic classes tend to be area-imbalanced within an image, *e.g.* eyes occupy less than 1% of a face while hair can take up larger than 40%, existing CNN-based networks may over-attend to larger semantic regions, while discounting smaller semantic regions that may be perceptually more salient. This will result in poor controllable editing in 3D space when we alter small semantic regions.

To mitigate these issues, we propose a novel framework, **Sem2NeRF**, that builds on NeRF [36] for 3D representation, by augmenting it with a semantic translation branch that conditionally generates high-quality 3D-consistent images. In particular, the framework is based on an encoder-decoder architecture that converts a singe-view 2D semantic mask to an embedded code, and then transfers it to a NeRF representation for rendering 3D-consistent images.

Our broad idea here is that, instead of directly learning to predict 3D structure from degenerate single-view 2D semantic masks, *the network can alternatively learn the 3D shape and appearance representation from large numbers of unstructured 2D RGB images.* This has achieved significant advances in NeRF-based generator [2,3,13,37,44], which transforms a random vector to a NeRF representation. In short, our scenario is thus: we have a well-trained 3D generator, but we aim to further control the generated content and appearance easily. The main idea is then to *learn a good mapping network* (like current methods for 2D GAN inversion [42,47]) *that can encode the semantic mask into the somewhat smaller latent space domain for 3D controllable translation and manipulation.* As for the second issue, we intriguingly discover that a *region-aware learning strategy* is of vital importance. We therefore aim to tame an encoder that is sensitive to image patches, and adopt a region-based sampling pattern for the decoder. Furthermore, augmenting the input semantic masks with extracted contours and distance field representations [5] also considerably helps to highlight the intended semantic changes, making them more easily perceptible.

Following the above analysis, we build our Sem2NeRF framework upon the Swin Transformer encoder [31] and the pre-trained π-GAN decoder [3]. To kick off the single-view Semantic-to-NeRF translation task, we pinpoint two suitable yet challenging datasets, including CelebAMask-HQ [25] and CatMask, where the latter contains cat faces rendered using π-GAN and labelled with 6-class semantic masks using DatasetGAN [63]. We showcase the superiority of our model over several strong baselines by considering SofGAN [4], pix2pixHD [51] with GAN-inversion [23], and pSp [42]. Our contributions are three-fold:

- We introduce a novel and challenging task, Semantic-to-NeRF translation, which converts a single-view 2D semantic mask to a 3D scene modelled by neural radiance fields.
- With the insight of needing a region-aware learning strategy, we propose a novel framework, Sem2NeRF, which is capable of achieving 3D-consistent free viewpoint image generation, semantic editing and multi-model synthesis, by taking as input only one single-view semantic mask of a specific category, *e.g.,* human face, cat face.
- We validate our insight regarding our region-aware learning strategy and the efficacy of Sem2NeRF via extensive ablation studies, and demonstrate that Sem2NeRF outperforms strong baselines on two challenging datasets.

2   Related Work

NeRF and Generative NeRF. Starting as an approach focused on modelling a single static scene, NeRF [36] had seen rapid development in different aspects. Several approaches managed to reduce the training [48] and inference time [29,33], while others improved visual quality [1]. Besides, it had also been extended in other ways, *e.g.*, dynamic scene [41], compositional scene [53], pose estimation [58], portrait generation [30], semantic segmentation [66].

Follow-up works that integrated NeRF with generative models were most relevant to ours. Schwarz *et al.* [44] proposed to learn a NeRF distribution by conditioning the input point positions with a sampled random vector. Niemeyer *et al.* [37] enabled multi-object generation by representing the whole scenes as a composition of different components. To improve the visual quality, π-GAN [3] adopted a SIREN-based [46] network structure with FiLM [40] conditioning. StyleNeRF [13] turned to embedding the volume rendering technique into Style-GAN [23]. More recently, VolumeGAN [57] relied on separately learning structure and texture features. MVCGAN [62] leveraged the underlying 3D geometry information. EG3D [2] proposed an efficient tri-plane hybrid 3D representation.

Our work belongs to the class of generative models, but unlike all existing methods that aimed to create a *random* scene, we aim to generate a *specific* scene that is conditioned by a given single-view semantic mask. Although there are concurrent works, *e.g.*, 3D-SGAN [59], FENeRF [49], exploring the similar condition settings, most of them purely focus on improving the quality of the generated images, while resort to existing GAN inversion [23] to do the mapping. In contrast, our work is more focused on improving the mapping from the mask to the NeRF-based scene.

Image-to-Image Translation is about converting an image from one source representation, *e.g.*, semantic masks, to another target representation, *e.g.*, photorealistic images. Since its introduction [18], progress has been made with regard to better image quality [6,51], multi-modal outputs [8,64,69], unsupervised learning [28,68], *etc.*. More recently, there is a new trend [42,45,56] of tackling this task by editing the latent space of a pre-trained generator, *e.g.*, StyleGAN.

In contrast to all mentioned work that aimed to map a semantic mask to an image, ours is focused on mapping to a 3D scene. We also notice that there are some recent approaches targeted at converting semantic masks to 3D scenes. Huang *et al.* [17] introduced rendering novel-view photorealistic images from a given semantic mask, by first applying semantic-to-image translation [38], then converting the single-view image to a 3D scene modelled by multiplane images (MPI) [67]. Hao *et al.* [14] proposed to learn a mapping from a semantically-labelled 3D block world to a NeRF-based 3D scene, using a scene-specific setting. Chen *et al.* [4] introduced a 3D-aware portrait generator by first mapping the given latent code to a semantic occupancy field (SOF) [7] for rendering novel view semantic masks, followed by applying image-to-image translation.

Unlike all mentioned attempts on learning semantic to 3D scene mappings, ours is the first to introduce the single-view semantic to NeRF translation task. Our work differs from theirs in: 1) We do not rely on any separate image-to-image translation stage, resulting in better multi-view consistency; 2) We do not require multi-view semantic masks for both training and testing phases, easing the data collection effort; 3) We pinpoint a solution for creating pseudo labels and demonstrate reasonable results beyond the human face domain.

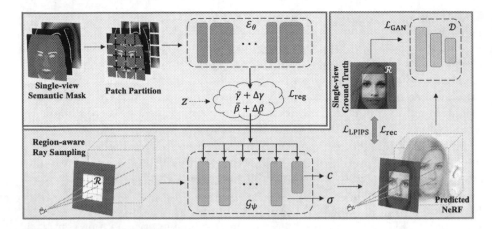

Fig. 2. Architecture of the Sem2NeRF framework. It aims to convert a single-view semantic mask to a 3D scene represented by NeRF. Specifically, a given semantic mask will be partitioned into patches, which will be further encoded by a patch-based encoder $\mathcal{E}_\theta$ into a latent style code (γ, β) of a pre-trained NeRF-based 3D generator $\mathcal{G}_\psi$. A region $\mathcal{R}$ will be randomly sampled to enforce awareness of differences among regions. And an optional latent vector $\mathbf{z}$ is included to enable multi-modal synthesis

3   Methodology

As shown in Fig. 1, our main goal is to train a Semantic-to-NeRF translation network $\Phi_{s\to\mathcal{V}}$, such that when presented with a single-view 2D semantic mask $\mathbf{s}$, it generates the corresponding NeRF representation $\mathcal{V}$, which can then be used to render realistic 3D-consistent images. This task is conceptually similar to the conventional semantic-to-image setting, except that here we opt to go beyond 2D image translation, and deal with the novel *controllable 3D translation*. More importantly, we can freely change the 3D content by *simply modifying the corresponding content in a single-view 2D semantic mask*.

In order to learn such a framework without enough supervision for arbitrary view appearances, we observed that 3D information can be learned from large image collections [2,3,20]. Therefore, our *key motivational insight* is this: instead of directly training $\Phi_{s\to\mathcal{V}}$ using *single-view* semantic-image pairs $(\mathbf{s}, \mathbf{I})$ (like current methods for 2D semantic-to-image translation [38,51]), we will train it as a two-stage pipeline shown in Fig. 2. Here, (A) we utilize a pre-trained 3D generator (lower portion $\mathcal{G}_\psi$) that learns 3D shape and appearance information from a large set of collected images; (B) we pose this challenging task as a *3D inversion* problem, where our main target is to design a front-end encoder (upper portion $\mathcal{E}_\theta$) that maps the semantic mask into the generator latent space accurately.

The two training stages are executed independently and can be separately implemented with different frameworks. There are at least two unique benefits of breaking down the entire controllable 3D translation into two-stages: 1) The training does *not* require copious views of semantic-image pairs for each instance, which are difficult to collect, or even impossible in some scenarios; 2) The compartmentalization of the 3D generator and the 2D encoder allows greater agility, where the 3D information can be previously learned on various tasks with a large collection of images and then be freely plugged into the 3D inversion pipeline.

3.1   3D Decoder with Region-Aware Ray Sampling

Preliminaries on NeRF. We first provide some preliminaries on NeRF before discussing how we exploit it for Sem2NeRF. NeRF [36] is one kind of implicit functions that represents a continuous 3D scene, which has achieved great successes in modeling 3D shape and appearance. A NeRF is a neural network that maps a 3D location $\mathbf{x} \in \mathbb{R}^3$ and a viewing direction $\mathbf{d} \in \mathbb{S}^2$ to a spatially varying volume density σ and a view-dependent emitted color $\mathbf{c} = (r, g, b)$. NeRFs trained on natural images are able to continuously render realistic images at arbitrary views. In particular, it requires to use the volume rendering [26], which computes the following integral to obtain the color of a pixel:

$$C(\mathbf{r}) = \int_{t_n}^{t_f} T(t)\sigma(\mathbf{r}(t))\mathbf{c}(\mathbf{r}(t), \mathbf{d})dt, \text{where } T(t) = \exp(-\int_{t_n}^{t} \sigma(\mathbf{r}(s))ds), \quad (1)$$

where $\boldsymbol{r}(t) = \boldsymbol{o} + t\boldsymbol{d}$ is the ray casting from the virtual camera located at $\boldsymbol{o}$, bounded by near t_n and far t_f, and $T(t)$ represents the accumulated transmittance of the ray traveling from t_n to t. The integral $C(\mathbf{r})$ is further implemented with a hierarchical volume sampling strategy [26,36], resulting in the optimization of a "coarse" network followed by a "fine" network.

NeRF-Based Generator. Our work is mainly based on a representative NeRF-based generator, π-GAN [3], which learns 3D representation using only 2D supervision. Inspired by StyleGAN2 [23], the architecture of π-GAN is mainly composed of two parts, a mapping network $\mathcal{F} : \mathcal{Z} \to \mathcal{W}$ that maps a latent vector z in the input latent space $\mathcal{Z}$ to an intermediate latent vector $w \to \mathcal{W}$, and a SIREN-based [46] synthesis network that maps w to the NeRF representation $\mathcal{V}$ that supports rendering 3D-consistent images from arbitrary camera poses.

Our Sem2NeRF framework can use various NeRF-based generators. Here, we choose π-GAN as the main decoder in our architecture for two main reasons. Firstly, among all *published* works related to NeRF-based generators, π-GAN achieves state-of-the-art performance in terms of rendered image quality and their underlying 3D consistency. Secondly and more importantly, similar to StyleGAN, the FiLM [40] conditioning used by π-GAN enables layer-wise control over the decoder and the mapping network decouples some high-level attributes, making it easier to perform *3D inversion* on top of NeRF, *i.e.*, searching for the desired latent code w that best reconstructs an ideal target. The similar observation has been previously explored in the latest 2D GAN inversion [9,23,42].

Region-Aware Ray Sampling. While π-GAN already provides high-quality view-consistent rendered images, our main goal is to accurately restore the NeRF from a single-view semantic mask, and even freely edit the 3D content via such a map. To achieve this, *the network should be sensitive to local small modifications.* However, this is not supported in the original π-GAN, which is trained on each entire image with a global perception. It stores scene-specific information in a latent code, which is shared across all points that are bounded by the rendering volume. As a result, a small change in an original latent code will easily cause a global modification in generation. This may not impact pure 3D generation, for which only the quality of global shape and appearance is paramount, but it has a large negative effect on recreating a 3D representation that accurately matches the corresponding semantic mask.

To mitigate this issue, we adopt a region-based ray sampling pattern [30, 44] in the π-GAN decoder, that *attempts to encourage latent codes to represent local regions at different scales and locations.* Suppose the rendered image $\mathbf{I}$ with a target size $h \times w$, a local region $\mathcal{R}$ used for training is randomly sampled as

$$\mathcal{R}(\alpha, (\Delta h, \Delta w)) = \{(\alpha h + \Delta h, \alpha w + \Delta w)\}, \tag{2}$$

where $(\alpha h + \Delta h, \alpha w + \Delta w)$ denotes the sampling coordinates of rays, with $\alpha \in (0, 1]$ being the scaling factor and $(\Delta h \in [0, (1 - \alpha)h], \Delta w \in [0, (1 - \alpha)w])$ being the translation factor. To obtain such training pairs between the NeRF rendered output and the local ground truth, we sample the original whole image using the same region coordinates $\mathcal{R}$ with bilinear interpolation. This strategy leads to large improvements on conditional generation as shown in the experiments.

3.2   3D Inversion Using Region-Aware 2D Encoder

3D Inversion. To inversely map a semantic mask $\mathbf{s}$ into the $\mathcal{W}$ latent space of the 3D generator $\mathcal{G}_\psi$ by an encoder $\mathcal{E}_\theta$, with respective parameters ψ and θ, we train $\mathcal{E}_\theta$ to minimize the reconstruction error between ground truth image $\mathbf{I}$ and output $\hat{\mathbf{I}}$. Specifically, Semantic-to-NeRF translation represents the mapping

$$\Phi_{\mathbf{s} \to \mathcal{V}}(\mathbf{x}, \mathbf{d}, \mathbf{z}; \mathbf{s}) = \mathcal{G}_\psi(\mathbf{x}, \mathbf{d}, \mathbf{z}; \mathcal{E}_\theta(\mathbf{s})) = \mathcal{V}(\sigma, \mathbf{c}) \tag{3}$$

where $\mathbf{x}, \mathbf{d}$ denotes point position and ray direction, while the derived density σ and color $\mathbf{c}$ can be used to calculate the corresponding pixel value via volume rendering as in Eq. (1). For *controllable* 3D generation, $\mathbf{s}$ is the input single-view semantic mask, embedded into $\mathcal{W}$ space to control the generated 3D content, while we also enable multi-modal synthesis by adding another latent vector $\mathbf{z}$ to model the generated appearance. Note that $\mathbf{s}$ only comes in a single view, which is not necessary the same as the output viewing direction. In short, *we use only single-view semantic-image pairs* $(\mathbf{s}, \mathbf{I})$ *for the Sem2NeRF training,* as the 3D view-consistent information has been captured by the *fixed* pre-trained 3D generator $\mathcal{G}_\psi$. Hence, we focus only on training the encoder network $\mathcal{E}_\theta$ to learn the posterior distribution $q(w|\mathbf{s})$ for 3D inversion.

Region-Aware 2D Encoder. A simple way Z is to directly apply an existing 2D GAN inversion framework. However, this straightforward idea does *not* work well as we originally discovered when using the state-of-the-art pSp encoder [42] in our setting, especially for small but perceptually important regions, such as eyes. Our conjecture is that the conventional CNN-based architecture integrates the neighboring information via overaggressive filtering, resulting in heavy loss of small details [60].

To mitigate this issue, we also deploy a region-aware learning strategy in the 2D encoder, which is inspired by the latest patch-based methods [10,65] that capture information in every patch with equal possibility. In other words, when we directly extract features from local patches, it will be *more sensitive to the semantic variation within each patch*, which can ameliorate the problem of imbalanced semantic distribution within an image. In particular, we adopt the Swin Transformer [31] as the encoder architecture. To embed the semantic mask $\mathbf{s}$ into the $\mathcal{W}$ latent space of the pre-trained 3D generator, we replace the final classification output size with the size of the latent vectors $\mathbf{w}$. Besides, to further stabilize the inversion training, we take inspiration from the truncation trick [22,42] and set the learned latent codes for the pre-trained decoder as

$$\gamma = \overline{\gamma} + \Delta\gamma, \beta = \overline{\beta} + \Delta\beta, \tag{4}$$

where γ and β represent the embedded vectors for the $\mathcal{W}$ latent space, *i.e.*, frequency and phase shift of π-GAN, respectively; $\Delta\gamma$ and $\Delta\beta$ are the outputs of the proposed encoder $\mathcal{E}_\theta$, while $\overline{\gamma}$ and $\overline{\beta}$ are the average latent codes extracted by the pre-trained π-GAN original mapping network $\mathcal{F} : \mathcal{Z} \to \mathcal{W}$.

Additional Inputs for the 2D Encoder. As mentioned, a semantic mask contains sparse information, where the changing of small regions may be imperceptible to the network, making the semantic-based controllable 3D editing very challenging. Considering that editing a semantic mask only effectively alters the boundaries between different semantic labels, we conjecture that explicitly augmenting the semantic input with *boundary information* will be useful for semantic editing. Therefore, we concatenate the semantic mask input with contours and distance field representations [5] for the region-aware encoder. These additional inputs further improve the semantic editing performance considerably as shown in the experiments. Note that contours and distance field representations are both directly calculated from the semantic masks (refer to the supplementary document for more details), which do *not involve any extra labels*.

3.3 Training Loss Functions

During the training phase, we use the single-view semantic mask $\mathbf{s}$, the corresponding viewing direction d_s, and the paired ground truth RGB image $\mathbf{I}$. Similar to Semantic-to-Image translation, we start by applying a pixel-level reconstruction loss,

$$\mathcal{L}_{\text{rec}}(\mathbf{I}, \mathbf{s}, d_s) = \|\mathbf{I} - \mathcal{G}_\psi(\mathcal{E}_\theta(\mathbf{s}), d_s)\|_2, \tag{5}$$

where $\mathcal{E}_\theta(\mathbf{s})$ denotes the latent codes mapped from $\mathbf{s}$ via the region-aware encoder $\mathcal{E}_\theta(\cdot)$, while $\mathcal{G}_\psi(\mathcal{E}_\theta(\mathbf{s}), d_s)$ represents the generated image rendered from direction d_s via the decoder $\mathcal{G}_\psi(\cdot)$. Unless otherwise specified, the aforementioned region-aware sampling strategy is applied to $\mathcal{G}_\psi$ and $\mathbf{I}$ before calculating any losses.

To further enforce the feature-level similarity between the generated image and the ground truth, the LPIPS loss [61] is leveraged,

$$\mathcal{L}_{\mathrm{LPIPS}}(\mathbf{I}, \mathbf{s}, d_s) = \|\mathcal{F}(\mathbf{I}) - \mathcal{F}(\mathcal{G}_\psi(\mathcal{E}_\theta(\mathbf{s}), d_s)))\|_2, \tag{6}$$

where $\mathcal{F}(\cdot)$ refers to the pre-trained feature extraction network.

Inspired by the truncation trick [22,42], we further encourage the decoder latent codes γ, β to be close to the average codes $\overline{\gamma}$, $\overline{\beta}$, which is achieved by regularizing the encoder with

$$\mathcal{L}_{\mathrm{reg}}(\mathbf{s}) = \|\mathcal{E}_\theta(\mathbf{s})\|_2. \tag{7}$$

To improve image quality, especially for novel views, we further apply a non-saturating GAN loss with R1 regularization [35],

$$\mathcal{L}_{\mathrm{GAN}}(\mathbf{I}, \mathbf{s}, d) = f(\mathcal{D}(\mathcal{G}_\psi(\mathcal{E}_\theta(\mathbf{I}), d))) + f(-\mathcal{D}(\mathbf{I})) + \lambda_{\mathrm{R1}} |\nabla \mathcal{D}(\mathbf{I})|^2,$$
$$\text{where } f(u) = -\log(1 + \exp(-u)). \tag{8}$$

Here $\mathcal{D}(\cdot)$ is a patch discriminator [18], aligned with our region-aware learning strategy for the decoder, and λ_{R1} is a hyperparameter that is set to 10. Note that here the viewing direction d is not required to be the same as the input semantic viewing direction d_s, and we randomly sample this viewing direction from a known distribution, *i.e.* Gaussian, following the settings of π-GAN [3].

Finally, the overall training objective for our framework is a weighted combination of the above loss functions as

$$\mathcal{L}_{\mathrm{Sem2NeRF}} = \lambda_{\mathrm{rec}}\mathcal{L}_{\mathrm{rec}} + \lambda_{\mathrm{LPIPS}}\mathcal{L}_{\mathrm{LPIPS}} + \lambda_{\mathrm{reg}}\mathcal{L}_{\mathrm{reg}} + \lambda_{\mathrm{GAN}}\mathcal{L}_{\mathrm{GAN}}. \tag{9}$$

3.4 Model Inference

For inference, our model takes as input a 2D single-view semantic mask, while d_s is optional, required only when rendering an image with the same viewing direction as the semantic mask. Different from the training phase, during inference the rays are cast to cover the whole image plane, rather than a local region.

Multi-view Generation. Since the employed decoder is a NeRF-based generator, Sem2NeRF inherently supports novel view generation. Specifically, given a semantic mask $\mathbf{s}$, it will first be mapped as an embedded vector in the $\mathcal{W}$ latent space that controls the "content" of the NeRF-based generator, whereupon a novel view image can then be generated by volume rendering the NeRF from an arbitrary viewing direction.

Multi-modal Synthesis. Similar to the diversified mapping in semantic-to-image [38], ideally a single semantic mask should be translated into multiple NeRFs consistent to it. Our Sem2NeRF framework inherently supports multi-modal synthesis in inference due to the usage of FiLM [40] conditioning on

π-GAN, without requiring any special customization in training. In practice, we additionally pass a random-sampled vector to the pre-trained π-GAN noise mapping module to obtain corresponding latent style codes z. Style mixing [22, 42] is then performed between z and $\mathcal{E}_\theta(\mathbf{s})$ to yield multi-modal outcomes.

4   Experiments

4.1   Settings

Datasets. To achieve Semantic-to-NeRF translation, we assume the training data to have single-view registered semantic masks and images, with the corresponding viewing directions. Two datasets were used for evaluation in our experiments. **CelebAMask-HQ** [25] contains images from CelebA-HQ [21,32], manually-labelled 19-class semantic masks, and head poses. We merged the left-right labels of symmetric parts, *i.e.*, eyes, eyebrows and ears, into one label per part. The dataset was randomly partitioned into training set with 28,000 samples and test set with 2,000 samples. **CatMask** is built using π-GAN and DatasetGAN [63] to further demonstrate the potential of Semantic-to-NeRF task and Sem2NeRF. Technical details are elaborated in the supplementary document.

Baselines. We identified the following three methods as baselines for comparison in our introduced Semantic-to-NeRF task. **SofGAN** [4] is an image translation approach. For a given single-view mask, we first apply inversion via iterative optimizations to find the corresponding latent vector for the preceding SOF [7] network, which can generate novel view semantic masks for further image-to-image mapping. Note that SofGAN requires training data to have high-quality multi-view semantic masks, which is not available nor needed in our task. **pix2pixHD** [51] is an image translation approach. We adopt it with general GAN-inversion techniques [3,23]. For a given mask, it is first mapped to a photorealistic image via pix2pixHD, which will then be mapped to the corresponding latent code in π-GAN via GAN-inversion. With the recovered codes, multi-view images can be directly obtained using π-GAN. **pSp** [42] is an image translation approach that is designed for encoding into StyleGAN2 [23]. We adapted it by using its ResNet [15]-based pSp encoder to replace the π-GAN mapping network, and we further trained the network with objective functions used by pSp.

Evaluation Metrics. We show qualitative results by rendering images with different viewing directions and FOV (Field of View). We also report Frechet Inception Distance (FID) [16] and Inception Score (IS) [43] using Inception-v3 [50] over the test sets. Average running time and model sizes are also compared.

Implementation Details. Swin-T is used in all experiments with input resolution 224×224. For the decoder, the size of local region $\mathcal{R}$ is set to 128×128. The step size of each ray is set to 28. Other miscellaneous settings of the pre-trained decoder, *e.g.*, ray depth ranges, are kept unchanged. Hyper-parameters in Eq. (9) are set as $\lambda_{rec} = 1$, $\lambda_{LPIPS} = 0.8$, $\lambda_{reg} = 0.005$, $\lambda_{GAN} = 0.08$. The implementation is done in PyTorch [39]. More details are provided in the supplementary document.

4.2   Results

Comparisons on CelebAMask-HQ. As shown in Fig. 3, compared to all other baseline models, **Sem2NeRF (1st and 5th columns)** achieved the best performance on both mapping accuracy and multi-view consistency. **pSp (2nd and 6th columns)** generated images with lower quality compared to ours, especially for novel views, mainly because our model is designed with a region-aware learning strategy and a GAN loss for random-posed images during training. The CNN-based encoder also failed to capture fine-grained details, *e.g.*, eyebrow shapes for the left face. Our method and the inversion-based pix2pixHD were better in matching semantics compared to pSp. **pix2pixHD (3rd and 7th columns)** can map semantic masks to high quality images in the same viewpoint (top row), but does not generate novel views well. Basic GAN-inversion is not an efficient or easy way to find the desired latent codes, since the current 3D generative models are still quite immature. Even though images with the same viewing direction as the masks are reasonable, those novel view outputs contain artifacts. **SofGAN (4th and 8th columns)** generated each single-view image with good quality; however, its results do not match with the given mask and lacked 3D consistency. The reason is that it is hard to map the given semantic mask to the desired latent codes of its semantic generator (SOF Net), whose

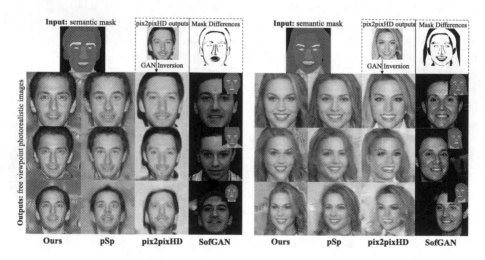

Fig. 3. Comparisons on CelebAMask. Images at each column are generated by the corresponding models mentioned at the bottom. Only SofGAN requires generation of multi-view semantic masks, shown at the top right corners of related images

sampling space is relatively small due to the lack of training data (only 122 subjects). The recovered mask did not match well with the given mask (top row). Besides, although the semantic masks show good multi-view consistency (top right corner of each image), conducting semantic-to-image mapping separately for each viewpoint does not guarantee that the consistency will be retained, since a semantic mask hardly contains any texture information and is geometrically ambiguous.

Quantitative results are give in Table 1. It can be seen that our Sem2NeRF method achieves the best performance, significantly outperforming the two baselines in both FID and IS scores. Note that we did not quantitatively compare single view image quality with SofGAN, considering that SofGAN for Semantic-to-NeRF is limited by its mask inversion quality and multi-view consistency, both of which cannot be measured by FID or IS scores. We also notice that scores of all models are lower than expected. The main reason is that π-GAN is initially trained on CelebA, but due to the requirement of semantic masks, our task conducted experiments using CelebA-HQ. The domain gap between CelebA and CelebA-HQ reduced the FID scores dramatically. Besides, our model also sees advantages in terms of running time and model size.

Table 1. Quantitative comparisons on CelebAMask @ 128×128

	FID ↓	IS ↑	Runtime (s) ↓	# Params(M) ↓
pix2pixHD [51] (with inversion)	67.32	1.72	161.59±0.859	~184.24
pSp [42]	55.56	1.74	0.25±0.004	~138.27
Sem2NeRF(Ours)	**41.52**	**2.03**	**0.18±0.003**	**~32.01**

Fig. 4. Editing 3D scenes by changing single-view semantic masks. Three viewpoints are shown for better comparison in each group

Mask Editing. As depicted in Fig. 4, our framework supports editing of 3D scenes by simply changing the given semantic mask, and is applicable to both labels associated with large regions, *e.g.*, hair, as well as small regions, *e.g.*, eyes, nose, mouth. This is not trivial since the semantic mask is not directly leveraged to control the 3D scene at the pixel level (if even possible), but is instead encoded into a sparse latent code, which may fail to preserve fine-grain editing. We address this challenge via the region-aware learning strategy.

Multi-modal Synthesis. Sem2NeRF supports multi-modal synthesis by simply changing the last few layers of the style codes. As shown in Fig. 5, we randomly sampled two style codes, and applied linear blending to continuously change the general styles of the 3D scenes generated by the given masks.

Ablation Studies. To further evaluate the efficacy of Sem2NeRF, we designed four ablation models, including 1) without region-aware encoder Φ_{wo_RE}, where the Swin-T encoder is replaced by the pSp encoder; 2) without region-aware decoder Φ_{wo_RD}, where the region-aware ray sampling strategy is discarded; 3) without input augmentation Φ_{wo_IA}, where contours and distance field representations are removed from the input; and 4) without random-pose GAN loss Φ_{wo_GAN}, where both Eq. (8) and the discriminator are removed.

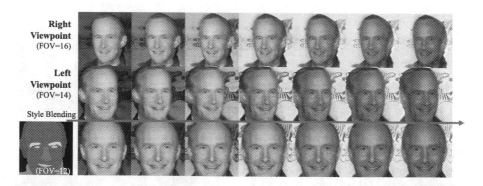

Fig. 5. Multi-modal synthesis. Styles are linearly blended from left to right. Three viewpoints are provided from top to bottom

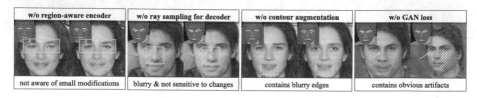

Fig. 6. Results of ablation studies. Each group (two views) is generated by a model without the component mentioned at the top. Main issues are described at the bottom

As shown in Fig. 6, compared to the full model (Fig. 4), Φ_{wo_RE} (1st group) is not sensitive to changes in small regions, *i.e.*, eyes, mainly because the CNN-based encoder tends to ignore small changes. Φ_{wo_RD} (2nd group) shows similar pattern (nose region) as the latent codes are not trained to be region-aware. It also has lower image quality, because the region-aware strategy enables denser sampling. Φ_{wo_IA} (3rd group) achieves comparable performance but with blurry edges for some regions, *e.g.*, mouth. This is because both contour and distance field representation help highlight the boundary information. Finally, images obtained by Φ_{wo_GAN} (4th group) have more artifacts in both views, demonstrating that GAN loss is important for improving the image quality of different poses.

Experiments beyond Human Faces. The introduced task can easily go beyond the human face domain by leveraging state-of-the-art weakly supervised semantic segmentation model to create pseudo labels. In this work, we present a Cat face example. Experimental results are shown in Fig. 7. Even when training with noisy pseudo labels, Sem2NeRF is robust enough to generate plausible results. For a given cat semantic mask, our model can map it to a 3D scene and render cat faces from arbitrary viewpoints, including different viewing directions (left part), and different FOV (right part). It also allows changing the 3D scenes by editing the single-view semantic masks, *e.g.*, changing the eye shape (left two rows). Multi-modal synthesis is also supported (right part in zigzag order).

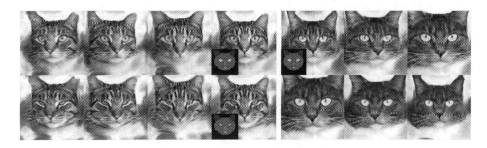

Fig. 7. Results on CatMask. Left part compares results of changing eyes shape. Right part showcases results of style linear blending (in zigzag order)

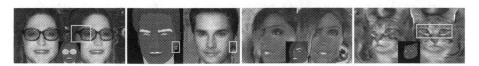

Fig. 8. Challenging cases of Sem2NeRF on Semantic-to-NeRF translation

Challenging Cases. Although Sem2NeRF addresses the Semantic-to-NeRF task in most cases, its advantages rely on an assumption, namely the generative capability of the pre-trained decoder. We show some challenging cases in Fig. 8. Accessories may have the wrong geometric shape (glasses in 1st case), or fail to render (earring in 2nd case), while masks with extreme poses might be converted to 3D scenes with abnormal texture or distorted contents (last two cases).

5   Conclusions

We have presented an initial step of extending the 2D image-to-image task to the 3D space, and introduced a new task called Semantic-to-NeRF translation. It aims to reconstruct a NeRF-based 3D scene, by taking as input only one single-view semantic mask. We further proposed Sem2NeRF model, which addresses the task via encoding the semantic mask into the latent space of a pre-trained 3D generative model. More importantly, we intriguingly found the importance of regional awareness for this new task, and tamed Sem2NeRF with a region-aware learning strategy. We demonstrated the capability of Sem2NeRF regarding free viewpoint generation, mask editing and multi-modal synthesis on two benchmark datasets, and showcased the superiority of our framework compared to three strong baselines. Future work will include adding more scenarios to the new task, and supporting changing styles for specific regions.

Acknowledgements. This research is partially supported by CRC Building 4.0 Project #44.

References

1. Barron, J.T., Mildenhall, B., Tancik, M., Hedman, P., Martin-Brualla, R., Srinivasan, P.P.: Mip-NeRF: a multiscale representation for anti-aliasing neural radiance fields. In: IEEE International Conference on Computer Vision, pp. 5855–5864 (2021)
2. Chan, E.R., et al.: Efficient geometry-aware 3D generative adversarial networks. IEEE Conference on Computer Vision and Pattern Recognition (2022)
3. Chan, E.R., Monteiro, M., Kellnhofer, P., Wu, J., Wetzstein, G.: pi-GAN: Periodic implicit generative adversarial networks for 3D-aware image synthesis. In: IEEE Conference on Computer Vision and Pattern Recognition, pp. 5799–5809 (2021)
4. Chen, A., Liu, R., Xie, L., Chen, Z., Su, H., Yu, J.: SofGAN: a portrait image generator with dynamic styling. ACM Trans. Graph. **41**(1), 1–26 (2022)
5. Chen, W., Hays, J.: Sketchygan: Towards diverse and realistic sketch to image synthesis. In: IEEE Conference on Computer Vision and Pattern Recognition, pp. 9416–9425 (2018)
6. Chen, Y., Huang, J., Wang, J., Xie, X.: Edge prior augmented networks for motion deblurring on naturally blurry images. arXiv preprint arXiv:2109.08915 (2021)
7. Chen, Z., Zhang, H.: Learning implicit fields for generative shape modeling. In: IEEE Conference on Computer Vision and Pattern Recognition, pp. 5939–5948 (2019)

8. Choi, Y., Uh, Y., Yoo, J., Ha, J.W.: StarGAN v2: diverse image synthesis for multiple domains. In: IEEE Conference on Computer Vision and Pattern Recognition, pp. 8188–8197 (2020)

9. Collins, E., Bala, R., Price, B., Susstrunk, S.: Editing in style: uncovering the local semantics of GANs. In: IEEE Conference on Computer Vision and Pattern Recognition, pp. 5771–5780 (2020)

10. Dosovitskiy, A., et al.: an image is worth 16x16 words: transformers for image recognition at scale. International Conference on Learning Representations (2021)

11. Goel, S., Kanazawa, A., Malik, J.: Shape and viewpoint without keypoints. In: Vedaldi, A., Bischof, H., Brox, T., Frahm, J.-M. (eds.) ECCV 2020. LNCS, vol. 12360, pp. 88–104. Springer, Cham (2020). https://doi.org/10.1007/978-3-030-58555-6_6

12. Goodfellow, I., et al.: Generative adversarial nets. Advances in Neural Information Processing Systems 27 (2014)

13. Gu, J., Liu, L., Wang, P., Theobalt, C.: StyleNeRF: a style-based 3D-aware generator for high-resolution image synthesis. In: International Conference on Learning Representations (2022)

14. Hao, Z., Mallya, A., Belongie, S., Liu, M.Y.: GANcraft: unsupervised 3D neural rendering of minecraft worlds. In: IEEE International Conference on Computer Vision, pp. 14072–14082 (2021)

15. He, K., Zhang, X., Ren, S., Sun, J.: Deep residual learning for image recognition. In: IEEE Conference on Computer Vision and Pattern Recognition, pp. 770–778 (2016)

16. Heusel, M., Ramsauer, H., Unterthiner, T., Nessler, B., Hochreiter, S.: GANs trained by a two time-scale update rule converge to a local nash equilibrium. In: Advances in Neural Information Processing Systems 30 (2017)

17. Huang, H.-P., Tseng, H.-Y., Lee, H.-Y., Huang, J.-B.: Semantic view synthesis. In: Vedaldi, A., Bischof, H., Brox, T., Frahm, J.-M. (eds.) ECCV 2020. LNCS, vol. 12357, pp. 592–608. Springer, Cham (2020), https://doi.org/10.1007/978-3-030-58610-2_35

18. Isola, P., Zhu, J.Y., Zhou, T., Efros, A.A.: Image-to-image translation with conditional adversarial networks. In: IEEE Conference on Computer Vision and Pattern Recognition, pp. 1125–1134 (2017)

19. Johnson, J., Alahi, A., Fei-Fei, L.: Perceptual losses for real-time style transfer and super-resolution. In: Leibe, B., Matas, J., Sebe, N., Welling, M. (eds.) ECCV 2016. LNCS, vol. 9906, pp. 694–711. Springer, Cham (2016). https://doi.org/10.1007/978-3-319-46475-6_43

20. Kanazawa, A., Tulsiani, S., Efros, A.A., Malik, J.: Learning category-specific mesh reconstruction from image collections. In: Ferrari, V., Hebert, M., Sminchisescu, C., Weiss, Y. (eds.) ECCV 2018. LNCS, vol. 11219, pp. 386–402. Springer, Cham (2018). https://doi.org/10.1007/978-3-030-01267-0_23

21. Karras, T., Aila, T., Laine, S., Lehtinen, J.: Progressive growing of GANs for improved quality, stability, and variation. In: International Conference on Learning Representations (2018)

22. Karras, T., Laine, S., Aila, T.: A style-based generator architecture for generative adversarial networks. In: IEEE Conference on Computer Vision and Pattern Recognition, pp. 4401–4410 (2019)

23. Karras, T., Laine, S., Aittala, M., Hellsten, J., Lehtinen, J., Aila, T.: Analyzing and improving the image quality of styleGAN. In: IEEE Conference on Computer Vision and Pattern Recognition, pp. 8110–8119 (2020)

24. Kingma, D.P., Ba, J.: Adam: a method for stochastic optimization. In: International Conference on Learning Representations (2014)
25. Lee, C.H., Liu, Z., Wu, L., Luo, P.: MaskGAN: towards diverse and interactive facial image manipulation. In: IEEE Conference on Computer Vision and Pattern Recognition, pp. 5549–5558 (2020)
26. Levoy, M.: Efficient ray tracing of volume data. ACM Trans. Graph. **9**(3), 245–261 (1990)
27. Ling, H., Kreis, K., Li, D., Kim, S.W., Torralba, A., Fidler, S.: EditGAN: high-precision semantic image editing. In: Advances in Neural Information Processing Systems (2021)
28. Lira, W., Merz, J., Ritchie, D., Cohen-Or, D., Zhang, H.: **GANHopper**: multi-hop GAN for unsupervised image-to-image translation. In: Vedaldi, A., Bischof, H., Brox, T., Frahm, J.-M. (eds.) ECCV 2020. LNCS, vol. 12371, pp. 363–379. Springer, Cham (2020). https://doi.org/10.1007/978-3-030-58574-7_22
29. Liu, L., Gu, J., Zaw Lin, K., Chua, T.S., Theobalt, C.: Neural sparse voxel fields. Adv. Neural Inform. Process. Syst. **33**, 15651–15663 (2020)
30. Liu, X., Xu, Y., Wu, Q., Zhou, H., Wu, W., Zhou, B.: Semantic-aware implicit neural audio-driven video portrait generation. arXiv preprint arXiv:2201.07786 (2022)
31. Liu, Z., et al.: Swin transformer: hierarchical vision transformer using shifted windows. In: IEEE Conference on Computer Vision, pp. 10012–10022 (2021)
32. Liu, Z., Luo, P., Wang, X., Tang, X.: Deep learning face attributes in the wild. In: IEEE Conference on Computer Vision and Pattern Recognition, pp. 3730–3738 (2015)
33. Lombardi, S., Simon, T., Schwartz, G., Zollhoefer, M., Sheikh, Y., Saragih, J.: Mixture of volumetric primitives for efficient neural rendering. ACM Trans. Graph. **40**(4), 1–13 (2021)
34. Luo, S., Hu, W.: Diffusion probabilistic models for 3d point cloud generation. In: IEEE Conference on Computer Vision and Pattern Recognition, pp. 2837–2845 (2021)
35. Mescheder, L., Geiger, A., Nowozin, S.: Which training methods for GANs do actually converge? In: International Conference on Machine Learning, pp. 3481–3490. PMLR (2018)
36. Mildenhall, B., Srinivasan, P.P., Tancik, M., Barron, J.T., Ramamoorthi, R., Ng, R.: NeRF: representing scenes as neural radiance fields for view synthesis. In: Vedaldi, A., Bischof, H., Brox, T., Frahm, J.-M. (eds.) ECCV 2020. LNCS, vol. 12346, pp. 405–421. Springer, Cham (2020). https://doi.org/10.1007/978-3-030-58452-8_24
37. Niemeyer, M., Geiger, A.: Giraffe: Representing scenes as compositional generative neural feature fields. In: IEEE Conference on Computer Vision and Pattern Recognition, pp. 11453–11464 (2021)
38. Park, T., Liu, M.Y., Wang, T.C., Zhu, J.Y.: Semantic image synthesis with spatially-adaptive normalization. In: IEEE Conference on Computer Vision and Pattern Recognition, pp. 2337–2346 (2019)
39. Paszke, A., et al.: PyTorch: an imperative style, high-performance deep learning library. In: Advances in Neural Information Processing Systems 32 (2019)
40. Perez, E., Strub, F., De Vries, H., Dumoulin, V., Courville, A.: Film: visual reasoning with a general conditioning layer. In: AAAI, vol. 32 (2018)
41. Pumarola, A., Corona, E., Pons-Moll, G., Moreno-Noguer, F.: D-NeRF: neural radiance fields for dynamic scenes. In: IEEE Conference on Computer Vision and Pattern Recognition, pp. 10318–10327 (2021)

42. Richardson, E., et al.: Encoding in style: a styleGAN encoder for image-to-image translation. In: IEEE Conference on Computer Vision and Pattern Recognition, pp. 2287–2296 (2021)

43. Salimans, T., Goodfellow, I., Zaremba, W., Cheung, V., Radford, A., Chen, X.: Improved techniques for training GANs. Advances in Neural Information Processing Systems 29 (2016)

44. Schwarz, K., Liao, Y., Niemeyer, M., Geiger, A.: GRAF: generative radiance fields for 3D-aware image synthesis. Adv. Neural Inform. Process. Syst. **33**, 20154–20166 (2020)

45. Shi, Y., Yang, X., Wan, Y., Shen, X.: SemanticStyleGAN: learning compositional generative priors for controllable image synthesis and editing. In: IEEE Conference on Computer Vision and Pattern Recognition, pp. 11254–11264 (2022)

46. Sitzmann, V., Martel, J., Bergman, A., Lindell, D., Wetzstein, G.: Implicit neural representations with periodic activation functions. Adv. Neural Inform. Process. Syst. **33**, 7462–7473 (2020)

47. Song, G., et al.: AgileGAN: stylizing portraits by inversion-consistent transfer learning. ACM Trans. Graph. **40**(4), 1–13 (2021)

48. Sun, C., Sun, M., Chen, H.T.: Direct voxel grid optimization: super-fast convergence for radiance fields reconstruction. In: IEEE Conference on Computer Vision and Pattern Recognition (2022)

49. Sun, J., et al.: FENeRF: face editing in neural radiance fields. In: IEEE Conference on Computer Vision and Pattern Recognition, pp. 7672–7682 (2022)

50. Szegedy, C., Vanhoucke, V., Ioffe, S., Shlens, J., Wojna, Z.: Rethinking the inception architecture for computer vision. In: IEEE Conference on Computer Vision and Pattern Recognition, pp. 2818–2826 (2016)

51. Wang, T.C., Liu, M.Y., Zhu, J.Y., Tao, A., Kautz, J., Catanzaro, B.: High-resolution image synthesis and semantic manipulation with conditional GANs. In: IEEE Conference on Computer Vision and Pattern Recognition, pp. 8798–8807 (2018)

52. Wu, J., Zhang, C., Xue, T., Freeman, B., Tenenbaum, J.: Learning a probabilistic latent space of object shapes via 3D generative-adversarial modeling. In: Advances in Neural Information Processing Systems 29 (2016)

53. Wu, Q., Liu, X., Chen, Y., Li, K., Zheng, C., Cai, J., Zheng, J.: Object-compositional neural implicit surfaces. arXiv preprint arXiv:2207.09686 (2022)

54. Wu, Z., Lischinski, D., Shechtman, E.: StyleSpace Analysis: disentangled controls for stylegan image generation. In: IEEE Conference on Computer Vision and Pattern Recognition, pp. 12863–12872 (2021)

55. Wu, Z., Nitzan, Y., Shechtman, E., Lischinski, D.: StyleAlign: analysis and applications of aligned styleGAN models. In: International Conference on Learning Representations (2022)

56. Xu, Y., et al.: TransEditor: transformer-based dual-space GAN for highly controllable facial editing. In: IEEE Conference on Computer Vision and Pattern Recognition, pp. 7683–7692 (2022)

57. Xu, Y., Peng, S., Yang, C., Shen, Y., Zhou, B.: 3D-aware image synthesis via learning structural and textural representations. In: IEEE Conference on Computer Vision and Pattern Recognition (2022)

58. Yen-Chen, L., Florence, P., Barron, J.T., Rodriguez, A., Isola, P., Lin, T.Y.: iNeRF: inverting neural radiance fields for pose estimation. In: IEEE International Conference on Intelligent Robots and Systems, pp. 1323–1330. IEEE (2021)

59. Zhang, J., Sangineto, E., Tang, H., Siarohin, A., Zhong, Z., Sebe, N., Wang, W.: 3D-aware semantic-guided generative model for human synthesis. arXiv preprint arXiv:2112.01422 (2021)
60. Zhang, R.: Making convolutional networks shift-invariant again. In: International Conference on Machine Learning, pp. 7324–7334. PMLR (2019)
61. Zhang, R., Isola, P., Efros, A.A., Shechtman, E., Wang, O.: The unreasonable effectiveness of deep features as a perceptual metric. In: IEEE Conference on Computer Vision and Pattern Recognition, pp. 586–595 (2018)
62. Zhang, X., Zheng, Z., Gao, D., Zhang, B., Pan, P., Yang, Y.: Multi-view consistent generative adversarial networks for 3D-aware image synthesis. In: IEEE IEEE Conference on Computer Vision and Pattern Recognition, pp. 18450–18459 (2022)
63. Zhang, Y., et al.: DatasetGAN: efficient labeled data factory with minimal human effort. In: IEEE Conference on Computer Vision and Pattern Recognition, pp. 10145–10155 (2021)
64. Zheng, C., Cham, T.J., Cai, J.: Pluralistic image completion. In: IEEE Conference on Computer Vision and Pattern Recognition, pp. 1438–1447 (2019)
65. Zheng, C., Cham, T.J., Cai, J.: TFill: image completion via a transformer-based architecture. In: IEEE Conference on Computer Vision and Pattern Recognition (2022)
66. Zhi, S., Laidlow, T., Leutenegger, S., Davison, A.J.: In-place scene labelling and understanding with implicit scene representation. In: International Conference on Computer Vision, pp. 15838–15847 (2021)
67. Zhou, T., Tucker, R., Flynn, J., Fyffe, G., Snavely, N.: Stereo magnification: learning view synthesis using multiplane images. ACM Transactions on Graphics (2018)
68. Zhu, J.Y., Park, T., Isola, P., Efros, A.A.: Unpaired image-to-image translation using cycle-consistent adversarial networks. In: International Conference on Computer Vision, pp. 2223–2232 (2017)
69. Zhu, J.Y., et al.: Toward multimodal image-to-image translation. In: Advances in Neural Information Processing Systems 30 (2017)

Author Index

Printed in the United States
by Baker & Taylor Publisher Services